Telecommunication Networks Handbook

Telecommunication Networks Handbook

Edited by **Bernhard Ekman**

CLANRYE INTERNATIONAL

New Jersey

Published by Clanrye International,
55 Van Reypen Street,
Jersey City, NJ 07306, USA
www.clanryeinternational.com

Telecommunication Networks Handbook
Edited by Bernhard Ekman

International Standard Book Number: 978-1-63240-482-4 (Hardback)

This book contains information obtained from authentic and highly regarded sources. Copyright for all individual chapters remain with the respective authors as indicated. A wide variety of references are listed. Permission and sources are indicated; for detailed attributions, please refer to the permissions page. Reasonable efforts have been made to publish reliable data and information, but the authors, editors and publisher cannot assume any responsibility for the validity of all materials or the consequences of their use.

The publisher's policy is to use permanent paper from mills that operate a sustainable forestry policy. Furthermore, the publisher ensures that the text paper and cover boards used have met acceptable environmental accreditation standards.

Trademark Notice: Registered trademark of products or corporate names are used only for explanation and identification without intent to infringe.

Printed in the United States of America.

Contents

Preface

This handbook provides a descriptive examination of telecommunications networks. It elucidates the basics of fast developing networks as well as the advanced concepts and future expectations of Telecommunications Networks. It recognizes and analyzes the most important research issues in Telecommunication and it consists of information contributed by the top researchers, industry and academic experts. This book also contains surveys of current publications that thoroughly examine important fields of interest like: simulation, optimization problems, etc. The book covers topics like Telecommunications, Traffic Engineering and Routing. It will serve as a good reference for both PhD and MA students.

The information contained in this book is the result of intensive hard work done by researchers in this field. All due efforts have been made to make this book serve as a complete guiding source for students and researchers. The topics in this book have been comprehensively explained to help readers understand the growing trends in the field.

I would like to thank the entire group of writers who made sincere efforts in this book and my family who supported me in my efforts of working on this book. I take this opportunity to thank all those who have been a guiding force throughout my life.

Editor

Part 1

Telecommunications

Quantum Secure Telecommunication Systems

Oleksandr Korchenko[1], Petro Vorobiyenko[2],
Maksym Lutskiy[1], Yevhen Vasiliu[2] and Sergiy Gnatyuk[1]
[1]National Aviation University
[2]Odessa National Academy of Telecommunication
named after O.S. Popov
Ukraine

Our scientific field is still in its embryonic stage. It's great that
we haven't been around for two thousands years. We are still at
a stage where very, very important results occur in front of our eyes
Michael Rabin

1. Introduction

Today there is virtually no area where information technology (IT) is not used in some way. Computers support banking systems, control the work of nuclear power plants, and control aircraft, satellites and spacecraft. The high level of automation therefore depends on the security level of IT.

The main features of information security are confidentiality, integrity and availability. Only providing these all gives availability for development secure telecommunication systems. *Confidentiality* is the basic feature of information security, which ensures that information is accessible only to authorized users who have an access. *Integrity* is the basic feature of information security indicating its property to resist unauthorized modification. *Availability* is the basic feature of information security that indicates accessible and usable upon demand by an authorized entity.

One of the most effective ways to ensure confidentiality and data integrity during transmission is cryptographic systems. The purpose of such systems is to provide key distribution, authentication, legitimate users authorisation, and encryption. *Key distribution is one of the most important problems of cryptography.* This problem can be solved with the help of (SECOQC White Paper on Quantum Key Distribution and Cryptography, 2007; Korchenko et al., 2010a):

- *Classical information-theoretic schemes* (requires channel with noise; efficiency is very low, 1–5%).
- *Classical public-key cryptography schemes* (Diffie-Hellman scheme, digital envelope scheme; it has computational security).

- *Classical computationally secure symmetric-key cryptographic schemes* (requires a pre-installed key on both sides and can be used only as scheme for increase in key size but not as key distribution scheme).
- *Quantum key distribution* (provides information-theoretic security; it can also be used as a scheme for increase in key length).
- *Trusted Couriers Key Distribution* (it has a high price and is dependent on the human factor).

In recent years, quantum cryptography (QC) has attracted considerable interest. Quantum key distribution (QKD) (Bennett, 1992; Bennett et al., 1992; Bennett et al., 1995; Bennett & Brassard, 1984; Bouwmeester et al., 2000; Gisin et al., 2002; Lütkenhaus & Shields, 2009; Scarani et al., 2009; Vasiliu & Vorobiyenko 2006; Williams, 2011) plays a dominant role in QC. The overwhelming majority of theoretic and practical research projects in QC are related to the development of QKD protocols. The number of different quantum technologies is increasing, but there is no comprehensive information about classification of these technologies in scientific literature (there are only a few works concerning different classifications of QKD protocols, for example (Gisin et al., 2002; Scarani, et al., 2009)). This makes it difficult to estimate the level of the latest achievements and does not allow using quantum technologies with full efficiency. The main purpose of this chapter is the systematisation and classification of up-to-date effective quantum technologies of data (transmitted via telecommunication channels) security, analysis of their strengths and weaknesses, prospects and difficulties of implementation in telecommunication systems.

The first of all *quantum technologies of information security* consist of (Korchenko et al., 2010b):

- Quantum key distribution.
- Quantum secure direct communication.
- Quantum steganography.
- Quantum secret sharing.
- Quantum stream cipher.
- Quantum digital signature, etc.

The theoretical basis of quantum cryptography is stated in set of books and review papers (see e.g. Bouwmeester et al., 2000; Gisin et al., 2002; Hayashi, 2006; Imre & Balazs, 2005; Kollmitzer & Pivk, 2010; Lomonaco, 1998; Nielsen & Chuang, 2000; Schumacher & Westmoreland, 2010; Vedral, 2006; Williams, 2011).

2. Main approaches to quantum secure telecommunication systems construction

2.1 Quantum key distribution

QKD includes the following protocols: protocols using single (non-entangled) qubits (two-level quantum systems) and qudits (d-level quantum systems, d>2) (Bennett, 1992; Bennett et al., 1992; Bourennane et al., 2002; Bruss & Macchiavello, 2002; Cerf et al., 2002; Gnatyuk et al., 2009); protocols using phase coding (Bennett, 1992); protocols using entangled states (Ekert, 1991; Durt et al., 2004); decoy states protocols (Brassard et al., 2000; Liu et al., 2010; Peng et al., 2007; Yin et al., 2008; Zhao et al., 2006a, 2006b); and some other protocols (Bradler, 2005; Lütkenhaus & Shields, 2009; Navascués & Acín, 2005; Pirandola et al., 2008).

The main task of QKD protocols is encryption key generation and distribution between two users connecting via quantum and classical channels (Gisin et al., 2002). In 1984 Ch. Bennett from IBM and G. Brassard from Montreal University introduced the first QKD protocol (Bennett & Brassard, 1984), which has become an alternative solution for the problem of key distribution. This protocol is called *BB84* (Bouwmeester et al., 2000) and it refers to QKD protocols using single qubits. The states of these qubits are the polarisation states of single photons. The BB84 protocol uses four polarisation states of photons (0°, 45°, 90°, 135°). These states refer to two mutually unbiased bases. Error searching and correcting is performed using classical public channel, which need not be confidential but only authenticated. For the detection of intruder actions in the BB84 protocol, an error control procedure is used, and for providing unconditionally security a privacy amplification procedure is used (Bennett et al., 1995). The efficiency of the BB84 protocol equals 50%. Efficiency means the ratio of the photons number which are used for key generation to the general number of transmitted photons.

Six-state protocol requires the usage of four states, which are the same as in the BB84 protocol, and two additional directions of polarization: right circular and left circular (Bruss, 1998). Such changes decrease the amount of information, which can be intercepted. But on the other hand, the efficiency of the protocol decreases to 33%.

Next, the *4+2 protocol* is intermediate between the BB84 and B92 protocol (Huttner et al., 1995). There are four different states used in this protocol for encryption: "0" and "1" in two bases. States in each base are selected non-orthogonal. Moreover, states in different bases must also be pairwise non-orthogonal. This protocol has a higher information security level than the BB84 protocol, when weak coherent pulses, but not a single photon source, are used by sender (Huttner et al., 1995). But the efficiency of the 4+2 protocol is lower than efficiency of BB84 protocol.

In the *Goldenberg-Vaidman protocol* (Goldenberg & Vaidman, 1995), encryption of "0" and "1" is performed using two orthogonal states. Each of these two states is the superposition of two localised normalised wave packets. For protection against intercept-resend attack, packets are sent at random times.

A modified type of Goldenberg-Vaidman protocol is called the *Koashi-Imoto protocol* (Koashi & Imoto, 1997). This protocol does not use a random time for sending packets, but it uses an interferometer's non-symmetrisation (the light is broken in equal proportions between both long and short interferometer arms).

The measure of QKD protocol security is Shannon's mutual information between legitimate users (Alice and Bob) and an eavesdropper (Eve): $I_{AE}(D)$ and $I_{BE}(D)$, where D is error level which is created by eavesdropping. For most attacks on QKD protocols, $I_{AE}(D) = I_{BE}(D)$, we will therefore use $I_{AE}(D)$. The lower $I_{AE}(D)$ in the extended range of D is, the more secure the protocol is.

Six-state protocol and BB84 protocol were generalised in case of using d-level quantum systems — qudits instead qubits (Cerf et al., 2002). This allows increasing the information capacity of protocols. We can transfer information using d-level quantum systems (which correspond to the usage of trits, quarts, etc.). It is important to notice that QKD protocols are intended for classical information (key) transfer via quantum channel.

The generalisation of BB84 protocol for qudits is called protocol using single qudits and two bases due to use of two mutually unbiased bases for the eavesdropping detection. Similarly, the generalisation of six-state protocol is called protocol using qudits and $d+1$ bases. These protocols' security against intercept-resend attack and non-coherent attack was investigated in a number of articles (see e.g. Cerf et al., 2002). Vasiliu & Mamedov have carried out a comparative analysis of the efficiency and security of different protocols using qudits on the basis of known formulas for mutual information (Vasiliu & Mamedov, 2008).

In fig. 1 dependences of $I_{AB}(D)$, $I_{AE}^{(d+1)}(D)$ and $I_{AE}^{(2)}(D)$ are presented, where $I_{AB}(D)$ is mutual information between Alice and Bob and $I_{AE}^{(d+1)}(D)$ and $I_{AE}^{(2)}(D)$ is mutual information between Alice and Eve for protocols using $d+1$ and two bases accordingly.

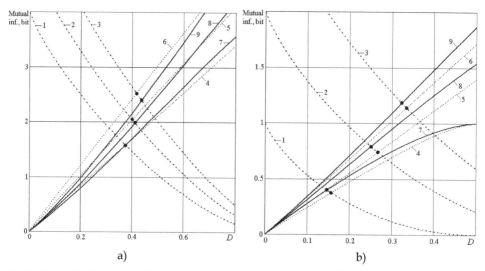

Fig. 1. Mutual information for non-coherent attack. 1, 2, 3 — $I_{AB}(D)$ for $d = 2, 4, 8$ (a) and $d = 16, 32, 64$ (b); 4, 5, 6 — $I_{AE}^{(d+1)}(D)$ for $d = 2, 4, 8$ (a) and $d = 16, 32, 64$ (b); 7, 8, 9 — $I_{AE}^{(2)}(D)$ for $d = 2, 4, 8$ (a) and $d = 16, 32, 64$ (b).

In fig. 1 we can see that at low qudit dimension (up to $d \sim 16$) the protocol's security against non-coherent attack is higher when $d+1$ bases are used (when $d = 2$ it corresponds as noted above to greater security of six-state protocol than BB84 protocol). But the protocol's security is higher when two bases are used in the case of large d, while the difference in Eve's information (using $d+1$ or two bases) is not large in the work region of the protocol, i.e. in the region of Alice's and Bob's low error level. That's why that the number of bases used has little influence on the security of the protocol against non-coherent attack (at least for the qudit dimension up to $d = 64$). The crossing points of curves $I_{AB}(D)$ and $I_{AE}(D)$ correspond to boundary values D, up to which one's legitimate users can establish a secret key by means of a privacy amplification procedure (even when eavesdropping occurs) (Bennett et al., 1995).

It is shown (Vasiliu & Mamedov, 2008) that the security of a protocol with qudits using two bases against intercept-resend attack is practically equal to the security of this protocol against non-coherent attack at any d. At the same time, the security of the protocol using $d+1$ bases against this attack is much higher. Intercept-resend attack is the weakest of all possible attacks on QKD protocols, but on the other hand, the efficiency of the protocol using $d+1$ bases rapidly decreases as d increases. A protocol with qudits using two bases therefore has higher security and efficiency than a protocol using $d+1$ bases.

Another type of QKD protocol is a *protocol using phase coding*: for example, the *B92 protocol* (Bennett, 1992) using strong reference pulses (Gisin et al., 2002). An eavesdropper can obtain more information about the encryption key in the B92 protocol than in the BB84 protocol for the given error level, however. Thus, the security of the B92 protocol is lower than the security of the BB84 protocol (Fuchs et al., 1997). The efficiency of the B92 protocol is 25%.

The *Ekert protocol (E91)* (Ekert, 1991) refers to QKD protocols using entangled states. Entangled pairs of qubits that are in a singlet state $\left|\psi^{-}\right\rangle = 1/\sqrt{2}\left(|0\rangle|1\rangle - |1\rangle|0\rangle\right)$ are used in this protocol. Qubit interception between Alice to Bob does not give Eve any information because no coded information is there. Information appears only after legitimate users make measurements and communicate via classical public authenticated channel (Ekert, 1991). But attacks with additional quantum systems (ancillas) are nevertheless possible on this protocol (Inamori et al., 2001).

Kaszlikowski et al. carried out the generalisation of the Ekert scheme for three-level quantum systems (Kaszlikowski et al., 2003) and Durt et al. carried out the generalisation of the Ekert scheme for d-level quantum systems (Durt et al., 2004): this increases the information capacity of the protocol a lot. Also the security of the protocol using entangled qudits is investigated (Durt et al., 2004). In the paper (Vasiliu & Mamedov, 2008), based on the results of (Durt et al., 2004), the security comparison of protocol using entangled qudits and protocols using single qudits (Cerf et al., 2002) against non-coherent attack is made. It was found that the security of these two kinds of protocols is almost identical. But the efficiency of the protocol using entangled qudits increases more slowly with the increasing dimension of qudits than the efficiency of the protocol using single qudits and two bases. Thus, from all contemporary QKD protocols using qudits, the most effective and secure against non-coherent attack is the protocol using single qudits and two bases (BB84 for qubits).

The aforementioned protocols with qubits are vulnerable to photon number splitting attack. This attack cannot be applied when the photon source emits exactly one photon. But there are still no such photon sources. Therefore, sources with Poisson distribution of photon number are used in practice. The part of pulses of this source has more than one photon. That is why Eve can intercept one photon from pulse (which contains two or more photons) and store it in quantum memory until Alice transfers Bob the sequence of bases used. Then Eve can measure stored states in correct basis and get the cryptographic key while remaining invisible. It should be noted that there are more advanced strategies of photon number splitting attack which allow Bob to get the correct statistics of the photon number in pulses if Bob is controlling these statistics (Lutkenhaus & Jahma, 2002).

In practice for realisation of BB84 and six-state protocols weak coherent pulses with average photon number about 0,1 are used. This allows avoiding small probability of two- and multi-photon pulses, but this also considerably reduces the key rate.

The *SARG04 protocol* does not differ much from the original BB84 protocol (Branciard et al., 2005; Scarani et al., 2004; Scarani et al., 2009). The main difference does not refer to the "quantum" part of the protocol; it refers to the "classical" procedure of key sifting, which goes after quantum transfer. Such improvement allows increasing security against photon number splitting attack. The SARG04 protocol in practice has a higher key rate than the BB84 protocol (Branciard et al., 2005).

Another way of protecting against photon number splitting attack is the use of *decoy states QKD protocols* (Brassard et al., 2000; Peng et al., 2007; Rosenberg et al., 2007; Zhao et al., 2006), which are also advanced types of BB84 protocol. In such protocols, besides information signals Alice's source also emits additional pulses (decoys) in which the average photon number differs from the average photon number in the information signal. Eve's attack will modify the statistical characteristics of the decoy states and/or signal state and will be detected. As practical experiments have shown for these protocols (as for the SARG04 protocol), the key rate and practical length of the channel is bigger than for BB84 protocols (Peng et al., 2007; Rosenberg et al., 2007; Zhao et al., 2006). Nevertheless, it is necessary to notice that using these protocols, as well as the others considered above, it is also impossible without users pre-authentication to construct the complete high-grade solution of the problem of key distribution.

As a conclusion, after the analysis of the first and scale quantum method, we must sum up and highlight the following *advantages of QKD protocols:*

1. These protocols always allow eavesdropping to be detected because Eve's connection brings much more error level (compared with natural error level) to the quantum channel. The laws of quantum mechanics allow eavesdropping to be detected and the dependence between error level and intercepted information to be set. This allows applying privacy amplification procedure, which decreases the quantity of information about the key, which can be intercepted by Eve. Thus, QKD protocols have unconditional (information-theoretic) security.
2. The information-theoretic security of QKD allows using an absolutely secret key for further encryption using well-known classical symmetrical algorithms. Thus, the entire information security level increases. It is also possible to synthesize QKD protocols with Vernam cipher (one-time pad) which in complex with unconditionally secured authenticated schemes gives a totally secured system for transferring information.

The disadvantages of quantum key distribution protocols are:

1. A system based only on QKD protocols cannot serve as a complete solution for key distribution in open networks (additional tools for authentication are needed).

2. The limitation of quantum channel length which is caused by the fact that there is no possibility of amplification without quantum properties being lost. However, the technology of quantum repeaters could overcome this limitation in the near future (Sangouard et al., 2011).
3. Need for using weak coherent pulses instead of single photon pulses. This decreases the efficiency of protocol in practice. But this technology limitation might be defeated in the nearest future.
4. The data transfer rate decreases rapidly with the increase in the channel length.
5. Photon registration problem which leads to key rate decreasing in practice.
6. Photon depolarization in the quantum channel. This leads to errors during data transfer. Now the typical error level equals a few percent, which is much greater than the error level in classical telecommunication systems.
7. Difficulty of the practical realisation of QKD protocols for d-level quantum systems.
8. The high price of commercial QKD systems.

2.2 Quantum secure direct communication

The next method of information security based on quantum technologies is the usage of *quantum secure direct communication (QSDC) protocols* (Boström & Felbinger, 2002; Chuan et al., 2005; Cai, 2004; Cai & Li, 2004a; Cai & Li, 2004b; Deng et al., 2003; Vasiliu, 2011; Wang et al., 2005a, 2005b). The main feature of QSDC protocols is that there are no cryptographic transformations; thus, there is no key distribution problem in QSDC. In these protocols, a secret message is coded by qubits' (qudits') – quantum states, which are sent via quantum channel. QSDC protocols can be divided into several types:

- *Ping-pong protocol (and its enhanced variants)* (Boström & Felbinger, 2002; Cai & Li, 2004b; Chamoli & Bhandari, 2009; Gao et al., 2008; Ostermeyer & Walenta, 2008;Vasiliu & Nikolaenko, 2009; Vasiliu, 2011).
- *Protocols using block transfer of entangled qubits* (Deng et al., 2003; Chuan et al., 2005; Gao et al., 2005; Li et al., 2006; Lin et al., 2008; Xiu et al., 2009; Wang et al., 2005a, 2005b).
- *Protocols using single qubits* (Cai, 2004; Cai & Li, 2004a).
- *Protocols using entangled qudits* (Wang et al., 2005b; Vasiliu, 2011).

There are QSDC protocols for two parties and for multi-parties, e.g. broadcasting or when one user sends message to another under the control of a trusted third party.

Most contemporary protocols require a transfer of qubits by blocks (Chuan et al., 2005; Wang et al., 2005). This allows eavesdropping to be detected in the quantum channel before transfer of information. Thus, transfer will be terminated and Eve will not obtain any secret information. But for storing such blocks of qubits there is a need for a large amount of quantum memory. The technology of quantum memory is actively being developed, but it is still far from usage in common standard telecommunication equipment. So from the viewpoint of technical realisation, protocols using single qubits or their non-large groups (for one cycle of protocol) have an advantage. There are few such protocols and they have only asymptotic security, i.e. the attack will be detected with high probability, but Eve can obtain some part of information before detection. Thus, the problem of privacy amplification appears. In other words, new pre-processing methods of transferring information are needed. Such methods should make intercepted information negligible.

One of the quantum secure direct communication protocols is the ping-pong protocol (Boström & Felbinger, 2002; Cai & Li, 2004b; Vasiliu, 2011), which does not require qubit transfer by blocks. In the first variant of this protocol, entangled pairs of qubits and two coding operations that allow the transmission of one bit of classical information for one cycle of the protocol are used (Boström & Felbinger, 2002). The usage of quantum superdense coding allows transmitting two bits for a cycle (Cai & Li, 2004b). The subsequent increase in the informational capacity of the protocol is possible by the usage instead of entangled pairs of qubits their triplets, quadruplets etc. in Greenberger-Horne-Zeilinger (GHZ) states (Vasiliu & Nikolaenko, 2009). The informational capacity of the ping-pong protocol with GHZ-states is equal to n bits on a cycle where n is the number of entangled qubits. Another way of increasing the informational capacity of ping-pong protocol is using entangled states of qudits. Thus, the corresponding protocol based on Bell's states of three-level quantum system (qutrit) pairs and superdense coding for qutrits is introduced (Wang et al., 2005; Vasiliu, 2011).

The advantages of QSDC protocols are a lack of secret key distribution, the possibility of data transfer between more than two parties, and the possibility of attack detection providing a high level of information security (up to information-theoretic security) for the protocols using block transfer. The main disadvantages are difficulty in practical realisation of protocols using entangled states (and especially protocols using entangled states for d-level quantum systems), slow transfer rate, the need for large capacity quantum memory for all parties (for protocols using block transfer of qubits), and the asymptotic security of the ping-pong protocol. Besides, QSDC protocols similarly to QKD protocols is vulnerable to man-in-the-middle attack, although such attack can be neutralized by using authentication of all messages, which are sent via the classical channel.

Asymptotic security of the ping-pong protocol (which is one of the simplest QSDC protocols from the technical viewpoint) can be amplified by using methods of classical cryptography. Security of several types of ping-pong protocols using qubits and qutrits against different attacks was investigated in series of papers (Boström & Felbinger, 2002; Cai, 2004; Vasiliu, 2011; Vasiliu & Nikolaenko, 2009; Zhang et al., 2005a).

The security of the ping-pong protocol using qubits against eavesdropping attack using ancilla states is investigated in (Boström & Felbinger, 2002; Chuan et al., 2005; Vasiliu & Nikolaenko, 2009).

Eve's information at attack with usage of auxiliary quantum systems (probes) on the ping-pong protocol with entangled n-qubit GHZ-states is defined by von Neumann entropy (Boström & Felbinger, 2002):

$$I_0 = S(\rho) \equiv -Tr\{\rho \log_2 \rho\} = -\sum_i \lambda_i \log_2 \lambda_i \tag{1}$$

where λ_i are the density matrix eigenvalues for the composite quantum system "transmitted qubits - Eve's probe".

For the protocol with Bell pairs and quantum superdence coding the density matrix ρ have size 4x4 and four nonzero eigenvalues:

$$\lambda_{1,2} = \frac{1}{2}(p_1 + p_2) \pm \frac{1}{2}\sqrt{(p_1 + p_2)^2 - 16p_1p_2d(1-d)},$$
$$\lambda_{3,4} = \frac{1}{2}(p_3 + p_4) \pm \frac{1}{2}\sqrt{(p_3 + p_4)^2 - 16p_3p_4d(1-d)}. \tag{2}$$

For the protocol with GHZ-triplets a density matrix size is 16x16, and a number of nonzero eigenvalues is equal to eight. At symmetrical attack their kind is (Vasiliu & Nikolaenko, 2009):

$$\lambda_{1,2} = \frac{1}{2}(p_1 + p_2) \pm \frac{1}{2}\sqrt{(p_1 + p_2)^2 - 16p_1p_2 \cdot \frac{2}{3}d\left(1 - \frac{2}{3}d\right)},$$
$$\lambda_{7,8} = \frac{1}{2}(p_7 + p_8) \pm \frac{1}{2}\sqrt{(p_7 + p_8)^2 - 16p_7p_8 \cdot \frac{2}{3}d\left(1 - \frac{2}{3}d\right)}. \tag{3}$$

For the protocol with n-qubit GHZ-states, the number of nonzero eigenvalues of density matrix is equal to 2^n, and their kind at symmetrical attack is (Vasiliu & Nikolaenko, 2009):

$$\lambda_{1,2} = \frac{1}{2}(p_1 + p_2) \pm \frac{1}{2}\sqrt{(p_1 + p_2)^2 - 16p_1p_2 \cdot \frac{2^{n-2}}{2^{n-1}-1}d\left(1 - \frac{2^{n-2}}{2^{n-1}-1}d\right)},$$
$$\lambda_{2^n-1,2^n} = \frac{1}{2}(p_{2^n-1} + p_{2^n}) \pm \frac{1}{2}\sqrt{(p_{2^n-1} + p_{2^n})^2 - 16p_{2^n-1}p_{2^n} \cdot \frac{2^{n-2}}{2^{n-1}-1}d\left(1 - \frac{2^{n-2}}{2^{n-1}-1}d\right)}, \tag{4}$$

where d is probability of attack detection by legitimate users at one-time switching to control mode; p_i are frequencies of n-grams in the transmitted message.

The probability of that Eve will not be detected after m successful attacks and will gain information $I = mI_0$ is defined by the equation (Boström & Felbinger, 2002):

$$s(I,q,d) = \left(\frac{1-q}{1-q(1-d)}\right)^{1/I_0}, \tag{5}$$

where q is a probability of switching to control mode.

In fig. 2 dependences of $s(I,q,d)$ for several n, identical frequencies $p_i = 2^{-n}$, $q = 0.5$ and $d = d_{max}$ are shown (Vasiliu & Nikolaenko, 2009). d_{max} is maximum probability of attack detection at one-time run of control mode, defined as

$$d_{max} = 1 - \frac{1}{2^{n-1}}. \tag{6}$$

At $d = d_{max}$ Eve gains the complete information about transmitted bits of the message. It is obvious from fig. 2 that the ping-pong protocol with many-qubit GHZ-states is asymptotically secure at any number n of qubits that are in entangled GHZ-states. A similar result for the ping-pong protocol using qutrit pairs is presented (Vasiliu, 2011).

A non-quantum method of security amplification for the ping-pong protocol is suggested in (Vasiliu & Nikolaenko, 2009; Korchenko et al., 2010c). Such method has been developed on the basis of a method of privacy amplification which is utilized in quantum key distribution protocols. In case of the ping-pong protocol this method can be some kind of analogy of the Hill cipher (Overbey et al., 2005).

Before the transmission Alice divides the binary message on l blocks of some fixed length r, we will designate these blocks as a_i ($i=1,...l$). Then Alice generates for each block separately random invertible binary matrix K_i of size $r \times r$ and multiplies these matrices by appropriate blocks of the message (multiplication is performed by modulo 2):

$$b_i = K_i a_i. \tag{7}$$

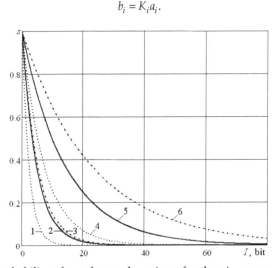

Fig. 2. Composite probability of attack non-detection s for the ping-pong protocol with many-qubit GHZ-states: $n=2$, original protocol (1); $n=2$, with superdense coding (2); $n=3$ (3); $n=5$ (4); $n=10$ (5); $n=16$ (6). I is Eve's information.

Blocks b_i are transmitted on the quantum channel with the use of the ping-pong protocol. Even if Eve, remained undetected, manages to intercept one (or more) from these blocks and without knowledge of used matrices K_i Eve won't be able to reconstruct source blocks a_i. To reach a sufficient security level the block length r and accordingly the size of matrices K_i should be selected so that Eve's undetection probability s after transmission of *one* block would be insignificant small. Matrices K_i are transmitted to Bob via usual (non-quantum) open authentic channel after the end of quantum transmission but only in the event when Alice and Bob were convinced lack of eavesdropping. Then Bob inverses the received matrices and having multiplied them on appropriate blocks b_i he gains an original message.

Let's mark that described procedure is not message enciphering, and can be named inverse hashing or hashing using two-way hash function, which role random invertible binary matrix acts.

It is necessary for each block to use individual matrix K_i which will allow to prevent cryptoanalytic attacks, similar to attacks to the Hill cipher, which are possible there at a multiple usage of one matrix for enciphering of several blocks (Eve could perform similar attack if she was able before a detection of her operations in the quantum channel to intercept several blocks, that are hashing with the same matrix). As matrices in this case are not a key and they can be transmitted on the open classical channel, the transmission of the necessary number of matrices is not a problem.

Necessary length r of blocks for hashing and accordingly necessary size $r \times r$ of hashing matrices should correspond to a requirement $r > I$, where I is the information which is gained by Eve. Thus, it is necessary for determination of r to calculate I at the given values of n, s, q and $d = d_{max}$.

Let's accept $s(I, q, d) = 10^{-k}$, then:

$$I = \frac{-kl_0}{\lg\left(\dfrac{1-q}{1-q(1-d)}\right)}. \tag{8}$$

The calculated values of I are shown in tab. 1:

n	q = 0,5; $d = d_{max}$	q = 0,5; $d = d_{max}/2$	q = 0,25; $d = d_{max}$	q = 0,25; $d = d_{max}/2$
2	69	113	180	313
3	74	122	186	330
4	88	145	216	387
5	105	173	254	458
6	123	204	297	537
7	142	236	341	620
8	161	268	387	706
9	180	302	434	793
10	200	335	481	881
11	220	369	529	970
12	240	403	577	1059
13	260	437	625	1149
14	279	471	673	1238
15	299	505	721	1328
16	319	539	769	1417
17	339	573	817	1507
18	359	607	865	1597
19	379	641	913	1686
20	399	675	961	1776

Table 1. Eve's information I at attack on the ping - pong protocol with n-qubit GHZ-states at $s = 10^{-6}$ (bit).

Thus, after transfer of hashed block, the lengths of which are presented in tab. 1, the probability of attack non-detection will be equal to 10^{-6}; there is thus a very high probability that this attack will be detected. The main disadvantage of the ping-pong protocol, namely its asymptotic security against eavesdropping attack using ancilla states, is therefore removed.

There are some others attacks on the ping-pong protocol, e.g. attack which can be performed when the protocol is executed in quantum channel with noise (Zhang, 2005a) or Trojan horse attack (Gisin et al., 2002). But there are some counteraction methods to these attacks (Boström & Felbinger, 2008). Thus, we can say that the ping-pong protocol (the security of which is amplified using method described above) is the most prospective QSDC protocol from the viewpoint of the existing development level of the quantum technology of information processing.

2.3 Quantum steganography

Quantum steganography aims to hide the fact of information transferral similar to classical steganography. Most current models of quantum steganography systems use entangled states. For example, modified methods of entangled photon pair detection are used to hide the fact of information transfer in patent (Conti et al., 2004).

A simple quantum steganographic protocol (stegoprotocol) with using four qubit entangled Bell states:

$$\left|\phi^+\right\rangle = \frac{1}{\sqrt{2}}\left(\left|0\right\rangle_1\left|0\right\rangle_2 + \left|1\right\rangle_1\left|1\right\rangle_2\right), \ \left|\phi^-\right\rangle = \frac{1}{\sqrt{2}}\left(\left|0\right\rangle_1\left|0\right\rangle_2 - \left|1\right\rangle_1\left|1\right\rangle_2\right),$$
$$\left|\psi^+\right\rangle = \frac{1}{\sqrt{2}}\left(\left|0\right\rangle_1\left|1\right\rangle_2 + \left|1\right\rangle_1\left|0\right\rangle_2\right), \ \left|\psi^-\right\rangle = \frac{1}{\sqrt{2}}\left(\left|0\right\rangle_1\left|1\right\rangle_2 - \left|1\right\rangle_1\left|0\right\rangle_2\right), \tag{9}$$

was proposed (Terhal et al., 2005). In this protocol n Bell states, including all four states (9) with equal probability is divided between two legitimate users (Alice and Bob) by third part (Trent). For all states the first qubit is sent to Alice and second to Bob. The secret bit is coded in the number of m singlet states $\left|\psi^-\right\rangle$ in the sequence of n states: even m represents "0" and odd represents "1". Alice and Bob perform local measurements each on own qubits and calculate the number of singlet states $\left|\psi^-\right\rangle$. That's why in this protocol Trent can secretly transmit information to Alice and Bob simultaneously.

Shaw & Brun proposed another one quantum stegoprotocol (Shaw & Brun, 2010). In this protocol the information qubit is hidden inside the error-correcting code. Thus, for intruder the qubits transmission via quantum channel looks like a normal quantum information transmission in the noise channel. For information qubit detection the receiver (Bob) must have a shared secret key with sender (Alice), which must be distributed before stegoprotocol starting. In the fig.3 the scheme of protocol proposed by Shaw & Brun is shown. Alice hides information qubit changing its places with qubit in her quantum codeword. She uses her secret key to determine which qubit in codeword must be replaced. Next, Alice uses key again to twirl (rotate) information qubit. This means that Alice uses one of the four single qubit operators (Pauli operators) I, σ_x, σ_y or σ_z for this qubit by determining a concrete operation using two current key bits.

For the intruder who hasn't a key, this qubit likes qubit in maximal mixed state (the rotation can be interpreted as quantum Vernam cipher). In the next stage Alice uses random depolarization mistakes (using the same Pauli operators σ_x, σ_y or σ_z) to some part of others qubits of codeword for simulating some level of noise in quantum channel. Next, she sent a codeword to Bob. For correct untwirl operation Bob use the shared secret key and then he uses a key again to find information qubit.

The security of this protocol depends on the security of previous key distribution procedure. When key distribution has information-theoretic security, and using information qubit twirl (equivalent to quantum Vernam cipher) all scheme can have information-theoretic security. It is known the information-theoretic security is provided by QKD protocols. But if an intruder continuously monitors the channel for a long time and he has a precise channel characteristics, in the final he discovers that Alice transmits information to Bob on quantum stegoprotocol. In addition, using quantum measurements of transmitted qubit states, an intruder can cancel information transmitting (Denial of Service attack).

Thus, in the present three basis methods of quantum steganography are proposed:

1. Hiding in the quantum noise;
2. Hiding using quantum error-correcting codes;
3. Hiding in the data formats, protocols etc.

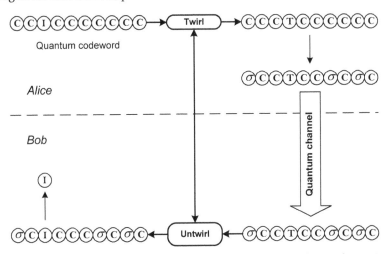

Fig. 3. The scheme of quantum stegoprotocol: C – qubit of codeword, I – information qubit, T – twirled information qubit, σ – qubit, to which Alice applies Pauli operator (qubit that simulate a noise).

The last method is the most promising direction of quantum steganography and also hiding using quantum error-correcting codes has some prospect in the future practice implementation.

It should be noted that theoretical research in quantum steganography has not reached the level of practical application yet, and it is very difficult to talk about the advantages and disadvantages of quantum steganography systems. Whether quantum steganography is superior to the classical one or not in practical use is still an open question (Imai & Hayashi, 2006).

2.4 Others technologies for quantum secure telecommunication systems construction

Quantum secret sharing (QSS). Most QSS protocols use properties of entangled states. The first QSS protocol was proposed by *Hillery, Buzek* and *Berthiaume* in 1998 (Hillery et al., 1998; Qin et al., 2007). This protocol uses GHZ-triplets (quadruplets) similar to some QSDC protocols. The sender shares his message between two (three) parties and only cooperation allows them to read this message. Semi-quantum secret sharing protocol using GHZ-triplets (quadruplets) was proposed by Li et al. (Li et al., 2009). In this protocol, users that receive a shared message have access to the quantum channel. But they are limited by some set of operation and are called "classical", meaning they are not able to prepare entangled states and perform any quantum operations or measurements. These users can measure qubits on a "classical" $\{|0\rangle, |1\rangle\}$ basis, reordering the qubits (via proper delay measurements), preparing (fresh) qubits in the classical basis, and sending or returning the qubits without disturbance. The sending party can perform any quantum operations. This protocol prevails over others QSS protocols in economic terms. Its equipment is cheaper because expensive devices for preparing and measuring (in GHZ-basis) many-qubit entangled states are not required. Semi-quantum secret sharing protocol exists in two variants: randomisation-based and measurement-resend protocols. Zhang et al. has been presented QSS using single qubits that are prepared in two mutually unbiased bases and transferred by blocks (Zhang et al., 2005b). Similar to the Hillery-Buzek-Berthiaume protocol, this allows sharing a message between two (or more) parties. The security improvement of this protocol against malicious acts of legitimate users is proposed (Deng et al., 2005). A similar protocol for multiparty secret sharing also is presented (Yan et al., 2008). QSS protocols are protected against external attackers and unfair actions of the protocol's parties. Both quantum and semi-quantum schemes allow detecting eavesdropping and do not require encryption unlike the classical secret-sharing schemes. The most significant imperfection of QSS protocols is the necessity for large quantum memory that is outside the capabilities of modern technologies today.

Quantum stream cipher (QSC) provides data encryption similar to classical stream cipher, but it uses quantum noise effect (Hirota et al., 2005) and can be used in optical telecommunication networks. QSC is based on the *Yuen-2000 protocol (Y-00, $\alpha\eta$ - scheme)*. Information-theoretic security of the Y-00 protocol is ensured by randomisation (based on quantum noise) and additional computational schemes (Nair & Yuen, 2007; Yuen, 2001). In a number of papers (Corndorf et al., 2005; Hirota & Kurosawa, 2006; Nair & Yuen, 2007) the high encryption rate of the Y-00 protocol is demonstrated experimentally, and a security analysis on the Yuen-2000 protocol against the fast correlation attack, the typical attack on stream ciphers, is presented (Hirota & Kurosawa, 2006). The next advantage is better security compared with usual (classical) stream cipher. This is achieved by quantum noise effect and by the impossibility of cloning quantum states (Wooters & Zurek, 1982). The complexity of practical implementation is the most important imperfection of QSC (Hirota & Kurosawa, 2006).

Quantum digital signature (QDS) can be implemented on the basis of protocols such as QDS protocols using single qubits (Wang et al., 2006) and QDS protocols using entangled states (authentic QDS based on quantum GHZ-correlations) (Wen & Liu, 2005). QDS is based on use of the quantum one-way function (Gottesman & Chuang, 2001). This function has better security than the classical one-way function, and it has information-theoretic security (its security does not depend on the power of the attacker's equipment). Quantum one-way function is defined by the following properties of quantum systems (Gottesman & Chuang, 2001):

1.   Qubits can exist in superposition "0" and "1" unlike classical bits.
2.   We can get only a limited quantity of classical information from quantum states according to the *Holevo theorem* (Holevo, 1977). Calculation and validation are not difficult but inverse calculation is impossible.

In the systems that use QDS, user identification and integrity of information is provided similar to classical digital signature (Gottesman & Chuang, 2001). The main advantages of QDS protocols are information-theoretic security and simplified key distribution system. The main disadvantage is the possibility to generate a limited number of public key copies and the leak of some quantities of information about incoming data of quantum one-way function (unlike the ideal classical one-way function) (Gottesman & Chuang, 2001).

Fig. 4 represents a general scheme of the methods of quantum secure telecommunication systems construction for their purposes and for using some quantum technologies.

2.5 Review of commercial quantum secure telecommunication systems

The world's first commercial quantum cryptography solution was *QPN Security Gateway (QPN-8505)* (QPN Security Gateway, 2011) proposed by *MagiQ Technologies (USA).* This system (fig. 5 a) is a cost-effective information security solution for governmental and financial organisations. It proposes VPN protection using QKD (up to 100 256-bit keys per second, up to 140 km) and integrated encryption. The QPN-8505 system uses BB84, 3DES (NIST, 1999) and AES (NIST, 2001) protocols.

The Swiss company *Id Quantique* (Cerberis, 2011) offers a systems called *Clavis²* (fig. 5 b) and *Cerberis.* Clavis² uses a proprietary auto-compensating optical platform, which features outstanding stability and interference contrast, guaranteeing low quantum bit error rate. Secure key exchange becomes possible up to 100 km. This optical platform is well documented in scientific publications and has been extensively tested and characterized. Cerberis is a server with automatic creation and secret key exchange over a fibre channel (FC-1G, FC-2G and FC-4G). This system can transmit cryptographic keys up to 50 km and carries out 12 parallel cryptographic calculations. The latter substantially improves the system's performance. The Cerberis system uses AES (256-bits) for encryption and BB84 and SARG04 protocols for quantum key distribution. Main features:

•   Future-proof security.

- Scalability: encryptors can be added when network grows.
- Versatility: encryptors for different protocols can be mixed.
- Cost-effectiveness: one quantum key server can distribute keys to several encryptors.

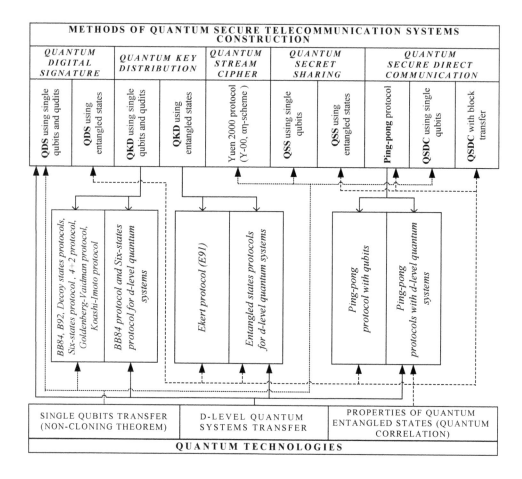

Fig. 4. Methods of quantum secure telecommunication systems construction.

Toshiba Research Europe Ltd (Great Britain) recently presented another QKD system named *Quantum Key Server* (QKS, 2011). This system (fig. 5 c) delivers digital keys for cryptographic applications on fibre optic based computer networks. Based on quantum cryptography it provides a failsafe method of distributing verifiably secret digital keys, with significant cost and key management advantages. The system provides world-leading performance. In particular, it allows key distribution over standard telecom fibre links exceeding 100 km in length and bit rates sufficient to generate 1 Megabit per second of key material over a distance of 50 km — sufficiently long for metropolitan coverage. Toshiba's system uses a simple "one-way" architecture, in which the photons travel from sender to receiver. This design has been rigorously proven as secure from most types of eavesdropping attack. Toshiba has pioneered active stabilisation technology that allows the system to distribute key material continuously, even in the most challenging operating conditions, without any user intervention. This avoids the need for recalibration of the system due to temperature-induced changes in the fibre lengths. Initiation of the system is also managed automatically, allowing simple turn-key operation. It has been shown to work successfully in several network field trials. The system can be used for a wide range of cryptographic applications, e.g., encryption or authentication of sensitive documents, messages or transactions. A programming interface gives the user access to the key material.

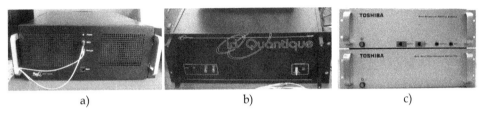

a)   b)   c)

Fig. 5. Some commercial quantum secure telecommunication systems.

Another British company, *QinetiQ*, realised the world's first network using quantum cryptography — *Quantum Net (Qnet)* (Elliot et al., 2003; Hughes et al., 2002). The maximum length of telecommunication lines in this network is 120 km. Moreover, it is a very important fact that Qnet is the first QKD system using more than two servers. This system has six servers integrated to the Internet.

In addition the world's leading scientists are actively taking part in the implementation of projects such as *SECOQC (Secure Communication based on Quantum Cryptography)* (SECOQC White Paper on Quantum Key Distribution and Cryptography, 2007), *EQCSPOT (European Quantum Cryptography and Single Photon Technologies)* (Alekseev & Korneyko, 2007) and *SwissQuantum* (Swissquantum, 2011).

SECOQC is a project that aims to develop quantum cryptography network. The European Union decided in 2004 to invest € 11 million in the project as a way of circumventing espionage attempts by ECHELON (global intelligence gathering system, USA). This project combines people and organizations in Austria, Belgium, the United Kingdom, Canada, the Czech Republic, Denmark, France, Germany, Italy, Russia, Sweden and Switzerland. On October 8, 2008 SECOQC was launched in Vienna.

Following no-cloning theorem, QKD only can provide point-to-point (sometimes called "1:1") connection. So the number of links will increase $N(N-1)/2$ as N represents the number of nodes. If a node wants to participate into the QKD network, it will cause some issues like constructing quantum communication line. To overcome these issues, SECOQC was started. SECOQC network architecture (fig. 6) can by divided by two parts. Trusted private network and quantum network consisted with QBBs (Quantum Back Bone). Private network is conventional network with end-nodes and a QBB. QBB provides quantum channel communication between QBBs. QBB is consisted with a number of QKD devices that are connected with other QKD devices in 1:1 connection. From this, SECOQC can provide easier registration of new end-node in QKD network, and quick recovery from threatening on quantum channel links.

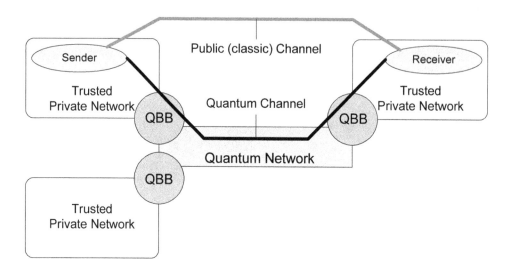

Fig. 6. Brief network architecture of SECOQC.

We also note that during the project SECOQC the seven most important QKD systems have been developed or refined (Kollmitzer & Pivk, 2010). Among these QKD systems are *Clavis²* and *Quantum Key Server* described above and also:

1. *The coherent one-way system (time-coding)* designed by GAP-Universite de Geneve and idQuantique realizes the novel distributed-phase-reference coherent one-way protocol.
2. *The entanglement-based QKD system* developed by an Austrian–Swedish consortium. The system uses the unique quantum mechanical property of entanglement for transferring the correlated measurements into a secret key.
3. *The free-space QKD system* developed by the group of H. Weinfurter from the University of Munich. It employs the BB84 protocol using polarization encoded attenuated laser pulses with photons of 850 nm wavelength. Decoy states are used to ensure key security even with faint pulses. The system is applicable to day and night operation using excessive filtering in order to suppress background light.
4. *The low-cost QKD system* was developed by John Rarity's team of the University of Bristol. The system can be applied for secure banking including consumer protection. The design philosophy is based on a future hand-held electronic credit card using free-space optics. A method is proposed to protect these transactions using the shared secret stored in a personal hand-held transmitter. Thereby Alice's module is integrated within a small device such as a mobile telephone, or personal digital

assistant, and Bob's module consists of a fixed device such as a bank asynchrone transfer mode.

The primary objective of EQCSPOT project is bringing quantum cryptography to the point of industrial application. Two secondary objectives exist to improve single photon technologies for wider applications in metrology, semiconductor characterisation, biosensing etc and to assess the practical use of future technologies for general quantum processors. The primary results will be in the tangible improvements in key distribution. The overall programme will be co-ordinated by British Defence Evaluation and Research Agency and the work will be divided into eight workparts with each workpart co-ordinated by one organisation. Three major workparts are dedicated to the development of the three main systems: NIR fibre, 1.3-1.55 µm fibre and free space key exchange. The other five are dedicated to networks, components and subsystems, software development, spin-off technologies and dissemination of results.

One of the key specificities of the SwissQuantum project is to aim at long-term demonstration of QKD and its applications. Although this is not the first quantum network to be deployed, it wills the first one to operate for months with real traffic. In this sense, the SwissQuantum network presents a major impetus for the QKD technology.

The SwissQuantum network consists of three layers:

- Quantum Layer. This layer performs Quantum Key Exchange.
- *Key Management Layer.* This layer manages the quantum keys in key servers and provides secure key storage, as well as advanced functions (key transfer and routing).
- *Application Layer.* In this layer, various cryptographic services use the keys distributed to provide secure communications.

There are many practical and theoretical research projects concerning the development of quantum technology in research institutes, laboratories and centres such as Institute for Quantum Optics and Quantum Information, Northwestern University, SmartQuantum, BBN Technologies of Cambridge, TREL, NEC, Mitsubishi Electric, ARS Seibersdorf Research and Los Alamos National Laboratory.

3. Conclusion

This chapter presents a classification and systematisation of modern quantum technology of information security. The characteristic of the basic directions of quantum cryptography from the point of view of the quantum technologies used is given. A qualitative analysis of the advantages and imperfections of concrete quantum protocols is made. Today the most developed direction of quantum secure telecommunication systems is QKD protocols. In research institutes, laboratories and centres, quantum cryptographic systems for secret key distribution for distant legitimate users are being developed. Most of the technologies used in these systems are patented in different countries (mainly in the U.S.A.). Such QKD systems can be combined with any classical cryptographic scheme, which provides information-theoretic security, and the entire cryptographic scheme will have information-theoretic security also. QKD protocols can generally provide higher information security level than appropriate classical schemes.

Other secure quantum technologies in practice have not been extended beyond laboratory experiments yet. But there are many theoretical cryptographic schemes that provide high information security level up to the information-theoretic security. QSDC protocols remove the secret key distribution problem because they do not use encryption. One of these is the ping-pong protocol and its improved versions. These protocols can provide high information security level of confidential data transmission using the existing level of technology with security amplification methods. Another category of QSDC is protocols with transfer qubits by blocks that have unconditional security, but these need a large quantum memory which is out of the capabilities of modern technologies today. It must be noticed that QSDC protocols are not suitable for the transfer of a high-speed flow of confidential data because there is low data transfer rate in the quantum channel. But when a high information security level is more important than transfer rate, QSDC protocols should find its application.

Quantum secret sharing protocols allow detecting eavesdropping and do not require data encryption. This is their main advantage over classical secret sharing schemes. Similarly, quantum stream cipher and quantum digital signature provide higher security level than classical schemes. Quantum digital signature has information-theoretic security because it uses quantum one-way function. However, practical implementation of these quantum technologies is also faced to some technological difficulties.

Thus, in recent years quantum technologies are rapidly developing and gradually taking their place among other means of information security. Their advantage is a high level of security and some properties, which classical means of information security do not have. One of these properties is the ability always to detect eavesdropping. Quantum technologies therefore represent an important step towards improving the security of telecommunication systems against cyber-terrorist attacks. But many theoretical and practical problems must be solved for wide practical use of quantum secure telecommunication systems.

4. Acknowledgment

Special thanks should be given to **Rector of National Aviation University (Kyiv, Ukraine) – Mykola Kulyk.** We would not have finished this chapter without his support.

5. References

Alekseev, D.A. & Korneyko, A.V. (2007). Practice reality of quantum cryptography key distribution systems, *Information Security*, No. 1, pp. 72–76.

Bennett, C. & Brassard, G. (1984). Quantum cryptography: public key distribution and coin tossing, *Proceedings of the IEEE International Conference on Computers, Systems and Signal Processing*. Bangalore, India, pp. 175–179.

Bennett, C. (1992). Quantum cryptography using any two non-orthogonal states, *Physical Review Letters*, Vol.68, No.21, pp. 3121–3124.

Bennett, C.; Bessette, F. & Brassard, G. (1992). Experimental Quantum Cryptography, *Journal of Cryptography*, Vol.5, No.1, pp. 3–28.

Bennett, C.; Brassard, G.; Crépeau, C. & Maurer, U. (1995). Generalized privacy amplification, *IEEE Transactions on Information Theory*, Vol.41, No.6, pp. 1915–1923.

Boström, K. & Felbinger, T. (2002). Deterministic secure direct communication using entanglement, *Physical Review Letters*, Vol.89, No.18, 187902.

Boström, K. & Felbinger, T. (2008). On the security of the ping-pong protocol, *Physics Letters A*, Vol.372, No.22, pp. 3953–3956.

Bourennane, M.; Karlsson, A. & Bjork, G. (2002). Quantum key distribution using multilevel encoding, *Quantum Communication, Computing, and Measurement 3*. N.Y.: Springer US, pp. 295–298.

Bouwmeester, D.; Ekert, A. & Zeilinger, A. (2000). *The Physics of Quantum Information. Quantum Cryptography, Quantum Teleportation, Quantum Computation*. Berlin: Springer-Verlag, 314 p.

Bradler K. (2005). Continuous variable private quantum channel, *Physical Review A*, Vol.72, No.4, 042313.

Branciard, C.; Gisin, N.; Kraus, B. & Scarani, V. (2005). Security of two quantum cryptography protocols using the same four qubit states, *Physical Review A*, Vol.72, No.3, 032301.

Brassard, G.; Lutkenhaus, N.; Mor, T. & Sanders, B. (2000). Limitations on practical quantum cryptography, *Physical Review Letters*, Vol.85, No.6, pp. 1330–1333.

Bruss, D. (1998). Optimal Eavesdropping in Quantum Cryptography with Six States, *Physical Review Letters*, Vol.81, No.14, pp. 3018–3021.

Bruss, D. & Macchiavello C. (2002). Optimal eavesdropping in cryptography with three-dimensional quantum states, *Physical Review Letters*, Vol.88, No.12, 127901.

Cai, Q.-Y. & Li, B.-W. (2004a). Deterministic Secure Communication Without Using Entanglement, *Chinese Physics Letters*, Vol.21 (4), pp. 601–603.

Cai, Q.-Y. & Li B.-W. (2004b). Improving the capacity of the Bostrom–Felbinger protocol, *Physical Review A*, Vol.69, No.5, 054301.

Cerberis. 01.10.2011, Available from: http://idquantique.com/products/cerberis.htm.

Cerf, N.J.; Bourennane, M.; Karlsson, A. & Gisin, N. (2002). Security of quantum key distribution using d-level systems, *Physical Review Letters*, Vol.88, No.12, 127902.

Chamoli, A. & Bhandari, C.M. (2009). Secure direct communication based on ping-pong protocol, *Quantum Information Processing*, Vol.8, No.4, pp. 347–356.

Chuan, W.; Fu Guo, D. & Gui Lu, L. (2005). Multi-step quantum secure direct communication using multi-particle Greenberg-Horne-Zeilinger state, *Optics Communications*, Vol.253, pp. 15–19.

Conti A.; Ralph, S.; Kenneth A. et al. *Patent* No 7539308 USA, H04K 1/00 (20060101). Quantum steganography, publ. 21.05.2004.

Corndorf, E., Liang, C. & Kanter, G.S. (2005). Quantum-noise randomized data encryption for wavelength-division-multiplexed fiber-optic networks, *Physical Review A*, Vol.71, No.6, 062326.

Deng, F.G.; Long, G.L. & Liu, X.S. (2003). Two-step quantum direct communication protocol using the Einstein–Podolsky–Rosen pair block. *Physical Review A*, 2003. Vol.68, No.4, 042317.

Deng, F. G.; Li, X. H.; Zhou, H. Y. & Zhang, Z. J. (2005). Improving the security of multiparty quantum secret sharing against Trojan horse attack, *Physical Review A*, Vol.72, No.4, 044302.

Desurvire, E. (2009). *Classical and Quantum Information Theory*. Cambridge: Cambridge University Press, 691 p.

Durt, T.; Kaszlikowski, D.; Chen, J.-L. & Kwek, L.C. (2004). Security of quantum key distributions with entangled qudits, *Physical Review A*, Vol.69, No.3, 032313.

Ekert, A. (1991). Quantum cryptography based on Bell's theorem, *Physical Review Letters*, Vol.67, No.6, pp. 661–663.

Elliot, C.; Pearson, D. & Troxel, G. (2003). Quantum Cryptography in Practice, *arXiv:quant-ph/0307049*.

Fuchs, C.; Gisin, N.; Griffits, R. et al. (1997). Optimal Eavesdropping in Quantum Cryptography. Information Bound and Optimal Strategy, *Physical Review A*, Vol.56, No.2, pp. 1163–1172.

Gao, T.; Yan, F.L. & Wang, Z.X. (2005). Deterministic secure direct communication using GHZ-states and swapping quantum entanglement. *Journal of Physics A: Mathematical and Theoretical*, Vol. 38, No.25, pp. 5761–5770.

Gao, F.; Guo, F.Zh.; Wen, Q.Y. & Zhu, F.Ch. (2008). Comparing the efficiencies of different detect strategies in the ping-pong protocol, *Science in China, Series G: Physics, Mechanics & Astronomy*, Vol.51, No.12. pp. 1853–1860.

Gisin, N.; Ribordy, G.; Tittel, W. & Zbinden, H. (2002). Quantum cryptography, *Review of Modern Physics*, Vol.74, pp. 145–195.

Gnatyuk, S.O.; Kinzeryavyy, V.M.; Korchenko, O.G. & Patsira, Ye.V. (2009). *Patent No 43779 UA, MPK H04L 9/08. System for cryptographic key transfer*, 25.08.2009.

Goldenberg, L. & Vaidman, L. (1995). Quantum Cryptography Based On Orthogonal States, *Physical Review Letters*, Vol.75, No.7, pp. 1239–1243.

Gottesman, D. & Chuang, I. (2001). Quantum digital signatures, *arXiv:quant-ph/0105032v2*.

Hayashi, M. (2006). *Quantum information. An introduction*. Berlin, Heidelberg, New York: Springer, 430 p.

Hillery, M.; Buzek, V. & Berthiaume, A. (1999). Quantum secret sharing, *Physical Review A*, Vol.59, No.3, pp. 1829–1834.

Hirota, O. & Kurosawa, K. (2006). An immunity against correlation attack on quantum stream cipher by Yuen 2000 protocol, *arXiv:quant-ph/0604036v1*.

Hirota, O.; Sohma, M.; Fuse, M. & Kato, K. (2005). Quantum stream cipher by the Yuen 2000 protocol: Design and experiment by an intensity-modulation scheme, *Physical Review A*, Vol.72, No.2, 022335.

Holevo, A.S. (1977). Problems in the mathematical theory of quantum communication channels, *Report of Mathematical Physics*, Vol.12, No.2, pp. 273–278.

Hughes, R.; Nordholt, J.; Derkacs, D. & Peterson, C. (2002). Practical free-space quantum key distribution over 10 km in daylight and at night, *New Journal of Physics*, Vol.4, 43 p.

Huttner, B.; Imoto, N.; Gisin, N. & Mor, T. (1995). Quantum Cryptography with Coherent States, *Physical Review A*, Vol.51, No.3, pp. 1863–1869.

Imai, H. & Hayashi, M. (2006). *Quantum Computation and Information. From Theory to Experiment*. Berlin: Springer-Verlag, Heidelberg, 235 p.

Imre, S. & Balazs, F. (2005). *Quantum Computing and Communications: An Engineering Approach*, John Wiley & Sons Ltd, 304 p.

Inamori, H.; Rallan, L. & Vedral, V. (2001). Security of EPR-based quantum cryptography against incoherent symmetric attacks, *Journal of Physics A*, Vol.34, No.35, pp. 6913–6918.

Kaszlikowski, D.; Christandl, M. et al. (2003). Quantum cryptography based on qutrit Bell inequalities, *Physical Review A*, Vol.67, No.1, 012310.

Koashi, M. & Imoto, N. (1997). Quantum Cryptography Based on Split Transmission of One-Bit Information in Two Steps, *Physical Review Letters*, Vol.79, No.12, pp. 2383–2386.

Kollmitzer, C. & Pivk, M. (2010). Applied Quantum Cryptography, *Lecture Notes in Physics 797*. Berlin, Heidelberg: Springer, 214 p.

Korchenko, O.G.; Vasiliu, Ye.V. & Gnatyuk, S.O. (2010a). Modern quantum technologies of information security against cyber-terrorist attacks, *Aviation*. Vilnius: Technika, Vol.14, No.2, pp. 58–69.

Korchenko, O.G.; Vasiliu, Ye.V. & Gnatyuk, S.O. (2010b). Modern directions of quantum cryptography, *"AVIATION IN THE XXI-st CENTURY"* – *"Safety in Aviation and Space Technologies": IV World Congress: Proceedings* (September 21–23, 2010), Kyiv, NAU, pp. 17.1–17.4.

Korchenko, O.G.; Vasiliu, Ye.V.; Nikolaenko, S.V. & Gnatyuk, S.O. (2010c). Security amplification of the ping-pong protocol with many-qubit Greenberger-Horne-Zeilinger states, *XIII International Conference on Quantum Optics and Quantum Information (ICQOQI'2010):* Book of abstracts (May 28 – June 1, 2010), pp. 58–59.

Li, Q.; Chan, W. H. & Long, D-Y. (2009). Semi-quantum secret sharing using entangled states, *arXiv:quant-ph/0906.1866v3*.

Li, X.H.; Deng, F.G. & Zhou, H.Y. (2006). Improving the security of secure direct communication based on the secret transmitting order of particles. *Physical Review A*, Vol.74, No.5, 054302.

Lin, S.; Wen, Q.Y.; Gao, F. & Zhu F.C. (2008). Quantum secure direct communication with chi-type entangled states, *Physical Review A*, Vol.78, No.6, 064304.

Liu, Y.; Chen, T.-Y.; Wang, J. et al. (2010). Decoy-state quantum key distribution with polarized photons over 200 km, *Optics Express*, Vol. 18, Issue 8, pp. 8587-8594.

Lomonaco, S.J. (1998). A Quick Glance at Quantum Cryptography, *arXiv:quant-ph/9811056*.

Lütkenhaus, N. & Jahma, M. (2002). Quantum key distribution with realistic states: photon-number statistics in the photon-number splitting attack, *New Journal of Physics*, Vol.4, pp. 44.1–44.9.

Lütkenhaus, N. & Shields, A. (2009). Focus on Quantum Cryptography: Theory and Practice, *New Journal of Physics*, Vol.11, No.4, 045005.

Nair, R. & Yuen, H. (2007). On the Security of the Y-00 (AlphaEta) Direct Encryption Protocol, *arXiv:quant-ph/0702093v2*.

Navascués, M. & Acín, A. (2005). Security Bounds for Continuous Variables Quantum Key Distribution, *Physical Review Letters*, Vol.94, No.2, 020505.

Nielsen, M.A. & Chuang, I.L. (2000). *Quantum Computation and Quantum Information.* Cambridge: Cambridge University Press, 676 p.

NIST. "FIPS-197: Advanced Encryption Standard." (2001). 01.10.2011, Available from: <http://csrc.nist.gov/publications/fips>.

NIST. "FIPS-46-3: Data Encryption Standard." (1999). 01.10.2011, Available from: <http://csrc.nist.gov/publications/fips>.

Ostermeyer, M. & Walenta N. (2008). On the implementation of a deterministic secure coding protocol using polarization entangled photons, *Optics Communications*, Vol. 281, No.17, pp. 4540–4544.

Overbey, J; Traves, W. & Wojdylo J. (2005). On the keyspace of the Hill cipher, *Cryptologia*, Vol.29, No.1, pp. 59–72.

Peng, C.-Z.; Zhang, J.; Yang, D. et al. (2007). Experimental long-distance decoy-state quantum key distribution based on polarization encoding, *Physical Review Letters*, Vol.98, No.1, 010505.

Pirandola, S.; Mancini, S.; Lloyd, S. & Braunstein S. (2008). Continuous-variable quantum cryptography using two-way quantum communication, *Nature Physics*, Vol.4, No.9, pp. 726–730.

Qin, S.-J.; Gao, F. & Zhu, F.-Ch. (2007). Cryptanalysis of the Hillery-Buzek-Berthiaume quantum secret-sharing protocol, *Physical Review A*, Vol.76, No.6, 062324.

QKS. Toshiba Research Europe Ltd. 01.10.2011, Available from: <http://www.toshiba-europe.com/research/crl/QIG/quantumkeyserver.html>.

QPN Security Gateway (QPN–8505). 01.10.2011, Available from: <http://www.magiqtech.com/MagiQ/Products.html>.

Rosenberg, D. et al. (2007). Long-distance decoy-state quantum key distribution in optical fiber, *Physical. Review Letters*, Vol.98, No.1, 010503.

Sangouard, N.; Simon, C.; de Riedmatten, H. & Gisin, N. (2011). Quantum repeaters based on atomic ensembles and linear optics, *Review of Modern Physics*, Vol.83, pp. 33–34.

Scarani, V.; Acin, A.; Ribordy, G. & Gisin, N. (2004). Quantum cryptography protocols robust against photon number splitting attacks for weak laser pulse implementations, *Physical Review Letters*, Vol.92, No.5, 057901.

Scarani, V.; Bechmann-Pasquinucci, H.; Nicolas J. Cerf et al. (2009). The security of practical quantum key distribution, *Review of Modern Physics*, Vol.81, pp. 1301–1350.

SECOQC White Paper on Quantum Key Distribution and Cryptography. (2007). *arXiv:quant-ph/0701168v1.*

Shaw, B. & Brun, T. (2010). Quantum steganography, *arXiv:quant-ph/1006.1934v1.*

Schumacher, B. & Westmoreland, M. (2010). *Quantum Processes, Systems, and Information.* Cambridge: Cambridge University Press, 469 p.

Terhal, B.M.; DiVincenzo, D.P. & Leung, D.W. (2001). Hiding bits in Bell states, *Physical review letters*, Vol.86, issue 25, pp. 5807–5810.

Vasiliu, E.V. (2011). Non-coherent attack on the ping-pong protocol with completely entangled pairs of qutrits, *Quantum Information Processing*, Vol.10, No.2, pp. 189–202.

Vasiliu, E.V. & Nikolaenko, S.V. (2009). Synthesis if the secure system of direct message transfer based on the ping–pong protocol of quantum communication, *Scientific works of the Odessa national academy of telecommunications named after O.S. Popov*, No.1, pp. 83–91.

Vasiliu, E.V. & Mamedov, R.S. (2008). Comparative analysis of efficiency and resistance against not coherent attacks of quantum key distribution protocols with transfer of multidimensional quantum systems, *Scientific works of the Odessa national academy of telecommunications named after O.S. Popov*, No.2, pp. 20–27.

Vasiliu, E.V. & Vorobiyenko, P.P. (2006). The development problems and using prospects of quantum cryptographic systems, *Scientific works of the Odessa national academy of telecommunications named after O.S. Popov*, No.1, pp. 3–17.

Vedral, V. (2006). *Introduction to Quantum Information Science*. Oxford University Press Inc., New York, 183 p.

Wang, Ch.; Deng, F.G. & Long G.L. (2005a). Multi – step quantum secure direct communication using multi – particle Greenberger – Horne – Zeilinger state, *Optics Communications*, Vol. 253, No.1, pp. 15–20.

Wang, Ch. et al. (2005b). Quantum secure direct communication with high dimension quantum superdense coding, *Physical Review A*, Vol.71, No.4, 044305.

Wang, J.; Zhang, Q. & Tang, C. (2006). Quantum signature scheme with single photons, *Optoelectronics Letters*, Vol.2, No.3, pp. 209–212.

Wen, X.-J. & Liu, Y. (2005). Quantum Signature Protocol without the Trusted Third Party, *arXiv:quant-ph/0509129v2*.

Williams, C.P. (2011). *Explorations in quantum computing, 2nd edition*. Springer-Verlag London Limited, 717 p.

Wooters, W.K. & Zurek, W.H. (1982). A single quantum cannot be cloned, *Nature*, Vol. 299, p. 802.

Xiu, X.-M.; Dong, L.; Gao, Y.-J. & Chi F. (2009). Quantum Secure Direct Communication with Four-Particle Genuine Entangled State and Dense Coding, *Communication in Theoretical Physics*, Vol.52, No.1, pp. 60–62.

Yan, F.-L.; Gao, T. & Li, Yu.-Ch. (2008). Quantum secret sharing protocol between multiparty and multiparty with single photons and unitary transformations, *Chinese Physics Letters*, Vol.25, No.4, pp. 1187–1190.

Yin, Z.-Q.; Zhao, Y.-B.; Zhou Z.-W. et al. (2008). Decoy states for quantum key distribution based on decoherence-free subspaces, *Physical Review A*, Vol.77, No.6, 062326.

Yuen, H.P. (2001). *In Proceedings of QCMC'00*, Capri, edited by P. Tombesi and O. Hirota New York: Plenum Press, p. 163.

Zhang, Zh.-J.; Li, Y. & Man, Zh.-X. (2005a). Improved Wojcik's eavesdropping attack on ping-pong protocol without eavesdropping-induced channel loss, *Physics Letters A*, Vol.341, No.5–6, pp. 385–389.

Zhang, Zh.-J.; Li, Y. & Man, Zh.-X. (2005b). Multiparty quantum secret sharing, *Physical Review A*, Vol.71, No.4, 044301.

Zhao, Y.; Qi, B.; Ma, X.; Lo, H.-K. & Qian, L. (2006a). Simulation and implementation of decoy state quantum key distribution over 60 km telecom fiber, *Proceedings of IEEE International Symposium on Information Theory*, pp. 2094–2098.

Zhao, Y.; Qi, B.; Ma, X.; Lo, H.-K. & Qian, L. (2006b). Experimental Quantum Key Distribution with Decoy States, *Physical Review Letters*, Vol.96, No.7, 070502.

2

Telecommunications Service Domain Ontology: Semantic Interoperation Foundation of Intelligent Integrated Services

Xiuquan Qiao, Xiaofeng Li and Junliang Chen
State Key Laboratory of Networking and Switching Technology
Beijing University of Posts and Telecommunications
China

1. Introduction

Network is the bearer of services and services are the soul of network. The convergent network extends the original communications service type and gradually forms new convergent services which integrate the traditional telecommunication services and a large number of value-added services or contents on Internet (Kolberg et al., 2010). The integrated service is essentially to handle the data and services across heterogeneous networks and service platforms. Facing the heterogeneity and diversity of service resources, integrated services need to run in a multi-terminal, multi-access network and multi-platform heterogeneous environment. These tremendous changes of service environment present a significant interoperability challenge for traditional service provisioning theory. Nowadays, the provision of context-awareness, adaptive personalized services is the development goal of future ubiquitous network (Park et al., 2009). It can enable seamless information exchange between humans, with humans and with entities (e.g., mobile devices), as well as entities and entities at any time, any place and in any way. To meet the development needs of adaptive personalized convergent services, dynamic service discovery and composition technologies are explored widely in the telecommunication service field (Bashah et al., 2010; Niazi & Mahmoud, 2009).

Today, semantic web service (McIlraith, 2001), as an establishing research paradigm, is defined as an augmentation of web service with semantic annotation, to facilitate the higher automation of service discovery, composition, invocation and monitoring in an open environment. Integration of the semantic web technology and telecommunications systems is explored widely in the telecommunication service field (Do & Jorstad, 2005; Vitvar & Viskova, 2005; Qiao et al., 2008a; Gutheim, 2011; Khan et al., 2011; Zander & Schandl, 2011). It is well known that ontology is the semantic interoperability and knowledge sharing foundation for semantic web services matching and context reasoning. Therefore, how to construct the telecommunications service domain ontology is an important factor of successfully applying semantic web services into telecommunication service systems (Veijalainen, 2007, 2008). However, telecommunication service field consists of a large number of concepts/terminologies and relations. How to abstract the sharing domain concepts and reasonably organize them is a big challenge. Some related work has been done mainly in applying ontology technology to the mobile service domain. Based on the need for a standardized ontology that describes semantic models of the domains relevant for scalable NGN (Next Generation Network) service delivery platforms, the (Villalonga et al., 2009; Su et al., 2009) provide an overview of Mobile Ontology which comprises a core ontology and several subontologies, and its application examples in the service delivery platform. This work, as a part of IST SPICE project (IST SPICE project, 2008), is a meaningful attempt to establish a standardized ontology for mobile service delivery in NGN. In addition, IST SIMS project explored the semantic interfaces as a new means to specify and design service components and to guarantee compatibility in static and dynamic component compositions. And they also defined a domain-specific ontology, and its main purpose of the ontology is to establish a common description of the SIMS-related concepts and their semantics (Rój, 2008). The (Zhu et al., 2010) introduces a mobile ontology construction and retrieval system architecture. However, there lacks a general domain ontology modelling methodology for telecommunications service and the corresponding engineering approach to support the development work for domain ontology. The (Li et al., 2010) briefly introduced the constructing method of telecommunications service domain ontology (TSDO) proposed by our research team. However, the approach is not perfect at that time and still needs to be further improved. In fact, telecommunication service domain ontology, as the important semantic interoperability foundation of telecom network, still has no significant progress up to now. This has become the biggest obstacle to hamper the applications of semantic web technology in telecom field.

In this chapter, we clearly presented a practical domain ontology modelling approach for telecommunications service field. Under the guidance of this approach, our research team has created an open telecommunications service domain ontology knowledge repository which consists of around 430 telecommunications services-related ontology concepts/terminologies and 245 properties. Based on this domain ontology, we described the telecom network capability services in the semantic level to validate its feasibility. The semantic annotation facilitated the accurate service description, discovery of telecommunication network services and addressed the semantic interoperability problem. The proposed model-driven domain ontology modelling approach separates domain conceptual model from the concrete ontology modelling languages, it enhances the reusability of domain conceptual model and greatly reduces the technical difficulty of domain ontology modelling.

The remainder of this chapter is structured as follows. In Section 2 we presented a general domain ontology modelling methodology for telecommunications service field, and also proposed a specific model-driven domain ontology modelling approach to support the above presented methodology. Section 3 introduced the experimental environment and the demo service to validate the feasibility of domain ontology. Finally, conclusions are drawn.

2. Domain ontology modelling methodology for telecommunications service

Here, technical modelling details for the proposed approach are described, namely telecommunications service domain ontology modeling methodology and a corresponding model-driven implementation mechanism.

2.1 Domain ontology modeling methodology

Based on our practical experiences in recent years, a concrete domain ontology modeling methodology is summarized as shown in Figure 1. The modelling process is illustrated in detail as follows.

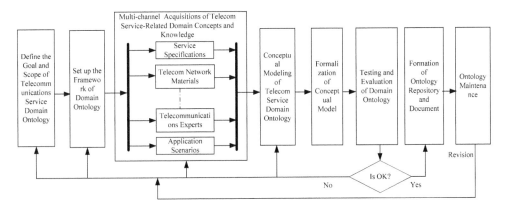

Fig. 1. Domain ontology modelling methodology.

2.1.1 Define the scope of telecommunications service domain ontology

The first step is mainly to define the scope and border of domain ontology. Telecommunications service domain ontology mainly addresses the semantic interoperability of telecommunications service. This domain ontology mainly provides the shared domain vocabularies and knowledge to support the semantic web applications in the telecommunication service field, such as semantic telecom service description, service discovery, and service context modelling. Therefore, TSDO should involve the service-related domain concepts and knowledge. For example, telecom services often involve network type, network carrier, billing policy, user terminal, service quality, service customer, service category, .etc. In fact, telecommunication service field consists of a large number of concepts/terminologies and relations. Some concepts have the higher sharing degree. However, some concepts are only related to concrete application field, such as service context ontology, service description ontology. Therefore, how to abstract the sharing domain concepts and reasonably organize them is a big challenge. The reusability and extensibility are two important ontology modeling factors considered. So an efficient ontology hierarchy modelling approach is needed.

In practice, we adopted a layered ontology modeling method to organize the domain concepts to improve the reusability and extensibility (see Figure 2). Common ontology, like time and space ontologies, can be shared in the different domains, like telecom, medical domain or any other domains. The concrete domain ontology can be shared by the different domain-related application ontologies. For example, TSDO may be used to create the service context ontology, network management ontology, etc. This method well distinguishes the border of TSDO, common ontology and telecom service-related application ontology.

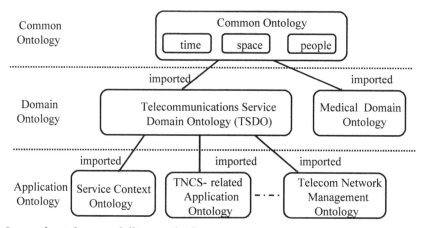

Fig. 2. Layered ontology modelling method.

2.1.2 Set the framework of telecommunications service domain ontology

When the goal and scope of TSDO are clear, the specific organization framework of TSDO should be set up. As TSDO involves a large number of telecom service domain concepts and relationships, how to reasonably classify and organize these terminologies is an important issue. Specifically, we adopted a modular modelling approach to construct TSDO. The principle of modular modelling is the "strong cohesion and loose coupling" way. The correlations among different concepts are the main reference of module division. The goal of modular modelling is to ensure that the correlation of concepts in the same module is stronger. Based on this modular design principle, TSDO is divided into several sub-ontologies as shown in Fig. 3.

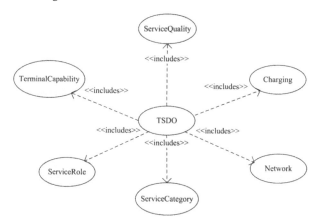

Fig. 3. The framework of telecommunications service domain ontology.

Specifically, TSDO mainly comprises six sub-ontologies, including Terminal Capability Ontology, Network Ontology, Service Role Ontology, Charging Ontology, Service Quality Ontology, and Service Category Ontology.

1. **Terminal Capability Ontology**: defines main concepts about terminal software, terminal hardware, terminal browser and network characteristics supported by terminal.
2. **Network Ontology**: specifies the network concepts, network category, network features, as well as the relationships of various networks, such as mobile network, internet, and fixed network, GSM, CDMA, UMTS, WCDMA, and WLAN.
3. **Service Role Ontology**: describes the stakeholders' concepts of the service supply chain, for example, service provider, content provider, network operator, service user.
4. **Service Category Ontology**: describes a telecommunications service classification. This ontology defines the relationship between various telecommunications services, like basic service, value-added service, voice service, data service, conference service, presence service, download service, browsing service, messaging service.
5. **Charging Ontology**: defines the charging-related concepts and rules about telecommunications services, including payment methods (such as prepaid and post-paid), charging types (such as time-based, volume-based, event-based, and content-based), billing rates, as well as account balances.
6. **Service Quality Ontology**: A telecommunication network must provide the services which have the end-to-end QoS guarantee. Depending on the technical characteristics, the QoS provided by different networks is varying. Service Quality Ontology mainly defines the QoS-related concepts about telecommunications service, including access network QoS, core network QoS and user's QoE, such as call delay, message size, call through rate, positioning accuracy, network bandwidth.

2.1.3 Multi-channel acquisitions of telecom service-related domain concepts and knowledge

After the framework of TSDO is set up, it needs to collect domain concepts and knowledge (including terminologies and their relationships) from multi-channel ways for each sub-ontology of TSDO. In general, the sources of knowledge acquisition include the released telecom service specifications, senior experts in the telecom field or some typical application

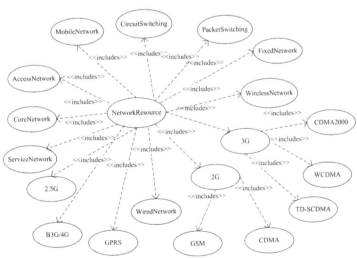

Fig. 4. Some collected domain concepts about telecom network.

scenarios. In this step, modellers need to list the collected concepts, relations and explanations as far as possible. It's unnecessary to care about the meaning overlap between the concepts and to consider how to express these concepts and their relation in class, property or instance ways. For example, Figure 4 briefly shows the concepts collection about network ontology.

2.1.4 Conceptual modelling of telecommunications service domain ontology

After the acquisition of a large number of telecom service related concepts, we need to make the concept classification, concept aggregation, and remove the duplicated concepts according to certain domain knowledge and logic. The goal of this step is to construct a conceptual model of TSDO. This concept model describes the involved domain concepts and their relations of each sub-ontology in detail. Note that, the relationships between the concepts not only involve the concepts of the same sub-ontology, may also be related to the concepts of different sub-ontologies. The concrete building of conceptual model is divided into three steps: (1) **Defining classes and class hierarchy**. In the process of defining the classes, we need to discover the inheritance hierarchy between the concepts and then distinguish the super-classes and sub-classes. (2) **Defining the properties of classes**. After the class is defined, its properties should be considered. There are two kinds of properties. One is datatype property, which is used to describe the features of the concept itself, such as name, age. The other is object property, which is used to depict the relationship between the concepts, like friendship relation between two people. (3) **The definition of domain axiom and knowledge**. When we use ontology to describe the real word things, there are often some contradictions or errors occurrences resulted by human negligence. For example, the range value of one person age property is negative, or a person has two biological fathers. To prevent these common-sense errors, some domain axiom and knowledge should be established. The axiom is to restrict the relationships of the concepts to ensure the consistency of domain knowledge, such as the range value or cardinality of properties.

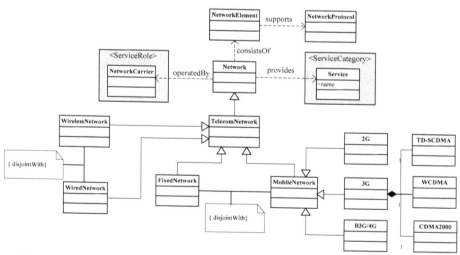

Fig. 5. Part conceptual model of network ontology.

Figure 5 shows the conceptual model of network ontology in part. Based on the terminologies collected in the above step, the class hierarchy and relationships are described. This conceptual model depicts the classification of network, the services provided by network and the operator of network. It can be seen that the ranges of object property "operatedBy" and "provides" are the concepts from ServiceRole and ServiceCategory sub-ontologies respectively. In addition, we define the domain axioms through the constraints way. For example, we define that "FixedNetwork" is disjointed with "MobileNetwork", i.e. if N1 is an instance of concept "FixedNetwork", then it will not be an instance of concept "MobileNetwork".

2.1.5 Formalization of conceptual model of telecommunications service domain ontology

As the conceptual model is one high-level abstract model and independent of any concrete ontology modelling languages, we need to formalize this conceptual model through a specific ontology modelling language like OWL (Web Ontology Language) (W3C, 2004a). In general, we can use the common ontology modelling tools to formally describe the terminologies, relationships and axioms in the conceptual model. Figure 6 shows the formalization description of part concepts and relationships of Figure 5 by OWL language. The concept is formally defined by "owl:Class", and the class hierarchy is organized by "owl:subClassOf". The "owl:ObjectProperty" is used to describe the relationships between the concepts and the "owl:disjointWith" clearly depicts the restrictions on the two disjointed concepts.

```
<owl:Class rdf:ID="Network"/>
<owl:Class rdf:about="#TelecomNetwork">
   <rdfs:subClassOf rdf:resource="#Network"/>
</owl:Class>
<owl:Class rdf:about="#MobileNetwork">
  <rdfs:subClassOf>
   <owl:Class rdf:about="#TelecomNetwork"/>
  </rdfs:subClassOf>
  <owl:disjointWith rdf:resource="#FixedNetwork"/>
</owl:Class>
<owl:Class rdf:ID="2G">
   <rdfs:subClassOf rdf:resource="#MobileNetwork"/>
  </owl:Class>
<owl:ObjectProperty rdf:ID="operatedBy">
   <rdfs:range rdf:resource="&ServiceRole;#NetworkCarrier"/>
   <rdfs:domain rdf:resource="#Network"/>
</owl:ObjectProperty>
```

Fig. 6. Part of network ontology formalized by OWL.

2.1.6 Evaluation of telecommunications service domain ontology

Ontology evaluation is an important issue that must be addressed if TSDO are to be widely adopted in the semantic related telecommunications applications. Ontology can be evaluated against many criteria: its coverage of a particular domain and the richness, complexity and granularity of that coverage; the specific use cases, scenarios, requirements, applications, and data sources it was developed to address; and formal properties such as the consistency and completeness of the ontology. We can test and validate whether the domain ontology satisfy the requirement or not. If yes, these ontologies will be added to the ontology repository; if no, we have to return back to previous steps to make some revisions until the requirement is satisfied.

In the specific use process, we often can find some existing shortcomings of domain ontology. The utilization of domain ontology to formally describe the concrete application scenario is a very effective evaluation approach. For example, when we defined the TSDO, we use network, service role and service category sub-ontologies to describe the network carrier resource (see Figure 7). We found that the operating scope of network carrier is an important characteristic. But the concept "NetworkOperator" of service role sub-ontology lacks this property. Actually, some carriers can provide services through out nation; however, some carriers can only provide services in a specific province or region. Therefore, the property "CoverageScope" should be added to the concept "NetworkOperator" of service role sub-ontology.

```
<ServiceRole:NetworkOperator rdf:ID="ChinaMobileCommunicationOperator"/>
    <ServiceRole:CoverageScope rdf:resource="&LocationSpace;#TroughOutNation"/>
</ServiceRole:NetworkOperator>

<TelecomNetwork:GSM rdf:ID="ChinaMobileNetwork">
    <TelecomNetwork:operatedBy rdf:resource="#ChinaMobileCommunicationOperator"/>
    <TelecomNetwork:provides rdf:resource="&ServiceCategory;#DataService"/>
    <TelecomNetwork:provides rdf:resource="&ServiceCategory;#VoiceService"/>
</TelecomNetwork:GSM>
```

Fig. 7. Ontology description of china mobile communication operator.

2.1.7 Maintenance of telecommunications service domain ontology

The construction of domain ontology is the basis of ontology applications. However, as the different domain experts or ontology modelers may have the different understandings of the same domain concepts or relationships, some created ontologies may need to be further revised or improved in the practical utilization process. In addition, the knowledge of real world is growing and updated continuously. This also results that regular maintenance is necessary after ontoloies have been constructed. Ontology maintenance refers to a series of amendments, corrections, improvements and adaptive maintenance for ontology, which mainly consists of improving maintenance and adaptive maintenance. The improving maintenance is to revise or correct some existing errors of domain ontology. However, the adaptive maintenance refers to the extensions of existing domain ontology with the external real world changes, such as the knowledge increase or technology advances.

In addition, with the maturity of ontology technology, there are some ontologies developed by different research teams or communities to satisfy their different application needs. The main advantage of ontology is the knowledge sharing and reuse. How to realize the interoperation with these existent distributed ontologies is a big problem of ontology maintenance. Therefore, sometimes, it needs to integrate several existent ontologies to address the reuse of different ontology knowledge. To implement the different ontology integration, the relationships among different ontologies should be analyzed. As the distributed feature and openness of WWW, knowledge ontologies maybe have the direct or indirect semantic relationships. For example, two ontologies maybe involve some same or similar concepts. The main relationships consist of two kinds: one is the repeat of terminologies definition. Some terminologies of this ontology might be equivalent to those defined in that ontology. It consists of the class equivalent and the property equivalent. For this equivalent relationship, we can use equivalent ontology mapping method to resolve as shown in Figure 8. The other is the subsumption of terminologies definition. It means that some terminologies of one ontology might subsume the semantic scope of those terminologies defined in other ontology. It also involves the class subsumption and property subsumption. For example, Figure 9 shows two independent ontologies: ontology 1 and ontology 2. In fact, the concept "Netowrk" of ontology 1 subsumes the concept "Internet" of ontology 2 in the semantic scope. Therefore, we can use the subsumption relationship to integrate these two ontologies into a new ontology.

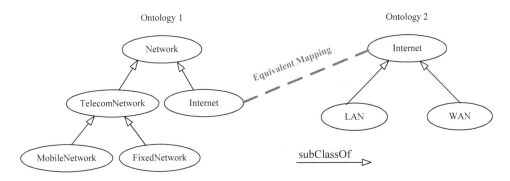

Fig. 8. Ontology integration based on the equivalent mapping.

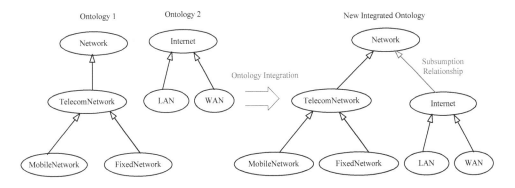

Fig. 9. Ontology integration based on the subsumption relationship.

2.2 A model-driven domain ontology modeling implementation approach

From the above descriptions in section 2.1, it can be seen that the construction of TSDO is a complex work, which involves not only several steps like terminology acquisition, concept modelling and formal description, but also different modellers like domain experts, formalization modeller. Currently, it lacks of a unified modelling tool to efficiently support this methodology. As the ontology modelling languages consists of a large number of logical symbols and formal description knowledge, it is not easy for general domain experts or software developers to understand and master. Although there are some visual modelling tools like Protege (Stanford, 2004) to support ontology modelling, the ontology modelling process still lacks the relation with mature software engineering method. For the general software developers, the current ontology modelling approach is not easy to master and it needs a strong professional background. Therefore, in the actual process of building domain ontology, domain experts often use UML (Unified Modelling Language) (OMG, 2005a) modelling tool or other office software to acquire domain terminologies or create concepts model, and then formalization modellers formalize the conceptual model by a specific ontology language through ontology modelling tool like Protege. As the existing UML modelling tool do not support the ontology modelling directly and the common ontology modelling tools also do not support the requirements and high-level conceptual modelling, the above proposed modelling process has to switch between different modelling tools. A key problem is that the high-level conceptual model cannot be automatically transformed into formal model encoded by a specific ontology language. This brings a lot of management and maintenance inconveniences of ontology modelling. The existing ontology modelling approach has limited the large-scale ontology development. Therefore, it needs a practical engineering approach and a unified modelling tool to support this modelling methodology completely.

Essentially, ontology engineering emphasizes the ontology modelling and knowledge reasoning; however, software engineering focuses on the complete system development methodology which mainly pays attention to requirement analysis, system design, implementation and dose not have the logical reasoning capability. So how to use mature software engineering theory and method to support the ontology development is very significant. Today, Model Driven Development (MDD) (Selic, 2003) is gaining significant momentum in both the software industry and the software engineering academic community. Model Driven Architecture (OMG, 2003), standardized by the Object Management Group (OMG), is a new strategy for designing software systems. Its main goal is to separate system function specification from specific implementation technique completely, enabling system's kernel function specification to be independent of the specific implementation platform technology. Therefore, MDA can retain the neutrality of programming languages, middleware platforms and vendors. In the face of heterogeneous and evolving technology, MDA is supposed to ensure: portability, increased application reuse and reduced development time. Thereby MDA minimizes the affection of technique changes.

Considering the development of domain ontology is a complex process and MDA is a new modeling approach which focuses on the model rather than the specific implementation technical details, we integrated MDA with ontology engineering together, and proposed a model driven domain ontology modeling approach to support the modelling methodology described in section 2.1. By this approach, domain experts or general software developers, who are familiar with UML, can conveniently build the domain conceptual model by UML modelling tools and then this conceptual model can be automatically transformed into the corresponding ontology model encoded by a specific ontology language. As this approach separates domain conceptual model from the concrete ontology modelling languages like OWL, it enhances the reusability of domain conceptual model and reduces the technical difficulty of domain ontology modelling. The implementation details are described in the following sections.

2.2.1 Overview of model-driven TSDO modeling approach

MDA adopts the model-based development mode (Miller & Mukerji, 2003) as shown in Figure 10. Computation Independent Model (CIM) mainly describes the requirements of software system, which specify the system function and boundary. Platform Independent Model (PIM) is the high level abstraction of system function, without any information related to implementation techniques; Platform Specific Model (PSM) is the model which contains specific implementation platform technique information. The MDA–based development process is: firstly, establishing CIM based on the system requirements; secondly, according to the specifications of CIM, creating PIM with the platform independent modeling language, such as UML; thirdly, transforming the PIM to PSM according to some specific mapping rules; lastly, generating platform specific code automatically or semi automatically. In this process, modeller can further refine the created models in CIM, PIM or PSM stage.

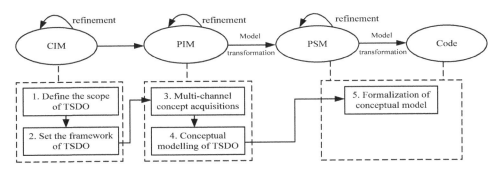

Fig. 10. Model-driven TSDO modelling approach.

According to the modelling idea of MDA, we presented a concrete model-driven TSDO modelling approach to provide a practical engineering implementation as shown in Figure 10. The definition of TSDO scope and the establishment of domain ontology framework belong to the CIM modelling stage. Modeller can employ use case diagram of UML to define the scope of TSDO and set up its framework. In this approach, PIM mainly focuses on the multi-channel domain concept acquisitions and the further conceptual integration and refinement, i.e. conceptual modelling. UML class diagram or use case diagram can be used to model the collected domain concepts and their relationships. After acquiring the domain terminologies, the following step is to integrate and refine these concepts and their relationships to form a high-level domain ontology model, which is independent of a specific ontology description language. The PSM and code steps are used to realize the formalization of high-level domain ontology conceptual model by a specific ontology language. By the model to model transformation technology, the high-level conceptual model (i.e. PIM) can be transformed into an ontology language specific model (i.e. PSM). And then by using model to code transformation technology, the concrete ontology description file encoded by a specific ontology language like OWL (i.e. code) can be generated from the ontology language specific model (i.e. PSM). When we need to revise or maintain the created ontology, we can return back to the CIM or PIM to modify the related models and then generated the corresponding code again. In this mode-driven ontology development approach, all processes adopt the standard UML model or UML extension mechanism (i.e. UML Profile). The technical details are described in the following sections.

2.2.2 CIM step: The scope and framework modeling of TSDO

In order to well organize the development of TSDO, this approach uses the UML use case diagram to model the scope and framework of TSDO. As is shown in Figure 11(a), the ontology hierarchy is represented by package *InfrastructureOfOntology*, which consists of three types of package: common ontology, domain ontology and application ontology. Each package contains the related ontology concepts and their relationships. The *Common Ontology* package contains some general concepts particularly designed for high reusability, where other different domain ontologies and application ontologies either import or specialize its specified concepts or relationships. This is illustrated in Figure 11(a), where it is shown how domain ontologies and application ontologies each depends on the common ontology. The common ontology is generally defined by some standard organizations or research communities. In this chapter, we mainly focus on the building of telecommunications service domain ontology. In order to facilitate reuse, the *Telecommunications Service Domain Ontology* package is further subdivided into a number of packages: *ServiceCategory, Netowrk, TerminalCapability, ServiceQuality, ServiceRole,* and *Charging,* as shown in Figure 11(b).

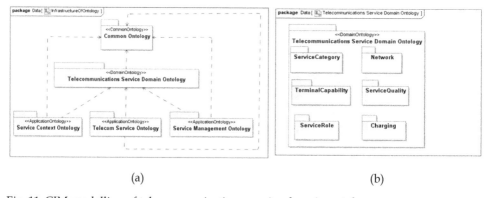

(a)   (b)

Fig. 11. CIM modelling of telecommunications service domain ontology.

2.2.3 PIM step: Terminology acquisitions and conceptual modeling of TSDO

After defining the CIM of TSDO, the following step is to construct the PIM of TSDO. It means that the telecom service related domain terminologies should be collected and then integrated into a high-level abstract domain ontology model which is independent of a specific ontology language like OWL. The collection of domain terminologies can be modeled by the UML Use Case diagram like Figure 4. However, the high-level domain conceptual modeling is the emphasis of PIM. How to model the conceptual model of domain ontology based on UML is needed to resolve. Fortunately, UML and ontology language have some common features, although sometimes represented differently. This provides a possible transformation from UML model to ontology model. For example, both ontology representation language and UML are based on *Class*. The *Generalization* elements of UML can represent the subClass or subProperty semantic of ontology. The ownedAttribute of UML Class can describe the DatatypeProperty of ontology language. The mapping example is illustrated in the Figure 12.

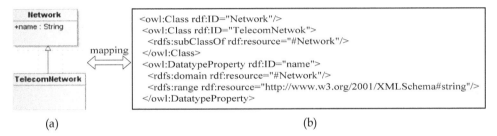

(a)   (b)

Fig. 12. The direct mapping example from UML to OWL.

However, although UML Class diagram has some constructs similar to the constructs of ontology representation language, there are still some ontology constructs which cannot be represented by UML constructs directly. We need to find the appropriate UML elements to represent some other ontology constructs, like objectProperty, equivalent class relation, and disjointing class relation. For instance, we can select the directedAssociation element of UML to represent the ObjectProperty and use the constraints anchored with association to represent the inverse, symmetric or transitive feature of ObjectProperty. An illustrated example is shown in Figure 13.

As a common software modeling language, most of software developers, system analysts and designers are familiar with UML. So, in order to decrease the technical threshold, it's a practical approach for the conceptual modeling of TSDO by UML. Although UML has some similar constructs with ontology language, however, the modeling goals and description capabilities of both languages have some differences. From the above analysis, in order to use UML to represent high-level ontology conceptual model, we need to define a specific tailored representation method to guide the modeler to build the conceptual model of domain ontology. Table 1 shows the main corresponding relation of UML elements with ontology elements. According to this semantic representation way, the modeler can use the UML elements to describe the semantic-enabled high-level ontology conceptual model like Figure 14.

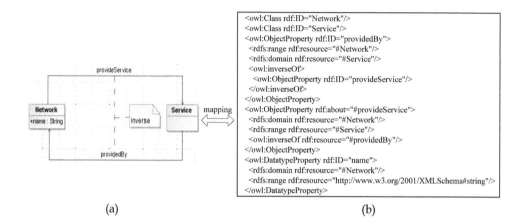

(a)   (b)

Fig. 13. The indirect mapping example from UML to OWL.

UML Elements	Ontology Elements	Comments
Class	Class	
Generalization	subClass, subProperty	
Instance	Individual	
Multiplicity	minCardinality maxCardinality	ontology cardinality declared only for range
ownedAttribute	Datatype Property	
directedAssociation	ObjectProperty	The value of "owned By" property of Association End A is the domain of ObjectProperty, the value of "owned By" property of Association End B is the range of ObjectProperty.
Constraint	Inverse Symmetric Transitive Functional	
Enumeration	oneOf	
Association Class	disjointWith equivalentClass	

Table 1. The defined UML representation method for high-level conceptual model of domain ontology.

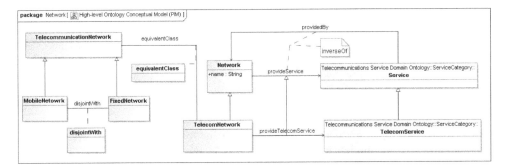

Fig. 14. PIM: A part of high-level conceptual model of network ontology.

2.2.4 PIM to PSM step: Formalization of ontology conceptual model

It can be seen that the high-level ontology conceptual model described by UML is independent of a specific ontology language. So, in order to generate the formal file encoded by a specific ontology language, we need to transform the PIM into PSM according to the concrete model transformation rules. Figure 15 shows the general model transformation mechanism of model driven architecture. Model transformation is essentially to map the source model elements to other elements of the target model. Models are usually the instantiation of its meta-model. The model transformation rules are generally defined in the metamodel level and then model transformation engine apply these rules to the model level to complete the model transformation.

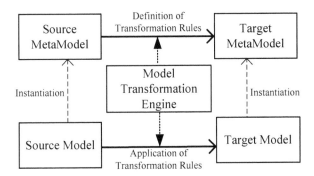

Fig. 15. The principle of model transformation.

Therefore, in order to transform the high-level ontology conceptual model (i.e. PIM) into platform specific model (i.e. PSM), we need to define the transformation rules according to the source and target metamodels. In our proposed approach, the high-level ontology conceptual model (i.e. PIM) is modeled by UML2.0, and the source metamodel is UML2.0 metamodel obviously. So we need a target metamodel relating to specific ontology language to describe the formal ontology model (i.e. PSM). In fact, OMG (Object Management Organization), which is the promoter of MDA, has considered this problem. In May 2009, OMG released the Ontology Definition Metamodel (ODM) v1.0 (OMG, 2009) based on the meta-modeling mechanism of MDA. This specification represents the foundation for an extremely important set of enabling capabilities for MDA based software engineering, namely the formal grounding for representation, management, interoperability, and application of business semantics. The ODM is applicable to knowledge representation, conceptual modeling, formal taxonomy development and ontology definition, and enables the use of a variety of enterprise models as starting points for ontology development. ODM is based on the Meta Object Facility (MOF) (OMG, 2006) meta-modeling architecture of MDA, illustrated by Fig.16, which is based on the traditional four layer metadata architecture. From top to bottom, meta-data is abstracted to 4 layers: M3 (meta-meta model), M2 (meta model), M1 (model) and M0 (object and instance). The under-layer is the instance of its up-layer in turn. M3 layer is the end of meta-layer, namely, MOF is self-described. MOF is a common, abstract language used to define meta-model. It defines some meta-modeling constructs, such as Class, DataType, Association, Package, and Constraint. So the meta-model of ODM or UML can be defined by MOF, whose power just lies in its capability to enable interoperability among different meta-models. Currently, there are 2 approaches to construct meta-models in M2 layer. One is to make use of MOF to define a completely new meta-model from syntax to semantics. Although this approach supports to define a new meta-model that will perfectly match the concepts and relation of the concrete domain, this need the underlying programming realization of corresponding new modeling tool. This is heavy-weight meta-modeling, such as UML and ODM. The other is to extend the existent UML meta-model and then construct a standard UML Profile through UML extension mechanism (Stereotype, TaggedValue, Constraints). This approach allows both defining domain specific conception and relation through UML extension mechanism and using the intrinsic UML elements. So it's a light-weight meta-modeling approach and most of existent MDA tools support this UML Profiling-based meta-modeling mechanism currently. There is no need to develop a new modeling tool. From the above analysis, the UML Profiling-based meta-modeling mechanism approach is adopted in our approach.

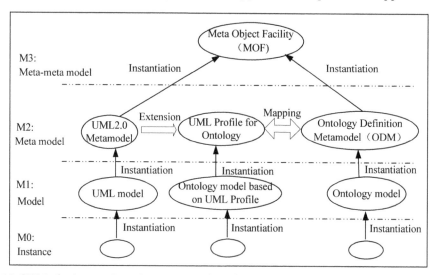

Fig. 16. ODM: the integration of semantic web and model driven architecture.

Therefore, in this approach, the metamodel of PSM employs the UML Profile for RDF and OWL defined in ODM specification. This profile is designed to support modelers developing vocabularies in Resource Description Framework (RDF) (W3C, 2004b) and richer ontologies in the Web Ontology Language (OWL) through reuse of UML notation using tools that support UML2 extension mechanisms. Table 2 specifies a part of stereotypes set that comprise the UML2 Profile for using UML to represent RDF/S and OWL vocabularies.

RDF, RDFS and OWL ontology	UML Base Class	UML Stereotype
rdfs:Resource	Class	《rdfsResource》
rdfs:Datatype	Class	《rdfsDatatype》
rdfs:domain	Association	《rdfsDomain》
rdfs:range	Association	《rdfsRange》
rdfs:subClassOf	Generalization	《rdfsSubClassOf》
rdfs:subPropertyOf	Generalization	《rdfsSubPropertyOf》
owl:Class	Class	《owlClass》
owl:Restriction	Class	《owlRestriction》
owl:ObjectPropert	Class AssociationClass Property Association	《objectProperty》
owl:DatatypeProperty	Class AssociationClass Property Association	《datatypeProperty》
owl:equivalentClass	Constraint	《equivalentClass》
owl:disjointWith	Constraint	《disjointWith》

Table 2. A part of UML Profile for RDF and OWL.

After the source and target metamodels are determined, we can define the model transformation rules from high-level ontology conceptual model (i.e. PIM) to ontology language related model (i.e. PSM). For example, based on the Table 1 and Table 2, we can define the following model transformation rules to support the model transformation like Figure 17. Notably, the source metamodel is UML2.0 metamodel and the target metamodel is UML Profile for RDF and OWL in this proposed approach.

When the transformation rules are defined, the model transformation engine can scan the elements of source model and then transform them into the corresponding elements of target model according to the transformation rules. As model transformation is a key technique used in model-driven architecture. In 2002, OMG issued a Request for proposal (RFP) on MOF Query/View/Transformation to seek a standard compatible with the MDA recommendation suite (UML, MOF, OCL, etc.). Several replies were given by a number of companies and research institutions that evolved during three years to produce a common proposal that was submitted and approved. QVT (Query/View/Transformation) (OMG,

2008) is a standard set of languages for model transformation defined by the Object Management Group. Currently, some MDA tools have declared to support the complete or part functions of QVT. For example, by using the transformation rules, the source model in Figure 14 is transformed into a target model in Figure 18.

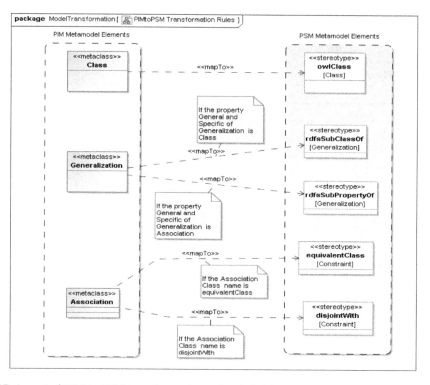

Fig. 17. A part of PIM to PSM transformation rules definitions.

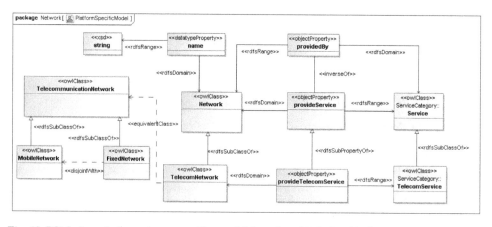

Fig. 18. PSM: A part of ontology specific model based on UML Profile for RDF and OWL.

2.2.5 PSM to code step: Formalization of ontology conceptual model

In order to generate the formal ontology file encoded by OWL, the PSM based on UML Profile for RDFS and OWL should be transformed into ontology file encoded by OWL, which is a model to code transformation process, which involves the model scanning technology.

2.2.5.1 Model to code transformation theory

Before we introduce the concrete transformation process, some related definitions are given firstly.

Definition 1. *Ontology model triples: UmlOnt (C, R, G).*

PSM based on the UML class diagram can be represented by a triple: UmlOnt(C, R, G). C is the class node set, and it is an ontology class definition of the concept in PSM. R is the relation node set, and it is the definition of the relations among ontology class. G is the relation set of C and R, which describes the relations among the nodes in C and R set.

Definition 2. *Relation Matrix (RM)*

Relation matrix is a n order square matrix, including elements a_{ij} in total of n * n, which

looks like $\begin{bmatrix} a_{11} & a_{12} & \cdots & a_{1n} \\ a_{21} & a_{22} & \cdots & a_{2n} \\ \cdots & \cdots & \cdots & \cdots \\ a_{n1} & a_{n2} & \cdots & a_{nn} \end{bmatrix}$, denoted as $A = (a_{ij})_{nn}$ (i,j = 1, 2, 3, ..., n) is used to

describe the relations between the class nodes and relation nodes in the PSM. And a_{ij} indicates whether there is relation between class node i and class node j, as well as the type of relation node.

Definition 3. *Connected subgraph*

Given a directed graph, which can be divided to several connected subgraphs {G1, G2, ⋯, Gn}, these subgraphs meet the following conditions:

- In any two connected subgraphs Gi and Gj, there is no such a node x which is both in Gi and Gj at the same time. That is, $\neg \exists x, x \in Gi \cap x \in Gj$.
- In a connected subgraph Gi, there always exist directed edge between any two nodes x and y (without regard to the direction of the edge).

Definition 4. *Model Transformation Automaton (MTA)*

Model transformation automaton is a quintuple: MTA = $(Q, \Sigma, \delta, q_0, F)$, including:

- Q: A nonempty and finite set of states and one state in it corresponds to an ontology class node. $\forall q \in Q$, q is called a state of MTA.
- Σ: Input events table, in which one input event corresponds to an ontology relation node.
- δ: Transfer function. One transfer function corresponds to a nonzero number a_{ij} ($a_{ij} \neq 0$), $\delta : Q \times \Sigma \rightarrow Q$ in relation matrix.
- q_0: The begin state of MTA, $q_0 \in Q$.

- F: The set of terminate states. F is included by Q. Any q∈F, q is called a terminate state of MTA.

According to the definitions given above, the transformation engine from PSM to formal file encoded by OWL can be described as: The model transformation engine firstly scans the PSM class graph, and the scanning result generates the ontology model triples UmlOnt(C, R, G). C is the set of all nodes in UML class graph, R is the set of all relations in UML class graph, G represents the structure relationship of the ontology class graph, which can be regarded as N connected subgraphs divided from a directed graph and these subgraphs correspond to N relation matrices {RM1, RM2, ⋯, RMn }. One nonzero number a_{ij} represents the relation type between the class node i and j, and these relations are all included in R.

When the transformation engine finishes scanning, it input the scan result to the model transformation automaton. In this MTA, the nonempty finite set of states corresponds to C in the ontology model triples; the input events table corresponds to R in the ontology model triples; the transfer function corresponds to G in the ontology model triples; q0 and F are elements in C. In the procedure of state transforming, the corresponding operations of model transformation are also performed in MTA. When the automaton arrives at the terminal, the transformation finishes.

2.2.5.2 The implementation mechanism of model to code transformation engine

In order to realize the model to code transformation according to the above mentioned theory, we design a model transformation engine based on Eclipse Plugin technology. In MDA, XML Metadata Interchange (XMI) (OMG, 2005b) is an Object Management Group (OMG) standard for exchanging metadata information via Extensible Markup Language (XML). As the most of MDA tools use XMI as an interchange format for UML models, the model transformation engine is responsible for scanning PSM encoded by XMI and then transforming PSM into ontology file encoded by OWL. The process of model to code transformation is indicated in Figure 19.

2.2.5.2.1 Model scanning module

When building ontology model, different modeling tool means different element label and different label structure in the model description file. Therefore, this chapter proposed a transitional model convert method, which adopts same data structure when describes different model format, i.e. the triples in *Definition 1*. This allows the model transformation is no longer constrained by the model structure. It thereby improves the versatility of transformation engine and is convenient to be maintained and updated.

The model scanning module in transformation engine scans the UML class graph encoded by XMI, and the scan result will generate two list sets in the transitional model. It is used to store the class nodes and relation nodes of UML graph, which corresponds to the C and R set in the UML ontology model triples.

2.2.5.2.2 Building relation matrix module

The function of relation matrix building module is used to generate the relation matrix of PSM, i.e. the G set in UML ontology model triples. There are mainly two kinds of nodes in

UML class graph: class node and relation node. Relation node connects class node and distinguishes them by direction, which is very similar to the directed graph. Hence directed graph is adopted to represent UML class graph. Usually matrix is used to represent the graph, and the values in matrix represent the type of relation. In ontology model class graph, there may be several independent subgraphs, which satisfy the description in definition 2. Therefore, it is necessary to handle the relation matrix to produce N independent sub relation matrices. This method can reduce the order or relation matrix and thereby reduce store space of the model, which also improves the efficiency of model transformation.

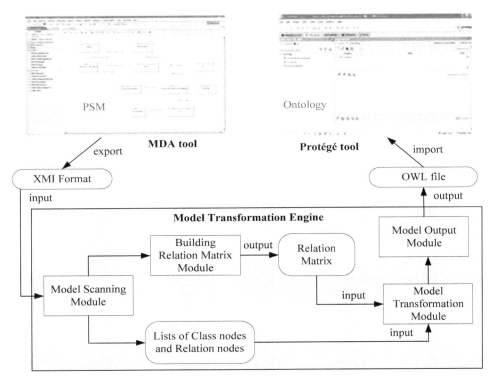

Fig. 19. Model to code transformation process.

2.2.5.2.3 Model transformation module

Model transformation module includes model transformation automaton and model transformation regulation table. In the process of states transition, the automaton performs transformation from PSM to OWL according to the corresponding transformation regulations. In this module, the automaton is separated with the model transformation regulations. Therefore, the changes of model transform regulation will not influence the running of automaton, and it is convenient to perform daily maintenance and update of the engine.

The model transformation automaton is a quintuple, and all information of this quintuple are included in the model transform transitional model, namely in the UML ontology description model triples (C, R, G). The states, input events and transformation function of the automaton correspond to the class nodes, relation nodes and relation matrix set respectively in PSM class graph. The begin state of automaton is the owlClass node or objectProperty/datatypeProperty node in class nodes and the terminate states set includes all class nodes whose out-degree are zero and all nodes have been transformed by the automaton.

The model transformation regulation table defines the transformation regulation from UML Profile for RDF and OWL to OWL language. In the states jump process of model transformation automaton, corresponding regulation is used to perform model transformation. In the process of formulating the transformation regulation, the relations between every label node should be unified, which makes the regulation can be formulated depending on the OWL label structure and the relations between UML model elements. And good regulation is easy to extend in the future.

2.2.5.2.4 Model output module

Model output module only stores the formalized result of the transformation to the appointed path. And in order to verify the validity of the OWL file transformed, user can import the generate code into protégé tool for verification. Protégé is an ontology editor developed by Stanford University, which represents the OWL structure in graphic interface and makes the verification of OWL code validity more quickly and conveniently.

By using the above mentioned model to code transformation approach, the PSM of Figure 18 is transformed into the corresponding ontology encoded by OWL like Figure 20.

3. Experimental environment, use cases and evaluation

In this section, we describe our experimental environment, the implemented service use case and present the obtained evaluation results to validate the semantic interoperability enabled by telecommunications service domain ontology.

3.1 Experimental environment

In order to support this model-driven domain ontology modelling approach, Borland Together (Borland, 2006), a famous MDA tool, is employed in our experiment. By using UML extension mechanism, we implemented the UML Profile for RDF and OWL in Borland Together. In addition, Borland Together tool enable the model-to-model transformation, and this facilitates the transformation from PIM to PSM.   Through the developed model-to-code transformation engine, we realize the transformation from PSM to ontology file encoded by OWL. At last, in order to verify whether the transformation is correct or not, the generated OWL file is imported in Protégé tool to test. By the experimental verification, the proposed model-driven ontology modelling approach can nicely support the constructing methodology of telecommunications service domain ontology.

Under the guidance of this approach, our research team has created a telecommunications service domain ontology knowledge repository which consists of around 430 telecommunications services-related ontology concepts/terminologies and 245 properties. Currently, these ontologies are published on our website (BUPT, 2009), see Figure 21.

```
<owl:Class rdf:ID="MobileNetwork">
 <rdfs:subClassOf>
  <owl:Class rdf:ID="TelecommunicationNetwork"/>
 </rdfs:subClassOf>
 <owl:disjointWith>
  <owl:Class rdf:ID="FixedNetwork"/>
 </owl:disjointWith>
</owl:Class>
<owl:Class rdf:ID="TelecomNetwork">
 <rdfs:subClassOf>
  <owl:Class rdf:ID="Network"/>
 </rdfs:subClassOf>
 <owl:equivalentClass>
  <owl:Class rdf:about="#TelecommunicationNetwork"/>
 </owl:equivalentClass>
</owl:Class>
<owl:Class rdf:about="#FixedNetwork">
 <owl:disjointWith rdf:resource="#MobileNetwork"/>
 <rdfs:subClassOf>
  <owl:Class rdf:about="#TelecommunicationNetwork"/>
 </rdfs:subClassOf>
</owl:Class>
<owl:Class rdf:about="#TelecommunicationNetwork">
 <rdfs:subClassOf rdf:resource="http://www.w3.org/2002/07/owl#Thing"/>
 <owl:equivalentClass rdf:resource="#TelecomNetwork"/>
</owl:Class>
<owl:ObjectProperty rdf:ID="provideTelecomService">
 <rdfs:range rdf:resource="&ServiceCategeory; #TelecomService"/>
 <rdfs:domain rdf:resource="#TelecomNetwork"/>
 <rdfs:subPropertyOf>
  <owl:ObjectProperty rdf:ID="provideService"/>
 </rdfs:subPropertyOf>
</owl:ObjectProperty>
<owl:ObjectProperty rdf:ID="prvidedBy">
 <rdfs:domain rdf:resource="&ServiceCategeory; #Service"/>
 <rdfs:range rdf:resource="#Network"/>
 <owl:inverseOf>
  <owl:ObjectProperty rdf:about="#provideService"/>
 </owl:inverseOf>
</owl:ObjectProperty>
<owl:ObjectProperty rdf:about="#provideService">
 <rdfs:range rdf:resource="&ServiceCategeory; #Service"/>
 <owl:inverseOf rdf:resource="#prvidedBy"/>
 <rdfs:domain rdf:resource="#Network"/>
</owl:ObjectProperty>
<owl:DatatypeProperty rdf:ID="name">
 <rdfs:range rdf:resource="http://www.w3.org/2001/XMLSchema#string"/>
 <rdfs:domain rdf:resource="#Network"/>
</owl:DatatypeProperty>
```

Fig. 20. A part of formal network ontology encoded by OWL.

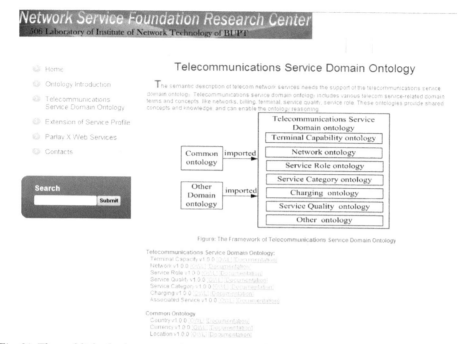

Fig. 21. The published telecommunications service domain ontology.

3.2 Use cases: Semantic telecommunications network capability services

In order to support the shift from traditional closed business model to open service ecosystem of telecom industry, NGN (Next Generation Network) and 3G network all adopt the open API (Application Programming Interface) technologies in the service layer, such as Parlay/OSA and Parlay X (Moerdijk & Klostermann, 2003). Thus, the telecommunication network services, such as call control, short messaging service, and location service, are available to the service developers in the form of APIs. This facilitates the value-added service development. With the development of distributed computing technology, Service-Oriented Architecture (SOA) is also imported into the telecommunications service domain by Parlay Web Service specifications. However, the open interface specifications of telecommunication networks are currently still in the syntactic level. As WSDL (Web Services Description Language)-based telecommunication network services lack the rich semantic annotation information, the keyword-based service matching cannot enable an accurate service discovery. So, currently value-added services often directly invoke the needed telecom network services provided by a specific network carrier. This results in the tight-coupling of application logic and service resources, which limits the provision of dynamically self-adaptive services. The applications cannot dynamically discover satisfied telecom network services and compose them according to the context environment. Facing the heterogeneous networks and personalized user demands, the self-adaptation has become a very important feature of future intelligent integrated service. Therefore, the semantic interoperability of telecom network and Internet in the service layer should be considered.

Based on this domain ontology, we described the telecom network capability services in the semantic level to validate its feasibility. We apply the semantic web service and ontology technologies to the telecommunications service domain, and present an infrastructure to enable the semantic interoperability of telecom network and Internet in the service layer (Qiao et al., 2008b). The proposed approach improves the accuracy of telecommunication network services description, discovery and matching, and unifies the semantic representations of telecommunication and Internet services.

3.3 Lessons learned

Currently, under the shift trend from Web2.0 to Web3.0 era, there have been some initial semantic web applications in Internet field. For example, the system of Twitter allows tweets to be tagged with information that will not appear in the message but can be read by computers (Twitter, 2010). Google is using structured data open standards such as microformats and RDFa to power the rich snippets feature. It's an experimental Semantic Web feature (Google, 2010). FOAF (Friend of a Friend) (FOAF, 2010) is a machine-readable ontology describing persons, their activities and their relations to other people and objects. As a "practical experiment" in the application of RDF and Semantic Web technologies to social networking, FOAF is becoming more and more popular now (FOAF, 2000). In addition, Linked Data (Linked Data, 2007) is a recommended best practice for exposing, sharing, and connecting pieces of data, information, and knowledge on the Semantic Web using URIs and RDF.

However, the semantic web applications in telecommunication services domain are still in an early research phase. Although RDF-based CC/PP (Composite Capability/Preference Profiles) (W3C, 2007) and UAProf (User Agent Profile) (OMA, 2001) are used to describe the terminal capability and user preference, other practical applications are very rare. Therefore, in order to eliminate the semantic gap between telecom network and Internet, the research on semantic web applications in telecommunications field still need to be further enhanced. Telecommunications service domain ontologies consist of various domain related concepts and knowledge, which is the base of semantic interoperability. The wide acceptance of standards and common practices of telecommunications service domain ontologies are still a way ahead. The promotion of the telecommunications service domain ontology by related standardization organizations would be in the foundation for the semantic interoperability of heterogeneous communications equipments and the industrial practical convergent service integration.

4. Conclusion

The network heterogeneity and service convergence are the main characteristics of future network. The provision of self-adaptive intelligent integrated services has become the pursuing goal of network carriers and value-added service providers. Dynamic discovery and composition of services are the important enabling technologies for self-adaptive integrated services. In the service discovery and composition process, semantic interoperability is a key issue. Actually, ontology, as a semantic interoperability and knowledge sharing foundation, has obtained more and more attentions. However, telecommunication service field consists of a large number of concepts/terminologies and relations. How to abstract the sharing domain concepts and reasonably organize them is a big challenge. In this chapter, we presented a practical domain ontology modelling approach for telecommunications service field. Based on this approach, we constructed an open telecommunication service domain ontology repository to support the knowledge sharing and reuse. This will partly facilitate the semantic interoperability of the telecommunications networks and the Internet in the service layer.

5. Acknowledgment

This work was supported by National Key Basic Research Program of China (973 Program) under Grant No. 2012CB315802, National Natural Science Foundation of China under Grant No. 60802034, No. 61171102 and No. 61132001, Beijing Nova Program under Grant No. 2008B50 and New generation broadband wireless mobile communication network Key Projects for Science and Technology Development under Grant No. 2011ZX03002-002-01. We thank Huawei Technologies Co., Ltd. for cooperation in promotion of this work. Thanks also to Dr. Anna Fensel, a Senior Researcher at FTW – Telecommunications Research Center Vienna and STI Innsbruck, University of Innsbruck, Austria, for her valuable comments and suggestions.

6. References

Bashah, N.S.K., Jorstad, I. & Thanh, D.V. (2010). Service Discovery in Future Open Mobile Environments. *Proceedings of ICDS 2010 Fourth International Conference on Digital Society*, pp.47-53, ISBN 978-1-4244-5805-9, St. Maarten, Netherlands Antilles, February 10-16, 2010

Borland, (2006). Together: Visual Modeling for Software Architecture Design. 26.09.2011, Available from http://www.borland.com/us/products/together/

BUPT, (2009). Semantic web application for telecommunications service. 26.09.2011, Available from http://www.int.bupt.cn/jsp/centers/bupt506/intro.htm

Do, T.V., Jorstad, I. (2005). A service-oriented architecture framework for mobile services. *Proceedings of Advanced Industrial Conference on Telecommunications/Service Assurance with Partial and Intermittent Resources Conference/E-Learning on Telecommunications Workshop AICT/SAPIR/ ELETE*, pp. 65 – 70, ISBN 0-7695-2388-9, Lisbon, Portugal, July 17-20, 2005

FOAF, (2010). FOAF Vocabulary Specification. 26.09.2011, Available from http://xmlns.com/foaf/spec/

FOAF project, (2000). The Friend of a Friend (FOAF) project. 26.09.2011, Available from http://www.foaf-project.org/

Google, (2010). Google's Semantic Web Push: Rich Snippets Usage Growing. 26.09.2011, Available from http://www.readwriteweb.com/archives/google_semantic_web_push_rich_snippets_usage_grow.php

Gutheim, P. (2011). An ontology-based context inference service for mobile applications in next-generation networks. *IEEE Communications Magazine*, Vol. 49, No. 1, (Jan. 2011), pp. 60 – 66, ISSN 0163-6804

IST SPICE project, (2008). 26.09.2011, Available from http://www.ist-spice.org/

Khan, A., Asghar, S. and Fong, S. (2011). Framework of integrated Semantic Web Services and Ontology Development for Telecommunication Industry. *Journal of Emerging Technologies in Web Intelligence*, Vol. 3, No. 2, (May 2011), pp.110-119, ISSN 1798-0461

Kolberg, M., Merabti, M. & Moyer, S. (2010). Consumer communication applications drive integration and convergence. *IEEE Communications Magazine*, Vol.48, No. 12, (December 2010), pp. 24 - 24, ISSN 0163-6804

Niazi, R., Mahmoud Q.H. (2009). An Ontology-Based Framework for Discovering Mobile Services. *Proceedings of CNSR 2009 Seventh Annual Communication Networks and Services Research Conference*, pp.178-184, ISBN 978-0-7695-3649-1, Moncton, New Brunswick, Canada, May 11-13, 2009

Li, X.F., Peng, H., Li, Y. & Qiao, X.Q. (2010). Research on Constructing Telecom Services Domain Ontology. *Proceedings of 2010 International Conference on Computational Intelligence and Software Engineering (CiSE)*, pp.1-4, ISBN 978-1-4244-5391-7, Wuhan, China, Dec. 10-12, 2010

Linked Data, (2007). Linked Data - Connect Distributed Data across the Web. 26.09.2011, Available from http://linkeddata.org/

McIlraith, S.A., Son, T.C. & Zeng H.L. (2001). Semantic Web services. *IEEE Intelligent Systems*, Vol. 16, No.2, (Mar.-Apr. 2001), pp. 46 – 53, ISSN 1541-1672

Miller, J. & Mukerji, J. (2003). MDA Guide V1.0.1. 26.09.2011, Available from http://www.omg.org/cgi-bin/doc?omg/03-06-01

Moerdijk, A.-J. & Klostermann, L. (2003). Opening the Networks with Parlay/OSA: Standards and Aspects Behind the APIs, *IEEE Network*, vol.17, no.3, (May-June 2003) pp: 58-64, ISSN 0890-8044

OMA, (2001). User Agent Profile. 26.09.2011, Available from http://www.openmobilealliance.org/tech/affiliates/wap/wap-248-uaprof-20011020-a.pdf

OMG. (2003). Model Driven Architecture. 26.09.2011, Available from http://www.omg.org/mda/

OMG. (2005). Unified Modeling Language. 26.09.2011, Available from http://www.omg.org/spec/UML/

OMG. (2005). XML Metadata Interchange. 26.09.2011, Available from http://www.omg.org/spec/XMI/

OMG. (2006). Meta Object Facility. 26.09.2011, Available from http://www.omg.org/mof/

OMG. (2008). Meta Object Facility (MOF) 2.0 Query/View/Transformation (QVT). 26.09.2011, Available from http://www.omg.org/spec/QVT/index.htm

OMG. (2009). Ontology Definition Metamodel (ODM). 26.09.2011, Available from http://www.omg.org/spec/ODM/1.0/

Park, K.L., Yoon, U.H. & Kim S.D. (2009). Personalized Service Discovery in Ubiquitous Computing Environments. *IEEE Pervasive Computing*, Vol.8, No.1, (Jan.-March 2009), pp.58 – 65, ISSN 1536-1268

Qiao, X.Q., Li X.F. and You, T. (2008). A Semantic Description Approach for Telecommunications Network Capability Services. *Proceedings of 11th Asia-Pacific Network Operations and Management Symposium*, pp.334-343, ISBN 978-3-540-88622-8, Beijing, China, October 22-24, 2008

Qiao, X.Q., Li, X.F., You, T., Sun, L.H. (2008). Semantic Telecommunications Network Capability Services. *Proceedings of the 3rd Asian Semantic Web Conference (ASWC 2008)*. pp. 508–523, ISBN 978-3-540-89703-3, Bangkok, Thailand, Dec. 8-11 2008

Rój, M. (2008). Recommendations for telecommunications services ontology, In: *IST SIMS project deliverables*, 26.09.2011, Available from http://www.ist-sims.org.

Selic, B. (2003). The pragmatics of model-driven development. *IEEE Software*, Vol. 20, No. 5, (Sept.-Oct. 2003), pp. 19 – 25, ISSN 0740-7459

Stanford Protégé team. (2004). Protégé. 26.09.2011, Available from http://protege.stanford.edu/

Su X.M., Alapnes, S. and Shiaa M.M. (2009). Mobile Ontology: Its Creation and Its Usage, In: *Constructing Ambient Intelligence*, H. Gerhauser, J. Hupp, C. Efstratiou and J. Heppner (Eds.), pp.75-79, Springer-Verlag Berlin Heidelberg, ISBN 3-642-10606-4, Germany

Twitter, (2010). Semantics, tagging and Twitter. 26.09.2011, Available from http://ml.sun.ac.za/2010/04/23/semantics-tagging-and-twitter/

Veijalainen, J. (2007)   Developing Mobile Ontologies; Who, Why, Where, and How? *Proceedings of 2007 International Conference on Mobile Data Management*, pp. 398 – 401, ISBN 1-4244-1241-2, Mannheim, Germany, May 1-1, 2007

Veijalainen, J. (2008). Mobile Ontologies: Concept, Development, Usage, and Business Potential. *International Journal on Semantic Web and Information Systems (IJSWIS)*, Vol. 4, No. 1, (Sep. 2008), pp.20-34, ISSN: 1552-6283

Villalonga, C., Strohbach, M., Snoeck, N., Sutterer, M., Belaunde, M., Kovacs, E., Zhdanova, A.V., Goix, L.W., Droegehorn, O. (2009). Mobile Ontology: Towards a Standardized Semantic Model for the Mobile Domain. *Proceedings of the 1st International Workshop on Telecom Service Oriented Architectures (TSOA 2007) at the 5th International Conference on Service-Oriented Computing*, pp. 248-257, ISBN, 978-3-540-93850-7. Vienna, Austria, September 17, 2007

Vitvar, T., Viskova, J. (2005). Semantic-enabled Integration of Voice and Data Services: Telecommunication Use Case, *Proceedings of 2005 IEEE European Conference on Web Services (ECOWS2005)*, pp. 138-151, ISBN 0-7695-2484-2, Vaxjo, Sweden, November14-16, 2005

W3C. (2004). OWL Web Ontology Language Overview. 26.09.2011, Available from http://www.w3.org/TR/owl-features/

W3C. (2004). Resource Description Framework (RDF). 26.09.2011, Available from http://www.w3.org/RDF/

W3C, (2007). Composite Capabilities/Preference Profiles: Structure and Vocabularies 2.0. 26.09.2011, Available from http://www.w3.org/Mobile/CCPP/

Zander, S., Schandl, B. (2011). Semantic Web-enhanced Context-aware Computing in Mobile Systems: Principles and Application. In: *Mobile Computing Techniques in Emerging Markets: Systems, Applications and Services*, Kumar, A.V. Senthil (Eds.), Retrieved from http://eprints.cs.univie.ac.at/2870/

Zhu, J.W., Li, B., Wang, F., Wang, S.CH. (2010). An Overview of CRS4MO Project: Construction and Retrieval System for Mobile Ontology. *Journal of Computational Information Systems*, Vol.6, No.6, (June, 2010), pp. 2009-2015, ISSN 1553-9105

3

Multicriteria Optimization in Telecommunication Networks Planning, Designing and Controlling

Valery Bezruk, Alexander Bukhanko,
Dariya Chebotaryova and Vacheslav Varich
Kharkov National University of Radio Electronics
Ukraine

1. Introduction

Modern telecommunication networks, irrespectively of their organization and type of the transmitted information, become more complex and possess many specific characteristics. The new generation of telecommunication networks and systems support a wide range of various communication-intensive real-time and non real-time various applications. All these net applications have their own different quality-of-service requirements in terms of throughput, reliability, and bounds on end-to-end delay, jitter, and packet-loss ratio etc. Thus, telecommunication network is a type of the information system considered as an ordered set of elements, relations and their properties. Their unique setting defines the goal searching system.

For such a type of information system as a telecommunication network it is necessary to perform a preliminary long-term planning (with structure designing and system relation defining) and a short-term operating control within networks functioning. The problem of the optimal planning, designing and controlling in the telecommunication networks involves: definition of an initial set of decisions, formation of a subset of system permissible variants, definition of an optimal criteria, and also a choice of the structure variants and network parameters, optimal by such a criteria. It is the task of a general decision making theory reduced to the implementation of some choice function of the best (optimal) system based on the set of valid variants. For the decision making tasks the following optimizing methods can be used: scalar and vector optimization, linear and nonlinear optimization, parametric and structure optimization, etc (Figueira, 2005; Taha, 1997; Saaty, 2005). We propose a method of the multicriteria optimization for optimum variants choice taking into account the set of quality indicators both in long-term and short-term planning and controlling.

The initial set of permissible variants of a telecommunication network is being formed through the definition of the different network topologies, transmission capacities of communication channels, various disciplines of service requests applied to different routing ways, etc. Obtained variants of the telecommunication network construction are estimated on a totality of given metrics describing the messages transmission quality. Thus, the formed set of the permissible design decisions is represented in the space of criteria ratings of quality indicators where, used of unconditional criteria of a preference, the subset of effective (Pareto-optimal) variants of the telecommunication network is selected. On a final stage of optimization any obtained effective variants of the network can be selected for usage. The unique variant choice of a telecommunication network with introducing some conventional criteria of preference as some scalar goal function is also possible.

In the present work some generalizations are made and all stages of solving multicriteria problems are analyzed with reference to telecommunication networks including the statement of a problem, finding the Pareto-optimal systems and selecting the only system variant. This chapter also considers the application particularities of multicriteria optimization methods at the operating control within telecommunication systems. The investigation results are provided on the example of solving of a particular management problem considering planning of cellular networks, optimal routing and choice of the speech codec, controlling network resources, etc.

2. Theoretical investigation in Pareto optimization

As far as the most general case is concerned, the system can be thought of as an ordered set of elements, relationships and their properties. The uniqueness of their assignment serves to define the system fully, notably, its structure and efficiency. The major objective of designing is to specify and define all the above-listed categories. The solution of this problem involves determining an initial set of solutions, generating a subset of pemissible solutions, assigning the criteria of the system optimality and selecting the system, which is optimal in terms of a criteria.

2.1 The problem statement in optimization system

It is assumed that the system $\varphi = (s, \vec{\beta}) \in \Phi_D$ is defined by the structure s (a set of elements and connections) and by the vector of parameters $\vec{\beta}$. A set of input actions X and output results Y should be assigned for an information system. This procedure defines the system as the mapping $\varphi : X \to Y$. The abstract determination of the system in the process of designing is considered to be exact. In particular, when formalizing the problem statement, a mathematical descripton of the working conditions (of signals, interferences) and of the functional purpose of a system (solutions obtained at the system output) are to be given, which, in fact, determine the variant of the system $\varphi \in \Phi$.

In particular, the limitations given on conditions of work, on the structure $s \in S_D$ and parameters $\beta \in B_D$, as well as on values of the system quality indicators define the subset of permissible project solutions $\Phi_a = S_a \times B_a$. Diverse ways of assigning a set of allowable are possible, in particular:

- implicit assignment using the limitations upon the operating conditions formulated in a rigorous mathematical form;
- enumeration of permissible variants of the system;
- determination of the formal mechanism for generating the system variants.

The choice of the optimal criteria is related to the formalization of the knowledge about an optimality. There exist two ways of describing the customer's preference of one variant to the other, i.e. ordinal and cardinal .

An ordinal approach is order-oriented (better-worse) and is based on introducing certain binary relations on a set of permissible alternatives. In this case the customer's preference is the binary relation R on the set Φ_D which reflects the customer's knowledge that the alternative φ' is better than the alternative: φ'': $\varphi'R\varphi''$.

Assume that a customer sticks to a certain rigorous preference $\succ$, which is asymmetric and transitive, as he decides on a set of permissible alternative Φ_D. The solution $\varphi_0 \in \Phi_D$ is called optimal with respect to $\succ$, unless there are other solution $\varphi \in \Phi_D$ for which $\varphi \succ \varphi^{(0)}$ holds true. A set of all optimal solutions in relation to $\succ$ is denoted by $\mathrm{opt}_\succ \Phi_D$. A set of optimal solutions can comprise the only element, a finite or infinite number of elements as a function of the structure of a permissible set or properties of the relation $\succ$. If the discernibility relation coincides with that of equality $=$, then the set $\mathrm{opt}_\succ \Phi_D$ (provided it is not empty) contains the only element.

A cardinal approach to describe the customer's preference assigns to each alternative $\varphi \in \Phi_D$, a certain number U being interpreted as the utility of the alternative φ. Each utility function determines a corresponding order (or a preference) R on die set $\Phi_D(\varphi'R\varphi')$ if and only if $U(\varphi') \geq U(\varphi'')$. In this case they say that the utility function $U(\cdot)$ is a preference indicator R. In point of fact this approach is related to assigning a certain scalar-objective function (a conventional preference criteria) whose optimization in a general case may result in the selection of the only optimal variant of the system.

The choice of the optimal criteria is based on formalizing the knowledge of a die system customer (i.e. a person who makes a decision) about its optimality. However, one often fails to formalize the knowledge of a decision-making person about the system optimality rigorously. Therefore, it appears impossible to assign the implicitly of the scalar optimal criteria resulting in the choice of the only decision variant $\varphi^{(0)} = \underset{\varphi \in \Phi_D}{\mathrm{extr}}\left[U(\varphi)\right]$, where $U(\varphi)$ is a certain objective function of the system utility (or usefulness). Therefore, at the initial design stages the system is characterized by a set of objective functions:

$$\vec{k}(\varphi) = (k_1(\varphi),...,k_i(\varphi),...,k_m(\varphi)), \tag{1}$$

which determines the influence of the structure s and the parameters $\vec{\beta}$ of the variant of the system $\varphi = (s,\vec{\beta})$ upon the system quality indicators. In this connection one has to deal with the newly emerged issues of optimizing approaches in terms of a collection of quality indicators, which likewise are called the problems of multicriteria or vector optimization. Basically, the statement and the solution of a multicriteria problems is related to replacing (approximation) customer's knowledge about the system optimality with a different optimality conception which can be formalized as a certain vector optimal criteria (1) and, consequently, the problem will be solved through the effective optimization procedure.

2.2 Forming a set of permissible variants of a system

When optimizing the information systems, as their decomposition into subsystems can be assigned, it would be judicious to proceed from the morphological approach which is widely applied in designing complicated systems. In this context it is assumed that any variant of a system has a definite structure, i.e. it consists of the finite number of elements (subsystems), and the distribution of system functions amongst them can be performed by the finite number of methods.

Now consider the peculiar features of generating the structural set of permissible variants of a system. Let us assume that the functional decomposition of the system into a set of elements is

$$\{\varphi_j, \ j = \overline{1,L}, \bigcup_{j=1}^{L} \varphi_j = \varphi\}.$$

What is considered to be assigned is as follows: a finite set of elements of the system E as well as the splitting of the set E into L morphological classes $\sigma(l), l = \overline{1,L}$ such as $\sigma(l) \cap \sigma(l') = \varnothing$ at $l \neq l'$.

A concept of the morphological space $\Lambda \subseteq 2^E$ is introduced, its elements being the morphological variant of the system $\varphi = (\varphi_1, \varphi_2, ..., \varphi_L)$. Each morphological variant φ is a certain set of representatives of the classes $\varphi(l) \in \sigma(l)$. Here for all $\varphi \in \Lambda$ and for any $l = \overline{1,L}$ the set $\varphi \in \Lambda$ contains a single element.

Under the assumption that there exist a multitude of alternative model of implementing each subsystem φ_{lk}, $k = \overline{1,L}$, $l = \overline{1,L}$, the following morphological table can be specified:

Morphological classes	Possible models of implementing the system elements	Number of modes of implementing the system
$\sigma(1)$	$\varphi_{11}[\varphi_{12}]\varphi_{13}\cdots\varphi_{1K_1}$	K_1
$\sigma(2)$	$\varphi_{21}\varphi_{22}\varphi_{13}\cdots[\varphi_{2K_2}]$	K_2
....	....	....
$\sigma(l)$	$\varphi_{11}\varphi_{12}[\varphi_{13}]\cdots\varphi_{1K_l}$	K_l
....	....	....
$\sigma(L)$	$[\varphi_{L1}]\varphi_{L2}\varphi_{L3}\cdots\varphi_{LK_L}$	K_L

Table 1. Morphological table.

As an example (see table 1), a q-th morphological variant of the system $\varphi^q = \langle \varphi_{12}, \varphi_{2K_2}, ..., \varphi_{13}, ..., \varphi_{L1} \rangle$ that determines the system structure is distinguished. The total number of all possible morphological variants of the system is generally determined as

$$Q = \prod_{l=1}^{L} K_l \ .$$

When generating a set of permissible variants Φ_D one has to allow for the constraints upon the structure, parameters and technical realization of elements and the system as a whole as well as for the permissible combination of elements connections and constraints up on the value of the quality indicators of the system as a whole.

Here, there exist conflicting requirements. On the one hand, it is desirable to present all conceivable variants of the system in their entirety so as not to leave out the potentially best variants. On the other hand, there are limitations specified by the permissible expenditures (of time and funds) on the designing of a system.

After a set of permissible variant of a system has been determined in terms of a particular structure, the value of the quality indicators is estimated, a set of Pareto-optimal variants is distinguished and gets narrowed down to the most preferable one.

2.3 Finding the system Pareto-optimal variants

As a collection of objective functions is being introduced, each variant of the system φ is mapped from a set of permissible variants Φ_D into the criteria space of estimates $V \in R^m$:

$$V = \vec{K}(\Phi_D) = \{\vec{v} \in R^m \mid \vec{v} = \vec{k}(\varphi),\ \varphi \in \Phi_D\}. \tag{2}$$

In this case to each approach φ corresponds its particular estimate of the selected quality indicators $\vec{v} = \vec{k}(\varphi)$ (2) and, vice versa, to each estimate corresponds an approach (in a general way, a single approach is not obligatory).

To the relation of the rigorous preference $\succ$ on the set Φ_D corresponds the relation $\succ$ in the criteria space of estimates V. According to the Pareto axiom, for any two estimates $\vec{v}, \vec{v}'' \in V$ satisfying the vector inequality $\vec{v}' \geq \vec{v}''$, the relation $\vec{v}' \succ \vec{v}''$ is always obeyed. Besides, according to the second Pareto action for any two approaches $\varphi', \varphi'' \in \Phi_D$, for which $\vec{k}(\varphi') \geq \vec{k}(\varphi'')$ is true, the relation $\varphi' \succ \varphi''$ always occurs. The Pareto axiom imposes definite limitations upon the character of the preference in multicriteria problem.

It is desirable for a customer to obtain the best possible value for each criteria. Yet in practice this case can be rarely found. Here, it should be emphasized that the quality indicators (objective function) of the system (1) may be of 3 types: neutral, consistent with one another and competing between one other. In the first two instances the system optimization can be performed separately in terms of each of indicators. In the third instance it appears impossible to arrive at a potential value of each of the individual indicators. In this case one can only attain the consistent optimum of introduced objective functions – the optimum according to the Pareto criteria which implies that each of the indicators can be further improved solly by lowering the remaining quality indicators of the system. To the Pareto optimum in the criteria space corresponds a set of Pareto-optimal estimates that satisfy the following expression:

$$P(V) = \text{opt}_\geq V = \{\vec{k}(\varphi^0) \in R^m \mid \forall \vec{k}(\varphi) \in V : \vec{k}(\varphi) \geq \vec{k}(\varphi^0)\}. \tag{3}$$

An optimum based on the Pareto criteria can be found either directly according to (3) by the exhaustive search of all permissible variants of the system Φ_D or with the use of special procedures such as the weighting method, methods of operating characteristics.

With the Pareto *weighting method* being employed. The optimal decisions are found by optimizing the weighted sum of objective functions

$$\operatorname*{extr}_{\varphi \in \Phi_D}[k_p(\varphi) = \lambda_1 k_1(\varphi) + \lambda_2 k_2(\varphi) + ... + \lambda_m k_m(\varphi)], \qquad (4)$$

in which the weighting coefficients $\lambda_1, \lambda_2, ..., \lambda_m$ are selected from the condition $\lambda_i > 0, \sum_{i=1}^{m} \lambda_i = 1$. The Pareto-optimal decisions are the system variants that satisfy eq. (4) with different permissible combination of the weighting coefficients $\lambda_1, \lambda_2, ..., \lambda_m$. When solving this problem one can observe the variation in the alternative systems $\varphi = (s, \vec{\beta}) \in \Phi_D$ within the limits of specified.

The method of operating characteristics consists all the objective functions, except for a single one, say, the first one, are transferred into a category of limitations of an inequality type, and its optimum is sought on a set of permissible alternatives

$$\operatorname*{extr}_{\varphi \in \Phi_D}[k_1(\varphi)], \; k_2(\varphi) = K_{2\varphi}; \; k_3(\varphi) = K_{3\varphi}, ..., \; k_m(\varphi) = K_{m\varphi}. \qquad (5)$$

Here $K_{2\varphi}, K_{3\varphi}, ..., K_{m\varphi}$ are the certain fixed, but arbitrary quality indicators values.

The optimization problem (5) is solved sequentially for all permissible combinations of the values $K_{2\varphi} \le K_{2D}, K_{3\varphi} \le K_{3D}, ..., K_{m\varphi} \le K_{mD}$. In each instance an optimal value of the indicator k_{1opt} is sought by variations $\varphi \in \Phi_D$. As a result a certain multidimensional working space in the criteria space is sought

$$k_{1opt} = f_p(K_{2\varphi}, K_{3\varphi}, ..., K_{m\varphi}). \qquad (6)$$

If the found relation (6) is monotonously decreasing in nature for each of the arguments, the working surface coincides with a Pareto-optimal surface. This surface can be connected, nonconnected and just a set of isolated points.

It should be pointed out that each point of the pareto-optimal surface offers the property of a m-fold optimum, i.e. this point checks with a potentially attainable (with variation $\varphi \in \Phi_D$) value of one of the indicators k_{1opt} at the fixed (corresponding to this point) value of other $(m-1)$ quality indicators. The Pareto-optimal surface can be described by any of the following relationships

$$k_{1opt} = f_{no}^1(k_2, k_3, ..., k_m), ..., k_{mopt} = f_{no}^m(k_1, k_2, ..., k_{m-1}), \qquad (7)$$

which represent the multidimensional diagram of the exchange between the quality indicators showing the way in which the potentially attainable value of the corresponding indicator depends upon the values of other indicators.

Thus, the Pareto-optimal surface connects the potentially attainable values of index is Pareto-optimum consistent, generally dependent and competing quality indicators Therefore, with the Pareto-optimal surface in the criteria space being obtained, the multidimensional potential characteristics of the system and related multidimensional exchange diagram are found.

It should be noted that they are different types of optimization problems depending upon the problem statement.

Discrete selection. The initial set Φ_D is specified by a finite number of variants of constructing the system $\{\phi_l, l = \overline{1, L_D}, \phi \in \Phi_D\}$. It is required that set of Pareto-optimal variants of the system $\text{opt}_{\succ}\Phi_D$. should be selected.

Parametric optimization. The structure of the system S_D is specified. It is necessary to find the magnitude of the vectors $\vec{\beta}^0 \in B_D$ at which $\phi = (s, \vec{\beta}) \in \text{opt}_{\succ}\Phi_D$.

Structural-parametric optimization. It is necessary to synthesize the structure $s \in S_D$ and to find the magnitude of the vector of the parameters $\vec{\beta} \in B_D$ at which $\phi = (s, \vec{\beta}) \in \text{opt}_{\succ}\Phi_D$.

The first two types of problems have been adequately developed in the theoiy of multicriteria optimization. The solution of the third-type problems is most complicated. To synthesize the Pareto-optimal structure and find the optimal parameters a set of functionals $k_1(s, \vec{\beta}), k_2(s, \vec{\beta}), ..., k_m(s, \vec{\beta})$ is to be optimized. Yet optimizing functionals even in a scalar case appears to be a rather challenging task from both the mathematical and some no less importants standpoints. In the case of a vector the solution to these types of problems becomes still more complicated. Therefore, in designing the systems with regard to a set of the quality indicators one has to simplify the optimization problem by decomposing the system into simpler subsystems, to reduce the number of quality indicators as the system structure is being synthesized.

If the set of Pareto-optimal systems variants, which has been found following the optimization procedure, turned out to be a narrow one, then any of them can be made use of as an optimal one. In this case the rigorous preference relation $\succ$ may be thought of as coinciding with the relation $\geq$ and, therefore, $\text{opt}_{\succ}V = P(V)$.

However, in practice the set $P(V)$ proves to be sufficiently wide. This implies that the relations $\succ$ and $\geq$ (although they are connected through the Pareto axiom) do not show a close agreement. Here, the inclusions $\text{opt}_{\succ}V \subset P(V)$ and $\text{opt}_{\succ}\Phi_D \subset P_{\overline{k}}(\Phi_D)$ are valid. Therefore, we will have to deal with an emerging problem of narrowing the found Pareto-optimal solutions involving additional information about the relation of the customer's rigorous preference. Yet the ultimate selection of optimal approaches should only be made within the limits of the found set of Pareto-optimal solution.

2.4 Narrowing of the set of Pareto-optimal solutions down to the only variant of a system

The formal model of the Pareto optimization problem does not contain any information to select the only alternative. In this particular instance a set of permissible variants gets narrowed only to a set of Pareto-optimal solution by eliminating the worse variants with respect to a precise variant.

However, the only variant of a system is normally to be chosen to ensure the subsequent designing stages. It is just for this reason why one feels it necessary to narrow the set of Pareto-optimal solutions down to the only variant of a system and to make use of some additional information about a customer's preference. This type of information is produced following the comprehensive analysis of Pareto-optimal variants of a system, particularly, of a structure, parameters, operating characteristics of the obtained variants of a system, a relative importance of input quality indicators, etc. Some additional information thus obtained concerning the customer's preferences is employed to construct choice function (an objective scalar function) whose optimization tends to select the sole variants of a system.

In order to solve the problem of narrowing a set of Pareto-optimal solution a diversity of approaches, especially those based on the theory of utility, the theory of fuzzy sets, etc. Now let us take a brief look at some of them.

The selection of optimal approaches using the scalar value function. One of the commonly used methods of narrowing a set of Pareto-optimal solution is constructing the scalar value function, which, if applied, gives rise to selecting one of the optimal variants of a system.

The numerical function $F(v_1, v_2, ..., v_m)$ of m variables is referred to as the value (utility) function for the relation $\succ$ if for the arbitrary estimates $\vec{v}', \vec{v}'' \in V$ the inequality $F(\vec{v}') > F(\vec{v}'')$ occurs if and only if $\vec{v}' \succ \vec{v}''$. If there exists the function of utility $F(\vec{v})$ for the relation $\succ$, then it is obvious that

$$\text{opt}_\succ V = \{\vec{v}^0 \in V : F(\vec{v}^0) = \max_{\vec{v} \in V} F(\vec{v})\}$$

and finding an optimal estimate boils down to solving the single-criteria problem of optimizing the function $F(\vec{v})$ on the set V. The value function of the type

$$F(v_1, v_2, ..., v_m) = \sum_{j=1}^{m} c_j f_j(v_j), \qquad (8)$$

where c_j is the scaling factor, $f_j(v_j)$ are the certain unidimensional value function which are the estimates of usefulnen of the system variant φ in terms of the index $k_j(\varphi)$.

The construction of the value function (8) consists in estimating the scale factors, forming unidimensional utility function $f_j(v_j)$ as well as in validating their independence and consistency. Here, use is made of the data obtained from interrogating a customer. Special interrogation procedures and program packages intended to acquire some additional information about the customer's preferences have been worked out.

The selection of optimal approaches based upon the theory of fuzzy sets. This procedure is based on the fact that due to the apriori uncertainty with regard to the customer's preference, the concept such as "the best variant of a system" cannot be accurately defined. This concept may be thought of as constituting a fuzzy set and in order to make an estimate of the system, the basic postulates of the fuzzy- set theory can be employed.

Let X be a certain set of possible magnitudes of a particular quality indicator of a system. The fuzzy set G on the set X is assigned by the membership function $\xi_G : X \rightarrow [0,1]$ which brings the real number ξ_G over the interval $[0,1]$ in line with each element of the set X. The value ξ_G defined the degree of membership of the set X elements to the fuzzy set G. The nearer is the value $\xi_G(x)$ to unity, the higher is the membership degree. The membership function $\xi_G(x)$ is the generalization of the characteristic function of sets, which takes two values only : 1 – at $x \in G$; 0 – at $x \notin G$. For discrete sets X the fuzzy set G is written as the set of pairs $G = \{x, \xi_G(x)\}$.

Thus, according to the theory of fuzzy sets each of the quality indicators can be assigned in the form of a fuzzy set

$$k_j = \{k_j, \xi_{k_j}(k_j)\},$$

where $\xi_{k_j}(\circ)$ is the membership function of the specific value of the j-th index to the optimal magnitude.

This type of writing is highly informative, since it gives an insight into its physical meaning and "worth" in relation to the optimal (extreme) value which is characterized by the membership function $\xi_{k_j}(\circ)$.

The main difficulty over the practical implementation of the considered approach consists in choosing the type of a membership function. In some sense the universal form of the membership function being interpreted in terms of the theory of fuzzy sets with regard to the collection of indicators is written as:

$$\xi_{\bar{k}}(k_1, k_2, ..., k_m) = \frac{1}{m} \left\{ \sum_{l=1}^{m} [\xi_{k_j}(k_j)]^\beta \right\}^{\frac{1}{\beta}}. \tag{9}$$

The advantage of this form is that depending upon the parameter β a wide class of functions is implemented. These functions range from the linear additive form at $\beta = 1$ to the particularly nonlinear relationships at $\beta \rightarrow \infty$.

It should be pointed out that with this particular approach it is essential that the information obtained from a customer by an expert estimates method be used to pick out a membership function and a variety of coefficients.

Selecting optimal approaches at quality indicators strictly ordered in terms of the level of their importance. Occasionally it appears desirable for a customer to obtain die maximum magnitude of one of the indicators, say, k_1 even at the expense of the "lasses" for the remaining indicators. This means that the indicator k_1 is found to be more important than other indicators.

In addition, there may be the case where the whole set of indicators $k_1, k_2, ..., k_m$ is strictly ordered in terms of their importance such k_1 is more important that other indicators $k_1, k_2, ..., k_m$; k_2 is more essential than all the indicators $k_1, k_2, ..., k_m$, etc. This corresponds the instance where the lexico-graphical relation lex is employed when a comparison is made between the estimates of approaches. Now we give the definition of the above relation.

Let there be two vectors of estimates $\vec{v}, \vec{v}' \in V \subset R^m$. The lexico-graphical relation lex is determined in the following way: the relation lex occurs if and only if one the following conditions is satisfied.

1) $v_1' > v_1''$;

2) $v_1' = v_1''; v_2' > v_2''$

....

m) $v_j' = v_j'', j = 1, 2, ..., m-1, v_m' > v_m''$,

$v' = (v_1', v_2', ..., v_m'); \quad v^n = (v_1^n, v_2^n, ..., v_m^n).$

In this case the components $v_1, v_2, ..., v_m$, i.e. the estimates of the system quality indicators $k_1(\varphi), k_2(\varphi), ..., k_m(\varphi)$ are said to be strictly order in terms of their importance. As the relation $\vec{v} \, lex \, \vec{v}''$ is satisfied they say that from the lexico-graphical stand point the vector $\vec{v}'$ is greater than the vector $\vec{v}'$. At $m-1$ the lexico-graphical relation coincides with the relation $\succ$ on the subset of real numbers.

In determining the lexico-graphical relation a major role is played by the order of enumerating quality indicators. The change in the numeration of quality indicators give rise to a different lexico-graphical relation.

3. Practical usage

Let us consider some practical peculiarities of an application of multicriteria optimization methods within a long-term and short-term planning, designing and controlling. In the examined examples of telecommunication networks operation and estimation of the quality indicators values is probed on mathematical models implemented on a computer using the packets of specific simulation modeling.

3.1 Telecommunication network variant choice

In particular, we considered features of an application of multicriteria optimization methods on the example of the packet switching network. For such a task the mathematical model of full-connected topology of a network was implemented. There was performed the simulation modeling of different variants of data transmission in the indicated network and the quality indicators estimates for each variant were obtained (Bezruk et al., 2008).

Pareto-optimal variants of the network were obtained with the methods of vector optimization and, among them, there was selected the single optimal variant of the network (fig. 1). The results of the optimization were used for the task of the network control when framing optimal control actions.

Thus, the control device collects the information on the current condition of the network and develops Pareto-optimal control actions which are directed to a variation of mechanisms of the arrival requests service and paths of packet transmission through the network.

The structure of the model, realized with a computer, includes simulators of the messages with a Poisson distribution and given intensities, procedures of the messages packing, their transmission through the communication channels. The procedures of the messages packing have simulated a batch data transmission with a mode of the window load control.

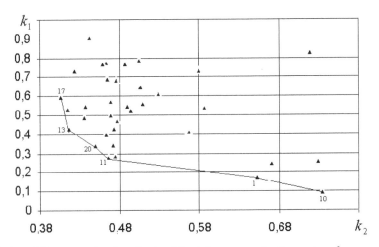

Fig. 1. Choice of Pareto-optimal variants of the telecommunication network.

The procedures of a packet transmission were simulated by the processes of transfer using duplex communication channels with errors. The simulation analysis of the transfer delays was stipulated at a packet transmission in the communication lines connected with final velocity of signals propagation in communication channels, fixed transmission channel capacity and packets arrival time in the queue for their transfer trough the communication channel.

Different variants of the telecommunication network functioning were realized at the simulation analysis, they differed in disciplines of service in the queues, ways of routing in a packet transmission and size of the window of the transport junction. In the considered example thirty six variants of the network functioning were obtained. Network functioning variants were estimated by the following quality indicators: average time of deliveries $k_1 = \overline{T}$ and average probability of message loss $k_2 = \overline{P}$. These quality indicators had contradictory character of interconnection. The obtained permissible set of network variants is presented in a criteria space (fig. 1). The subset of the Pareto-optimal network operation is selected by the exclusion of the inferior variants. The left low bound set of the valid variants corresponds to Pareto-optimal variants. Among Pareto-optimal variants of the network Φ_0 was selected a single variant from the condition of a minimum of the introduced resulting quality indicator $k_{pn} = C_1 k_1 + C_2 k_2$. For the case $C_1 = 0,4$, $C_2 = 0,6$ the single variant 11 was selected; the discipline service of the requests (in the random order) was established for it as well as the way of routing (weight method) and size of the "transmission window" (equal 8).

The given task is urgent for practical applications being critical to the delivery time (in telecommunication systems of video and voice intelligences, systems of the banking terminals, alarm installations, etc).

3.2 Multicriteria optimization in radio communication networks designing

Let us consider some practical aspects of multicriteria optimization methods when planning radio communication networks, on an example of cellular communication network (CCN). The process of finding CCN optimal variants includes such stages:

- setting the initial set of the system variants differed in the following terms: radio standards, the engaged frequency band, the number and activity of subscribers, covered territory, sectoring and the height of antennas, the power of base station transmitters, the parameter of radio wave attenuation, etc;
- separation of the permissible set of variants with regard of limitations on the network structure and parameters, limitation on the value of the quality indicators;
- choice of the subset of Pareto-optimal CCN variants;
- analysis of obtained Pareto-optimal CCN variants;
- choice of a single CCN variant.

In the considered example there was formed a set of permissible variants of CCN (GSM standard), which were defined by different initial data including the following ones: the planned number of subscribers in the network; dimensions of the covered territory (an area); the activity of subscribers at HML (hour of maximum load); the frequency bandwidth authorized for the network organization; sizes of clusters; the permissible probability of call blocking and percentage of the time of the communication quality deterioration.

The following technical parameters of CCN were calculated by a special technique.

1. The general number of frequency channels authorized for deployment of CCN in the given town, is defined as

$$N_k = \text{int}(_\Delta F / F_k),$$

where F_k is the frequency band.

2. The number of radio frequencies needed for service of subscribers in one sector of each cell, is defined as

$$n_s = \text{int}(N_k / C \cdot M).$$

3. A value of the permissible telephone load in one sector of one cell or in a cell (for base stations incorporating antennas with the circular pattern) is defined by the following relationships

$$A = n_O \left[1 - \sqrt{1 - \left(P_{sl} \sqrt{\pi n_O / 2} \right)^{1/n_O}} \right] \text{ at } P_{sl} \leq \sqrt{\frac{2}{\pi n_o}};$$

$$A = n_O + \sqrt{\frac{\pi}{2} + 2 n_O \ln \left(P_{sl} \sqrt{\pi n_O / 2} \right)} - \sqrt{\frac{\pi}{2}} \text{ at } P_{sl} > \sqrt{\frac{2}{\pi n_o}},$$

where $n_0 = n_s \cdot n_a$; n_a is the number of subscribers which can use one frequency channel simultaneously. The value is defined by standard.

4. The number of subscribers under service of the base station, depending on the number of sectors, permissible telephone load and activity of subscribers

$$N_{aBTS} = M\,int(A\,/\,\beta).$$

5. The necessary number of the base stations at the given territory of covering, is defined as

$$N_{BTS} = int(N_a\,/\,N_{aBTS}).$$

where N_a is the given number of subscribers to be under service of the cellular communication network.

6. The cell radius, under condition that the load is uniformly distributed over the entire zone, is defined by the formula

$$R = \sqrt{\frac{1,21 \cdot S_0}{\pi N_{BTS}}}.$$

7. The value of the protective distance between BTS with equal frequency channels, is defined as

$$D = R\sqrt{3C},$$

and other parameters such as the necessary power at the receiver input, the probability of error in the process of communication session, the efficiency of radio spectrum use, etc.

Finding the subset of Pareto-optimal network variants is performed in criteria space of the quality indicators estimates. A single variant of CCN was chosen with the use of the conditional criteria of preference by finding the extreme of the scalar criteria function as $c_i = \frac{1}{7}$, $i = \overline{1,7}$.

For a choice of optimal design solutions on the basis of multicriteria optimization methods, there was developed the program complex. It includes two parts solving the following issues.

1. Setting initial data and calculation of technical parameters for some permissible set of variants of CCN.
2. A choice of Pareto-optimal network variants and narrowing them to a single one.

Fig. 2 shows, as an example, the program complex interface. Here is shown part of table with values of 14 indicators for 19 CCN variants. There is the possibility to choose («tick off») concrete quality indicators to be taken into account at the multicriteria optimization. Besides, here are given values of coefficients of relative importance of chosen quality indicators.

There was selected a subset of Pareto-optimal variants including 71 network variants. Therewith 29 certainly worst variants are rejected. From the condition of minimum conditional criteria of preference as of the Pareto subset, a single variant is chosen (№72). It is characterized by the following initial and calculated parameters: the number of subscribers is 30000; the area under service is 320 km²; activity of subscribers is 0.025 Erl; the frequency bandwidth is 4 MHz; the permissible probability of call blocking is 0.01; percentage of the connection quality deterioration time is 0.07; the density of service is 94 active subscribers per km²; the cluster size is 7; the number of base stations in the network is 133; the number of subscribers serviced by one BS is 226; the efficiency of radio frequency spectrum is $1.614 \cdot 10^{-4}$ active subscribers per Hz; the telephone load is 3.326 Erl; the probability of error is $5.277 \cdot 10^{-7}$; the angle of antenna radiation pattern is 120 degrees.

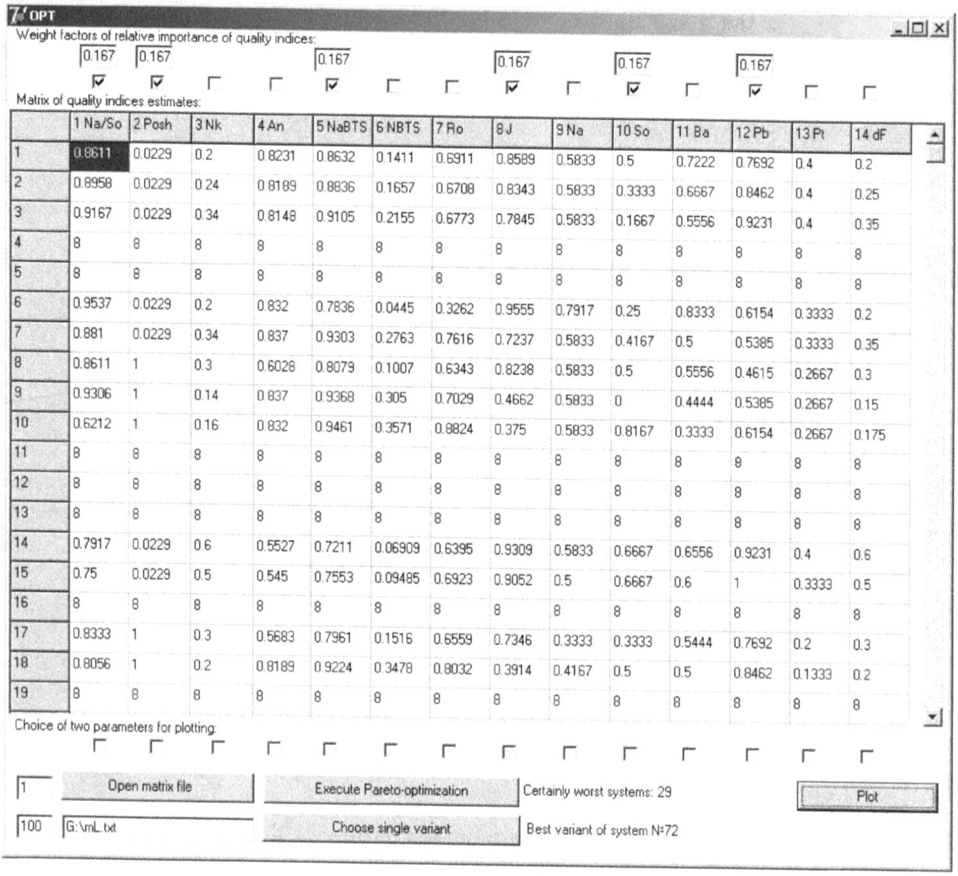

	1 Na/So	2 Posh	3 Nk	4 An	5 NaBTS	6 NBTS	7 Ro	8 J	9 Na	10 So	11 Ba	12 Pb	13 Pt	14 dF
1	0.8611	0.0229	0.2	0.8231	0.8632	0.1411	0.6911	0.8589	0.5833	0.5	0.7222	0.7692	0.4	0.2
2	0.8958	0.0229	0.24	0.8189	0.8836	0.1657	0.6708	0.8343	0.5833	0.3333	0.6667	0.8462	0.4	0.25
3	0.9167	0.0229	0.34	0.8148	0.9105	0.2155	0.6773	0.7845	0.5833	0.1667	0.5556	0.9231	0.4	0.35
4	8	8	8	8	8	8	8	8	8	8	8	8	8	8
5	8	8	8	8	8	8	8	8	8	8	8	8	8	8
6	0.9537	0.0229	0.2	0.832	0.7836	0.0445	0.3262	0.9555	0.7917	0.25	0.8333	0.6154	0.3333	0.2
7	0.881	0.0229	0.34	0.837	0.9303	0.2763	0.7616	0.7237	0.5833	0.4167	0.5	0.5385	0.3333	0.35
8	0.8611	1	0.3	0.6028	0.8079	0.1007	0.6343	0.8238	0.5833	0.5	0.5556	0.4615	0.2667	0.3
9	0.9306	1	0.14	0.837	0.9368	0.305	0.7029	0.4662	0.5833	0	0.4444	0.5385	0.2667	0.15
10	0.6212	1	0.16	0.832	0.9461	0.3571	0.8824	0.375	0.5833	0.8167	0.3333	0.6154	0.2667	0.175
11	8	8	8	8	8	8	8	8	8	8	8	8	8	8
12	8	8	8	8	8	8	8	8	8	8	8	8	8	8
13	8	8	8	8	8	8	8	8	8	8	8	8	8	8
14	0.7917	0.0229	0.6	0.5527	0.7211	0.06909	0.6395	0.9309	0.5833	0.6667	0.6556	0.9231	0.4	0.6
15	0.75	0.0229	0.5	0.545	0.7553	0.09485	0.6923	0.9052	0.5	0.6667	0.6	1	0.3333	0.5
16	8	8	8	8	8	8	8	8	8	8	8	8	8	8
17	0.8333	1	0.3	0.5683	0.7961	0.1516	0.6559	0.7346	0.3333	0.3333	0.5444	0.7692	0.2	0.3
18	0.8056	1	0.2	0.8189	0.9224	0.3478	0.8032	0.3914	0.4167	0.5	0.5	0.8462	0.1333	0.2
19	8	8	8	8	8	8	8	8	8	8	8	8	8	8

Fig. 2. Interface of program complex.

As results of Pareto-optimization, there were obtained multivariate patterns of exchange (MPE) of the quality indicators, being of antagonistic character. For illustration, some MPE are presented at fig. 3. Each MPE point defines the potentially best values of each indicator which can be attained at fixed but arbitrary values of other quality indicators. MPE also show how the improvement of some quality indicators is achieved at the expense of other.

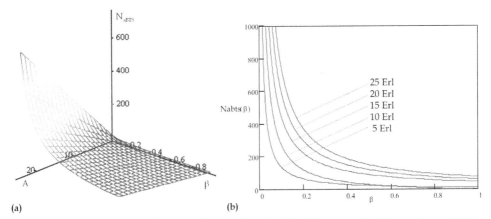

Fig. 3. MPE of the quality indicators (the number of subscribers serviced by one base station (a), the load, the activity of subscribers (b)) for CCN of GSM standard.

3.3 Features of a choice of Pareto-optimal routes

We have a set of permissible solutions (routes) on the finite network graph $G = (V, X)$, where $V = \{v\}$ – set of nodes, $E = \{e\}$ – set of network lines. Each route X is defined by a subset of the nodes and links. The goal task is presented by the model $\{X, F\} \rightarrow x^*$, where $X = \{x\}$ – set of permissible solutions (routes) on the network graph $G = (V, E)$; $F(x)$ – objective function of choice of the routes; x^* – optimal solution of the routing problem. The multicriteria approach of a choice of the best routes relies to perform decomposition of the function $F(x)$ to set (vector) partial choice functions. In this case on the set X it is given the vector of the objective function (Bezruk & Varich, 2011):

$$F(x) = \left(W_1(x), ..., W_j(x), ..., W_m(x)\right),$$

where components determine the values of quality routes indicators.

The route variant $x^* \in X$ is Pareto-optimal route if another route $x \in X$ doesn't exist, order to perform inequality $F_j(x^*) \le F_j(\tilde{x}), j = 1, ..., m$, where at least one of the inequalities is strict. We propose to solve the problem of finding Pareto-optimal routes by using weight method. It is used for finding extreme values of the objective route function as a weighted sum of the partial choice functions for all possible values of the weighting coefficients λ_j:

$$\underset{\text{var } x \in X}{\text{extr}} = \left(F(x)\right) = \sum_{j=1}^{m} \lambda_j W_j(x).$$

Pareto-optimal routes have some characteristic features. Particularly, Pareto-optimal alternative routes corresponds to the Pareto coordinated optimum partial objective functions $W_1(x), ..., W_j(x), ..., W_m(x)$. When selecting a subset of the Pareto-optimal routes there was dropped a certainly worst variant in terms of the absolute criteria of preference.

Pareto-optimal alternatives of the routes are equivalent to the Pareto criteria and could be used for organizing multipath routing in the multi-service telecommunication networks.

Network model consists of twelve nodes; they are linked by communication lines with losses (fig. 4).

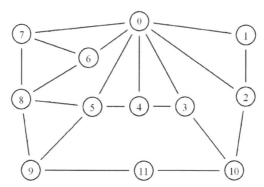

Fig. 4. The structure of the investigated network.

The quality indicators normalized to maximum values are presented in table 2.

The link	The delay time of packets transmission k_1	The level of packet loss k_2	The cost of using the line k_3
0-1	0.676	1	0.333
0-2	1	0.25	1
0-3	0.362	1	0.333
0-4	0.381	0.25	1
0-5	0.2	1	0.333
0-6	0.19	1	0.333
0-7	0.571	0.25	1
7-6	0.4	0.25	0.333
7-8	0.362	0.25	0.667
8-6	0.314	0.5	0.5
8-5	0.438	0.25	0.333
8-9	0.248	0.5	0.333
9-5	0.257	0.25	1
9-11	0.571	0.25	0.667
11-10	0.762	0.25	0.333
5-4	0.381	0.25	0.667
2-10	0.457	0.25	0.333
3-10	0.79	0.25	0.333
4-3	0.286	0.25	0.333
1-2	0.448	0.25	0.333

Table 2. Network quality indicators.

Network analysis shows that for each destination node there are many options to choose the route directly. For example, between node 0 and node 8 there are 22 routes.

Fig. 5 shows the set of the alternative routes between nodes 0 and 8 in the space of the quality indicators k_1 and k_2. Subset of the Pareto-optimal alternatives routes corresponds to the left lower border which includes three variants, they are marked (▲). This subset corresponds to be coordinated in Pareto optimum of the quality indicators.

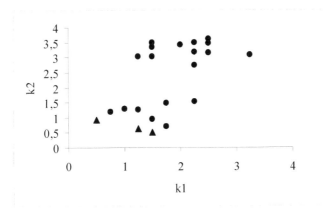

Fig. 5. Set of the routes between nodes 0 and 8.

The resulting subset of the Pareto-optimal alternative routes can be used for organizing multipath routing when using MPLS technology. It will allow to provide a load balancing and a traffic management and to provide given quality-of-service taking into account the set of the quality indicators.

3.4 Pareto-optimal choice of the speech codec

Proposed theoretical investigations can be used for Pareto-optimal choice of the speech codec used in IP-telephony systems (Bezruk & Skorik, 2010).

For carrying out the comparative analysis of basic speech codec and the optimal codec variant choice there have been used the data about 23 speech codecs described by the set of the technical and economic indicators: coding rate, quality of the speech coding, complexity of the realization, frame size, total time delay, etc. The initial values of the quality indicators are presented in table 3. It is easy to see that presented quality indicators are connected between each other with competing interconnections.

The time delay is increasing with frame size increasing as well as with complexity of the coding algorithm realization. Then, when transferring speech the permissible delay can not be bigger than 250 ms in one direction.

A frame size influences on the quality of a reproduced speech: the bigger is the frame, the more effective is the speech modeled. On other hand, the big frames increase an influence of the time delay on processing the information transferring. A frame size is defined by the compromise amongst these requirements.

№	Codec	Speech coding, kbps	Coding quality, MOS (1-5)	Complexity of the realization, MIPS	Frame size, ms	Total delay, ms
1	G 711	64	3,83	11,95	0,125	60
2	G 721	32	4,1	7,2	0,125	30
3	G 722	48	3,83	11,95	0,125	31,5
4	G 722(a)	56	4,5	11,95	0,125	31,5
5	G 722(b)	64	4,13	11,95	0,125	31,5
6	G 723.1(a)	5,3	3,6	16,5	30	37,5
7	G 723.1	6,4	3,9	16,9	30	37,5
8	G 726	24	3,7	9,6	0,125	30
9	G 726(a)	32	4,05	9,6	0,125	30
10	G 726(b)	40	3,9	9,6	0,125	30
11	G 727	24	3,7	9,9	0,125	30
12	G 727(a)	32	4,05	9,9	0,125	30
13	G 727(b)	40	3,9	9,9	0,125	30
14	G 728	16	4	25,5	0,625	30
15	G 729	8	4,05	22,5	10	35
16	G 729a	8	3,95	10,7	10	35
17	G 729b	8	4,05	23,2	10	35
18	G 729ab	8	3,95	11,5	10	35
19	G 729e	8	4,1	30	10	35
20	G 729e(a)	11,8	4,12	30	10	35
21	G 727(c)	16	4	9,9	0,125	30
22	G 728(a)	12,8	4,1	16	0,625	30
23	G 729d	6,4	4	20	10	35

Table 3. Codecs characteristics.

Complexity of the realization is connected with providing necessary calculations in real time. The coding algorithm complexity influences on the physical size of coding, decoding or combined devices, and also on its cost and power consumption.

In table 4 are presented some transformations results of the initial values of the quality indicators. In particular, there were performed the rationing operations of the indicators to their maximum values $k_{iн} = \dfrac{k_i}{k_{imax}}$. These indicators were transformed to a comparable kind where all indicators had the same character depending on the technical codecs characteristics. In particular, for indicators k_{3n} and k_{5n} the transformations $k'_{3н} = \dfrac{1}{k_{3н}}$, $k'_{5н} = \dfrac{1}{k_{5н}}$ were done.

№	Codec	K_{1n}	K_{2n}	K'_{3n}	K_{4n}	K'_{5n}	Pareto-optimal choice
1	G 711	1	0,851	0,604	0,004	0,515	-
2	G 721	0,5	0,911	1	0,004	1	+
3	G 722	0,75	0,851	0,604	0,004	0,969	-
4	G 722(a)	0,875	1	0,604	0,004	0,969	+
5	G 722(b)	1	0,918	0,604	0,004	0,969	+
6	G 723.1(a)	0,083	0,8	0,439	1	0,818	+
7	G 723.1	0,1	0,867	0,424	1	0,818	+
8	G 726	0,375	0,822	0,748	0,004	1	-
9	G 726(a)	0,5	0,9	0,748	0,004	1	-
10	G 726(b)	0,625	0,866	0,748	0,004	1	+
11	G 727	0,375	0,822	0,727	0,004	1	-
12	G 727(a)	0,5	0,9	0,727	0,004	1	-
13	G 727(b)	0,625	0,866	0,727	0,004	1	-
14	G 728	0,25	0,889	0,281	0,021	1	+
15	G 729	0,125	0,9	0,317	0,333	0,879	+
16	G 729a	0,125	0,878	0,669	0,333	0,879	+
17	G 729b	0,125	0,9	0,309	0,333	0,879	-
18	G 729ab	0,125	0,878	0,626	0,333	0,879	-
19	G 729e	0,125	0,911	0,237	0,333	0,879	-
20	G 729e(a)	0,184	0,915	0,237	0,333	0,879	+
21	G 727(c)	0,25	0,889	0,727	0,004	1	-
22	G 728(a)	0,2	0,911	0,453	0,021	1	+
23	G 729d	0,1	0,889	0,359	0,333	0,879	+

Table 4. Transformed quality indicators.

On the base of received results there were considered the practical application features examined methods of the allocation of the Pareto-optimal speech codec variant set taking into account a set of the quality indicators as well as the unique design decision choice. From the initial set of the 23 speech codecs variants there was allocated the Pareto subset included 12 codecs variants (marked + in table 4).

The only one project decision was chosen from the condition of the scalar goal function extreme (9) with two different values of β defined characters of this function changing. In table 5 are presented the values of the given function for Pareto-optimal speech codecs variants at $\beta = 2$ and $\beta = 3$. It was obtained that an extreme goal function value, depending on β, is reached for the same speech codec G 722 (b).

Within statement of a problem we have chosen the codec of series G.722b which has following values of the quality indicators: speech coding – 64 kbps, coding quality – 4,13 MOS, complexity of the realization – 11,95 MIPS, the frame size – 0,125 ms, total delay – 31,5 ms.

№	Codec	Values $\xi_{\bar{k}}$ for diffrent β	
		$\beta = 2$	$\beta = 3$
2	G 721	0,35099	0,24688
4	G 722(a)	0,35039	0,28188
5	G 722(b)	0,35476	0,28532
6	G 723.1(a)	0,31677	0,25791
7	G 723.1	0,32312	0,26308
10	G 726(b)	0,32863	0,26445
14	G 728	0,27801	0,24056
15	G 729	0,26904	0,22785
16	G 729a	0,29103	0,23837
20	G 729e(a)	0,26912	0,22898
22	G 728(a)	0,28812	0,24582
23	G 729d	0,26927	0,22716

Table 5. Results of multicriteria optimization.

3.5 Network resources controlling

Let us consider some features of the short-term planning issues in the telecommunication system. There was shown the important place of multi-service network occupied with models, methods and facilities of network resources controlling in modern and perspective technologies. To the basic network resources facilities belong: channel resources control facilities (channels throughput, buffers size, etc), information resources control (user traffic).

Considered system was presented as the model of a distributed telecommunication system, consisting from a set of operating agents, for each autonomous system (fig 6).

In this model the process of network resources control was carried out by finding the distribution streams vector of the following type (Bezruk & Bukhanko, 2010):

$$\vec{K} = (k_1, k_2, ..., k_l), \quad \sum_i^l k_i = 1,$$

with next limitation

$$0 \le k_i \le 1, \quad i = \overline{1...l};$$

$$\lambda_i^{out} k_i \le c_i, \quad i = \overline{1...l}.$$

Each element of this vector characterizes a part of outgoing user traffic from autonomous system operating agent transferred by using a corresponding channel. Within a given model, the task of network resources controlling comes to solving the optimization problem connected to function minimization.

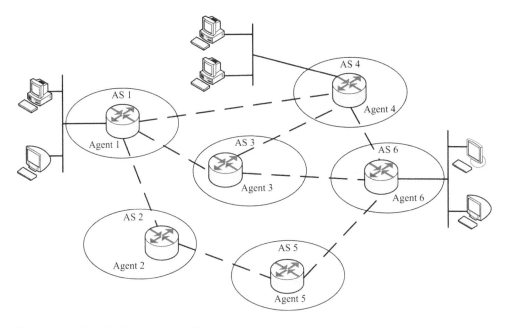

Fig. 6. Considered telecommunication system.

$$\varepsilon(\vec{K}) = \min(q_1\Phi + q_2\sigma_1(\vec{K}) + q_3\sigma_2(\vec{K})),$$ (10)

where $\sigma_1(\vec{K})$ – standard deviation of channels loading x_i, $i = \overline{1...I}$;

$$\sigma_1(\vec{K}) = \sqrt{\frac{1}{I-1}\sum_{i=1}^{I}\left(x_i - \overline{x}\right)^2};$$

$\sigma_2(\vec{K})$ – standard deviation of agents loading Z_i, $i = \overline{1...I}$;

$$\sigma_2(\vec{K}) = \sqrt{\frac{1}{I-1}\sum_{i=1}^{I}\left(Z_i - \overline{Z}\right)^2};$$

Φ – used routing protocol metric;

$$\Phi = \sum_{i=1}^{I}\varphi_i x_i;$$

φ_i – cost of full used channel ($\sum\lambda_i = c_i$);

q_1, q_2, q_3 – weight coefficients characterized the traffic balancing cost using standard metric, agents and channels loading.

The considered mathematical model of the distributed network resources controlling uses specific criteria of optimality included standard routing protocol metrics, a measure of channels and agents loading in given telecommunication network.

Obviously, under condition of $\sigma_1(\vec{K})$ and $\sigma_2(\overline{K})$ absence, function (10) becomes the model of the load balancing under the routes with equal or non-equal metric. However, absence of the decentralized control behind the autonomous system of telecommunication network can finally result in an uncontrollable overload. That fact is defined by the presence of additional minimized indicators leading to the practical value of the proposed model. Thus a choice of the relation of weight coefficients q_1, q_2 and q_3 is an independent problem demanding some future investigations and formalizations. In this model this task was dared with expert's estimations.

The proposed imitation model included up to 18 agents (fig. 7). Researches for different variants of connectivity between agents have been carried out.

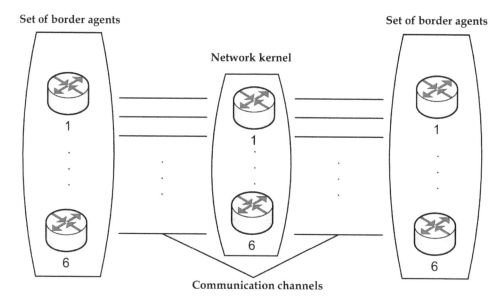

Fig. 7. Used imitation model.

During practical investigation there were analyzed several models of multipath routing and load balancing. These models are listed below:

M1 – model of routing by RIP;
M2 – model of multipath routing by an equal metric;
M3 – model of multipath routing by an non-equal metric (IGRP);
M4 – Gallagher stream model;
M5 – considered model with multicriteria account of two indicators (10);
M6 – considered model with multicriteria account of three indicators (10).

Below are presented some results of the analytic and imitation modeling within comparative analysis of considered existing and proposed models. These results are shown as dependences of the blocking probability and average delay time from the network loading (fig. 8).

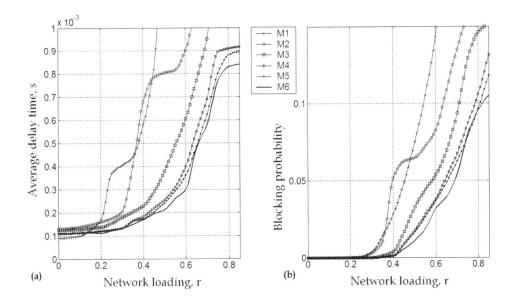

(a)   Network loading, r

(b)   Network loading, r

Fig. 8. Received dependences of average delay time (a) and blocking probability (b).

The use of the proposed models allows to:

-   lower the average delay time (a) in comparison with the best known model (M4), for 3 – 12% (M5) and for 6 – 25% (M6);
-   lower the general blocking probability (b) for 6 – 11% (M5) and 6 – 20% (M6).

4. Conclusion

The present work deals with the methodology of generating and selecting the variants of information systems when they are optimized in terms of the set of quality indicators. The multicriteria system-optimization problems are solved in three stages. By using the morphological approach a structural set of permissible variants of a system is initially generated. This set is mapped into the space of vector estimates. In this space a subset of Pareto-optimal estimates is selected, defining the potential characteristics of the system on the basis of the set of quality indicators. At the conclusive stage the only variant is selected amongst the Pareto-optimal variants of the system provided there exists an extreme of a certain scalar functional whose form is determined with the use of some additional information obtained from a customer.

Multicriteria optimization issues and methods based on Pareto conclusions are introduced for the long-term and short-term practical planning, designing and controlling within different types of telecommunication networks. In the process of solving the optimization problems we consider the set of network quality indicators as the different network topologies, transmission capacities of communication channels, various disciplines of service requests applied to different routing ways, etc.

Peculiarities of the long-term multicriteria optimization methods used for solving problems of the cellular networks planning are considered. As an example, the Pareto-optimization solution within planning of the cellular communication networks is also presented.

Practical features of the multicriteria approach in solving the optimal routing problem in the multi-service networks are considered within organizing multipath routing as well as speech codec choice based on a set of the quality indicators. The model of the information resources balancing on a basis of the decentralized operating agents system with a multicriteria account of chosen quality indicators is also offered. Considered adaptive balancing traffic algorithm improves the basic characteristics of the telecommunication network in a process of the short-term controlling for chosen cases of topologies.

5. Acknowledgment

The research described in this work was made possible in part by the scientific direction "Telecommunication and information networks optimization", headed by prof. Bezruk V., of the Communication Network Department within Kharkov National University of Radio Electronics, Ukraine.

6. References

Bezruk, V. & Skorik, Y. (2010). Optimization of speech codec on set of indicators of quality. *Proceedings of TCSET'2010 Modern problems of radio engineering, telecommunications and computer science*, p. 212, ISBN 978-966-553-875-2, Lviv – Slavske, Ukraine, February 23 – 27, 2010

Bezruk, V. & Bukhanko, O. (2010). Control mode of network resources in multiservice telecommunication systems on basis of distributed system of agents. *Proceedings of CriMiCo'2010 Microwave and Telecommunication Technology*, pp. 526-527, ISBN 978-966-335-329-6, Sevastopol, Crimea, Ukraine, September 13 – 17, 2010

Bezruk, V. & Varich, V. (2011). The multicriteria routing problem in multiservice networks with use composition quality indicators. *Proceedings of CriMiCo'2011 Microwave and Telecommunication Technology*, pp. 519 – 520, ISBN 978-966-335-254-8, Sevastopol, Crimea, Ukraine, September 12 – 16, 2011

Figueira, J. (Ed(s).). (2005). *Multiple Criteria Decision Analysis: State of the Art Surveys*, Springer Science + Business Media, Inc, ISBN 978-0-387-23081-8, Boston, USA

Saaty, P. (2005). *Theory and Applications of the Analytic Network Process: Decision Making with Benefits, Opportunities, Costs and Risks*, RWS Publications, ISBN 1-888603-06-2, Pittsburgh, USA

Taha, H. (1997). *Operations Research: An Introduction*, Prentice Hall Inc., ISBN 0-13-272915-6, New Jersey, USA

Part 2

Traffic Engineering

4

Modelling a Network Traffic Probe Over a Multiprocessor Architecture

Luis Zabala, Armando Ferro,
Alberto Pineda and Alejandro Muñoz
University of the Basque Country (UPV/EHU)
Spain

1. Introduction

The need to monitor and analyse data traffic grows with increasing network usage by businesses and domestic users. Disciplines such as security, quality of service analysis, network management, billing and even routing require traffic monitoring and analysis systems with high performance. Thus, the increasing bandwidth in data networks and the amount and variety of network traffic have increased the functional requirements for applications that capture, process or store monitored traffic. Besides, the availability of capture hardware (monitoring cards, taps, etc.) and mass storage solutions at a reasonable cost makes the situation better in the field of network traffic monitoring. For these reasons, several research groups are studying how to monitor heterogeneous network environments, such as wired broadband backbone networks, next generation cellular networks, high-speed access networks or WLAN in campus-like environments. In keeping with this line, our research group NQaS (Networking, Quality and Security) aims to contribute in this challenge and presents theoretical and experimental research to study the behaviour of a probe (Ksensor) that can perform traffic capturing and analysis tasks in Gigabit Ethernet networks. Not only do we intend to progress in the design of traffic analysis systems, but we also want to obtain mathematical models to study the performance of these devices.

The widespread of 1/10 Gigabit Ethernet networks, emphasizes the problems related to system losses which invalidate the results for certain analyses. New Gigabit networks, even at 40 and 100 Gbps, are already being implemented and the problem becomes accentuated. On top of that, commodity systems are not optimized for monitoring [Wang&Liu, 2004] and, as a result, processing resources are often wasted on inefficient tasks. Because of this, new research works have arisen focusing on the development of analysis systems that are able to process all the information carried by actual networks.

Taking all this into account, we would like to develop analytical models that represent traffic monitoring systems in order to provide solutions to the problems mentioned before. Modelling helps to predict the system's performance when it is subjected to a variety of network traffic load conditions. Designers and administrators can identify bottlenecks, deficiencies and key system parameters that impact its performance, and thereby the system can be properly tuned to give the optimal performance. By means of modelling technique, it is possible to draw qualitative and, in many cases, also quantitative conclusions about features related to modelled systems even without having to develop them. The impact of developing costs, which is a determining factor in some cases, can be dramatically reduced by using modelling.

Having this in mind, and considering the experience of our group, we present our original design (Ksensor) that improves system performance, as well as a mathematical model based on a closed queueing network which represents the behaviour of a multiprocessor traffic monitoring and analysis system. Both things are considered together in the validation of the model, where Ksensor is used as well as a testing platform developed by NQaS. All these aspects are presented throughout this chapter.

A number of papers has addressed the issue of modelling traffic monitoring systems. However, there are more related to the hardware and software involved in this type of systems.

Regarding hardware proposals, one of the most relevant was the development of the high-performance DAG capture cards [Cleary et al., 2000] at the University of Waikato (New Zealand). Several research works and projects have made use of these cards for traffic analysis system design. Some other works proposed the use of Network Processors (NP) [Intel, 2002]. Conventional hardware also showed bottlenecks and new input/output architectures were proposed, such as Intel's CSA (Communication Streaming Architecture).

At the software level, Mogul and Ramakrishnan [Mogul&Ramakrishnan, 1996] identified the most important performance issues on interrupt-driven capture systems. Zero-copy architectures are also remarkable [Zhu et al, 2006]. They try to omit the path followed by packets through the system kernel to the user-level applications, providing a direct access to captured data or mapping memory spaces (mmap). Biswas and Sinha proposed a DMA ring architecture [Biswas&Sinha, 2006] shared by user and kernel levels. Luca Deri suggests a passive traffic monitoring system over general purpose hardware at Gbps speeds (nProbe). Deri has also suggested improvements for the capture subsystem of GNU/Linux, such as a driver-level ring [Deri, 2004], and a user-level library, nCap [Deri, 2005a]. Recently, Deri has proposed a method for speeding up network analysis applications running on Virtual Machines [Cardigliano, 2011], and has presented a framework [Fusco&Deri, 2011] that can be exploited to design and implement this kind of applications.

Other proposals focus on parallel systems. Varenni et al. described the logic architecture of a multiprocessor monitoring system based on a circular capture buffer [Varenni et al.,2003] and designed an SMP driver for DAG cards. We must also remark the KNET module [Lemoine et al., 2003], a packet classifying system at the NIC to provide independent per connection queues for processors. In addition, Schneider and Wallerich studied the performance challenges over general purpose architectures and described a methodology [Schneider, 2007] for evaluating and selecting the ideal hardware/software in order to monitor high-speed networks.

Apart from the different proposals about architectures for capture and analysis systems, there are analytical studies which aim at the performance evaluation of these computer systems. Among them, we want to underline the works done by the group led by Salah

[Salah, 2006][Salah et al., 2007]. They analyse the performance of the capturing system considering CPU consumptions in a model based on queuing theory. Their last contributions explain the evolution of their models towards applications like Snort or PC software routers. Another work in the same line was developed by Wu [Wu et al., 2007], where a mathematical model based on the 'token bucket' algorithm characterized Linux packet reception process.

We also have identified more complex models whose application to traffic capturing and analysis systems can be very beneficial. They are models based on queuing systems with vacations. In this field, we want to underline the contributions from Lee [Lee, 1989], Takagi [Takagi, 1994, 1995] and Fiems [Fiems, 2004].

Most of the previous approaches are for single processor architectures. However, it is clear interest in the construction of analytical models for multiprocessor architectures, in order to evaluate their performance. This paper contributes in this sense from a different point of view, given that the model is based on a closed queueing network. Furthermore, the analytical model and the techniques presented in this paper can be considerably useful not only to model traffic monitoring systems, but also to characterize similarly-behaving queueing systems, particularly those of multiple-stage service. These systems may include intrusion detection systems, network firewalls, routers, etc.

The rest of the chapter is organized as follows: in Section 2 we introduce the framework of our traffic and analysis system called 'Ksensor'. Section 3 presents the analytical model for evaluating the performance of the traffic monitoring system. Section 4 provides details on the analytical solution of the model. Section 5 deals with the validation and obtained results are discussed. Finally, Section 6 remarks the conclusions and future work.

2. Ksensor: Multithreaded kernel-level probe

In a previous work [Muñoz et al., 2007], our research group, NQaS, proposed a design for an architecture able to cope with high-speed traffic monitoring using commodity hardware. This kernel-level framework is called Ksensor and its design is based on the following elements:

- Migration to the kernel which consists in migrating the processing module from user-level to the kernel of the operating system.
- Execution threads defined to take advantage of multiprocessor architectures at kernel-level and solve priority problems. Independent instances are defined for capture and analysis phases. There are as many analysing instances as processors, and as many capturing instances as capturing NICs.
- A single packet queue, shared by all the analysing instances, omitting the filtering module and so saving processing resources for the analysis.

This section explains the main aspects of Ksensor, because of its importance in the validation of the mathematical model which will be explained in a subsequent section.

2.1 Architecture of Ksensor

The kernel-level framework, called Ksensor, intended to exploit the parallelism in QoS algorithms, improving the overall performance.

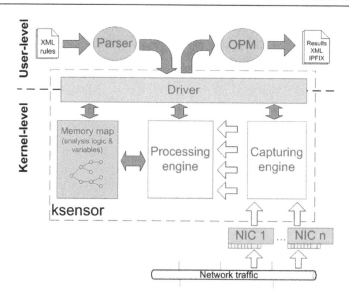

Fig. 1. Architecture of Ksensor.

Fig. 1 shows the architecture of Ksensor. As we can see, only the system configuration (parser) and the result management (Offline Processing Module, OPM) modules are at user-level. Communication between user and Kernel spaces is offered by a module called driver. The figure also shows a module called memory map. This module is shared memory where the analysis logic and some variables are stored.

The definition of execution threads is aimed to take advantage of multiprocessor architectures at kernel-level and solve priority problems, minimizing context and CPU switching. Kernel threads are scheduled at the same level than other processes, so the Kernel's scheduler is responsible for this task.

Ksensor executes two tasks. On one hand, it has to capture network traffic. On the other hand, it has to analyse those captured packets. In order to do that, we define independent instances for capture and analysis phases. Each thread belongs to an execution instance of the system and is always linked with the same processor. All threads share information through the Kernel memory.

In Fig. 2 we can see the multithreaded execution instances in Ksensor. There are as many analysing instances as processors (ksensord#n) and as many capturing instances as capturing NICs (ksoftirqd#n). For example, if the system has two processors, one of them is responsible for capturing packets and analysing some of them and the other one is responsible for analysing packets. This way an analysis task could fill the 100% of one processor's resources if necessary.

The capturing instance takes the packets that the networking subsystem captures and stores them in the packet queue. There is only one packet queue. Processing instances take packets from that queue in order to analyse them.

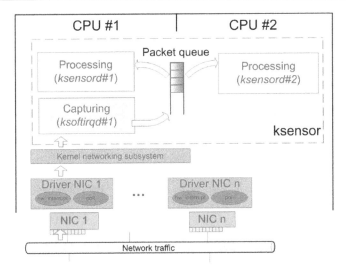

Fig. 2. Multithreaded execution instances in Ksensor.

It does not matter what processing thread analyses a packet because all of them use the same analysis logic. As we said before, there is a shared memory (memory map module) that stores the analysis logic. All the processing threads can access this memory.

2.2 Capturing mechanism in Linux

Ksensor is integrated into the Linux Kernel. In order to capture the packets of the net, Ksensor uses the Kernel networking subsystem. The capturing interface of this subsystem is called NAPI (New API). Nowadays, all the devices have been upgraded to NAPI. Because of that it is important to explain how this interface works [Benvenuti, 2006].

When the first packet arrives to the NIC, it is stored on the card's internal buffer. When the PCI bus is free, the packets are copied from the NIC's buffer to a ring buffer through DMA. The ring buffer is also known as DMA buffer. Once this copy has finished, a hardware interrupt (hardirq) is generated. All of these actions have been executed without consuming any processor's resources.

If the network interface copies a lot of packets in the ring buffer and the Kernel does not take them out, the ring buffer fills up. In this case, unless the interrupts are disabled, another interrupt is generated in order to notify this situation. Then, while the ring buffer is full, the new captured packets will be stored on the NIC's buffer. When this buffer fills up too, the arriving packets will be dropped.

In any case, when the kernel detects the network card interrupt, its handler is executed. In this handler, the NIC driver registers the network interface in an especial list called poll list. This means that this interface has captured packets and needs the Kernel to take them out of the ring buffer. In order to do that, a new software interrupt (softIRQ) is scheduled. Finally, hardIRQs are disabled. From now on, the NIC will not notify new packet arrivals or overload of the ring buffer.

2.3 Network interfaces polling

The softIRQ handler takes out packets from the ring buffer. In Ksensor, after taking out a packet from the ring buffer, the handler stores it in a special queue called packet queue, as we can see in Fig. 2.

The system decides when a softIRQ handler is executed. When its execution starts, the handler polls the first interface in the poll list and starts taking out packets from its ring buffer. In each poll, the softIRQ handler can only pull out packets up to a maximum number called quota. When it reaches the quota it has to poll the next interface in the poll list. If an interface does not have more packets it is deleted from the poll list. Besides, in a softIRQ, the handler can only take out a maximum number of packets called budget. When the handler reaches this maximum, the softIRQ finishes. If there are interfaces left in the poll list, a new softIRQ is scheduled. Furthermore, a softIRQ may take one jiffy (4 ms) at most. If it consumes this time and there are still packets to pull out, the softIRQ finishes and a new one is scheduled.

There is only one poll list in each processor. When the hardIRQ handler is called it registers the network interface in the poll list of the processor that is executing the handler. The softIRQ handler is executed in the same processor. At any given time, a network interface can only be registered in one poll list.

Ksensor has a system to improve the performance in case of congestion. When the packet queue reaches a maximum number of stored packets, this system forces NAPI to stop capturing packets. This means that all the resources of all the processors are dedicated to analysing instances. When the number of packets in the packet queue reaches a fixed threshold value the system starts capturing again.

3. Model for a traffic monitoring system

This section introduces an analytical model which works out some characteristics of network traffic analysis systems. There are several alternatives to model theoretically this type of system. For example, you can use models of queuing theory, Petri nets and, even, mixed models. The ultimate goal is to have a theoretical model that allows us to study the performance of a network traffic analysis system, considering those parameters that are the most representative: throughput, number of processors, analysis load and so on.

We have chosen a theoretical model based on closed queuing networks. It is able to represent accurately the behaviour of a system in charge of analysing network traffic loaded in a multiprocessor architecture. Queuing theory allows us to develop models in order to study the performance of computer's systems [Kobayashi, 1978]. Proposed model consists in a closed queue network where CPU consumptions are related to the service capacity of the queues.

It is worth mentioning that both the flowing traffic and the processing capacity at the nodes are modelled by Poisson arrival rates and exponential service rates. Poisson's distributions are considered to be acceptable for modelling incoming traffic [Barakat et al., 2002]. This assumption can be relaxed to more general processes such as MAPs (Markov Arrival Processes) [Altman et al., 2000], or non homogeneous Poisson processes, but we will keep working with it for simplicity of the analysis. Regarding service rate modelling, although program's code has a quite deterministic behaviour, some randomness is introduced by Poisson incoming traffic, variable length of packets and kernel scheduler uncertainty.

3.1 Description of the model

The proposed queuing network for modelling a traffic monitoring system is showed in Fig. 3. It consists of two parts; the upper one has a set of multi-server queues which represents the processing ability of the traffic analysis system. The lower part models the injection of network traffic with λ rate with a simple queue. The number of packets that are permitted in the closed queue network is fixed and its value is N.

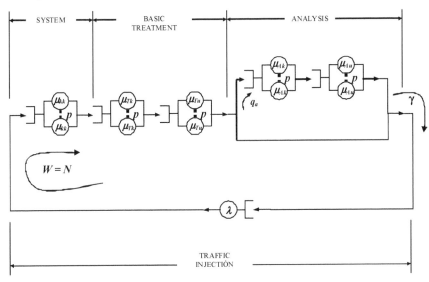

Fig. 3. General model for the traffic analysis system.

Some stages are divided into multiple queues, due to the need to differentiate the processing done in the Kernel and the processing done at user level. Although the process code is usually running on the user level, system calls that require Kernel services are also used.

Four different stages have been distinguished for the closed network, each one with a specific function:

- System stage (system queue): it consists in a queue of μ_{kk} (measured in packets per second) capacity. This stage represents the time spent on the Kernel level of the operating system by the traffic analysis system. It comprises treatments of device controllers and attention paid by kernel to interruptions (hardIRQ and softIRQ) due to packet arrival.
- Basic treatment stage (treatment queues): it is modelled by two queues with μ_{Tk} and μ_{Tu} capacities. This stage represents the amount of time consumed by the system to perform basic treatment to packets captured from the net. This is mainly accomplished by studying control headers of the packets and by determining through a decision tree whether a packet need to be further analysed or not.

- Analysis stage (analysis queues): it is integrated by two queues with μ_{Ak} and μ_{Au}. This stage simulates the analysis treatment that the system does to packets that need further analysis. Not all the packets need to be analysed in this stage. For this reason, a rate called q_a has been defined to represent the proportion of received packet that has to be analysed.
- Traffic injection stage (injection queue): it is a simple queue of λ capacity. This stage simulates the arrival of packets to the system with a λ rate. Since the number of packets in the closed network is fixed to N, the traffic injection queue can be empty. This situation simulates the blocking and new packets will not be introduced on the system.

Each service queue has p servers that represent the p processors of a multiprocessor system. Multiple server representation has been chosen to emphasize the possibility of parallelizing every stage of processing. However, all stages may not be necessarily parallelizable. For example, only one processor can access NIC at the same time, so the packet capturing process will not be parallelizable in different instances.

Another aspect to consider is that packets cannot flow freely in the closed network, because the sum of packets attended in the servers that represent the traffic monitoring system never exceeds the maximum number of processors available. Therefore, we have to assure that, at any time, the maximum number of packets in the upper queues of Fig 3 is not greater than p (the number of processors).

Considering an arrival rate of λ packets per second, the traffic analysis system will be able to keep pace with a part of that traffic, defined as $q \cdot \lambda$. Remaining traffic $((1-q) \cdot \lambda)$ will be lost because the platform is not capable of dealing with all the packets. Captured traffic, $q \cdot \lambda$, goes through the system and basic treatment stages. Nevertheless, all traffic will not be subject of further analysis because of features of the modelled system. For example, a system in charge of calculating QoS parameters of all connections that arrive to a server will discard the packets with other destination address or monitoring systems which use sampling techniques will discard a percentage of packets or intrusion detection will apply further detection techniques only to suspicious packets. Therefore, q_a coefficient has been defined to represent the rate of captured packets liable of being further analysed (analysis stage) than treated only (treatment stage). Thus, $q_a \cdot q \cdot \lambda$ of the initial flow will go through the analysis stage.

3.2 Simplifications of the model

The model presented in Fig. 3 is very general, but if we observe it, some simplifications are possible. Simplifications allow us to group different service rates to identify parameters that may be analysed easily. Among the possible simplifications, we highlight two: one related to CPU consumption and another one, to the equivalent traffic monitoring system.

3.2.1 Model of CPU consumption

This simplification proposes to group all the kernel consumptions in a simple queue, whereas user processes consumptions are represented in a multi-queue. It considers that kernel services are hardly parallelizable.

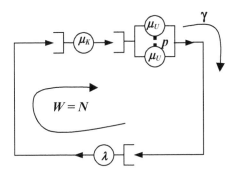

Fig. 4. Model of CPU consumption.

The equivalent service rates can be calculated as follows.

$$\frac{1}{\mu_K} = \frac{1}{\mu_{pk}} + \frac{1}{\mu_{Kk}} = \frac{q_a}{\mu_{Ak}} + \frac{1}{\mu_{Tk}} + \frac{1}{\mu_{Kk}} \tag{1}$$

$$\frac{1}{\mu_U} = \frac{1}{\mu_{pu}} = \frac{q_a}{\mu_{Au}} + \frac{1}{\mu_{Tu}} \tag{2}$$

3.2.2 Model of the equivalent traffic monitoring system

The main feasible simplification preserving the identity of the system is to replace the whole system with an equivalent multi-server queue applying the Norton equivalence [Chandy et al., 1975]. The Norton theorem establishes that in networks with solution in product form, any subnetwork can be replaced by a queue with a state-dependent service capacity. Our theoretical model has exponential service rates in all stages, so applying the Norton equivalence, the new equivalent queue will have a state-dependent service capacity $\mu_{eq}(n,q_a)$.

The simple queue μ_S of the Fig. 5 represents non-parallelizable processes of the system and the multiple queue μ_M represents parallelizable ones.

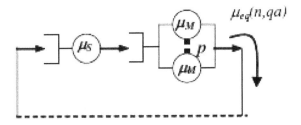

Fig. 5. Traffic monitoring system that Norton equivalence is applied to.

This model adapts perfectly to Ksensor, because we identify a non-parallelizable process that corresponds with the packet capture and parallelizable processes that are related to analysis. Both μ_S and μ_M (in packets per second) can be measured in the laboratory.

4. Analytical study of the model

This section presents the analytical study of the model. It can be directly addressed by analytical calculation, assuming Poisson arrivals and exponential service times. Perhaps the greatest difficulty lies in determining the abstractions that are necessary to adapt the model to the actual characteristics of the traffic monitoring system. Likewise, we propose a method of calculation based on mean value analysis which allows us to solve systems with more elements, where the analytical solution may be more complex to develop.

4.1 Equations of the general model

Viewing the simplifications that have been developed, we might observe that, in the study of this model, a topology is repeated at different levels of abstraction. This topology corresponds with a closed network model with two queues in series; first, a simple one, and second, another one with multiple servers, as shown in Fig. 6. This structure usually occurs in every processing stage. Processing at Kernel level is usually not parallelizable, and therefore, the model is represented as a simple queue. On the other hand, the user processing is usually parallelizable and it is represented by a multiple queue with p servers, being p the number of processor of the platform. The appearance of this topology allows us to define a simple model that we can solve analytically.

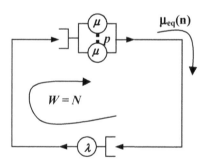

Fig. 6. Closed queue network simplified for the general model.

In order to get the total throughput of the system, first, we calculate the state probabilities for the network, putting N packets in circulation through the closed network, but assuming that the upper multiple queue can have at most p packets being served and the rest waiting in the queue. We also assume that the service capacity in every state of the multiple queue is not proportional to the number of packets. Thus, we will consider μ_i as the service capacity for the state i. The state diagram for this topology is presented in Fig. 7. In this model we are representing the state i of the multiple queue. N packets are flowing through the closed network and we refer to the state i when there are i packets in the multiple queue and the rest, N-i, in the simple queue. The probability of that state is represented as p_i. Finally, the simple queue with rate λ is the packet injection queue.

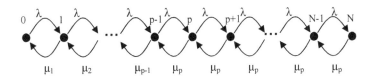

Fig. 7. State diagram for the multiple queue.

It is possible to deduce the balance equations from the diagram of states and, subsequently, the expression of the probability of any state i as a function of the probability of zero state p_0:

$$\forall i = 1, \cdots, p \Rightarrow \begin{cases} p_0 \cdot \lambda = p_1 \cdot \mu_1 \\ p_1 \cdot \lambda = p_2 \cdot \mu_2 \\ \cdots\cdots\cdots\cdots \\ p_{p-1} \cdot \lambda = p_p \cdot \mu_p \end{cases} \Rightarrow p_i = \frac{\lambda}{\mu_i} \cdot p_{i-1} \tag{3}$$

$$\Rightarrow p_i = \overbrace{\frac{\lambda}{\mu_i} \cdot \frac{\lambda}{\mu_{i-1}} \cdots \frac{\lambda}{\mu_1}}^{i \text{ terms}} \cdot p_0 = \frac{\lambda^i}{\prod\limits_{j=1}^{i} \mu_j} \cdot p_0 \tag{4}$$

From this equation, we deduce p_p, the probability of the state p:

$$\Rightarrow p_p = \frac{\lambda^p}{\prod\limits_{j=1}^{p} \mu_j} \cdot p_0 \tag{5}$$

For the states with i>p, their probabilities can be expressed as:

$$\forall i = p + 1, \cdots, N \Rightarrow \begin{cases} p_p \cdot \lambda = p_{p+1} \cdot \mu_p \\ p_{p+1} \cdot \lambda = p_{p+2} \cdot \mu_p \\ \cdots\cdots\cdots\cdots \\ p_{N-1} \cdot \lambda = p_N \cdot \mu_p \end{cases} \Rightarrow p_i = p_{i-1} \cdot \frac{\lambda}{\mu_p} \tag{6}$$

$$p_i = \overbrace{\frac{\lambda}{\mu_p} \cdot \frac{\lambda}{\mu_p} \cdots \frac{\lambda}{\mu_p}}^{(i-p) \text{ terms}} \cdot p_p = \left(\frac{\lambda}{\mu_p} \right)^{i-p} \cdot p_p \tag{7}$$

From this equation we can also derive the expression of the probability p_N, which is interesting because it indicates the probability of having all the packets in the multiple queue and there is none in the simple queue. This probability defines the blocking probability (P_B) of the simple queue.

$$P_N = P_B = \frac{\lambda^N}{\mu_p^{N-P} \cdot \prod_{j=1}^{p} \mu_j} \cdot p_0 \tag{8}$$

Applying the normalization condition (the sum of all probabilities must be equal to 1), we can obtain the general expression for p_0 and, then, we get every state probabilities.

$$\sum_{i=0}^{N} p_i = 1 = p_0 + \sum_{i=1}^{P} p_i + \sum_{i=p+1}^{N} p_i \tag{9}$$

$$1 = p_0 + p_0 \sum_{i=1}^{P} \frac{\lambda^i}{\prod_{j=1}^{i} \mu_j} + p_0 \frac{\lambda^P}{\prod_{j=1}^{p} \mu_j} \cdot \sum_{i=p+1}^{N} \frac{\lambda^{i-p}}{\mu_p^{i-p}} \tag{10}$$

$$\Rightarrow p_0 = \left(1 + \sum_{i=1}^{P} \frac{\lambda^i}{\prod_{j=1}^{i} \mu_j} + \frac{\lambda^P}{\prod_{j=1}^{p} \mu_j} \cdot \sum_{i=p+1}^{N} \frac{\lambda^{i-p}}{\mu_p^{i-p}} \right)^{-1} \tag{11}$$

Considering equations (8) and (11), we have the following blocking probability p_N.

$$P_N = \frac{\lambda^N / \mu_p^{N-p}}{\left(\prod_{j=1}^{p} \mu_j + \sum_{i=1}^{P} \left(\lambda^i \cdot \prod_{j=i}^{p} \mu_j \right) + \lambda^P \cdot \sum_{i=p+1}^{N} \frac{\lambda^{i-p}}{\mu_p^{i-p}} \right)} \tag{12}$$

P_N is the probability of having N packets in the multiple queue (traffic analysis system queue) of Fig. 6 , so there is not any packet in the injection queue. This situation describes the loss of the system. In order to calculate the throughput γ of the system, (13) is used.

$$\gamma = \lambda \cdot (1 - P_N) \tag{13}$$

Taking into account these expressions, which are valid for the general case, we can develop the equations of the model for some particular cases that will be detailed below: the calculation of the equivalence for the traffic monitoring system and the solution for the closed network with incoming traffic load.

4.2 Calculation of the equivalence for the traffic monitoring system

In general, multiprocessor platforms that implement traffic monitoring systems have certain limitations to parallelize some parts of the processing they do. In particular, Kernel services are not usually parallelizable. This means that, despite having a multiprocessor architecture with p processors that can work in parallel, some services will be performed sequentially and we will lose some of the potential of the platform. For all this, in order to calculate the

Norton equivalence for a traffic monitoring system, one must begin with a model that contains a simple queue and a multi-server queue. This is a particular case of the general model studied before.

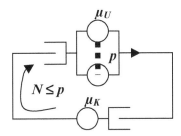

Fig. 8. Equivalence for the traffic monitoring system.

The simple queue with service rate μ_K models non-parallelizable Kernel services, whereas the multiple queue with p servers and service rate μ_U models the system capacity to parallelize certain services. The particularity of this model with regard to the general model is that, at most, only p packets can circulate on the closed network maximum. We are interested in solving this model to work out the equivalent service rate of the traffic monitoring system for every state in the network.

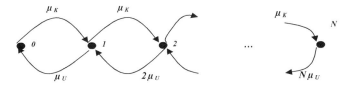

Fig. 9. State diagram for the traffic monitoring system equivalence.

The state diagram makes sense for values of N that are less or equal to the highest number of processors. The service rate of the traffic monitoring system will be different for every value of N and, given that some services are not parallelizable, in general, it does not follow a linear evolution. Following a similar approach to the general case, we can calculate the probability of the highest state, p_N, which is useful to estimate the effective service rate of the equivalence.

$$
\left.
\begin{aligned}
p_0 \cdot \mu_K &= p_1 \cdot \mu_U \\
p_1 \cdot \mu_K &= p_2 \cdot 2\mu_U \\
&\cdots \\
p_{i-1} \cdot \mu_K &= p_i \cdot i \cdot \mu_U
\end{aligned}
\right\}
\Rightarrow p_i = \frac{\mu_K}{i \cdot \mu_U} \cdot p_{i-1}
\tag{14}
$$

$$
p_i = \frac{\mu_K}{i \cdot \mu_U} \cdot p_{i-1} = \frac{\mu_K^2}{\mu_U^2 \cdot i \cdot (i-1)} \cdot p_{i-2} = \cdots = \frac{\mu_K^i}{\mu_U^i \cdot i!} \cdot p_0
\tag{15}
$$

After considering the normalization condition, we can determine the expression for p_N:

$$p_0 + \sum_{i=1}^{N} p_i = 1 = p_0 + \sum_{i=1}^{N} \frac{\mu_K^i}{\mu_U^i \cdot i!} \cdot p_0 = p_0 \cdot \left(1 + \sum_{i=1}^{N} \frac{\rho^i}{i!}\right) \qquad (16)$$

$$\Rightarrow p_0 = \frac{1}{\left(1 + \sum_{i=1}^{N} \frac{\rho^i}{i!}\right)} \qquad (17)$$

$$P_N = \frac{\mu_K^N}{\mu_U^N \cdot N!} \cdot \frac{1}{\left(1 + \sum_{i=1}^{N} \frac{\rho^i}{i!}\right)} = \frac{\rho^N}{N! + \sum_{i=1}^{N} \frac{N! \cdot \rho^i}{i!}} = \frac{\rho^N}{\sum_{i=0}^{N} \frac{N! \cdot \rho^i}{i!}} \qquad (18)$$

Thus, taking into account that the throughput of the closed network is the equivalent service rate, we have the following expression:

$$\mu_{eq}(n) = \mu_K \cdot (1 - p_n) \qquad (19)$$

$$\mu_{eq}(n) = \mu_K \cdot \left(1 - \frac{\rho^n}{\sum_{i=0}^{n} \frac{n! \cdot \rho^i}{i!}}\right) \quad // \quad \rho = \frac{\mu_K}{\mu_U} \qquad (20)$$

Note that this case is really a particular case of the general case where $\lambda = \mu_K$ and $\mu_i = i \cdot \mu_U$.

4.3 Solution for the closed network model with incoming traffic

The previously explained Norton equivalence takes into consideration the internal problems of the traffic monitoring system related to the non-parallelizable tasks. Now we will complete the model adding the traffic injection queue to the equivalent system calculated before.

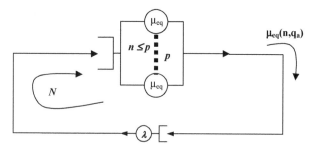

Fig. 10. General model with incoming traffic.

The entire system under traffic load is modelled as a closed network with an upper multiple queue, which is the Norton equivalent queue of the traffic analysis system, and a lower simple queue, simulating the injection of network traffic with rate λ. In this closed network, a finite number N of packets circulate. In general, this number N is greater than p, the number of available processors.

The analytical solution of this model is similar to that proposed for the general model taking into account that the service rates μ_1, μ_2..., μ_p will correspond with the calculation of the Norton equivalent model $\mu_{eq}(n, q_a)$ with values of n from 1 to p. This model allows us to calculate the theoretical throughput of the traffic monitoring system for different loads of network traffic.

$$\gamma = \lambda \cdot (1 - p_N) \tag{21}$$

The value of N will allow us to estimate the system losses. There will be losses when the N packets of the closed network are located in the upper queue. At that time, the traffic injection queue will be empty and, therefore, it will simulate the blocking of the incoming traffic. That will be less likely, the higher the value of N is.

4.4 Mean value analysis

Apart from the analytic solution explained above, we have also considered an iterative method based on the mean value analysis (MVA), in order to simplify the calculations even more. This theorem states that 'when one customer in an N-customer closed system arrives at a service facility he/she observes the rest of the system to be in the equilibrium state for a system with N−1 customers' [Reiser&Lavengerg, 1980]. The application of this theorem to our case requires taking into account the dependencies between some states and others in a complex state diagram, where the state transitions can be also performed with different probabilities, because there are state dependent service rates.

4.4.1 Probability flows between adjacent states

The mean value analysis is based on the iterative dependency between the probability of a certain state with regard to the probabilities of the closest states. The state transitions will not be possible between any two states, they can only occur between adjacent states.

$$p(i,j) = f(p(i-1,j), p(i,j-1)) \tag{22}$$

It is necessary to do a balance of probability flows between states considering the service rates that are dependent on the state of each queue.

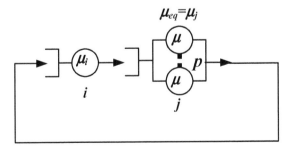

Fig. 11. General model for the closed queue network.

To begin with, we consider the general model for the closed queue network. We call queue i to the simple queue of the model. We assume that this simple queue is in state i and its service rate is μ_i. Likewise, we call queue j to the multi-server queue which is in state j with a state dependent equivalent service rate μ_j. A fixed number of packets (N) are circulating in the closed network, so that there is a dependence between the state i and j.

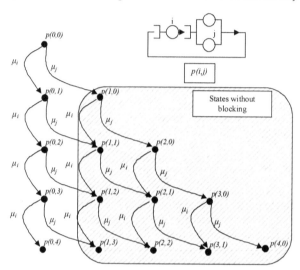

Fig. 12. Probability flows between adjacent states with two processors.

Fig.12 shows the dependencies of the probability of a given state with regard to the closer states in the previous stage with one packet less.

4.4.2 Iterative calculation method

Little's law [Little, 1961] can help us to interpret the relationship between the state probabilities at different stages of the closed queue network.

$$E(T) = \frac{E(n)}{\gamma} \qquad (23)$$

This formula is applied to any queue system that is in equilibrium in which there are users who enter the system, consume time to be attended and then leave. In the formula, γ can be understood as the throughput of the system, E(T) as the average time spent in the system and E (n) as the average number of users.

The iterative method applied to the closed queue network is based on solving certain interesting statistics of the network at every stage, using the data obtained in the previous stage. You go from one stage with N packets to the next with N+1 packets, adding one packet to the closed queue network once the system is in stable condition. Knowing the state probability distribution in stage N, we can calculate the average number of users on each server.

$$E(n_i) = \sum_{i=1}^{N} i \cdot p_N(i, N-i) \qquad E(n_j) = \sum_{j=1}^{N} j \cdot p_N(N-j, j) \qquad (24)$$

We can calculate every state probability in the stage N as the ratio of the average stay time in this state, $t_N(i,j)$ and the total time for that stage T_{TN}. The total time T_{TN} can be calculated as the sum of all the partial times $t_N(i, j)$ of each state at that stage.

$$p_N(i,j) = \frac{t_N(i,j)}{T_{TOTAL,N}} \qquad (25)$$

$$T_{TOTAL,N} = \sum_{i=0}^{N} t_N(i, N-i) \qquad (26)$$

If we consider Reiser's theorem [Reiser, 1981], it is possible to set a relation between the state probabilities of a certain state with regard to the ones which are adjacent in the previous stage. In particular, in equilibrium, when we have N packets, the state probability distribution is equal to the distribution at the moment of a new packet arrival at the closed network. In the state diagram of our model, in general, every state depends on two states of the previous stage. We will have the following probability flows:

Transition $(i-1,j) \rightarrow (i,j)$ a new packet arrives at queue i

$$p_N'(i,j) = p_{N-1}(i-1,j) \qquad (27)$$

Transition $(i,j-1) \rightarrow (i,j)$ a new packet arrives at queue j

$$p_N''(i,j) = p_{N-1}(i,j-1) \qquad (28)$$

Knowing the iterative relations of the probabilities between different stages and basing on Little's formula, we can calculate the average stay time $t_N(i, j)$ in the system in a given state, accumulating the average time in queue i, $t^i_n(i, j)$ and the average time in queue j, $t^j_n(i, j)$.

$$t_N(i,j) = t_N^i(i,j) + t_N^j(i,j) \qquad (29)$$

Applying Little's law:

$$t_N^i(i,j) = \frac{E_N^i(i)}{\mu_i(i)} = \frac{p_N'(i,j) \cdot i}{\mu_i(i)} = \frac{p_{N-1}(i-1,j) \cdot i}{\mu_i(i)} \qquad (30)$$

$$t_N^j(i,j) = \frac{E_N^j(j)}{\mu_j(j)} = \frac{p_N''(i,j) \cdot j}{\mu_j(j)} = \frac{p_{N-1}(i,j-1) \cdot j}{\mu_j(j)} \qquad (31)$$

Considering the probability distribution of the previous stage:

$$t_N(i,j) = \frac{p_{N-1}(i-1,j) \cdot i}{\mu_i(i)} + \frac{p_{N-1}(i,j-1) \cdot j}{\mu_j(j)} \qquad (32)$$

Taking into account that, for a given state (i, j), the average stay time of a packet in the queues i and j is given by t_i and t_j respectively, we can express the probability of that state as:

$$\tau_i = \frac{i}{\mu_i(i)} \qquad \tau_j = \frac{j}{\mu_j(j)} \tag{33}$$

$$t_N(i,j) = \frac{p_{N-1}(i-1,j) \cdot i}{\mu_i(i)} + \frac{p_{N-1}(i,j-1) \cdot j}{\mu_j(j)} \tag{34}$$

$$p_N(i,j) = \frac{t_N(i,j)}{T_{TN}} = p_{N-1}(i-1,j) \cdot \frac{\tau_i}{T_{TN}} + p_{N-1}(i,j-1) \cdot \frac{\tau_j}{T_{TN}} \tag{35}$$

Eq. 35 allows us to calculate a certain state probability of the stage with N packets, having the probabilities of the adjacent states in the stage N. Using this equation, we can iteratively calculate the state probability distribution for every stage.

4.4.3 Adjusting losses depending on N

The losses of the traffic monitoring system can be measured assessing the blocking probability of the injection queue. If we consider the general model with an incoming traffic of λ, we can calculate (Eq. 21) the volume of traffic processed by the traffic monitoring system (γ) and also the caused losses (δ).

$$\gamma = \lambda \cdot \left(1 - p(0,N)\right) \tag{36}$$

$$\delta = \lambda - \gamma = \lambda \cdot p(0,N) \tag{37}$$

If we look at the evolution of the blocking probability of the injection queue with increasing number of packets N in the closed network, we can see how that probability stage is reduced in each stage. The same conclusion can be derived from Eq. 18.

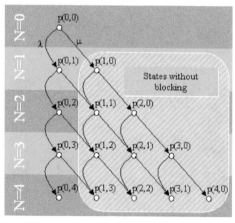

Fig. 13. Evolution of probability flows as a function of N.

A parameter that can be difficult to assess is N, the number of packets that are circulating in the closed network. In general, this parameter depends on specific features of the platform, such as the number of available processors and the ability of the Kernel to accept packets in transit regardless of whether they have processors available at that time.

One conclusion to be drawn from the model, is that it is possible to estimate the value of the parameter N by adjusting the losses that the model has with regard to those which actually occur in a traffic monitoring system.

5. Model validation

This section presents the validation tests to verify the correctness of our analytical model. The aim is to compare theoretical results with those obtained by direct measurement in a real traffic monitoring system, in particular, in the Ksensor prototype developed by NQaS which is integrated into a testing architecture. It is also worth mentioning that, prior to obtaining the theoretical performance results, it is necessary to introduce some input parameters for the model. These initial necessary values will also be extracted from experimental measurements in Ksensor and the testing platform, making use of an appropriate methodology. With all this, we report experimental and analysis results of the traffic monitoring system in terms of two key measures, which are the mean throughput and the CPU utilization. These measures are plotted against incoming packet arrival rate. Finally, we discuss the results obtained.

5.1 Test setup

In this section, we describe the hardware and software setup that we use for our evaluation. Our hardware setup (see Fig. 14) consists of four computers: one for traffic generation (injector), a second one for capturing and analysing the traffic (sensor or Ksensor), a third one for packet reception (receiver) and the last one for managing, configuring and launching the tests (manager). All they are physically connected to the same Gigabit Ethernet switch.

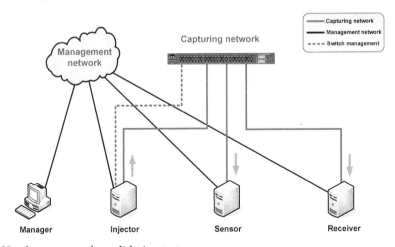

Fig. 14. Hardware setup for validation tests.

However, two virtual networks are distinguished: the first one is the capturing network that connects the elements that play some role during the tests; the second one is the management network which contains the elements that are responsible for the management tasks that can be needed before or after doing tests. The use of two separate networks is necessary, so that the information exchange between the management elements does not interfere with the test results.

The basic idea is to overwhelm Ksensor (sensor) with high traffic generated from the injector. Despite the fact that we do not have 10 Gigabit Ethernet hardware for our tests available, we can achieve our goal of studying the behaviour of the traffic capturing and analysis software at high rates. In addition, we can compare the results with the analytical model and also identify the possible bottlenecks of all analysed systems.

Regarding software, we use a testing architecture [Beaumont et al., 2005] designed by NQaS that allows the automation of tasks like configuration, running and gathering results related to validation tests. The manager, the injector and the sensor that appear in Fig. 14 are part of this testing architecture. They have installed the necessary software to perform the functions of manager, agent, daemon or formatter as we will explain in the next subsection. On the other hand, the receiver is simply the destination of the traffic entered into the network by the injector and it does not have any other purpose.

5.2 Architecture to automatically test a traffic capturing and analysis system

As mentioned previously, in this section, we use a testing architecture for experimental performance measures and, also, to estimate the values of certain input parameters required for the analytical model. It is, therefore, advisable to explain, albeit briefly, the main elements of this platform.

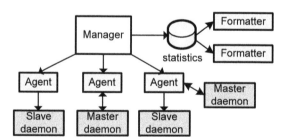

Fig. 15. Logical elements of the testing architecture used in validation tests.

The testing architecture consists of four types of logical elements as Fig. 15 shows. Each of them implements a perfectly defined function:

• Manager is the interface with the user. This element, in the infrastructure shown in Fig. 14, is located on the machine with the same name. It is in charge of managing the rest of the logical elements (agents, daemons and formatters) according to the configuration received from the administrator. After introducing the test setup, it is distributed from the manager to the other elements and the test is launched when the manager sends the start command. At the end of every test, the manager receives and stores the results obtained by the rest of the elements.

- Agents are responsible for attending manager's requests and acting on different devices. Agents are always listening and they have to start and stop the daemons, as well as to collect the performance results. During a test in the infrastructure, one agent is executed in the injector and another one, in the sensor.
- Daemons are in charge of acting on the different physical elements which are involved in each test. Its function can be very variable. For example, the injection of network traffic according to the desired parameterization, the configuration of the capturing buffers, the execution of control programs in the sensor, the acquisition of information or some element's statics, etc. Depending on the relationship with the agent two different types of daemons can be distinguished: master and slave. Master daemons have got some intelligence. The agent will start them but they will indicate when their work has finished. On the other hand, slave daemons do not determine the end of its execution. In each test, to do all the tasks, as many daemons as necessary are executed in the injector and in the sensor.
- Formatters are the programs which select and translate the information stored by the manager to more appropriate formats for its representation. They are executed in the machine called manager, at the end of every test.

5.3 Experimental estimation for certain parameters of the model

In section 3, we have defined an analytical model which functionally responds to a traffic monitoring system. In order to perform an assessment of the model, first we need some values for certain input parameters. We are referring to some service rates that appear in the model based on closed queue networks and are necessary to obtain theoretical performance results. Then we can compare these analytical results with those obtained in the laboratory.

In general, we talk about μ service rates, but, in this subsection, it is easier to talk about mean service times. For this reason, we use the nomenclature based on average processing time in which an average time t_{ij} can be expressed as the inverse of its service rate $1/\mu_{ij}$.

We want to adapt the theoretical model to Ksensor, a real network traffic probe. The best approach is to consider the model of the equivalent traffic monitoring system (see Fig.5) where we distinguish a non-parallelizable process and a parallelizable one. In Ksensor, this separation corresponds with the packet capturing process and the analysis process.

The packet capturing process is not parallelizable because the softIRQ is responsible for the capture and it only runs in one CPU. Fig. 16 shows experimental measurements about average packet capturing times. They have been obtained running tests with Ksensor under different conditions: variable packet injection rate in packets per second and traffic analysis load in number of cycles (null, 1K, 5K or 25K). The inverse of the average softIRQ times shown in Fig. 16 will be the service rate μ_s that appears in the model.

On the other hand, the analysis process is parallelizable in Ksensor. In the same way that softIRQ times have been obtained, we experimentally get average analysis processing times that are shown in Fig. 17. The inverse of the average times shown in Fig.17 will be the service rate μ_M that appears in the multi-queue of the model. It is necessary to comment that, in Fig. 16, the average softIRQ times are not constant. This is because neither all the injected packets are captured by the system, nor all the captured packets are analysed and this causes different computational flow balances.

The values μ_s and μ_M, derived from these experimental measurements, will be taken to the performance evaluation of the model that will be explained later. In addition to the two parameters mentioned, there is another one which is q_a, but it is always $q_a=1$ in our test configuration.

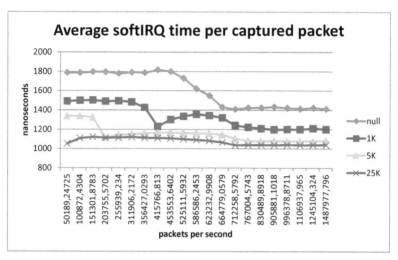

Fig. 16. Average softIRQ per captured packet.

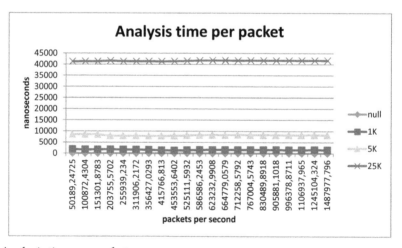

Fig. 17. Analysis time per packet.

5.4 Performance measurements - Evaluation and discussion

The analytical model has been tested with Ksensor under different conditions: packet injection rate (packets per second) varies between 0 and 1.5 million, packet length is 64-1500 bytes and traffic analysis load (at present we simulate QoS algorithm processing times, from 0 to 25000 cycles). The number of processors has been 2 in every test.

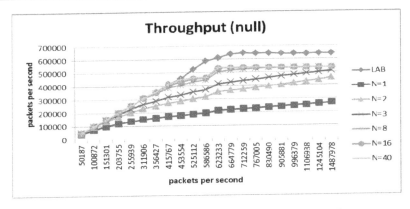

Fig. 18. Theoretical and experimental throughputs without analysis load.

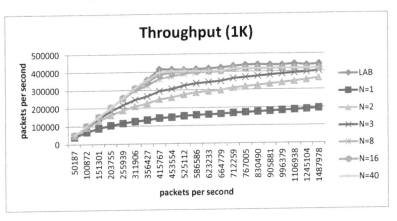

Fig. 19. Theoretical and experimental throughputs with 1Kcycle analysis load.

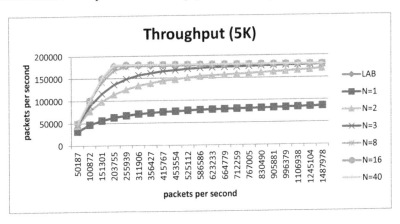

Fig. 20. Theoretical and experimental throughputs with 5Kcycle analysis load.

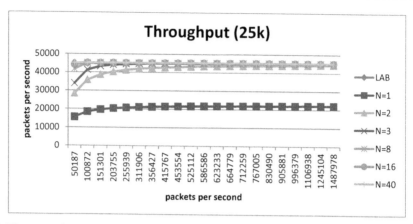

Fig. 21. Theoretical and experimental throughputs with 25Kcycle analysis load.

Fig. 18, Fig. 19, Fig. 20 and Fig. 21 show the comparison between the theoretical model's throughput for different values of N and the real probe's throughput measured experimentally (marked as LAB in the graph). 64 byte-length packets have been used in the lab test and its corresponding service rates in the theoretical calculation. The service rates has been calculated according to the method explained in subsection 5.3.

In all the cases, the throughput grows until a maximum is reached (saturation point). We also observe in these graphs that, with increasing N, the theoretical throughput is close to the real one. It shows, therefore, that the analytical model fits the real system.

6. Conclusion

In this chapter we have presented an analytical model that represents a multiprocessor traffic monitoring system. This model analyses and quantifies the system performance and it can be useful to improve aspects related to hardware and software design. Even, the model can be extended to more complex cases which have not been treated in the laboratory.

Thus, the major contribution of this chapter is the development of a theoretical model based on a closed queuing network that allows to study the behaviour of a multiprocessor network probe. A series of simplifications and adaptations is proposed for the closed network, in order to fit it better to the real system. We obtain the model's analytic solution and we also propose a recurrent calculation method based on the mean value analysis. The model has been validated comparing theoretical results with experimental measures. In the validation process we have made use of a testing architecture that not only has measured the performance, it has also provided values for some necessary input parameters of the mathematical model. Moreover, the architecture helps to setup tests faster as well as to collect and plot results easier. Ksensor, a real probe, is part of the testing architecture and, therefore, it is directly involved in the validation process. As has been seen in the validation section, Ksensor's throughput is acceptably calculated by the model proposed in this chapter. The conclusions obtained have been satisfactory with regard to the behaviour of the model.

This paper has also come in useful to explain the main aspects of Ksensor, a multithreaded kernel-level probe developed by NQaS research group. It is remarkable that this system introduces performance improving design proposals into traffic analysis systems for passive QoS monitoring.

As a future work, we suggest two main lines: the first one is related to Ksensor and it is about a new hardware-centered approach whose objective is to embed our proposals onto programmable network devices like FPGAs. The second research line aims at completing and adapting the model to the real system in a more accurate way. We are already making progress on new mathematical scenarios which can represent, in detail, aspects such as packet capturing process, congestion avoidance mechanisms between capturing and analysis stages, specific analysis algorithms applied in QoS monitoring and packet filtering.

Finally, it is worth mentioning that the test setup, which has been used to validate the model, will be improved acquiring network hardware at 10 Gbps and installing Ksensor over a server with more than two processors. The model will be tested under these new conditions and we hope to obtain satisfactory results, too.

Thus, further work is necessary to analyse this type of systems with a higher precision, compare their results, in certain conditions, better and prevent us from developing high-cost prototypes.

7. References

Altman, E.; Avratchenkov,K. & Barakat, C.. (2000). A stochastic model for TCP/IP with stationary random losses. *ACM SIGCOMM 2000*.

Barakat, C.; Thiran, P.; Iannaccone, G.; Diot, C. & Owezarski, P. (2002). A flow-based model for Internet backbone traffic, *Proceedings of the 2nd ACM SIGCOMM Workshop on Internet measurement, 2002*.

Beaumont, A.; Fajardo, J.; Ibarrola, E. & Perfecto, C. (2005). Arquitectura de red para la automatización de pruebas. *VI Jornadas de Ingeniería Telemática.*, Vigo, Spain.

Benvenuti, C. (2006). *Understanding Linux Network Internals*, O' Reilly Media.

Biswas, A.; Sinha, P. (2006). Efficient real-time Linux interface for PCI devices: A study on hardening a Network Intrusion Detection System. *SANE 2006*, Delft, The Netherlands.

Cardigliano, A. (2011). Towards wire-speed network monitoring using Virtual Machines. *Master Thesis*, University of Pisa, Italy.

Chandy, K.M.; Herzog, U. & Woo, L.S. (1975). Parametric Analysis of Queueing Networks Learning Techniques, *IBM J. Research and Development*, vol. 19, no. 1, pp. 43-49, January 1975.

Cleary, J.; Donnelly, S.; Graham, I.; McGregor, A. & Pearson, M. (2000). Design principles for accurate passive measurement. *Passive and Active Measurement. PAM 2000*, Hamilton, New Zealand.

Deri, L. (2004). Improving Passive Packet Capture: Beyond Device Polling. *SANE 2004*, Amsterdam, The Netherlands.

Deri, L. (2005). nCap: Wire-speed Packet Capture and Transmission. *E2EMON 2005*, Nice, France.

Fiems, D. (2004). Analysis of discrete-time queueing systems with vacations. *PhD Thesis,* Ghent University, Belgium.

Fusco, F. & Deri, L. (2010). High Speed Network Traffic Analysis with Commodity Multi-core Systems. *Internet Measurement Conference 2010,* Melbourne, Australia.

Intel-CSA. (2002). Communication Streaming Architecture: Reducing the PCI Network Bottleneck.

Kobayashi, H. (1978). *Modeling and Analysis: An Introduction to System Performance Evaluation Methodology,* Ed. Wiley-Interscience, ISBN 0-201-14457-3.

Lee, T. (1989). M M/G/1/N queue with vacation time and limited service discipline. *Performance Evaluation,* vol. 9, no. 3, pp. 181-190.

Lemoine, E.; Pham, C. & Lefèvre, L. (2003). Packet classification in the NIC for improved SMP-based Internet servers. *ICN'04,* Guadeloupe, French Caribbean.

Little, J. D. C. (1961). A proof of the queueing formula: L=λ·W, *Operations Research,* vol. 9, no. 3, pp. 383-386, 1961.

Mogul, J.C. & Ramakrishnan, K.K. (1996). Eliminating Receive Livelock in an Interrupt-driven Kernel. *USENIX 1996 Annual Technical Conference,* San Diego, California.

Muñoz, A.; Ferro, A.; Liberal, F. & López, J. (2007). A Kernel-Level Monitor over Multiprocessor Architectures for High-Performance Network Analysis with Commodity Hardware. *SensorComm 2007,* Valencia, Spain.

Reiser, M. (1981). Mean value analysis and convolution method for queue-dependent servers in closed queueing networks, *Performance Evaluation,* vol. 1, no. 1, pp. 7-18, January 1981.

Reiser, M. & Lavengerg, S.S. (1980). Mean Value Analysis of Closed Multichain Queueing Networks, *Journal of the ACM,* vol. 27, no. 2, pp. 313-322, April 1980.

Salah, K. (2006). Two analytical models for evaluating performance of Gigabit Ethernet hosts with finite buffer. *AEU - International Journal of Electronics and Communications,* vol. 60, no. 8, pp. 545-556.

Salah, K.; El-Badawi, K. & Haidari, F. (2007). Performance analysis and comparison of interrupt-handling schemes in gigabit networks. *Computer Communications,* vol. 30, no. 17, pp. 3425-3441.

Schneider, F. (2007). Packet Capturing with Contemporary Hardware in 10 Gigabit Ethernet Environments. *Passive and Active Measurement. PAM 2007,* Louvain-la-Neuve, Belgium.

Takagi, H. (1991). *Queueing Analysis, A Foundation of Performance Evaluation Volume 1: Vacation and Priority Systems (Part 1),* North-Holland, Amsterdam, The Netherlands.

Takagi, H. (1994). M/G/1/N Queues with Server Vacations and Exhaustive Service. *Operations Research,* pp. 926-939.

Varenni, G.; Baldi, M.; Degioanni, L. & Risso, F. (2003). Optimizing Packet Capture on Symmetric Multiprocessing Machines. *15th Symposium on Computer Architecture and High Performance Computing,* Sao Paulo, Brazil.

Wang, P. & Liu, Z. (2004). Operating system support for high performance networking, a survey. *The Journal of China Universities of Posts and Telecommunications,* vol. 11, no. 3, pp. 32-42.

Wu, W; Crawford, M. & Bowden, M. (2007). The performance analysis of linux networking - Packet receiving. *Computer Communications,* vol. 30, no. 5, pp. 1044-1057.

Zhu, H.; Liu, T.; Zhou, C. & Chang, G. (2006). Research and Implementation of Zero-Copy Technology Based on Device Driver in Linux. *IMSCCS'06.*

5

Optical Burst-Switched Networks Exploiting Traffic Engineering in the Wavelength Domain

João Pedro[1,2] and João Pires[2]
[1]Nokia Siemens Networks Portugal S.A.
[2]Instituto de Telecomunicações, Instituto Superior Técnico
Portugal

1. Introduction

In order to simplify the design and operation of telecommunications networks, it is common to describe them in a layered structure constituted by a service network layer on top of a transport network layer. The service network layer provides services to its users, whereas the transport network layer comprises the infrastructure required to support the service networks. Hence, transport networks should be designed to be as independent as possible from the services supported, while providing functions such as transmission, multiplexing, routing, capacity provisioning, protection, and management. Typically, a transport network includes multiple network domains, such as access, aggregation, metropolitan and core, ordered by decreasing proximity to the end-users, increasing geographical coverage, and growing level of traffic aggregation.

Metropolitan and, particularly, core transport networks have to transfer large amounts of information over long distances, consequently demanding high capacity and reliable transport technologies. Multiplexing of lower data rate signals into higher data rate signals appropriate for transmission is one of the important tasks of transport networks. Time Division Multiplexing (TDM) is widely utilized in these networks and is the fundamental building block of the Synchronous Digital Hierarchy (SDH) / Synchronous Optical Network (SONET) technologies. The success of SDH/SONET is mostly due to the utilization of a common time reference, improving the cost-effectiveness of adding/extracting lower order signals from the multiplexed signal, the augmented reliability and interoperability, and the standardization of optical interfaces. SDH/SONET networks also generalized the use of optical fibre as the transmission medium of metropolitan and core networks. Essentially, when compared to twisted copper pair and coaxial cable, optical fibre benefits from a much larger bandwidth and lower attenuation, as well as being almost immune to electromagnetic interferences. These features are key to transmit information at larger bit rates over longer distances without signal regeneration.

Despite the proved merits of SDH/SONET systems, augmenting the capacity of transport networks via increasing their data rates is only cost-effective up to a certain extent, whereas adding parallel systems by deploying additional fibres is very expensive. The prevailing solution to expand network capacity was to rely on Wavelength Division Multiplexing (WDM) to transmit parallel SDH/SONET signals in different wavelength channels of the same fibre. Nevertheless, since WDM was only used in point-to-point links, switching was performed in the electrical domain, demanding Optical-Electrical (OE) conversions at the input and Electrical-Optical (EO) conversions at the output of each intermediate node, as well as electrical switches. Both the OE and EO converters and the electrical switches are expensive and they represent a large share of the network cost.

Nowadays, transport networks already benefit from optical switching, thereby alleviating the use of expensive and power consuming OE and EO converters and electrical switching equipment operating at increasingly higher bit rates (Korotky, 2004). The main ingredients to support optical switching are the utilization of reconfigurable nodes, like Reconfigurable Optical Add/Drop Multiplexers (ROADMs) and Optical Cross-Connects (OXCs), along with a control plane, such as the Generalized Multi-Protocol Label Switching (GMPLS), (IETF, 2002), and the Automatically Switched Optical Network (ASON), (ITU-T, 2006). The control plane has the task of establishing/terminating optical paths (lightpaths) in response to connection requests from the service network. As a result, the current type of dynamic optical networks is designated as Optical Circuit Switching (OCS).

In an OCS network, bandwidth is allocated between two nodes by setting up one or more lightpaths (Zang et al., 2001). Consequently, the capacity made available for transmitting data from one node to the other can only be incremented or decremented in multiples of the wavelength capacity, which is typically large (e.g., 10 or 40 Gb/s). Moreover, the process of establishing a lightpath can be relatively slow, since it usually relies on two-way resource reservation mechanisms. Therefore, although the deployment of OCS networks only makes use of already mature optical technologies, these networks are inefficient in supporting bursty data traffic due to their coarse wavelength granularity and limited ability to adapt the allocated wavelength resources to the traffic demands in short time-scales, which can also increase the bandwidth waste due to capacity overprovisioning.

Diverse solutions have been proposed to overcome the limitations of OCS networks and improve the bandwidth utilization efficiency of future optical transport networks. The less disruptive approach consists of an optimized combination of optical and electrical switching at the network nodes. In this case, entire wavelength channels are switched optically at a node if the carried traffic flows, originated at upstream nodes, approximately occupy the entire wavelength capacity. Alternatively, traffic flows with small bandwidth requirements can be groomed (electrically) into one wavelength channel with enough spare capacity (Zhu et al., 2005). This hybrid switching solution demands costly OE/EO converters and electrical switches, albeit in/of smaller numbers/sizes than those needed in opaque implementations relying only on electrical switching. However, OCS networks with electrical grooming only become attractive when it is possible to estimate in advance the fractions of traffic to be groomed and switched transparently at each node, enabling to accurately dimension both the optical and electrical switches needed to accomplish an optimized trade-off between maximizing the bandwidth utilization and minimizing the electrical switching and OE/EO conversion equipment. Otherwise, when the traffic pattern cannot be accurately predicted, this trade-off can become difficult to attain and both optical and electrical switches may have to be overdimensioned, hampering the cost-effectiveness of this hybrid approach.

The most advanced all-optical switching paradigm for supporting data traffic over optical transport networks is Optical Packet Switching (OPS). Ideally, OPS would replicate current store-and-forward packet-switched networks in the optical domain, thereby providing statistical multiplexing with packet granularity, rendering the highest bandwidth utilization when supporting bursty data traffic. In the full implementation of OPS, both data payload and their headers are processed and routed in the optical domain. However, the logical operations needed to perform address lookup are difficult to realize in the optical domain with state-of-the-art optics. Similarly to MPLS, Optical Label Switching (OLS) simplifies these logical operations through using label switching as the packet forwarding technique (Chang et al., 2006). In their simplest form, OPS networks can even rely on processing the header/label of each packet in the electrical domain, while the payload is kept in the optical domain. Nevertheless, despite the complexity differences of the implementations proposed in the literature, the deployment of any variant of OPS networks is always hampered by current limitations in optical processing technology, namely the absence of an optical equivalent of electronic Random-Access Memory (RAM), which is vital both for buffering packets while their header/label is being processed and for contention resolution (Tucker, 2006; Zhou & Yang, 2003), and the difficulty to fabricate large-sized fast optical switches, essential for per packet switching at high bit rates (Papadimitriou et al., 2003).

The above discussion highlighted that OCS networks are relatively simple to implement but inefficient for transporting bursty data traffic, whereas OPS networks are efficient for transporting this type of traffic but very difficult to implement with state-of-the-art optical technology. Next-generation optical networks would benefit from an optical switching approach whose bandwidth utilization and optical technology requirements lie between those of OCS and OPS. In order to address this challenge, an intermediate optical switching paradigm has been proposed and studied in the literature – Optical Burst Switching (OBS).

The basic premise of OBS is the development of a novel architecture for next-generation optical WDM networks characterized by enhanced flexibility to accommodate rapidly fluctuating traffic patterns without requiring major technological breakthroughs. A number of features have been identified as key to attain this objective (Chen et al., 2004). In order to overview some of them, consider an optical network comprising edge nodes, interfacing with the service network, and core nodes, as illustrated in Fig. 1. OBS networks grant intermediate switching granularity (between that of circuits and packets) via: assembling multiple packets into larger data containers, designated as data bursts, at the ingress edge nodes, enforcing per burst switching at the core nodes, and disassembling the packets at the egress edge nodes. Noteworthy, data bursts are only assembled and transmitted into the OBS network when data from the service network arrives at an edge node. This circumvents the stranded capacity problem of OCS networks, where the bandwidth requirements from the service network evolve throughout the lifetime of a lightpath and during periods of time can be considerably smaller than the provisioned capacity. Furthermore, the granularity at which the OBS network operates can be controlled through varying the number of packets contained in the data bursts, enabling to regulate the control and switching overhead.

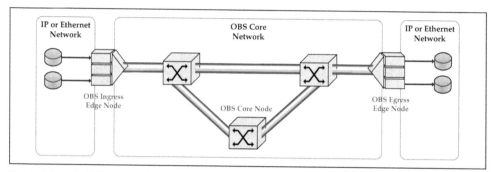

Fig. 1. Generic OBS network architecture.

In OBS networks, similarly to OCS networks, control information is transmitted in a separate wavelength channel and processed in the electronic domain at each node, avoiding complex optical processing functions inherent to OPS networks. More precisely, a data burst and its header packet are decoupled in both the wavelength and time domains, since they are transmitted in different wavelengths and the header precedes the data burst by an offset time. Channel separation of headers and data bursts, a distinctive feature of out-of-band signalling schemes, is suitable to efficiently support electronic processing of headers while preserving data in the optical domain, because OE/EO converters at the core nodes are only needed for the control channel. The offset time has a central role in OBS networks, since it is dimensioned to guarantee the burst header is processed and resources are reserved for the upcoming data burst before the latter arrives to the node. Accordingly, a data burst can cut through the core nodes all-optically, avoiding being buffered at their input during the time needed for header processing. Moreover, since the transmission of data bursts can be asynchronous, complex synchronization schemes are not mandatory. Combined, these features ensure OBS networks can be implemented without making use of optical buffering.

The prospects of deploying OBS in future transport networks can be improved provided that the bandwidth utilization achievable with OBS networks can be enhanced without significantly increasing their complexity or, alternatively, by easing their implementation without penalizing network performance. Noteworthy, OBS networks are technologically more demanding than OCS networks in several aspects. Firstly, although OBS protocols avoid optical buffering, OBS networks still demand some technology undergoing research, namely all-optical wavelength converters (Poustie, 2005) and fast optical switches scalable to large port counts (Papadimitriou et al., 2003). Secondly, the finer granularity of OBS is accomplished at the expense of a control plane more complex than the one needed for OCS networks (Barakat & Darcie, 2007). Nevertheless, the expected benefits of adopting a more bandwidth efficient optical switching paradigm fuelled significant research efforts in OBS, which even resulted in small network demonstrators (Sahara et al., 2003; Sun et al., 2005).

The performance of OBS networks is mainly limited by data loss due to contention for the same transmission resources between multiple data bursts (Chen et al., 2004). The lack of optical RAM limits the effectiveness of contention resolution in OBS networks. Wavelength conversion is usually assumed to be available to resolve contention for the same wavelength channel. In view of the complexity and immaturity of all-optical wavelength converters, decreasing the number of converters utilized or using simpler ones without degrading performance would enhance the cost-effectiveness of OBS networks. Nevertheless, even if wavelength conversion is available, contention occurs when the number of bursts directed to the same link exceeds the number of wavelength channels. Moreover, the asynchronous transmission of data bursts creates voids between consecutive data bursts scheduled in the same wavelength channel, further contributing to contention. Consequently, minimizing these voids and smoothing burst traffic without resorting to complex contention resolution strategies would also improve the cost-effectiveness of OBS networks.

In alternative or as a complement to contention resolution strategies, such as wavelength conversion, the probability of resource contention in an OBS network can be proactively reduced using contention minimization strategies. Essentially, these strategies optimize the resources allocated for transmitting data bursts in such way that the probability of multiple data bursts contending for the same network resources is reduced. Contention minimization strategies for OBS networks mainly consist of optimizing the wavelength assignment at the ingress edge nodes to decrease contention for the same wavelength channel (Wang et al., 2003), mitigating the performance degradation from unused voids between consecutive data bursts scheduled in the same wavelength channel (Xiong et al., 2000), and selectively smoothing the burst traffic entering the network (Li & Qiao, 2004). Albeit the utilization of these strategies can entail additional network requirements, namely augmenting the (electronic) processing capacity in order to support more advanced algorithms, it is expected that the benefits in terms of performance or complexity reduction will justify their support.

This chapter details two contention minimization strategies, which when combined provide traffic engineering in the wavelength domain for OBS networks. The utilization of this approach is shown to significantly improve network performance and reduce the number of wavelength converters deployed at the network nodes, enhancing their cost-effectiveness.

The remaining of the chapter is organized as follows. The second section introduces the problem of wavelength assignment in OBS networks whose nodes have no wavelength converters or have a limited number of wavelength converters. A heuristic algorithm for optimizing the wavelength assignment in these networks is described and exemplified. The third section addresses the utilization of electronic buffering at the ingress edge nodes of OBS networks, highlighting its potential for smoothing the input burst traffic and describing how it can be combined with the heuristic algorithm detailed in the previous section to attain traffic engineering in the wavelength domain. The performance improvements and node complexity reduction made possible by employing these strategies in an OBS network are evaluated via network simulation in the fourth section. Finally, the fifth and last section presents the final remarks of the work presented in this chapter.

2. Priority-based wavelength assignment

OBS networks utilize one-way resource reservation, such as the Just Enough Time (JET) protocol (Qiao & Yoo, 1999). The principles of burst transmission are as follows. Upon assembling a data burst from multiple packets, the ingress node generates a Burst Header Packet (BHP) containing the offset time between itself and the data burst, as well as the length of the data burst. This node also sets a local timer to the value of the offset time.

The BHP is transmitted via a control wavelength channel and processed at the control unit of each node along the routing path of the burst. The control unit uses the information in the BHP to determine the resources (e.g., wavelength channel in the designated output fibre link) to be allocated to the data burst during the time interval it is expected to be traversing the core node. This corresponds to a delayed resource reservation, since the resources are not immediately set up, but instead are only set up just before the arrival time of the data burst. Furthermore, the resources are allocated to the burst during the time strictly necessary for it to successfully pass through the node. This minimizes the bandwidth waste because these resources can be allocated to other bursts in non-overlapping time intervals. Before forwarding the BHP to the next node, the control unit updates the offset time, reducing it by the amount of time spent by the BHP at the node. Meanwhile, the data burst buffered at the ingress node is transmitted after the timer set to the offset time expires. In case of successful resource reservation by its BHP at all the nodes of the routing path, the burst cuts through the core nodes in the optical domain until it arrives to the egress node. Otherwise, when resource reservation is unsuccessful at a node, both BHP and data burst are dropped at that node and the failed burst transmission is signalled to the ingress node.

As a result of using one-way resource reservation, there is a large probability that data bursts arrive at a core node on the same wavelength channel from different input fibre links and being directed to the same output fibre link of that node. This leads to contention for the same wavelength channel at the output fibre link. These contention events can be efficiently resolved using wavelength converters and/or minimized in advance through an optimized assignment of wavelengths at the ingress nodes. In view of the immaturity of all-optical wavelength converters, strategies for minimizing the probability of wavelength contention become of paramount importance in order to design cost-effective OBS core nodes.

2.1 Problem statement

Consider an OBS network modelled as a directed graph $G = (V, E)$, where $V = \{v_1, v_2, ..., v_N\}$ is the set of nodes, $E = \{e_1, e_2, ..., e_L\}$ is the set of unidirectional fibre links and the network has a total of N nodes and L fibre links. Each fibre link supports a set of W data wavelength channels, $\{\lambda_1, \lambda_2, ..., \lambda_{W-1}, \lambda_W\}$. Let $\Pi = \{\pi_1, \pi_2, ..., \pi_{|\Pi|-1}, \pi_{|\Pi|}\}$ denote the set of routing paths used to transmit data bursts in the network, E_i denote the set of fibre links traversed by path $\pi_i \in \Pi$, and γ_i denote the average traffic load offered to path π_i. It is assumed that the average offered traffic load values are obtained empirically or based on long-term predictions of the network load. Ideally, this input information would be used to formulate a combinatorial optimization problem for determining a wavelength search ordering, that is, an ordered list of all W wavelength channels, for each routing path such that a relevant performance metric, like the average burst blocking probability, is minimized. However, blocking probability performance metrics can only be computed via network simulation or, in particular cases, estimated by solving a set of non-linear equations (Pedro et al., 2006a). As a result, the objective function cannot be expressed in terms of the problem variables in an analytical closed-form manner (Teng & Rouskas, 2005). Moreover, even if this was possible, the size of the solutions search space would grow steeply with the number of wavelength channels W and the number of routing paths

$|\Pi|$, since there are $(W!)^{|\Pi|}$ combinations of wavelength channel orderings. Consequently, for OBS networks of realistic size, this would prevent computing the optimum wavelength search orderings in a reasonable amount of time.

In view of the aforementioned limitations in both problem formulation and resolution, the wavelength search orderings must be computed without knowing the resulting average burst blocking probability and by relying on heuristic algorithms. Notably, when the core nodes have limited or no wavelength conversion capabilities, burst blocking probability is closely related with the expected amount of unresolved wavelength contention. Consider two routing paths, π_1 and π_2, that traverse a common fibre link. Clearly, the chances of data bursts going through these paths and contending for the same wavelength channel at the common fibre link are minimized if their ingress nodes search for an available wavelength using opposite orderings of the wavelengths, that is, the ingress node of π_1 uses, for instance, $\lambda_1, \lambda_2, ..., \lambda_{W-1}, \lambda_W$, whereas the ingress node of π_2 uses $\lambda_W, \lambda_{W-1}, ..., \lambda_2, \lambda_1$. This simple scenario is illustrated is Fig. 2 for $W = 4$, where most of the burst traffic on π_1 (π_2) will go through λ_1, λ_2 (λ_4, λ_3). However, in realistic network scenarios, each routing path shares fibre links with several other paths and, consequently, it is not feasible to have opposite wavelength search orderings for each pair of overlapping paths. Still, as long as it is possible for two overlapping paths to have two different wavelength channels ranked as the highest priority wavelengths, the probability of wavelength contention among data bursts going through these paths is expected to be reduced. This observation constitutes the foundation of the heuristic traffic engineering approaches described in the following.

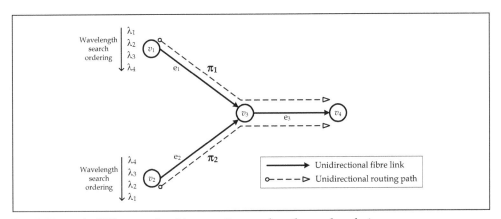

Fig. 2. Example OBS network with opposite wavelength search orderings.

2.2 Heuristic minimum priority interference

Intuitively, the chances of wavelength contention between data bursts going through different routing paths are expected to increase with both the average traffic load offered to the paths and with the number of common fibre links. Bearing this in mind, it is useful to define the concept of interference level of routing path π_i on routing path π_j with $i \neq j$ as,

$$I(\pi_i, \pi_j) = \gamma_i \,|\, E_i \cap E_j\,|, \tag{1}$$

where $|E_i \cap E_j|$ denotes the number of fibre links shared by both paths, and to define the combined interference level between routing paths π_i and π_j with $i \neq j$ as,

$$I^c(\pi_i, \pi_j) = I(\pi_i, \pi_j) + I(\pi_j, \pi_i) = (\gamma_i + \gamma_j) |E_i \cap E_j|. \tag{2}$$

The higher the combined interference level between two routing paths, the higher the likelihood that data bursts going through those paths will contend for the same fibre link resources. Consequently, routing paths with higher combined interference level should use wavelength search orderings as opposed as possible. This constitutes the basic principle exploited by First Fit-Traffic Engineering (FF-TE) (Teng & Rouskas, 2005), which was the first offline algorithm proposed to determine wavelength search orderings that are expected to reduce the probability of wavelength contention. However, this algorithm oversimplifies the problem resolution by computing a single wavelength search ordering for all the routing paths with the same ingress node. A detailed discussion of the limitations of the FF-TE algorithm is presented in (Pedro et al., 2006b). To overcome these shortcomings, the more advanced Heuristic Minimum Priority Interference (HMPI) algorithm, which computes an individual wavelength search ordering per routing path, is described below.

2.2.1 Algorithm description

The algorithm proposed in (Pedro et al., 2006b) for minimizing wavelength contention aims to determine an individual wavelength search ordering for each routing path with a reduced computational effort. The HMPI algorithm uses as input information the network topology, the routing paths and the average traffic load offered to the routing paths.

In order to determine the wavelength search ordering of a routing path, a unique priority must be assigned to each of the wavelengths. The wavelength ranked with the highest priority, called the primary wavelength, is expected to carry the largest amount of burst traffic going through the routing path. The other wavelengths, ordered by decreasing priority, expectedly carry diminishing amounts of burst traffic. In view of the importance of the primary wavelengths, the HMPI algorithm comprises a first stage dedicated to optimize them, consisting of the following three steps.

(S1)   Reorder the routing paths of Π such that if $i < j$ one of the following conditions holds,

$$\sum_{\substack{\pi_k \in \Pi, \\ k \neq i}} I(\pi_i, \pi_k) > \sum_{\substack{\pi_k \in \Pi, \\ k \neq j}} I(\pi_j, \pi_k); \tag{3}$$

$$\sum_{\substack{\pi_k \in \Pi, \\ k \neq i}} I(\pi_i, \pi_k) = \sum_{\substack{\pi_k \in \Pi, \\ k \neq j}} I(\pi_j, \pi_k) \text{ and } |E_i| > |E_j|. \tag{4}$$

(S2)   Consider W sub-sets of the routing paths, one per wavelength, initially empty, that is, $|\Pi_j| = 0$ for $j = 1, ..., W$. Following the routing path ordering defined for Π, include path π_i in the sub-set Π_j such that for any $k \neq j$ one of the subsequent conditions holds,

$$\sum_{\substack{\pi_l \in \Pi_j, \\ l \neq i}} I^c(\pi_i, \pi_l) < \sum_{\substack{\pi_l \in \Pi_k, \\ l \neq i}} I^c(\pi_i, \pi_l); \tag{5}$$

$$\sum_{\substack{\pi_l \in \Pi_j, \\ l \neq i}} I^c(\pi_i, \pi_l) = \sum_{\substack{\pi_l \in \Pi_k, \\ l \neq i}} I^c(\pi_i, \pi_l) \text{ and } |\Pi_j| > |\Pi_k|. \tag{6}$$

(S3) Select wavelength channel λ_j as the primary wavelength of all the paths in sub-set Π_j, that is,

$$P(\lambda_j, \pi_i) = \begin{cases} W, & \text{if } \pi_i \in \Pi_j \\ 0, & \text{otherwise} \end{cases}. \tag{7}$$

The first step of this stage of the HMPI algorithm is used to order the routing paths by decreasing interference level on the remaining paths. Ties are broken by giving preference to the longer routing paths. Considering W sub-sets of routing paths, the second step sequentially includes each routing path on the sub-set with minimum combined interference level between the routing path and the paths already included in the sub-set. Ties are broken by preferring the sub-set with larger number of paths. Finally, the third step assigns to all routing paths of a sub-set the primary wavelength associated with that sub-set. As a result of this stage, the routing paths with minimum combined interference level, carrying data bursts that are less prone to contend with each other for the same wavelength channel, will share the same primary wavelength.

In the second stage of the algorithm, the non-primary wavelengths for all routing paths are determined sequentially, starting with the second preferred wavelength channel and ending with the least preferred wavelength. When determining for each routing path the wavelength with priority $p < W$, it is intuitive to select one to which has been assigned, so far in the algorithm execution, the lowest priorities on routing paths that share fibre links with the routing path being considered. This constitutes the basic rule used in the second stage of the HMPI algorithm.

The following steps are executed for priorities $1 \leq p \leq W - 1$ in decreasing order and considering, for each priority p, all the routing paths according to the path ordering defined in the first stage of the algorithm.

(S1) Let $\Lambda = \{\lambda_j : P(\lambda_j, \pi_i) = 0, 1 \leq j \leq W\}$ denote the initial set of candidate wavelengths, containing all wavelengths that have been assigned a priority of zero on routing path π_i. If $|\Lambda| = 1$, go to **(S7)**.

(S2) Let $P = \{k : \exists \pi_l, l \neq i, P(\lambda_j, \pi_l) = k, |E_l \cap E_i| > 0, \lambda_j \in \Lambda\}$ be the set of priorities that have already been assigned to candidate wavelengths on paths that overlap with π_i.

(S3) Let $\psi = \min_{\lambda_j \in \Lambda} \max_{\pi_l \in \Pi} \{P(\lambda_j, \pi_l) : l \neq i, |E_l \cap E_i| > 0, P(\lambda_j, \pi_l) \in P\}$ be the lowest priority among the set containing the highest priority assigned to each candidate wavelength on paths that share links with π_i. Update the set of candidate wavelengths as follows,

$$\Lambda \leftarrow \Lambda \setminus \{\lambda_j : \exists \pi_l, l \neq i, P(\lambda_j, \pi_l) > \psi, |E_l \cap E_i| > 0\}; \tag{8}$$

If $|\Lambda| = 1$, go to **(S7)**.

(S4) Define $C(\lambda_j, e_m) = \sum \{\gamma_l : E_l \supset e_m, |E_l \cap E_i| > 0, P(\lambda_j, \pi_l) = \psi\}$ as the cost associated with wavelength channel $\lambda_j \in \Lambda$ on link $e_m \in E_i$ and $\alpha_e = \min_{\lambda_j \in \Lambda} \max_{e_m \in E_i} C(\lambda_j, e_m)$ as the minimum cost among the set containing the highest cost associated with each candidate wavelength on the fibre links of π_i. Update the set of candidate wavelengths as follows,

$$\Lambda \leftarrow \Lambda \setminus \{\lambda_j : \exists e_m, C(\lambda_j, e_m) > \alpha_e, e_m \in E_i\} ; \tag{9}$$

If $|\Lambda| = 1$, go to **(S7)**.

(S5) Define $C(\lambda_j, \pi_i) = \sum_{e_m \in E_i} C(\lambda_j, e_m)$ as the cost associated with wavelength λ_j on path π_i and $\alpha_\pi = \min_{\lambda_j \in \Lambda} C(\lambda_j, \pi_i)$ as the minimum cost among the costs associated with the candidate wavelengths on π_i. Update the set of candidate wavelengths as follows,

$$\Lambda \leftarrow \Lambda \setminus \{\lambda_j : C(\lambda_j, \pi_i) > \alpha_\pi\} ; \tag{10}$$

If $|\Lambda| = 1$, go to **(S7)**.

(S6) Update the set of priorities assigned to the candidate wavelengths as follows,

$$P \leftarrow P \setminus \{k : k \geq \psi\} ; \tag{11}$$

If $|P| > 0$, go to **(S3)**. Else, randomly select a candidate wavelength $\lambda \in \Lambda$.

(S7) Assign priority p to the candidate wavelength $\lambda \in \Lambda$ on path π_i, that is, $P(\lambda, \pi_i) = p$.

The first step of the second stage of the HMPI algorithm is used to define the candidate wavelength channels by excluding the ones that have already been assigned a priority larger than zero on the routing path, whereas the second step determines the priorities assigned to these wavelengths on paths that overlap with the routing path under consideration. The third, fourth and fifth step are used to reduce the number of candidate wavelengths. As soon as there is only one candidate wavelength, it is assigned to it the priority p on path π_i, concluding the iteration. In the third step, the highest priority already assigned to each of the candidate wavelength channels on paths that overlap with π_i is determined. Only the wavelengths with the lowest of these priorities are kept in the set of candidates. If needed, the fourth step tries to break ties by associating a cost with each candidate wavelength on each fibre link of π_i. This cost is given by the sum of the average traffic load offered to paths that traverse the fibre link and use the wavelength with priority ψ. The wavelengths whose largest link cost, among all links of π_i, is the smallest one (α_e) are kept as candidates. When there are still multiple candidate wavelengths, the fifth step associates a cost with each wavelength on path π_i, which is simply given by the sum of the cost associated to the wavelength on all links of the routing path. The candidate wavelengths with smallest path cost (α_π) are kept. If necessary, the sixth step removes the priorities equal or larger than ψ from the set of priorities assigned to candidate wavelengths on paths that overlap with the path being considered and repeats the iteration. Finally, if all priorities have been removed and there are still multiple candidate wavelengths, one of them is randomly selected.

As the outcome of executing the HMPI algorithm, each wavelength channel λ_j is assigned a unique priority on routing path π_i, $1 \leq P(\lambda_j, \pi_i) \leq W$. Equivalently, this solution for the priority assignment problem can be represented as an ordering of the W wavelengths, $\{\lambda^1(\pi_i), \lambda^2(\pi_i), ..., \lambda^j(\pi_i), ..., \lambda^W(\pi_i)\}$, where $\lambda^j(\pi_i)$ denotes the jth wavelength channel to be searched when assigning a wavelength to data bursts directed to routing path π_i. In order to enforce these search orderings, each of these lists must be uploaded from the point where they are computed to the ingress nodes of the routing paths. Hence, assuming single-path routing, each ingress node will have to maintain at most $N - 1$ lists of ordered wavelengths.

The computational complexity of the HMPI algorithm, as derived in (Pedro et al., 2009c), is given by $O(W^2 \cdot |\Pi|^2)$, that is, in the worst case it scales with the square of the number of wavelength channels times the square of the number of routing paths.

2.2.2 Illustrative example

In order to give a better insight into the HMPI algorithm, consider the example OBS network of Fig. 3, which has 6 nodes and 8 fibre links (Pedro et al., 2009c). The number of routing paths used to transmit bursts in the network is $|\Pi| = 6$ and each fibre link supports a number of wavelength channels $W = 4$. Moreover, the average traffic load offered to each routing path is 1, except for routing path π_4, which has an average offered traffic load of 1.2, that is, $\gamma_i = 1$ for $i = 1, 2, 3, 5, 6$ and $\gamma_4 = 1.2$.

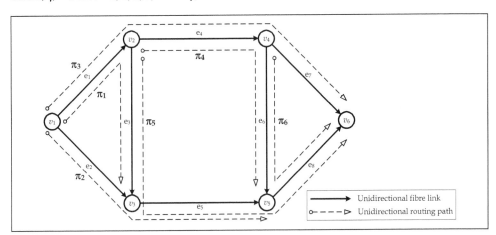

Fig. 3. OBS network used to exemplify the HMPI algorithm (Pedro et al., 2009c).

The HMPI algorithm starts by computing the interference level of all pairs of routing paths, as shown in Table 1. Step (**S1**) of the first stage of the algorithm orders the routing paths by decreasing order of their interference level over other paths, which results in the path order $\{\pi_5, \pi_4, \pi_3, \pi_1, \pi_6, \pi_2\}$. The path with the highest interference level over other paths is π_5, which overlaps with three paths, and the path with the second highest interference level over other paths is π_4, which overlaps with two paths. Although π_3, π_1 and π_6 also overlap with two paths, π_4 is offered more traffic load and consequently can cause more contention. In addition, π_3 precedes π_1 and π_6 because it is longer than the later paths. Since paths π_1 and

π_6 are tied, the path with the smallest index was given preference. Finally, the path with the lowest interference level over other paths is π_2.

$I(\pi_i, \pi_j)$	π_1	π_2	π_3	π_4	π_5	π_6
π_1	--	0	1	0	1	0
π_2	0	--	0	0	1	0
π_3	1	0	--	1	0	0
π_4	0	0	1.2	--	0	1.2
π_5	1	1	0	0	--	1
π_6	0	0	0	1	1	--

Table 1. Interference level of the routing paths.

Step (S2) starts by creating one sub-set of routing paths per wavelength, that is, Π_1, Π_2, Π_3, Π_4. Following the determined path order, π_5 is included in the first empty sub-set, Π_1. Path π_4 is also included in Π_1, because $I^C(\pi_4, \pi_5) = 0$ and Π_1 has more paths than the remaining sub-sets. Since path π_3 overlaps with π_4, $I^C(\pi_3, \pi_4) = 2.2$, and π_4 is already included in Π_1, π_3 is included in the empty sub-set Π_2. Moreover, path π_1 overlaps with both π_5 and π_3 and thus it is included in empty sub-set Π_3. Path π_6 can be included in sub-sets Π_2 and Π_3 because it only overlaps with the paths of Π_1. The tie is broken by selecting the sub-set with smallest index, that is, Π_2. Similarly, path π_2 is also included in this sub-set as it does not overlap with the paths in Π_2 and Π_3 and $|\Pi_2| > |\Pi_3|$. Since every path has been included in one sub-set, $\Pi_1 = \{\pi_4, \pi_5\}$, $\Pi_2 = \{\pi_2, \pi_3, \pi_6\}$ and $\Pi_3 = \{\pi_1\}$, step (S3) concludes the first stage of the algorithm by making λ_1 the primary wavelength of paths π_4 and π_5, λ_2 the primary wavelength of paths π_2, π_3 and π_6, and λ_3 the primary wavelength of path π_1. The other wavelengths are temporarily assigned priority 0 on the routing paths. Table 2 shows the priorities assigned to the wavelengths on the routing paths after the entire HMPI algorithm has been executed.

$P(\lambda_j, \pi_i)$	π_1	π_2	π_3	π_4	π_5	π_6
λ_1	1	1	1	4	4	1
λ_2	2	4	4	1	1	4
λ_3	4	3	2	2	2	3
λ_4	3	2	3	3	3	2

Table 2. Wavelengths priority on the routing paths.

The second stage of the algorithm is initiated with $p = 3$ and proceeds path by path according to the order already defined. For path π_5, the algorithm starts by creating the initial set of candidate wavelengths, $\Lambda = \{\lambda_2, \lambda_3, \lambda_4\}$, in (S1). Since this path overlaps with π_1, π_2 and π_6, the set of priorities assigned to wavelengths of Λ on these paths, determined in (S2), is P = {0, 4}. Wavelength λ_4 is assigned priority 0 on all paths that overlap with π_5 and thus $\rho = 0$. Accordingly, in (S3) the set of candidate wavelengths is updated, $\Lambda = \{\lambda_4\}$, and λ_4 is assigned priority 3 on path π_5. For path π_4, $\Lambda = \{\lambda_2, \lambda_3, \lambda_4\}$, P = {0, 4}, and $\rho = 0$. The set of candidate wavelengths is updated to $\Lambda = \{\lambda_3, \lambda_4\}$, because both λ_3 and λ_4 are assigned priority 0 on paths that overlap with π_4. In this particular case, the algorithm cannot break the tie and in (S7) randomly selects wavelength λ_4 to be assigned priority 3 on path π_4. For the remaining paths, there is only one candidate wavelength whose priority on other paths equals ρ. Wavelength λ_4 is assigned priority 3 on paths π_3 and π_1 and wavelength λ_3 is assigned this priority on paths π_6 and π_2.

The second stage of the algorithm is executed again, but with $p = 2$. For path π_5, the initial set of candidate wavelengths is $\Lambda = \{\lambda_2, \lambda_3\}$. Both wavelengths are assigned priority 4 on at least one of the paths that overlaps with π_5 ($\rho = 4$), λ_2 on π_2 and π_6 and λ_3 on π_1. Paths π_1, π_2, and π_6 share with π_5 links e_3, e_5 and e_8, respectively, and the average traffic load offered to these paths is 1. Thus, according to (S4), the cost associated with λ_2 and λ_3 on each link is at most 1 ($\alpha_e = 1$). However, λ_2 has this link cost on two links, which in (S5) results in a cost $C(\lambda_2, \pi_5) = 2$, whereas λ_3 has this link cost on a single link, $C(\lambda_3, \pi_5) = 1$. Consequently, $\alpha_\pi = 1$ and the set of candidate wavelengths is updated to $\Lambda = \{\lambda_3\}$. For path π_4, $\Lambda = \{\lambda_2, \lambda_3\}$, $P = \{0, 3, 4\}$, and $\rho = 3$. Only wavelength λ_3 is used with a priority smaller or equal than 3 in all links, which reduces the set of candidates to λ_3. In the case of path π_3, $\Lambda = \{\lambda_1, \lambda_3\}$ and λ_1 is assigned priority 4 on π_4, whereas λ_3 is assigned this priority on π_1. Since $\gamma_4 > \gamma_1$, the highest link cost associated to λ_1 is larger than that for λ_3, and the candidate wavelengths are reduced to λ_3. For path π_1, $\Lambda = \{\lambda_1, \lambda_2\}$ and both these wavelengths observe $\rho = 4$, $\alpha_e = 1$ and $\alpha_\pi = 1$. The algorithm has to randomly select one of the wavelengths (λ_2). For both π_6 and π_2, $\Lambda = \{\lambda_1, \lambda_4\}$, $\rho = 3$, but only λ_4 is assigned a priority smaller or equal to 3 in all of the links. The set of candidate wavelengths is reduced to $\Lambda = \{\lambda_4\}$.

Finally, for $p = 1$ the wavelength assignment is trivial, because there is only one wavelength still assigned priority 0 on each path. The complete wavelength search ordering of each path can be obtained from Table 2. The following observations show that these orderings should effectively reduce contention. Firstly, overlapping paths do not share the same primary wavelength. Instead, primary wavelengths are reused by link-disjoint routing paths (e.g., λ_2 is the primary wavelength of π_2, π_3 and π_6). Secondly, paths use with smallest possible priority the primary wavelengths of overlapping paths (e.g., π_1, π_2 and π_6 overlap with π_5 and use the primary wavelength of this path with priority 1).

3. Traffic engineering in the wavelength domain

Noteworthy, at the ingress edge nodes of an OBS network, data bursts are kept in electronic buffers before a wavelength channel is assigned to them and they are transmitted optically towards the egress edge nodes. Clearly, the flexibility of scheduling data bursts in the wavelength channels is considerably higher when the bursts are still buffered at the ingress nodes than when they have already been converted to the optical domain. For instance, a data burst can be delayed at one of the ingress buffers by the exact amount of time required for a wavelength channel to become available in the designated output fibre link. This procedure is not possible at the core nodes due to the lack of optical RAM. The capability of delaying data bursts at an ingress node by a random amount of time, not only increases the chances of successfully scheduling bursts at the output fibre link of their ingress nodes, but also enables implementing strategies that reduce in advance the probability of contention at the core nodes.

The Burst Overlap Reduction Algorithm proposed in (Li & Qiao, 2004) exploits the additional degree of freedom provided by delaying data bursts at the electronic buffers of the ingress nodes to shape the burst traffic departing from these nodes in such way that the probability of contention at the core nodes can be reduced. The principle underlying BORA is that a decrease on the number of different wavelength channels allocated to the data bursts assembled at an ingress node can smooth the burst traffic at the input fibre links of the core nodes and, as a result, reduce the probability that the number of overlapping data bursts directed to the same output fibre link exceeds the number of wavelength channels. In its simpler implementation, BORA relies on using the same wavelength search ordering at all the ingress nodes of the network and utilizing the buffers in these nodes to transmit the maximum number of bursts in the first wavelength channels according to such ordering. In order to limit the extra transfer delay incurred by data bursts, as well as the added buffering and processing requirements, the ingress node can impose a maximum ingress burst delay, Δt_{max}^{RAM}, defined as the maximum amount of time a data burst can be kept at an electronic buffer of its ingress node excluding the time required to assemble the burst and the offset time between the data burst and its correspondent BHP.

The concept of BORA is appealing in OBS networks with wavelength conversion, since these algorithms have not been designed to mitigate wavelength contention. Moreover, BORA algorithms do not account for the capacity fragmentation of the wavelength channels, which is also a performance limiting factor in OBS networks. These limitations have motivated the development of a novel strategy in (Pedro et al., 2009b) that also exploits the electronic buffers of the ingress edge nodes to selectively delay data bursts, while providing a twofold advantage over BORA: enhanced contention minimization at the core nodes and support of core node architectures with relaxed wavelength conversion capabilities.

The first principle of the proposed strategy is related with the availability of RAM at the ingress nodes. In the process of judiciously delaying bursts to schedule them using the smallest number of different wavelength channels, the delayed bursts can be scheduled with minimum voids between them and the preceding bursts already scheduled on the same wavelength channel. This is only possible because the bursts assembled at the node can be delayed by a random amount of time. The serialization of data bursts not only smoothes the burst traffic, with the consequent decrease of the chances of contention at the core nodes, but also reduces the fragmentation of the wavelengths capacity at the output fibre links of the ingress nodes. These serialized data bursts traverse the core nodes, where some of them must be converted to other wavelength channels to resolve contention. The wavelength conversions break the series of data bursts and, as a result, create voids between a burst converted to another wavelength channel and the bursts already scheduled on this wavelength. A large number of these voids lead to wasting bandwidth, as the core nodes will not be able to use them to carry data.

In essence, the first key principle consists of serializing data bursts at the ingress nodes to mitigate the voids between them. Noticeably, if these bursts traverse a set of common fibre links without experiencing wavelength conversion, the formation of unusable voids is reduced at those links. Hence, the second key principle of the proposed strategy consists of improving the probability that serialized bursts routed via the same path are kept in the same wavelength channel for as long as possible. This can reduce the number of unusable voids created in the fibre links traversed before wavelength conversion is used, improving network performance.

The task of keeping the data bursts, which are directed to the same routing path and have been serialized at the ingress node, in the same wavelength channel requires minimizing the chances that bursts on overlapping routing paths contend for the same wavelength channel and, as a result, demand wavelength conversion. This objective is the same as that of the HMPI algorithm presented in Section 2. For that reason, the strategy proposed in (Pedro et al., 2009b), which is designated as Traffic Engineering in the wavelength domain with Delayed Burst Scheduling (TE-DBS), combines the wavelength contention minimization capability of HMPI with selectively delaying data bursts at the electronic buffers of their ingress nodes not only to smooth burst traffic, but also to maximize the amount of data bursts carried in the wavelength channels ranked with the highest priorities by HMPI.

The key principles of the TE-DBS strategy can be illustrated with the example of Fig. 4. The OBS network depicted comprises six nodes and five fibre links. Three paths, π_1, π_2, and π_3,

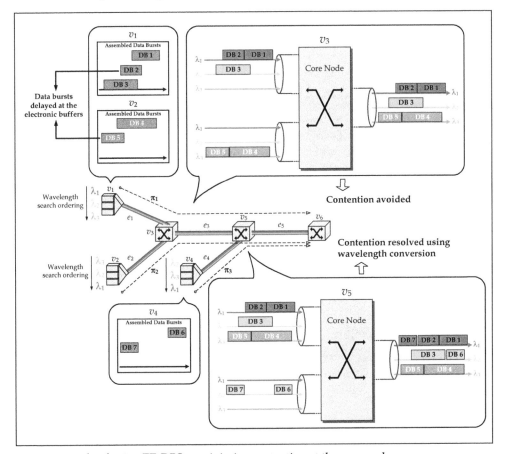

Fig. 4. Example of using TE-DBS to minimize contention at the core nodes.

are used to transmit bursts between one of the three ingress nodes, v_1, v_2, and v_4, and node v_6. Contention between bursts from different input fibre links and directed to the same output fibre link can occur at core nodes v_3 and v_5. Each ingress node uses its own wavelength search ordering and selectively delays bursts with the purpose of transmitting them on the wavelength channels which have been ranked with the highest priorities by an algorithm for minimizing contention in the wavelength domain. Similarly to what occurs with BORA, a maximum ingress burst delay, Δt_{max}^{RAM}, is imposed at each ingress node.

As can be seen, v_1 has assembled three data bursts (DB 1, DB 2, and DB 3), which overlap in time, and v_2 has assembled two data bursts (DB 4 and DB 5), which also overlap in time. The first two bursts assembled by v_1 are transmitted in wavelength channel λ_1, whereas the third cannot be transmitted in this wavelength without infringing the maximum ingress burst delay and, therefore, has to be transmitted in λ_2. The two bursts assembled by v_2 are transmitted in the wavelength ranked with highest priority, λ_3. These bursts traverse v_3, where contention is avoided since the bursts arrive in different wavelengths. Meanwhile, the ingress node v_4 has assembled two data bursts (DB 6 and DB 7) and transmits them in the wavelength ranked with highest priority, λ_2. All seven data bursts traverse core node v_5, where DB 7 must be converted to another wavelength in order to resolve contention.

The major observations provided by this example are as follows. Similarly to using BORA, the burst traffic is smoothed at the ingress nodes, reducing contention at the core nodes from an excessive number of data bursts directed to the same output fibre link. Moreover, since the burst traffic of routing paths π_1, π_2, and π_3 is mostly carried in different wavelengths, contention for the same wavelength channel is also reduced. As a result, the pairs of bursts serialized at the ingress nodes, DB 1 and DB 2 in routing path π_1 and DB 4 and DB5 in routing path π_2, can be kept in the same wavelength channel until they reach node v_6, mitigating the fragmentation of the capacity of wavelengths λ_1 and λ_3 in the fibre links traversed by routing paths π_1 and π_2. Since this is accomplished through minimizing the probability of wavelength contention, it can also relax the wavelength conversion capabilities of the core nodes without significantly degrading network performance.

The TE-DBS strategy requires the computation of one wavelength search ordering, $\{\lambda^1(\pi_i), \lambda^2(\pi_i), ..., \lambda^W(\pi_i)\}$, for each routing path π_i. The HMPI algorithm is used to optimize offline the wavelength search orderings. These orderings are stored at the ingress nodes and the control unit of these nodes uses them for serializing data bursts on the available wavelength channel ranked with the highest priority on the routing path the bursts will follow.

4. Results and discussion

This section presents a performance analysis of the framework for traffic engineering in the wavelength domain TE-DBS, described in the Section 3, and assuming the HMPI algorithm, detailed in Section 2, is employed offline to optimize the wavelength search ordering for each routing path in the network.

The results are obtained via network simulation using the event-driven network simulator described in (Pedro et al., 2006a). The network topology used in the performance study is a 10-node ring network. All of the network nodes have the functionalities of both edge and core nodes and the resource reservation is made using the JET protocol. It is also assumed that all the wavelength channels in a fibre link have a capacity μ = 10 Gb/s, the time required to configure an optical space switch matrix is t_g = 1.6 μs, each node can process the BHP of a data burst in t_p = 1 μs and the offset time between BHP and data burst is given by $t_g + h_i \cdot t_p$, where h_i is the number of hops of burst path $\pi_i \in \Pi$. The switch matrix of each node is assumed to be strictly non-blocking. Unless stated otherwise, the simulation results were obtained assuming W = 32 wavelength channels per fibre link.

The traffic pattern used in the simulations is uniform, in the sense that a burst generated at an ingress node is randomly destined to one of the remaining nodes. Bursts are always routed via the shortest path. Both the data burst size and the burst interarrival time are negative-exponentially distributed. An average burst size of 100 kB is used, which results in an average burst duration of 80 μs. In the network simulations, increasing the average offered traffic load is obtained through reducing the average burst interarrival time. The average offered traffic load normalized to the network capacity is given by,

$$\Gamma = \frac{\sum_{\pi_i \in \Pi} \gamma_i \cdot h_i^{SP}}{L \cdot W \cdot \mu}, \tag{12}$$

where h_i^{SP} is the number of links traversed between the edge nodes of $\pi_i \in \Pi$.

In OBS networks, the most relevant performance metric is the average burst blocking probability, which measures the average fraction of burst traffic that is discarded by the network. The network performance can also be evaluated via the average offered traffic load that results in an objective average burst blocking probability B_{obj}. This metric is estimated by performing simulations with values of Γ spaced by 0.05, determining the load values between which the value with blocking probability B_{obj} is located and then using linear interpolation (with logarithmic scale for the average burst blocking probability). All of the results presented in this section were obtained through running 10 independent simulations for calculating the average value of the performance metric of interest, as well as a 95% confidence interval on this value. However, these confidence intervals were found to be so narrow that have been omitted from the plots for improving readability.

The majority of OBS proposals assumes the utilization of full-range wavelength converters deployed in a dedicated configuration, that is, one full-range wavelength converter is used at each output port of the switch matrix, as illustrated in Fig. 5. Each full-range wavelength converter must be capable of converting any wavelength at its input to a fixed wavelength at its output and if a node has M output fibres, its total number of converters is $M \cdot W$.

Fig. 6 plots the average burst blocking probability as a function of the maximum ingress burst delay for different values of the offered traffic load and considering both TE-DBS and the previously described BORA strategy. It also displays the blocking performance that corresponds to delaying bursts at the ingress nodes whenever a free wavelength channel is not immediately found. More precisely, the DBS strategy consists of delaying a data burst at its ingress node by the minimum amount of time, upper-bounded to the maximum ingress burst delay, such that one wavelength becomes available in the output fibre link.

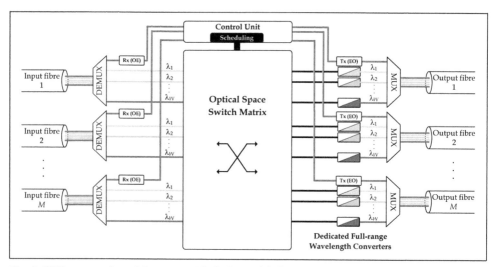

Fig. 5. OBS core node architecture with dedicated full-range wavelength converters.

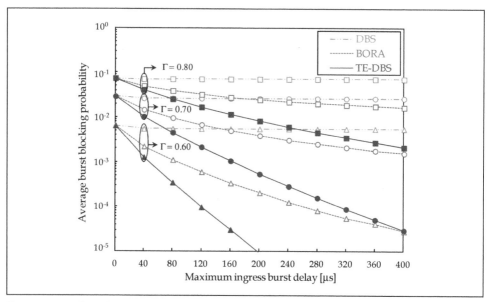

Fig. 6. Network performance with dedicated full-range wavelength converters for different values of the average offered traffic (Pedro et al., 2009a).

The curves for DBS show that exploiting the electronic buffers at the ingress nodes only for contention resolution does not improve blocking performance. On the contrary, with both BORA and TE-DBS the average burst blocking probability is decreased as the maximum ingress burst delay is increased, confirming that these strategies proactively reduce the probability of contention by selectively delaying bursts at their ingress nodes.

The results also indicate TE-DBS is substantially more efficient than BORA in exploiting larger maximum ingress burst delays to reduce the burst blocking probability. The proposed strategy outperforms BORA for the same maximum ingress burst delay or, alternatively, requires a smaller maximum ingress burst delay to attain the same blocking performance of BORA. Particularly, the decrease rate of the burst losses with increasing the maximum ingress burst delay is considerably larger with TE-DBS than that with BORA. In addition, with TE-DBS the slope of the curves of the burst blocking probability is much steeper for smaller values of the average offered traffic load, a trend less pronounced with BORA.

Table 3 presents the average traffic load that can be offered to the network as to support an objective average burst blocking probability, B_{obj}, of 10^{-3} and 10^{-4}. The results include two values of the maximum ingress burst delay for BORA and TE-DBS, $\Delta t_{max}^{RAM} = 200$ μs and $\Delta t_{max}^{RAM} = 400$ μs, and the case of immediate burst scheduling at the ingress nodes, $\Delta t_{max}^{RAM} = 0$.

B_{obj}	$\Delta t_{max}^{RAM} = 0$	$\Delta t_{max}^{RAM} = 200$ μs		$\Delta t_{max}^{RAM} = 400$ μs	
		BORA	TE-DBS	BORA	TE-DBS
10^{-3}	0.522	0.654	0.723	0.689	0.782
10^{-4}	0.453	0.584	0.659	0.632	0.729

Table 3. Average offered traffic load for an objective average burst blocking probability of 10^{-3} and 10^{-4} (Pedro et al., 2009a).

The OBS network supports more offered traffic load for the same average burst blocking probability when using the TE-DBS and BORA strategies instead of employing immediate burst scheduling. In addition, the former strategy provides the largest improvements in supported offered traffic load. For instance, with $B_{obj} = 10^{-3}$, the network supports 32% more offered traffic load when using BORA with a maximum ingress burst delay of 400 μs instead of immediate burst scheduling, whereas when using the TE-DBS strategy the performance improvement is more expressive, enabling an increase of 50% in offered traffic load.

In order to provide evidence of the principles underlying contention minimization with BORA and TE-DBS, the first set of results differentiates the burst blocking probability at the ingress nodes (ingress bursts) and at the core nodes (transit bursts). Fig. 7 plots the average burst blocking probability, discriminated in terms of ingress bursts and transit bursts, as a function of the maximum ingress burst delay for $\Gamma = 0.70$.

The plot shows that without additional delays at the ingress nodes, the blocking probability of ingress bursts and of transit bursts are of the same order of magnitude. However, as the maximum ingress burst delay is increased, the blocking probability of ingress bursts is rapidly reduced, as a result of the enhanced ability of ingress nodes to buffer bursts during longer periods of time. This holds for the three channel scheduling algorithms. Therefore, the average burst blocking probability of transit bursts becomes the dominant source of blocking. Notably, using DBS does not reduce burst losses at the core nodes, rendering this strategy useless, whereas BORA and TE-DBS strategies exploit the selective ingress delay to reduce blocking of transit bursts. Moreover, TE-DBS is increasingly more effective than BORA in reducing these losses, which supports its superior performance displayed in Fig. 6.

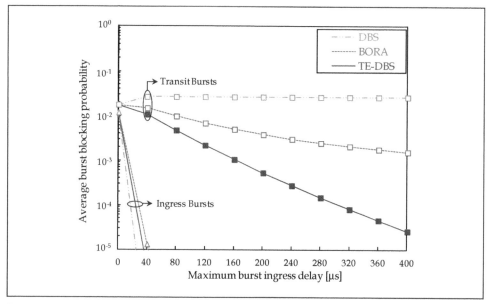

Fig. 7. Average burst blocking probability of ingress and transit bursts (Pedro et al., 2009a).

The major dissimilarity between the TE-DBS and BORA strategies is the order by which free wavelength channels are searched to schedule the data bursts assembled at the ingress nodes. Particularly, the TE-DBS strategy exploits the selective delaying of data bursts at the electronic buffers of these nodes not only to smooth the burst traffic entering the core network, similarly to BORA, but also to proactively reduce the unusable voids formed between consecutive data bursts scheduled in the same wavelength channel. As described in Section 3, complying with the latter objective demands enforcing that the serialized data bursts are kept in the same wavelength for as long as possible along their routing path, which means that contention for the same wavelength among bursts on overlapping paths must be minimized. Intuitively, the success of keeping the serialized data bursts in the same wavelength channel for as long as possible should be visible in the form of a reduced number of bursts experiencing wavelength conversion at the core nodes. In order to observe this effect, Fig. 8 presents the average wavelength conversion probability, defined as the fraction of transit data bursts that undergo wavelength conversion, as a function of the maximum ingress burst delay for different values of the average offered traffic load.

The curves for TE-DBS exhibit a declining trend as the maximum ingress burst delay increases, with this behaviour being more pronounced for smaller average offered traffic load values. These observations confirm that the probability of the data bursts serialized at the ingress nodes being kept in the same wavelength channel, as they go through the core nodes, is higher for larger values of the maximum ingress burst delay and smaller values of offered traffic load. Conversely, with BORA the wavelength conversion probability remains insensitive to variations in both the maximum ingress burst delay and offered traffic load, corroborating the fact that it cannot reduce the utilization of wavelength conversion at the core nodes. The reduced wavelength contention characteristic of the TE-DBS strategy, which is absent in BORA, is critical to mitigate the fragmentation of the wavelengths capacity, resulting in the smaller transit burst losses reported with TE-DBS in Fig. 7 and ultimately explaining the enhanced blocking performance provided by this strategy.

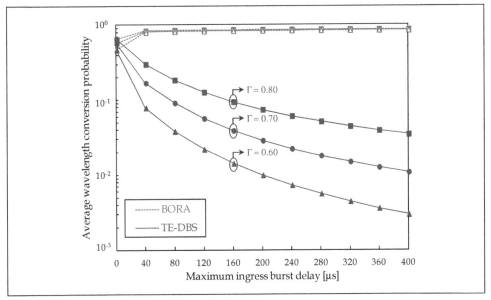

Fig. 8. Average wavelength conversion probability (Pedro et al., 2009b).

Fig. 9 shows the blocking performance as a function of the maximum ingress burst delay for different numbers of wavelength channels and Γ = 0.80. The results indicate that the slope of the average burst blocking probability curves for TE-DBS increases with the number of wavelength channels, augmenting the performance gain of using this strategy instead of BORA. This behaviour is due to the fact that when the number of wavelength channels per fibre link increases the effectiveness of the HMPI algorithm in determining appropriate wavelength search orderings improves, enhancing the isolation degree of serialized burst traffic from overlapping routing paths on different wavelength channels.

In principle, only a fraction of transit bursts experience wavelength contention, demanding the use of a wavelength converter. Consequently, the deployment of a smaller number of converters, in a shared configuration, has been proposed in the literature. Converter sharing at the core nodes can be implemented on a per-link or per-node basis, depending on whether each converter can only be used by bursts directed to a specific output link or can be used by bursts directed to any output link of the node (Chai et al., 2002). The latter sharing strategy enables to deploy a smaller number of converters. Fig. 10 exemplifies the architecture of a core node with C full-range wavelength converters shared per-node, where $C \leq M \cdot W$. In this core node architecture, each wavelength converter must be capable of converting any wavelength channel at its input to any wavelength channel at its output and the switch matrix has to be augmented with C input ports and C output ports.

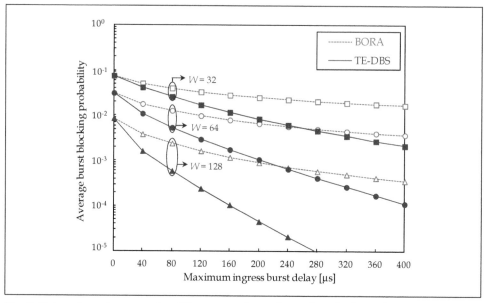

Fig. 9. Network performance for different numbers of wavelength channels (Pedro et al., 2009b).

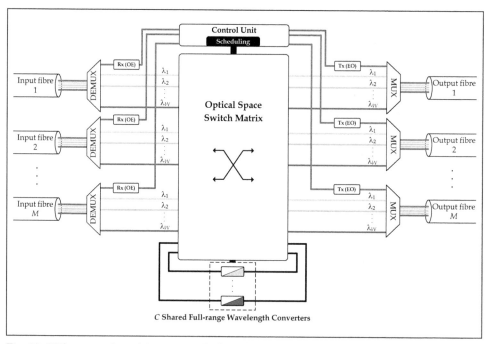

Fig. 10. OBS core node architecture with shared full-range wavelength converters.

The minimization of wavelength contention experienced by transit bursts is a key enabler for TE-DBS to improve the loss performance of OBS networks. Particularly, the simulation results presented in Fig. 8 confirm that the utilization of this strategy reduces the probability of wavelength conversion, and consequently the utilization of the wavelength converters, as the maximum ingress burst delay is increased. This attribute can extend the usefulness of TE-DBS to OBS networks with shared full-range wavelength converters because in this network scenario the lack of available converters at the core nodes can become the major cause of unresolved contention, specially for small values of C.

In order to illustrate the added-value of the TE-DBS strategy in OBS networks whose core nodes have shared full-range wavelength converters, consider the 10-node ring network with $W = 32$. When using wavelength converters in a dedicated configuration, each node of this network needs $M \cdot W = 64$ converters. Fig. 11 plots the average burst blocking probability as a function of the number of shared full-range wavelength converters at the nodes, C, for different values of the average offered traffic load and using BORA and TE-DBS strategies with $\Delta t_{max}^{RAM} = 160$ μs.

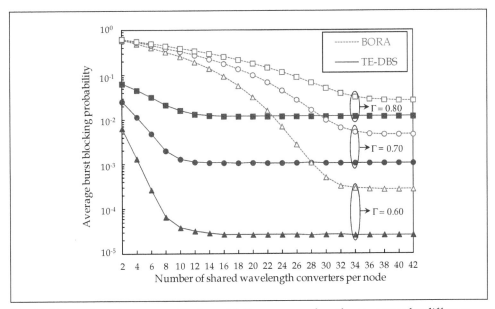

Fig. 11. Network performance with shared full-range wavelength converters for different values of the average offered traffic load (Pedro et al., 2009a).

The blocking performance curves clearly show that the OBS network using TE-DBS can benefit not only in terms of enhanced blocking performance, but also from enabling using simplified core node architectures. More precisely, the burst loss curves indicate that for very small numbers of shared wavelength converters, the utilization of TE-DBS results in a burst blocking probability that can be multiple orders of magnitude lower than that obtained using BORA. Furthermore, using TE-DBS demands a much smaller number of shared wavelength converters to match the blocking performance of a network using core nodes with dedicated wavelength converters. Particularly, with TE-DBS around 16 shared converters per node are enough to match the loss performance obtained with 64 dedicated converters, whereas with BORA this number more than doubles, since around 36 shared converters are required. The larger savings in the number of wavelength converters enabled by TE-DBS also mean that the expansion of the switch matrix to accommodate the shared converters is smaller, leading to an even more cost-effective network solution.

5. Conclusions

Optical burst switching is seen as a candidate technology for next-generation transport networks. This chapter has described and analyzed the performance benefits of a strategy to enforce traffic engineering in the wavelength domain in OBS networks. The TE-DBS strategy is based on using the HMPI algorithm to optimize offline the order by which wavelength channels are searched for each routing path and employing at the ingress nodes a selective delaying of data bursts as a way to maximize the amount of burst traffic sent via the wavelength channels ranked with highest priority. Both the HMPI offline algorithm and the online selective delaying of bursts were revisited and exemplified.

A network simulation study has highlighted the performance improvements attained by using TE-DBS in an OBS network with dedicated full-range wavelength converters and with shared full-range wavelength converters. It was shown that the utilization of the TE-DBS strategy enables to reduce the average burst blocking probability for a given average offered traffic load, or augment the average offered traffic load for an objective burst blocking probability, when compared to utilizing a known contention minimization strategy. The simulation results shown that increasing the maximum delay a burst can experience at the ingress node and augmenting the number of wavelength channels per link can improve the effectiveness of the TE-DBS strategy and also provided evidence of the burst serialization and traffic isolation in different wavelengths inherent to this strategy. Finally, the analysis confirms that the utilization of TE-DBS in OBS networks with shared full-range wavelength converters can provide noticeable savings in the number of expensive all-optical wavelength converters and a smaller increase in the size of the switch matrix of the core nodes.

6. References

Barakat, N. & Darcie, T. (2007). The Control-Plane Stability Constraint in Optical Burst Switching Networks. *IEEE Communications Letters*, Vol. 11, No. 3, (March 2007), pp. 267-269, ISSN 1089-7798

Chai, T.; Cheng, T. ; Shen, G.; Bose, S. & Lu, C. (2002). Design and Performance of Optical Cross-Connect Architectures with Converter Sharing. *Optical Networks Magazine*, Vol. 3, No. 4, (July/August 2002), pp. 73-84, ISSN 1572-8161

Chang, G.; Yu, J. ; Yeo, Y.; Chowdhury, A. & Jia, Z. (2006). Enabling Technologies for Next-Generation Packet-Switching Networks. *Proceedings of the IEEE*, Vol. 94, No. 5, (May 2006), pp. 892-910, ISSN 0018-9219

Chen, Y.; Qiao, C. & Yu, X. (2004). Optical Burst Switching: A New Area in Optical Networking Research. *IEEE Network*, Vol. 18, No. 3, (May/June 2004), pp. 16-23, ISSN 0890-8044

IETF (2002). *RFC 3945: Generalized Multi-Protocol Label Switching (GMPLS) Architecture,* Internet Engineering Task Force, September 2002

ITU-T (2006). *Recommendation G.8080: Architecture for the Automatically Switched Optical Network (ASON),* International Telecommunication Union – Telecommunication Standardization Sector, June 2006

Korotky, S. (2004). Network Global Expectation Model: A Statistical Formalism for Quickly Quantifying Network Needs and Costs. *IEEE/OSA Journal of Lightwave Technology,* Vol. 22, No. 3, (March 2004), pp. 703-722, ISSN 0733-8724

Li, J. & Qiao, C. (2004). Schedule Burst Proactively for Optical Burst Switched Networks. *Computer Networks,* Vol. 44, (2004), pp. 617-629, ISSN 1389-1286

Papadimitriou, G.; Papazoglou, C. & Pomportsis, A. (2003). Optical Switching: Switch Fabrics, Techniques, and Architectures. *IEEE/OSA Journal of Lightwave Technology,* Vol. 21, No. 2, (February 2003), pp. 384-405, ISSN 0733-8724

Pedro, J.; Castro, J.; Monteiro, P. & Pires, J. (2006a). On the Modelling and Performance Evaluation of Optical Burst-Switched Networks, *Proceedings of IEEE CAMAD 2006 11th International Workshop on Computer-Aided Modeling, Analysis and Design of Communication Links and Networks,* pp. 30-37, ISBN 0-7803-9536-0, Trento, Italy, June 8-9, 2006

Pedro, J.; Monteiro, P. & Pires, J. (2006b). Wavelength Contention Minimization Strategies for Optical-Burst Switched Networks, *Proceedings of IEEE GLOBECOM 2006 49th Global Telecommunications Conference,* paper OPNp1-5, ISBN 1-4244-0356-1, San Francisco, USA, November 27-December 1, 2006

Pedro, J.; Monteiro, P. & Pires, J. (2009a). On the Benefits of Selectively Delaying Bursts at the Ingress Edge Nodes of an OBS Network, *Proceedings of IFIP ONDM 2009 13th Conference on Optical Network Design and Modelling,* ISBN 978-1-4244-4187-7, Braunschweig, Germany, February 18-20, 2009

Pedro, J.; Monteiro, P. & Pires, J. (2009b). Contention Minimization in Optical Burst-Switched Networks Combining Traffic Engineering in the Wavelength Domain and Delayed Ingress Burst Scheduling. *IET Communications,* Vol. 3, No. 3, (March 2009), pp. 372-380, ISSN 1751-8628

Pedro, J.; Monteiro, P. & Pires, J. (2009c). Traffic Engineering in the Wavelength Domain for Optical Burst-Switched Networks. *IEEE/OSA Journal of Lightwave Technology,* Vol. 27, No. 15, (August 2009), pp. 3075-3091, ISSN 0733-8724

Poustie, A. (2005). Semiconductor Devices for All-Optical Signal Processing, *Proceedings of ECOC 2005 31st European Conference on Optical Communication,* Vol. 3, pp. 475-478, ISBN 0-86341-543-1, Glasgow, Scotland, September 25-29, 2005

Qiao, C. & Yoo, M. (1999). Optical Burst Switching (OBS) – A New Paradigm for an Optical Internet. *Journal of High Speed Networks,* Vol. 8, No. 1, (January 1999), pp. 69-84, ISSN 0926-6801

Sahara, A.; Shimano, K.; Noguchi, K.; Koga, M. & Takigawa, Y. (2003). Demonstration of Optical Burst Data Switching using Photonic MPLS Routers operated by GMPLS Signalling, *Proceedings of OFC 2003 Optical Fiber Communications Conference,* Vol. 1, pp. 220-222, ISBN 1-55752-746-6, Atlanta, USA, March 23-28, 2003

Sun, Y.; Hashiguchi, T. ; Minh, V.; Wang, X.; Morikawa, H. & Aoyama, T. (2005). Design and Implementation of an Optical Burst-Switched Network Testbed. *IEEE*

Communications Magazine, Vol. 43, No. 11, (November 2005), pp. s48-s55, ISSN 0163-6804

Teng, J. & Rouskas, G. (2005). Wavelength Selection in OBS Networks using Traffic Engineering and Priority-Based Concepts. *IEEE Journal on Selected Areas in Communications*, Vol. 23, No. 8, (August 2005), pp. 1658-1669, ISSN 0733-8716

Tucker, R. (2006). The Role of Optics and Electronics in High-Capacity Routers. *IEEE/OSA Journal of Lightwave Technology*, Vol. 24, No. 12, (December 2006), pp. 4655-4673, ISSN 0733-8724

Wang, X.; Morikawa, H. & Aoyama, T. (2003). Priority-Based Wavelength Assignment Algorithm for Burst Switched WDM Optical Networks. *IEICE Transactions on Communications*, Vol. E86-B, No. 5, (2003), pp. 1508-1514, ISSN 1745-1345

Xiong, Y.; Vandenhoute, M. & Cankaya, H. (2000). Control Architecture in Optical Burst-Switched WDM Networks. *IEEE Journal on Selected Areas in Communications*, Vol. 18, No. 10, (October 2000), pp. 1838-1851, ISSN 0733-8716

Zang, H.; Jue, J. ; Sahasrabuddhe, L.; Ramamurthy, R. & Mukherjee, B. (2001). Dynamic Lightpath Establishment in Wavelength-Routed WDM Networks. *IEEE Communications Magazine*, Vol. 39, No. 9, (September 2001), pp. 100-108, ISSN 0163-6804

Zhou, P. & Yang, O. (2003). How Practical is Optical Packet Switching in Core Networks?, *Proceedings of IEEE GLOBECOM 2003 49th Global Telecommunications Conference*, pp. 2709-2713, ISBN 0-7803-7974-8, San Francisco, USA, December 1-5, 2003

Zhu, K.; Zhu, H. & Mukherjee, B. (2005). *Traffic Grooming in Optical WDM Mesh Networks*, Springer, ISBN 978-0-387-25432-6, New York, USA

6

Routing and Traffic Engineering in Dynamic Packet-Oriented Networks

Mihael Mohorčič and Aleš Švigelj
Jožef Stefan Institute
Slovenia

1. Introduction

Spurred by the vision of seamless connectivity anywhere and anytime, ubiquitous and pervasive communications are playing increasingly important role in our daily lives. New types of applications are also affecting behaviour of users and changing their habits, essentially reinforcing the need for being always connected. This clearly represents a challenge for the telecommunications community especially for operating scenarios characterised by high dynamics of the network requiring appropriate routing and traffic engineering.

Routing and traffic engineering are cornerstones of every future telecommunication system, thus, this chapter is concerned with an adaptive routing and traffic engineering in highly dynamic packet-oriented networks such as mobile ad hoc networks, mobile sensor networks or non-geostationary satellite communication systems with intersatellite links (ISL). The first two cases are recently particularly popular for smaller scale computer or data networks, where scarce energy resources represent the main optimisation parameter both for traffic engineering and routing. However, they require a significantly different approach, typically based on clustering, which exceeds the scope of this chapter. The third case, on the other hand, is particularly interesting from the aspect of routing and traffic engineering in large scale telecommunication networks. Even more so, since it exhibits a high degree of regularity, predictability and periodicity. It combines different segments of communication network and generally requires distinction between different types of traffic. Different restrictions and requirements in different segments typically require separate optimization of resource management.

So, in order to explain all routing functions and different techniques used for traffic engineering in highly dynamic networks we use as an example the ISL network, characterized by highly dynamic conditions. Nonetheless, wherever possible the discussion is intentionally kept independent of the type of underlying network or particular communication protocols and mechanisms (e.g. IP, RIP, OSPF, MPLS, IntServ, DiffServ, etc.), although some presented techniques are an integral part of those protocols. Thus, this chapter is focusing on general routing and traffic engineering techniques that are suitable for the provision of QoS in packet-oriented ISL networks. Furthermore, most concepts, described techniques, procedures and algorithms, even if explained on an example of ISL network, can be generalised and used also in other types of networks exhibiting high level of dynamics (Liu et al., 2011; Long et al., 2010; Rao & Wang, 2010, 2011). The modular approach allows easy (re)usage of presented procedures and techniques, thus, only particular or entire procedures can be used.

ISL network exhibits several useful properties which support the development of routing procedures. These properties include (Wood et al., 2001):

- Predictability – motion of satellites around the earth is deterministic, thus the position of satellites and their connectivity can be computed in advance, taking into account the parameters of the satellite orbit and constellation. Consequently, in an ISL network only undeterministic parameters need to be monitored and distributed through the network, thus minimizing the signalling load.
- Periodicity – satellite positions and thus the configuration of the space segment, repeats with the orbit period, which is defined uniquely by the selected orbit altitude. Taking into account also the terrestrial segment, an ISL network will experience a quasi-periodic behaviour on a larger scale, defined as the smallest common integer multiple of the orbit period and the traffic intensity period, referred to as the system period.
- Regularity – a LEO constellation with an ISL network is characterized by a regular mesh topology, enabling routing procedures to be considered independently of the actual serving satellite (i.e. concealing the motion of satellites with respect to the earth from the routing procedure). Furthermore, the high level of node connectivity (typically between 2 and 6 links to the neighbouring nodes) provides several alternative paths between a given pair of satellites.
- Constant number of network nodes – routing procedures in ISL networks are based typically on the explicit knowledge of the network topology which, in the case of satellite constellation, has a constant, predefined number of network nodes in the space (satellites) and terrestrial (gateways) segments (except in the case of a node or a link failure). This property has a direct influence on the calculation of routing tables.

The above properties are incorporated in the described routing and traffic modelling techniques and procedures. Special attention is given to properties which support the development of efficient, yet not excessively complex, adaptive routing and traffic engineering techniques.

However, for the verification, validation and performance evaluation of algorithms, protocols, or whole telecommunication systems, the development of suitable traffic models, which serve as a vital input parameter in any simulation model, is of paramount importance. Thus, at the end of the chapter we are presenting the methodology for modelling global aggregate traffic comprising of four main modules. It can be used as a whole or only selected modules can be used for particular purposes connected with simulation of particular models.

Routing and traffic engineering on one side require good knowledge of the type of network and its characteristics and on the other side also of the type of traffic in the network. This is needed not only for adapting particular techniques, procedures and algorithms to the network and traffic conditions but also for their simulation, testing and benchmarking. To this end this chapter is complemented by description of a methodology for developing a global traffic model suitable for the non-geostationary ISL networks, which consists of modules describing distribution of sources, their traffic intensity and its temporal variation, as well as traffic flow patterns.

2. Routing functions

The main task of any routing is to find suitable paths for user traffic from the source node to destination node in accordance with the traffic's service requirements and the network's service restrictions. Paths should accommodate all different types of services using different optimisation metrics (e.g. delay, bandwidth, etc.). Thus, different types of traffic can be routed over different routes. Routing functionality can be in general split in four core routing functions, (i) acquiring information about the network and user traffic state, and link cost calculation, (ii) distributing the acquired information, (iii) computing routes according to the traffic state information and chosen optimization criteria, and (iv) forwarding the user traffic along the routes to the destination node.

For each of these functions, several policies exist. Generally speaking, the selection of a given policy will impact (i) the performance of the routing protocol and (ii) the cost of running the protocol. These two aspects are dual and a careful design in the routing algorithm must achieve a suitable balance between the two. The following sub-sections will discuss the four core routing functions.

2.1 Acquiring information about the network and link cost calculation

The parameters of the link-cost metric should directly represent the fundamental network characteristics and the changing dynamics of the network status. Furthermore, they should be orthogonal to each other, in order to eliminate unnecessary redundant information and inter-dependence among the variables (Wang & Crowcroft, 1996). Depending on the composition rule we distinguish additive, multiplicative, concave and convex link-cost metrics (Wang, 1999). In additive link-cost metrics the total cost of the path is a sum of costs on every hop. Additive link costs include delay, jitter, cost and hop-count. Total cost of the path in the case of multiplicative link-cost metrics is a product of individual costs of links. A typical example of multiplicative link cost is link reliability. In concave and convex link-cost metrics the total cost of the path equals the cost on the hop with the minimum and maximum link cost respectively, and a typical example of link-cost metric is the available bandwidth.

2.1.1 Link cost for delay sensitive traffic

We show the use of the additive link-cost metric as an example for the link-cost function for the delay sensitive traffic, considering two dynamically changing parameters. The first is the intersatellite distance between neighbouring satellites, while the second is the traffic load on a particular satellite. They have a significant effect on the routing performance and are scalable with the network load and link capacity, thus being well suited for link-cost metric. (Mohorcic et al., 2004; Svigelj et al., 2004a).

The distance between satellite pairs in a non-geostationary satellite system is deterministic and can be calculated in advance. We consider this distance of a particular link l through propagation delay (T_{Pl}). Propagation delay in satellite communications is proportional to the number of hops between source and destination satellites, which could be used as a simplified cost metric or an additional criterion.

The traffic load on a particular satellite and its outgoing links is constantly changing in a random fashion, thus it needs to be estimated in real-time. To estimate the traffic load on particular link we can use two parameters. It can be estimated through the queuing delay, which reflects the past values of traffic load, or expected queuing delay, which estimates the future value of queuing delay in a given outgoing queue. In addition, both parameters can be improved with additional functions (i.e. exponential forgetting function, exponential smoothing function), which are described in the following subsections. Thus, in general the link costs (LC_l) for delay sensitive traffic on the link l at time t_i are calculated using Equation (1) at the end of each routing table update interval. It includes the propagation delay (T_{Pl}) and traffic load represented by (T_{Ql}).

$$LC_l(t_i) = T_{Pl}(t_i) + T_{Ql}(t_i) \tag{1}$$

2.1.1.1 Link cost based on the queuing delay enhanced with Exponential forgetting function EFF

In this case we monitor the traffic load on a satellite through the packet queuing delay (T_{ql}) at the respective port of the node, which is directly proportional to the traffic load on the selected outgoing link l as shown in Equation (2), where L_r denotes the length of the rth packet in outgoing queue and C_l is the capacity of the link l

$$T_{ql} = \frac{\sum_r L_r}{C_l} \tag{2}$$

Due to variation of these queuing delays, the queuing delay value T_{Ql}, considered in the link-cost function, is periodically estimated using a fixed-size window exponential forgetting function $EFF(n, \chi, T_{ql})$ on a set of the last n values of packet queuing delay collected in a given time interval (i.e. $T_{ql}[n]$ being the last collected value, and the other values considered being $T_{ql}[n-1],...,T_{ql}[1]$). In the EFF function, n (the depth of the function) denotes the number of memory cells in the circular register. If the number of collected T_{ql} values m is smaller than n, then only these values are considered in the EFF function. Furthermore, as shown in Equation (3), a forgetting factor, $\chi \in (0, 1)$, is introduced to make the more recent T_{ql} values more significant in calculating T_{Ql}.

$$T_{Ql} = EFF(n, \chi, T_{ql}) = \begin{cases} (1-\chi) \cdot \sum_{r=0}^{m-1} \chi^r \cdot T_{ql}[m-r] & \text{for } m < n \\ \\ (1-\chi) \cdot \sum_{r=0}^{m-1} \chi^r \cdot T_{ql}[n-r] & \text{for } m \geq n \end{cases} \tag{3}$$

2.1.1.2 Link cost based on expected queuing delay enhanced with Exponential Smoothing Link-Cost Function

In the case of using expected queuing delay in the assessment of the traffic load, we monitor the outgoing queues of particular traffic. A packet entering a given output queue at time t will have the expected queuing delay, T_{exp}, given by Equation (4), where L_{av} is the average packet length, C the link capacity, and $n(t)$ the number of packets in the queue.

$$T_{exp}(t) = n(t) \cdot \frac{L_{av}}{C} \tag{4}$$

Calculation of the expected queuing delay does not require any distribution of link status between neighbouring nodes, and has the advantage of fast response to congestions on the link. However, for calculation of pre-computed routing tables the average expected queuing delay T_{exp_av} has to be determined using Equation (5) at the end of each update interval T_I starting at time t_S. This average expected queuing delay could subsequently be already used as a link-cost metric parameter T_{Ql}, as shown in Equation (6), which expresses traffic load on the link.

$$T_{exp\_av}(t_S + T_I) = \frac{1}{T_I} \cdot \int_{t_S}^{t_S+T_I} n(t) \cdot \frac{L_{av}}{C} \cdot dt \tag{5}$$

$$T_{Ql}(t_i) = T_{exp\_av}(t_S + T_I) \tag{6}$$

The consideration of link load in the link cost calculation, and consequently in route computation, may cause traffic load oscillations between alternative paths in the network (Bertsekas & Gallager, 1987). In particular, routing of packets along a given path increases the cost of used links. At the end of routing update interval this information is fed back to the routing algorithm, which chooses for the next routing update interval an alternative path. In extreme cases this may result in complete redirection of traffic load to alternative paths, eventually leading to traffic load oscillation between the two alternative paths in consecutive routing tables and hence routing instability. In ISL networks for instance traffic load oscillations impose a particular effect on delay sensitive traffic, as there are many alternative paths between a given pair of satellites with similar delays. Oscillations are especially inconvenient under heavy traffic load conditions, where the impact of traffic load parameter on the link cost is much higher than that of the propagation delay T_P. Under such conditions oscillations lead to congestion on particular links, which significantly degrades routing performance. In addition, the oscillations of traffic load have also a great impact on triggered signalling, where the signalling load depends on a significant change of link cost. In order to introduce the triggered signalling, the reduction of the oscillation of traffic load and consequently the oscillation of link cost is inevitable. Smoothing of the link cost on a particular link can be done in two ways:

- Directly by modifying the link cost on particular link with a suitable smoothing function.

- Indirectly by using advanced forwarding policies, which send traffic also along the alternative paths and distribute traffic more evenly on the first and the second shortest paths and consequently smooth-out the link cost. (see section 2.4.)

To reduce the oscillations one can use an exponential smoothing link-cost function, which iteratively calculates the traffic load parameter T_{Ql} from its previous values according to Equation (7). The influence of the previous value is regulated with a parameter k, defined between 0 and 1 ($k \in [0,1]$), while the initial value for the parameter T_{Ql} is set to 0.

$$T_{Ql}(t_0) = 0$$
$$T_{Ql}(t_i) = \left[T_{exp\_av}(t_i) - T_{Ql}(t_{i-1}) \right] \cdot k + T_{Ql}(t_{i-1}) = \qquad (7)$$
$$= k \cdot T_{exp\_av}(t_i) + (1-k) \cdot T_{Ql}(t_{i-1})$$

Taking into account this parameter, the cost of a given link l is calculated using Equation (1). If k equals 1, there is no influence of previous values on current link cost and Equation (7) transforms to Equation (6). On the other hand, if k equals 0, only propagation delay is considered in link cost calculation, which leads to traffic insensitive routing.

One of the drawbacks of the exponential smoothing link-cost function is that it takes into account in each iteration all previous values of parameter T_{exp_av}. The value of T_Q as a function of n previous values of T_{exp_av} is given in Equation (8). It can be seen that the impact of previous values of T_{exp_av} decreases exponentially with increasing value of n.

$$T_{Ql}(t_n) = k \cdot ((1-k)^0 \cdot T_{exp\_av}(t_n) + (1-k)^1 \cdot T_{exp\_av}(t_{n-1}) +$$
$$+ (1-k)^2 \cdot T_{exp\_av}(t_{n-2}) + ... + (1-k)^{n-1} \cdot T_{exp\_av}(t_1)) \qquad (8)$$

The main goal of the exponential smoothing link-cost function, which tends to suppress the traffic load oscillations, is that the link cost should reflect the actual traversing traffic flow and the traffic intensity of the region served by the satellite, and not the instantaneous fluctuations of traffic load due to oscillation. In such manner exponential smoothing algorithm promises more evenly distribution of traffic load between links and consequently a better performance for different traffic types. Furthermore it ensures, that in a lightly loaded network, the routing performance is not decreased, while it is notably enhanced in heavily loaded network. A more exhaustive explanation of exponential smoothing link cost function and optimum definition of parameter k is given in (Svigelj et al., 2004a).

2.1.1.3 Weighted delay calculation

The relative impacts of traffic load and propagation delay on the link cost are linearly regulated with a traffic weight factor (TWF_l) and a propagation delay weight factor ($PDWF_l$), respectively, as shown in Equation (9) defining weighted delay (WD_l) on the link l. This allows biasing of link cost towards shortest-path routes ($PDWF_l > TWF_l$) or towards least loaded but slightly longer routes ($PDWF_l < TWF_l$).

$$WD_l = PDWF_l \cdot T_{Pl} + TWF_l \cdot T_{Ql} \qquad (9)$$

In general, as indicated in Equation (9), different weights can be used on different links. In a non-geostationary satellite system, however, satellites are continuously revolving around the rotating earth, so weights cannot be optimized for the traffic load of certain regions but should either be fixed or should adapt to the conditions in a given region. The later gives opportunity for further optimisation using some traffic aware heuristic approach.

Weighted delay on the link, as given by Equation (9), can already be used as a simple continuous link-cost function with a linear relation between both metrics. In general, however, a more sophisticated link-cost function should be able to control the relative cost of heavily loaded links with respect to lightly loaded links. This can be accomplished by a non-linear link-cost function, such as an exponentially growing function with exponent α, as given in Equation (10), where WD_L and WD_U represent lower and upper boundary values of weighted delay on the links respectively.

$$LC_l = \left(\frac{WD_l - WD_L}{WD_U - WD_L} \right)^\alpha + \frac{WD_L}{WD_U} \tag{10}$$

The first term in Equation (10) represents the normalised dynamically changing link cost according to variation of propagation delay (e.g. ISL length) and traffic load (e.g. queuing delay). Since it is not suitable that link cost be zero, which can cause high oscillations, a small constant (WD_L/WD_U) is added to the normalised term of the link-cost function. This constant represents the normalised cost of the shortest link without any traffic load. When $a = 0$ a link-cost function has no influence on the routing algorithm, and path selection reduces to cost-independent routing (i.e. minimum hop count routing), while with $a = 1$ it selects a path with the minimum sum of link costs. Exponent values larger than 1 ($a > 1$) tend to eliminate heavily loaded (high cost) links from consideration, while exponent values smaller than 1 ($a < 1$) tend to preserve lightly loaded links. Combining Equations (9) and(10), the link cost for the delay sensitive traffic, which takes into consideration delay on the link, is calculated as given by Equation (11).

$$LC_l = \left(\frac{PDWF_l \cdot T_{Pl} + TWF_l \cdot T_{Ql} - WD_L}{WD_U - WD_L} \right)^\alpha + \frac{WD_L}{WD_U} \tag{11}$$

2.1.1.4 Discretization

Regardless of the selected link-cost function the calculated link cost needs to be distributed throughout the network and stored in nodes for the subsequent calculation of new routing tables. In order to reduce computation effort and memory requirements, routing algorithms have been proposed that perform path selection on a small set of discrete link-cost levels. In these algorithms the appropriate number of link-cost levels needs to be defined to balance between the accuracy and computational complexity.

Equation (13) represents a suitable function, which converts the continuous link-cost function, given in Equation (12), to L discrete levels denoted as C_{Dl} in the range between 0 and 1. In this link-cost function the minimum and maximum value for weighted delay are used, WD_{min} and WD_{max}. Any link with weighted delay below WD_{min} is assigned the minimum cost $1/L$, while links with weighted delay higher than WD_{max} have link cost set to 1.

$$C_i = WD_i^{\alpha} \tag{12}$$

$$C_{DI} = \begin{cases} \dfrac{1}{L} & WD_l < WD_{min} \\[2em] \dfrac{\left[\left(\dfrac{WD_l - WD_{min}}{WD_{max} - WD_{min}}\right)^a \cdot (L-1)\right] + 1}{L} & WD_{min} < WD_l < WD_{max} \\[2em] 1 & WD_l \geq WD_{max} \end{cases} \tag{13}$$

2.1.2 Link cost function for the throughput sensitive traffic

The most suitable optimization parameter for the throughput sensitive traffic, on the other hand, is the available bandwidth on the link. Thus, on each link the lengths of the traversing packets are monitored between consecutive routing table updates, and the link utilization (LU_l) is calculated according to Equation (14), where L_r denotes the length of the rth traversing packet. The selected time interval between consecutive calculations of the sum of the packet lengths was equal to the routing table update interval T_l starting at time t_S.

$$LU_l(t_S + T_l) = \frac{\sum_r L_r}{T_l \cdot C_l} \tag{14}$$

The link-cost metric for the throughput sensitive traffic is a typical concave metric. The optimization problem is to find the paths with the maximum available bandwidth and, as an additional constraint, with minimum hop count, which minimizes the use of resources in the network. Thus, the link cost for throughput sensitive traffic is the normalized available bandwidth on the link, calculated at the end of the routing table update interval according to Equation (15).

$$LC_l(t_i) = 1 - LU_l(t_S + T_l) \tag{15}$$

2.2 Distributing the acquired information – signalling

Before the routes are calculated the information about network state should be distributed between nodes. An effective signalling scheme must achieve a trade-off between (a) bandwidth consumed for signalling information (b) computing and memory capacity dedicated to signalling processing and (c) improvement of the routing decisions due to the presence of signalling information (Franck & Maral, 2002a). Signalling is subdivided in two families: unsolicited and on-demand signalling. The following subsections detail these two families.

2.2.1 Unsolicited signalling

Unsolicited signalling is similar to unsolicited mail ads. Nodes receive at given time intervals information about the state of the other nodes. Conversely, nodes broadcast in the network information about their own state. Because a node has no control of the time it receives state information, the information might be non-topical once used for route computation. Non topical information is undesirable since it introduces a discrepancy between what is known and what the reality is. This is of particular importance for those systems which incorporate non-permanent links. Non topical information results in inaccurate and possibly poor routing decisions. Unsolicited signalling is further subdivided into periodic and triggered signalling.

Periodic signalling works by having each node broadcasting state information every p units of time, p being the broadcast period. It is not required for the broadcast period be equal for all nodes, however, it is practical to do so because (a) all nodes run the same software (b) it avoids discrepancies in the topicality of state information. Since the quality of routing decisions depends on how topical the state information is, it is expected that increasing the broadcast period results in increasing the connection blocking probability. On the other hand, increasing the broadcast period helps to keep the signalling traffic low. Periodic signalling supports easy dimensioning since the amount of signalling traffic does not depend on the amount of traffic flowing in the network and therefore can be quantified analytically. Unfortunately, this interesting characteristic is also a drawback: if the state of a node does not change during the whole broadcast period, the next broadcast will take place, regardless of whether it is useful or not. Likewise, some important state change might occur in the middle of the broadcast period without any chance for these changes to be advertised prior to the next broadcast. For these reasons, triggered signalling is worth investigating.

Instead of broadcasting periodically, the node using **triggered signalling** permanently monitors its state and initiates a broadcast upon a significant change of its state (threshold function). This approach is supposed to alleviate signalling traffic, holding down useless broadcasts. Triggered updates for instance are used for Routing Information Protocol (RIP). Unfortunately, triggered signalling has two down sides. First, while periodic signalling does not depend on the actual content of state information, triggered signalling must be aware of the semantics of the state information to define what a significant state change is. Second, the amount of signalling traffic generated depends on the characteristics of the traffic load in the constellation. It does not depend on the amount of data traffic but rather on the traffic variations in the nodes and links. Since routing impacts how traffic is distributed in the network, the behaviours of routing and triggered signalling are tightly interlaced. Triggered signalling can be further sub-divided in additional versions depending on the chosen threshold function.

In networks there are two changing parameters, which have the impact on the link cost: propagation delay between neighbouring nodes and traffic load. The first can be computed in advance in each node, so it can be eliminated from signalling information. For delay sensitive traffic the new value of T_{Ql} is broadcasted only if the value exceeds predefined threshold (Svigelj et al., 2012). If T_{Ql} does not exceed the threshold, only the propagation delay is used as a link cost in routes calculation. In the case of throughput sensitive traffic the link cost is broadcasted only if LC_l is lower than threshold (i.e. the available bandwidth is lower than threshold), otherwise value 1 (i.e. empty link) is used in routes calculation.

With an appropriate selection of thresholds the signalling load can be significantly reduced, especially for nodes, which has no intensive traffic. To omit the impact of oscillations of the link costs the triggered signalling can be used in a combination with exponential smoothing link-cost function or adaptive forwarding.

2.2.2 On-demand signalling

Compared to unsolicited signalling, on-demand signalling works the other way around. When a node (called the requesting node) requires state information, it queries the other nodes (called the serving nodes) for this information. Thus, on-demand signalling yields the state information as recent as possible, with expected benefit for the routing decisions. Furthermore, the type of state information which is queried (e.g. capacity or buffer occupancy) may vary according to the type of route that must be computed. On the other hand, since the signalling procedure is triggered for each route computation, the amount of traffic generated by on-demand signalling is likely to be higher than with unsolicited signalling. Additionally, the requesting node has to gather complete information before initiating the route computation. On-demand signalling is more convenient for connection oriented networks, where the source node requests the network state information from other nodes before setting up a connection and then the route to destination node is computed. As the number of packets during a signalling session is high, additional mechanisms (caching, snooping) have to be devised, in order to limit the number of signalling packets (Franck & Maral, 2002a).

2.3 Computing routes

In the case of per-hop packet-switched routing routes cannot be computed on demand. Instead, routing tables are pre-computed for all nodes periodically or in response to a significant change in link costs, thus defining routing update intervals. Link-cost metrics for the delay sensitive traffic are typical additive metrics, and thus the shortest routes are typically calculated using the Dijkstra algorithm. The main feature of an additive metric is that the total cost for any path is a sum of costs of individual links.

On the other hand, the link cost for the throughput sensitive traffic is a concave metric. Thus, the total cost for any path equals the one on the link with minimum cost. A typical optimization criterion for the throughput sensitive traffic is to find the paths within minimum hop count with the maximum available bandwidth. Minimum hop count is an additional constraint, which is used to minimize the use of resources. The Bellman-Ford shortest path algorithm is well suited to compute paths of the maximum available bandwidths within a minimum hop count. It is a property of the Bellman-Ford algorithm that, at its hth iteration, it identifies the optimal path (in our context the path with the maximum available bandwidth) between the source and each destination not more than h hops away. In other words, because the Bellman-Ford algorithm progresses by increasing the hop count, it provides the hop count of a path as a side result, which can be used as a second optimization criterion.

Regardless of the type of traffic the second shortest path with disjoint first link can be calculated by eliminating the first link on the shortest route (i.e. LC_l is set to infinity for delay sensitive traffic and to 0 in the case of throughput sensitive traffic) and using Dijkstra and Bellman Ford algorithm on such modified network. The alternative paths are used in the case of adaptive forwarding.

2.4 Forwarding the user traffic

In the route execution phase packets are forwarded on outgoing links to the next node along the path according to most recently calculated routing tables. In particular, packets are placed into an appropriate first in first out (FIFO) queue with a suitable scheduler according to the traffic type they belong to and according to the selected forwarding policy.

2.4.1 Static forwarding

Two representatives of static forwarding policies originally developed for regular network topologies, such as exhibited by ISL networks, are alternate link routing with deflection in the source node (ALR-S) and alternate link routing with deflection in all nodes (ALR-A) (Mohorcic et al., 2000, 2001). Both policies are based on an iterative calculation of routing algorithm for determining alternative routes between satellite pairs. An additional restriction considered in static forwarding policies is that the alternative routes must consist of the same (i.e., minimum) number of hops, with a different link for the first hop. Such alternative routes with the same number of hops guarantee that the propagation delay increase for the second-choice route is kept within a well-defined limit.

After determination of alternative routes with the same number of hops between each pair of nodes (satellites) the selected forwarding policy decides which packets are forwarded along each of these routes. Different forwarding policies are depicted in Fig. 1

According to the routing table given in Table 1, the SPR policy is only forwarding user traffic along the shortest routes. This leads to very non-uniform traffic load particularly on links (A-D, B-E, and C-F).

	Next hops on the route to satellite F and the cost of the route					
From	Shortest route		Second shortest route		Third shortest route	
Satellite A	D, E, F	14	B, E, F	15	B, C, F	16
Satellite B	E, F	10	C, F	11	/	/
Satellite C	F	6	/	/	/	/
Satellite D	E, F	10	/	/	/	/
Satellite E	F	5	/	/	/	/

Table 1. Alternative paths to Satellite F with the same minimum number of hops.

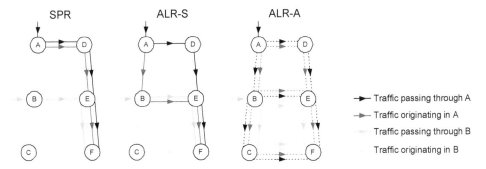

Fig. 1. Path selection with different forwarding policies.

The ALR-S policy ensures a more uniform distribution of traffic load over the network, as it distinguishes between the packets passing through a particular node and the packets that are originating in that node. Packets originating in a particular node are forwarded on the link of the second shortest route (e.g. from A to F via B, from B to F via C), while packets passing through the node are forwarded on the link of the shortest route (e.g. through A to F via D, through B to F via E). By using the second-choice route only for originating packets, the delay is increased with respect to the shortest route only on the first hop, hence the increase in delay does not accumulate for the packets with a large number of hops. Between the consecutive updates of routing tables, all packets between a given pair of nodes follow the same route. Thus, ALR-S policy maintains the correct sequence of the packets within the routing interval, the same as the SPR forwarding policy.

The ALR-A policy promises an even more uniform distribution of traffic load and thus further improvement of link utilisation by alternating between the shortest and the second shortest route regardless of the packet origination node (this is denoted in Fig. 1 by dashed lines). However, packets belonging to the same session can be forwarded along different routes even within one routing table update interval, thus additional buffering is required in the destination nodes to re-order terminated packets and obtain the correct sequence.

The static forwarding policies, such as ALR-S and ALR-A, distribute packets according to a pre-selected rule. They allow significant reduction of traffic load fluctuation between links, however they do not adapt to the actual traffic load on alternative routes.

2.4.2 Adaptive forwarding

In contrast to static forwarding an adaptive forwarding policy has to take into account the link status information to support the selection of the most appropriate between the alternative outgoing links on the route to the destination. An example of such approach is adaptive forwarding policy based on local information about the link load (Svigelj et al, 2003, 2004b; Mohorcic et al. 2004). This policy selects the most suitable outgoing link taking into account routing tables with alternative routes, calculated using link costs obtained during the previous routing update interval, and current local information on the link status.

In particular, for delay sensitive traffic local information can be based on the expected queuing delay as defined in Equation (7). The expected queuing delay for a particular link can be calculated locally and does not require any information distribution between neighbouring nodes, thus enabling a very fast response to congestion on the link. Depending on this local information, packets are forwarded on the shortest or on the alternative second shortest path. The alternative second shortest path is used only if it has the same or a smaller number of hops (h) to the destination and if the expected queuing delay in the outgoing queue on the shortest path (T_{exp1}) is more than a given threshold $\Delta_{tr}{}^D$ (where D is denoting delay sensitive traffic) higher than the expected queuing delay in the outgoing queue on the second shortest path (T_{exp2}). This condition for selecting the alternative second shortest path is given in Equation (16). Different threshold values can be used for different traffic types.

$$(h_2 \le h_1) \wedge (T_{\exp 1}(t) - T_{\exp 2}(t) < \Delta_{tr}^D \qquad (16)$$

For the throughput sensitive traffic we monitor the number of packets in outgoing queues (n). The alternative second shortest path is used only if it has the same or a smaller number of hops (h) to the destination and if the number of packets (n) in the outgoing queue on the shortest path (n_1) is more than a given threshold $\Delta_{tr}{}^T$ (where T is denoting throughput sensitive traffic) higher than the number of packets in the outgoing queue on the alternative path (n_2), as given in Equation (17).

$$(h_2 \leq h_1) \wedge (n_1(t) - n_2(t) < \Delta_{tr}^T \tag{17}$$

The significance of the threshold is that it regulates distribution of traffic between alternative paths based on local information about the link status, and thus differentiates between lightly and heavily loaded nodes. The higher the threshold value the more congested the shortest path needs to be to allow forwarding along the alternative second shortest path. In the extreme, setting the threshold value to infinity prevents forwarding along the second shortest path (i.e. adaptive forwarding deteriorates to SPR), while no threshold (i.e. $\Delta_{tr}{}^T = 0$) means that packets are forwarded along the second shortest path as soon as the expected queuing delay for the corresponding link is smaller than the one on the shortest path.

Routing with the proposed adaptive forwarding promises more uniform distribution of traffic load between links and the possibility to react quickly to link failure. However, packets belonging to the same session can be forwarded along different routes, even within the same routing update interval, so additional buffering is required in destination nodes to reorder terminated packets and obtain the correct sequence.

3. Traffic modelling for global networks

As we have shown in previous section, the general routing and traffic engineering functions consist of many different algorithms, methods and policies that need to be carefully selected and adapted to the particular network characteristics as well as types of traffic to be used in the network. Clearly, the more dynamic and non-regular the network and the more different types of traffic, the more demanding is the task of optimising network performance, requiring good understanding of the fundamental network operating conditions and the traffic characteristics. The later largely affect the performance of routing and traffic engineering, typically requiring appropriate traffic models to be used in simulating, testing and benchmarking different routing and traffic engineering solutions. In the following a methodology is described for developing a global traffic model suitable for supporting the dimensioning and computer simulations of various procedures in the global networks but focusing in particular on the non-geostationary ISL networks, which are well suited for supporting asymmetric applications such as data, audio and video streaming, bulk data transfer, and multimedia applications with limited interactivity, as well as the broadband access to Internet services beyond densely populated areas. Such traffic models are an important input to network dimensioning tasks (Werner et al., 2001) as well as to simulators devoted to the performance evaluation of particular network functions such as routing and traffic engineering (Mohorcic et al., 2001, 20021, Svigelj et al., 2004a).

A typical multimedia application contains a mix of packets from various sources. Purely mathematical traffic generators cannot capture the traffic characteristics of such applications in real networks to the extent that would allow detailed performance evaluation of the network. Hence, the applicability of traffic analysis based on mathematical tractability is diminishing, while the importance of computer simulation has grown considerably, but poses different requirements for traffic source models (Ryu, 1999). A suitable traffic source model should represent real traffic, while the possibility of mathematical description is less important. In global non-geostationary satellite network traffic source model needs to be complemented by a suitable model of other elementary phenomena causing traffic dynamics, i.e. geographical distribution of traffic sources and destinations, temporal variation of traffic load and traffic flow patterns between different geographical regions.

In the following the approach to modelling global aggregate traffic intensity is described, in particular useful for the dimensioning of satellite networks and computer simulations of various procedures in the ISL network segment, including routing and traffic engineering.

The model is highly parameterized and consists of four main modules:

- module for global distribution of traffic sources and destinations;
- module for temporal variations of traffic sources' intensity;
- module describing the traffic flow patterns between regions; and
- module describing statistical behaviour of aggregated traffic sources.

3.1 Module for global distribution of traffic sources and destinations

The module for global distribution of traffic sources and destinations should support the representation of an arbitrary distribution.

A simple representative of a geographically dependent source/ destination distribution assumes homogeneous distribution over the landmasses, considering continents and major islands (called landmass distribution), while traffic intensity above the oceans equals 0 (Mohorcic et al., 2002b). More realistic source/destination distributions should reflect the geographic distribution of traffic intensity, which is related to several techno-economic factors including the population density and distribution, the existing telecommunication infrastructure, industrial development, service penetration and acceptance level, gross domestic product (GDP) in a given region, and pricing of services and terminals (Werner & Maral, 1997, Hu & Sheriff, 1997, Werner & Lutz 1998). Thus, the estimation of traffic distribution in the yet non-existing system demands a good understanding of the types of services and applications that will be supported by the network. Furthermore, it should also consider attractiveness of particular services for potential users, which in turn depends also on different socio-economic factors.

The methodology for estimating the market distribution for different terminal classes, i.e. lap-top, briefcase and hand-held, is reported in (Hu & Sheriff, 1998) Essentially, countries over the globe are categorized into three different bands according to their annual GDP per capita: low (less than 6 kEuro), medium (between 6 kEuro and 22 kEuro) and high (greater than 22 kEuro). A yearly growth for GDP per capita for each country is then predicted by linearly extrapolating historical data. This, together with the tariff of a particular service and a predicted market saturation value, is used to determine the yearly service take-up for each country via the logistic model. The yearly service penetration for each country is estimated by multiplying the predicted yearly gross potential market with the yearly take-up (Mohorcic et al., 2003).

Taking into account techno-economic and socio-economic factors and the above methodology, we can define different non-homogeneous geographic-dependent distributions taking into account a more realistic distribution of sources and destinations for provisioning of the particular types of service. Such geographic-dependent distributions are typically based on statistical data provided on the level of countries, and only for some larger countries also on the level of states and territories. In addition to limitations of data availability, we also face the problem of the accuracy of its representation, which depends on the granularity of the model and on the assumption regarding the source/destination distribution within the smallest geographical unit (i.e. country). The simplest approach in country-based non-homogeneous geographic-dependent distributions assumes that a nation's subscribers are evenly distributed over the country. The weakness of this approach is representation of traffic demand in large countries spanning several units of geographical granularity. In determining the distribution, different levels of geographical granularity may be adopted; however, in order to be able to individually represent also small countries, the geographical granularity should be in the range of those small countries. In (Mohorcic et al., 2003), a traffic grid of dimension $180° \times 360°$ has been generated in steps of $1°$ in both latitude and longitude directions.

3.2 Module for temporal variations of traffic sources' intensity

Temporal variation of traffic load in a non-geostationary satellite system is caused by daily variation of traffic load due to the local time of day and geographical variation of this daily load behaviour according to geographical time zones. Both are considered in the module for temporal variation of traffic load, which actually mimics the geographically dependent daily behaviour of users. Daily variation can be taken into account with an appropriate daily user profile curve (for average or for local users). An example of such a daily user profile curve is shown in Fig. 2. For geographical time zones a simplified model can be considered, which increments the local hour every 15 degrees longitude eastward from the GMT.

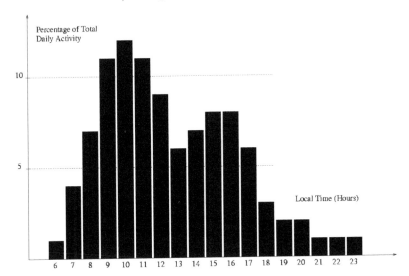

Fig. 2. Daily user profile curve.

An alternative approach defines temporal variation of traffic load in conjunction with the global distribution of traffic sources and destinations, which inherently takes into account geographical time zones. An example of relative traffic intensity considering distribution of traffic sources and destinations combined with temporal variation of traffic load is depicted in Fig. 3, where traffic intensity is normalised to the highest value (i.e. the maximum value of normalized traffic load equals 1, but for better visualization we bounded the z-axis in Fig. 3 to 0.3). The traffic intensity is generated by assuming that a single session is established per day per user and that each session on average lasts for about 2 minutes.

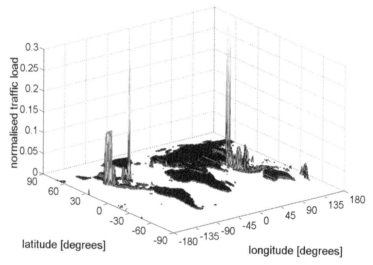

Fig. 3. Global distribution and activity of traffic sources and destinations at midnight GMT.

Another contribution to temporal variation of traffic load in non-geostationary ISL networks in addition to user activity dynamics is the rapidly changing satellite visibility, and consequently active users' coverage, on the ground. To a certain extent this temporal variation as well as multiple visibility of satellites can be captured with a serving satellite selection scheme. Implementing a satellite selection scheme in case of multiple visibility has two aspects. For fixed earth stations line-of-sight conditions are assumed, so that the serving satellite can be determined according to a simple deterministic rule, e.g., maximum elevation satellite. For mobile earth stations, the stochastic feature of unexpected handover situations due to propagation impairments can be considered through the shares of traffic on alternative satellites also estimated according to a simple rule (e.g., equal sharing between all satellites above the minimum elevation) or using a simple formula (e.g., shares are a function of the elevation angle of each alternative satellite as one main indicator for channel availability).

3.3 Module describing the traffic flow patterns between regions

This module assigns traffic flow destinations using a traffic flow pattern resembling the flow characteristic between different regions. Interregional patterns should be defined at least on the level of the Earth's six continental regions shown in Fig. 4, similarly as in (Werner & Maral,

1997), but preferably on a smaller scale between countries/territories. In a destination region, the traffic can be divided among the satellites proportionally to their coverage of that region.

Customized traffic flow patterns should be based on the density distribution of sources and/or destinations for the selected type of service.

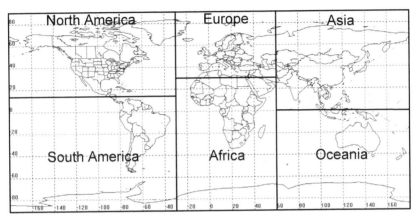

Fig. 4. Geographical division of six source/destination regions.

3.4 Module describing statistical behaviour of aggregated traffic sources

The fourth module concerns modelling of the aggregated traffic sources. In particular, the module comprises of suitable aggregate traffic source generator, which is modulated by the normalized cumulative traffic on each satellite obtained from distribution of traffic sources and destinations and temporal variation of traffic sources' intensity. Thus data packets are actually generated considering the relative traffic intensity experienced by a particular satellite in its coverage area, while taking into account the statistics of the selected aggregate traffic source model.

Ideally, the traffic source model should capture the essential characteristics of traffic that have significant impact on network performance with only a small number of parameters, and should allow fast generation of packets. Among the most important traffic characteristics for circuit switched networks are the connection duration distribution and the average number of connection requests per time unit. By contrast, in the case of packet switched networks, traffic characteristics are given typically by packet lengths and packet inter-arrival times (in the form of distributions or histograms), burstiness, moments, autocorrelations, and scaling (including long-range dependence, self-similarity, and multifractals). For generating cumulative traffic load on a particular satellite, the traffic source generator should model an aggregate traffic of many sources overlaid with the effect of a multiple access scheme, which is expected to significantly shape source traffic originating from single or multiplexed ground terminal applications due to the uplink resource management and traffic scheduling.

One approach for modelling aggregate traffic sources is by using traces of real traffic. Trace-driven traffic generators are recommended for model validation, but suffer from two drawbacks: firstly, the traffic generator can only reproduce something that has happened in the past, and secondly, there is seldom enough data to generate all possible scenarios, since the extreme situations are particularly hard to capture. In the case of satellite networks with no appropriate system to obtain the traffic traces, the use of traces is even more inconvenient.

An alternative approach, increasingly popular in the field of research, is to base the modelling of traffic sources on empirical distributions obtained by measurement from real traffic traces. The measurements can be performed on different segments of real networks, i.e. in the backbone network or in the access segment. In order to generate cumulative traffic load representing an aggregate of many individual traffic sources in the coverage area of the satellite, the traffic properties have to be extracted from a representative aggregate traffic trace (Svigelj at al., 2004a), such as a real traffic trace captured on the 622 Mbit/s backbone Internet link carrying 80 Mbit/s traffic (Micheel, 2002). The selected traffic trace comprises aggregate traffic from a large number of individual sources. Such traffic trace resembles the traffic load experienced by a satellite, both from numerous traffic sources within its coverage area, and from aggregate flows transferred over broadband intersatellite links. A suitable traffic source model, which resembles IP traffic in the backbone network, can already be built by reproducing some of the first order statistical properties of the real traffic trace that have major impact on network performance, e.g. inter-arrival time and packet length distribution. A simple traffic generator can be developed using a look-up table with normalized values, which allows packet inter-arrival time and packet length values to be scaled, so as to achieve the desired total traffic load. Distributions of packet inter-arrival time and packet length obtained with such a traffic generator are depicted in Fig. 5 and Fig. 6 respectively. The main advantage of traffic sources, whose distributions conform to those obtained by measurements of real traffic, is that they are relatively simple to implement and allow high flexibility.

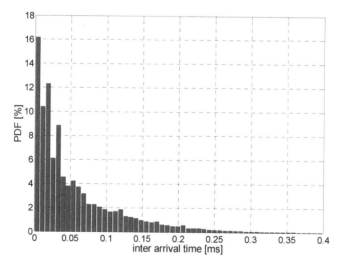

Fig. 5. Packet inter-arrival time distribution obtained with empirical traffic generator.

For the more accurate prediction of the behaviour of the traffic source exhibiting long-range dependence, the traffic model requires detailed modelling of also the second order statistics of the packet arrival process. The accurate fitting of modelled traffic to the traffic trace can be achieved using modelling process with a discrete-time batch Markovian arrival process that jointly characterizes the packet arrival process and the packet length distribution (Salvador et al., 2004). Such modelling allows very close fitting of the auto-covariance, the marginal distribution and the queuing behaviour of measured traces.

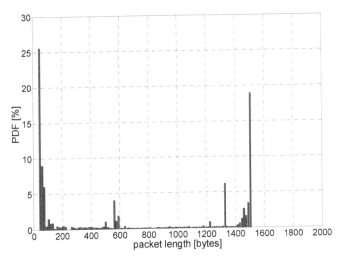

Fig. 6. Packet length distribution obtained with empirical traffic generator.

The potential drawback of traffic sources based on real traffic traces is that the empirically obtained traffic properties (i.e. obtained from the aggregated traffic on the backbone Internet link in this particular example) may not be suitably representative for the system under consideration, so it can sometimes deviate considerably from real situations and lead to incorrect conclusions.

In addition to traffic sources based on traffic traces (directly or via statistical distributions) traffic sources can also be implemented in classical way with pure mathematical distributions such as Poisson, Uniform, Self-Similar, etc. Although such mathematically tractable traffic sources never fully resemble the characteristics of real traffic, they can serve as a reference point to compare simulation results obtained with different scenarios, however they should exhibit the same values of first order statistic (i.e. mean inter-arrival time and average packet length) as obtained from traces.

In the case of supporting different levels of services, packets belonging to different types of traffic (e.g. real time, high throughput, best effort) should be generated using different traffic source models, which should reproduce statistical properties of that particular traffic. However, as different services and applications will generate different traffic intensity depending on regions and users' habits, also separate traffic flow patterns will have to be developed for different types of traffic, to be used in conjunction with different traffic source generators.

3.5 Global aggregate traffic intensity model

Integration of individual modules in the global aggregate traffic intensity model is schematically illustrated in Fig. 7. Instead of simulating individual sources and destinations, a geographic distribution of relative traffic source intensity is calculated for any location on the surface of the Earth. The cumulative traffic intensity of sources within its coverage area are mapped to the currently serving satellite. Satellite footprint coverage areas on the Earth, overlaid over geographic distribution of traffic sources and destinations, are identified from the satellite positions in a given moment.

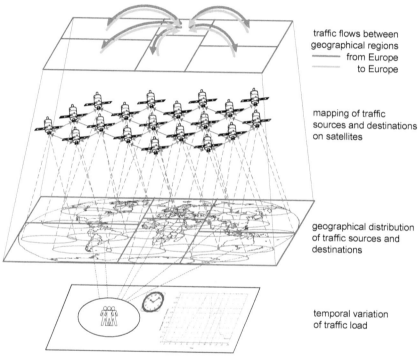

Fig. 7. Global aggregate traffic intensity model.

With the normalized cumulative traffic on each satellite, which is proportional to the intensity of traffic sources in the satellite's coverage area, it is possible to modulate the selected traffic source generator (not shown in Fig. 7). Thus data packets are actually generated considering the relative traffic intensity experienced by a particular satellite.

The destination satellite is selected for each packet in accordance with the traffic flow pattern. The probability of selecting a given satellite as a destination node is proportional to its coverage share in the destination region divided by the sum of all coverage shares in that region. Thus, although in a simplified manner, the model is taking into consideration also multiple coverage. In the case of using different traffic source models to generate distinct types of traffic by global aggregate traffic intensity model, one should also consider different, service specific traffic flow patterns.

4. Summary

Traffic engineering involves adapting the routing of traffic to the network conditions with two main goals: (i) providing sufficient quality of service, which is important from user's point of view, and (ii) efficient use of network resources, which is important for operators of telecommunication's network. The presented routing and traffic engineering issues addressed both goals that are explained using the ISL network as a concrete example of highly dynamic telecommunication network with several useful properties, which can be exploited by developing of routing procedures. However, the presented work is not limited to ISL networks, but can be used also in other networks as described in (Liu et al., 2011; Long et al., 2010; Rao & Wang, 2010, 2011). Routing and traffic engineering functions are presented in modular manner for easier reuse of particular procedures.

Adaptation of routing requires, in addition to good understanding of the fundamental network operating conditions, also good knowledge of the characteristics of different types of traffic in the network. In order to support better modelling of traffic characteristics a modular methodology is described for developing a global aggregate traffic intensity model suitable for supporting the dimensioning and computer simulations of various procedures in the global networks. It is based on the integration of modules describing traffic characteristics on four different levels of modelling, i.e. geographical distribution of traffic sources and destinations, temporal variations of traffic sources' intensity, traffic flows patterns and statistical behaviour of aggregated traffic sources.

5. References

Bertsekas D. & Gallager R. (1987). *Data Networks*, Englewood Cliffs: Prentice-Hall International.

Franck L. & Maral G. (2002a). Signaling for inter satellite link routing in broadband non GEO satellite systems. *Computer Networks*, Vol. 39, No. 1, pp. 79-92.

Franck L. & Maral G. (2002b). Routing in Networks of Intersatellite Links. *IEEE Transaction on Aerospace and Electronic Systems,* Vol. 38, No. 3, pp. 902-917.

Hu Y. F. & Sheriff R. E. (1997). The Potential Demand for the Satellite Component of the Universal Mobile Telecommunication System. *Electronics and Communication Engineering Journal*, April 1997, pp. 59-67.

Hu Y. F. & Sheriff R. E. (1999). Evaluation of the European Market for Satellite-UMTS Terminals. *International Journal of Satellite Communications*, Vol. 17, pp. 305-323.

Liu, X.; Ma, J. & Hao, X. (2011). Self-Adapting Routing for Two-Layered Satellite Networks. *China Communications*, Volume 8, Issue 4, July 2011, pp. 116-124.

Long F; Xiong N.; Vasilakos A.V.; Yang L.T. & Sun, F. (2010). A sustainable heuristic QoS routing algorithm for pervasive multi-layered satellite wireless networks. *Wireless Networks*, Volume 16, Issue 6, August 2010, Pages 1657-1673.

Micheel, 2002. National Laboratory for Applied Network Research), Passive Measurement and Analysis. http://pma.nlanr.net/PMA/, 22 October, 2002.

Mohorcic M.; Svigelj A. & Kandus G. 2004. Traffic Class Dependent Routing in ISL Networks. *IEEE Transaction on Aerospace and Electronic Systems,* Vol. 39, pp. 1160-1172.

Mohorcic M.; Svigelj A.; Kandus G. & Werner M. (2000). Comparison of Adaptive Routing Algorithms in ISL Networks Considering Various Traffic Scenarios. In: *Proc. of 4th*

European Workshop on Mobile and Personal Satellite Communications (EMPS 2000), pp. 72-81, London, UK; September 18, 2000.

Mohorcic M.; Svigelj A.; Kandus G. & Werner M. (2002a). Performance Evaluation of Adaptive Routing Algorithms in Packet Switched Intersatellite Link Networks. *International Journal of Satellite Communications*, Vol. 20, pp. 97-120.

Mohorcic M.; Svigelj A.; Kandus G.; Hu Y. F. & Sheriff R. E. (2003). Demographically weighted traffic flow models for adaptive routing in packet switched non-geostationary satellite meshed networks. *Computer Networks*, No. 43, pp. 113-131.

Mohorcic M.; Svigelj A.; Werner M. & Kandus G. (2001). Alternate link routing for traffic engineering in packet oriented ISL networks. *International Journal of Satellite Communications*, No. 19, pp. 463-480.

Mohorcic M.; Werner M.; Svigelj A. & Kandus G. (2002b). Adaptive Routing for Packet-Oriented Inter Satellite Link Networks: Performance in various Traffic Scenarios. *IEEE Transactions on Wireless Communications*, Vol. 1, No. 4, pp. 808-818.

Rao Y. & Wang R. (2011). Performance of QoS routing using genetic algorithm for Polar-orbit LEO satellite networks. *AEU - International Journal of Electronics and Communications*, Vol. 65 (6), pp. 530-538.

Rao, Y. & Wang, R. (2010). Agent-based load balancing routing for LEO satellite networks. *Computer Networks*, Volume 54, Issue 17, 3 December 2010, pp. 3187-3195.

Ryu B., (1999). Modeling and Simulation of Broadband Satellite Networks: Part II - Traffic Modeling, *IEEE Communication Magazine*, July 1999.

Salvador P.; Pacheco A. & Valadas R. (2004). Modeling IP traffic: joint characterization of packet arrivals and packet sizes using BMAPs. *Computer Networks*, No. 44, pp. 335-352.

Svigelj A.; Mohorcic M. & Kandus G. (2004b) Traffic class dependent routing in ISL networks with adaptive forwarding based on local link load information. *Space communications*, Vol. 19, pp. 158-170.

Svigelj A.; Mohorcic M.; Franck L. & Kandus G. (2012). Signalling Analysis for Traffic Class Dependent Routing in Packet Switched ISL Networks. To appear in: *Space communications*, Vol. 22:2.

Svigelj A.; Mohorcic M.; Kos A.; Pustisek M.; Kandus G. & Bester J. (2004a). Routing in ISL networks Considering Empirical IP Traffic. *IEEE Journal on Selected Areas in Communications*, Vol. 22, No. 2, pp. 261-272.

Wang Z & Crowcroft J.(1996). Quality-of-Service Routing for Supporting Multimedia Applications. *IEEE Journal on Selected Areas in Communications*, Vol. 14, No. 7, pp. 1228-1234.

Wang Z. (1999). On the complexity of quality of service routing. *Information Processing Letters*, Vol. 69, pp. 111-114.

Werner M. & Lutz E. (1998). Multiservice Traffic Model and Bandwidth Demand for Broadband Satellite Systems. In proceedings: *M. Ruggieri (Ed.), Mobile and Personal Satellite Communications 3*, pp. 235-253, Venice, Italy, November 1998.

Werner M. & Maral G. (1997). Traffic Flows and Dynamic Routing in LEO Intersatellite Link Networks. In: *Proc. IMSC '97*, pp. 283-288, Pasadena, California, USA, June 1997.

Werner M; Frings J.; Wauquiez F. & Maral G. (2001). Topological Design, Routing and Capacity Dimensioning for ISL Networks in Broadband LEO Satellite Systems. *International Journal of Satellite Communications*, No. 19, pp. 499-527.

Wood, L.; Clerget, A.; Andrikopoulos I.; Pavlou G. & Dabbous W. (2001). IP Routing Issues in Satellite Constellation Networks. *International Journal of Satellite Communications*, Vol. 19, No. 1, pp. 69 92.

7

Modeling and Simulating the Self-Similar Network Traffic in Simulation Tool

Matjaž Fras[1], Jože Mohorko[2] and Žarko Čučej[2]
[1]Margento R&D, Maribor,
[2]University of Maribor, Faculty of Electrical Engineering
and Computer Science, Maribor,
Slovenia

1. Introduction

Telecommunication networks are growing very fast. The user's needs, in regards to new services and applications that have a higher bandwidth requirement, are becoming bigger every day. A telecommunication network requires early design, planning, maintenance, continuous development and updating, as demand increases. In that respect we are forced to incessantly evaluate the telecommunication network's efficiency by utilizing methods such as measurement, analysis modeling and simulations of these networks.

Measuring, analyses and the modeling of self-similar traffic has still been one of the main research challenges. Several studies have been carried-out over the last fifteen years on: analysis of network traffic on the Internet [30], [31], traffic measurements in the high speed networks [32], and also measurement in the next generation networks [33]. Also, a lot of research works exist, where attention had been given to analysis of the network traffic caused by different applications, such as P2P [34], [35], network games [36] and VoIP application Skype [37]. Analyses of the measured network traffic help us to understand the basic behavior of network traffic. Various have showed that traffic in contemporary communication networks is well described with a self-similar statistical traffic model, which is based on fractal theory [6]. The pioneers in this field are: Leland, Willinger, and many others [1], [5], [6]. They introduced the new network traffic description in 1994. New description appeared as an alternative to traditional models, as were Poisson and Markov, which were used as a good approximation for telephone networks (PSNT networks) when describing the process of call durations and time between calls [5], [20]. These models do not allow descriptions of bursts, which are distinctive in today's network traffic. Such bursts can be described by a self-similarity model [5], [6], because it shows bursts over a wide-range of time scales. This contrasts with the traditional traffic model (Poisson model), which became very smooth during the aggregation process. The measure of bursts and also self-similarity present the Hurst parameter [1]-[4], which is correlated with another very important property called long-range dependence [5]-[8]. This property is also manifested with heavy-tailed probability of density distributions [5], [6], such as Pareto [43] or Weibull [44]. So Pareto's and Weibull's heavy-tailed distributions became the most frequently used distributions to describe self-similar network traffic in communication networks.

During past years another aspect of network traffic studying has also appeared. In this case, the network traffic is researched from application or data source point of view, especially focused on statistics of file sizes and inter-arrival times between files [19]. These research works are very important for describing a relation between packet network traffic on lower ISO/OSI layers and data source network traffic on higher layers of ISO/OSI model. Based on the research of WWW network traffic, it has been shown that file sizes of such traffic are best described by Pareto distribution with shape parameter $a = 1$ [38]. That was also shown for the FTP traffic, where the shape parameter of Pareto distribution is in the range $0.9 < a < 1.1$ [20]. In [6], [39], and [40] it is shown that inter-arrival time of TCP connections are self-similar processes, which can be described by Weibull heavily tailed distribution.

With expansion of simulation tools, which are used for simulation of communication networks, the knowledge about simulating the network traffic also becomes very important. One of the important tasks in simulations is also knowledge about modeling and simulating of network traffic. Network traffic is usually modeled in simulation tools from an application point of view [42], [45]. It is usually supposed that the file size statistics and file inter-arrival times are known [39], [40]. Such kinds of traffic models are supported by most commercial telecommunication simulation tools such as the OPNET Modeler [10], [11], [24], used in our simulations and experiments. Consequently, for using the measured data of packet traffic, when modeling file statistics, it is necessary to transform packets' statistics into files' statistics [9, 10]. This transformation contains opposite operations in relation to the fragmentation and encapsulation process. Extensive research and investigation about traffic sources in contemporary networks show that this approach requires an in-depth analysis of packet's traffic (which needs specialized, very powerful and consequently, expensive instruments). This approach, in the case of encrypted packets and non-standard application protocols, is not completely possible. In such cases, capture of entire packets is also necessary, which can be problematic in contemporary high-speed networks. Another approach estimates distribution parameters of file data sources from measured packets' network traffic. For such approach, we have developed and tested different methods [42], [45]. Estimated distribution parameters are used for modeling of the measured network traffic for simulation purposes. Through the use of these methods we want to minimize discrepancies between the measured and simulated traffic in regards to an average bit rate and bursts, which are characteristic of self-similar traffic.

2. Network traffic

2.1 Packet network traffic measuring

The measuring and analyzing of real network traffic provide us with a very important knowledge about computer network states. In analyzing process, we need statistical mathematical tools. These tools are crucial for accuracy of a derived mathematical model, described by stochastic parameters for packet size and inter-arrival time [9]. Using this simulation model, we want to acquire information about telecommunication network's performances for:

- improvement of the current network,
- bottleneck searching,
- building and development of new network devices and protocols,

- and for ensuring quality of service (QoS) for real-time streaming multimedia applications.

Using this information, network administrators can make the network more efficient.

The simplest tools that measure and capture the packets of network traffic are packet sniffers. Packet sniffers, also known as protocol or network analyzers, are tools that monitor and capture network traffic with all content of network traffic. We can use sniffers to obtain the main information about network traffic, such as packet size, inter-arrival time and the type and structure of IP protocol. Sniffers have become very important and indispensable tools for network administrators. Figure 1 shows traffic captured by a packet sniffer.

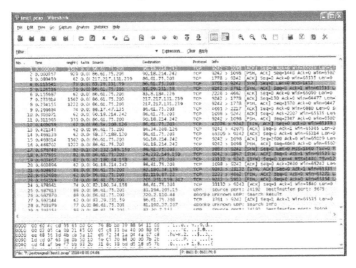

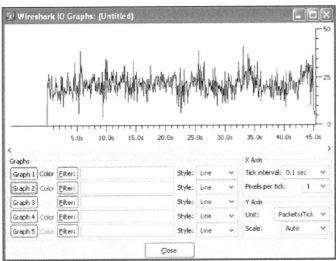

Fig. 1. User interface of WireShark sniffer during the network capturing.

Any sniffers are able to extract this data from the IP headers. Knowing them, it is then simple to calculate a length of IP PDU (Protocol Data Unit), which also contains a header of higher layer protocols. Using an in-depth header analysis, it is possible, in the similar way to the IP header, to calculate the lengths of all these headers.

An analytical description of network traffic does not exist, because we cannot predict the size and arrival time of the next packet. Therefore, we can only describe network traffic as a stochastic process. Hence, we have tried to describe these two stochastic processes (arrival time and packet size) with the use of Hurst parameter and probability distributions.

2.2 Self-similarity

In the 1990s, new descriptions and models of network's traffic were developed, which then replaced the traditional traffic models, such as Poisson and Markov [5], [20]. The Poisson process was widely used in the past, because it gave a good approximation of telephone network (PSNT networks), especially when describing times between each call and call durations. This model is usually described by exponential probability distribution, which is characterized by the parameter λ (number of events per second). However, these models do not allow for descriptions of bursts, which are distinctive in today's network traffic. Such

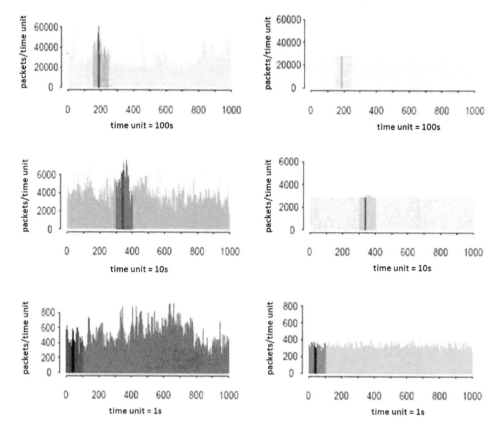

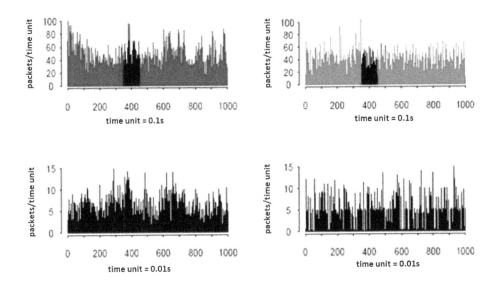

Fig. 2. Comparison of self-similar network traffic (left) and synthetic traffic created by Poisson model (right) on different time scales (100, 10, 1, 0.1 and 0.01s). Self similar traffic contains bursts on all time scales in contrast to the generated synthetic traffic, based on the Poisson model, which tends to average on longer time [1].

bursts can be described by a self-similarity model, because it shows bursts over a wide-range of time scales [1]-[4]. This contrasts the traditional traffic model (Poisson model), which becomes very smooth during the aggregation process.

2.3 Self-similarity

The definition of self-similarity is usually based on fractals for the standard stationary time series [5], [6], [21].

Let $X = (X_t, t = 0, 1, 2,...)$ be a covariance stationary stochastic process; that is a process with a constant mean, finite variance $\sigma^2 = E[(X_t - \mu)2]$, with auto-covariance function $\gamma(k) = E[(X_t - \mu)(X_t+k - \mu)]$, that depends only on k. Then the autocorrelation function $r(k)$ is:

$$r(k) = \frac{\gamma(k)}{\sigma^2} = \frac{E[(X_t - \mu)(X_{t+k} - \mu)]}{E[(X_t - \mu)^2]}, \qquad k = 0,1,2,... \tag{1}$$

Assume X has an autocorrelation function, which is asymptotically equal to:

$$r(k) \approx k^{-\beta} L_1(k), \quad k \to \infty, \quad 0 < \beta < 1, \tag{2}$$

where $L_1(k)$ slowly varies at infinity, that is $\lim_{t \to \infty}(L_1(tx) / L_1(t)) = 1$ for all $x > 0$. Such functions are for example $L_1(t) = const.$ and $L_1(t) = \log(t))$ [5], [6].

The measure of self-similarity is the Hurst parameter (H), which is in a relationship with the parameter β in equation (3).

$$H = 1 - \frac{\beta}{2} \qquad (3)$$

Let's define the aggregation process for the time series [5], [6]:

For each $m = 1, 2, 3, \ldots$ let $X^{(m)} = (X_k^{(m)}, k = 1,2,..m)$ denote a new time series obtained by averaging the original series X over a non-overlapping block of size m. That is, for $m=1, 2, 3, \ldots$, $X^{(m)}$ is given by:

$$X_k^{(m)} = \frac{1}{m}(X_{km-m+1} + \ldots + X_{km}), \qquad k = 1,2,3,\ldots \qquad (4)$$

$X_k^{(m)}$ is the process with average mean and autocorrelation function $r^{(m)}(k)$ [6].

The process X is called an exactly second order with parameter H, which represents the measure of self-similarity if the corresponding aggregated $X^{(m)}$ has the same correlation structures as X and $\operatorname{var}(X^{(m)}) = \sigma^2 m^{-\beta}$ for all $m = 1, 2, \ldots$:

$$r^{(m)}(k) = r(k), \text{ for all } m = 1,2,\ldots \quad k = 1,2,\ldots \qquad (5)$$

The process X is called an asymptotically second order with parameter $H = 1 - \beta/2$, if for all k it is large enough,

$$r^{(m)}(k) \rightarrow r(k), \qquad m \rightarrow \infty \qquad (6)$$

It follows from definitions that the process is the second order self-similar in the exact or asymptotical sense, if their corresponding aggregated process $X^{(m)}$ is the same as X or becomes indistinguishable from X-at least with respect to their autocorrelation function. The most striking property in both cases, exact and asymptotical self-similar processes, is that their aggregated processes $X^{(m)}$ possess a no degenerate correlation structure as $m \rightarrow \infty$. This contrasts with the Poisson stochastic models, where their aggregated processes tend to second order pure noise as $m \rightarrow \infty$:

$$r^{(m)}(k) \rightarrow 0, \quad m \rightarrow \infty, \quad k = 0,1,2,\ldots \qquad (7)$$

Network traffic with bursts is self-similar, if it shows bursts over many time scales, or it can be also said over a wide-range of time scales. This contrasts with traditional models such as Poisson and Markov, where their aggregation processes become very smooth.

2.4 Long-range dependence

The self-similar process can also contain a property of long-range dependence [5]-[8]. Long range dependence describes the memory effect, where a current value strongly depends upon the past values, of a stochastic process, and it is characterized by its autocorrelation function. This property has a stochastic process, which satisfies relation (2), order with relation $r(k) = \gamma(k)/\sigma^2$.

For $0 < H < 1, H \neq 1/2$ it holds [6]

$$r(k) \approx H(2H-1)k^{-2H-2}, \qquad r \to \infty \tag{8}$$

For values $0.5 < H < 1$ autocorrelation function $r(k)$ behavior, in an asymptotic mean, as $ck^{-\beta}$ for values $0 < \beta < 1$, where c is constant $c > 0$, $\beta = 2 - 2H$, and we have:

$$\sum_{k=-\infty}^{\infty} r(k) = \infty \,. \tag{9}$$

The autocorrelation function decays hyperbolically, as the k increases, which means that autocorrelation function is non-summable. This is opposite to the property of short-range dependence (SRD), where the autocorrelation function decays exponentially and the equation (9) has a finite value. Short and long-range dependence have a common relationship with the value of the Hurst parameter of the self-similar process [6], [21]:

- $0 < H < 0.5 \to$SRD - Short Range Dependence
- $0.5 < H < 1 \to$LRD - Long Range Dependence

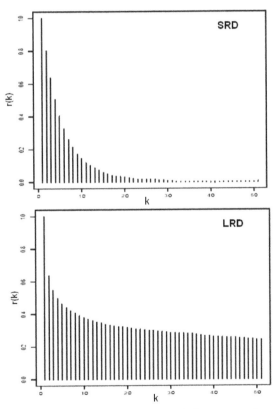

Fig. 3. Comparison between autocorrelation function of short range dependence process (left) and autocorrelation function of long range dependence process (right) [15].

2.5 Heavy-tailed distributions

Self-similar processes can be described by heavy-tailed distributions [5], [6], [9]. The main property of heavy-tailed distributions is that they decay hyperbolically, which is opposite to the light-tailed distribution, which decays exponentially. The simplest heavy-tailed distribution is Pareto. The probability density function of Pareto distribution is given by [43]:

$$p(x) = \frac{\alpha k^{\alpha}}{x^{\alpha+1}}, \quad k \le x, \quad \alpha, k > 0 \tag{10}$$

where parameter a represents the shape parameter, and k represents the local parameter of distribution (also a minimum possible positive value of the random variable x).

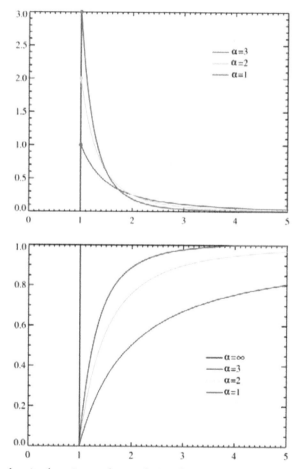

Fig. 4. Probability density function and cumulative distribution function of Pareto distribution for various shape parameters α and constant location parameter $k = 1$ [43].

Another very important heavy-tailed distribution is Weibull distribution, which is described by [44]:

$$p(x) = \frac{\alpha}{k} \cdot \left(\frac{x}{k}\right)^{\alpha-1} \cdot e^{-\left(\frac{x}{k}\right)^{\alpha}}, \quad x \geq 0, \quad \alpha, k > 0 \tag{11}$$

where parameter a presents the shape parameter, and k presents the local parameter of distribution.

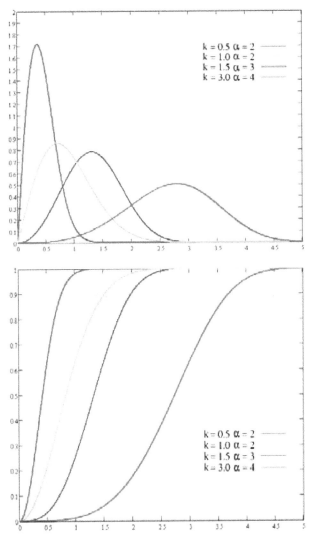

Fig. 5. Probability density function and cumulative distribution function of Weibull distribution for various shape parameters a and constant location parameter k [44].

2.6 Network traffic definitions

The network traffic can be observed on different layers of ISO/OSI model, for that reason we define different kinds of network traffics. The network traffic can be represented as a stochastic process, which can be interpreted as the traffic volume – measured in packets, bytes or bits per time unit, and it is consequent on data or packets, which are sent through the network in time unit. If we observe network traffic on the low level of ISO/OSI model, then define the packet network traffic [45] $Z_p[n]$:

Let define the packet network traffic $Z_p[n]$ as a stochastic process interpreted as the traffic volume, measured in packets per time unit. $Z_p[n]$ can be described as a composite of two stochastic processes:

$$Z_p[n] = X_p[n] \circ Y_p[n], \quad n \in \mathbb{R}. \tag{12}$$

where $X_p[n]$ represents packet size process and $Y_p[n]$ represents the packet inter-arrival time.

Packet-size process $X_p[n]$ is defined as a series of packet sizes l_{Pi} measured in bits (b) or bytes (B).

$$X_p[n] = \{l_{P1}, l_{P2}, \dots l_{Pi}, \dots, l_{Pn}\}, \quad 1 \le i \le n \tag{13}$$

where sizes of packets' l_{Pi} are limited by the shortest l_m and the longest l_{MTU} packet size (MTU - Maximum Transmission Unit).

$$l_m \le l_{Pi} \le l_{MTU} \tag{14}$$

Packet inter-arrival time process $Y_p[n]$ is defined as a series of times between packet arrivals t_{Pi} (time stamps).

$$Y_p[n] = \{t_{P2} - t_{P1}, \dots, t_{Pi} - t_{pi-i}, \dots, t_{Pn} - t_{Pn-i}\}, \quad 1 \le i \le n$$
$$= \{\Delta t_{p1}, \Delta t_{p2}, \dots, \Delta t_{pi}, \dots, \Delta t_{pn-1}\}, \quad 1 \le i \le n \tag{15}$$

The measured network traffic is packet network traffic, which can be captured using special software program or hardware devices. For that reason, the measured network traffic is marked as $Z_{pm}[n]$. We also define modeled (simulated) network traffic as $Z_{ps}[n]$. We suppose, that the measured and modeled traffic is statistically equal, denoted by the symbol $\approx$,

$$Z_{pm}[n] \approx Z_{ps}[n] \tag{16}$$

if there are also statistical equalities between a packet size and inter-arrival time processes of measured, and modeled traffic.

$$X_{pm}[n] \approx X_{ps}[n]$$

and

$$Y_{pm}[n] \approx Y_{ps}[n] \qquad (17)$$

Let's define network traffic on higher layers (application) of ISO/OSI model. Data source network traffic $Z_d[n]$ can be described as a composite of data source lengths $X_d[n]$ and data inter-arrival times $Y_d[n]$ processes:

$$Z_d[n] = X_d[n] \circ Y_d[n], \quad n \in \mathbb{R} \qquad (18)$$

To provide statistical equality between packet network traffic $Z_p[n]$ and data sources network traffic $Z_d[n]$, we have performed a transformation between packet size process $X_p[n]$ and the process of data length $X_d[n]$ as well as transformation between packet inter-arrival time $Y_p[n]$ and data inter-arrival time $Y_d[n]$.

$$X_{pm}[n] \xrightleftharpoons{transformation} X_d[n] \qquad (19)$$

$$Y_{pm}[n] \xrightleftharpoons{transformation} Y_d[n] \qquad (20)$$

Transformation (19) and (20) allows estimation of packet traffic processes from data source traffic processes or vice verse.

3. Network traffic analysis and modeling

3.1 Hurst parameter estimations

Hurst's parameter represents the measure of self-similarity. There are several methods for estimating Hurst's parameter (H) [1]-[4] of stochastic self-similar processes. However, there are no criteria as to which method gives the best results. There are several different methods for estimating the Hurst parameter which can lead to diverse results [9], [10]. This is the reason why Hurst's parameter cannot be calculating but can be estimated. The most often used methods for Hurst's parameter estimation are [6], [8], [21]:

- Variance method is a graphical method, which is based on the property of slowly decaying variance. In a log-log scale plot, a sample variance versus a non-overlapping block of size m is drawn for each aggregation level. From the line with slope β we can estimate Hurst's parameter as a relationship, from equation (3).
- R/S method is also a graphical method. It is based on a range of partial sums regarding data series deviations from mean value, rescaled by its standard deviation. The slope in the log-log plot of the R/S statistic versus aggregated points is the estimation for Hurst's parameter.
- Periodogram method plots spectral density in a logarithm scale versus frequency (also in logarithm scale). The slope in periodogram allows the estimation of parameter H.

Figure 6 presents an example of test traffic and estimations of Hurst's parameter through different methods.

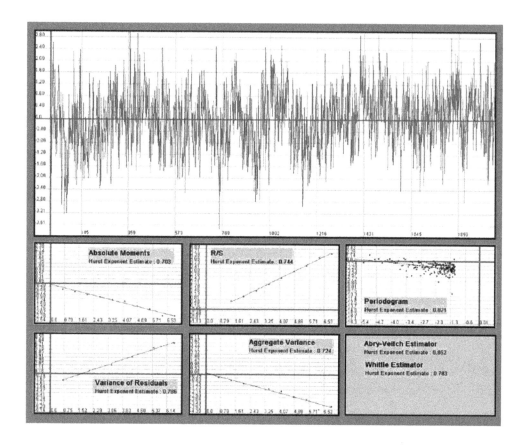

Fig. 6. Estimating parameter H for self-similar traffic (upper-left) with the variances method (lower left), R/S method (upper-right) and periodogram method (lower-right) using SELFIS tool [8].

3.2 Distribution parameter estimation for stochastic process of network traffic

Network traffic can be described by two stochastic processes, one for packet/data sizes and one for packet/data inter-arrival time. All processes are usually described by probability distributions. Self-similar process can be described by heavy tailed distributions. The main task for modeling the stochastic process with probability distribution is to choose the right distribution, which would be a good representation of our network traffic stochastic process. The statistic distribution parameters of data sources are then estimated by fitting tools [9], [25], [26] or other known methods, such as CCDF [6] or Hill estimator [17], [18]. Mathematical fitting tools are used (EasyFit), which allow us to automatically include the fit distribution of the stochastic process, and also estimate parameters of distribution from the captured traffic [9], [29].

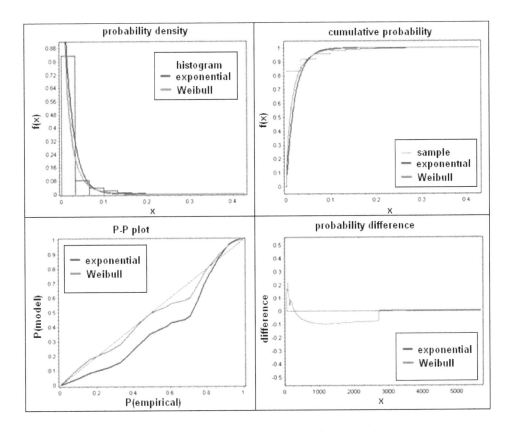

Fig. 7. For the stochastic process of inter-arrival time, distribution and estimate parameters of these distributions are chosen based on the histogram (upper left), and cumulative distribution function (upper right). Differences between empirical and theoretical distributions in P-P plot (lower left), and deferential distribution (lower right).

4. Simulation of network traffic in simulation tools

One of the very important tasks in simulation is modeling the real network parameters and network elements for simulation purposes. The main goal in successful modeling of network traffic is to minimize discrepancies between the measured simulations and by simulations statistically-modeled and generated traffic. This means, that both traffics are similar within the different criteria, such as bit and packet-rate, bursts (Hurst's parameter), variance, etc.

Network traffic simulations are usually based on modeling of data sources or applications. One of the most known simulation tools is OPNET Modeler [22], [23]. A simulation of network traffic in this tool is based on the "on/off" models [41] or more often used traffic generators. Difference between these manners is in a modeling manner. In the first case, the arrival process is described by Hurst's parameter (H) and the data length process is described by probability density function (*pdf*). In the second case, processes of data length and data inter-arrival time are both described by *pdf*.

In OPNET Modeler, two standard node models appear [9]:

- Raw Packet Generator (RPG)
- IP station

Raw Packet Generator (RPG) is a traffic source model [16], [27] implemented specially to generate self-similar traffic, which is based on different fractal point processes (FPP) [41]. Self similar traffic is modeled with an arrival process, which is described by Hurst's parameter and the distribution probability for packet sizes. This arrival process can be based on many different parameters, such as Hurst parameter, average arrival rate, fractal onset time scale, source activity ratio and peak to mean ratio [16]. There are several different fractal point processes (FPP). In our case, we used the superposition of the fractal renewal process (Sub-FRP) model, which is defined as the superposition of M independent and probably identical renewal fractal processes. Each FRP stream is a point renewal processes and M numbers of independent sources compose the Sub-FRP model. Common inter-arrival probability density function $p(t)$ of this process is:

$$p(t) = \begin{cases} \gamma A^{-1} e^{-\gamma t/A} & 0 \le t \le A \\ \gamma e^{-\gamma} A^{\gamma} t^{-(\gamma+1)} & t \ge A \end{cases} \tag{21}$$

where $1 < \gamma < 2$. Process FRP can be defined as Sup-FRP process, when the number of independent identical renewal processes (M) is equal to 1. A model Sub-FRP is described by three parameters: γ, A and M. γ represents the fractal exponent, A is the location parameter, and M is the number of sources. These three parameters are in relationship with three OPNET parameters. These parameters are Hurst's average arrival-rate λ, and fractal onset time-scale (FOTS). The relationships between these three parameters of Sub-FRP and parameters in OPNET model are:

$$H = (3 - \gamma)/2$$

$$\lambda = M\gamma[1 + (\gamma-1)^{-1}e^{-\gamma}]^{-1}A^{-1} \tag{22}$$

$$T^{\alpha} = 2^{-1}\gamma^{-2}e^{-\gamma}(\gamma-1)^{-1}(2-\gamma)(3-\gamma)[1+(\gamma-1)e^{\gamma}]^2 A^{\alpha},$$

where $\gamma = 2 - \beta$. Hurst parameter H is defined by equation (3). In the Sub-FRP model from OPNET, we can set Hurst's parameter (H), average arrival-rate (λ) and fractal onset time-scale (FOTS) in seconds. The recommended value for the parameter FOTS in OPNET is 1 second.

The IP station [16] can contain an arbitrary number of independent simultaneous working-traffic generators. Each generator enables the use of heavy-tailed distributions, such as Pareto or Weibull, for the generation of a self-similar network traffic by two distributions, one for length of a data source process and another for data inter-arrival time process. In our research, a traffic generator contained in an Ethernet IP station model of the OPNET Modeler simulation tool is used, as shown in the Figure 8.

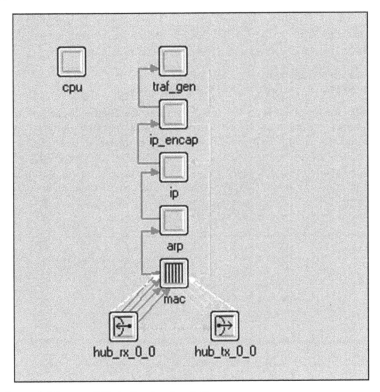

Fig. 8. Node model for used IP station in simulation.

In the IP station model, the traffic generator is placed above the IP encapsulation layer, which takes care of packets' formations and fragmentation. This is the process of segmentation of long data into the shorter packets, or vice versa, according to the RFC 793 [12]. Padding of the packet data payload with additional bits is also performed when data is shorter than a predefined minimal payload. Because the traffic is modeled, above IP level of the TCP/IP model, to the lengths of the generated data, 20 bytes of IP header are added. 18 bytes of information for MAC (14bytes) and CRC (4 bytes) are also further added. Structure of Ethernet frame used in the IP station model. Using this model, the applications' protocol does not impact the generated traffic. The model is suitable for the simulation cases, when we want to statistically model the network traffic, which can be caused by many arbitrary communications' applications. Using this approach, we can model such network traffic by single traffic source.

5. Estimation of simulation parameters of measured network traffic

The main problem of measured packet network traffic modeling is to estimate the parameter, which is needed for modeling measured network traffic in simulation tools. It has already been mentioned that the parameters of data source traffic processes are needed. We already described that transformation from packet network traffic $Z_p[n]$ to data source network traffic $Z_d[n]$ is needed (section 2.6) [45]. There are many possibilities to make a transformation from $Z_p[n]$ to $Z_d[n]$, which allows estimation of parameters of data source network traffic processes. We investigated two algorithms [28]:

1. algorithm with an in-depth analysis of all packet headers,
2. algorithm with a coarse inspection of IP header only.

The main differences between them are complexity and the needed execution time. The first algorithm mimics a complete decapsulation process, and defragmentation in higher layers of the communication model. Any sniffers are able to extract this data from the IP header. Knowing them, it is then simple to calculate a length of IP PDU (Protocol Data Unit) which also contains a header of higher layer protocols. Through the use of an in-depth header analysis, it is possible, in the similar way as the IP header, to calculate the lengths of all these headers. Each packed IP header has four the so-called fragmentation fields that contain information about data fragmentation, which is shown on Figure 9.

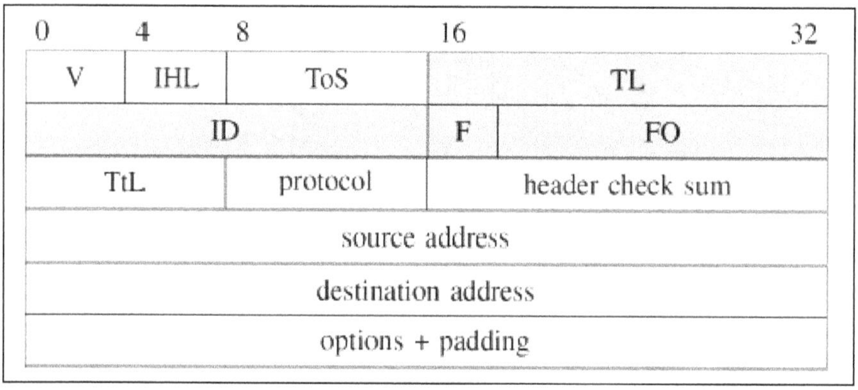

0	4	8	16	32
V	IHL	ToS	TL	
ID			F	FO
TtL		protocol	header check sum	
source address				
destination address				
options + padding				

Fig. 9. IP header. Shadowed fields are used in the defragmentation process. Legend:
V: protocol version; IHL: Internet Header Length; ToS: Type of Service; TL: Total Length; ID: Identification Data; F: Flags; FO: Fragment Offset; TTL: Time to Live.

Extensive research and investigation about traffic sources in contemporary networks show that this approach requires an in-depth analysis of packets (where need specialized, very powerful and consequently, expensive instruments), which in case of encrypted packets and non-standard application protocols, is not completely possible. In such cases, it is also necessary to capture the entire packets, which can be problematic in the high-speed networks. For these reasons, a simple algorithm has been developed, where only information of packets sizes, packet time stamps and IP addresses are needed.

The second algorithm skips decapsulation by considering the average lengths of packet headers and then uses only packet lengths and inter-arrival times. In the second case, the algorithm offers the estimation of data source network traffic, not the exact reconstructed data source traffic. The second algorithm represents the main part of method by mimic defragmentation process, which is described in detail in [45]. The main idea of mimic defragmentation process method is to compose data from the captured packet traffic, which is previously fragmented at the transmitter. The data source traffic estimation is carried out by finding and summing fragmented packets' sequences without an in-depth analysis of packets. Fragmented sequence is defined as a sequence of l_{MTU} sized packets associated with the same source and destination addresses and terminated by packet shorter than l_{MTU}.

6. Simulation results

In real networks, we have captured packets of different network traffic through a Wireshark sniffer. The two different types of measured traffic are used for analysis, modeling and simulation purposes. These two test traffics are shown in Figure 10.

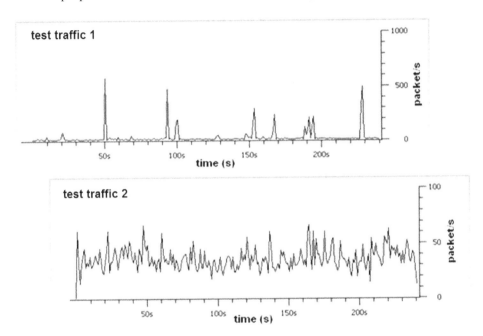

Fig. 10. Measured test traffic 1 and 2 captured by Wireshark sniffer.

measured test traffics	packet rate (p/s)	bit rate (kb/s)	variance method	R/S method	periodogram method
test traffic 1	24.02	108.90	0.630	0.723	0.843
test traffic 2	35.612	114.51	0.592	0.580	0.477

Table 1. The main properties of captured traffics. On the right side of the table the Hurst parameter is estimated using different methods for both test traffics.

For each of test traffics, the Hurst parameter has been estimated through different methods. The Hurst parameters for both cases are bigger than 0.5, so we can classify these test traffics as a self-similar network traffic. Table 1 contains the estimated parameters H for both traffics, which are estimated by variance, R/S and periodogram methods. We also conducted tests about short and long-range dependence. In the case of the first test traffic, the autocorrelation function decayed hyperbolically, which means, that this traffic can have the property of a long-range dependence. For the second test traffic autocorrelation, function decayed exponentially towards 0. For this case, the sum of autocorrelations has finite results and, therefore, the test traffic 2 has the property of short-range dependence.

For both test traffics (test traffic 1 and test traffic 2) we estimate distribution and its parameters for data source traffic processes for simulation purpose. For that reason, we made an estimation of data source traffic from the captured packet traffic through the mimic defragmentation process method [45]. For both test traffics, the suitably heavy (Pareto or Weibull) and also light-tailed (exponential) distributions are chosen.

Based on the estimated distribution parameters for both measured test traffic (test traffic 1 and test traffic 2), we generated self-similar traffic in the OPNET simulation tool with two different station types – RPG and IP stations. We have created six different scenarios for each of test traffic. In the first two scenarios, the network traffic is generated by an RPG station, where a self-similarity is described by Hurst parameter. During the first scenario, we use heavy-tailed distribution for the data size process, while in the second a light-tailed distribution (exponential) is used. In the next four scenarios, network traffic is generated using the IP station, where we use different combination's distributions for the data size process and data inter-arrival time. One of the criterions, for successful modeling, is the difference between bit and packet-rates of the test traffic and modeled traffic in OPNET simulation tool. Besides the average values of bit and packet-rates, the more important criteria are also bursts' intensity within the network traffic. For each of test traffics (test traffic 1 and test traffic 2), the traffic which best represents the measured test traffic is chosen from six modeled traffics.

Test traffic 1 poses the property of long-range dependence, so there are a lot of bursts in the traffic. We model this measured-test traffic over six different scenarios. The results are shown in Figure 6 and Table 2. Table 2 shows the main properties of measured test traffic 1 and estimated distribution parameters which were used in OPNET simulation tool for simulating network traffic (the left side of Table 2). Table 2 (the right side) also shows main properties of simulated network traffics (six different scenarios) in OPNET simulation tool based on estimated distributions.

Table 2 shows modeling results for test traffic 1 over six different scenarios in OPNET simulation tool. There are estimated statistical parameters such as Hurst parameters and distributions used in models and simulation results using these models. Figure 11 shows all six modeled traffic traffics generated by OPNET, with estimated distributions and parameters from Table 2.

The best approximation for test traffic 1 is modeled traffic 5 from Table 2, which is described by Pareto distribution for data size process and Weibull distribution for data inter-arrival time. Figure 12 shows a comparison between the second test traffic and the modeled traffic 5 for bit rates. From all critera after comparison, we can say that the modeled traffic 5 is a good approximation of measured test traffic 1.

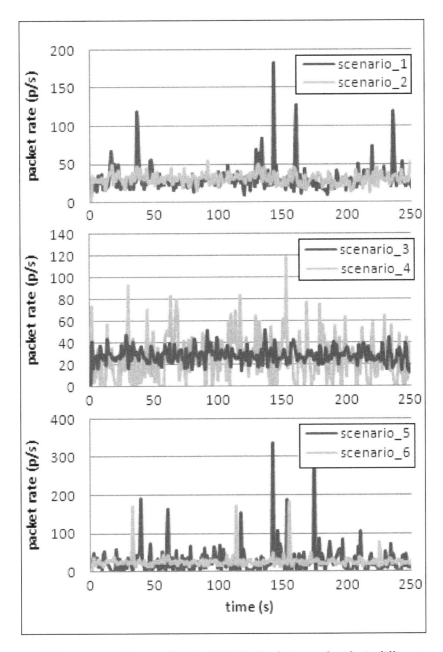

Fig. 11. Modeling measured test traffic 1 in OPNET simulation tool with six different estimated parameters from Table 2 (scenario 1 and 2 with RPG station, scenario 3, 4, 5, 6 with IP station).

parameters for modeling			parameters of measured and modeled traffic in OPNET		
traffic	data inter-arrival process	data size process	packet rate (p/s)	bite rate (kb/s)	H
measured test traffic 1	X	X	24	108.90	0.73
modeled 1	$H = 0.732$	Pareto $a = 0.9835$ $\beta = 432$	33.82	128.75	0.59
modeled 2	$H = 0.732$	exponential $\lambda = 7547.2$	29.18	181.44	0.59
modeled 3	exponential $\lambda = 0.0458$	exponential $\lambda = 933.4$	27.56	168.94	0.51
modeled 4	Weibull $a = 0.304$ $\beta = 0.00578$	exponential $\lambda = 933.4$	25.14	153.71	0.62
modeled 5	Weibull $a = 0.304$ $\beta = 0.00578$	Pareto $a = 0.9835$ $\beta = 34$	25.32	88.70	0.66
modeled 6	exponential $\lambda = 0.0458$	Pareto $a = 0.9835$ $\beta = 34$	26.63	81.30	0.55

Table 2. The left side of table shows the estimated distributions and parameters for measured test traffic 1 (six different distribution combinations). The right side of table shows main properties of modeled network traffic in OPNET simulation tool (six scenarios), where estimated distributions were used.

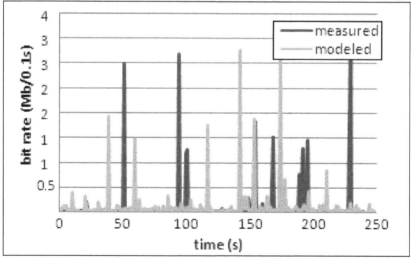

Fig. 12. Comparison between the modeled traffic 5 generated in OPNET simulation tool and the measured test traffic 1 in bits per second (kb/s).

Test traffic 2 is also modeled over six different scenarios, just like in the first case. Table 3 shows the main properties of measured test traffic 2 and estimated distribution parameters which were used in OPNET simulation tool for simulating network traffic (left side of Table 3). Table 3 (right side) also shows main properties of simulated network traffics (six different scenarios) in OPNET simulation tool.

As the best modeled traffic of test traffic 2 from all six cases (Table 3), we choose the case where simulated traffic is described by the exponential distribution for packet sizes and Weibull heavy-tailed distribution for inter-arrival time (modeled traffic 4). The bit-rate of this traffic is 33.27 (p/s) and packet-rate is 126.79 (kb/s), which are very close to the measured values. The Hurst parameter of the simulated traffic is 0.58, which is also close to the estimated values of the measured traffic. Figure 13 shows the comparison between the measured test traffic 2 and the best-modeled traffic (modeled traffic 4) for bit rates. From all critera after comparison, we can say that the simulated traffic is a good approximation of the measured traffic 2.

parameters for modeling			parameters of measured and modeled traffic in OPNET		
traffic	data inter-arrival process	data size process	packet rate (p/s)	bite rate (kb/s)	H
measured test traffic 2	X	X	35.61	114.51	0.55
modeled 1	$H = 0.55$	Pareto $a = 0.8373$ $\beta = 272$	49.46	231.98	0.62
modeled 2	$H = 0.55$	exponential $\lambda = 3619$	36.66	140.72	0.58
modeled 3	exponential $\lambda = 0.029$	exponential $\lambda = 452.48$	35.66	135.89	0.53
modeled 4	Weibull $a = 0.57$ $\beta = 0.01894$	exponential $\lambda = 452.48$	33.27	126.79	0.58
modeled 5	Weibull $a = 0.57$ $\beta = 0.01894$	Pareto $a = 0.8373$ $\beta = 34$	52.27	298.25	0.62
modeled 6	exponential $\lambda = 0.029$	Pareto $a = 0.8373$ $\beta = 34$	55.12	315.61	0.53

Table 3. The left side of table shows the estimated distributions and parameters for measured test traffic 2 (six different distribution combinations). The right side of table shows main properties of modeled network traffic in OPNET simulation tool (six scenarios), where estimated distributions and its parameters were used.

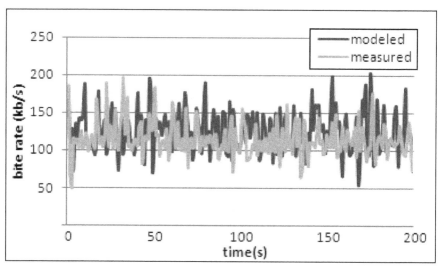

Fig. 13. Comparison between modeled traffic 4 generated in OPNET simulation tool and measured test traffic 2 in bits per second (kb/s).

7. Conclusion

In this chapter, we present our research in the area of measurements, modeling and simulations of the self-similar network traffic. Firstly, the state of the art method for modeling and simulating of self-similar network traffic is presented. We also describe a number of facts about self-similarity, long range dependences and probability, which are used to describe such stochastic processes. Described as well are the mechanism and models to simulate network traffic in the OPNET Modeler simulation tool. The main goal of our research is to simulate measured network traffic, where we tend to minimize discrepancies between the measured and the simulated network traffic in the sense of packet-rate, bit-rate, bursts intensity, and variances. One of the big challenges in our research work was to find appropriate method to estimate parameters of data source network traffic processes that are based on measured network packet's traffic. The estimated parameters are needed during the modeling of the measured network traffic in the simulation tool. For those reasons, we have developed different methods, which allow estimation of the parameters of data source network traffic processes, based on the measured network packet's traffic.

At the end of the chapter, all phases needed for simulating the measured network traffic in the OPNET simulation tool are presented. During the analysis phase we pay attention to the self-similar property, which has become the basic model for describing today's network traffic. In the network traffic theory, the properties of short and long-range dependence are directly prescribed by the values of estimated parameter H. In our network traffic analysis, we prove that network traffic (test traffic 2) can exist where Hurst parameter is bigger than 0.5, but this process does not have the property of a long-range dependence.

For the purpose of parameters estimation of data source network traffic processes, we have used a method that mimics packet defragmentation. Through the use of this method we offer estimated parameters, used in simulations, where six traffics are simulated by different distributions for each of the measured test traffic. It can be seen from simulations that in the case of modeling self-similar traffic, short-range dependence is more appropriate for choosing exponential distribution to describe a packet-size process. The exponential distribution does not impact the extreme peaks in the modeled traffic. Pareto distribution is unsuitable for this purpose.

Heavy-tailed distributions, especially Pareto, are suitable for modeling a packet-size process of the measured network traffic, which are self-similar and also have the property of a long-range dependence (test traffic 1).

There are discrepancies between the measured and the modeled traffics in the sense of packet-rate, bit-rate, bursts intensity, and variances. With a method which mimics defragmentation, a good approximation of the measured network traffic is obtained. We cannot claim that this is the optimal method for all situations, because there are some limitations, although it shows good results through simulation in OPNET Modeler. We have noticed that estimating the shape-parameter of Pareto is very delicate, because a small deviation in the parameter causes large discrepancies regarding the network traffic's average values, which is one of the important criteria for traffic modeling.

8. Acknowledgment

This work has been partly financed by the Slovenian Ministry of Defense as part of the target research program "Science for Peace and Security": M2-0140 - Modeling of Command and Control information systems, and partly by the Slovenian Ministry of Higher Education and Science, research program P2-0065 "Telematics".

9. References

[1] W. E. Leland, M. S. Taqqu, W. Willinger and D. V. Wilson, On the self-similar nature of Ethernet traffic (Extended version), IEEE/ACM Transactions on Networking, Vol.2, pp.1-15, 1994.

[2] W. Willinger and V. Paxson, Where mathematics meets the Internet, Notices of the American Mathematical Society, 45(8): 961–970, 1998.

[3] K. Park, G. Kim and M. E. Crovella, On the Relationship Between File Sizes Transport Protocols, and Self-Similar Network Traffic, International Conference on Network Protocols, 171–180, Oct 1996.

[4] M. E. Crovella and A. Bestavros, Self-Similarity in World Wide Web Traffic Evidence and Possible Causes, IEEE/ACM Transactions on Networking, 1997.

[5] O. Sheluhin, S. Smolskiy and A. Osin, Self-Similar Processes in Telecommunications, John Wiley & Sons, 2007.

[6] K. Park and W. Willinger, Self-Similar Network Traffic and Performance Evaluation, John Wiley & Sons, 2000.

[7] T. Karagiannis, M. Molle and M. Faloutos, Understanding the limitations of estimation methods for long-range dependence, University of California.

[8] T. Karagiannis and M. Faloutos, Selfis: A tool for self-similarity and long range dependence analysis, University of California.

[9] M. Fras, J. Mohorko and Ž. Čucej, Estimating the parameters of measured self similar traffic for modeling in OPNET, IWSSIP Conference, 27.-30 June 2007, Maribor, Slovenia.

[10] J. Mohorko and M. Fras, Modeling of IRIS Replication Mechanism in a Tactical Communication network, using OPNET, Computer Networks, v 53, n 7, p 1125-36, 13 May 2009.

[11] J. Mohorko, M. Fras and Ž. Čucej: Modeling of IRIS replication mechanism in tactical communication network with OPNET, OPNETWORK 2007 - the eleventh annual OPNET technology Conference, August 27th-31st, Washington, D.C., 2007.

[12] RFC 793 - Transmission Control Protocol:. [Online]. Available: http://www.faqs.org/rfcs/rfc793.html

[13] M. Chakravarti, R. G. Laha and J. Roy, Handbook of Methods of Applied Statistics, Volume I, John Wiley and Sons, pp. 392-394, 1967.

[14] W. T. Eadie, D. Drijard, F. E. James, M. Roos and B. Sadoulet, Statistical Methods in Experimental Physics, Amsterdam, North-Holland, 269-271, 1971.

[15] A. Adas, Traffic Models in Broadband Telecommunication Networks, Communications Magazine, IEEE , vol 35/7, 82–89, 1997.

[16] J. Potemans, B. Van den Broeck, Y. Guan, J. Theunis, E. Van Lil and A. Van de Capelle, Implementation of an Advanced Traffic Model in OPNET Modeler, OPNETWORK 2003, Washington D.C., USA, 2003.

[17] B. Hill, A Simple Approach to Inference About tbc Tail of a Distribution, Annals of Statistics, Vol. 3, No. 5, 1975, pp.1163-1174.

[18] J. Judge, H. W. Beadle and J. Chicharo, Sampling HTTP response packets for prediction of web traffic volume statistics, IEEE Global Communications Conference (GLOBECOM'98), Sydney, Australia, Nov. 8-12, 1998.

[19] K. Park, G. Kim and M. E. Crovella, On the Relationship Between File Sizes Transport Protocols, and Self-Similar Network Traffic, International Conference on Network Protocols, 171–180, Oct 1996.

[20] V. Paxon and S. Floyd, Wide area traffic: the failure of Poisson modeling, IEEE/ACM Transactions on Networking, 3(3): 226–244, 1995.

[21] H. Yõlmaz, IP over DVB: Managment of self-similarity, Master of Science, Boğaziçi University, 2002.

[22] B. Vujičić, Modeling and Characterization of Traffic in Public Safety Wireless Networks, Master of Applied science, Simon Fraser University, Vancouver, 2006.

[23] M. Jiang, S. Hardy in Lj. Trajkovic, Simulating CDPD networks using OPNET, OPNETWORK 2000, Washington D.C., August 2000.

[24] J. Mohorko, M. Fras and Ž. Čučej, Modeling methods in OPNET simulations of tactical command and control information systems, IWSSIP Conference, 27.-30 June 2007, Maribor, Slovenia.

[25] A. M. Law and M. G. McComas, How the Expertfit distribution fitting software can make simulation models more valid, Proceedings of the 2001 Winter Simulation Conference.

[26] Free (demo) fitting tool EasyFit software [Online]. Available: www.mathwave.com/.

[27] F. Xue and S. J. Ben Yoo, On the Generation and Shaping Self-similar Traffic in Optical Packet-switched Networks, OPNETWORK 2002, Washington D.C., USA, 2002.

[28] Ž. Čučej and M.Fras, Data source statistics modeling based on measured packet traffic : a case study of protocol algorithm and analytical transformation approach, TELSIKS 2009, 9th International Conference on Telecommunications in Modern Satellite, Cable and Broadcasting Services, Serbia, Niš, 7-9 October, 2009.

[29] M. Fras, J. Mohorko and Ž. Čučej, Analysis, modeling and simulation of P2P file sharing traffic impact on networks' performances. Inf. MIDEM, 38(2):117–123, 2008.

[30] H. Abrahamsson, Traffic measurement and analysis, Swedish Institute of Computer Science, 1999.

[31] C. Williamson, Internet traffic measurement, IEEE internet computing, vol. 5, no. 6, pp. 70–74, 2001.

[32] P. Celeda, High-speed network traffic acquisition for agent systems, in Proc. IEEE/WIC/ACM International Conference on High-Speed Network Traffic Acquisition for Agent Systems, Intelligent Agent Technology, November 2-5, 2007, pp. 477–480.

[33] D. Pezaros, Network Traffic Measurement for the Next Generation Internet. Computing Department Lancaster University, 2005.

[34] D. Epema, J. Pouwelse, P. Garbacki and H. Sips, The bittorrent P2P filesharing system: Measurements and analysis. Peer-to-Peer Systems IV, 2005.

[35] S. Saroiu, P. K. Gummadi and S. D. Gribble, A Measurement Study of Peer-to-Peer File Sharing Systems, in Proc. of the Multimedia Computing and Networking (MMCN), January 2-5, San Jose, Ca, USA, 2002.

[36] E. Asensio, J. M. Orduna and P. Morillo, Analyzing the Network Traffic Requirements of Multiplayer Online Games, in Proc. 2nd International Conference on Advanced Engineering Computing and Applications in Sciences: ADVCOMP'08, 2008, pp. 229–234.

[37] Y. Yu, D. Liu, J. Li and C. Shen, Traffic Identification and Overlay Measurement of Skype, in Proc. International Conference on Computational Intelligence and Security, November 3-6, vol. 2, 2006, p. 1043 – 1048.

[38] M. E. Crovella and L. Lipsky, Long-lasting transient conditions in simulations with heavy-tailed workloads, in Proc. 1997 Winter Simulation Conference, December 7-10, vol. Atlanta, GA, USA, Edmonton, Canada, 1997.

[39] A. Feldmann, A. C. Gilbert, P. Huang and W. Willinger, Dynamics of IP traffic: a study of the role of variability and the impact of control, in Proc. Applications, technologies, architectures, and protocols for computer communication, August 30-September 03, Cambridge, Massachusetts, USA, 1999, pp. 301–313.

[40] C. Nuzman, I. Saniee, W. Sweldens and A. Weiss, A compound model for TCP connection arrivals for LAN and WAN applications, Computer Networks: The International Journal of Computer and Telecommunications Networking, vol. 40, no. 3, pp. 319–337, 2002.

[41] B. Ryu and S. Lowen. Fractal Traffic Model for Internet Simulation. In Proc. 5th IEEE Symposium on Computers and Communications (ISCC 2000), 2000.

[42] M. Fras, Methods for the statistical modeling of measured network traffic for simulation purposes, Ph.D. thesis, 2009, Maribor, Slovenia.

[43] http://en.wikipedia.org/wiki/Pareto_distribution.

[44] http://en.wikipedia.org/wiki/Weibull_distribution.

[45] M. Fras, J. Mohorko and Ž. Čučej, Modeling of captured network traffic by the mimic defragmentation process, Simulation: Transactions of The Society for Modeling and Simulation International, San Diego, USA, Published online 20 September 2010.

[46] M. Fras, J. Mohorko and Ž. Čučej, Modeling of measured self-similar network traffic in OPNET simulation tool, Inf. MIDEM, 40(3): 224-231, September 2010.

Part 3

Routing

8

On the Fluid Queue Driven by an Ergodic Birth and Death Process

Fabrice Guillemin[1] and Bruno Sericola[2]
[1]Orange Labs, Lannion
[2]INRIA Rennes - Bretagne Atlantique, Campus de Beaulieu,
35042 Rennes Cedex
France

1. Introduction

Fluid models are powerful tools for evaluating the performance of packet telecommunication networks. By masking the complexity of discrete packet based systems, fluid models are in general easier to analyze and yield simple dimensioning formulas. Among fluid queuing systems, those with arrival rates modulated by Markov chains are very efficient to capture the burst structure of packet arrivals, notably in the Internet because of bulk data transfers. By exploiting the Markov property, very efficient numerical algorithms can be designed to estimate performance metrics such as the overflow probability, the delay of a fluid particle or the duration of a busy period.

In the last decade, stochastic fluid models and in particular Markov driven fluid queues, have received a lot of attention in various contexts of system modeling, e.g. manufacturing systems (see Aggarwal et al. (2005)), communication systems (in particular TCP modeling; see vanForeest et al. (2002)) or more recently peer to peer file sharing process (see Kumar et al. (2007)) and economic systems (risk analysis; see Badescu et al. (2005)). Many techniques exist to analyze such systems.

The first studies of such queuing systems can be dated back to the works by Kosten (1984) and Anick et al. (1982), who analyzed fluid models in connection with statistical multiplexing of several identical exponential on-off input sources in a buffer. The above studies mainly focused on the analysis of the stationary regime and have given rise to a series of theoretical developments. For instance, Mitra (1987) and Mitra (1988) generalize this model by considering multiple types of exponential on-off inputs and outputs. Stern & Elwalid (1991) consider such models for separable Markov modulated rate processes which lead to a solution of the equilibrium equations expressed as a sum of terms in Kronecker product form. Igelnik et al. (1995) derive a new approach, based on the use of interpolating polynomials, for the computation of the buffer overflow probability.

Using the Wiener-Hopf factorization of finite Markov chains, Rogers (1994) shows that the distribution of the buffer level has a matrix exponential form, and Rogers & Shi (1994) explore algorithmic issues of that factorization. Ramaswami (1999) and da Silva Soares & Latouche (2002), Ahn & Ramaswami (2003) and da Silva Soares & Latouche (2006) respectively exhibit and exploit the similarity between stationary fluid queues in a finite Markovian environment and quasi birth and death processes.

Following the work by Sericola (1998) and that by Nabli & Sericola (1996), Nabli (2004) obtained an algorithm to compute the stationary distribution of a fluid queue driven by a finite Markov chain. Most of the above cited studies have been carried out for finite modulating Markov chains.

The analysis of a fluid queue driven by infinite state space Markov chains has also been addressed in many research papers. For instance, when the driving process is the M/M/1 queue, Virtamo & Norros (1994) solve the associated infinite differential system by studying the continuous spectrum of a key matrix. Adan & Resing (1996) consider the background process as an alternating renewal process, corresponding to the successive idle and busy periods of the M/M/1 queue. By renewal theory arguments, the fluid level distribution is given in terms of integral of Bessel functions. They also obtain the expression of Virtamo and Norros via an integral representation of Bessel functions. Barbot & Sericola (2002) obtain an analytic expression for the joint stationary distribution of the buffer level and the state of the M/M/1 queue. This expression is obtained by writing down the solution in terms of a matrix exponential and then by using generating functions that are explicitly inverted.

In Sericola & Tuffin (1999), the authors consider a fluid queue driven by a general Markovian queue with the hypothesis that only one state has a negative drift. By using the differential system, the fluid level distribution is obtained in terms of a series, which coefficients are computed by means of recurrence relations. This study is extended to the finite buffer case in Sericola (2001). More recently, Guillemin & Sericola (2007) considered a more general case of infinite state space Markov process that drives the fluid queue under some general uniformization hypothesis.

The Markov chain describing the number of customers in the M/M/1 queue is a specific birth and death process. Queueing systems with more general modulating infinite Markov chain have been studied by several authors. For instance, van Dorn & Scheinhardt (1997) studied a fluid queue fed by an infinite general birth and death process using spectral theory.

Besides the study of the stationary regime of fluid queues driven by finite or infinite Markov chains, the transient analysis of such queues has been studied by using Laplace transforms by Kobayashi & Ren (1992) and Ren & Kobayashi (1995) for exponential on-off sources. These studies have been extended to the Markov modulated input rate model by Tanaka et al. (1995). Sericola (1998) has obtained a transient solution based on simple recurrence relations, which are particularly interesting for their numerical properties. More recently, Ahn & Ramaswami (2004) use an approach based on an approximation of the fluid model by the amounts of work in a sequence of Markov modulated queues of the quasi birth and death type. When the driving Markov chain has an infinite state space, the transient analysis is more complicated. Sericola et al. (2005) consider the case of the M/M/1 queue by using recurrence relations and Laplace transforms.

In this paper, we analyze the transient behavior of a fluid queue driven by a general ergodic birth and death process using spectral theory in the Laplace transform domain. These results are applied to the stationary regime and to the busy period analysis of that fluid queue.

2. Model description

2.1 Notation and fundamental system

Throughout this paper, we consider a queue fed by a fluid traffic source, whose instantaneous transmitting bit rate is modulated by a general birth and death process (Λ_t) taking values in $\mathbb{N} = \{0, 1, 2, \ldots\}$. The input rate is precisely $r(\Lambda_t)$, where r is a given increasing function from $\mathbb{N}$ into $\mathbb{R}$.

The birth and death process (Λ_t) is characterized by the infinitesimal generator given by the infinite matrix

$$A = \begin{pmatrix} -\lambda_0 & \lambda_0 & 0 & \cdot & \cdot \\ \mu_1 & -(\lambda_1 + \mu_1) & \lambda_1 & \cdot & \cdot \\ 0 & \mu_2 & -(\lambda_2 + \mu_2) & \lambda_2 & \cdot \\ \cdot & & \cdot & & \cdot & \cdot \end{pmatrix}, \tag{1}$$

where $\lambda_i > 0$ for $i \geq 0$ is the transition rate from state i to state $i + 1$ and $\mu_j > 0$ for $j \geq 1$ is the transition rate from state j to state $j - 1$.

We assume that the birth and death process (Λ_t) is ergodic, which amounts to assuming (see Asmussen (1987) for instance) that

$$\sum_{i=0}^{\infty} \frac{1}{\lambda_i \pi_i} = \infty \quad \text{and} \quad \sum_{i=0}^{\infty} \pi_i < \infty, \tag{2}$$

where the quantities π_i are defined by:

$$\pi_0 = 1 \quad \text{and} \quad \pi_i = \frac{\lambda_0 \ldots \lambda_{i-1}}{\mu_1 \ldots \mu_i}, \quad \text{for } i \geq 1.$$

Under the above assumption, the birth and death process (Λ_t) has a unique invariant probability measure: in steady state, the probability of being in state i is

$$p(i) = \frac{\pi_i}{\displaystyle\sum_{j=0}^{\infty} \pi_j}.$$

Let $p_0(i)$ denote, for $i \geq 0$, the probability that the birth and death process (Λ_t) is in state i at time 0, i.e., $\mathbb{P}(\Lambda_0 = i) = p_0(i)$. Note that if $p_0(i) = p(i)$ for all $i \geq 0$, then $\mathbb{P}(\Lambda_t = i) = p(i)$ for all $t \geq 0$ and $i \geq 0$.

We assume that the queue under consideration is drained at constant rate $c > 0$. Furthermore, we assume that $r(i) > c$ when i is greater than a fixed $i_0 > 0$ and that $r(i) < c$ for $0 \leq i \leq i_0$. (It is worth noting that we assume that $r(i) \neq c$ for all $i \geq 0$ in order to exclude states with no drift and thus to avoid cumbersome special cases.) In addition, the parameters c and $r(i)$ are such that

$$\rho = \sum_{i=0}^{\infty} \frac{r(i)}{c} p(i) < 1 \tag{3}$$

so that the system is stable. The quantity $r_i = r(i) - c$ is either positive or negative and is the net input rate when the modulating process (Λ_t) is in state i.

Let X_t denote the buffer content at time t. The process (X_t) satisfies the following evolution equation: for $t \geq 0$,

$$\frac{dX_t}{dt} = \begin{cases} r(\Lambda_t) - c & \text{if } X_t > 0 \text{ or } r(\Lambda_t) > c, \\ 0 & \text{if } X_t = 0 \text{ and } r(\Lambda_t) \leq c. \end{cases} \tag{4}$$

Let $f_i(t,x)$ denote the joint probability density function defined by

$$f_i(t,x) = \frac{\partial}{\partial x} \mathbb{P}(\Lambda_t = i, X_t \leq x).$$

As shown in Sericola (1998), on top of its usual jump at point $x = 0$, when $X_0 = x_0 \geq 0$, the distribution function $\mathbb{P}(\Lambda_t = i, X_t \leq x)$ has a jump at points $x = x_0 + r_i t$, for t such that $x_0 + r_i t > 0$, which corresponds to the case when the Markov chain $\{\Lambda_t\}$ starts and remains during the whole interval $[0,t)$ in state i.

We focus in the rest of the paper on the probability density function $f_i(t,x)$ for $x > 0$ along with its usual jump at point $x = 0$. A direct consequence of the evolution equation (4) is the forward Chapman-Kolmogorov equations satisfied by $(f_i(t,x), \ x \geq 0, i \in \mathbb{N})$, which form the fundamental system to be solved.

Proposition 1 (Fundamental system). *The functions $(x,t) \rightarrow f_i(t,x)$ for $i \in \mathbb{N}$ satisfy the differential system (in the sense of distributions):*

$$\frac{\partial f_i}{\partial t} = -r_i \frac{\partial}{\partial x} \left(\left(\mathbb{1}_{\{i > i_0\}} + \mathbb{1}_{\{i \leq i_0\}} \mathbb{1}_{\{x > 0\}} \right) f_i \right) - (\lambda_i + \mu_i) f_i + \lambda_{i-1} f_{i-1} + \mu_{i+1} f_{i+1}, \tag{5}$$

with the convention $\lambda_{-1} = 0, f_{-1} \equiv 0$ and $f_i(t,x) = 0$ for $x < 0$.

Note that the differential system (5) holds for the density probability functions $f_i(t,x)$. The differential system considered in Parthasarathy et al. (2004) and van Dorn & Scheinhardt (1997) governs the probability distribution functions $\mathbb{P}(X_t \leq x, \Lambda_t = i), i \geq 0$. The differential system (5) is actually the equivalent of Takács' integro-differential formula for the $M/G/1$ queue, see Kleinrock (1975). The resolution of this differential system is addressed in the next section.

2.2 Basic matrix Equation

Introduce the double Laplace transform

$$F_i(s,\xi) = \int_{0^-}^{\infty} \int_{0^-}^{\infty} e^{-st - \xi x} f_i(t,x) dt dx = \int_0^{\infty} e^{-st} \mathbb{E}\left(e^{-\xi X_t} \mathbb{1}_{\{\Lambda_t = i\}} \right) dt$$

and define the functions $f_i^{(0)}(\xi)$ and $h_i(s)$ for $i \in \mathbb{N}$ as follows

$$f_i^{(0)}(\xi) = \int_0^{\infty} e^{-x\xi} \mathbb{P}\{\Lambda_0 = i, X_0 \in dx\}$$

$$h_i(s) = \int_0^{\infty} e^{-st} \mathbb{P}\{\Lambda_t = i, X_t = 0\} dt.$$

The functions $f_i^{(0)}$ are related to the initial conditions of the system and are known functions. For $i > i_0$, we have $P\{\Lambda_t = i, X_t = 0\} = 0$, which implies that $h_i(s) = 0$, for $i > i_0$. On the contrary, for $i \leq i_0$, the functions h_i are unknown and have to be determined by taking into account the dynamics of the system.

By taking Laplace transforms in Equation (5), we obtain the following result.

Proposition 2. *Let $F(s,\xi)$, $f^{(0)}$, and $h(s)$ be the infinite column vectors, which components are $F_i(s,\xi)/\pi_i$, $f_i^{(0)}/\pi_i$, and $h_i(s)/\pi_i$ for $i \geq 0$, respectively. Then, these vectors satisfy the matrix equation*

$$(s\mathbb{I} + \xi R - A)F(s,\xi) = f^{(0)}(\xi) + \xi Rh(s), \tag{6}$$

where $\mathbb{I}$ is the identity matrix, A is the infinitesimal generator of the birth and death process $\{\Lambda_t\}$ defined by Equation (1), and R is the diagonal matrix with diagonal elements r_i, $i \geq 0$.

Proof. Taking the Laplace transform of $\partial f_i/\partial t$ gives rise to the term $sF_i - f_i^{(0)}$. In the same way, taking the Laplace transform of $\partial(\mathbb{1}_{\{x>0\}}f_i)/\partial x$ yields the term $\xi F_i - \xi h_i$. Hence, taking Laplace transforms in Equation (5) and dividing all terms by π_i gives, for $i \geq 0$,

$$s\frac{F_i}{\pi_i} - \frac{f_i^{(0)}}{\pi_i} = -r_i\xi\frac{F_i}{\pi_i} + r_i\xi\frac{h_i}{\pi_i} - (\lambda_i + \mu_i)\frac{F_i}{\pi_i} + \lambda_i\frac{F_{i+1}}{\pi_{i+1}} + \mu_i\frac{F_{i-1}}{\pi_{i-1}},$$

which can be rewritten in matrix form as Equation (6) □

When we consider the stationary regime of the fluid queue, we have to set $f^{(0)}(\xi) \equiv 0$ and eliminate the term $s\mathbb{I}$ in Equation (6), which then becomes

$$(\xi R - A)F(\xi) = \xi Rh, \tag{7}$$

where h is the vector, which ith component is $h_i = \lim_{t\to\infty} P\{\Lambda_t = i, X_t = 0\}/\pi_i$ and $F(\xi)$ is the vector, which ith component is $\mathbb{E}\left[e^{-\xi X_t}\mathbb{1}_{\{\Lambda_t=i\}}\right]/\pi_i$. This is the Laplace transform version of Equation (12) by van Dorn & Scheinhardt (1997), which addresses the resolution of Equation (7).

3. Resolution of the fundamental system

In this section, we show how Equation (6) can be solved. For this purpose, we analyze the structure of this equation and in a first step, we prove that the functions $F_i(s,\xi)$ can be expressed in terms of the function $F_{i_0}(s,\xi)$. (Recall that the index i_0 is the greatest integer such that $r(i) - c < 0$ and that for $i \geq i_0 + 1$, $r(i) > c$.). The proof greatly relies on the spectral properties of some operators defined in adequate Hilbert spaces.

3.1 Basic orthogonal polynomials

In the following, we use the orthogonal polynomials $Q_i(s;x)$ defined by recursion: $Q_0(s;x) \equiv 1$, $Q_1(s;x) = (s + \lambda_0 - r_0x)/\lambda_0$ and for $i \geq 1$,

$$\frac{\lambda_i}{|r_i|}Q_{i+1}(s;x) + \left(x - \frac{s+\lambda_i+\mu_i}{|r_i|}\right)Q_i(s;x) + \frac{\mu_i}{|r_i|}Q_{i-1}(s;x) = 0. \tag{8}$$

By suing Favard's criterion (see Askey (1984) for instance), it is easily checked that the polynomials $Q_i(s;x)$ for $i \geq 0$ form an orthogonal polynomial system.

The polynomials $\frac{\lambda_0...\lambda_{i-1}}{|r_0...r_{i-1}|}Q_i(s;-z)$, $i \geq 0$ are the successive denominators of the continued fraction

$$\mathcal{F}^e(s;z) = \cfrac{1}{z + \frac{s+\lambda_0}{|r_0|} - \cfrac{\frac{\mu_1\lambda_0}{|r_0 r_1|}}{z + \frac{s+\lambda_1+\mu_1}{|r_1|} - \cfrac{\frac{\mu_2\lambda_1}{|r_2 r_1|}}{z + \frac{s+\lambda_2+\mu_2}{|r_2|} - \ddots}}}$$

which is itself the even part of the continued fraction

$$\mathcal{F}(s;z) = \cfrac{\alpha_1(s)}{z + \cfrac{\alpha_2(s)}{1 + \cfrac{\alpha_3(s)}{z + \cfrac{\alpha_4(s)}{1 + \ddots}}}}, \tag{9}$$

where the coefficients $\alpha_k(s)$ are such that $\alpha_1(s) = 1$, $\alpha_2(s) = (s+\lambda_0)/|r_0|$, and for $k \geq 1$,

$$\alpha_{2k}(s)\alpha_{2k+1}(s) = \frac{\lambda_{k-1}\mu_k}{|r_{k-1}r_k|}, \quad \alpha_{2k+1}(s) + \alpha_{2(k+1)}(s) = \frac{s+\lambda_k+\mu_k}{|r_k|}. \tag{10}$$

We have the following property, which is proved in Appendix A.

Lemma 1. *The continued fraction $\mathcal{F}(s;z)$ defined by Equation (9) is a converging Stieltjes fraction for all $s \geq 0$.*

As a consequence of the above lemma, there exists a unique bounded, increasing function $\psi(s;x)$ in variable x such that

$$\mathcal{F}(s;z) = \int_0^\infty \frac{1}{z+x}\psi(s;dx).$$

The polynomials $Q_n(s;x)$ are orthogonal with respect to the measure $\psi(s;dx)$ and satisfy the orthogonality relation

$$\int_0^\infty Q_i(s;x)Q_j(s;x)\psi(s;dx) = \frac{|r_0|}{|r_i|\pi_i}\delta_{i,j} \tag{11}$$

As a consequence, it is worth noting that the polynomial $Q_i(s;x)$ has i real, simple and positive roots.

It is possible to associate with the polynomials $Q_i(s,x)$ a new class of orthogonal polynomials, referred to as associated polynomials and denoted by $Q_i(i_0 + 1; s; x)$ and satisfying the recurrence relations: $Q_0(i_0 + 1; s; x) = 1$, $Q_1(i_0 + 1; s; x) = (s + \lambda_{i_0+1+i} + \mu_{i_0+1+i} - r_{i_0+1+i}x)/\lambda_{i_0+1+i}$ and, for $i \geq 0$,

$$\frac{\lambda_{i_0+1+i}}{r_{i_0+1+i}} Q_{i+1}(i_0 + 1; s; x) + \left(x - \frac{s + \lambda_{i_0+1+i} + \mu_{i_0+1+i}}{r_{i_0+1+i}}\right) Q_i(i_0 + 1; s; x)$$

$$+ \frac{\mu_{i_0+1+i}}{r_{i_0+1+i}} Q_{i-1}(i_0 + 1; s; x) = 0. \quad (12)$$

The polynomials $Q_i(i_0 + 1; s; z)$ are related to the denominator of the continued fraction

$$F_{i_0}^e(z) = \cfrac{1}{z + \cfrac{s+\lambda_{i_0+1}+\mu_{i_0+1}}{r_{i_0+1}} - \cfrac{\frac{\lambda_{i_0+1}\mu_{i_0+2}}{r_{i_0+1}r_{i_0+2}}}{z + \cfrac{s+\lambda_{i_0+2}+\mu_{i_0+2}}{r_{i_0+2}} - \cfrac{\frac{\lambda_{i_0+2}\mu_{i_0+3}}{r_{i_0+2}r_{i_0+3}}}{z + \cfrac{s+\lambda_{i_0+3}+\mu_{i_0+3}}{|r_{i_0+3}|} - \cdots}}}$$

which is the even part of the continued fraction $\mathcal{F}_{i_0}(z)$ defined by

$$\mathcal{F}_{i_0}(s; z) = \cfrac{\beta_1(s)}{z + \cfrac{\beta_2(s)}{1 + \cfrac{\beta_3(s)}{z + \cfrac{\beta_4(s)}{1 + \cdots}}}}, \quad (13)$$

where the coefficients $\beta_k(s)$ are such that

$$\beta_1(s) = 1, \quad \beta_2(s) = (s + \lambda_{i_0+1} + \mu_{i_0+1})/|r_{i_0+1}|,$$

and for $k \geq 1$,

$$\beta_{2k}(s)\beta_{2k+1}(s) = \frac{\lambda_{i_0+k}\mu_{i_0+k+1}}{r_{i_0+k}r_{i_0+1+k}},$$

$$\beta_{2k+1}(s) + \beta_{2(k+1)}(s) = \frac{s + \lambda_{i_0+1+k} + \mu_{i_0+1+k}}{r_{i_0+1+k}}. \quad (14)$$

Since the continued fraction $\mathcal{F}(s; z)$ is a converging Stieltjes fraction, it is quite clear that the continued fraction $\mathcal{F}_{i_0}(s; z)$ defined by Equation (13) is a converging Stieltjes fraction for all $s \geq 0$. There exists hence a unique bounded, increasing function $\psi^{[i_0]}(s; x)$ in variable x such that

$$\mathcal{F}_{i_0}(s; z) = \int_0^\infty \frac{1}{z + x} \psi^{[i_0]}(s; dx).$$

The polynomials $Q_i(i_0 + 1; s; x)$ are orthogonal with respect to the measure $\psi^{[i_0]}(s; dx)$ and satisfy the orthogonality relation

$$\int_0^\infty Q_i(i_0 + 1; s; x)Q_j(i_0 + 1; s; x)\psi^{[i_0]}(s; dx) = \frac{r_{i_0+1}\pi_{i_0+1}}{r_{i_0+1+i}\pi_{i_0+1+i}}\delta_{i,j}.$$

3.2 Resolution of the matrix equation

We show in this section how to solve the matrix Equation (6). In a first step, we solve the $i_0 + 1$ first linear equations.

Lemma 2. *The functions $F_i(s, \xi)$, for $i \leq i_0$, are related to function $F_{i_0+1}(s, \xi)$ as follows: for $\xi \neq \zeta_k(s), k = 0, \ldots, i_0,$*

$$F_i(s, \xi) = \frac{\pi_i}{r_0} \sum_{j=0}^{i_0} (f_j^{(0)}(\xi) + r_j \xi h_j(s)) \int_0^\infty \frac{Q_j(s; x) Q_i(s; x)}{\xi - x} \psi_{[i_0]}(s; dx)$$

$$+ \mu_{i_0+1} \frac{\pi_i}{r_0} F_{i_0+1}(s, \xi) \int_0^\infty \frac{Q_{i_0}(s; x) Q_i(s; x)}{\xi - x} \psi_{[i_0]}(s; dx), \quad (15)$$

where the $\zeta_k(s)$ are the roots of the polynomial $Q_{i_0+1}(s; x)$ defined by Equation (8) and the measure $\psi_{[i_0]}(s; dx)$ is defined by Equation (45) in Appendix A.

Proof. Let $\mathbb{I}_{[i_0]}$, $A_{[i_0]}$ and $R_{[i_0]}$ denote the matrices obtained from the infinite identity matrix, the infinite matrix A defined by Equation (1) and the infinite diagonal matrix R by deleting the rows and the columns with an index greater than i_0, respectively. Denoting by $F_{[i_0]}$, $h_{[i_0]}$ and $f_{[i_0]}$ the finite column vectors which ith components are F_i / π_i, h_i / π_i and $f_i^{(0)} / \pi_i$, respectively for $i = 0, \ldots, i_0$, Equation (6) can be written as

$$(s\mathbb{I}_{[i_0]} + \xi R_{[i_0]} - A_{[i_0]}) F_{[i_0]} = f_{[i_0]} + \xi R_{[i_0]} h_{[i_0]} + \frac{\lambda_{i_0}}{\pi_{i_0+1}} F_{i_0+1} e_{i_0},$$

where e_{i_0} is the column vector with all entries equal to 0 except the i_0th one equal to 1.

Since $r(i) < c$ for all $i \leq i_0$, the matrix $R_{[i_0]}$ is invertible and the above equation can be rewritten as

$$\left(\xi \mathbb{I}_{[i_0]} + R_{[i_0]}^{-1} (s\mathbb{I}_{[i_0]} - A_{[i_0]}) \right) F_{[i_0]} = R_{[i_0]}^{-1} f_{[i_0]} + \xi h_{[i_0]} + \frac{\lambda_{i_0}}{r_{i_0} \pi_{i_0+1}} F_{i_0+1} e_{i_0}.$$

From Lemma 6 proved in Appendix B, we know that the operator associated with the finite matrix $(\xi \mathbb{I}_{[i_0]} + R_{[i_0]}^{-1} (s\mathbb{I}_{[i_0]} - A_{[i_0]}))$ is selfadjoint in the Hilbert space $H_{i_0} = \mathbb{C}^{i_0+1}$ equipped with the scalar product

$$(c, d)_{i_0} = \sum_{k=0}^{i_0} c_k \overline{d_k} |r_k| \pi_k.$$

The eigenvalues of the operator $(\xi \mathbb{I}_{[i_0]} + R_{[i_0]}^{-1} (s\mathbb{I}_{[i_0]} - A_{[i_0]}))$ are the quantities $\xi - \zeta_k(s)$ for $k = 0, \ldots, i_0$, where the $\zeta_k(s)$ are the roots of the polynomial $Q_{i_0+1}(s; x)$ defined by Equation (8). Hence, for $\xi \notin \{\zeta_0(s), \ldots, \zeta_{i_0}(s)\}$, we have

$$F_{[i_0]} = \left(\xi \mathbb{I}_{[i_0]} + R_{[i_0]}^{-1} (s\mathbb{I}_{[i_0]} - A_{[i_0]}) \right)^{-1} R_{[i_0]}^{-1} f_{[i_0]} + \xi \left(\xi \mathbb{I}_{[i_0]} + R_{[i_0]}^{-1} (s\mathbb{I}_{[i_0]} - A_{[i_0]}) \right)^{-1} h_{[i_0]}$$

$$+ \frac{\lambda_{i_0}}{r_{i_0} \pi_{i_0+1}} F_{i_0+1} \left(\xi \mathbb{I}_{[i_0]} + R_{[i_0]}^{-1} (s\mathbb{I}_{[i_0]} - A_{[i_0]}) \right)^{-1} e_{i_0}.$$

By introducing the vectors $Q_{[i_0]}(s, \zeta_k(s))$ for $k = 0, \ldots, i_0$ defined in Appendix B, the column vector e_i with all entries equal to 0 except the ith one equal to 1 can be written as

$$e_j = \frac{|r_j|\pi_j}{|r_0|} \int_0^\infty Q_j(s, x) Q_{[i_0]}(s, x) \psi_{[i_0]}(s; dx)$$

where the measure $\psi_{[i_0]}(s; dx)$ is defined by Equation (45). Since the vectors $Q_{[i_0]}(s, \zeta_k(s))$ are such that

$$\left(\zeta \mathbb{I}_{[i_0]} + R_{[i_0]}^{-1}(s\mathbb{I}_{[i_0]} - A_{[i_0]})\right)^{-1} Q_{[i_0]}(s, \zeta_k(s)) = \frac{1}{\zeta - \zeta_k(s)} Q_{[i_0]}(s, \zeta_k(s)),$$

we deduce that

$$\left(\zeta \mathbb{I}_{[i_0]} + R_{[i_0]}^{-1}(s\mathbb{I}_{[i_0]} - A_{[i_0]})\right)^{-1} e_j = \frac{|r_j|\pi_j}{|r_0|} \int_0^\infty \frac{Q_j(s, x)}{\zeta - x} Q_{[i_0]}(s, x) \psi_{[i_0]}(s; dx)$$

Hence, if $f = \sum_{j=0}^{i_0} f_j e_j$, then

$$\left(\zeta \mathbb{I}_{[i_0]} + R_{[i_0]}^{-1}(s\mathbb{I}_{[i_0]} - A_{[i_0]})\right)^{-1} f = \sum_{j=0}^{i_0} f_j \frac{|r_j|\pi_j}{|r_0|} \int_0^\infty \frac{Q_j(s, x)}{\zeta - x} Q_{[i_0]}(s, x) \psi_{[i_0]}(s; dx)$$

and the ith component of the above vector is

$$\left(\left(\zeta \mathbb{I}_{[i_0]} + R_{[i_0]}^{-1}(s\mathbb{I}_{[i_0]} - A_{[i_0]})\right)^{-1} f\right)_i = \sum_{j=0}^{i_0} f_j \frac{|r_j|\pi_j}{|r_0|} \int_0^\infty \frac{Q_j(s, x) Q_i(s, x)}{\zeta - x} \psi_{[i_0]}(s; dx)$$

Applying the above identity to the vectors $R_{[i_0]}^{-1} f_{[i_0]}$, $h_{[i_0]}$ and e_{i_0}, Equation (15) follows. □

We now turn to the analysis of the second part of Equation (6).

Lemma 3. *For $s \geq 0$, the functions $F_i(s, \zeta)$ are related to function $F_{i_0}(s, \zeta)$ by the relation: for $i \geq 0$,*

$$F_{i_0+i+1}(s, \zeta) = \lambda_{i_0} \frac{\pi_{i_0+i+1}}{r_{i_0+1}\pi_{i_0+1}} F_{i_0}(s, \zeta) \int_0^\infty \frac{Q_i(i_0 + 1; s; x)}{\zeta + x} \psi^{[i_0]}(s; dx)$$

$$+ \frac{\pi_{i_0+i+1}}{r_{i_0+1}\pi_{i_0+1}} \sum_{j=0}^\infty f_{i_0+j+1}^{(0)}(\zeta) \int_0^\infty \frac{Q_j(i_0 + 1; s; x) Q_i(i_0 + 1; s; x)}{x + \zeta} \psi^{[i_0]}(s; dx), \quad (16)$$

where the measure $\psi^{[i_0]}(s; dx)$ is the orthogonality measure of the associated polynomials $Q_i(i_0 + 1; s; x)$, $i \geq 0$.

Proof. Let $\mathbb{I}^{[i_0]}$, $A^{[i_0]}$ and $R^{[i_0]}$ denote the matrices obtained from $\mathbb{I}$, A and R by deleting the first $(i_0 + 1)$ lines and columns, respectively. The infinite matrix $(R^{[i_0]})^{-1}(s\mathbb{I}^{[i_0]} - A^{[i_0]})$ induces in the Hilbert space H^{i_0} defined by

$$H^{i_0} = \left\{ (f_n) \in \mathbf{C}^{\mathbb{N}} : \sum_{n=0}^{\infty} |f_n|^2 r_{i_0+n+1} \pi_{i_0+n+1} < \infty \right\}$$

and equipped with the scalar product

$$(f,g) = \sum_{n=0}^{\infty} f_n \overline{g}_n r_{i_0+n+1} \pi_{i_0+n+1},$$

where $\overline{g}_n$ is the conjugate of the complex number g_n, an operator such that for $f \in H^{i_0}$

$$((R^{[i_0]})^{-1}(s\mathbb{I}^{[i_0]} - A^{[i_0]})f)_n =$$
$$-\frac{\mu_{i_0+1+n}}{r_{i_0+n+1}} f_{n-1} + \frac{s + \lambda_{i_0+n+1} + \mu_{i_0+1+n}}{r_{i_0+n+1}} f_n - \frac{\lambda_{i_0+n+1}}{r_{i_0+n+1}} f_{n+1}.$$

The above operator is symmetric in H^{i_0}. To show that this operator is selfadjoint, we have to prove that the domains of this operator and its adjoint coincide. In Guillemin (2012), it is shown that given the special form of the operator under consideration, this condition is equivalent to the convergence of the Stieltjes fraction defined by Equation (13) and if this is the case, the spectral measure is the orthogonality measure $\psi^{[i_0]}(s; dx)$. Since the continued fraction $\mathcal{F}_{i_0}(s; z)$ is a converging Stieltjes fraction, the above operator is hence selfadjoint.

Let $Q^{[i_0]}(s; x)$ the column vector which ith entry is $Q_i(i_0 + 1; s; x)$. This vector is in H^{i_0} if and only if $\|Q^{[i_0]}(s; x)\|^2 \overset{def}{=} (Q^{[i_0]}(s; x), Q^{[i_0]}(s; x)) < \infty$. If it is the case, then the measure $\psi^{[i_0]}(s; dx)$ has an atom at point x with mass $1/\|Q^{[i_0]}(s; x)\|^2$. Otherwise, the vector $Q^{[i_0]}(s; x)$ is not in H^{i_0} but from the spectral theorem we have

$$H^{i_0} = \int^{\oplus} H_x^{i_0} \psi^{[i_0]}(s; dx)$$

where $H_x^{i_0}$ is the vector space spanned by the vector $Q^{[i_0]}(s; x)$ for x in the support of the measure $\psi^{[i_0]}(s; dx)$. In addition, we have the resolvent identity: For $f, g \in H^{i_0}$ and $\xi \in \mathbf{C}$ such that $-\xi$ is not in the support of the measure $\psi^{[i_0]}(s; dx)$,

$$\left(\left(\xi\mathbb{I}^{[i_0]} + (R^{[i_0]})^{-1}(s\mathbb{I}^{[i_0]} - A^{[i_0]}) \right)^{-1} f, g \right) = \int_0^{\infty} \frac{(f_x, g)}{\xi + x} \psi^{[i_0]}(s; dx). \tag{17}$$

where f_x is the projection on $H_x^{i_0}$ of the vector f.

For $i \geq 0$, let e_i denote the column vector, which ith entry is equal to 1 and the other entries are equal to 0. Denoting by $F^{[i_0]}$ and $\hat{f}^{[i_0]}$ the column vectors which ith components are $F_{i_0+1+i}/\pi_{i_0+1+i}$ and $f_{i_0+1+i}^{(0)}/\pi_{i_0+1+i}$, respectively, Equation (6) can be written as

$$(s\mathbb{I}^{[i_0]} + \xi R^{[i_0]} - A^{[i_0]})F^{[i_0]} = f^{[i_0]} + \frac{\mu_{i_0+1}}{\pi_{i_0}} F_{i_0} e_0,$$

since $h_i(s) \equiv 0$ for $i > i_0$.

Given that $r_i > 0$ for $i > i_0$, the matrix $R^{[i_0]}$ is invertible and the above equation can be rewritten as

$$\left(\xi \mathbb{I}^{[i_0]} + (R^{[i_0]})^{-1} (s\mathbb{I}^{[i_0]} - A^{[i_0]}) \right) F^{[i_0]} = (R^{[i_0]})^{-1} f^{[i_0]} + \frac{\mu_{i_0+1}}{\pi_{i_0}} F_{i_0} \hat{R}^{-1} e_0,$$

The operator $\left(\xi \mathbb{I}^{[i_0]} + (R^{[i_0]})^{-1} (s\mathbb{I}^{[i_0]} - A^{[i_0]}) \right)$ is invertible for ξ such that $-\xi$ is not in the support of the measure $\psi^{[i_0]}(s, dx)$, and we have

$$F^{[i_0]} = \left(\xi \mathbb{I}^{[i_0]} + (R^{[i_0]})^{-1} (s\mathbb{I}^{[i_0]} - A^{[i_0]}) \right)^{-1} (R^{[i_0]})^{-1} f^{[i_0]}$$

$$+ \frac{\mu_{i_0+1}}{r_{i_0+1} \pi_{i_0}} F_{i_0} \left(\xi \mathbb{I}^{[i_0]} + (R^{[i_0]})^{-1} (s\mathbb{I}^{[i_0]} - A^{[i_0]}) \right)^{-1} e_0.$$

By using the spectral identity (17), we can compute F_i for $i > i_0$ as soon as F_{i_0} is known. Indeed, we have

$$F^{[i_0]} = \sum_{j=0}^{\infty} \frac{F_{i_0+1+j}}{\pi_{i_0+1+j}} e_j,$$

and then, for $i \geq i_0 + 1$, by using the fact that $r_{i_0+1+i} F_{i_0+1+i} = (F^{[i_0]}, e_i)$, we have

$$r_{i_0+1+i} F_{i_0+1+i} = \left(\left(\xi \mathbb{I}^{[i_0]} + (R^{[i_0]})^{-1} (s\mathbb{I}^{[i_0]} - A^{[i_0]}) \right)^{-1} (R^{[i_0]})^{-1} f^{[i_0]}, e_i \right)$$

$$+ \frac{\mu_{i_0+1}}{r_{i_0+1} \pi_{i_0}} F_{i_0} \left(\left(\xi \mathbb{I}^{[i_0]} + (R^{[i_0]})^{-1} (s\mathbb{I}^{[i_0]} - A^{[i_0]}) \right)^{-1} e_0, e_i \right).$$

By using the fact that for $j \geq 0$,

$$(e_j)_x = \frac{r_{i_0+j+1} \pi_{i_0+j+1}}{r_{i_0+1} \pi_{i_0+1}} Q_j(i_0 + 1; s; x) Q^{[i_0]}(s; x),$$

Equation (16) follows by using the resolvent identity (17). $\square$

From the two above lemmas, it turns out that to determine the functions $F_i(s, \xi)$ it is necessary to compute the function $h_i(s)$ for $i = 0, \ldots, i_0 + 1$. For this purpose, let us introduce the non negative quantities $\eta_\ell(s)$, $\ell = 0, \ldots, i_0$, which are the $(i_0 + 1)$ solution to the equation

$$1 - \frac{\lambda_{i_0} \mu_{i_0+1} \pi_{i_0}}{r_{i_0+1} r_0} F_{i_0}(s; \xi) \int_0^{\infty} \frac{Q_{i_0}(s; x)^2}{\xi - x} \psi_{[i_0]}(s; dx) = 0. \tag{18}$$

Then, we can state the following result, which gives a means of computing the unknown functions $h_j(s)$ for $j = 0, \ldots, i_0$.

Proposition 3. *The functions $h_j(s)$, $j = 0, \ldots, i_0$, satisfy the linear equations: for $\ell = 0, \ldots, i_0$,*

$$\frac{\lambda_{i_0} \mathcal{F}_{i_0}(s; \eta_\ell(s)) \eta_\ell(s)}{r_{i_0}} \left(\left(\eta_k(s) \mathbb{I}_{[i_0]} + R_{[i_0]}^{-1}(s \mathbb{I}_{[i_0]} - A_{[i_0]}) \right)^{-1} e_{i_0}, h(s) \right)_{i_0}$$

$$= \left(\left(\eta_k(s) \mathbb{I}^{[i_0]} + (R^{[i_0]})^{-1}(s \mathbb{I}^{[i_0]} - A^{[i_0]}) \right)^{-1} e_0, (R^{[i_0]})^{-1} f^{[i_0]}(\eta_k(s)) \right)$$

$$- \frac{\lambda_{i_0} \mathcal{F}_{i_0}(s; \eta_\ell(s))}{r_{i_0}} \left(\left(\eta_k(s) \mathbb{I}_{[i_0]} + R_{[i_0]}^{-1}(s \mathbb{I}_{[i_0]} - A_{[i_0]}) \right)^{-1} e_{i_0}, R_{[i_0]}^{-1} f_{[i_0]}(\eta_k(s)) \right)_{i_0}, \quad (19)$$

where $\mathcal{F}_{i_0}(s; z)$ is the continued fraction (13) and $f^{[i_0]}((\xi)$ and $f_{[i_0]}(\xi)$ are the vectors, which ith components are equal to $f_{i_0+i+1}^{(0)}(\xi)/\pi_{i_0+i+1}$ and $f_i^{(0)}(\xi)/\pi_i$, respectively.

Proof. From Equation (16) for $i = i_0 + 1$ and Equation (15) for $i = i_0$, we deduce that

$$\left(1 - \frac{\lambda_{i_0} \mu_{i_0+1} \pi_{i_0}}{r_0 r_{i_0+1}} \mathcal{F}_{i_0}(s; \xi) \int_0^\infty \frac{Q_{i_0}(s; x)^2}{\xi - x} \psi_{[i_0]}(s; dx) \right) F_{i_0+1}(s, \xi) =$$

$$\frac{\lambda_{i_0} \pi_{i_0}}{r_0 r_{i_0+1}} \mathcal{F}_{i_0}(s; \xi) \sum_{j=0}^{i_0} (f_j^{(0)}(\xi) + r_j \xi h_j(s)) \int_0^\infty \frac{Q_j(s; x) Q_{i_0}(s; x)}{\xi - x} \psi_{[i_0]}(s; dx)$$

$$+ \frac{1}{r_{i_0+1}} \sum_{j=0}^\infty f_{i_0+j+1}^{(0)}(\xi) \int_0^\infty \frac{Q_j(i_0+1; s; x)}{x + \xi} \psi^{[i_0]}(s; dx). \quad (20)$$

From equation (15), since the Laplace transform $F_i(s, \xi)$ should have no poles for $\xi \geq 0$, the roots $\zeta_k(s)$ for $k = 0, \ldots, i_0$ should be removable singularities and hence for all $i, j, k = 0, \ldots, i_0$

$$Q_i(s; \zeta_k(s)) \left(\left(f_j^{(0)}(\zeta_k(s)) + r_j \zeta_k(s) h_j(\zeta_k(s)) \right) Q_j(s; \zeta_k(s)) \right.$$

$$\left. + \mu_{i_0+1} F_{i_0+1}(s, \zeta_k(s)) Q_{i_0}(s, \zeta_k(s)) \right) = 0.$$

By using the interleaving property of the roots of successive orthogonal polynomials, we have $Q_i(s; \zeta_k(s)) \neq 0$ for all $i, k = 0, \ldots, i_0$. Hence, the term between parentheses in the above equation is null and we deduce that the points $\zeta_k(s)$, $k = 0, \ldots, i_0$, are removable singularities in expression (20). The quantities $h_j(s)$, $j = 0, \ldots, i_0$, are then determined by using the fact that the r.h.s. of equation (20) must cancel at points $\eta_k(s)$ for $k = 0, \ldots, i_0$. This entails that for $k = 0, \ldots, i_0$, the terms

$$\sum_{j=0}^\infty f_{i_0+j+1}^{(0)}(\eta_k(s)) \int_0^\infty \frac{Q_j(i_0+1; s; x)}{x + \eta_k(s)} \psi^{[i_0]}(s; dx)$$

$$+ \frac{\lambda_{i_0} \pi_{i_0} \mathcal{F}_{i_0}(s; \eta_k(s))}{r_0} \sum_{j=0}^{i_0} v_j(s) \int_0^\infty \frac{Q_j(s; x) Q_{i_0}(s; x)}{\eta_k(s) - x} \psi_{[i_0]}(s; dx) \quad (21)$$

must cancel, where

$$v_j(s) = f_j^{(0)}(\eta_k(s)) + \eta_k(s) r_j h_j(s).$$

By using the fact that

$$\int_0^\infty \frac{Q_j(s;x)Q_{i_0}(s;x)}{\eta_k(s)-x}\psi_{[i_0]}(s;dx) =$$

$$\frac{|r_0|}{|r_{i_0}|\pi_{i_0}|r_j|\pi_j}\left(\left(\eta_k(s)\mathbb{I}_{[i_0]}+R_{[i_0]}^{-1}(s\mathbb{I}_{[i_0]}-A_{[i_0]})\right)^{-1}e_{i_0},e_j\right)_{i_0}$$

and

$$\int_0^\infty \frac{Q_j(i_0+1;s;x)}{x+\eta_k(s)}\psi^{[i_0]}(s;dx) =$$

$$\frac{1}{r_{i_0+1+j}\pi_{i_0+j+1}}\left(\left(\eta_k(s)\mathbb{I}^{[i_0]}+(R^{[i_0]})^{-1}(s\mathbb{I}^{[i_0]}-A^{[i_0]})\right)^{-1}e_0,e_j\right),$$

Equation (19) follows.   □

By solving the system of linear equations (19), we can compute the unknown functions $h_j(s)$ for $j=0,\ldots,i_0$. The function $F_{i_0+1}(s,\xi)$ is then given by

$$\left(1-\frac{\lambda_{i_0}\mu_{i_0+1}\pi_{i_0}}{r_{i_0+1}r_0}\mathcal{F}_{i_0}(s;\xi)\int_0^\infty \frac{Q_{i_0}(s;x)^2}{\xi-x}\psi_{[i_0]}(s;dx)\right)F_{i_0+1}(s,\xi) =$$

$$=\frac{1}{r_{i_0+1}}\left(\left(\xi\mathbb{I}^{[i_0]}+(R^{[i_0]})^{-1}(s\mathbb{I}^{[i_0]}-A^{[i_0]})\right)^{-1}e_0,(R^{[i_0]})^{-1}f^{[i_0]}(\xi)\right)$$

$$-\frac{\lambda_{i_0}\mathcal{F}_{i_0}(s;\xi)}{r_{i_0}r_{i_0+1}}\left(\left(\xi\mathbb{I}_{[i_0]}+R_{[i_0]}^{-1}(s\mathbb{I}_{[i_0]}-A_{[i_0]})\right)^{-1}e_{i_0},R_{[i_0]}^{-1}f_{[i_0]}(\xi)+\xi h(s)\right)_{i_0},\quad(22)$$

The function $F_{i_0}(s,\xi)$ is computed by using equation (22) and equation (15) for $i=i_0$. The other functions $F_i(s,\xi)$ are computed by using Lemmas 2 and 3.

The above procedure can be applied for any value i_0 but expressions are much simpler when $i_0=0$, i.e., when there is only one state with negative net input rate. In that case, we have the following result, when the buffer is initially empty and the birth and death process is in state 1.

Proposition 4. *Assume that $r_0<0$ and $r_i>0$ for $i>0$. When the buffer is initially empty and the birth and death process is in the state 1 at time 0 (i.e., $p_0(i)=\delta_{1,i}$ for all $i\geq0$), the Laplace transform $h_0(s)$ is given by*

$$h_0(s)=\frac{r_0\eta_0(s)+s+\lambda_0}{\lambda_0\eta_0(s)|r_0|}=\frac{\mu_1\mathcal{F}_0(s;\eta_0(s))}{r_1|r_0|\eta_0(s)}.\qquad(23)$$

where $\eta_0(s)$ is the unique positive solution to the equation

$$1-\frac{\lambda_0\mu_1\mathcal{F}_0(s;\xi)}{r_1(s+\lambda_0+r_0\xi)}=0.$$

In addition,

$$F_1(s, \xi) = \frac{\dfrac{1}{r_1}\left(1 + \dfrac{\lambda_0 \xi r_0 h_0(s)}{s + \lambda_0 + r_0 \xi}\right) \mathcal{F}_0(s; \xi)}{1 - \dfrac{\lambda_0 \mu_1}{r_1(s + \lambda_0 + r_0 \xi)}\mathcal{F}_0(s; \xi)}. \tag{24}$$

Proof. In the case $i_0 = 0$, the unique root to the equation $Q_1(s; x)$ is $\zeta_0(s) = (s + \lambda_0)|r_0|$. The measure $\psi_{[0]}(s; dx)$ is given by

$$\psi_{[0]}(s; dx) = \delta_{\zeta_0(s)}(dx)$$

and Equation (18) reads

$$1 - \frac{\lambda_0 \mu_1}{r_1}\mathcal{F}_0(s; \xi)\frac{1}{s + \lambda_0 + r_0 \xi} = 0$$

which has a unique solution $\eta_0(s) > 0$. When the buffer is initially empty and the birth and death process is in the state 1 at time 0, we have $f_i^{(0)}(\xi) = \delta_{1,j}$. Then,

$$\left(\left(\eta_0(s)\mathbb{I}^{[0]} + (R^{[0]})^{-1}(s\mathbb{I}^{[0]} - A^{[0]})\right)^{-1}e_0, (R^{[0]})^{-1}f^{[0]}(\eta_0(s))\right)$$

$$= \frac{1}{r_1 \pi_1}\left(\left(\eta_0(s)\mathbb{I}^{[0]} + (R^{[0]})^{-1}(s\mathbb{I}^{[0]} - A^{[0]})\right)^{-1}e_0, e_0\right) = \int_0^\infty \frac{1}{\eta_0(s) + x}\psi^{[0]}(s; dx)$$

$$= \mathcal{F}_0(s; \eta_0(s)),$$

where we have used the resolvent identity (17) and the fact that $(e_0)_x = Q^{[0]}(s; x)$. Moreover,

$$\left(\left(\eta_0(s)\mathbb{I}_{[0]} + R_{[0]}^{-1}(s\mathbb{I}_{[0]} - A_{[0]})\right)^{-1}e_0, R_{[0]}^{-1}f_{[0]}(\eta_0(s)) + h(s)\right)_0$$

$$= \frac{h_0(s)}{\eta_0(s) + \frac{s + \lambda_0}{r_0}}(e_0, e_0)_0 = \frac{h_0(s)|r_0|}{\eta_0(s) + \frac{s + \lambda_0}{r_0}}.$$

By using Equation (19) for $i_0 = 0$, Equation (23) follows. Finally, Equation (24) is obtained by using Equation (22). $\qquad\square$

4. Analysis of the stationary regime

In this section, we analyze the stationary regime. In this case, we have to take $s = 0$ and $f^{(0)} \equiv 0$. To alleviate the notation, we set $\psi_{[i_0]}(0; dx) = \psi_{[i_0]}(dx)$, $\psi^{[i_0]}(0; dx) = \psi^{[i_0]}(dx)$ and $Q_j(0; x) = Q_j(x)$ and $Q_j(i_0 + 1; 0; x) = Q_j(i_0 + 1; x)$. Equation (20) then reads

$$\left(1 - \frac{\lambda_{i_0}\mu_{i_0+1}\pi_{i_0}}{r_{i_0+1}r_0}\mathcal{F}_{i_0}(\xi)\int_0^\infty \frac{Q_{i_0}(x)^2}{\xi - x}\psi_{[i_0]}(dx)\right)\mathcal{F}_{i_0+1}(\xi)$$

$$= \frac{\lambda_{i_0}\pi_{i_0}\xi \mathcal{F}_{i_0}(\xi)}{r_0 r_{i_0+1}}\sum_{j=0}^{i_0} r_j h_j \int_0^\infty \frac{Q_j(x)Q_{i_0}(x)}{\xi - x}\psi_{[i_0]}(dx), \tag{25}$$

where $h_j = \lim_{t \to \infty} \mathbb{P}(\Lambda_t = j, X_t = 0)$, $\mathcal{F}_{i_0}(\xi) = \mathcal{F}_{i_0}(0; \xi)$ and $\mathcal{F}_{i_0+1}(\xi) = \mathcal{F}_{i_0+1}(0; \xi)$.

The continued fraction $\mathcal{F}_{i_0}(\xi)$ has the following probabilistic interpretation:

$$\mu_{i_0+1}\mathcal{F}_{i_0}(\xi)/r_{i_0+1} = \mathbb{E}\left(e^{-\xi\theta_{i_0}}\right)$$

where θ_{i_0} is the passage time of the birth and death process with birth rates $\lambda_n/|r_n|$ and death rates $\mu_n/|r_n|$ from state $i_0 + 1$ to state i_0 (see Guillemin & Pinchon (1999) for details). This entails in particular that $\mathcal{F}_{i_0}(0) = r_{i_0+1}/\mu_{i_0+1}$.

Let us first characterize the measure $\psi_{[i_0]}(dx)$. For this purpose, let us introduce the polynomials of the second kind associated with the polynomials $Q_i(x)$. The polynomials of the second kind $P_i(x)$ satisfy the same recursion as the polynomials $Q_i(x)$ but wit the initial conditions $P_0(x) = 0$ and $P_1(x) = |r_0|/\lambda_0$. The even numerators of the continued fraction $F(z) \overset{def}{=} F(0; z)$, where $F(s; z)$ is defined by Equation (9), are equal to $\frac{\lambda_0...\lambda_{n-1}}{|r_0...r_{n-1}|} P_n(-z)$ and the even denominators to $\frac{\lambda_0...\lambda_{n-1}}{|r_0...r_{n-1}|} Q_n(-z)$.

Lemma 4. The spectral measure $\psi_{[i_0]}(dx)$ of the non negative selfadjoint operator $R_{[i_0]}^{-1}A_{[i_0]}$ in the Hilbert space H_{i_0} is such that

$$\int_0^\infty \frac{1}{z - x}\psi_{[i_0]}(dx) = -\frac{P_{i_0+1}(z)}{Q_{i_0+1}(z)}. \tag{26}$$

The measure $\psi_{[i_0]}(dx)$ is purely discrete with atoms located at the zeros ζ_k, $k = 0, \ldots, i_0$, of the polynomial $Q_{i_0+1}(z)$.

Proof. Let $P_{[i_0]}(z)$ (resp. $Q_{[i_0]}(z)$) denote the column vector, which ith component for $0 \leq i \leq i_0$ is $P_i(z)$ (resp. $Q_i(z)$). For any $x, z \in \mathbb{C}$, we have

$$\left(z\mathbb{I}_{[i_0]} - R_{[i_0]}^{-1}A_{[i_0]}\right)\left(P_{[i_0]}(z) + xQ_{[i_0]}(z)\right) = e_0 - \frac{\lambda_{i_0}}{|r_{i_0+1}|}\left(P_{i_0+1}(z) + xQ_{i_0+1}(z)\right)e_{i_0}.$$

Hence, if $z \neq \zeta_i$ for $0 \leq i \leq i_0$, where ζ_i is the ith zero of the polynomial $Q_{i_0+1}(x)$, and if we take $x = -P_{i_0+1}(z)/Q_{i_0+1}(z)$, we see that

$$\left(z\mathbb{I}_{[i_0]} - R_{[i_0]}^{-1}A_{[i_0]}\right)^{-1}e_0 = P_{[i_0]}(z) - \frac{P_{i_0+1}(z)}{Q_{i_0+1}(z)}Q_{[i_0]}(z).$$

From the spectral identity for the operator $R_{[i_0]}^{-1}A_{[i_0]}$ (similar to Equation (17)), we have

$$\left(\left(z\mathbb{I}_{[i_0]} - R_{[i_0]}^{-1}A_{[i_0]}\right)^{-1}e_0, e_0\right)_{i_0} = \int_0^\infty \frac{((e_0)x, e_0)_{i_0}}{z - x}\psi_{[i_0]}(dx) = -\frac{P_{i_0+1}(z)}{Q_{i_0+1}(z)}|r_0|.$$

Since $(e_0)x = Q_{[i_0]}(x)$ because of the orthogonality relation (11), Equation (26) immediately follows. $\square$

By using the above lemma, we can show that the smallest solution to the equation

$$1 - \frac{\lambda_{i_0}\mu_{i_0+1}\pi_{i_0}}{r_{i_0+1}r_0}\mathcal{F}_{i_0}(\xi)\int_0^\infty \frac{Q_{i_0}(x)^2}{\xi - x}\psi_{[i_0]}(dx) = 0 \tag{27}$$

is $\eta_0 = 0$. The above equation is the stationary version of Equation (18).

Lemma 5. *The solutions η_j, $j = 0, \ldots, i_0$, to Equation (27) are such that $\eta_0 = 0 < \eta_1 < \cdots < \eta_{i_0}$. For $\ell = 1, \ldots, i_0$, η_ℓ is solution to equation*

$$1 = \frac{\mu_{i_0+1}}{r_{i_0+1}}\mathcal{F}_{i_0}(\xi)\frac{Q_{i_0}(\xi)}{Q_{i_0+1}(\xi)}. \tag{28}$$

Proof. The fraction $P_{i_0+1}(z)/Q_{i_0+1}(z)$ is a terminating fraction and from Equation (26), we have

$$\frac{P_{i_0+1}(-z)}{Q_{i_0+1}(-z)} = \int_0^\infty \frac{1}{z+x}\psi_{[i_0]}(dx).$$

On the one hand, by applying Theorem 12.11d of Henrici (1977) to this fraction, we have

$$\frac{P_{i_0+1}(-z)}{Q_{i_0+1}(-z)} - \frac{P_{i_0}(-z)}{Q_{i_0}(-z)} = \int_0^\infty \frac{Q_{i_0}(x)^2}{Q_{i_0}(-z)^2}\frac{\psi_{[i_0]}(dx)}{z+x}. \tag{29}$$

On the other hand, by using the fact that

$$\frac{P_{i_0+1}(-z)}{Q_{i_0+1}(-z)} - \frac{P_{i_0}(-z)}{Q_{i_0}(-z)} = \frac{|r_0|}{\lambda_{i_0}\pi_{i_0}Q_{i_0+1}(-z)Q_{i_0}(-z)}, \tag{30}$$

we deduce that

$$\int_0^\infty \frac{Q_{i_0}(x)^2}{x}\psi_{[i_0]}(dx) = \frac{|r_0|}{\lambda_{i_0}\pi_{i_0}},$$

since $Q_i(0) = 1$ for all $i \geq 0$. In addition, by using the fact that $\mathcal{F}_{i_0}(0) = r_{i_0+1}/\mu_{i_0+1}$, we deduce that the smallest root of Equation (27) is $\eta_0 = 0$. The other roots are positive. Equation (27) can be rewritten as Equation (28) by using Equations (29) and (30). $\square$

Note that by using the same arguments as above, we can simplify Equation (18). As a matter of fact, we have

$$\frac{P_{i_0+1}(s,-z)}{Q_{i_0+1}(s,-z)} - \frac{P_{i_0}(s,-z)}{Q_{i_0}(s,-z)} = \frac{|r_0|}{\lambda_{i_0}\pi_{i_0}Q_{i_0+1}(s,-z)Q_{i_0}(s,-z)},$$

so Equation (18) becomes

$$1 = \frac{\mu_{i_0+1}}{r_{i_0+1}}\mathcal{F}_{i_0}(s,\xi)\frac{Q_{i_0}(s,\xi)}{Q_{i_0+1}(s,\xi)}. \tag{31}$$

The quantities h_i are evaluated by using the normalizing condition $\sum_{i=0}^{i_0} h_i = 1 - \rho$, where ρ is defined by Equation (3), and by solving the i_0 linear equations

$$\ell = 1, \ldots, i_0, \quad \left(\left(\eta_\ell\mathbb{I} - R_{[i_0]}^{-1}A_{[i_0]}\right)^{-1}e_{i_0}, h\right)_{i_0} = 0, \tag{32}$$

where h is the vector which ith component is h_i/π_i. Once the quantities h_i, $i = 0, \dots, i_0$ are known, the function $F_{i_0+1}(\xi)$ is computed by using relation (25). The function $F_{i_0}(\xi)$ is computed by using the relation

$$F_{i_0+1}(\xi) = \frac{\lambda_{i_0}}{r_{i_0+1}} F_{i_0}(\xi) \mathcal{F}_{i_0}(\xi).$$

This allows us to determine the functions $F_{i_0+1}(\xi)$ and $F_{i_0}(\xi)$. The functions $F_i(\xi)$ for $i = 0, \dots, i_0$ are computed by using Equation (15) for $s = 0$ and $f^{(0)} \equiv 0$. The functions $F_i(\xi)$ for $i > i_0$ are computed by using Equation (16) for $s = 0$ and $f^{(0)} \equiv 0$. This leads to the following result.

Proposition 5. *The Laplace transform of the buffer content X in the stationary regime is given by*

$$
\mathbb{E}\left(e^{-\xi X}\right) = \sum_{i=0}^{\infty} F_i(\xi) = \frac{1}{r_0} \sum_{j=0}^{i_0} r_j \xi h_j \int_0^{\infty} \frac{Q_j(x)\Pi(x)}{\xi - x} \psi_{[i_0]}(dx)
$$
$$
+ \frac{\lambda_{i_0}}{r_{i_0+1}} F_{i_0}(\xi) \left(\frac{\mu_{i_0+1}}{r_0} \mathcal{F}_{i_0}(\xi) \int_0^{\infty} \frac{Q_{i_0}(x)\Pi(x)}{\xi - x} \psi_{[i_0]}(dx) + \frac{1}{\pi_{i_0+1}} \int_0^{\infty} \frac{\Pi_{i_0}(x)}{x+\xi} \psi^{[i_0]}(dx) \right) \quad (33)
$$

with

$$
\Pi(x) = \sum_{i=0}^{i_0} \pi_i Q_i(x),
$$

$$
\Pi_{i_0}(x) = \sum_{i=0}^{\infty} \pi_{i_0+1+i} Q_i(i_0+1;x),
$$

$$
F_{i_0}(\xi) = \frac{\dfrac{\pi_{i_0}}{r_0} \sum_{j=0}^{i_0} r_j \xi h_j \int_0^{\infty} \dfrac{Q_j(x)Q_{i_0}(x)}{\xi - x} \psi_{[i_0]}(dx)}{1 - \dfrac{\lambda_{i_0}\mu_{i_0+1}\pi_{i_0}}{r_0 r_{i_0+1}} \mathcal{F}_{i_0}(\xi) \int_0^{\infty} \dfrac{Q_{i_0}(x)^2}{\xi - x} \psi_{[i_0]}(dx)}.
$$

In the case when there is only one state with negative drift, the above result can be simplified as follows.

Corollary 1. *When there is only one state with negative drift, the Laplace transform of the buffer content is given by*

$$
\mathbb{E}\left(e^{-\xi X}\right) = \frac{\xi(1-\rho)r_0}{r_0\xi + \lambda_0 - \frac{\lambda_0 \mu_1}{r_1} F_0(\xi)} \left(1 + \frac{\lambda_1}{r_1} \int_0^{\infty} \frac{\Pi_0(x)}{x+\xi} \psi^{[0]}(dx) \right). \quad (34)
$$

Proof. Since $\psi_{[0]}(dx) = \delta_{\zeta_0}(dx)$ with $\zeta_0 = \lambda_0/|r_0|$ and $\Pi(x) = 1$, we have

$$
\int_0^{\infty} \frac{\Pi(x)}{\xi - x} \psi_{[i_0]}(dx) = \frac{r_0}{r_0\xi + \lambda_0}.
$$

Moreover, we have $h_0 = 1 - \rho$ and then

$$F_0(\xi) = \frac{(1-\rho)\xi r_0}{r_0\xi + \lambda_0 - \frac{\lambda_0\mu_1}{r_1}F_0(\xi)}.$$

Simple algebra then yields equation (34). $\qquad\qquad\qquad\qquad\qquad\qquad\qquad\square$

By examining the singularities in Equation (34), it is possible to determine the tail of the probability distribution of the buffer content in the stationary regime. The asymptotic behavior greatly depends on the properties of the polynomials $Q_i(x)$ and their associated spectral measure.

5. Busy period

In this section, we are interested in the duration of a busy period of the fluid reservoir. At the beginning of a busy period, the buffer is empty and the modulating process is in state $i_0 + 1$. More generally, let us introduce the occupation duration B which is the duration the server is busy up to an idle period. The random variable B depends on the initial conditions and we define the conditional probability distribution

$$H_i(t,x) = \mathbb{P}(B \leq t \mid \Lambda_0 = i, X_0 = x).$$

The probability distribution function of a busy period β of the buffer is clearly given by

$$\mathbb{P}(\beta \leq t) = H_{i_0+1}(t,0). \tag{35}$$

It is known in Barbot et al. (2001) that for $t > 0$ and $x > 0$, $H_i(t,x)$ satisfies the following partial differential equations

$$\frac{\partial}{\partial t}H_i(t,x) - r_i\frac{\partial}{\partial x}H_i(t,x) = -\mu_i H_{i-1}(t,x) + (\lambda_i + \mu_i)H_i(t,x) - \lambda_i H_{i+1}(t,x) \tag{36}$$

with the boundary conditions

$$H_i(t,0) = 1 \quad \text{if} \quad t \geq 0,\ r_i \leq 0,$$

$$H_i(0,x) = 0 \quad \text{if} \quad x > 0,$$

$$H_i(0,0) = 0 \quad \text{if} \quad r_i > 0.$$

Define then conditional Laplace transform

$$\theta_i(u,x) = \mathbb{E}\left(e^{-uB} \mid \Lambda_0 = i, Q_0 = x\right).$$

By taking Laplace transforms in Equation (36), we have

$$r_i\frac{\partial}{\partial x}\theta_i(u,x) = u\theta_i(u,x) - \mu_i\theta_{i-1}(u,x) + (\lambda_i + \mu_i)\theta_i(u,x) - \lambda_i\theta_{i+1}(u,x)$$

By introducing the conditional double Laplace transform

$$\tilde{\theta}_i(u,\xi) = \int_0^\infty e^{-\xi x}\theta_i(u,x)dx.$$

we obtain for $i \geq 0$

$$r_i\xi\tilde{\theta}_i(u,\xi) - r_i\theta_i(u,0) = u\tilde{\theta}_i(u,\xi) - \mu_i\tilde{\theta}_{i-1}(u,\xi) + (\lambda_i + \mu_i)\tilde{\theta}_i(u,\xi) - \lambda_i\tilde{\theta}_{i+1}(u,\xi)$$

By introducing the infinite vector $\Theta(u,\xi)$, which ith component is $\tilde{\theta}_i(u,\xi)$, the above equations can be rewritten in matrix form as

$$\xi R\Theta(u,\xi) = RT(u) + (u\mathbb{I} - A)\Theta(u,\xi), \tag{37}$$

where $T(u)$ is the vector which ith component is equal to $\theta_i(u,0)$. We clearly have $\theta_i(u,0) = 1$ for $i = 0,\ldots,i_0$. For the moment, the functions $\theta_i(u,0)$ for $i > i_0$ are unknown functions.

Equation (37) can be solved by using the same technique as in Section 3. In the following, we assume that the measure $\psi^{[i_0]}(s;dx)$ has a discrete spectrum with atoms located at points $\chi_k(s) > 0$ for $k \geq 0$. This assumption is satisfied for instance when the measure $\psi(s;dx)$ has a discrete spectrum (see Guillemin & Pinchon (1999) for details). Under this assumption, let $\chi_k(s) > 0$ for $k \geq 0$ be the solutions to the equation

$$\frac{\mu_{i_0+1}}{r_{i_0+1}}\frac{Q_{i_0}(u;-\xi)}{Q_{i_0+1}(u;-\xi)}\mathcal{F}_{i_0}(u,-\xi) = 1.$$

Proposition 6. *The Laplace transforms $\theta_{i_0+1+j}(u,0)$ for $j \geq 0$ satisfy the following linear equations:*

$$\frac{1}{r_{i_0+1}\pi_{i_0+1}}\frac{Q_{i_0}(u;-\xi)}{Q_{i_0+1}(u;-\xi)}\sum_{j=0}^\infty r_{i_0+1+j}\pi_{i_0+1+j}\theta_{i_0+1+j}(u,0)\int_0^\infty \frac{Q_j(i_0+1;u;x)}{\xi-x}\psi^{[i_0]}(u;dx)$$

$$+ \frac{1}{|r_0|}\sum_{j=0}^{i_0}|r_j|\pi_j\int_0^\infty \frac{Q_{i_0}(u;x)Q_j(u;x)}{\xi+x}\psi_{[i_0]}(u;dx) = 0 \tag{38}$$

for $\xi \in \{\chi_k(s), k \geq 0\}$.

Proof. Equation (37) can be split into two parts. The first part reads

$$\left(\xi\mathbb{I}_{[i_0]} - R_{[i_0]}^{-1}\left(u\mathbb{I}_{[i_0]} - A_{[i_0]}\right)\right)\Theta_{[i_0]} = e_{[i_0]} - \frac{\lambda_{i_0}}{r_{i_0}}\tilde{\theta}_{i_0+1}(u,\xi)e_{i_0}, \tag{39}$$

where $e_{[i_0]}$ is the finite vector with all entries equal to 1 for $i = 0,\ldots,i_0$ and $\Theta_{[i_0]}$ is the finite vector, which ith entry is $\tilde{\theta}_i(u,\xi)$ for $i = 0,\ldots,i_0$. The second part of the equation is

$$\left(\xi\mathbb{I}^{[i_0]} - \left(R^{[i_0]}\right)^{-1}\left(u\mathbb{I}^{[i_0]} - A^{[i_0]}\right)\right)\Theta^{[i_0]} = T^{[i_0]} - \frac{\mu_{i_0+1}}{r_{i_0+1}}\tilde{\theta}_{i_0}(u,\xi)e_0, \tag{40}$$

where the vector $T^{[i_0]}$ (resp. $\Theta^{[i_0]}$) has entries equal to $\theta_{i_0+1+i}(u,0)$ (resp. $\tilde{\theta}_{i_0+1+i}(u,\xi)$) for $i \geq 0$.

By adapting the proofs in Section 3, we have for $i = 0, \ldots, i_0$

$$\tilde{\theta}_i(u, \xi) = \frac{1}{|r_0|} \sum_{j=0}^{i_0} |r_j| \pi_j \int_0^\infty \frac{Q_i(u; x) Q_j(u; x)}{\xi + x} \psi_{[i_0]}(u; dx)$$

$$+ \frac{\mu_{i_0+1} \pi_{i_0+1}}{|r_0|} \tilde{\theta}_{i_0+1}(u, \xi) \int_0^\infty \frac{Q_{i_0}(u; x) Q_i(s; x)}{\xi + x} \psi_{[i_0]}(u; dx), \quad (41)$$

and for $i \geq 0$

$$\tilde{\theta}_{i_0+i+1}(u, \xi) = -\frac{\mu_{i_0+1+i}}{r_{i_0+1}} \tilde{\theta}_{i_0}(u, \xi) \int_0^\infty \frac{Q_i(i_0+1; u; x)}{\xi - x} \psi^{[i_0]}(u; dx)$$

$$+ \frac{1}{r_{i_0+1} \pi_{i_0+1}} \sum_{j=0}^\infty r_{i_0+1+j} \pi_{i_0+1+j} \theta_{i_0+1+j}(u, 0) \int_0^\infty \frac{Q_j(i_0+1; u; x) Q_i(i_0+1; u; x)}{\xi - x} \psi^{[i_0]}(u; dx)$$

$$(42)$$

By using Equation 41 for $i = i_0$ and Equation (42) for $i = 0$, we obtain

$$\left(1 - \frac{\mu_{i_0+1}}{r_{i_0+1}} \frac{Q_{i_0}(u; -\xi)}{Q_{i_0+1}(u; -\xi)} \mathcal{F}_{i_0}(u, -\xi)\right) \tilde{\theta}_{i_0}(u, \xi) =$$

$$\frac{1}{|r_0|} \sum_{j=0}^{i_0} |r_j| \pi_j \int_0^\infty \frac{Q_{i_0}(u; x) Q_j(u; x)}{\xi + x} \psi_{[i_0]}(u; dx)$$

$$+ \frac{1}{r_{i_0+1} \pi_{i_0+1}} \frac{Q_{i_0}(u; -\xi)}{Q_{i_0+1}(u; -\xi)} \sum_{j=0}^\infty r_{i_0+1+j} \pi_{i_0+1+j} \theta_{i_0+1+j}(u, 0) \int_0^\infty \frac{Q_j(i_0+1; u; x)}{\xi - x} \psi^{[i_0]}(u; dx)$$

where we have used the fact

$$\int_0^\infty \frac{Q_{i_0}(u; x)^2}{\xi + x} \psi_{[i_0]}(u; dx) = \frac{|r_0|}{\lambda_{i_0} \pi_{i_0}} \frac{Q_{i_0}(u; -\xi)}{Q_{i_0+1}(u; -\xi)}$$

and

$$\int_0^\infty \frac{1}{\xi - x} \psi^{[i_0]}(u; dx) = -\mathcal{F}_{i_0}(u; -\xi).$$

Since the function $\tilde{\theta}_{i_0}(u; \xi)$ shall have no poles in $[0, \infty)$, the result follows.   □

6. Conclusion

We have presented in this paper a general method for computing the Laplace transform of the transient probability distribution function of the content of a fluid reservoir fed with a source, whose transmission rate is modulated by a general birth and death process. This Laplace transform can be evaluated by solving a polynomial equation (see equation (18)). Once the zeros are known, the quantities $h_i(s)$ for $i = 0, \ldots, i_0$ are computed by solving the system of linear equations (19). These functions then completely determined the two critical functions F_{i_0} and F_{i_0+1}, which are then used for computing the functions F_i for $i > i_0 + 1$ and F_i for $i < i_0$

by using equations (16) and (15), respectively. Moreover, we note that the theory of orthogonal polynomials and continued fractions plays a crucial role in solving the basic equation (6).

The above method can be used for evaluating the Laplace transform of the duration of a busy period of the fluid reservoir as shown in Section 5. The results obtained in this section can be used to study the asymptotic behavior of the busy period when the service rate of the buffer becomes very large. Occupancy periods of the buffer then become rare events and one may expect that buffer characteristics converge to some limits. This will be addressed in further studies.

7. Appendix

A. Proof of Lemma 1

From the recurrence relations (10), the quantities $A_k(s)$ defined by $A_0(s) = 1$ and for $k \geq 1$

$$A_k(s) = |r_0 \ldots r_{k-1}| \prod_{j=1}^{k} \alpha_{2j}(s)$$

satisfy the recurrence relation for $k \geq 1$

$$A_{k+1}(s) = (s + \lambda_k + \mu_k) A_k(s) - \lambda_{k-1} \mu_k A_{k-1}(s).$$

It is clear that $A_k(s)$ is a polynomial in variable s. In fact, the polynomials $A_k(s)$ are the successive denominators of the continued fraction

$$\mathcal{G}^e(z) = \cfrac{1}{s + \lambda_0 - \cfrac{\mu_1 \lambda_0}{s + \lambda_1 + \mu_1 - \cfrac{\mu_2 \lambda_1}{s + \lambda_2 + \mu_2 - \ddots}}}$$

which is itself the even part of the continued fraction

$$G(s) = \cfrac{\alpha_1}{z + \cfrac{\alpha_2}{1 + \cfrac{\alpha_3}{z + \cfrac{\alpha_4}{1 + \ddots}}}}, \tag{43}$$

where the coefficients α_k are such that $\alpha_1 = 1$, $\alpha_2 = \lambda_0$, and for $k \geq 1$,

$$\alpha_{2k} \alpha_{2k+1} = \lambda_{k-1} \mu_k, \quad \alpha_{2k+1} + \alpha_{2(k+1)} = \lambda_k + \mu_k.$$

It is straightforwardly checked that $\alpha_{2k} = \lambda_{k-1}$ and $\alpha_{2k+1} = \mu_k$ for $k \geq 1$. The continued fraction $G(s)$ is hence a Stieltjes fraction and is converging for all $s > 0$ if and only if $\sum_{k=0}^{\infty} a_k =$

∞ where the coefficients a_k are defined by

$$\alpha_1 = \frac{1}{a_1}, \quad \alpha_k = \frac{1}{a_{k-1}a_k} \text{ for } k \geq 1.$$

(See Henrici (1977) for details.) It is easily checked that for $k \geq 1$

$$a_{2k} = \frac{1}{\lambda_{k-1}\pi_{k-1}} \quad \text{and} \quad a_{2k+1} = \pi_k.$$

Since the process (Λ_t) is assumed to be ergodic, $\sum_{k \geq 1} a_k = \infty$, which shows that the continued fraction $\mathcal{G}(s)$ is converging for all $s > 0$ and that there exists a unique measure $\varphi(dx)$ such that $\mathcal{G}(s)$ is the Stieltjes transform of $\varphi(dx)$, that is, for all $s \in \mathbb{C} \setminus (-\infty, 0]$

$$\mathcal{G}(s) = \int_0^\infty \frac{1}{z+x}\varphi(dx).$$

The support of $\varphi(dx)$ is included in $[0, \infty)$ and this measure has a mass at point $x_0 \geq 0$ if and only if

$$\sum_{k=0}^\infty \frac{A_k(-x_0)^2}{\lambda_0 \ldots \lambda_{k-1}\mu_1 \ldots \mu_k} < \infty.$$

Since the continued fraction $\mathcal{G}(s)$ is converging for all $s > 0$, we have

$$\sum_{k=0}^\infty \frac{A_k(s)^2}{\lambda_0 \ldots \lambda_{k-1}\mu_1 \ldots \mu_k} = \infty. \tag{44}$$

Since the polynomials $A_k(s)$ are the successive denominator of the fraction $\mathcal{G}^e(s)$, the polynomials $A_k(-s)$, $k \geq 1$, are orthogonal with respect to some orthogonality measure, namely the measure $\varphi(dx)$. From the general theory of orthogonal polynomials Askey (1984); Chihara (1978), we know that the polynomial $A_k(-s)$ has k simple, real, and positive roots. Since the coefficient of the leading term of $A_k(-s)$ is $(-1)^k$, this implies that $A_k(s)$ can be written as $A_k(s) = (s + s_{1,k}) \ldots (s + s_{k,k})$ with $s_{i,k} > 0$ for $i = 1, \ldots, k$. Hence, $A_k(s) \geq 0$ for all $s \geq 0$ and then, for all $k \geq 0$, $\alpha_k(s) \geq 0$ for all $s \geq 0$ and hence the continued fraction $\mathcal{F}(s, z)$ defined by Equation (9) is a Stieltjes fraction.

The continued fraction $\mathcal{F}(s, z)$ is converging if and only if $\sum_{k=0}^\infty a_k(s) = \infty$ where the coefficients $a_k(s)$ are defined by

$$\alpha_1(s) = \frac{1}{a_1(s)}, \quad \alpha_k(s) = \frac{1}{a_{k-1}(s)a_k(s)} \text{ for } k \geq 1.$$

(See Henrici (1977) for details.)

It is easily checked that

$$a_{2k+1}(s) = \frac{|r_k|}{|r_0|}\frac{A_k(s)^2}{\lambda_{k-1} \ldots \lambda_0\mu_k \ldots \mu_1} \quad \text{and} \quad a_{2k} = |r_0|\frac{\lambda_0 \ldots \lambda_{k-2}\mu_1 \ldots \mu_{k-1}}{A_k(s)A_{k-1}(s)}.$$

For $k > i_0$, $r_k \geq r_{i_0+1}$ and then by taking into account Equation (44), we deduce that for all $s > 0$, $\sum_{k=0}^{\infty} a_k(s) = \infty$ and the continued fraction $\mathcal{F}(s;z)$ is then converging for all $s > 0$. For $s = 0$, we have

$$a_{2k}(0) = \frac{|r_0|}{\lambda_{k-1}\pi_{k-1}}$$

and then $\sum_{k=0}^{\infty} a_k(0) = \infty$ since the process (Λ_t) is ergodic (see Condition (2)). This shows that the Stieltjes fraction $\mathcal{F}(s;z)$ is converging for all $s \geq 0$.

B. Selfadjointness properties

We consider in this section the Hilbert space $H_{i_0} = \mathbb{C}^{i_0+1}$ equipped with the scalar product

$$(c,d)_{i_0} = \sum_{k=0}^{i_0} c_k \overline{d_k} |r_k| \pi_k.$$

The main result of this section is the following lemma.

Lemma 6. For $s \geq 0$, the finite matrix $-R_{[i_0]}^{-1}(s\mathbb{I}_{[i_0]} - A_{[i_0]})$ defines a selfadjoint operator in the Hilbert space H_{i_0}; the spectrum is purely point-wise and composed by the (positive) roots of the polynomial $Q_{i_0+1}(s;x)$ defined by Equation (8), denoted by $\zeta_k(s)$ for $k = 0,\ldots,i_0$.

Proof. The finite matrix $-R_{[i_0]}^{-1}(s\mathbb{I}_{[i_0]} - A_{[i_0]})$ is given by

$$\begin{pmatrix} -\frac{s+\lambda_0}{|r_0|} & \frac{\lambda_0}{|r_0|} & 0 & \cdot & \cdot \\ \frac{\mu_1}{|r_1|} & -\frac{(s+\lambda_1+\mu_1)}{|r_1|} & \frac{\lambda_1}{|r_1|} & \cdot & \cdot \\ 0 & \frac{\mu_2}{|r_2|} & -\frac{(s+\lambda_2+\mu_2)}{|r_2|} & \frac{\lambda_2}{|r_2|} & \cdot \\ \cdot & & \cdot & \cdot & \cdot \\ & & & \frac{\mu_{i_0}}{|r_{i_0}|} & -\frac{s+\lambda_{i_0}+\mu_{i_0}}{|r_{i_0}|} \end{pmatrix}.$$

The symmetry of the matrix with respect to the scalar product $(.,.)_{i_0}$ is readily verified by using the relation $\lambda_k\pi_k = \mu_{k+1}\pi_{k+1}$. Since the dimension of the Hilbert space H_{i_0} is finite, the operator associated with the matrix $-R_{[i_0]}^{-1}(s\mathbb{I}_{[i_0]} - A_{[i_0]})$ is selfadjoint and its spectrum is purely point-wise.

If f is an eigenvector for the matrix $-R_{[i_0]}^{-1}(s\mathbb{I}_{[i_0]} - A_{[i_0]})$ associated with the eigenvalue x, then under the hypothesis that $f_0 = 1$, the sequence f_n verifies the same recurrence relation as $Q_k(s;x)$ for $k = 0,\ldots,i_0 - 1$. This implies that x is an eigenvalue of the above matrix if an only if $Q_{i_0+1}(s;x) = 0$, that is, x is one of the (positive) zeros of the polynomial $Q_{i_0+1}(s;x)$, denoted by $\zeta_k(s)$ for $k = 0,\ldots,i_0$. □

Let us introduce the column vector $Q_{[i_0]}(s,\zeta_k(s))$ for $k = 0,\ldots,i_0$, whose ℓth component is $Q_\ell(s,\zeta_k(s))$. The vector $Q_{[i_0]}(s,\zeta_k(s))$ is the eigenvector associated with the eigenvalue $\zeta_k(s)$ of the operator $-R_{[i_0]}^{-1}(s\mathbb{I}_{[i_0]} - A_{[i_0]})$. From the spectral theorem, the vectors $Q_{[i_0]}(s,\zeta_k(s))$ for

$k = 0, \ldots, i_0$ form an orthogonal basis of the Hilbert space H_{i_0}. The vectors e_j for $j = 0, \ldots, i_0$ such that all entries are equal to 0 except the jth one equal to 1 form the natural orthogonal basis of the space H_{i_0}. We can moreover write for $j = 0, \ldots, i_0$

$$e_j = \sum_{k=0}^{i_0} \alpha_k^{(j)} Q_{[i_0]}(s, \zeta_k(s)).$$

By using the orthogonality of the vectors $Q_{[i_0]}(s, \zeta_k(s))$ for $k = 0, \ldots, i_0$, we have

$$(e_j, Q_{[i_0]}(s, \zeta_k(s)))_{i_0} = |r_j| \pi_j Q_j(s, \zeta_k(s)) = \|Q_{[i_0]}(s, \zeta_k(s))\|_{i_0}^2 \alpha_k^{(j)}$$

where for $f \in H_{i_0}$, $\|f\|_{i_0}^2 = (f, f)_{i_0}$. We hence deduce that

$$|r_j| \pi_j \sum_{k=0}^{i_0} \frac{Q_j(s, \zeta_k(s)) Q_\ell(s, \zeta_k(s))}{\|Q_{[i_0]}(s, \zeta_k(s))\|_{i_0}^2} = \delta_{j,\ell},$$

where $\delta_{j,\ell}$ is the Kronecker symbol. It follows that if we define the measure $\psi_{[i_0]}(s; dx)$ by

$$\psi_{[i_0]}(s; dx) = |r_0| \sum_{k=0}^{i_0} \frac{1}{\|Q_{[i_0]}(s, \zeta_k(s))\|_{i_0}^2} \delta_{\zeta_k(s)}(dx) \tag{45}$$

the polynomials $Q_k(s, x)$ for $k = 0, \ldots, i_0$ are orthogonal with respect to the above measure, that is, they verify

$$\int_0^\infty Q_j(s, x) Q_\ell(s, x) \psi_{[i_0]}(s; dx) = \frac{|r_0|}{|r_j| \pi_j} \delta_{j,\ell},$$

and the total mass of the measure $\psi_{[i_0]}(s; dx)$ is equal to 1, i.e,

$$\int_0^\infty \psi_{[i_0]}(s; dx) = 1.$$

8. References

Adan, I. & Resing, J. (1996). Simple analysis of a fluid queue driven by an M/M/1 queue, *Queueing Systems - Theory and Applications*, Vol. 22, pp. 171–174.

Aggarwal, V., Gautam, N., Kumara, S. R. T. & Greaves, M. (2005). Stochastic fluid flow models for determining optimal switching thresholds, *Performance Evaluation*, Vol. 59, pp. 19–46.

Ahn, S. & Ramaswami, V. (2003). Fluid flow models and queues - a connection by stochastic coupling, *Stochastic Models*, Vol. 19, No. 3, pp. 325–348.

Ahn, S. & Ramaswami, V. (2004). Transient analysis of fluid flow models via stochastic coupling to a queue, *Stochastic Models*, Vol. 20, No. 1, pp. 71–101.

Anick, D., Mitra, D. & Sondhi, M. M. (1982). Stochastic theory of a data-handling system with multiple sources, *Bell System Tech. J.*, Vol. 61, No. 8, pp. 1871–1894.

Askey, R. & Ismail, M. (1984). Recurrence relations, continued fractions, and orthogonal polynomials, *Memoirs of the American Mathematical Society*, Vol. 49, No. 300.

Asmussen, S. (1987). Applied probability and queues, *J. Wiley and Sons*.

Badescu, A., Breuer, L., da Silva Soares, A., Latouche, G., Remiche, M.-A. & Stanford, D. (2005). Risk processes analyzed as fluid queues, *Scandinav. Actuar. J.*, Vol. 2, pp. 127–141.

Barbot, N., Sericola, B. & Telek, M. (2001). Dsitribution of busy period in stochastic fluid models, *Stochastic Models*, Vol. 17, No. 4, pp. 407–427.

Barbot, N. & Sericola, B. (2002). Stationary solution to the fluid queue fed by an M/M/1 queue, *Journ. Appl. Probab.*, Vol. 39, pp. 359–369.

Chihara, T. S. (1978). *An introduction to orthogonal polynomials*. Gordon and Breach, New York, 1978.

da Silva Soares, A. & Latouche, G. (2002). Further results on the similarity between fluid queues and QBDs, In G. Latouche and P. Taylor, editors, *Proc. of the 4th Int. Conf. on Matrix-Analytic Methods (MAM'4)*, Adelaide, Australia, 89–106, World Scientific.

da Silva Soares, A. & Latouche, G. (2006). Matrix-analytic methods for fluid queues with finite buffers, *Performance Evaluation*, Vol. 63, No. 4, pp. 295–314.

Guillemin, F. (2012). Spectral theory of birth and death processes, *Submitted for publication*.

Guillemin, F. & Pinchon, D. (1999). Excursions of birth and death processes, orthogonal polynomials, and continued fractions, *J. Appl. Prob.*, Vol. 36, pp. 752–770.

Guillemin, F. & Sericola, B. (2007). Stationary analysis of a fluid queue driven by some countable state space Markov chain, *Methodology and Computing in Applied Probability*, Vol. 9, pp. 521–540.

Henrici, P. (1977). Applied and computational complex analysis, *Wiley, New York*, Vol. 2.

Igelnik, B., Kogan, Y., Kriman, V. & Mitra, D. (1995). A new computational approach for stochastic fluid models of multiplexers with heterogeneous sources, *Queueing Systems - Theory and Applications*, Vol. 20, pp. 85–116.

Kleinrock, L. (1975). Queueing Systems, *J. Wiley*, Vol. 1.

Kobayashi, H. & Ren, Q. (1992). A mathematical theory for transient analysis of communication networks, *IEICE Trans. Communications*, Vol. 75, No. 12, pp. 1266–1276.

Kosten, L. (1984). Stochastic theory of data-handling systems with groups of multiple sources, In *Proceedings of the IFIP WG 7.3/TC 6 Second International Symposium on the Performance of Computer-Communication Systems*, Zurich, Switzerland, pp. 321–331.

Kumar, R., Liu, Y. & Ross, K. W. (2007). Stochastic Fluid Theory for P2P Streaming Systems, In *Proceedings of INFOCOM*, Anchorage, Alaska, USA, pp. 919–927.

Mitra, D. (1987). Stochastic fluid models, In *Proceedings of Performance'87, P. J. Courtois and G. Latouche Editors*, Brussels, Belgium, pp. 39–51.

Mitra, D. (1988). Stochastic theory of a fluid model of producers and consumers coupled by a buffer, *Advances in Applied Probability*, Vol. 20, pp. 646–676.

Nabli, H. & Sericola, B. (1996). Performability analysis: a new algorithm, *IEEE Trans. Computers*, Vol. 45, pp. 491–494.

Nabli, H. (2004). Asymptotic solution of stochastic fluid models, *Performance Evaluation*, Vol. 57, pp. 121–140.

Parthasarathy, P. R., Vijayashree, K. V. & Lenin, R. B. (2004). Fluid queues driven by a birth and death process with alternating flow rates, *Mathematical Problems in Engineering*, Vol. 5, pp. 469–489.

Ramaswami, V. (1999). Matrix analytic methods for stochastic fluid flows, In D. Smith and P. Hey, editors, *Proceedings of the 16th International Teletraffic Congress : Teletraffic Engineering in a Competitive World (ITC'16)*, Edinburgh, UK, Elsevier, pp. 1019–1030.

Ren, Q. & Kobayashi, H. (1995). Transient solutions for the buffer behavior in statistical multiplexing, *Performance Evaluation*, Vol. 23, pp. 65–87.

Rogers, L. C. G. (1994). Fluid models in queueing theory and wiener-hopf factorization of Markov chains, *Advances in Applied Probability*, Vol. 4, No. 2.

Rogers, L. C. G. & Shi, Z. (1994). Computing the invariant law of a fluid model, *Journal of Applied Probability*, Vol. 31, No. 4, pp. 885–896.

Sericola, B. (1998). Transient analysis of stochastic fluid models, *Performance Evaluation*, Vol. 32, pp. 245–263.

Sericola, B. & Tuffin, B. (1999). A fluid queue driven by a Markovian queue, *Queueing Systems - Theory and Applications*, Vol. 31, pp. 253–264.

Sericola, B. (2001). A finite buffer fluid queue driven by a Markovian queue, *Queueing Systems - Theory and Applications*, Vol. 38, pp. 213–220.

Sericola, B., Parthasarathy, P. R. & Vijayashree, K. V. (2005). Exact transient solution of an M/M/1 driven fluid queue, *Int. Journ. of Computer Mathematics*, Vol. 82, No. 6.

Stern, T. E. & Elwalid, A. I. (1991). Analysis of separable Markov-modulated rate models for information-handling systems, *Advances in Applied Probability*, Vol. 23, pp. 105–139.

Tanaka, T., Hashida, O. & Takahashi, Y. (1995). Transient analysis of fluid models for ATM statistical multiplexer, *Performance Evaluation*, Vol. 23, pp. 145–162.

van Dorn, E. A. & Scheinhardt, W. R. (1997). A fluid queue driven by an infinite-state birth and death process, In V. Ramaswami and P. E. Wirth, editors, *Proceedings of the 15th International Teletraffic Congress : Teletraffic Contribution for the Information Age (ITC'15)*, Washington D.C., USA, Elsevier, pp. 465–475.

vanForeest, N., Mandjes, M. & Scheinhardt, W. R. (2003). Analysis of a feedback fluid model for TCP with heterogeneous sources, *Stochastic Models*, Vol. 19, pp. 299–324.

Virtamo, J. & Norros, I. (1994). Fluid queue driven by an M/M/1 queue, *Queueing Systems - Theory and Applications*, Vol. 16, pp. 373–386.

9

Simulation and Optimal Routing of Data Flows Using a Fluid Dynamic Approach

Ciro D'Apice[1], Rosanna Manzo[1] and Benedetto Piccoli[2]

[1]*Department of Electronic and Information Engineering, University of Salerno,*
Fisciano (SA)
[2]*Department of Mathematical Sciences, Rutgers University, Camden, New Jersey*
[1]*Italy*
[2]*USA*

1. Introduction

There are various approaches to telecommunication and data networks (see for example Alderson et al. (2007), Baccelli et al. (2006), Baccelli et al. (2001), Kelly et al. (1998), Tanenbaum (1999), Willinger et al. (1998)). A first model for data networks, similar to that used for car traffic, has been proposed in D'Apice et al. (2006), where two algorithms for dynamics at nodes were considered and existence of solutions to Cauchy Problems was proved. Then in D'Apice et al. (2008), following the approach of Garavello et al. (2005) for road networks (see also Coclite et al. (2005); Daganzo (1997); Garavello et al. (2006); Holden et al. (1995); Lighthill et al. (1955); Newell (1980); Richards (1956)), sources and destinations have been introduced, thus taking care of the packets paths inside the network.

In this Chapter we deal with the fluid-dynamic model for data networks together with optimization problems, reporting some results obtained in Cascone et al. (2010); D'Apice et al. (2006; 2008; 2010).

A telecommunication network consists in a finite collection of transmission lines, modelled by closed intervals of $\mathbb{R}$ connected by nodes (routers, hubs, switches, etc.). Taking the Internet network as model, we assume that:

1) Each packet seen as a particle travels on the network with a fixed speed and with assigned final destination;

2) Nodes receive, process and then forward packets which may be lost with a probability increasing with the number of packets to be processed. Each lost packet is sent again.

Since each lost packet is sent again until it reaches next node, looking at macroscopic level, it is assumed that the packets number is conserved. This leads to a conservation law for the packets density ρ on each line:

$$\rho_t + f(\rho)_x = 0. \tag{1}$$

The flux $f(\rho)$ is given by $v(\rho) \cdot \rho$ where v is the average speed of packets among nodes, derived considering the amount of packets that may be lost.

The key point of the model is the loss probability, used to define the flux function. Indeed the choice of a non reasonable loss probability function could invalidate the model. To achieve the goal of the validation of the model assumptions, the loss probability function has been compared with the behaviour of the packet loss derived from known models used in literature to infer network performance and the shape of the velocity and flux functions has been discussed. All the comparisons confirm the validity of the assumptions underlying the fluid-dynamic model (see D'Apice et al. (2010)).

To describe the evolution of networks in which many lines intersect, Riemann Problems (RPs) at junctions were solved in D'Apice et al. (2006) proposing two different routing algorithms:

(RA1) Packets from incoming lines are sent to outgoing ones according to their final destination (without taking into account possible high loads of outgoing lines);

(RA2) Packets are sent to outgoing lines in order to maximize the flux through the node.

One of the drawback of (RA2) is that it does not take into account the global path of packets, therefore leading to possible cycling to bypass congested nodes. These cyclings are avoided if we consider that the packets originated from a source and with an assigned destination have paths inside the network.

Taking this in mind the model was refined in D'Apice et al. (2008). On each transmission line a vector π describing the traffic types, i.e. the percentages of packets going from a source to a destination, has been introduced. Assuming that packets velocity is independent from the source and the destination, the evolution of π follows a semilinear equation

$$\pi_t + v(\rho)\pi_x = 0, \tag{2}$$

hence inside transmission lines the evolution of π is influenced by the average speed of packets.

Different distribution traffic functions describing different routing strategies have been analysed:

- at a junction the traffic started at source s and with d as final destination, coming from the transmission line i, is routed on an assigned line j;
- at a junction the traffic started at source s and with d as final destination, coming from the transmission line i, is routed on every outgoing lines or on some of them.

In particular two ways according to which the traffic at a junction is splitted towards the outgoing lines have been defined. Starting from the distribution traffic function, and using the vector π, the traffic distribution matrix, which describes the percentage of packets from an incoming line that are addressed to an outgoing one, has been assigned. Then, methods to solve RPs according to the routing algorithms (RA1) and (RA2) have been proposed. Optimizations results have been obtained for the model consisting of the conservation law (1). In particular priority parameters and traffic distribution coefficients have been considered as controls and two functionals to measure the efficiency of the network have been defined in Cascone et al. (2010):

1) The velocity of packets travelling through the network.
2) The travel time taken by packets from source to destination.

Due to the nonlinear relation among cost functionals, the optimization of velocity and travel time can give different control parameters.

The analytical treatment of a complex network is very hard due to the high nonlinearity of the dynamics and discontinuities of the I/O maps. For these reasons, a decentralized strategy has been adapted as follows:

Step 1. The optimal controls for asymptotic costs in the case of a single node with constant initial data is computed.

Step 2. For a complex network, the (locally) optimal parameters at every node are used. Thus, the optimal control is determined at each node independently.

The optimization problem for nodes of 2×2 type, i.e. with two entering and two exiting lines, and traffic distribution coefficient α and priority parameter p as control parameters, constant initial data and asymptotic functionals has been completely solved.

Then a test telecommunication network, consisting of 24 nodes, each one of 2×2 type has been studied. Three different choices have been tested for the traffic distribution coefficients and priority parameters: (locally) optimal, static random and dynamic random. The first choice is given by Step 1. By static random parameters, we mean a random choice done at the beginning of the simulation and then kept constant. Finally, dynamic random coefficients are chosen randomly at every instant of time for every node.

The results present some interesting features: the performances of the optimal coefficients are definitely superior with respect to the other two. Then, how the dynamic random choice, which sometimes is equal in performance to the optimal ones, may be not feasible for modelling and robustness reasons has been discussed.

The Chapter is organized as follows. Section 2 reports the model for data networks. Then, in Section 3, we consider possible choices of the traffic distribution functions, and how to compute the traffic distribution matrix from the latter functions and the traffic-type function. We describe two routing algorithms, giving explicit unique solutions to RPs. In Section 4, we discuss the validity of the assumption on the loss probability function, the velocity and flux. The subsequent Section 5 is devoted to the analysis of the optimal control problem introducing the cost functionals. Simulations for three different choices of parameters (optimal, static and dynamic random) in the case of a complex network are presented. The paper ends with conclusions in Section 6.

2. Basic definitions

A telecommunication network is a finite collection of transmission lines connected together by nodes, some of which are sources and destinations. Formally we introduce the following definition:

Definition 1. *A telecommunication network is given by a 7-tuple* $(N, \mathcal{I}, \mathcal{F}, \mathcal{J}, \mathcal{S}, \mathcal{D}, \mathcal{R})$ *where*

Cardinality N *is the cardinality of the network, i.e. the number of lines in the network;*

Lines $\mathcal{I}$ *is the collection of lines, modelled by intervals* $I_i = [a_i, b_i] \subseteq \mathbb{R}, i = 1, ..., N;$

Fluxes $\mathcal{F}$ *is the collection of flux functions* $f_i : [0, \rho_i^{\max}] \mapsto \mathbb{R}, i = 1, ..., N;$

Nodes $\mathcal{J}$ *is a collection of subsets of* $\{\pm 1, ..., \pm N\}$ *representing nodes. If* $j \in J \in \mathcal{J}$, *then the transmission line* $I_{|j|}$ *is crossing at* J *as incoming line (i.e. at point* b_j) *if* $j > 0$ *and as outgoing line* *(i.e. at point a_i) if $j < 0$. For each junction $J \in \mathcal{J}$, we indicate by $\mathrm{Inc}(J)$ the set of incoming lines, that are I_i's such that $i \in J$, while by $\mathrm{Out}(J)$ the set of outgoing lines, that are I_i's such that $-i \in J$. We assume that each line is incoming for (at most) one node and outgoing for (at most) one node;*

Sources $\mathcal{S}$ is the subset of $\{1, ..., N\}$ representing lines starting from traffic sources. Thus, $j \in \mathcal{S}$ if and only if j is not outgoing for any node. We assume that $\mathcal{S} \neq \emptyset$;

Destinations $\mathcal{D}$ is the subset of $\{1, ..., N\}$ representing lines leading to traffic destinations, Thus, $j \in \mathcal{D}$ if and only if j is not incoming for any node. We assume that $\mathcal{D} \neq \emptyset$;

Traffic distribution functions $\mathcal{R}$ is a finite collection of functions (also multivalued) $r_J : \mathrm{Inc}(J) \times \mathcal{S} \times \mathcal{D} \to \mathrm{Out}(J)$. For every J, $r_J(i, s, d)$ indicates the outgoing direction of traffic that started at source s has d as final destination and reached J from the incoming road i.

2.1 Dynamics on lines

Following D'Apice et al. (2008), we recall the model used to define the dynamics of packet densities along lines. We make the following hypothesis:

(H1) Lines are composed of consecutive processors N_k, which receive and send packets. The packets number at N_k is indicated by $R_k \in [0, R_{max}]$;

(H2) There are two time-scales: Δt_0, the physical travel time of a single packet from node to node (assumed to be independent of the node for simplicity); T, the processing time, during which each processor tries to operate the transmission of a given packet;

(H3) Each processor N_k tries to send all packets R_k at the same time. Packets are lost according to a loss probability function $p : [0, R_{max}] \to [0, 1]$, computed at R_{k+1}, and lost packets are sent again for a time slot of length T;

(H4) The number of packets not transmitted for a whole processing time slot is negligible.

Since the packet transmission velocity on the line is assumed constant, it is possible to compute an average velocity function and thus an average flux function.

Let us focus on two consecutive nodes N_k and N_{k+1}, assume a static situation, i.e. R_k and R_{k+1} are constant. Indicate by δ the distance between the nodes, Δt_{av} the packets average transmission time, $\bar{v} = \frac{\delta}{\Delta t_0}$ the packet velocity without losses and $v = \frac{\delta}{\Delta t_{av}}$ the average packets velocity. Then, we can compute:

$$\Delta t_{av} = \sum_{n=1}^{M} n \Delta t_0 (1 - p(R_{k+1})) p^{n-1}(R_{k+1}),$$

where $M = [T/\Delta t_0]$ (here $[\cdot]$ indicates the floor function) represents the number of attempts of sending a packet and T is the length of a processing time slot. The hypothesis (H4) corresponds to assume $\Delta t_0 \ll T$ or, equivalently, $M \sim +\infty$. Making the identification, $M = +\infty$, we get:

$$\Delta t_{av} = \frac{\Delta t_0}{1 - p(R_{k+1})},$$

and

$$v = \bar{v}(1 - p(R_{k+1})). \tag{3}$$

Let us call now ρ the averaged density and ρ_{max} its maximum. We can interpret the probability loss function p as a function of ρ and, using (3), determine the corresponding flux function, given by the averaged density times the average velocity. A possible choice of p is the following:

$$p(\rho) = \begin{cases} 0, & 0 \le \rho \le \sigma, \\ \frac{\rho_{max}(\rho - \sigma)}{\rho(\rho_{max} - \sigma)}, & \sigma \le \rho \le \rho_{max}, \end{cases} \tag{4}$$

from which

$$v(\rho) = \begin{cases} \bar{v}, & 0 \le \rho \le \sigma, \\ \bar{v}\frac{\sigma(\rho_{max} - \rho)}{\rho(\rho_{max} - \sigma)}, & \sigma \le \rho \le \rho_{max}, \end{cases} \tag{5}$$

and

$$f(\rho) = \begin{cases} \bar{v}\rho, & 0 \le \rho \le \sigma, \\ \frac{\bar{v}\sigma(\rho_{max} - \rho)}{\rho_{max} - \sigma}, & \sigma \le \rho \le \rho_{max}. \end{cases} \tag{6}$$

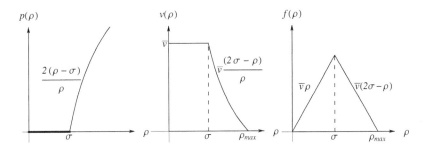

Fig. 1. Loss probability, average velocity and flux behaviours for $\rho_{max} = 1, \sigma = \frac{1}{2}, \bar{v} = 1$.

To simplify the treatment of the corresponding conservation laws, we will assume the following:

(F) Setting $\rho_{max} = 1$, on each line the flux $f : [0, 1] \to R$ is concave, $f(0) = f(1) = 0$ and there exists a unique maximum point $\sigma \in]0, 1[$.

Notice that the "tent" function

$$f(\rho) = \begin{cases} \rho, & 0 \le \rho \le \frac{1}{2}, \\ 1 - \rho, & \frac{1}{2} \le \rho \le 1, \end{cases} \tag{7}$$

and the parabolic flux

$$f(\rho) = \rho(1 - \rho), \rho \in [0, 1], \tag{8}$$

satisfy the assumption (F).

2.2 Dynamics on the network

On each transmission line I_i we consider the evolution equation

$$\partial_t \rho_i + \partial_x f_i(\rho_i) = 0, \tag{9}$$

where we use the assumption (F). Therefore, the network load evolution is described by a finite set of functions $\rho_i : [0, +\infty[\times I_i \mapsto [0, \rho_i^{max}]$.

Moreover, inside each line I_i we define a traffic-type function π_i, which measures the portion of the whole density coming from each source and travelling towards each destination:

Definition 2. *A traffic-type function on a line I_i is a function*

$$\pi_i : [0, \infty[\times [a_i, b_i] \times \mathcal{S} \times \mathcal{D} \mapsto [0, 1]$$

such that, for every $t \in [0, \infty[$ and $x \in [a_i, b_i]$

$$\sum_{s \in \mathcal{S}, d \in \mathcal{D}} \pi_i(t, x, s, d) = 1.$$

In other words, $\pi_i(t, x, s, d)$ specifies the density fraction $\rho_i(t, x)$ that started from source s and is moving towards the final destination d.

Assuming, on the discrete model, that a FIFO policy is used at nodes, it is natural that the averaged velocity, obtained in the limit procedure, is independent from the original sources of packets and their final destinations. In other words, we make the following hypothesis:

(H5) On each line I_i, the average velocity of packets depends only on the value of the density ρ_i and not on the values of the traffic-type function π_i.

As a consequence of hypothesis (H5), we deduce the semilinear equation

$$\partial_t \pi_i(t, x, s, d) + \partial_x \pi_i(t, x, s, d) \cdot v_i(\rho_i(t, x)) = 0. \tag{10}$$

This equation is coupled with equation (9) on each line I_i. More precisely, equation (10) depends on the solution of (9), while in turn at junctions the values of π_i will determine the traffic distribution on outgoing lines as explained below.

For simplicity and without loss of generality, we assume from now on that the fluxes f_i are all the same and we indicate them with f. Thus, the model for a single transmission line, consists in the system of equations:

$$\begin{cases} \rho_t + f(\rho)_x = 0, \\ \pi_t + \pi_x \cdot v(\rho) = 0. \end{cases}$$

To treat the evolution at junctions, let us introduce some notations. Fix a junction J with n incoming transmission lines, say $I_1, ..., I_n$, and m outgoing transmission lines, say $I_{n+1}, ..., I_{n+m}$ (junction of $n \times m$ type). The basic ingredient for the solution of Cauchy Problems by Wave Front Tracking method is the solution of Riemann Problems (RPs) (see Bressan (2000), Dafermos (1999), Serre (1999)).

We call RP for a junction the Cauchy Problem corresponding to an initial data $\rho_{1,0}, ..., \rho_{n+m,0} \in [0, 1]$, and $\pi_1^{s,d}, ..., \pi_{n+m}^{s,d} \in [0, 1]$ which are constant on each transmission line.

Definition 3. *A Riemann Solver (RS) for the junction J is a map that associates to Riemann data $\rho_0 = (\rho_{1,0}, ..., \rho_{n+m,0})$ and $\Pi_0 = (\pi_{1,0}, ..., \pi_{n+m,0})$ at J the vectors $\hat{\rho} = (\hat{\rho}_1, ..., \hat{\rho}_{n+m})$ and $\hat{\Pi} = (\hat{\pi}_1, ..., \hat{\pi}_{n+m})$ so that the solution on an incoming transmission line I_i, $i = 1, ..., n$, is given by the wave $(\rho_{i,0}, \hat{\rho}_i)$ and on an outgoing one I_j, $j = n + 1, ..., n + m$, is given by the waves $(\hat{\rho}_j, \rho_{j,0})$ and $(\hat{\pi}_j, \pi_{j,0})$. We require the following consistency condition:*

(CC) $RS(RS(\rho_0, \Pi_0)) = RS(\rho_0, \Pi_0)$.

Once a RS is defined and the solution of the RP is obtained, we can define admissible solutions at junctions.

3. Riemann Solvers at junctions

Consider a junction J of $n \times m$ type. We denote with $\rho_i(t,x), i = 1, ..., n$ and $\rho_j(t,x), j = n+1, ..., n+m$ the traffic densities, respectively, on the incoming transmission lines and on the outgoing ones and by $(\rho_{1,0}, ..., \rho_{n+m,0})$ the initial datum.

Define the maximum flux that can be obtained by a single wave solution on each transmission line as follows:

$$\gamma_i^{max} = \begin{cases} f(\rho_{i,0}), & \text{if } \rho_{i,0} \in [0,\sigma], \\ f(\sigma), & \text{if } \rho_{i,0} \in]\sigma, 1], \end{cases} i = 1, ..., n, \tag{11}$$

and

$$\gamma_j^{max} = \begin{cases} f(\sigma), & \text{if } \rho_{j,0} \in [0,\sigma], \\ f(\rho_{j,0}), & \text{if } \rho_{j,0} \in]\sigma, 1], \end{cases} j = n+1, ..., n+m. \tag{12}$$

Finally denote with

$$\Omega_i = [0, \gamma_i^{max}], i = 1, ..., n,$$
$$\Omega_j = [0, \gamma_j^{max}], j = n+1, ..., n+m,$$

and with $\hat{\gamma}_{inc} = (f(\hat{\rho}_i), ..., f(\hat{\rho}_n)), \hat{\gamma}_{out} = (f(\hat{\rho}_{n+1}), ..., f(\hat{\rho}_{n+m}))$ where $\hat{\rho} = (\hat{\rho}_1, ..., \hat{\rho}_{n+m})$ is the solution of the RP at the junction.

Now, we discuss some possible choices for the traffic distribution function:

1) $r_J : \text{Inc}(J) \times \mathcal{S} \times \mathcal{D} \to \text{Out}(J)$;
2) $r_J : \text{Inc}(J) \times \mathcal{S} \times \mathcal{D} \hookrightarrow \text{Out}(J)$, i.e. r_J is a multifunction.

If r_J is of type 1), then each packet has a deterministic route, it means that, at the junction J, the traffic that started at source s and has d as final destination, coming from the transmission line I_i, is routed on an assigned line I_j ($r_J(i,s,d) = j$).

Instead if r_J is of type 2), at the junction J, the traffic with source s and destination d coming from a line I_i is routed on every line $I_j \in \text{Out}(J)$ or on some lines $I_j \in \text{Out}(J)$. We can define $r_J(i,s,d)$ in two different ways:

2a) $r_J : \text{Inc}(J) \times \mathcal{S} \times \mathcal{D} \hookrightarrow \text{Out}(J)$,
$r_j(i,s,d) \subseteq \text{Out}(J)$;

2b) $r_J : \text{Inc}(J) \times \mathcal{S} \times \mathcal{D} \to [0,1]^{\text{Out}(J)}$,
$$r_J(i,s,d) = (\alpha_J^{i,s,d,n+1}, ..., \alpha_J^{i,s,d,n+m})$$
with $0 \le \alpha_J^{i,s,d,j} \le 1, j \in \{n+1, ..., n+m\}, \sum_{j=n+1}^{n+m} \alpha_J^{i,s,d,j} = 1$.

In case 2a) we have to specify in which way the traffic at junction J is splitted towards the outgoing lines.

The definition 2b) means that, at the junction J, the traffic with source s and destination d coming from line I_i is routed on the outgoing line $I_j, j = n+1, ..., n+m$ with probability $\alpha_J^{i,s,d,j}$.

Let us analyze how the distribution matrix A is constructed using π and r_J.

Definition 4. *A distribution matrix is a matrix*

$$A \doteq \left\{ \alpha_{j,i} \right\}_{j=n+1,\dots,n+m, i=1,\dots,n} \in \mathbb{R}^{m \times n}$$

such that

$$0 < \alpha_{j,i} < 1, \quad \sum_{j=n+1}^{n+m} \alpha_{j,i} = 1,$$

for each $i = 1, \dots, n$ and $j = n+1, \dots, n+m$, where $\alpha_{j,i}$ is the percentage of packets arriving from the i-th incoming transmission line that take the j-th outgoing transmission line.

In case 1) we can define the matrix A in the following way. Fix a time t and assume that for all $i \in \mathrm{Inc}(J), s \in \mathcal{S}$ and $d \in \mathcal{D}, \pi_i(t, \cdot, s, d)$ admits a limit at the junction J, i.e left limit at b_j. For $i \in \{1, \dots, n\}, j \in \{n+1, \dots, n+m\}$, we set

$$\alpha_{j,i} = \sum_{\substack{s \in \mathcal{S}, d \in \mathcal{D}, \\ r_J(i,s,d)=j}} \pi_i(t, b_i-, s, d).$$

The fluxes $f_i(\rho_i)$ to be consistent with the traffic-type functions must satisfy the following relation:

$$f_j(\rho_j(\cdot, a_j+)) = \sum_{i=1}^{n} \alpha_{j,i} f_i(\rho_i(\cdot, b_i-)),$$

for each $j = n+1, \dots, n+m$.

Let us analyze how to define the matrix A in the case 2a). We may assign $\varphi(i,s,d) \in r_J(i,s,d)$ and set

$$\alpha_{j,i} = \sum_{\substack{s \in \mathcal{S}, d \in \mathcal{D}, \\ i:\varphi(i,s,d)=j}} \pi_i(t, b_i-, s, d),$$

$$\alpha_{j,i} = 0, \text{ if } j \notin r_J(i,s,d).$$

However, it is more natural to assign a flexible strategy defining a set of admissible matrices A in the following way

$$\mathcal{A} = \left\{ \begin{array}{l} A : \exists \alpha_J^{i,s,d,j} \in [0,1], \ \sum_{j=n+1}^{n+m} \alpha_J^{i,s,d,j} = 1, \alpha_J^{i,s,d,j} = 0, \text{ if } j \notin r_J(i,s,d) : \\ \alpha_{j,i} = \sum_{\substack{s \in \mathcal{S}, d \in \mathcal{D}, \\ j \in r_J(i,s,d)}} \pi_i(t, b_i-, s, d) \alpha_J^{i,s,d,j} \end{array} \right\}.$$

Finally, we treat now the case 2b). In this case the matrix A is unique and is defined by

$$\alpha_{j,i} = \sum_{s \in \mathcal{S}, d \in \mathcal{D}} \pi_i(t, b_i-, s, d) \alpha_J^{i,s,d,j}. \tag{13}$$

We describe two different RSs at a junction that represent two different routing algorithms:

(RA1) We assume that

(A) the traffic from incoming transmission lines is distributed on outgoing transmission lines according to fixed coefficients;

(B) respecting (A) the router chooses to send packets in order to maximize fluxes (i.e., the number of packets which are processed).

(RA2) We assume that the number of packets through the junction is maximized both over incoming and outgoing lines.

3.1 Algorithm (RA1)

We have to distinguish case 2a) and 2b).

In case 2a) first we observe that the set $\mathcal{A}$ is convex. The admissible region given by

$$\Omega_{adm} = \{\hat{\gamma} : \hat{\gamma} \in \Omega_1 \times \dots \times \Omega_n, \exists A \in \mathcal{A} \ t.c. A\hat{\gamma} \in \Omega_{n+1} \times \dots \times \Omega_{n+m}\},$$

is convex at least for the case of junctions of 2×2.

If the region Ω_{adm} is convex than rules (A) and (B) amount to the Linear Programming problem:

$$\max_{\hat{\gamma} \in \Omega_{adm}} (\hat{\gamma}_1 + \hat{\gamma}_2).$$

This problem has clearly a solution, which may not be unique.

Let us consider the case 2b). We need some more notations.

Definition 5. *Let* $\tau : [0,1] \to [0,1]$ *be the map such that* $f(\tau(\rho)) = f(\rho)$ *for every* $\rho \in [0,1]$ *and* $\tau(\rho) \neq \rho$ *for every* $\rho \in [0,1]\backslash\{\sigma\}$.

We need some assumption on the matrix A (satisfied under generic conditions for $m = n$). Let $\{e_1, \dots, e_n\}$ be the canonical basis of $\mathbb{R}^n$ and for every subset $V \subset \mathbb{R}^n$ indicate by $V^{\perp}$ its orthogonal. Define for every $i = 1, \dots, n$, $H_i = \{e_i\}^{\perp}$, i.e. the coordinate hyperplane orthogonal to e_i and for every $j = n+1, \dots, n+m$ let $\alpha_j = \{\alpha_{j1}, \dots, \alpha_{jn}\} \in \mathbb{R}^n$ and define $H_j = \{\alpha_j\}^{\perp}$. Let $\mathcal{K}$ be the set of indices $k = (k_1, \dots, k_l), 1 \leq l \leq n - 1$, such that $0 \leq k_1 < k_2 < \dots < k_l \leq n + m$ and for every $k \in \mathcal{K}$ set $H_k = \bigcap\limits_{h=1}^{l} H_h$. Letting $\mathbf{1} = (1, \dots, 1) \in \mathbb{R}^n$, we assume

(C) for every $k \in \mathcal{K}, \mathbf{1} \notin H_k^{\perp}$.

In case 2b) the following result holds

Theorem 6. *(Theorem 3.1 in Coclite et al. (2005) and 3.2 in Garavello et al. (2005)) Let* $(N, \mathcal{I}, \mathcal{F}, \mathcal{J}, \mathcal{S}, \mathcal{D}, \mathcal{R})$ *be an admissible network and* J *a junction of* $n \times m$ *type. Assume that the flux* $f : [0,1] \to \mathbb{R}$ *satisfies (F) and the matrix* A *satisfies condition (C). For every* $\rho_{1,0}, \dots, \rho_{n+m,0} \in [0,1]$, *and for every* $\pi_1^{s,d}, \dots \pi_{n+m}^{s,d} \in [0,1]$, *there exist densities* $\hat{\rho}_1, \dots, \hat{\rho}_{n+m}$ *and a unique admissible centered weak solution,* $\rho = (\rho_1, \dots, \rho_{n+m})$ *at* J *such that*

$$\rho_1(0, \cdot) \equiv \rho_{1,0}, \dots, \rho_{n+m}(0, \cdot) \equiv \rho_{n+m,0},$$
$$\pi^1(0, \cdot s, d) = \pi_1^{s,d}, \dots, \pi^{n+m}(0, \cdot, s, d) = \pi_{n+m}^{s,d}, (s \in \mathcal{S}, d \in \mathcal{D}).$$

We have

$$\hat{\rho}_i \in \begin{cases} \{\rho_{i,0}\} \cup]\tau(\rho_{i,0}), 1], & \text{if } 0 \leq \rho_{i,0} \leq \sigma, \\ [\sigma, 1], & \text{if } \sigma \leq \rho_{i,0} \leq 1, \end{cases} \quad i = 1, \dots, n, \tag{14}$$

$$\hat{\rho}_j \in \begin{cases} [0, \sigma], & if\ 0 \leq \rho_{j,0} \leq \sigma, \\ \{\rho_{j,0}\} \cup \left[0, \tau(\rho_{j,0})\right[, & if\ \sigma \leq \rho_{j,0} \leq 1, \end{cases} \quad j = n+1, ..., n+m, \tag{15}$$

and on each incoming line I_i, $i = 1, ..., n$, the solution consists of the single wave $(\rho_{i,0}, \hat{\rho}_i)$, while on each outgoing line I_j, $j = n+1, ..., n+m$, the solution consists of the single wave $(\hat{\rho}_j, \rho_{j,0})$. Moreover $\hat{\pi}_i(t, \cdot, s, d) = \pi_i^{s,d}$ for every $t \geq 0, i \in \{1, ..., n\}, s \in S, d \in D$ and

$$\hat{\pi}_j(t, a_j+, s, d) = \frac{\sum_{i=1}^{n} \alpha_j^{i,s,d,j} \pi_i^{s,d}(t, b_i-, s, d) f(\hat{\rho}_i)}{f(\hat{\rho}_j)}$$

for every $t \geq 0, j \in \{n+1, ..., n+m\}, s \in S, d \in D$.

3.2 Algorithm (RA2)

To solve RPs according to (RA2) we need some additional parameters called priority and traffic distribution parameters. For simplicity of exposition, consider, junction J of 2×2 type. In this case we have only one priority parameter $q \in]0,1[$ and one traffic distribution parameter $\alpha \in]0,1[$. We denote with $(\rho_{1,0}, \rho_{2,0}, \rho_{3,0}, \rho_{4,0})$ and $(\pi_{1,0}^{s,d}, \pi_{2,0}^{s,d}, \pi_{3,0}^{s,d}, \pi_{4,0}^{s,d})$ the initial data.

In order to maximize the number of packets through the junction over incoming and outgoing lines we define

$$\Gamma = \min\{\Gamma_{in}^{max}, \Gamma_{out}^{max}\},$$

where $\Gamma_{in}^{max} = \gamma_1^{max} + \gamma_2^{max}$ and $\Gamma_{out}^{max} = \gamma_3^{max} + \gamma_4^{max}$. Thus we want to have Γ as flux through the junction.

One easily see that to solve the RP, it is enough to determine the fluxes $\hat{\gamma}_i = f(\hat{\rho}_i), i = 1, 2$. In fact, to have simple waves with the appropriate velocities, i.e. negative on incoming lines and positive on outgoing ones, we get the constraints (14), (15). Observe that we compute $\hat{\gamma}_i = f(\hat{\rho}_i), i = 1, 2$ without taking into account the type of traffic distribution function.

We have to distinguish two cases:

I $\Gamma_{in}^{max} = \Gamma$,
II $\Gamma_{in}^{max} > \Gamma$.

In the first case we set $\hat{\gamma}_i = \gamma_i^{max}, i = 1, 2$.
Let us analyze the second case in which we use the priority parameter q. Not all packets can enter the junction, so let C be the amount of packets that can go through: qC packets come from first incoming line and $(1 - q)C$ packets from the second. In the space (γ_1, γ_2), define the following lines:

$$r_q : \gamma_2 = \frac{1-q}{q}\gamma_1, \quad r_\Gamma : \gamma_1 + \gamma_2 = \Gamma,$$

and P the point of intersection of r_q and r_Γ. Recall that the final fluxes should belong to the region:

$$\Omega_{in} = \{(\gamma_1, \gamma_2) : 0 \leq \gamma_i \leq \gamma_i^{max}, i = 1, 2\}.$$

We distinguish two cases:

a) P belongs to Ω_{in},

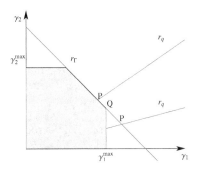

Fig. 2. P belongs to Ω_{in} and P is outside Ω_{in}.

b) P is outside Ω_{in}.

In the first case we set $(\hat{\gamma}_1, \hat{\gamma}_2) = P$, while in the second case we set $(\hat{\gamma}_1, \hat{\gamma}_2) = Q$, with $Q = proj_{\Omega_{in} \cap r_\Gamma}(P)$ where $proj$ is the usual projection on a convex set, see Figure 2.

As for the algorithm (RA1) $\hat{\pi}_i^{s,d} = \pi_{i,0}^{s,d}, i = 1, 2.$

Let us now determine $\hat{\gamma}_j, j = 3, 4$. We have to distinguish again two cases :

I $\Gamma_{out}^{max} = \Gamma,$

II $\Gamma_{out}^{max} > \Gamma.$

In the first case $\hat{\gamma}_j = \gamma_j^{max}, j = 3, 4$. Let us determine $\hat{\gamma}_j$ in the second case, using the traffic distribution parameter α. Since not all packets can go on the outgoing transmission lines, we let C be the amount that goes through. Then αC packets go on the outgoing line I_3 and $(1 - \alpha)C$ on the outgoing line I_4. Consider the space (γ_3, γ_4) and define the following lines:

$$r_\alpha : \gamma_4 = \frac{1 - \alpha}{\alpha} \gamma_3,$$

$$r_\Gamma : \gamma_3 + \gamma_4 = \Gamma.$$

We have to distinguish case 2a) and 2b) for the traffic distribution function.

3.2.1 Case 2a)

Let us introduce the connected set

$$\mathcal{G} = \left\{ A \hat{\gamma}_{inc}^T : A \in \mathcal{A} \right\},$$

and G_1 and G_2 its endpoints. Since in case 2a) we have an infinite number of matrices A, each of one determines a line r_α, we choose the most "natural" line r_α, i.e. the one nearest to the statistic line determined by measurements on the network.

Recall that the final fluxes should belong to the region:

$$\Omega_{out} = \left\{ (\gamma_3, \gamma_4) : 0 \leq \gamma_j \leq \gamma_j^{max}, j = 3, 4 \right\}.$$

Define $P = r_\alpha \cap r_\Gamma, R = (\Gamma - \gamma_4^{max}, \gamma_4^{max}), Q = (\gamma_3^{max}, \Gamma - \gamma_3^{max})$. We distinguish 3 cases:

a) $\mathcal{G} \cap \Omega_{out} \cap r_\Gamma \neq \varnothing$,

b) $\mathcal{G} \cap \Omega_{out} \cap r_\Gamma = \varnothing$ and $\gamma_3(G_1) < \gamma_3(R)$,

c) $\mathcal{G} \cap \Omega_{out} \cap r_\Gamma = \varnothing$ and $\gamma_3(G_1) > \gamma_3^{max}$.

If the set $\mathcal{G}$ has a priority over the line r_Γ we set $(\hat{\gamma}_3, \hat{\gamma}_4)$ in the following way. In case a) we define $(\hat{\gamma}_3, \hat{\gamma}_4) = proj_{\mathcal{G} \cap \Omega_{out} \cap r_\Gamma}(P)$, in case b) $(\hat{\gamma}_3, \hat{\gamma}_4) = R$, and finally in case c) $(\hat{\gamma}_3, \hat{\gamma}_4) = Q$.

Otherwise, if r_Γ has a priority over $\mathcal{G}$ we set $(\hat{\gamma}_3, \hat{\gamma}_4) = \min\limits_{\gamma \in \Omega_{out}} \mathcal{F}(\gamma, r_\alpha, \mathcal{G})$ where $\mathcal{F}$ is a convex functional which depends on γ, r_α and on the set $\mathcal{G}$ of the routing standards.

The vector $\hat{\pi}_i^{s,d}$, $j = 3, 4$ are computed in the same way as for the algorithm (RA1).

3.2.2 Case 2b)

In case 2b) we have a unique matrix A. The fluxes on outgoing lines are computed as in the case without sources and destinations.

We distinguish two cases:

a) P belongs to Ω,

b) P is outside Ω.

In the first case we set $(\hat{\gamma}_3, \hat{\gamma}_4) = P$, while in the second case we set $(\hat{\gamma}_3, \hat{\gamma}_4) = Q$, where $Q = proj_{\Omega_{adm}}(P)$. Again, we can extend to the case of m outgoing lines.

Finally we define $\hat{\pi}_i^{s,d}$, $j = 3, 4$ as in the case 2a):

$$\hat{\pi}_j(t, a_j+, s, d) = \frac{\sum\limits_{i=1}^{n} \alpha_j^{i,s,d,j} \pi_i^{s,d}(t, b_i-, s, d) f(\hat{\rho}_i)}{f(\hat{\rho}_j)}$$

for every $t \geq 0, j \in \{n+1, ..., n+m\}, s \in \mathcal{S}, d \in \mathcal{D}$.

Once solutions to RPs are given, one can use a Wave Front Tracking algorithm to construct a sequence of approximate solutions.

4. Model assumptions

The aim of this section is to verify that the assumptions underlying the data networks fluid-dynamic model (shortly FD model) are correct. Here we focus on the fixed-point models to describe TCP, and considering various set-ups with TCP traffic in a single bottleneck topology, we investigate queueing models for estimating packet loss rate. In what follows we suppose $\rho_{max} = 1$ and $\sigma = \frac{1}{2}$.

4.1 Loss probability function

It is reasonable to assume that the loss probability function p is null for some interval, which is a right neighborhood of zero. This means that at low densities no packet is lost. Then p should be increasing, reaching the value 1 at the maximal density, the situation of complete stuck. With the above assumptions the loss probability function in (4) can be written as:

$$p(\rho) = \begin{cases} 0, & 0 \leq \rho \leq 1/2, \\ \frac{2\rho-1}{\rho}, & 1/2 \leq \rho \leq 1. \end{cases} \tag{16}$$

We analyze some models used in literature to evaluate the packets loss rate with the aim to compare its behaviour with the function depicted in Figure 1.

4.1.1 The proportional-excess model

Let us consider the transmission of two consecutive routers. The node that transmits packets is called *sender*, while the receiving one is said *receiver*. Among the nodes, there is a link or channel, with limited capacity. Assume that the sender and the receiver are synchronized each other, i.e. the receiver is able to process in real time all packets, sent by the sender. In few words, no packets are lost. The packets loss can occur only on the link, due to its finite capacity. Under the zero buffer hypotheses the loss rate is defined as the proportional excess of offered traffic over the available capacity. If R is the sender bit rate and C is the link capacity, we have a loss if $R > C$. The model is said *proportional-excess* or briefly P/E and suppose deterministic arrivals. The packets bit rate is:

$$p = \begin{cases} 0, & R < C, \\ \frac{R-C}{C}, & R > C. \end{cases} \tag{17}$$

In Figure 3, loss probability for P/E model (continuous curve) and FD model (dashed curve) are shown, assuming $C = \sigma = 1/2$. For values $C < \rho < 2C$, the FD model overestimates the loss probability.

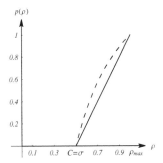

Fig. 3. Loss probabilities. Dashed line: FD model. Continuous line: P/E model.

Observe that the P/E model is not realistic. In fact, the sender and the receiver are never synchronized each other and whatever transmission protocol is used by the transport layer, the receiver has a finite length buffer, where the packets wait to be processed and eventually sent to the next node. Thus queueing models are needed, to infer about network performance.

4.1.2 Models with finite capacity

Queueing models are good at predicting loss in a network with many independent users, probably using different applications. Consider the traffic from TCP sources that send packets through a bottleneck link. The traffic is aggregated and used as an arrival process for the link. The arrival process, being the aggregation of independent sources, is approximated as a Poisson process, and the aggregated throughput is used as the rate of the Poisson process (see Wierman et al. (2003)). These considerations justify the assumption that the times between the packets arrivals are exponentially distributed. Depending on the hypothesis on the length of packets arriving to the queue the data transmission can be modelled with different queueing models, as $M/D/1/B$ and $M/M/1/B$, characterized by deterministic and exponentially distributed lengths, respectively, and a buffer with capacity $B - 1$. From the queue length distribution, known in closed formulas or iteratively in the finite buffer case, expected time in queue and in the system, as well as packet loss rate can be derived. In what follows we denote the arrival intensity by λ, the service intensity by μ and define the load as $\rho = \lambda/\mu$.

4.1.2.1 Fixed packets dimension

In a scenario where all senders use the same data packets size, the queueing model $M/D/1/B$ is the most natural choice. The probability that the buffer is full gives the loss rate:

$$p(\rho) = \frac{1 + (\rho - 1)\,\alpha_B\,(\rho)}{1 + \rho\alpha_B\,(\rho)}, \tag{18}$$

where

$$\alpha_B\,(\rho) = \sum_{k=0}^{B-2} \frac{e^{\rho(B-k-1)}\,(-1)^k\,(B-k-1)^k\,\rho^k}{k!}, \quad B \geq 2.$$

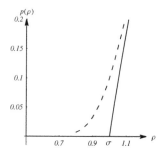

Fig. 4. Loss rates. Dashed line: $M/D/1/B$ model. Continuous line: FD model.

Figure 4 shows a comparison among the loss rate (16) and (18), assuming $B = 10$. However, an $M/D/1/B$ queue predicts a lower loss rate and higher throughput than is seen in the true network. This is due to fact that in real routers packet sizes are not always fixed to the maximum segment size, therefore packet sizes are more variable than a deterministic distribution.

4.1.2.2 Exponentially distributed packets size

Assume the packet size is exponentially distributed. This assumption is true if we consider the total amount of traffic as the superposition of traffic fluxes, coming from different TCP sources, each configured to use its own packet size. The $M/M/1/B$ queue is a good approximation of the simulated bottleneck link shared among TCP sources under any traffic load (Wierman et al. (2003)). The loss rate for the $M/M/1/B$ queueing model is:

$$p(\rho) = \frac{\rho^B\,(1 - \rho)}{1 - \rho^{B+1}}. \tag{19}$$

In Figure 5, left, the loss bit rate for different values of the buffer ($B = 10, 20, 30$) is reported. Notice that, increasing the B values, dashed lines tend to the continuous one.

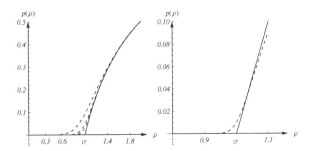

Fig. 5. Left: Loss bit rate for different values of the buffer. Right: Loss probability function. Dashed lines: $M/M/1/B$. Continuous line: P/E model.

In fact, the loss probability of the FD model represents for $\sigma = 1$ (up to a scale factor equal to 2) a limit case of (19):

$$\lim_{B \to \infty} \frac{\rho^B (1 - \rho)}{1 - \rho^{B+1}} = \begin{cases} 0, & 0 < \rho \leq 1, \\ \frac{\rho-1}{\rho}, & \rho > 1. \end{cases}$$

The loss probability for the queueing model (dashed line) and the P/E one (continuous line) is shown in Figure 5, right. The two curves almost match for small bit rate values, i.e. in the load range $0.9\sigma < \rho < 1.1\sigma$. For greater loads values, the P/E model overestimates the loss probability.

Theoretical and simulative studies pointed out that $M/D/1/B$ and $M/M/1/B$ queueing models give good prediction of the loss rate in network with many independent users performing short file transfers (shorts FTP). In literature other queueing models have been considered to describe different scenarios, as bach arrivals. For a comparison among different models see Figure 6, where the packet loss rate for $M/D/1/B$, $M/M/1/B$, $M^2/M/1/B$, $M^5/M/1/B$ and the P/E models are reported for the case $B = 100$ and loads in the interval $0.8 < \rho < 1.1$. Observe that $M^r/M/1/B$ denotes a queue with Poisson batch arrivals of size r and describes the fact that TCP traffic is likely to be quite bursty due to synchronized loss events that are experienced by multiple users.

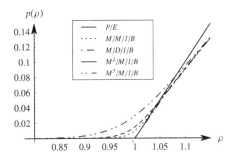

Fig. 6. Comparison of different queueing models.

Significant difference are restricted to the range $0.9\sigma < \rho < 1.1\sigma$. As the load increases above 1.1 the loss estimates become very close in the different queueing models. Any of these models predict the loss rate equally well. However, under low loss environments, the best queueing model depends on the type of transfers by TCP sources, i.e. persistent or transient. It is shown in Olsen (2003) that $M/D/1/B$ queues estimations of the loss rate can be used for transient sources. However, for sources with a slightly longer on and off periods, $M/M/1/B$ queues best predict the loss rate, and for (homogeneous) persistent sources, $M^r/M/1/B$ queues give better performance inferences, due to the traffic burstiness stemming from the TCP slow-start and source synchronization effect. Even if some models are more appropriate in situations of low load, others when the load is heavy, Figure 6 shows that the assumption on the loss probability function of the FD model is valid.

4.2 Velocity

The loss probability, influencing the average transmission time, has effects on the average velocity of packets:

$$v(\rho) = \bar{v}\left(1 - p(\rho)\right).$$

The behaviour of the average velocity in the FD model

$$v(\rho) = \begin{cases} \bar{v}, & 0 \leq \rho \leq /2, \\ \bar{v}\frac{1-\rho}{\rho}, & 1/2 \leq \rho \leq 1, \end{cases} \tag{20}$$

is depicted in Figure 1. Notice that the velocity is constant if the system is free (no losses). Over the threshold, losses occur, and the average travelling time increasing reduces the velocity. The average packet velocity for the P/E model and the $M/M/1/B$ model is plotted in Figure 7. Such two curves fit the curve of the FD model, confirming the goodness of its assumptions.

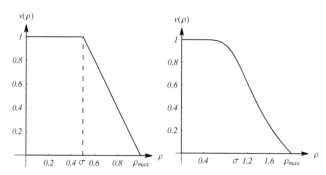

Fig. 7. Average velocity. Left: P/E model. Right: $M/M/1/B$ model.

4.3 Flux

Once the velocity function is known, the flux is given by:

$$f(\rho) = v(\rho)\rho.$$

In case of the FD model

$$f(\rho) = \begin{cases} \bar{v}\rho, & 0 \leq \rho \leq 1/2, \\ \bar{v}(1 - \rho), & 1/2 \leq \rho \leq 1, \end{cases} \tag{21}$$

see Figure 1. For the P/E model, we get

$$f(\rho) = \begin{cases} \rho \bar{v}, & 0 \leqslant \rho \leqslant \sigma, \\ \frac{(2\sigma - \rho)\bar{v}\rho}{\sigma}, & \sigma \leqslant \rho \leqslant \rho_{max}. \end{cases} \tag{22}$$

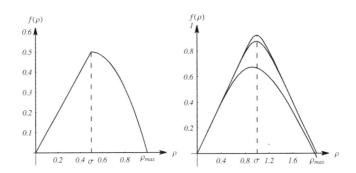

Fig. 8. Flux. Left: P/E model. Right: $M/M/1/B$ (for $B = 5$, $B = 15$, $B = 25$).

The flux in the P/E model and $M/M/1/B$ model are depicted in Figure 8. Note the effects of a finite buffer on the maximal value of the flux. If B tends to infinity, the flux best approximates the FD model flux. For small B values, the maximal flux decreases and the load value in which the maximum is attained is shifted on the right due to the fact that packets are lost for load values smaller than the threshold.

5. Optimal control problems for telecommunication networks

Now we state optimal control problems on the network.

We have a network $(\mathcal{I}, \mathcal{J})$, with nodes of at most 2×2 type, and an initial data $\rho_0 = (\rho_{i,0})_{i=1,...,N}$. The evolution is determined by equation (9) on each line I_i and by Riemann Solvers RS_J, depending on priority and traffic distribution parameters, q and α, respectively. For the definition of RS_J see the case when the traffic distribution function is of type 2b).

We now consider α and q as controls. To measure the efficiency of the network, it is natural to consider two quantities:

1) The average velocity at which packets travel through the network.
2) The average time taken by packets from source to destination.

Clearly, to optimize 1) and 2) is the same if we refer to a single packet, but the averaged values may be very different (since there is a nonlinear relation among the two quantities). As the model consider macroscopic quantities, we can estimate the averages integrating over time and space the average velocity and the reciprocal of average velocity, respectively. We thus define the following:

$$J_1(t) = \sum_i \int_{I_i} v(\rho_i(t, x)) \, dx,$$

$$J_2(t) = \sum_i \int_{I_i} \frac{1}{v(\rho_i(t, x))} \, dx,$$

and, to obtain finite values, we assume that the optimization horizon is given by $[0, T]$ for some $T > 0$.

Notice that this corresponds to the following operation:

- average in time and then w.r.t packets, to compute the probability loss function;
- average in space, to pass to the limit and get model (9);
- integrate in space and time to get the final value.

The value of such functionals depends on the order in which averages and integrations are taken.

Summarizing, we get the following optimal control problems:

Data. Network $(\mathcal{I}, \mathcal{J})$; initial data $\bar{\rho} = (\bar{\rho}_i)_{i=1,\dots,N}$; optimization horizon $[0, T]$, $T > 0$.

Dynamics. Equation (9) on each line $I \in \mathcal{I}$ and Riemann Solver RS_J for each $J \in \mathcal{J}$, depending on controls α and q.

Control Variables. Traffic distribution parameter $t \mapsto \alpha_J(t)$ and priority parameter $t \mapsto q_J(t)$, i.e. two controls for every node $J \in \mathcal{J}$.

Control Space. $\{(\alpha_J, q_J) : J \in \mathcal{J}, \ \alpha_J, q_J \in L^\infty([0, T], [0, 1])\}$.

Cost functions. Integrated functionals:

$$\max \int_0^T J_1(t)\, dt, \qquad \min \int_0^T J_2(t)\, dt.$$

Definition 7. *We call* **(P$_i$)** *the optimal control problem referred to the functional J_i:*

$$(\mathbf{P_1}) \quad \max_{(\alpha, q)} J_1, \text{ subject to (9).}$$

$$(\mathbf{P_2}) \quad \min_{(\alpha, q)} J_2, \text{ subject to (9).}$$

The direct solution of problems **(P$_i$)** corresponds to a centralized approach. We propose the alternative approach of decentralized algorithm more precisely:

Step 1 For every node J and Riemann Solver RS_J, solve the simplified optimal control problem:

$$\max (or \ min) \ J_i(T),$$

for T sufficiently big, on the network formed only by J with constant initial data, taking approximate solutions when there is lack of existence.

Step 2 Apply the obtained optimal control at every time t in the optimization horizon and at every node J, taking the value at J on each line as initial data.

Notice that, for T sufficiently big, we can assume that the datum is constant on each line: this strongly simplifies the approach.

We consider a single node J with incoming lines, labelled by 1 and 2, and with outgoing lines, labelled by 3 and 4.

Since $\check{\rho} = \check{\gamma}$, $0 \leq \check{\rho} \leq \frac{1}{2}$, and $\hat{\rho} = 1 - \hat{\gamma}$, $\frac{1}{2} \leq \hat{\rho} \leq 1$, we have that $v(\hat{\rho}_\varphi) = H(-s_\varphi) +$

$\frac{1-\widehat{\rho}_\varphi}{\widehat{\rho}_\varphi} H\left(s_\varphi\right), \varphi = 1, 2, v\left(\widehat{\rho}_\psi\right) = H\left(-s_\psi\right) + \frac{1-\widehat{\rho}_\psi}{\widehat{\rho}_\psi} H\left(s_\psi\right), \psi = 3, 4$, where $H(x)$ is the Heavyside function and s_φ and s_ψ are determined by the solution to the RP at J:

$$s_\varphi = \begin{cases} -1, \text{ if } \rho_{\varphi,0} \leq \frac{1}{2} \text{ and } \Gamma = \Gamma_{in}, \text{ or } \rho_{\varphi,0} \leq \frac{1}{2}, q_\varphi \Gamma = \gamma_\varphi^{max} \text{ and } \Gamma = \Gamma_{out}, \\ +1 \text{ if } \rho_{\varphi,0} > \frac{1}{2}, \text{ or } \rho_{\varphi,0} \leq \frac{1}{2}, q_\varphi \Gamma < \gamma_\varphi^{max} \text{ and } \Gamma = \Gamma_{out}, \end{cases} \varphi = 1, 2,$$

$$s_\psi = \begin{cases} -1, \text{ if } \rho_{\psi,0} < \frac{1}{2}, \text{ or } \rho_{\psi,0} \geq \frac{1}{2}, \alpha_\psi \Gamma < \gamma_\psi^{max} \text{ and } \Gamma = \Gamma_{in}, \\ +1 \text{ if } \rho_{\psi,0} \geq \frac{1}{2} \text{ and } \Gamma = \Gamma_{out}, \text{ or } \rho_{\psi,0} \geq \frac{1}{2}, \alpha_\psi \Gamma = \gamma_\psi^{max} \text{ and } \Gamma = \Gamma_{in}. \end{cases} \psi = 3, 4,$$

with:

$$q_\varphi = \begin{cases} q, & \text{if } \varphi = 1, \\ 1 - q, \text{ if } \varphi = 2, \end{cases} \qquad \alpha_\psi = \begin{cases} \alpha, & \text{if } \psi = 3, \\ 1 - \alpha, \text{ if } \psi = 4. \end{cases}$$

Then, for T sufficiently big,

$$J_1(T) = 2\left[v\left(\widehat{\rho}_1\right) + v\left(\widehat{\rho}_2\right) + v\left(\widehat{\rho}_3\right) + v\left(\widehat{\rho}_4\right)\right]; \tag{23}$$

$$J_2(T) = t\left(\widehat{\rho}_1\right) + t\left(\widehat{\rho}_2\right) + t\left(\widehat{\rho}_3\right) + t\left(\widehat{\rho}_4\right), \tag{24}$$

with

$$t\left(\widehat{\rho}_x\right) = \frac{\widehat{\rho}_x}{H\left(s_x\right) + \widehat{\rho}_x\left[H\left(-s_x\right) - H\left(s_x\right)\right]}.$$

We want to maximize the cost $J_1(T)$ and to minimize the cost $J_2(T)$ with respect to the parameters α and q. In Marigo (2006) and Cascone et al. (2007), you can find a similar approach for telecommunication networks and road networks, respectively, modelled with flux function (8). Let

$$\beta^- = \frac{\Gamma - \gamma_3^{max}}{\gamma_3^{max}}, \quad \beta^+ = \frac{\gamma_4^{max}}{\Gamma - \gamma_4^{max}},$$

$$p^- = \frac{\Gamma - \gamma_1^{max}}{\gamma_1^{max}}, \quad p^+ = \frac{\gamma_2^{max}}{\Gamma - \gamma_2^{max}}.$$

Theorem 8. *Consider a junction J of 2×2 type. If $\Gamma = \Gamma_{in} = \Gamma_{out}$ and T is sufficiently big, the cost functionals $J_1(T)$ and $J_2(T)$ depend neither on α nor q. If $\Gamma = \Gamma_{in}$, the cost functionals $J_1(T)$ and $J_2(T)$ depend only on α. The optimal values for $J_1(T)$ are the following:*

(i) *if $s_3 = s_4 = +1$, and $\beta^- \leq 1 \leq \beta^+$, $\beta^- \beta^+ > 1$, or $1 \leq \beta^- \leq \beta^+$, $\alpha \in \left[0, \frac{1}{1+\beta^+}\right]$;*

(ii) *if $s_3 = s_4 = +1$, and $\beta^- \leq 1 \leq \beta^+$, $\beta^- \beta^+ = 1$, $\alpha \in \left[0, \frac{1}{1+\beta^+}\right] \cup \left[\frac{1}{1+\beta^-}, 1\right]$;*

(iii) *if $s_3 = s_4 = +1$, and $\beta^- \leq 1 \leq \beta^+$, $\beta^- \beta^+ < 1$, or $\beta^- \leq \beta^+ \leq 1$, $\alpha \in \left[\frac{1}{1+\beta^-}, 1\right]$;*

(iv) *if $s_3 = -s_4 = -1$, $\alpha \in \left[0, \frac{1}{1+\beta^+}\right]$ in the cases: $\beta^- \leq 1 \leq \beta^+$, $1 \leq \beta^- \leq \beta^+$, or $\beta^- \leq \beta^+ \leq 1$;*

(v) *if $s_3 = -s_4 = +1$, $\alpha \in \left]\frac{1}{1+\beta^-}, 1\right]$ in the cases: $\beta^- \leq 1 \leq \beta^+$, $1 \leq \beta^- \leq \beta^+$, or $\beta^- \leq \beta^+ \leq 1$.*

If $\Gamma = \Gamma_{in}$, the optimal values for $J_2(T)$ are the following:

(i) *if $s_3 = s_4 = +1$ or $s_c = -s_d = -1$, and $\beta^- \leq 1 \leq \beta^+$, $\alpha = \frac{1}{2}$;*

(ii) *if $s_3 = s_4 = +1$, and $\beta^- \leq \beta^+ \leq 1$, $\alpha \in \left[0, \frac{1}{1+\beta^+}\right]$;*

(iii) *if $s_3 = s_4 = +1$, and $1 \leq \beta^- \leq \beta^+$, $\alpha \in \left[\frac{1}{1+\beta^-}, 1\right]$;*

(iv) *if $s_3 = -s_4 = -1$, and $1 \leq \beta^- \leq \beta^+$, or $\beta^- \leq \beta^+ \leq 1$, $\alpha \in \left[0, \frac{1}{1+\beta^+}\right]$;*

(v) *if $s_3 = -s_4 = +1$, and $\beta^- \leq 1 \leq \beta^+$, or $1 \leq \beta^- \leq \beta^+$, or $\beta^- \leq \beta^+ \leq 1$, $\alpha \in \left]\frac{1}{1+\beta^-}, 1\right]$.*

If $\Gamma = \Gamma_{out}$, the cost functionals $J_1(T)$ and $J_2(T)$ depend only on q. The optimal values for $J_1(T)$ and $J_2(T)$ are the same for α when $\Gamma = \Gamma_{in}$, if we substitute α with q, β^- with p^-, and β^+ with p^+.

5.1 A case study

In what follows, we report the simulation results of a test telecommunication network, that consists of nodes of 2×2 type. The network, represented in Figure 9, is characterized by:

- 24 nodes;
- 12 incoming lines: $1, 2, 5, 8, 9, 16, 19, 20, 31, 32, 45, 46$;
- 12 outgoing lines: $6, 17, 29, 43, 48, 50, 52, 54, 56, 58, 59, 60$;
- 36 inner lines: $3, 4, 7, 10, 11, 12, 13, 14, 15, 18, 21, 22, 23, 24, 25, 26, 27, 28, 30, 33, 34, 35, 36, 37, 38, 39, 40, 41, 42, 44, 47, 49, 51, 53, 55, 57$.

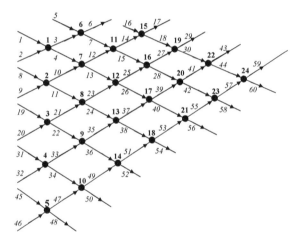

Fig. 9. Network with 24 nodes.

We distinguish three case studies, that can be called, case A, B, and C. In Table 1, we report the initial conditions $\rho_{i,0}$ and the boundary data (if necessary) $\rho_{bi,0}$ for case A.

As for case B, instead, we consider the same initial conditions of case A, but boundary data equal to 0.75.

Table 2 contains initial and boundary conditions for case C. An initial condition of 0.75 is assumed for the inner lines of the network, that are not present in Table 2.

As in Bretti et al. (2006), we consider approximations obtained by the numerical method of Godunov (Godunov (1959)), with space step $\Delta x = 0.0125$ and time step determined by the CFL condition (Godlewsky et al. (1996)). The telecommunication network is simulated in a time interval $[0, T]$, where $T = 50$ min. We study four simulation cases, choosing the flux function (7) or the flux function (8):

Line	$\rho_{i,0}$	$\rho_{bi,0}$	Line	$\rho_{i,0}$	$\rho_{bi,0}$	Line	$\rho_{i,0}$	$\rho_{bi,0}$
1	0.4	0.4	21	0.3	/	41	0.1	/
2	0.35	0.35	22	0.2	/	42	0.1	/
3	0.3	/	23	0.1	/	43	0.25	0
4	0.2	/	24	0.1	/	44	0.3	/
5	0.35	0.35	25	0.2	/	45	0.4	0.4
6	0.2	0	26	0.1	/	46	0.3	0.3
7	0.25	/	27	0.2	/	47	0.2	/
8	0.4	0.4	28	0.25	/	48	0.4	0
9	0.35	0.35	29	0.2	0	49	0.35	/
10	0.3	/	30	0.4	/	50	0.3	0
11	0.2	/	31	0.35	0.35	51	0.2	/
12	0.1	/	32	0.3	0.3	52	0.1	0
13	0.1	/	33	0.2	/	53	0.1	/
14	0.25	/	34	0.35	/	54	0.2	0
15	0.3	/	35	0.2	/	55	0.1	/
16	0.4	0.4	36	0.25	/	56	0.2	0
17	0.3	0	37	0.4	/	57	0.25	/
18	0.2	/	38	0.35	/	58	0.2	0
19	0.4	0.4	39	0.3	/	59	0.15	0
20	0.35	0.35	40	0.2	/	60	0.15	0

Table 1. Initial conditions and boundary data for the lines of the network for case A.

Line	$\rho_{i,0}$	$\rho_{bi,0}$	Line	$\rho_{i,0}$	$\rho_{bi,0}$	Line	$\rho_{i,0}$	$\rho_{bi,0}$
1	0.4	0.4	19	0.4	0.4	48	0.5	0.7
2	0.5	0.5	20	0.5	0.5	50	0.5	0.7
5	0.5	0.5	29	0.4	0.7	52	0.4	0.7
6	0.4	0.7	31	0.4	0.4	54	0.5	0.7
8	0.4	0.4	32	0.4	0.4	56	0.4	0.7
9	0.5	0.5	43	0.4	0.7	58	0.5	0.7
16	0.4	0.4	45	0.4	0.4	59	0.5	0.7
17	0.4	0.7	46	0.5	0.5	60	0.5	0.7

Table 2. Initial conditions and boundary data for the lines of the network for case C.

1. at each node parameters, that optimize the cost functionals J_1 and J_2 (optimal case);
2. random α and q parameters (static random case) chosen in a random way at the beginning of the simulation process (for each simulation case, 100 static random simulations are made);
3. dynamic random parameters (dynamic random case) which change randomly at every step of the simulation process.

In the following pictures, we show the values of the functionals J_1 and J_2, computed on the whole network, as function of time. A legend for every picture indicates the different simulation cases.

The algorithm of optimization, which is of local type, can be applied to complex networks, without compromising the possibility of a global optimization. This situation is evident if we

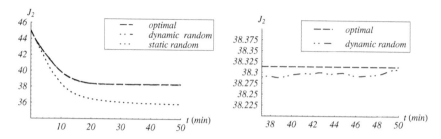

Fig. 10. J_1 for flux function (8), case A, and zoom around the optimal and dynamic random case (right).

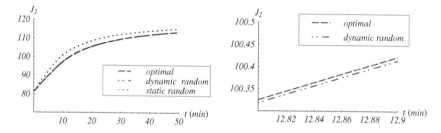

Fig. 11. J_2 for flux function (8), case B, and zoom around the optimal and dynamic random case (right).

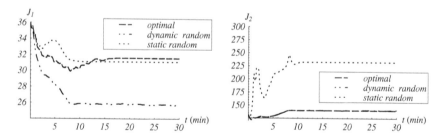

Fig. 12. J_1 and J_2 for flux function (7), case C.

consider the behaviour of J_1 for case A and J_2 for case B. For cases A and B, the cost functionals simulated with flux function (7) are constant, which is not surprising since the initial data on the lines is less than $\frac{1}{2}$. In case C, we present the behaviour of the cost functionals J_1 and J_2 for flux function (7). Boundary data are of Dirichlet type (unlike case A and B where we have considered Neumann boundary conditions) and the network is simulated with high incoming fluxes for the incoming lines and high initial conditions for inner lines. We can see, from Figure 12, that J_1 and J_2 are not constant as in cases A and B. Moreover, we have to take in mind that we have two different optimization algorithms for J_1 and J_2. Notice that the dynamic random case follows the optimal case for J_2 and not for J_1. Indeed, the optimal algorithm for J_1 presents an interesting aspect. When simulation begins, it is worst than the static random configuration. In the steady state, instead, the optimal configuration is the highest.

As for the dynamic random simulation, its behaviour looks very similar to the optimal one for cases A and B (for case C, only J_2 presents optimal and dynamic random configurations, that are very similar). Hence, we could ask if it is possible to avoid the optimization of the network, and operate in dynamic random conditions. Indeed, this last case originates strange phenomena, that cannot be modelled, hence it is preferred to avoid such a situation for telecommunication network design. To give a confirmation of this intuition, focus the attention on line 13, that is completely inside the network and it is strongly influence by the dynamics at various nodes. In Figure 13, we see that, using optimal parameters, the density on line 13 shows a smoother profile than the one obtained through a dynamic random simulation.

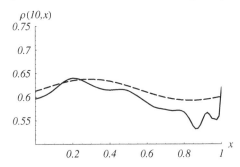

Fig. 13. Behaviour of the density on line 13 of the network of Figure 9, for $t = 10$, flux function (7), case C, in optimal and dynamic random simulations. Dashed line: optimal simulation for J_2; solid line: dynamic random simulation.

6. Conclusions

A fluid-dynamic model for data networks has been described. The main advantages of this approach, with respect to existing ones, can be summarized as follows. The fluid-dynamic models are completely evolutive, thus they are able to describe the traffic situation of a network every instant of time, overcoming the difficulties encountered by many static models. An accurate description of queues formation and evolution on the network is possible. The theory permits the development of efficient numerical schemes for very large networks. The model is based on packets conservation at intermediate time scales, whose flux is determined via a loss probability function (at fast time scales) and on a semilinear equation for the evolution of the percentage of packets going from an assigned source to a given destination. The choice of the loss probability function is of paramount importance in order to achieve a feasible model. The fluid dynamic model has been compared with those obtained using various queueing paradigms, from proportional/excess to models with finite capacity, including different distributions for packet sizes. The final result is that such models give rise to velocity profiles and flux functions which are quite similar to the fluid dynamic ones. In order to solve dynamics at node, Riemann Solvers have been defined considering different traffic distribution functions (which indicate for each junction J the outgoing direction of traffic that started at source s, has d as final destination and reached J from an assigned incoming road) and rules RA1 and RA2. The algorithm RA1, already used for road traffic models, requires the definition of a traffic distribution matrix, whose coefficients describe the percentage of packets, forwarded from incoming lines to outgoing ones. Using the algorithm

RA2, not considered for urban traffic as redirections are not expected from modelling point of view (except in particular cases, as strong congestions or road closures), priority parameters, indicating priorities among flows of incoming lines, and distribution coefficients have to be assigned.

The main differences between the two algorithms are the following. The first one simply sends each packet to the outgoing line which is naturally chosen according to the final packet destination. The algorithm is blind to possible overloads of some outgoing lines and, by some abuse of notation, is similar to the behaviour of a "switch". The second algorithm, on the contrary, sends packets to outgoing lines in order to maximize the flux both on incoming and outgoing lines, thus taking into account the loads and possibly redirecting packets. Again by some abuse of notation, this is similar to a "router" behaviour. Hence, RA1 forwards packets on outgoing lines without considering the congestion phenomena, unlike RA2. Observe that a routing algorithm of RA1 type working through a routing table, according to which flows are sent with prefixed probabilities to the outgoing links, is of "distance vector" type. Reverse, an algorithm of RA2 type can redirect packets on the basis of link congestions, so it works on the link states (hence on their congestions) and so it is of "link-state" type.

The performance analysis of the networks was made through the use of different cost functionals, measuring average velocity and average travelling time, using the model consisting of the conservation law. The optimization is over parameters, which assign priority among incoming lines and traffic distribution among outgoing lines. A complete solution is provided in a simple case, and then used as local optimal choice for a complex test network. Three different choices of parameters have been considered: locally optimal, static random, and dynamic random (changing in time). The local optimal outperforms the others. Then, the behaviour of packets densities on the lines, that permits to rule out the dynamic random case has been analyzed.

All the optimization results have been obtained using a decentralized approach, i.e. an approach which sets local optimal parameters for each junction of the network. The cooperative aspect of such decentralized approach is the following. When a router optimizes the (local) functionals, it takes into considerations entering and exiting lines. Such lines reach other nodes, which benefit from the optimal choice. This in fact reflects in good global behavior as showed by simulations, described below. In future we aim to extend the optimization results to more general junctions and to explore global optimization techniques.

7. References

Alderson, D.; Chang, H.; Roughan, M.; Uhlig, S. & Willinger, W. (2007). The many facets of internet topology and traffic, *Networks and Heterogenous Media*, Vol. 1, Issue 4, 569–600, ISSN 1556-1801.

Baccelli, F.; Chaintreau, A.; De Vleeschauwer, D. & McDonald, D. (2006). HTTP turbulence, *Networks and Heterogeneous Media*, Vol. 1, 1–40, ISSN 1556-1801.

Baccelli, F; Hong, D. & Liu, Z. (2001). Fixed points methods for the simulation of the sharing of a local loop by large number of interacting TCP connections, *Proceedings of the ITC Specialist Conference on Local Loop*, 1–27, Barcelona, Spain, (also available in Technical Report RR-4154, INRIA, Le Chesnay Cedex, France), ISBN 0249-6399.

Bretti, G.; Natalini, R. & Piccoli, B. (2006). Numerical approximations of a traffic flow model on networks, *Networks and Heterogeneous Media*, Vol. 1, 57–84, ISSN 1556-1801.

Bressan, A. (2000). *Hyperbolic Systems of Conservation Laws - The One-dimensional Cauchy Problem*, Oxford University Press, ISBN 0198507003, Oxford.

Cascone, A.; D'Apice, C.; Piccoli, B. & Raritá, L. (2007). Optimization of traffic on road networks, *Mathematical Models in Applied Sciences*, Vol. 17, 1587–1617, ISSN 0218-2025.

Cascone, A.; Marigo, A.; Piccoli, B. & Rarità, L. (2010). Decentralized optimal routing for packets flow on data networks, *Discrete and Continuous Dynamical Systems - Series B (DCDS - B)*, Vol. 13, No. 1, 59–78, ISSN 15313492.

Coclite, G.; Garavello, M. & and Piccoli, B. (2005). Traffic Flow on a Road Network, *SIAM Journal on Mathematical Analysis*, Vol. 36, 1862–1886, ISNN 0036-1410.

Dafermos, C. (1999). *Hyperbolic Conservation Laws in Continuum Physics*, Springer-Verlag, ISBN 354064914X, New York.

Daganzo, C. (1997). *Fundamentals of Transportation and Traffic Operations*, Pergamon-Elsevier, ISBN 0080427855, Oxford.

D'Apice, C.; Manzo, R. & Piccoli, B. (2006). Packet flow on telecommunication networks, *SIAM Journal on Mathematical Analysis*, Vol. 38, No. 3, 717–740, ISNN 0036-1410.

D'Apice, C.; Manzo, R. & Piccoli, B. (2008). A fluid dynamic model for telecommunication networks with sources and destinations, *SIAM Journal on Applied Mathematics (SIAP)*, Vol. 68, No. 4, 981–1003, ISSN 0036-1399.

D'Apice, C.; Manzo, R. & Piccoli, B. (2010). On the validity of fluid-dynamic models for data networks, *Journal of Networks*, submitted, ISSN 1796-2056.

Garavello, M. & Piccoli, B. (2006). *Traffic flow on networks*, AIMS Series on Applied Mathematics, vol. 1, American Institute of Mathematical Sciences, ISBN 1601330006, United States.

Godlewsky E. & Raviart, P. (1996). *Numerical Approximation of Hyperbolic Systems of Conservation Laws*, Springer Verlag, ISBN 978-0-387-94529-3, Heidelberg.

Garavello, M. & Piccoli, B. (2005). Source-Destination Flow on a Road Network, *Communication in Mathematical Sciences*, Vol. 3, 261–283, ISSN 1539-6746.

Godunov, S. K. (1959). A difference method for numerical calculation of discontinuous solutions of the equations of hydrodynamics, *Mat. Sb.*, Vol. 47, 271–306, ISSN 0368-8666.

Holden, H. & Risebro, N. H. (1995). A mathematical model of traffic flow on a network of unidirectional roads, *SIAM Journal on Mathematical Analysis*, Vol. 26, 999–1017, ISSN 0036-1410.

Marigo, A. (2006). Optimal distribution coefficients for telecommunication networks, *Networks and Heterogeneous Media*, Vol. 1, 315–336, ISNN 1556-1801.

Kelly, F.; Maulloo, A. K. & Tan, D. K. H. (1998). Rate control in communication networks: shadow prices, proportional fairness and stability, *Journal of the Operational Research Society*, Vol. 49, 237–252, ISSN 0160-5682.

Lighthill, M. J. & Whitham, G. B. (1955). On kinetic waves. II. Theory of Traffic Flows on Long Crowded Roads, *Proc. R. Soc. Lond. Ser. A Math. Phys. Eng. Sci.*, Vol. 229, 317–345, doi: 10.1098/rspa.1955.0089.

Newell, G. F. (1980). *Traffic Flow on Transportation Networks*, MIT Press, ISBN 0262140322, Cambridge (MA,USA).

Olsén, J. (2003). On Packet Loss Rates used for TCP Network Modeling, *Technical Report*, Uppsala University.

Richards, P. I. (1956). Shock Waves on the Highway, *Oper. Res.*, Vol. 4, 42–51, ISSN 0030-364X.

Serre, D. (1999). *Systems of conservation laws I and II*, Cambridge University Press, ISBN 521582334, 521633303, Cambridge.

Tanenbaum, A. S. (2003). *Computer Networks*, Prentice Hall, ISBN 0130661023, Upper Saddle River.

Wierman, A.; Osogami, T. & Olsén, J. (2003). A Unified Framework for Modeling TCP-Vegas, TCP-SACK, and TCP-Reno, *Proceedings of the IEEE/ACM International Symposium on modeling, Analysis and Simulation of Computer and Telecommunication Systems (MASCOTS)*, 269–278, ISBN 0-7695-2039-1, Orlando, Florida, October 2003, Los Alamitos, California, Washington.

Willinger, W. & Paxson, V. (1998). Where Mathematics meets the Internet, *Notices of the AMS*, Vol. 45, 961–970, ISSN 0002-9920.

10

Optimal Control Strategies for Multipath Routing: From Load Balancing to Bottleneck Link Management

C. Bruni, F. Delli Priscoli, G. Koch, A. Pietrabissa and L. Pimpinella
Dipartimento di Informatica e Sistemistica "A. Ruberti",
"Sapienza" Università di Roma, Roma,
Italy

1. Introduction

In this work we face the Routing problem defined as an optimal control problem, with control variables representing the percentages of each flow routed along the available paths, and with a cost function which accounts for the distribution of traffic flows across the network resources (multipath routing). In particular, the scenario includes the load balancing problem already dealt with in a previous work (Bruni et al., 2010) as well as the bottleneck minimax control problem. The proposed approaches are then compared by evaluating the performances of a sample network.

In a given network, the resource management problem consists in taking decisions about handling the traffic amount which is carried by the network, while respecting a set of Quality of Service (QoS) constraints.

As stated in Bruni et al., 2009a, b, the resource management problem is hardly tackled by a single procedure. Rather, it is currently decomposed in a number of subproblems (Connection Admission Control (CAC), traffic policing, routing, dynamic capacity assignment, congestion control, scheduling), each one coping with a specific aspect of such problem. In this respect, the present work is embedded within the general approach already proposed by the authors in Bruni et al., 2009a, b, according to which each of the various subproblems is given a separate formulation and solution procedure, which strives to make the other sub-problems easier to be solved. More specifically, the above mentioned approach consists in charging the CAC with the task of deciding, on the basis of the network congestion state, new connection admission/blocking and possible forced dropping of the in-progress connections with the aim of maximizing the number of accepted connections, whilst satisfying the QoS requirements.

According to the proposed approach, the role of the other resource management procedures is the one of keeping the network as far as possible far from the congestion state. Indeed, the more the network is kept far from congestion, the higher is the number of new connection set-up attempts that can be accepted by the CAC without infringing the QoS constraints, and hence the traffic carried by the network increases. By so doing, the CAC and the other resource management procedures can work in a consistent way, while being kept independent.

This work deals with the multipath routing problem. Multipath routing is a widespread topic in the literature. For example, Cidon et al., 1999, and Banner and Orda, 2007, demonstrate the advantages of multipath routing with respect to single-path routing in terms of network performances; Chen et al., 2004, considers the multipath routing problem under bandwidth and delay constraints; Lin and Shroff, 2006, formulate the multipath routing problem as a utility maximization problem with bandwidth constraints; Guven et al., 2008, extend the multipath routing to multicast flows; Jaffe, 1981, Tsai et al., 2006, Tsai and Kim, 1999 deal with the multipath routing as a minimax optimization problem.

In this work we face the multipath routing problem formulated as an optimal control problem, with control variables representing the percentages of each flow routed along the available paths. As a matter of fact, in the most advanced networks each flow can be simultaneously routed over more than one path: the routing procedure has to decide the percentages of the traffic belonging to the considered flow which have to be routed over the paths associated to the flow in question. According to the above mentioned vision, we assume that other resource management control units (specifically the CAC) already dealt with and decided about issues such as how many, which ones, when and for how long connections have to be admitted in the network, with specific QoS constraints (related to losses and delays) to be satisfied. Therefore, the routing control unit has to deal with an already defined offered traffic. Thus, the admissible set for the routing control variables turns out to be closed, bounded and non-empty, and the existence of (at least) an optimal solution of the routing problem is guaranteed.

The goal of an optimal routing policy aims the routing problem solution towards a network traffic pattern which should make QoS requirements and consequently the CAC task (implicitly) easier to be satisfied. The quality of the routing solution will be evaluated by different performance indices, which take a nominal capacity for each link into account.

As far as the dynamical aspects of a routing problem, we first note that explicitly accounting for them would call for a reliable and sufficiently general dynamical model for the offered traffic. However it is widely acknowledged that such a model is not available and hard to design, due to unpredictable features of Internet traffic. And, in any case, the requested dynamical characters are committed to the CAC procedures, where the more reliable connection dynamics model along with the feedback structure may properly handle the issue.

In addition, a non-dynamical set up for the routing problem makes it much easier to be dealt with. Moreover, this approach could be justified by assuming that the time scale for changes in the routing policy is surely slower than the bit rate fluctuations in the in-progress connections, but it is reasonably faster than the evolution of traffic statistical features. Thus, the routing policy has to be periodically computed to fit the most likely traffic pattern at each given period of time.

In this work, we consider the possibility/opportunity of splitting the given network into sub-networks as detailed in Bruni et al., 2010 each one controlled by a separate subset of variables.

This work is organized as follows. In Section 2, a definition for a reference communication network and its decomposition is given, which is useful for the routing problem; in Sections 3, we in depth study the optimal routing control problem with reference to a number of different cost functions; Section 4 shows some results in order to evaluate the performance and to compare the found optimal solutions for traffic balancing and bottleneck link management; finally, concluding remarks in Section 5 end the work.

2. Reference telecommunication network definition and decomposition

At any fixed time, the telecommunication network can be defined in terms of its topological description as well as in terms of its traffic pattern. As far as network topology is concerned, we consider the network nodes $n \in N = \{n_1, n_2, ..., n_N\}$ and the network links defined as ordered pairs of nodes $l \in \Lambda = \{l_1, l_2, ..., l_L\}$. To describe the network traffic request we first define a path $v \in \Omega = \{v_1, v_2, ..., v_V\}$ as a collection of consecutive links, denoted by Λ_v, from an ingoing node i to an outgoing node j (where $i,j \in N$). Moreover a certain set of different Service Classes $k \in K = \{k_1, k_2, ..., k_K\}$, is defined, each one characterized by a set of Quality of Service (QoS) parameters. According to the most recent trends, the QoS control is performed on a per flow basis, where a flow $f \in \Phi = \{f_1, f_2, ..., f_F\}$ is defined as the triple $f = (n_i, n_j, k_p)$, with n_i denoting the ingoing node, n_j denoting the outgoing node and k_p denoting the service class. The traffic associated with a given flow f may possibly be routed on a set Ω_f of one or more paths. We further introduce the set of indices $\{a(l,v), l \in \Lambda, v \in \Omega\}$, defined as follows:

$$a(l,v) = \begin{cases} 1, & \text{if } l \in v \\ 0, & \text{otherwise} \end{cases} \qquad (1)$$

For each link $l \in \Lambda$, at the given time, we may consider its occupancy level $c(l)$ defined as the sum of all contributions to the occupancy due to the flows routed on the link itself. Each contribution of this type will be quantified by the bit rate $R(l,f)$ which, in turn, is the sum of bit rates of all in-progress connections going through the link l and relevant to the flow f, possibly weighted by a coefficient $\alpha(l,f)$ which accounts for the specific need of the flow itself. Therefore we have:

$$c(l) = \sum_{f \in \Phi} \alpha(l,f) R(l,f) \qquad (2)$$

where $\alpha(l,f)$ are positive known coefficients which take into account the fact that some technologies differentiate the classes of service by varying modulation, coding, and so on. For each link l, we consider the so-called nominal capacity $c_{NOM}(l)$, that is the value of the occupancy level suggested for a proper behaviour of the link (typically in terms of QoS)[1].

---

[1] $c(l)$ and $c_{NOM}(l)$ can be interpreted as generalizations of "load factor" and "Noise Rise" in UMTS (see Holma and Toskala, 2002).

that indicates the fraction of $R(f)$ to be routed on path $v \in \Omega_f$. Then, due to the bit conservation law, we have:

$$R(l,f) = \sum_{v \in \Omega_f} \alpha(l,f) R(f) u(f,v) \qquad (3)$$

where obviously:

$$u(f,v) \in [0,1], \ \forall f \in \Phi, \ \forall v \in \Omega_f$$

$$\sum_{v \in \Omega_f} u(f,v) = 1, \ \forall f \in \Phi \qquad (4)$$

As shown in Bruni *et al.*, 2010, with reference to the routing control problem, the link set Λ might be decomposed into separated subclasses $\Lambda^{(j)}$, $j = 0,1,2,...,P$, each of them involving separate subsets of control variables, where $\Lambda^{(0)}$ is the set, possibly empty, of links that cannot be controlled by any control variable and which therefore they are not involved in any routing control problem.

For every communicating class of links $\Lambda^{(j)} \subset \Lambda$, there exists the (uniquely) corresponding communicating class of flows $\Phi^{(j)} \subset \Phi$ defined as the set of flows such that, for each $f \in \Phi^{(j)}$, there exists (at least) a link $l \in \Lambda^{(j)}$, and therefore a pair of links (generally depending on f itself), which are controllable with respect to f. Clearly, the set $\Phi^{(0)}$ coincides with the empty set. We now observe that the set $\{\Phi^{(j)}, \ j \geq 0\}$ of flow communicating classes forms a partition of Φ, corresponding to the fact that the set $\{\Lambda^{(j)}, \ j \geq 0\}$ of link communicating classes forms a partition of Λ. This partition for Λ and Φ immediately induces a partition of the network. Note that each j-th part of the network is controlled by a corresponding subvector of control variables, later defined as $u^{(j)}$ independently of the other parts; the components of the vector $u^{(j)}$ are the variables $u(f,v)$, $f \in \Phi^{(j)}$, $v \in \Omega_f$. In the following $\{\Lambda^{(j)}, \Phi^{(j)}\}$ will denote a sub-network. We will use the detailed network decomposition procedure described in Bruni *et al.*, 2010, facing the routing control problem in each sub-network (but in $\Lambda^{(0)}$).

3. A rationale for the network loading

In the following, we will focus attention on the routing problem for any given sub-network $\{\Lambda^{(j)}, \Phi^{(j)}\}$. As mentioned above, any such problem is characterized by a set $u^{(j)}$ of control variables, which may be (optimally) selected independently of the other ones. As stated in Bruni *et al.*, 2010, the admissible set for $u^{(j)}$ is defined by the constraints:

$$u(f,v) \in [0,1], \ \forall f \in \Phi^{(j)}, \ \forall v \in \Omega_f \qquad (5)$$

$$\sum_{v \in \Omega_f} u(f,v) = 1, \ \forall f \in \Phi^{(j)} \qquad (6)$$

so that the set itself turns out to be convex. From here on, for sake of simplicity the apices j will be dropped.

The optimal choice for u within its (convex) admissible set may be performed according to a cost function which assesses the network loading. In a previous work Bruni et al., 2010, the control goal was the normalized load balancing in the sub-network, evaluated by the function:

$$J(u) = \sum_{l \in \Lambda} \left(\frac{c(l)}{c_{NOM}(l)} - k \right)^2 \tag{7}$$

with k a given constant. If, for any given u, we optimize (7) with respect to k, we get:

$$k = \frac{1}{L} \sum_{l \in \Lambda} \frac{c(l)}{c_{NOM}(l)} \tag{8}$$

with L denoting the cardinality of Λ. In Bruni et al., 2010, and Bruni et al., 2010 (to appear), a shortcoming of (7) was enlightened, which is due to the partial controllability property (therein defined) of some of the links. These links, in the following referred to as "ballast", are such that they are bound to accept traffic flows not controlled by the components of the control vector u. Thus other choices of the cost function might be considered which more explicitly account for the network overloading.

One first possibility is to assess the link overflow setting $k = 0$ in (7), thus more generally arriving at the functions:

$$J(u) = \sum_{l \in \Lambda} \left(\frac{c(l)}{c_{NOM}(l)} \right)^m \tag{9}$$

for some integer $m \geq 1$. If the target is to give more importance to the links belonging to several paths the function (9) can be rewritten as follows:

$$J(u) = \sum_{v \in \Omega} \sum_{l \in \Lambda_v} \left(\frac{c(l)}{c_{NOM}(l)} \right)^m \tag{10}$$

According to (9), (10) we try to distribute the total load in the network in such a way that the higher the normalized load for a link is, the stronger is the effort in reducing it. This selective attention to the most heavily loaded links progressively increases with m. As m keeps increasing, then function (10) is approximated by:

$$J(u) = \sum_{v \in \Omega} (G_v)^m \tag{11}$$

where:

$$G_v = \max_{l \in \Lambda_v} \frac{c(l)}{c_{NOM}(l)} \tag{12}$$

Thus for each path v the optimization attention is just focused on the most heavily loaded link of the path itself (bottleneck). Eventually we can consider the worst bottleneck load over the whole sub-network:

$$J(u) = \max_{v \in \Omega} G_v \qquad (13)$$

Remark. Some methods are proposed in the literature to solve the above minimax optimization problem (see Warren *et al.*, 1967, Osborne and Wetson, 1969, Blander *et al.*, 1972, Blander and Charambous, 1972). The original minimax problem (11) is equivalent to the following:

$$\min_{u,g \in U} J(u,g) \qquad (14)$$

$$J(u,g) = \sum_{v \in \Omega} [g(v)]^m \qquad (15)$$

$$U = \left\{ (u,g) \in R^{V(F+1)} : u(f,v) \geq 0, \ \sum_{v \in \Omega_f} u(f,v) = 1, \right.$$

$$\left. \frac{c(l)}{c_{NOM}(l)} \leq g(v), \forall f \in \Phi, \forall v \in \Omega_f \right\} \qquad (16)$$

where g is the vector of auxiliary variables $g(v)$, $v \in \Omega$. This is a nonlinear (linear if $m = 1$, quadratic if $m = 2$) programming problem that can be solved by well-established methods. We observe that the equivalence lies in the fact that, once (14) (15) (16) is solved, the optimal value assumed by $g(v)$ coincides with G_v in (12), for $v \in \Omega$, i.e., it represents the normalized bottleneck link load of path v.

The load balancing problem (7) (8), with constraints (5) (6) and the bottleneck load management problem (14) (15) (16) are easily seen to be convex. This allows standard minimization routines to be used for its solution, such as MatLab simulation tools.

Remark. The cost function (13) enlightens a further advantage of network decomposition. Indeed, in case the decomposition had not been performed, then (13) would describe an ill-posed optimal control problem whenever the worst bottleneck over the whole network happens to be an uncontrollable link. Similar considerations hold for cost function (11).

4. Evaluation and comparison of optimal routing procedures

4.1 Network structure and decomposition

The considered scenario is composed by 16 nodes and 19 links (see Fig. 1 a)). The traffic pattern involves 4 traffic flows of the same service class k, from 4 source nodes n_i, $i = 1,..,4$, to 4 different destination nodes n_j, $j = 11,..,14$. The traffic pattern is described by the set of traffic flows $\Phi = \{f_1, f_2, f_3, f_4\}$, where each traffic flow is identified by the following triples: $f_1 = (n_1, n_{11}, k)$, $f_2 = (n_2, n_{12}, k)$, $f_3 = (n_3, n_{13}, k)$, $f_4 = (n_4, n_{14}, k)$. After performing the network decomposition as in Bruni *et al.*, 2010, we recognize three sub-networks (see, Fig. 1 b), c) and d)). The network topology is summarized in Table 1, where the network decomposition is reported as well.

	l_1	l_2	l_3	l_4	l_5	l_6	l_7	l_8	l_9	l_{10}	l_{11}	l_{12}	l_{13}	l_{14}	l_{15}	l_{16}	l_{17}	l_{18}	l_{19}	f_1	f_2	f_3	f_4
c_{NOM} [kbps]	10	10	5.4	5.4	5.4	5.4	10	5.4	5.4	5.4	5.4	5.4	5.4	5.4	5.4	5.4	5.4	5.4	5.4				
v_1	x									x											x		
v_2		x				x				x											x		
v_3			x			x						x										x	
v_4			x			x							x			x						x	
v_5				x		x							x				x						x
v_6				x				x						x									x
v_7				x	x									x									x
v_8						x									x		x						x
$\Lambda^{(0)}$			x	x																			
$\Lambda^{(1)}$	x	x			x	x	◊			x	x			x		x			x	x	x		x
$\Lambda^{(2)}$								x	x		x	x		x		x	x			x	x		

Table 1. Network Topology and Decomposition; the first row shows the nominal link capacities in [Mbps]; the generic entry (l_i,v_j) is denoted by 'x' if $l_i \in \Lambda^{(j)}v$; the generic entry (f_i,v_j) is denoted by 'x' if it is possible to route f_i on path v_j; the generic entry $(l_i,\Lambda^{(j)})$ is denoted by 'x' if $l_i \in \Lambda^{(j)}$, or by '◊' if $l_i \in \Lambda^{(j)}$ and l_i is a ballast link; the generic entry $(f_i,\Lambda^{(j)})$ is denoted by 'x' if $f_i \in \Phi^{(j)}$.

The considered scenario has been simulated with MatLab. In particular we have tested two simuation sets reported in subsection 4.2 and 4.3respectively. In subsection 4.2 we considered the Bottleneck Link Management by varying the weights of the bottleneck loads, while in subsection 4.3 we made comparisons between Load Balancing and Bottleneck Link Management.

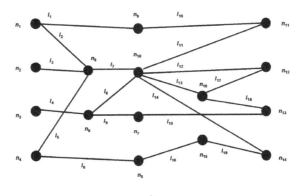

a)

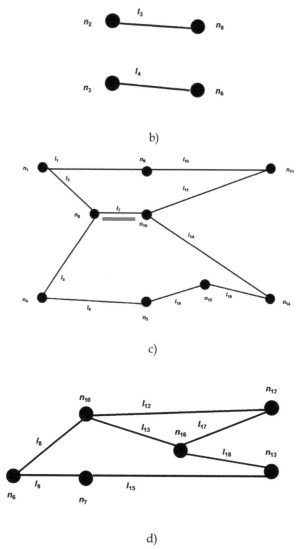

b)

c)

d)

Fig. 1. a) Global Network, b) Sub-network 0 ($\Lambda^{(0)}$), c) Sub-network 1 ($\Lambda^{(1)}$), d) Sub-network 2 ($\Lambda^{(2)}$).

4.2 Optimal routing for different weights of bottleneck loads

In this simulation set we consider that the bit rate of traffic flows f_1, f_3, f_4 is equal to 5 Mbps whilst the bit rate of traffic flow f_2 is equal to 5.4 Mbps. Fig. 2 and 3 show the dependence of the optimal solutions on index m of the Bottleneck Link Management problem.

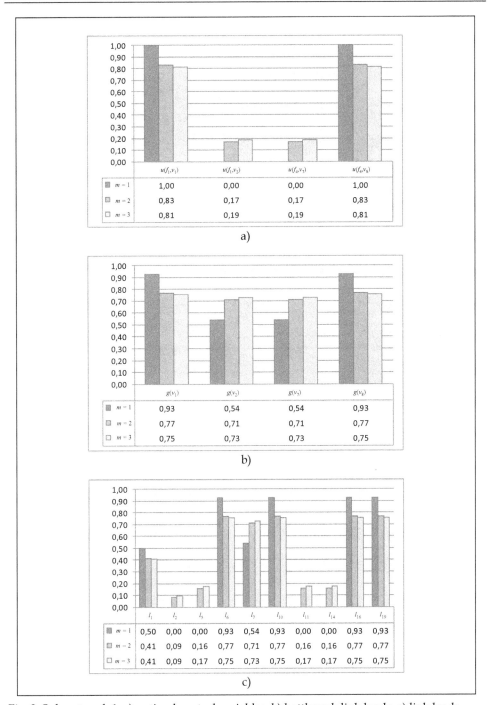

Fig. 2. Sub-network 1: a) optimal control variables, b) bottleneck link loads, c) link loads.

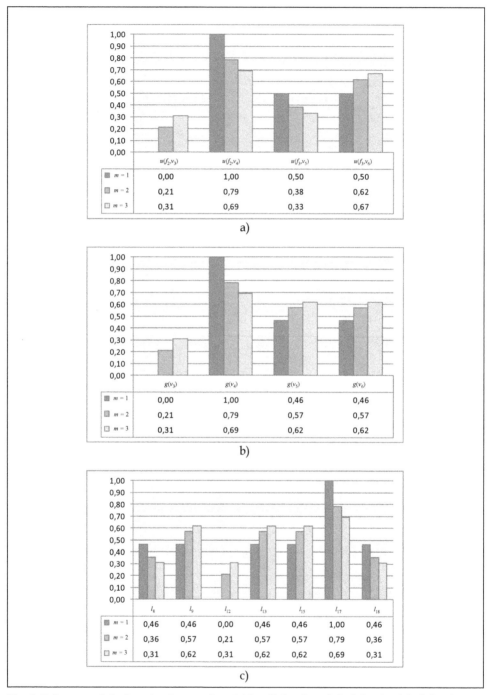

Fig. 3. Sub-network 2: a) optimal control variables, b) bottleneck link loads, c) link loads.

4.3 Comparisons between load balancing and bottleneck link management

In this simulation set we consider that all the traffic sources transmit with an increasing trend from 4.5 Mbps to 8.5 Mbps. Tables 2-5 show the network load as the sources bit rate increase, and compares the optimal bottleneck control solutions for $m = 1, 2, 3$ with the load balancing optimal solution.

In Tables 2-5, we denote by bold characters the normalized link loads exceeding 1; hereinafter, the corresponding links will be denoted as overloaded links.

The bottleneck control for $m \geq 2$ manages a higher network load than the load balancing approach. In fact, the tables show that the solutions of the bottleneck control problem are such that no link is overloaded until the flow rates exceed 5 Mbps, 6.5 Mbps and 6.5 Mbps for $m = 1,2,3$, respectively; on the other hand, the load balancing solutions are such that no link is overloaded until the flow rates exceed 5 Mbps. Similar results are obtained for sub-network 2.

Rate [Mbps]	4	4.5	5	5.5	6	6.5	7	7.5	8	8.5
$u(f_1,v_1)$						1,00				
$u(f_1, v_2)$						0,00				
$u(f_4, v_7)$						0,00				
$u(f_4, v_8)$						1,00				
$g(v_1)$	0,74	0,83	0,93	**1,02**	**1,11**	**1,20**	**1,30**	**1,39**	**1,48**	**1,57**
$g(v_2)$	0,40	0,45	0,50	0,55	0,60	0,65	0,70	0,75	0,80	0,85
$g(v_7)$	0,40	0,45	0,50	0,55	0,60	0,65	0,70	0,75	0,80	0,85
$g(v_8)$	0,74	0,83	0,93	**1,02**	**1,11**	**1,20**	**1,30**	**1,39**	**1,48**	**1,57**
l_1	0,40	0,45	0,50	0,55	0,60	0,65	0,70	0,75	0,80	0,85
l_2	0,00	0,00	0,00	0,00	0,00	0,00	0,00	0,00	0,00	0,00
l_5	0,00	0,00	0,00	0,00	0,00	0,00	0,00	0,00	0,00	0,00
l_6	0,74	0,83	0,93	**1,02**	**1,11**	**1,20**	**1,30**	**1,39**	**1,48**	**1,57**
l_7	0,40	0,45	0,50	0,55	0,60	0,65	0,70	0,75	0,80	0,85
l_{10}	0,74	0,83	0,93	**1,02**	**1,11**	**1,20**	**1,30**	**1,39**	**1,48**	**1,57**
l_{11}	0,00	0,00	0,00	0,00	0,00	0,00	0,00	0,00	0,00	0,00
l_{14}	0,00	0,00	0,00	0,00	0,00	0,00	0,00	0,00	0,00	0,00
l_{16}	0,74	0,83	0,93	**1,02**	**1,11**	**1,20**	**1,30**	**1,39**	**1,48**	**1,57**
l_{19}	0,74	0,83	0,93	**1,02**	**1,11**	**1,20**	**1,30**	**1,39**	**1,48**	**1,57**

Table 2. Sub-network 1: Optimal Solutions under bottleneck control, $m = 1$.

Rate [Mbps]	4	4.5	5	5.5	6	6.5	7	7.5	8	8.5
$u(f_1,v_1)$						0,81				
$u(f_1, v_2)$						0,19				
$u(f_4, v_7)$						0,19				
$u(f_4, v_8)$						0,81				
$g(v_1)$	0,60	0,67	0,75	0,82	0,90	0,97	1,05	1,12	1,20	1,27
$g(v_2)$	0,55	0,62	0,69	0,76	0,83	0,90	0,97	1,04	1,11	1,18
$g(v_7)$	0,55	0,62	0,69	0,76	0,83	0,90	0,97	1,04	1,11	1,18
$g(v_8)$	0,60	0,67	0,75	0,82	0,90	0,97	1,05	1,12	1,20	1,27
l_1	0,32	0,36	0,40	0,44	0,48	0,52	0,57	0,61	0,65	0,69
l_2	0,08	0,09	0,10	0,11	0,12	0,13	0,13	0,14	0,15	0,16
l_5	0,14	0,16	0,18	0,20	0,21	0,23	0,25	0,27	0,29	0,30
l_6	0,60	0,67	0,75	0,82	0,90	0,97	**1,05**	**1,12**	**1,20**	**1,27**
l_7	0,55	0,62	0,69	0,76	0,83	0,90	0,97	**1,04**	**1,11**	**1,18**
l_{10}	0,60	0,67	0,75	0,82	0,90	0,97	**1,05**	**1,12**	**1,20**	1,27
l_{11}	0,14	0,16	0,18	0,20	0,21	0,23	0,25	0,27	0,29	0,30
l_{14}	0,14	0,16	0,18	0,20	0,21	0,23	0,25	0,27	0,29	0,30
l_{16}	0,60	0,67	0,75	0,82	0,90	0,97	**1,05**	**1,12**	**1,20**	**1,27**
l_{19}	0,60	0,67	0,75	0,82	0,90	0,97	**1,05**	**1,12**	**1,20**	**1,27**

Table 3. Sub-network 1: Optimal Solutions under bottleneck control, $m = 2$.

Rate [Mbps]	4	4.5	5	5.5	6	6.5	7	7.5	8	8.5
$u(f_1,v_1)$						0,79				
$u(f_1, v_2)$						0,21				
$u(f_4, v_7)$						0,21				
$u(f_4, v_8)$						0,79				
$g(v_1)$	0,59	0,66	0,73	0,81	0,88	0,95	1,03	1,10	1,18	1,25
$g(v_2)$	0,57	0,64	0,71	0,78	0,85	0,92	0,99	1,06	1,13	1,20
$g(v_7)$	0,57	0,64	0,71	0,78	0,85	0,92	0,99	1,06	1,13	1,20
$g(v_8)$	0,59	0,66	0,73	0,81	0,88	0,95	1,03	1,10	1,18	1,25
l_1	0,32	0,36	0,40	0,44	0,48	0,52	0,56	0,59	0,63	0,67
l_2	0,08	0,09	0,10	0,11	0,12	0,13	0,14	0,16	0,17	0,18
l_5	0,15	0,17	0,19	0,21	0,23	0,25	0,27	0,29	0,31	0,33
l_6	0,59	0,66	0,73	0,81	0,88	0,95	**1,03**	**1,10**	**1,18**	**1,25**
l_7	0,57	0,64	0,71	0,78	0,85	0,92	0,99	**1,06**	**1,13**	**1,20**
l_{10}	0,59	0,66	0,73	0,81	0,88	0,95	**1,03**	**1,10**	**1,18**	**1,25**
l_{11}	0,15	0,17	0,19	0,21	0,23	0,25	0,27	0,29	0,31	0,33
l_{14}	0,15	0,17	0,19	0,21	0,23	0,25	0,27	0,29	0,31	0,33
l_{16}	0,59	0,66	0,73	0,81	0,88	0,95	**1,03**	**1,10**	**1,18**	**1,25**
l_{19}	0,59	0,66	0,73	0,81	0,88	0,95	**1,03**	**1,10**	**1,18**	**1,25**

Table 4. Sub-network 1: Optimal Solutions under bottleneck control, $m = 3$.

Rate [Mbps]	4	4.5	5	5.5	6	6.5	7	7.5	8	8.5
$u(f_1,v_1)$						0,60				
$u(f_1, v_2)$						0,40				
$u(f_4, v_7)$						0,45				
$u(f_4, v_8)$						0,55				
$g(v_1)$	0,24	0,27	0,30	0,33	0,36	0,39	0,42	0,45	0,48	0,51
$g(v_2)$	0,16	0,18	0,20	0,22	0,24	0,26	0,28	0,30	0,32	0,34
$g(v_7)$	0,33	0,37	0,41	0,45	0,50	0,54	0,58	0,62	0,66	0,70
$g(v_8)$	0,41	0,46	0,51	0,56	0,62	0,67	0,72	0,77	0,82	0,87
l_1	0,74	0,83	0,92	**1,01**	**1,11**	**1,20**	**1,29**	**1,38**	**1,48**	**1,57**
l_2	0,45	0,50	0,56	0,61	0,67	0,72	0,78	0,84	0,89	0,95
l_5	0,30	0,33	0,37	0,41	0,44	0,48	0,52	0,55	0,59	0,63
l_6	0,33	0,37	0,41	0,45	0,50	0,54	0,58	0,62	0,66	0,70
l_7	0,41	0,46	0,51	0,56	0,62	0,67	0,72	0,77	0,82	0,87
l_{10}	0,41	0,46	0,51	0,56	0,62	0,67	0,72	0,77	0,82	0,87
l_{11}	0,24	0,27	0,30	0,33	0,36	0,39	0,42	0,45	0,48	0,51
l_{14}	0,16	0,18	0,20	0,22	0,24	0,26	0,28	0,30	0,32	0,34
l_{16}	0,33	0,37	0,41	0,45	0,50	0,54	0,58	0,62	0,66	0,70
l_{19}	0,41	0,46	0,51	0,56	0,62	0,67	0,72	0,77	0,82	0,87

Table 5. Sub-network 1: Optimal Solutions under load balancing control.

4.4 Decomposition evaluation

With the purpose of evaluating the decomposition strategy, in this simulation set we consider randomly generated networks, flows and paths, and use the decomposition algorithm to partition the network in sub-networks. The networks were generated starting from a grid of nodes; in particular, the considered network width is 10 nodes. Each column of the grid can be assigned a number of nodes; in the considered network, the number of nodes per column is [18, 18, 18, 16, 10, 10, 16, 18, 18, 18]. 30 flows were considered, starting from a random node of the first column of the network and directed to a random node of the last column. Similarly, each network path is directed from a node of the first column of the network and directed to a node of the last column Fig. 4 a) shows an example of randomly generated network, whereas Fig. 4 a) shows an example of sub-network. The results were obtained by averaging 20 simulations. The average number of variables of the original problem (i.e., the non-decomposed one) is 1984.8, whereas the decomposition manages to decompose the network in 10.2 sub-network (in the average): each sub-network optimization problem has therefore 194.6 variables, i.e., each sub-network problem is reduced by about one order of magnitude.

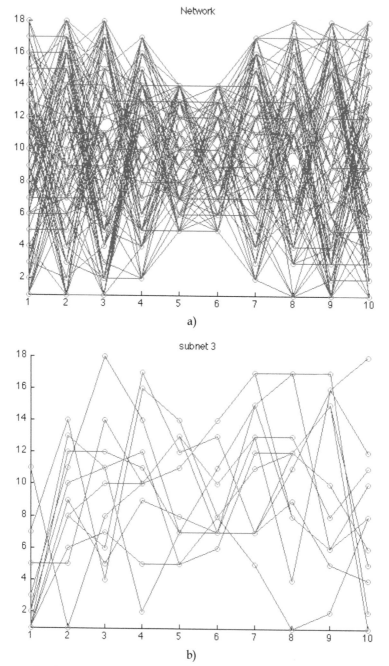

Fig. 4. a) example of a network (width=10, height=18), b) one of the sub-networks resulting from the decomposition of the network in Fig. 4 a).

5. Conclusion

In this work we formulate the multipath routing problem as an optimal control problem considering various performance indices. In particular, the scenario includes the load balancing problem already dealt with in a previous work Bruni *et al.*, 2010, as well as the bottleneck minimax control problem, in which the traffic load of the bottleneck (raised to a given power *m*) is minimized. The mathematical structure of the problem might easily suggest some issues which are evidentiated by the results of Section 4, simply intended to provide a numerical example of more general behaviours. On one side, the load balancing performance index obviously allows to achieve a higher uniformity in the loading of the various links, but it cannot prevent overloading of possible ballast links (apart from *ad hoc* modifications suggested in Bruni *et al.*, 2010).

On the other side, the minimax (bottleneck) approach succeeds in keeping the bottleneck loads (including the ones of the ballast links), as low as possible, with an effort which happens to be more successful the higher the value of *m* is. This allows accommodating for a higher traffic flow.

Moreover, we stress the fact that the choice of the proper performance index is a matter left to the network manager in charge of the routing control problem, who will have to take into account at the same time the network structure and capacity, as well as the admitted traffic flow and the possible presence of ballast links.

As a final conclusion, we have considered several cost functions for the multipath routing which are suitable for a certain network load situation. Those cost functions can be properly switched during the operations according to the network needs. In that way our approach is strongly oriented with the most innovative vision of the Future Internet perspective (see Delli Priscoli, 2010), in which the core idea is to take consistent and coordinated decisions according to the present contest.

6. References

Bruni, C., Delli Priscoli, F., Koch, G., Marchetti, I. (2009). Resource management in network dynamics: An optimal approach to the admission control problem, Computers & Mathematics with Applications, article in press, available at www.sciencedirect.com, 8 September 2009,doi:10.1016/j.camwa.2009.01.046

Bruni, C., Delli Priscoli, F., Koch, G., Marchetti, I. (2009,b)"Optimal Control of Connection Admission in Telecommunication Networks", European Conrol Conference (ECC) 09, Budapest (Hungary), pp. 2929-2935.

Bruni, C., Delli Priscoli, F., Koch, G., Pietrabissa, A., Pimpinella, L., (2010) "Multipath Routing by Network Decomposition and Traffic Balancing", Proceedings Future Network and Mobile Summit.

Holma H., Toskala A., (2002) WCDMA for UMTS, 2nd Edition.

Warren, A. D., Lasdon, L. S., Suchman, D. F., (1967) Optimization in engineering design, Proc. IEEE, 1885-1897.

Osborne, M. R., Watson, G. A., (1969) An Algorithm for minimax approximation in the non-linear case, Comput. J., 12, pp. 63-68.

Bandler, J. W., Srinivasan, T.V., Charalambous, (1972) Minimax Optimization of networks by Grazor Search, IEEE Trans. Microwave Theory Tech., MTT-20, 596-604.

Bandler, J. W., Charalambous, C. (1972), Practical least pth optimization of networks, IEEE Trans. Microwave Theory Tech., MTT-20, 834-840.

Brayton, R.K., S.W. Director, G.D. Hachtel, and L.Vidigal, (1979), A New Algorithm for Statistical Circuit Design Based on Quasi-Newton Methods and Function Splitting, IEEE Trans. Circuits and Systems, Vol. CAS-26, pp. 784-794. Demyanov, V. F., Malozemov, V. N., (1974) Introduction to minimax, John Wiley & Sons.

Cidon, I., Rom R., Shavitt Y., (1999) Analysis of Multipath Routing, IEEE/ACM Transactions ON Networking, Vol. 7, No. 6, pp. 885-896

Banner, R., Orda A., (2007), Multipath Routing Algorithms for Congestion Minimization, IEEE/ACM Transactions on Networking, Vol. 15, No. 2, pp. 413-424

Chen, J., Chan, S.-H. Gary, Li V. O. K., (2004) Multipath Routing for Video Delivery Over Bandwidth-Limited Networks, IEEE Journal on Selected Areas in Communications, Vol. 22, No. 10, pp. 1920-1932

Lin, X., Shroff, N. B., (2006)"Utility Maximization for Communication Networks With Multipath Routing", IEEE Transactions on Automatic Control, Vol. 51, No. 5, pp. 766-781

Güven, T., La, R. J., Shayman, M. A., Bhattacharjee, B., (2008), A Unified Framework for Multipath Routing for Unicast and Multicast Traffic, IEEE/ACM Transactions on Networking, Vol. 16, No. 5, pp. 1038-1051

Jaffe J. M., "Bottlneck Flow Control", IEEE Transactions on Communications, Vol 29, No 7, July 1981, pp. 954-962

Tsai, D., Liau, T. C., Tsai, Wei K., (2006) Least Square Approach to Multipath Maxmin Rate Allocation, 14th IEEE International Conference on Networks, 2006 (ICON '06), Vol. 1, pp. 1-6.

Tsai, Wei K., Kim, Y., (1999) Re-Eximining Maxmin Protocols: A Fundamental Study on Convergence, Complexity, Variations and Performance, 18th Annual Joint Conference of the IEEE Computer and Communications Societies (INFOCOM '99), Vol. 2, March 1999, pp. 811-818

Delli Priscoli, F., (2010) A Fully Cognitive Approach for Future Internet, Future Internet ISSN 1999-5903 available at www.mdpi.com/journal/futureinternet.

Permissions

The contributors of this book come from diverse backgrounds, making this book a truly international effort. This book will bring forth new frontiers with its revolutionizing research information and detailed analysis of the nascent developments around the world.

We would like to thank Professor Jesús Hamilton Ortiz, for lending his expertise to make the book truly unique. He has played a crucial role in the development of this book. Without his invaluable contribution this book wouldn't have been possible. He has made vital efforts to compile up to date information on the varied aspects of this subject to make this book a valuable addition to the collection of many professionals and students.

This book was conceptualized with the vision of imparting up-to-date information and advanced data in this field. To ensure the same, a matchless editorial board was set up. Every individual on the board went through rigorous rounds of assessment to prove their worth. After which they invested a large part of their time researching and compiling the most relevant data for our readers. Conferences and sessions were held from time to time between the editorial board and the contributing authors to present the data in the most comprehensible form. The editorial team has worked tirelessly to provide valuable and valid information to help people across the globe.

Every chapter published in this book has been scrutinized by our experts. Their significance has been extensively debated. The topics covered herein carry significant findings which will fuel the growth of the discipline. They may even be implemented as practical applications or may be referred to as a beginning point for another development. Chapters in this book were first published by InTech; hereby published with permission under the Creative Commons Attribution License or equivalent.

The editorial board has been involved in producing this book since its inception. They have spent rigorous hours researching and exploring the diverse topics which have resulted in the successful publishing of this book. They have passed on their knowledge of decades through this book. To expedite this challenging task, the publisher supported the team at every step. A small team of assistant editors was also appointed to further simplify the editing procedure and attain best results for the readers.

Our editorial team has been hand-picked from every corner of the world. Their multi-ethnicity adds dynamic inputs to the discussions which result in innovative outcomes. These outcomes are then further discussed with the researchers and contributors who give their valuable feedback and opinion regarding the same. The feedback is then collaborated with the researches and they are edited in a comprehensive manner to aid the understanding of the subject.

Apart from the editorial board, the designing team has also invested a significant amount of their time in understanding the subject and creating the most relevant covers. They scrutinized every image to scout for the most suitable representation of the subject and create an appropriate cover for the book.

The publishing team has been involved in this book since its early stages. They were actively engaged in every process, be it collecting the data, connecting with the contributors or procuring relevant information. The team has been an ardent support to the editorial, designing and production team. Their endless efforts to recruit the best for this project, has resulted in the accomplishment of this book. They are a veteran in the field of academics and their pool of knowledge is as vast as their experience in printing. Their expertise and guidance has proved useful at every step. Their uncompromising quality standards have made this book an exceptional effort. Their encouragement from time to time has been an inspiration for everyone.

The publisher and the editorial board hope that this book will prove to be a valuable piece of knowledge for researchers, students, practitioners and scholars across the globe.

List of Contributors

Oleksandr Korchenko, Maksym Lutskiy and Sergiy Gnatyuk
National Aviation University, Ukraine

Petro Vorobiyenko and Yevhen Vasiliu
Odessa National Academy of Telecommunication named after O.S. Popov, Ukraine

Xiuquan Qiao, Xiaofeng Li and Junliang Chen
State Key Laboratory of Networking and Switching Technology, Beijing University of Posts and Telecommunications, China

Valery Bezruk, Alexander Bukhanko, Dariya Chebotaryova and Vacheslav Varich
Kharkov National University of Radio Electronics, Ukraine

Luis Zabala, Armando Ferro, Alberto Pineda and Alejandro Muñoz
University of the Basque Country (UPV/EHU), Spain

João Pedro
Nokia Siemens Networks Portugal S.A., Portugal

João Pedro and João Pires
Instituto de Telecomunicações, Instituto Superior Técnico, Portugal

Mihael Mohorčič and Aleš Švigelj
Jožef Stefan Institute, Slovenia

Matjaž Fras
Margento R&D, Maribor, Slovenia

Jože Mohorko and Žarko Čučej
University of Maribor, Faculty of Electrical Engineering and Computer Science, Maribor, Slovenia

Fabrice Guillemin
Orange Labs, Lannion, France

Bruno Sericola
INRIA Rennes - Bretagne Atlantique, Campus de Beaulieu, 35042 Rennes Cedex, France

Ciro D'Apice and Rosanna Manzo
Department of Electronic and Information Engineering, University of Salerno, Fisciano (SA), Italy

Benedetto Piccoli

Department of Mathematical Sciences, Rutgers University, Camden, New Jersey, USA

C. Bruni, F. Delli Priscoli, G. Koch, A. Pietrabissa and L. Pimpinella

Dipartimento di Informatica e Sistemistica "A. Ruberti", "Sapienza" Università di Roma, Roma, Italy

Printed in the USA
CPSIA information can be obtained
at www.ICGtesting.com
JSHW011441221024
72173JS00004B/894

Concepts and Design of Embedded Systems

Edited by **Alan Moore**

LANRYE
INTERNATIONAL

New Jersey

Published by Clanrye International,
55 Van Reypen Street,
Jersey City, NJ 07306, USA
www.clanryeinternational.com

Concepts and Design of Embedded Systems
Edited by Alan Moore

International Standard Book Number: 978-1-63240-116-8 (Hardback)

This book contains information obtained from authentic and highly regarded sources. Copyright for all individual chapters remain with the respective authors as indicated. A wide variety of references are listed. Permission and sources are indicated; for detailed attributions, please refer to the permissions page. Reasonable efforts have been made to publish reliable data and information, but the authors, editors and publisher cannot assume any responsibility for the validity of all materials or the consequences of their use.

The publisher's policy is to use permanent paper from mills that operate a sustainable forestry policy. Furthermore, the publisher ensures that the text paper and cover boards used have met acceptable environmental accreditation standards.

Trademark Notice: Registered trademark of products or corporate names are used only for explanation and identification without intent to infringe.

Printed in the United States of America.

Contents

Preface

This book aims to highlight the current researches and provides a platform to further the scope of innovations in this area. This book is a product of the combined efforts of many researchers and scientists, after going through thorough studies and analysis from different parts of the world. The objective of this book is to provide the readers with the latest information of the field.

These days, embedded systems, i.e. the computer systems that are embedded in different types of devices, play a crucial role in particular control functions, and have led to different aspects of industry. Hence, we can hardly talk about our life and society without mentioning embedded systems. A large number of high-quality fundamental and applied researches are required for the development of wide-ranging embedded systems; only diversified technologies together can make these systems. This book attempts to cover a broad range of research topics on embedded systems, inclusive of basic researches, theoretical studies, and practical work that would be helpful to researchers and engineers around the world.

I would like to express my sincere thanks to the authors for their dedicated efforts in the completion of this book. I acknowledge the efforts of the publisher for providing constant support. Lastly, I would like to thank my family for their support in all academic endeavors.

Editor

Part 1

Real-Time Property, Task Scheduling, Predictability, Reliability, and Safety

Ways for Implementing Highly-Predictable Embedded Systems Using Time-Triggered Co-Operative (TTC) Architectures

Mouaaz Nahas and Ahmed M. Nahhas
Department of Electrical Engineering, College of Engineering and Islamic Architecture,
Umm Al-Qura University, Makkah,
Saudi Arabia

1. Introduction

Embedded system is a special-purpose computer system which is designed to perform a small number of dedicated functions for a specific application (Sachitanand, 2002; Kamal, 2003). Examples of applications using embedded systems are: microwave ovens, TVs, VCRs, DVDs, mobile phones, MP3 players, washing machines, air conditions, handheld calculators, printers, digital watches, digital cameras, automatic teller machines (ATMs) and medical equipments (Barr, 1999; Bolton, 2000; Fisher et al., 2004; Pop et al., 2004). Besides these applications, which can be viewed as "noncritical" systems, embedded technology has also been used to develop "safety-critical" systems where failures can have very serious impacts on human safety. Examples include aerospace, automotive, railway, military and medical applications (Redmill, 1992; Profeta et al., 1996; Storey, 1996; Konrad et al., 2004).

The utilization of embedded systems in safety-critical applications requires that the system should have real-time operations to achieve correct functionality and/or avoid any possibility for detrimental consequences. Real-time behavior can only be achieved if the system is able to perform *predictable* and *deterministic* processing (Stankovic, 1988; Pont, 2001; Buttazzo, 2005; Phatrapornnant, 2007). As a result, the correct behavior of a real-time system depends on the time at which these results are produced as well as the logical correctness of the output results (Avrunin et al., 1998; Kopetz, 1997). In real-time embedded applications, it is important to predict the timing behavior of the system to guarantee that the system will behave correctly and consequently the life of the people using the system will be saved. Hence, predictability is the key characteristic in real-time embedded systems.

Embedded systems engineers are concerned with all aspects of the system development including hardware and software engineering. Therefore, activities such as specification, design, implementation, validation, deployment and maintenance will all be involved in the development of an embedded application (Fig. 1). A design of any system usually starts with ideas in people's mind. These ideas need to be captured in requirements specification documents that specify the basic functions and the desirable features of the system. The system design process then determines how these functions can be provided by the system components.

Fig. 1. The system development life cycle (Nahas, 2008).

For successful design, the system requirements have to be expressed and documented in a very clear way. Inevitably, there can be numerous ways in which the requirements for a simple system can be described.

Once the system requirements have been clearly defined and well documented, the first step in the design process is to design the overall system *architecture*. Architecture of a system basically represents an overview of the system components (i.e. sub-systems) and the interrelationships between these different components. Once the software architecture is identified, the process of implementing that architecture should take place. This can be achieved using a lower-level system representation such as an operating system or a *scheduler*. Scheduler is a very simple operating system for an embedded application (Pont, 2001). Building the scheduler would require a *scheduling algorithm* which simply provides the set of rules that determine the order in which the tasks will be executed by the scheduler during the system operating time. It is therefore the most important factor which influences predictability in the system, as it is responsible for satisfying timing and resource requirements (Buttazzo, 2005). However, the actual implementation of the scheduling algorithm on the embedded microcontroller has an important role in determining the functional and temporal behavior of the embedded system.

This chapter is mainly concerned with so-called "Time-Triggered Co-operative" (TTC) schedulers and how such algorithms can be implemented in highly-predictable, resource-constrained embedded applications.

The layout of the chapter is as follows. Section 2 provides a detailed comparison between the two key software architectures used in the design of real-time embedded systems, namely "time-triggered" and "event-triggered". Section 3 introduces and compares the two most known scheduling policies, "co-operative" and "pre-emptive", and highlights the advantages of co-operative over pre-emptive scheduling. Section 4 discusses the relationship between scheduling algorithms and scheduler implementations in practical embedded systems. In Section 5, Time-Triggered Co-operative (TTC) scheduling algorithm is introduced in detail with a particular focus on its strengths and drawbacks and how such drawbacks can be addressed to maintain its reliability and predictability attributes. Section 6 discusses the sources and impact of timing jitter in TTC scheduling algorithm. Section 7 describes various possible ways in which the TTC scheduling algorithm can be implemented on resource-constrained embedded systems that require highly-predictable system behavior. In Section 8, the various scheduler implementations are compared and contrasted in terms of jitter characteristics, error handling capabilities and resource requirements. The overall chapter conclusions are presented in Section 9.

2. Software architectures of embedded systems

Embedded systems are composed of hardware and software components. The success of an embedded design, thus, depends on the right selection of the hardware platform(s) as well as the software environment used in conjunction with the hardware. The selection of hardware and software architectures of an application must take place at early stages in the development process (typically at the design phase). Hardware architecture relates mainly to the type of the processor (or microcontroller) platform(s) used and the structure of the various hardware components that are comprised in the system: see Mwelwa (2006) for further discussion about hardware architectures for embedded systems.

Provided that the hardware architecture is decided, an embedded application requires an appropriate form of software architecture to be implemented. To determine the most appropriate choice for software architecture in a particular system, this condition must be fulfilled (Locke, 1992): *"The [software] architecture must be capable of providing a provable prediction of the ability of the application design to meet all of its time constraints."*

Since embedded systems are usually implemented as collections of *real-time tasks*, the various possible system architectures may then be determined by the characteristics of these tasks. In general, there are two main software architectures which are typically used in the design of embedded systems:

Event-triggered (ET): tasks are invoked as a response to aperiodic events. In this case, the system takes no account of time: instead, the system is controlled purely by the response to external events, typically represented by interrupts which can arrive at anytime (Bannatyne, 1998; Kopetz, 1991b). Generally, ET solution is recommended for applications in which sporadic data messages (with unknown request times) are exchanged in the system (Hsieh and Hsu, 2005).

Time-triggered (TT): tasks are invoked periodically at specific time intervals which are known in advance. The system is usually driven by a global clock which is linked to a hardware timer that overflows at specific time instants to generate periodic interrupts (Bennett, 1994). In distributed systems, where multi-processor hardware architecture is used, the global clock is distributed across the network (via the communication medium) to synchronise the local time base of all processors. In such architectures, time-triggering mechanism is based on time-division multiple access (TDMA) in which each processor-node is allocated a periodic time slot to broadcast its periodic messages (Kopetz, 1991b). TT solution can suit many control applications where the data messages exchanged in the system are periodic (Kopetz, 1997).

Many researchers argue that ET architectures are highly flexible and can provide high resource efficiency (Obermaisser, 2004; Locke, 1992). However, ET architectures allow several interrupts to arrive at the same time, where these interrupts might indicate (for example) that two different faults have been detected at the same time. Inevitably, dealing with an occurrence of several events at the same time will increase the system complexity and reduce the ability to predict the behavior of the ET system (Scheler and Schröder-Preikschat, 2006). In more severe circumstances, the system may fail completely if it is heavily loaded with events that occur at once (Marti, 2002). In contrast, using TT architectures helps to ensure that only a single event is handled at a time and therefore the behavior of the system can be highly-predictable.

Since highly-predictable system behavior is an important design requirement for many embedded systems, TT software architectures have become the subject of considerable attention (e.g. see Kopetz, 1997). In particular, it has been widely accepted that TT architectures are a good match for many safety-critical applications, since they can help to improve the overall safety and reliability (Allworth, 1981; Storey, 1996; Nissanke, 1997; Bates; 2000; Obermaisser, 2004). Liu (2000) highlights that TT systems are easy to validate, test, and certify because the times related to the tasks are deterministic. Detailed comparisons between the TT and ET concepts were performed by Kopetz (1991a and 1991b).

3. Schedulers and scheduling algorithms

Most embedded systems involve several tasks that share the system resources and communicate with one another and/or the environment in which they operate. For many projects, a key challenge is to work out how to schedule tasks so that they can meet their timing constraints. This process requires an appropriate form of *scheduler*[1]. A scheduler can be viewed as a very simple operating system which calls tasks periodically (or aperiodically) during the system operating time. Moreover, as with desktop operating systems, a scheduler has the responsibility to manage the computational and data resources in order to meet all temporal and functional requirements of the system (Mwelwa, 2006).

According to the nature of the operating tasks, any real-time scheduler must fall under one of the following types of scheduling policies:

Pre-emptive scheduling: where a multi-tasking process is allowed. In more details, a task with higher priority is allowed to pre-empt (i.e. interrupt) any lower priority task that is currently running. The lower priority task will resume once the higher priority task finishes executing. For example, suppose that – over a particular period of time – a system needs to execute four tasks (Task A, Task B, Task C, Task D) as illustrated in Fig. 2.

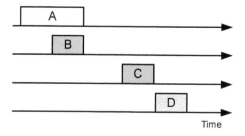

Fig. 2. A schematic representation of four tasks which need to be scheduled for execution on a single-processor embedded system (Nahas, 2008).

Assuming a single-processor system is used, Task C and Task D can run as required where Task B is due to execute before Task A is complete. Since no more than one task can run at the same time on a single-processor, Task A or Task B has to relinquish control of the CPU.

---

[1] Note that schedulers represent the core components of "Real-Time Operating System" (RTOS) kernels. Examples of commercial RTOSs which are used nowadays are: VxWorks (from Wind River), Lynx (from LynxWorks), RTLinux (from FSMLabs), eCos (from Red Hat), and QNX (from QNX Software Systems). Most of these operating systems require large amount of computational and memory resources which are not readily available in low-cost microcontrollers like the ones targeted in this work.

In pre-emptive scheduling, a higher priority might be assigned to Task B with the consequence that – when Task B is due to run – Task A will be interrupted, Task B will run, and Task A will then resume and complete (Fig. 3).

Fig. 3. Pre-emptive scheduling of Task A and Task B in the system shown in Fig. 2: Task B, here, is assigned a higher priority (Nahas, 2008).

Co-operative (or "non-pre-emptive") scheduling: where only a single-tasking process is allowed. In more details, if a higher priority task is ready to run while a lower priority task is running, the former task cannot be released until the latter one completes its execution. For example, assume the same set of tasks illustrated in Fig. 2. In the simplest solution, Task A and Task B can be scheduled co-operatively. In these circumstances, the task which is currently using the CPU is implicitly assigned a high priority: any other task must therefore wait until this task relinquishes control before it can execute. In this case, Task A will complete and then Task B will be executed (Fig. 4).

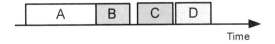

Fig. 4. Co-operative scheduling of Task A and Task B in the system shown in Fig. 2 (Nahas, 2008).

Hybrid scheduling: where a limited, but efficient, multi-tasking capabilities are provided (Pont, 2001). That is, only one task in the whole system is set to be pre-emptive (this task is best viewed as "highest-priority" task), while other tasks are running co-operatively (Fig. 5). In the example shown in the figure, suppose that Task B is a short task which has to execute immediately when it arrives. In this case, Task B is set to be pre-emptive so that it acquires the CPU control to execute whenever it arrives and whether (or not) other task is running.

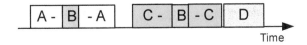

Fig. 5. Hybrid scheduling of four-tasks: Task B is set to be pre-emptive, where Task A, Task C and Task D run co-operatively (Nahas, 2008).

Overall, when comparing co-operative with pre-emptive schedulers, many researchers have argued that co-operative schedulers have many desirable features, particularly for use in safety-related systems (Allworth, 1981; Ward, 1991; Nissanke, 1997; Bates, 2000; Pont, 2001). For example, Bates (2000) identified the following four advantages of co-operative scheduling over pre-emptive alternatives:

- The scheduler is simpler.
- The overheads are reduced.
- Testing is easier.
- Certification authorities tend to support this form of scheduling.

Similarly, Nissanke (1997) noted: *"[Pre-emptive] schedules carry greater runtime overheads because of the need for context switching - storage and retrieval of partially computed results. [Co-operative] algorithms do not incur such overheads. Other advantages of co-operative algorithms include their better understandability, greater predictability, ease of testing and their inherent capability for guaranteeing exclusive access to any shared resource or data."*

Many researchers still, however, believe that pre-emptive approaches are more effective than co-operative alternatives (Allworth, 1981; Cooling, 1991). This can be due to different reasons. As in (Pont, 2001), one of the reasons why pre-emptive approaches are more widely discussed and considered is because of confusion over the options available. Pont gave an example that the basic cyclic scheduling, which is often discussed by many as an alternative to pre-emptive, is not a representative of the wide range of co-operative scheduling architectures that are available.

Moreover, one of the main issues that concern people about the reliability of co-operative scheduling is that long tasks can have a negative impact on the responsiveness of the system. This is clearly underlined by Allworth (1981): *"[The] main drawback with this co-operative approach is that while the current process is running, the system is not responsive to changes in the environment. Therefore, system processes must be extremely brief if the real-time response [of the] system is not to be impaired."*

However, in many practical embedded systems, the process (task) duration is extremely short. For example, calculations of one of the very complicated algorithms, the "proportional integral differential" (PID) controller, can be carried out on the most basic (8-bit) 8051 microcontroller in around 0.4 ms: this imposes insignificant processor load in most systems – including flight control – where 10 ms sampling rate is adequate (Pont, 2001). Pont has also commented that if the system is designed to run long tasks, *"this is often because the developer is unaware of some simple techniques that can be used to break down these tasks in an appropriate way and – in effect – convert long tasks called infrequently into short tasks called frequently"*: some of these techniques are introduced and discussed in Pont (2001).

Moreover, if the performance of the system is seen slightly poor, it is often advised to update the microcontroller hardware rather than to use a more complex software architecture. However, if changing the task design or microcontroller hardware does not provide the level of performance which is desired for a particular application, then more than one microcontroller can be used. In such cases, long tasks can be easily moved to another processor, allowing the host processor to respond rapidly to other events as required (for further details, see Pont, 2001; Ayavoo et al., 2007).

Please note that the very wide use of pre-emptive schedulers can simply be resulted from a poor understanding and, hence, undervaluation of the co-operative schedulers. For example, a co-operative scheduler can be easily constructed using only a few hundred lines of highly portable code written in a high-level programming language (such as 'C'), while the resulting system is highly-predictable (Pont, 2001).

It is also important to understand that sometimes pre-emptive schedulers are more widely used in RTOSs due to commercial reasons. For example, companies may have commercial benefits from using pre-emptive environments. Consequently, as the complexity of these environments increases, the code size will significantly increase making 'in-house' constructions of such environments too complicated. Such complexity factors lead to the sale of commercial RTOS products at high prices (Pont, 2001). Therefore, further academic research has been conducted in this area to explore alternative solutions. For example, over the last few years, the Embedded Systems Laboratory (ESL) researchers have considered various ways in which simple, highly-predictable, non-pre-emptive (co-operative) schedulers can be implemented in low-cost embedded systems.

4. Scheduling algorithm and scheduler implementation

A key component of the scheduler is the *scheduling algorithm* which basically determines the order in which the tasks will be executed by the scheduler (Buttazzo, 2005). More specifically, a scheduling algorithm is the set of rules that, at every instant while the system is running, determines which task must be allocated the resources to execute.

Developers of embedded systems have proposed various scheduling algorithms that can be used to handle tasks in real-time applications. The selection of appropriate scheduling algorithm for a set of tasks is based upon the capability of the algorithm to satisfy all timing constraints of the tasks: where these constraints are derived from the application requirements. Examples of common scheduling algorithms are: Cyclic Executive (Locke, 1992), Rate Monotonic (Liu & Layland, 1973), Earliest-Deadline-First (Liu & Layland, 1973; Liu, 2000), Least-Laxity-First (Mok, 1983), Deadline Monotonic (Leung, 1982) and Shared-Clock (Pont, 2001) schedulers (see Rao et al., 2008 for a simple classification of scheduling algorithms). This chapter outlines one key example of scheduling algorithms that is widely used in the design of real-time embedded systems when highly-predictable system behavior is an essential requirement: this is the Time Triggered Co-operative scheduler which is a form of cyclic executive.

Note that once the design specifications are converted into appropriate design elements, the system implementation process can take place by translating those designs into software and hardware components. People working on the development of embedded systems are often concerned with the software implementation of the system in which the system specifications are converted into an executable system (Sommerville, 2007; Koch, 1999). For example, Koch interpreted the implementation of a system as the way in which the software program is arranged to meet the system specifications.

The implementation of schedulers is a major problem which faces designers of real-time scheduling systems (for example, see Cho et al., 2005). In their useful publication, Cho and colleges clarified that the well-known term *scheduling* is used to describe the process of finding the optimal schedule for a set of real-time tasks, while the term *scheduler implementation* refers to the process of implementing a physical (software or hardware) scheduler that enforces – at run-time – the task sequencing determined by the designed schedule (Cho et al., 2007).

Generally, it has been argued that there is a wide gap between scheduling theory and its implementation in operating system kernels running on specific hardware, and for any meaningful validation of timing properties of real-time applications, this gap must be bridged (Katcher et al., 1993). The relationship between any scheduling algorithm and the number of possible implementation options for that algorithm – in practical designs – has generally been viewed as 'one-to-many', even for very simple systems (Baker & Shaw, 1989; Koch; 1999; Pont, 2001; Baruah, 2006; Pont et al., 2007; Phatrapornnant, 2007). For example, Pont et al. (2007) clearly mentioned that if someone was to use a particular scheduling architecture, then there are many different implementation options which can be available. This claim was also supported by Phatrapornnant (2007) by noting that the TTC scheduler (which is a form of cyclic executive) is only an algorithm where, in practice, there can be many possible ways to implement such an algorithm.

The performance of a real-time system depends crucially on implementation details that cannot be captured at the design level, thus it is more appropriate to evaluate the real-time properties of the system after it is fully implemented (Avrunin et al., 1998).

5. Time-triggered co-operative (TTC) scheduling algorithm

A key defining characteristic of a time-triggered (TT) system is that it can be expected to have highly-predictable patterns of behavior. This means that when a computer system has a time-triggered architecture, it can be determined in advance – before the system begins executing – exactly what the system will do at every moment of time while the system is operating. Based on this definition, completely defined TT behavior is – of course – difficult to achieve in practice. Nonetheless, approximations of this model have been found to be useful in a great many practical systems. The closest approximation of a "perfect" TT architecture which is in widespread use involves a collection of periodic tasks which operate co-operatively (or "non-pre-emptively"). Such a time-triggered co-operative (TTC) architecture has sometimes been described as a cyclic executive (e.g. Baker & Shaw, 1989; Locke, 1992).

According to Baker and Shaw (1989), the cyclic executive scheduler is designed to execute tasks in a sequential order that is defined prior to system activation; the number of tasks is fixed; each task is allocated an execution slot (called a *minor cycle* or a *frame*) during which the task executes; the task – once interleaved by the scheduler – can execute until completion without interruption from other tasks; all tasks are periodic and the deadline of each task is equal to its period; the worst-case execution time of all tasks is known; there is no context switching between tasks; and tasks are scheduled in a repetitive cycle called *major cycle*. The major cycle can be defined as the time period during which each task in the scheduler executes – at least – once and before the whole task execution pattern is repeated. This is numerically calculated as the lowest common multiple (LCM) of the periods of the scheduled tasks (Baker & Shaw, 1989; Xu & Parnas, 1993). Koch (1999) emphasized that cyclic executive is a "proof-by-construction" scheme in which no schedulability analysis is required prior to system construction.

Fig. 6 illustrates the (time-triggered) cyclic executive model for a simple set of four periodic tasks. Note that the final task in the task-group (i.e. Task D) must complete execution before the arrival of the next timer interrupt which launches a new (major) execution cycle.

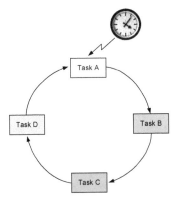

Fig. 6. A time-triggered cyclic executive model for a set of four periodic tasks (Nahas, 2011b).

In the example shown, each task is executed only once during the whole major cycle which is, in this case, made up of four minor cycles. Note that the task periods may not always be identical as in the example shown in Fig. 6. When task periods vary, the scheduler should define a sequence in which each task is repeated sufficiently to meet its frequency requirement (Locke, 1992).

Fig. 7 shows the general structure of the time-triggered cyclic executive (i.e. time-triggered co-operative) scheduler. In the example shown in this figure, the scheduler has a minor cycle of 10 ms, period values of 20, 10 and 40 ms for the tasks A, B and C, respectively. The LCM of these periods is 40 ms, therefore the length of the major cycle in which all tasks will be executed periodically is 40 ms. It is suggested that the minor cycle of the scheduler (which is also referred to as the tick interval: see Pont, 2001) can be set equal to or less than the greatest common divisor value of all task periods (Phatrapornnant, 2007). In the example shown in Fig. 7, this value is equal to 10 ms. In practice, the minor cycle is driven by a periodic interrupt generated by the overflow of an on-chip hardware timer or by the arrival of events in the external environment (Locke, 1992; Pont, 2001). The vertical arrows in the figure represent the points at which minor cycles (ticks) start.

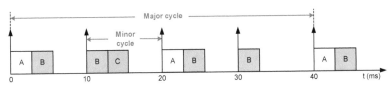

Fig. 7. A general structure of the time-triggered co-operative (TTC) scheduler (Nahas, 2008).

Overall, TTC schedulers have many advantages. A key recognizable advantage is its simplicity (Baker & Shaw, 1989; Liu, 2000; Pont, 2001). Furthermore, since pre-emption is not allowed, mechanisms for context switching are, hence, not required and, as a consequence, the run-time overhead of a TTC scheduler can be kept very low (Locke, 1992; Buttazzo, 2005). Also, developing TTC schedulers needs no concern about protecting the integrity of shared data structures or shared resources because, at a time, only one task in the whole system can exclusively use the resources and the next due task cannot begin its execution until the running task is completed (Baker & Shaw, 1989; Locke, 1992).

Since all tasks are run regularly according to their predefined order in a deterministic manner, the TTC schedulers demonstrate very low levels of task jitter (Locke, 1992; Bate, 1998; Buttazzo, 2005) and can maintain their low-jitter characteristics even when complex techniques, such as dynamic voltage scaling (DVS), are employed to reduce system power consumption (Phatrapornnant & Pont, 2006). Therefore, as would be expected (and unlike RM designs, for example), systems with TTC architectures can have highly-predictable timing behavior (Baker & Shaw, 1989; Locke, 1992). Locke (1992) underlines that with cyclic executive systems, *"it is possible to predict the entire future history of the state of the machine, once the start time of the system is determined (usually at power-on). Thus, assuming this future history meets the response requirements generated by the external environment in which the system is to be used, it is clear that all response requirements will be met. Thus it fulfills the basic requirements of a hard real time system."*

Provided that an appropriate implementation is used, TTC architectures can be a good match for a wide range of low-cost embedded applications. For example, previous studies have described – in detail – how these techniques can be applied in various automotive applications (e.g. Ayavoo et al., 2006; Ayavoo, 2006), a wireless (ECG) monitoring system (Phatrapornnant & Pont, 2004; Phatrapornnant, 2007), various control applications (e.g. Edwards et al., 2004; Key et al., 2004; Short & Pont, 2008), and in data acquisition systems, washing-machine control and monitoring of liquid flow rates (Pont, 2002). Outside the ESL group, Nghiem et al. (2006) described an implementation of PID controller using TTC scheduling algorithm and illustrated how such architecture can help increase the overall system performance as compared with alternative implementation methods.

However, TTC architectures have some shortcomings. For example, many researchers argue that running tasks without pre-emption may cause other tasks to wait for some time and hence miss their deadlines. However, the availability of high-speed, COTS microcontrollers nowadays helps to reduce the effect of this problem and, as processor speeds continue to increase, non-pre-emptive scheduling approaches are expected to gain more popularity in the future (Baruah, 2006).

Another issue with TTC systems is that the task schedule is usually calculated based on estimates of Worst Case Execution Time (WCET) of the running tasks. If such estimates prove to be incorrect, this may have a serious impact on the system behavior (Buttazzo, 2005).

One recognized disadvantage of using TTC schedulers is the lack of flexibility (Locke, 1992; Bate, 1998). This is simply because TTC is usually viewed as 'table-driven' static scheduler (Baker & Shaw, 1989) which means that any modification or addition of a new functionality, during any stage of the system development process, may need an entirely new schedule to be designed and constructed (Locke, 1992; Koch, 1999). This reconstruction of the system adds more time overhead to the design process: however, with using tools such as those developed recently to support "automatic code generation" (Mwelwa et al., 2006; Mwelwa, 2006; Kurian & Pont, 2007), the work involved in developing and maintaining such systems can be substantially reduced.

Another drawback of TTC systems, as noted by Koch (1999), is that constructing the cyclic executive model for a large set of tasks with periods that are prime to each other can be unaffordable. However, in practice, there is some flexibility in the choice of task periods (Xu & Parnas, 1993; Pont, 2001). For example, Gerber et al. (1995) demonstrated how a feasible solution for task periods can be obtained by considering the period harmonicity relationship of each task with all its successors. Kim et al. (1999) went further to improve and automate this period calibration method. Please also note that using a table to store the task schedule is only one way of implementing TTC algorithm where, in practice, there can be other implementation methods (Baker & Shaw, 1989; Pont, 2001). For example, Pont (2001) described an alternative to table-driven schedule implementation for the TTC algorithm which has the potential to solve the co-prime periods problem and also simplify the process of modifying the whole task schedule later in the development life cycle or during the system run-time.

Furthermore, it has also been reported that a long task whose execution time exceeds the period of the highest rate (shortest period) task cannot be scheduled on the basic TTC scheduler (Locke, 1992). One solution to this problem is to break down the long task into multiple short tasks that can fit in the minor cycle. Also, possible alternative solution to this problem is to use a Time-Triggered Hybrid (TTH) scheduler (Pont, 2001) in which a limited degree of pre-emption is supported. One acknowledged advantage of using TTH scheduler is that it enables the designer to build a static, fixed-priority schedule made up of a collection of co-operative tasks and a single (short) pre-emptive task (Phatrapornnant, 2007). Note that TTH architectures are not covered in the context of this chapter. For more details about these scheduling approaches, see (Pont, 2001; Maaita & Pont, 2005; Hughes & Pont, 2008; Phatrapornnant, 2007).

Please note that later in this chapter, it will be demonstrated how, with extra care at the implementation stage, one can easily deal with many of the TTC scheduler limitations indicated above.

6. Jitter in TTC scheduling algorithm

Jitter is a term which describes variations in the timing of activities (Wavecrest, 2001). The work presented in this chapter is concerned with implementing highly-predictable embedded systems. Predictability is one of the most important objectives of real-time embedded systems which can simply be defined as the ability to determine, in advance, exactly what the system will do at every moment of time in which it is running. One way in which predictable behavior manifests itself is in low levels of task jitter.

Jitter is a key timing parameter that can have detrimental impacts on the performance of many applications, particularly those involving period sampling and/or data generation (e.g. data acquisition, data playback and control systems: see Torngren, 1998). For example, Cottet & David (1999) show that – during data acquisition tasks – jitter rates of 10% or more can introduce errors which are so significant that any subsequent interpretation of the sampled signal may be rendered meaningless. Similarly, Jerri (1977) discusses the serious impact of jitter on applications such as spectrum analysis and filtering. Also, in control systems, jitter can greatly degrade the performance by varying the sampling period (Torngren, 1998; Marti et al., 2001).

When TTC architectures (which represent the main focus of this chapter) are employed, possible sources of task jitter can be divided into three main categories: scheduling overhead variation, task placement and clock drift.

The overhead of a conventional (non-co-operative) scheduler arises mainly from context switching. However, in some TTC systems the scheduling overhead is comparatively large and may have a highly variable duration due to code branching or computations that have non-fixed lengths. As an example, Fig. 8 illustrates how a TTC system can suffer release jitter as a result of variations in the scheduler overhead (this relates to DVS system).

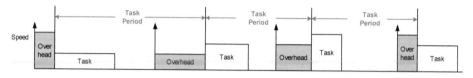

Fig. 8. Release jitter caused by variation of scheduling overhead (Nahas, 2011a).

Even if the scheduler overhead variations can be avoided, TTC designs can still suffer from jitter as a result of the task placement. To illustrate this, consider Fig. 9. In this schedule example, Task C runs sometimes after A, sometimes after A and B, and sometimes alone. Therefore, the period between every two successive runs of Task C is highly variable. Moreover, if Task A and B have variable execution durations (as in Fig. 8), then the jitter levels of Task C will even be larger.

Fig. 9. Release jitter caused by task placement in TTC schedulers (Nahas, 2011a).

For completeness of this discussion, it is also important to consider clock drift as a source of task jitter. In the TTC designs, a clock "tick" is generated by a hardware timer that is used to trigger the execution of the cyclic tasks (Pont, 2001). This mechanism relies on the presence of a timer that runs at a fixed frequency. In such circumstances, any jitter will arise from variations at the hardware level (e.g. through the use of a low-cost frequency source, such as a ceramic resonator, to drive the on-chip oscillator: see Pont, 2001). In the TTC scheduler implementations considered in this study, the software developer has no control over the clock source. However, in some circumstances, those implementing a scheduler must take such factors into account. For example, in situations where DVS is employed (to reduce CPU power consumption), it may take a variable amount of time for the processor's phase-locked loop (PLL) to stabilize after the clock frequency is changed (see Fig. 10).

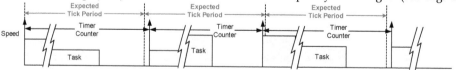

Fig. 10. Clock drift in DVS systems (Nahas, 2011a).

As discussed elsewhere, it is possible to compensate for such changes in software and thereby reduce jitter (see Phatrapornnant & Pont, 2006; Phatrapornnant, 2007).

7. Various TTC scheduler implementations for highly-predictable embedded systems

In this section, a set of "representative" examples of the various classes of TTC scheduler implementations are reviewed. In total, the section reviews six TTC implementations.

7.1 Super loop (SL) scheduler

The simplest practical implementation of a TTC scheduler can be created using a "Super Loop" (SL) (sometimes called an "endless loop: Kalinsky, 2001). The super loop can be used as the basis for implementing a simple TTC scheduler (e.g. Pont, 2001; Kurian & Pont, 2007). A possible implementation of TTC scheduler using super loop is illustrated in Listing 1.

```
int main(void)
    {
    ...
    while(1)
        {
        TaskA();
        Delay_6ms();
        TaskB();
        Delay_6ms();
        TaskC();
        Delay_6ms();
        }

    // Should never reach here
    return 1
    }
```

Listing 1. A very simple TTC scheduler which executes three periodic tasks, in sequence.

By assuming that each task in Listing 1 has a fixed duration of 4 ms, a TTC system with a 10 ms "tick interval" has been created using a combination of super loop and delay functions (Fig. 11).

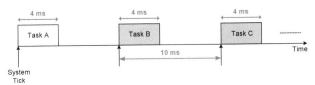

Fig. 11. The task executions resulting from the code in Listing 1 (Nahas, 2011b).

In the case where the scheduled tasks have variable durations, creating a fixed tick interval is not straightforward. One way of doing that is to use a "Sandwich Delay" (Pont et al., 2006) placed around the tasks. Briefly, a Sandwich Delay (SD) is a mechanism – based on a hardware timer – which can be used to ensure that a particular code section always takes approximately the same period of time to execute. The SD operates as follows: [1] A timer is set to run; [2] An activity is performed; [3] The system waits until the timer reaches a pre-determined count value.

In these circumstances – as long as the timer count is set to a duration that exceeds the WCET of the sandwiched activity – SD mechanism has the potential to fix the execution period. Listing 2 shows how the tasks in Listing 1 can be scheduled – again using a 10 ms tick interval – if their execution durations are not fixed

```
int main(void)
   {
   ...
   while(1)
      {
      // Set up a Timer for sandwich delay
      SANDWICH_DELAY_Start();
      // Add Tasks in the first tick interval
      Task_A();
      // Wait for 10 millisecond sandwich delay
      // Add Tasks in the second tick interval
      SANDWICH_DELAY_Wait(10);
      Task_B();
      // Wait for 20 millisecond sandwich delay
      // Add Tasks in the second tick interval
      SANDWICH_DELAY_Wait(20);
      Task_C();
      // Wait for 30 millisecond sandwich delay
      SANDWICH_DELAY_Wait(30);
      }
   // Should never reach here
   return 1
   }
```

Listing 2. A TTC scheduler which executes three periodic tasks with variable durations, in sequence.

Using the code listing shown, the successive function calls will take place at fixed intervals, even if these functions have large variations in their durations (Fig. 12). For further information, see (Nahas, 2011b).

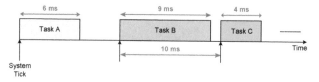

Fig. 12. The task executions expected from the TTC-SL scheduler code shown in Listing 2 (Nahas, 2011b).

7.2 A TTC-ISR scheduler

In general, software architectures based on super loop can be seen simple, highly efficient and portable (Pont, 2001; Kurian & Pont, 2007). However, these approaches lack the provision of accurate timing and the efficiency in using the power resources, as the system always operates at full-power which is not necessary in many applications.

An alternative (and more efficient) solution to this problem is to make use of the hardware resources to control the timing and power behavior of the system. For example, a TTC scheduler implementation can be created using "Interrupt Service Routine" (ISR) linked to the overflow of a hardware timer. In such approaches, the timer is set to overflow at regular "tick intervals" to generate periodic "ticks" that will drive the scheduler. The rate of the tick interval can be set equal to (or higher than) the rate of the task which runs at the highest frequency (Phatrapornnant, 2007).

In the TTC-ISR scheduler, when the timer overflows and a tick interrupt occurs, the ISR will be called, and awaiting tasks will then be activated from the ISR directly. Fig. 13 shows how such a scheduler can be implemented in software. In this example, it is assumed that one of the microcontroller's timers has been set to generate an interrupt once every 10 ms, and thereby call the function `Update()`. This `Update()` function represents the scheduler ISR. At the first tick, the scheduler will run Task A then go back to the while loop in which the system is placed in the idle mode waiting for the next interrupt. When the second interrupt takes place, the scheduler will enter the ISR and run Task B, then the cycle continues. The overall result is a system which has a 10 ms "tick interval" and three tasks executed in sequence (see Fig. 14)

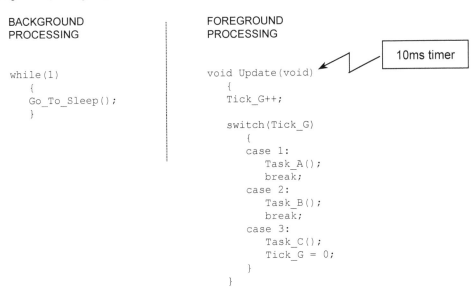

Fig. 13. A schematic representation of a simple TTC-ISR scheduler (Nahas, 2008).

Whether or not the idle mode is used in TTC-ISR scheduler, the timing observed is largely independent of the software used but instead depends on the underlying timer hardware (which will usually mean the accuracy of the crystal oscillator driving the microcontroller). One consequence of this is that, for the system shown in Fig. 13 (for example), the successive function calls will take place at precisely-defined intervals, even if there are large variations in the duration of tasks which are run from the Update() function (Fig. 14). This is very useful behavior which is not easily obtained with implementations based on super loop.

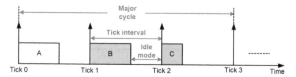

Fig. 14: The task executions expected from the TTC-ISR scheduler code shown in Fig. 13 (Nahas, 2008).

The function call tree for the TTC-ISR scheduler is shown in Fig. 15. For further information, see (Nahas, 2008).

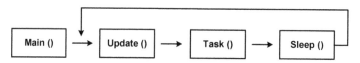

Fig. 15: Function call tree for the TTC-ISR scheduler (Nahas, 2008).

7.3 TTC-dispatch scheduler

Implementation of a TTC-ISR scheduler requires a significant amount of hand coding (to control the task timing), and there is no division between the "scheduler" code and the "application" code (i.e. tasks). The TTC-Dispatch scheduler provides a more flexible alternative. It is characterized by distinct and well-defined scheduler functions.

Like TTC-ISR, the TTC-Dispatch scheduler is driven by periodic interrupts generated from an on-chip timer. When an interrupt occurs, the processor executes an Update() function. In the scheduler implementation discussed here, the Update() function simply keeps track of the number of ticks. A Dispatch() function will then be called, and the due tasks (if any) will be executed one-by-one. Note that the Dispatch() function is called from an "endless" loop placed in the function Main(): see Fig. 16. When not executing the Update() or Dispatch() functions, the system will usually enter the low-power idle mode.

In this TTC implementation, the software employs a SCH_Add_Task() and a SCH_Delete_Task() functions to help the scheduler add and/or remove tasks during the system run-time. Such scheduler architecture provides support for "one shot" tasks and dynamic scheduling where tasks can be scheduled online if necessary (Pont, 2001). To add a task to the scheduler, two main parameters have to be defined by the user in addition to the task's name: task's *offset*, and task's *period*. The offset specifies the time (in ticks) before the task is first executed. The period specifies the interval (also in ticks) between repeated executions of the task. In the Dispatch() function, the scheduler checks these parameters for each task before running it. Please note that information about tasks is stored in a user-defined scheduler data structure. Both the "sTask" data type and the "SCH_MAX_TASKS" constant are used to create the "Task Array" which is referred to throughout the scheduler as "sTask SCH_tasks_G[SCH_MAX_TASKS]". See (Pont, 2001) for further details. The function call tree for the TTC-Dispatch scheduler is shown in Fig. 16.

Fig. 16. Function call tree for the TTC-Dispatch scheduler (Nahas, 2011a).

Fig. 16 illustrates the whole scheduling process in the TTC-Dispatch scheduler. For example, it shows that the first function to run (after the startup code) is the Main() function. The Main() calls Dispatch() which in turn launches any tasks which are currently scheduled to execute. Once these tasks are complete, the control will return back to Main() which calls Sleep() to place the processor in the idle mode. The timer interrupt then occurs which will wake the processor up from the idle state and invoke the ISR Update(). The function call then returns all the way back to Main(), where Dispatch() is called again and the whole cycle thereby continues. For further information, see (Nahas, 2008).

7.4 Task Guardians (TG) scheduler

Despite many attractive characteristics, TTC designs can be seriously compromised by tasks that fail to complete within their allotted periods. The TTC-TG scheduler implementation described in this section employs a Task Guardian (TG) mechanism to deal with the impact of such task overruns. When dealing with task overruns, the TG mechanism is required to shutdown any task which is found to be overrunning. The proposed solution also provides the option of replacing the overrunning task with a backup task (if required).

The implementation is again based on TTC-Dispatch (Section 7.3). In the event of a task overrun with ordinary Dispatch scheduler, the timer ISR will interrupt the overrunning task (rather than the Sleep() function). If the overrunning task keeps executing then it will be periodically interrupted by Update() while all other tasks will be blocked until the task finishes (if ever): this is shown in Fig. 17. Note that (a) illustrates the required task schedule, and (b) illustrates the scheduler operation when Task A overrun by 5 tick interval.

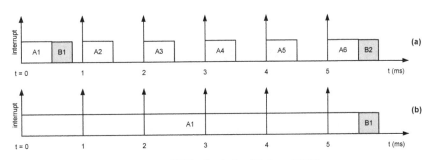

Fig. 17. The impact of task overrun on a TTC scheduler (Nahas, 2008).

In order for the TG mechanism to work, various functions in the TTC-Dispatch scheduler are modified as follows:

- `Dispatch()` indicates that a task is being executed.
- `Update()` checks to see if an overrun has occurred. If it has, control is passed back to `Dispatch()`, shutting down the overrunning task.
- If a backup task exists it will be executed by `Dispatch()`.
- Normal operation then continues.

In a little more detail, detecting overrun in this implementation uses a simple, efficient method employed in the `Dispatch()` function. It simply adds a "Task_Overrun" variable which is set equal to the task index before the task is executed. When the task completes, this variable will be assigned the value of (for example) 255 to indicate a successful completion. If a task overruns, the `Update()` function in the next tick should detect this since it checks the Task_overrun variable and the last task index value. The `Update()` then changes the return address to an `End_Task()` function instead of the overrunning task. The `End_Task()` function should return control to Dispatch. Note that moving control from `Update()` to `End_Task()` is a nontrivial process and can be done by different ways (Hughes & Pont, 2004).

The `End_Task()` has the responsibility to shutdown the overrunning task. Also, it determines the type of function that has overrun and begins to restore register values accordingly. This process is complicated which aims to return the scheduler back to its normal operation making sure the overrun has been resolved completely. Once the overrun is dealt with, the scheduler replaces the overrunning task with a backup task which is set to run immediately before running other tasks. If there is no backup task defined by the user, then the TTC-TG scheduler implements a mechanism which turns the priority of the task that overrun to the lowest so as to reduce the impact of any future overrunning by this task. The function call tree for the TTC-TTG scheduler can be shown in Fig. 18.

Fig. 18. Function call tree for the TTC-TG scheduler (Nahas, 2008).

Note that the scheduler structure used in TTC-TG scheduler is same as that employed in the TTC-Dispatch scheduler which is simply based on ISR Update linked to a timer interrupt and a Dispatch function called periodically from the Main code (Section 7.3). For further details, see (Hughes & Pont, 2008).

7.5 Sandwich Delay (SD) scheduler

In Section 6, the impact of task placement on "low-priority" tasks running in TTC schedulers was considered. The TTC schedulers described in Sections 7.1 - 7.4 lack the ability to deal with jitter in the starting time of such tasks. One way to address this issue is to place "Sandwich Delay" (Pont et al., 2006) around tasks which execute prior to other tasks in the same tick interval.

In the TTC-SD scheduler described in this section, sandwich delays are used to provide execution "slots" of fixed sizes in situations where there is more than one task in a tick interval. To clarify this, consider the set of tasks shown in Fig. 19. In the figure, the required SD prior to Task C – for low jitter behavior – is equal to the WCET of Task A plus the WCET of Task B. This implies that in the second tick (for example), the scheduler runs Task A and then waits for the period equals to the WCET of Task B before running Task C. The figure shows that when SDs are placed around the tasks prior to Task C, the periods between successive runs of Task C become equal and hence jitter in the release time of this task is significantly reduced.

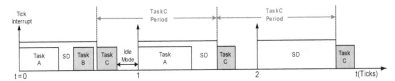

Fig. 19: Using Sandwich Delays to reduce release jitter in TTC schedulers (Nahas, 2011a).

Note that – with this implementation – the WCET for each task is input to the scheduler through a `SCH_Task_WCET()` function placed in the `Main` code. After entering task parameters, the scheduler employs `Calc_Sch_Major_Cycle()` and `Calculate_Task_RT()` functions to calculate the scheduler major cycle and the required release time for the tasks, respectively. The release time values are stored in the "Task Array" using the variable `SCH_tasks_G[Index].Rls_time`. Note that the required release time of a task is the time between the start of the tick interval and the start time of the task "slot" plus a little safety margin. For further information, see (Nahas, 2011a).

7.6 Multiple Timer Interrupts (MTI) scheduler

An alternative to the SD technique which requires a large computational time, a "gap insertion" mechanism that uses "Multiple Timer Interrupts" (MTIs) can be employed.

In the TTC-MTI scheduler described in this section, multiple timer interrupts are used to generate the predefined execution "slots" for tasks. This allows more precise control of timing in situations where more than one task executes in a given tick interval. The use of interrupts also allows the processor to enter an idle mode after completion of each task, resulting in power saving. In order to implement this technique, two interrupts are required:

- Tick interrupt: used to generate the scheduler periodic tick.
- Task interrupt: used – within tick intervals – to trigger the execution of tasks.

The process is illustrated in Fig. 20. In this figure, to achieve zero jitter, the required release time prior to Task C (for example) is equal to the WCET of Task A plus the WCET of Task B plus scheduler overhead (i.e. ISR `Update()` function). This implies that in the second tick (for example), after running the ISR, the scheduler waits – in idle mode – for a period of time equals to the WCETs of Task A and Task B before running Task C. Fig. 20 shows that when an MTI method is used, the periods between the successive runs of Task C (the lowest priority task in the system) are always equal. This means that the task jitter in such implementation is independent on the task placement or the duration(s) of the preceding task(s).

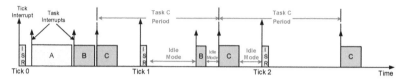

Fig. 20. Using MTIs to reduce release jitter in TTC schedulers (Nahas, 2011a).

In the implementation considered in this section, the WCET for each task is input to the scheduler through SCH_Task_WCET() function placed in the Main() code. The scheduler then employs Calc_Sch_Major_Cycle() and Calculate_Task_RT() functions to calculate the scheduler major cycle and the required release time for the tasks, respectively. Moreover, there is no Dispatch() called in the Main() code: instead, "interrupt request wrappers" – which contain Assembly code – are used to manage the sequence of operation in the whole scheduler. The function call tree for the TTC-MTI scheduler is shown in Fig. 21 (compare with Fig. 16).

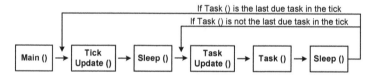

Fig. 21. Function call tree for the TTC-MTI scheduler (in normal conditions) (Nahas, 2011a).

Unlike the normal Dispatch schedulers, this implementation relies on two interrupt Update() functions: Tick Update() and Task Update(). The Tick Update() – which is called every tick interval (as normal) – identifies which tasks are ready to execute within the current tick interval. Before placing the processor in the idle mode, the Tick Update() function sets the match register of the task timer according to the release time of the first due task running in the current interval. Calculating the release time of the first task in the system takes into account the WCET of the Tick Update() code.

When the task interrupt occurs, the Task Update() sets the return address to the task that will be executed straight after this update function, and sets the match register of the task timer for the next task (if any). The scheduled task then executes as normal. Once the task completes execution, the processor goes back to Sleep() and waits for the next task interrupt (if there are following tasks to execute) or the next tick interrupt which launches a new tick interval. Note that the Task Update() code is written in such a way that it always has a fixed execution duration for avoiding jitter at the starting time of tasks.

It is worth highlighting that the TTC-MTI scheduler described here employs a form of "task guardians" which help the system avoid any overruns in the operating tasks. More specifically, the described MTI technique helps the TTC scheduler to shutdown any overrunning task by the time the following interrupt takes place. For example, if the overrunning task is followed by another task in the same tick, then the task interrupt –

which triggers the execution of the latter task – will immediately terminate the overrun. Otherwise, the task can overrun until the next tick interrupt takes place which will terminate the overrun immediately. The function call tree for the TTC-MTI scheduler – when a task overrun occurs – is shown in Fig. 22. The only difference between this process and the one shown in Fig. 21 is that an ISR will interrupt the overrunning task (rather than the `Sleep()` function). Again, if the overrunning task is the last task to execute in a given tick, then it will be interrupted and terminated by the `Tick Update()` at the next tick interval: otherwise, it will be terminated by the following `Task Update()`. For further information, see (Nahas, 2011a).

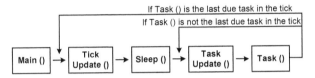

Fig. 22. Function call tree for the TTC-MTI scheduler (with task overrun) (Nahas, 2008).

8. Evaluation of TTC scheduler implementations

This section provides the results of the various TTC implementations considered in the previous section. The results include jitter levels, error handling capabilities and resource (i.e. CPU and memory) requirements. The section begins by briefing the experimental methodology used in this study.

8.1 Experimental methodology

The empirical studies were conducted using Ashling LPC2000 evaluation board supporting Philips LPC2106 processor (Ashling Microsystems, 2007). The LPC2106 is a modern 32-bit microcontroller with an ARM7 core which can run – under control of an on-chip PLL – at frequencies from 12 MHz to 60 MHz.

The compiler used was the GCC ARM 4.1.1 operating in Windows by means of Cygwin (a Linux emulator for windows). The IDE and simulator used was the Keil ARM development kit (v3.12).

For meaningful comparison of jitter results, the task-set shown in Fig. 23 was used to allow exploring the impact of schedule-induced jitter by scheduling Task A to run every two ticks. Moreover, all tasks were set to have variable execution durations to allow exploring the impact of task-induced jitter.

For jitter measurements, two measures were recorded: Tick Jitter: represented by the variations in the interval between the release times of the periodic tick, and Task Jitter: represented by the variations in the interval between the release times of periodic tasks. Jitter was measured using a National Instruments data acquisition card 'NI PCI-6035E' (National Instruments, 2006), used in conjunction with appropriate software LabVIEW 7.1 (LabVIEW, 2007). The "difference jitter" was reported which is obtained by subtracting the minimum period (between each successive ticks or tasks) from the maximum period obtained from the measurements in the sample set. This jitter is sometimes referred to as "absolute jitter" (Buttazzo, 2005).

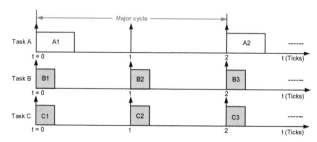

Fig. 23. Graphical representation of the task-set used in jitter test (Nahas, 2011a).

The CPU overhead was measured using the performance analyzer supported by the Keil simulator which calculates the time required by the scheduler as compared to the total runtime of the program. The percentage of the measured CPU time was then reported to indicate the scheduler overhead in each TTC implementation.

For ROM and RAM memory overheads, the CODE and DATA memory values required to implement each scheduler were recorded, respectively. Memory values were obtained using the ".map" file which is created when the source code is compiled. The STACK usage was also measured (as DATA memory overhead) by initially filling the data memory with 'DEAD CODE' and then reporting the number of memory bytes that had been overwritten after running the scheduler for sufficient period.

8.2 Results

This section summarizes the results obtained in this study. Table 1 presents the jitter levels, CPU requirements, memory requirements and ability to deal with task overrun for all schedulers. The jitter results include the tick and tasks jitter. The ability to deal with task overrun is divided into six different cases as shown in Table 2. In the table, it is assumed that Task A is the overrunning task.

Scheduler	Tick Jitter (µs)	Task A Jitter (µs)	Task B Jitter (µs)	Task C Jitter (µs)	CPU %	ROM (Bytes)	RAM (Bytes)	Ability to deal with task overrun
TTC-SL	1.2	1.5	4016.2	5772.2	100	2264	124	1b
TTC-ISR	0.0	0.1	4016.7	5615.8	39.5	2256	127	1a
TTC Dispatch	0.0	0.1	4022.7	5699.8	39.7	4012	325	1b
TTC-TG	0.0	0.1	4026.2	5751.9	39.8	4296	446	2b
TTC-SD	0.0	0.1	1.5	1.5	74.0	5344	310	1b
TTC-MTI	0.0	0.1	0.0	0.0	39.6	3620	514	3a

Table 1. Results obtained in the study detailed in this chapter.

From the table, it is difficult to obtain zero jitter in the release time of the tick in the TTC-SL scheduler, although the tick jitter can still be low. Also, the TTC-SL scheduler always requires a full CPU load (~ 100%). This is since the scheduler does not use the low-power "idle" mode when not executing tasks: instead, the scheduler waits in a "while" loop. In the TTC-ISR scheduler, the tick interrupts occur at precisely-defined intervals with no measurable delays or jitter and the release jitter in Task A is equal to zero. Inevitably, the memory values in the TTC-Dispatch scheduler are somewhat larger than those required to implement the TTC-SL and TTC-ISR schedulers. The results from the TTC-TG scheduler are very similar to those obtained from the TTC-Dispatch scheduler except that it requires slightly more data memory. When the TTC-SD scheduler is used, the low-priority tasks are executed at fixed intervals. However, there is still a little jitter in the release times of Tasks B and Task C. This jitter is caused by variation in time taken to leave the software loop – which is used in the SD mechanism to check if the required release time for the concerned task is matched – and begin to execute the task. With the TTC-MTI scheduler, the jitter in the release time of all tasks running in the system is totally removed, causing a significant increase in the overall system predictability.

Regarding the ability to deal with task overrun, the TTC-TG scheduler detects and hence terminates the overrunning task at the beginning of the tick following the one in which the task overruns. Moreover, the scheduler allows running a backup task in the same tick in which the overrun is detected and hence continues to run the following tasks. This means that one tick shift is added to the schedule. Also, the TTC-MTI scheduler employs a simple TG mechanism and – once an interrupt occurs – the running task (if any) will be terminated. Note that the implementation employed here did not support backup tasks.

Schedule	Shut down time (after Ticks)	Backup task	Comment
1a	---	Not applicable	Overrunning task is not shut down. The number of elapsed ticks – during overrun – is not counted and therefore tasks due to run in these ticks are ignored.
1b	---	Not applicable	Overrunning task is not shut down. The number of elapsed ticks – during overrun – is counted and therefore tasks due to run in these ticks are executed immediately after overrunning task ends.
2a	1 Tick	Not available	Overrunning task is detected at the time of the next tick and shut down.
2b	1 Tick	Available – BK(A)	Overrunning task is detected at the time of the next tick and shut down: a replacement (backup) task is added to the schedule.
3a	WCET(Ax)	Not available	Overrunning task is shut down immediately after it exceeds its estimated WCET.
3b	WCET(Ax)	Available – BK(A)	Overrunning task is shut down immediately after it exceeds its estimated WCET. A backup task is added to the schedule.

Table 2. Examples of possible schedules obtained with task overrun (Nahas, 2008).

9. Conclusions

The particular focus in this chapter was on building embedded systems which have severe resource constraints and require high levels of timing predictability. The chapter provided necessary definitions to help understand the scheduling theory and various techniques used to build a scheduler for the type of systems concerned with in this study. The discussions indicated that for such systems, the "time-triggered co-operative" (TTC) schedulers are a good match. This was mainly due to their simplicity, low resource requirements and high predictability they can offer. The chapter, however, discussed major problems that can affect the performance of TTC schedulers and reviewed some suggested solutions to overcome such problems.

Then, the discussions focused on the relationship between scheduling algorithm and scheduler implementations and highlighted the challenges faced when implementing software for a particular scheduler. It was clearly noted that such challenges were mainly caused by the broad range of possible implementation options a scheduler can have in practice, and the impact of such implementations on the overall system behavior.

The chapter then reviewed six various TTC scheduler implementations that can be used for resource-constrained embedded systems with highly-predictable system behavior. Useful results from the described schedulers were then provided which included jitter levels, memory requirements and error handling capabilities. The results suggested that a "one size fits all" TTC implementation does not exist in practice, since each implementation has advantages and disadvantages. The selection of a particular implementation will, hence, be decided based on the requirements of the application in which the TTC scheduler is employed, e.g. timing and resource requirements.

10. Acknowledgement

The research presented in this chapter was mainly conducted in the Embedded Systems Laboratory (ESL) at University of Leicester, UK, under the supervision of Professor Michael Pont, to whom the authors are thankful.

11. References

Allworth, S.T. (1981) "An Introduction to Real-Time Software Design", Macmillan, London.

Ashling Microsystems (2007) "LPC2000 Evaluation and Development Kits datasheet", available online (Last accessed: November 2010)
http://www.ashling.com/pdf_datasheets/DS266-EvKit2000.pdf

Avrunin, G.S., Corbett, J.C. and Dillon, L.K. (1998) "Analyzing partially-implemented real-time systems", IEEE Transactions on Software Engineering, Vol. 24 (8), pp.602-614.

Ayavoo, D. (2006) "The Development of Reliable X-by-Wire Systems: Assessing The Effectiveness of a 'Simulation First' Approach", PhD thesis, Department of Engineering, University of Leicester, UK.

Ayavoo, D., Pont, M.J. and Parker, S. (2006) "Does a 'simulation first' approach reduce the effort involved in the development of distributed embedded control systems?", 6th UKACC International Control Conference, Glasgow, Scotland, 2006.

Ayavoo, D., Pont, M.J., Short, M. and Parker, S. (2007) "Two novel shared-clock scheduling algorithms for use with CAN-based distributed systems", Microprocessors and Microsystems, Vol. 31(5), pp. 326-334.

Baker, T.P. and Shaw, A. (1989) "The cyclic executive model and Ada. Real-Time Systems", Vol. 1 (1), pp. 7-25.

Bannatyne, R. (1998) "Time triggered protocol-fault tolerant serial communications for real-time embedded systems", WESCON/98 Conference Proceedings, Anaheim, CA, USA, pp. 86-91.

Barr, M. (1999) "Programming Embedded Systems in C and C++", O'Reilly Media.

Baruah S.K. (2006) "The Non-preemptive Scheduling of Periodic Tasks upon Multiprocessors", Real-Time Systems, Vol. 32, pp. 9-20.

Bate, I.J. (1998), "Scheduling and Timing Analysis for Safety Critical Real-Time Systems", PhD thesis, Department of Computer Science, University of York.

Bates, I. (2000) "Introduction to scheduling and timing analysis", in The Use of Ada in Real-Time System, IEE Conference Publication 00/034.

Bolton, W. (2000) "Microprocessor Systems", Longman.

Buttazzo, G. (2005), "Hard real-time computing systems: predictable scheduling algorithms and applications", Second Edition, Springer.

Cho, Y., Yoo, S., Choi, K., Zergainoh, N.E. and Jerraya, A. (2005) "Scheduler implementation in MPSoC Design", In: Asia South Pacific Design Automation Conference (ASPDAC'05), pp. 151-156.

Cho, Y., Zergainoh, N-E., Yoo, S., Jerraya, A.A. and Choi, K. (2007) "Scheduling with accurate communication delay model and scheduler implementation for multiprocessor system-on-chip", Design Automation for Embedded Systems, Vol. 11 (2-3), pp. 167-191.

Cooling, J.E. (1991) "Software design for real time systems", Chapman and Hall.

Cottet, F. (2002) "Scheduling in Real-time Systems", Wiley.

Fisher, J.A., Faraboschi, P. and Young, C. (2004) "Embedded Computing: A VLIW Approach to Architecture, Compilers and Tools", Morgan Kaufmann.

Hsieh, C-C. and Hsu, P-L. (2005) "The event-triggered network control structure for CAN-based motion system",Proceeding of the 2005 IEEE conference on Control Applications, Toronto, Canada, August 28 – 31, 2005.

Hughes, Z.M. and Pont, M.J. (2008) "Reducing the impact of task overruns in resource-constrained embedded systems in which a time-triggered software architecture is employed", Trans Institute of Measurement and Control.

Jerri, A.J. (1977), "The Shannon sampling theorem: its various extensions and applications a tutorial review", Proc. of the IEEE, Vol. 65, pp. 1565-1596.

Kalinsky, D. (2001) " Context switch, Embedded Systems Programming", Vol. 14(1), 94-105.

Kamal, R. (2003) "Embedded Systems: Architecture, Programming and Design", McGraw-Hill.

Katcher, D., Arakawa, H. and Strosnider, J. (1993) "Engineering and analysis of fixed priority schedulers", IEEE Transactions on Software Engineering, Vol. 19 (9), pp. 920-934.

Kim, N., Ryu, M., Hong, S. and Shin, H. (1999) "Experimental Assessment of the Period Calibration Method: A Case Study", Real-Time Systems, Vol. 17 (1), pp. 41-64.

Koch, B. (1999) "The Theory of Task Scheduling in Real-Time Systems: Compilation and Systematization of the Main Results", Studies thesis, University of Hamburg.

Konrad, S., Cheng, B.H. C. and Campbell, L.A. (2004) "Object analysis patterns for embedded systems", IEEE Transactions on Software Engineering, Vol. 30 (12), pp. 970- 992.

Kopetz, H. (1991a) "Event-triggered versus time-triggered real-time systems", In: Proceedings of the InternationalWorkshop on Operating Systems of the 90s and Beyond, London, UK, Springer-Verlag, pp. 87-101.

Kopetz, H. (1991b), "Event-triggered versus time-triggered real-time systems", Technical Report 8/91, Technical University of Vienna, Austria.

Kopetz, H. (1997) "Real-time systems: Design principles for distributed embedded applications", Kluwer Academic.

Kurian, S. and Pont, M.J. (2007) "Maintenance and evolution of resource-constrained embedded systems created using design patterns", Journal of Systems and Software, Vol. 80 (1), pp. 32-41.

LabVIEW (2007) "LabVIEW 7.1 Documentation Resources", WWW website (Last accessed: November   2010) http://digital.ni.com/public.nsf/allkb/06572E936282C0E486256EB0006B70B4

Leung J.Y.T. and Whitehead, J. (1982) "On the Complexity of Fixed-Priority Scheduling of Periodic Real-Time Tasks", Performance Evaluation, Vol. 2, pp. 237-250.

Liu, C.L. and Layland, J.W. (1973), "Scheduling algorithms for multi-programming in a hard real-time environment", Journal of the AVM 20, Vol. 1, pp. 40-61.

Liu, J.W.S. (2000), "Real-time systems", Prentice Hall.

Locke, C.D. (1992), "Software architecture for hard real-time applications: cyclic executives vs. fixed priority executives", Real-Time Systems, Vol. 4, pp. 37-52.

Maaita, A. and Pont, M.J. (2005) "Using 'planned pre-emption' to reduce levels of task jitter in a time-triggered hybrid scheduler". In: Koelmans, A., Bystrov, A., Pont, M.J., Ong, R. and Brown, A. (Eds.), Proceedings of the Second UK Embedded Forum (Birmingham, UK, October 2005), pp. 18-35. Published by University of Newcastle upon Tyne

Marti, P. (2002), "Analysis and design of real-time control systems with varying control timing constraints", PhD thesis, Automatic Control Department, Technical University of Catalonia.

Marti, P., Fuertes, J.M., Villa, R. and Fohler, G. (2001), "On Real-Time Control Tasks Schedulability", European Control Conference (ECC01), Porto, Portugal, pp. 2227-2232.

Mok, A.K. (1983) "Fundamental Design Problems of Distributed Systems for the Hard Real-Time Environment", Ph.D Thesis, MIT, USA.

Mwelwa, C. (2006) "Development and Assessment of a Tool to Support Pattern-Based Code Generation of Time-Triggered (TT) Embedded Systems", PhD thesis, Department of Engineering, University of Leicester, UK.

Mwelwa, C., Athaide, K., Mearns, D., Pont, M.J. and Ward, D. (2006) "Rapid software development for reliable embedded systems using a pattern-based code generation tool", Paper presented at the Society of Automotive Engineers (SAE) World Congress, Detroit, Michigan, USA, April 2006. SAE document number: 2006-01-1457. Appears in: Society of Automotive Engineers (Ed.) "In-vehicle software and hardware systems", Published by Society of Automotive Engineers.

Nahas, M. (2008) "Bridging the gap between scheduling algorithms and scheduler implementations in time-triggered embedded systems", PhD thesis, Department of Engineering, University of Leicester, UK.

Nahas, M. (2011a) "Employing two 'sandwich delay' mechanisms to enhance predictability of embedded systems which use time-triggered co-operative architectures", International Journal of Software Engineering and Applications, Vol. 4, No. 7, pp. 417-425

Nahas, M. (2011b) "Implementation of highly-predictable time-triggered cooperative scheduler using simple super loop architecture", International Journal of Electrical and Computer Sciences,Vol. 11, No. 4, pp. 33-38.

National Instruments (2006) "Low-Cost E Series Multifunction DAQ – 12 or 16-Bit, 200 kS/s, 16 Analog Inputs", available online (Last accessed: November 2010) http://www.ni.com/pdf/products/us/4daqsc202-204_ETC_212-213.pdf

Nghiem, T., Pappas, G.J., Alur, R. and Girard, A. (2006) "Time-triggered implementations of dynamic controllers", Proceedings of the 6th ACM & IEEE International conference on Embedded software, Seoul, Korea, pp. 2-11.

Nissanke, N. (1997) "Real-time Systems", Prentice-Hall.

Obermaisser, R (2004) "Event-Triggered and Time-Triggered Control Paradigms", Kluwer Academic.

Phatrapornnant, T. (2007) "Reducing Jitter in Embedded Systems Employing a Time-Triggered Software Architecture and Dynamic Voltage Scaling", PhD thesis, Department of Engineering, University of Leicester, UK.

Phatrapornnant, T. and Pont, M.J. (2004) "The application of dynamic voltage scaling in embedded systems employing a TTCS software architecture: A case study", Proceedings of the IEE / ACM Postgraduate Seminar on "System-On-Chip Design, Test and Technology", Loughborough, UK, 15 September 2004. Published by IEE. ISBN: 0 86341 460 5 (ISSN: 0537-9989), pp. 3-8.

Phatrapornnant, T. and Pont, M.J. (2006), "Reducing jitter in embedded systems employing a time-triggered software architecture and dynamic voltage scaling", IEEE Transactions on Computers, Vol. 55 (2), pp. 113-124.

Pont, M.J. (2001) "Patterns for time-triggered embedded systems: Building reliable applications with the 8051 family of microcontrollers", ACM Press / Addison-Wesley.

Pont, M.J. (2002) "Embedded C", Addison-Wesley.

Pont, M.J., Kurian, S. and Bautista-Quintero, R. (2006) "Meeting real-time constraints using 'Sandwich Delays'", In: Zdun, U. and Hvatum, L. (Eds) Proceedings of the Eleventh European conference on Pattern Languages of Programs (EuroPLoP '06), Germany, July 2006: pp. 67-77. Published by Universitätsverlag Konstanz.

Pont, M.J., Kurian, S., Wang, H. and Phatrapornnant, T. (2007) "Selecting an appropriate scheduler for use with time-triggered embedded systems", Paper presented at the twelfth European Conference on Pattern Languages of Programs (EuroPLoP 2007).

Pop et al., 2002

Pop, P., Eles, P. and Peng, Z. (2004) "Analysis and Synthesis of Distributed Real-Time Embedded Systems", Springer.

Profeta III, J.A., Andrianos, N.P., Bing, Yu, Johnson, B.W., DeLong, T.A., Guaspart, D. and Jamsck, D. (1996) "Safety-critical systems built with COTS", IEEE Computer, Vol. 29 (11), pp. 54-60.

Rao, M.V.P, Shet, K.C, Balakrishna, R. and Roopa, K. (2008) "Development of Scheduler for Real Time and Embedded System Domain", 22nd International Conference on Advanced Information Networking and Applications - Workshops, 25-28 March 2008, AINAW, pp. 1-6.

Redmill, F. (1992) "Computers in safety-critical applications", Computing & Control Engineering Journal, Vol. 3 (4), pp.178-182.

Sachitanand, N.N. (2002). "Embedded systems - A new high growth area". The Hindu. Bangalore.

Scheler, F. and Schröder-Preikschat, W. (2006) "Time-Triggered vs. Event-Triggered: A matter of configuration?", GI/ITG Workshop on Non-Functional Properties of Embedded Systems (NFPES), March 27 – 29, 2006, Nürnberg, Germany.

Sommerville, I. (2007) "Software engineering", 8th edition, Harlow: Addison-Wesley.

Stankovic, J.A. (1988) "Misconceptions about real-time computing", IEEE Computers, Vol. 21 (10).

Storey, N. (1996) "Safety-critical computer systems", Harlow, Addison-Wesley.

Torngren, M. (1998), "Fundamentals of implementing real-time control applications in distributed computer systems", Real-Time Systems, Vol. 14, pp. 219-250.

Ward, N.J. (1991) "The static analysis of a safety-critical avionics control systems", Air Transport safety: Proceedings of the Safety and Reliability Society Spring Conference, In: Corbyn D.E. and Bray, N.P. (Eds.)

Wavecrest (2001), "Understanding Jitter: Getting Started", Wavecrest Corporation.

Xu , J. and Parnas, D.L. (1993) "On satisfying timing constraints in hard - real - time systems", IEEE Transactions on Software Engineering, Vol. 19 (1), pp. 70-84.

2

Real-Time Operating Systems and Programming Languages for Embedded Systems

Javier D. Orozco and Rodrigo M. Santos
Universidad Nacional del Sur - CONICET
Argentina

1. Introduction

Real-time embedded systems were originally oriented to industrial and military special purpose equipments. Nowadays, mass market applications also have real-time requirements. Results do not only need to be correct from an arithmetic-logical point of view but they also need to be produced before a certain instant called deadline (Stankovic, 1988). For example, a video game is a scalable real-time interactive application that needs real-time guarantees; usually real-time tasks share the processor with other tasks that do not have temporal constraints. To organize all these tasks, a scheduler is typically implemented. Scheduling theory addresses the problem of meeting the specified time requirements and it is at the core of a real-time system.

Paradoxically, the significant growth of the market of embedded systems has not been accompanied by a growth in well-established developing strategies. Up to now, there is not an operating system dominating the market; the verification and testing of the systems consume an important amount of time.

A sign of this is the contradictory results between two prominent reports. On the one hand, The Chaos Report (*The Chaos Report*, 1994) determined that about 70 % had problems; 60 % of those projects had problems with the statement of requirements. On the other hand, a more recent evaluation (Maglyas et al., 2010) concluded that about 70% of them could be considered successful. The difference in the results between both studies comes from the model adopted to analyze the collected data. While in The Chaos Report (1994) a project is considered to be successful if it is completed on time and budget, offering all features and functions as initially specified, in (Maglyas et al., 2010) a project is considered to be successful even if there is a time overrun. In fact, in (Maglyas et al., 2010) only about 30% of the projects were finished without any overruns, 40% have time overrun and the rest of the projects have both overruns (budget and time) or were cancelled. Thus, in practice, both studies coincide in that 70 % of the projects had some kind of overrun but they differ in the criteria used to evaluate a project as successful.

In the literature there is no study that conducts this kind of analysis for real time projects in particular. The evidence from the reports described above suggests that while it is difficult to specify functional requirements, specifying non functional requirements such as temporal constraints, is likely to be even more difficult. These usually cause additional redoes and errors motivated by misunderstandings, miscommunications or mismanagement. These errors could be more costly on a time critical application project than on a non real time one given that not being time compliant may cause a complete re-engineering of the system. The introduction of non-functional requirements such as temporal constraints makes the design and implementation of these systems increasingly costly and delays the introduction of the final product into the market. Not surprisingly, development methodologies for real-time frameworks have become a widespread research topic in recent years.

Real-time software development involves different stages: modeling, temporal characterization, implementation and testing. In the past, real-time systems were developed from the application level all the way down to the hardware level so that every piece of code was under control in the development process. This was very time consuming. Given that the software is at the core of the embedded system, reducing the time needed to complete these activities reduces the time to market of the final product and, more importantly, it reduces the final cost. In fact, as hardware is becoming cheaper and more powerful, the actual bottleneck is in software development. In this scenario, there is no guarantee that during the software life time the hardware platform will remain constant or that the whole system will remain controlled by a unique operating system running the same copy of the operating embedded software. Moreover, the hardware platform may change even while the application is being developed. Therefore, it is then necessary to introduce new methods to extend the life time of the software (Pleunis, 2009).

In this continuously changing environment it is necessary to introduce certainty for the software continuity. To do such a thing, in the last 15 years the paradigm Write Once Run Anywhere (WORA) has become dominant. There are two alternatives for this: Java and .NET. The first one was first introduced in the mid nineties and it is supported by Sun Microsystems and IBM among others (Microsystems, 2011). Java introduces a virtual machine that eventually runs on any operating system and hardware platform. .NET was released at the beginning of this century by Microsoft and is oriented to Windows based systems only and does not implement a virtual machine but produces a specific compilation of the code for each particular case. (Zerzelidis & Wellings, 2004) analyze the requirements for a real-time framework for .NET.

Java programming is well established as a platform for general purpose applications. Nevertheless, hardware independent languages like Java are not used widely for the implementation of control applications because of low predictability, no real-time garbage collection implementation and cumbersome memory management (Robertz et al., 2007). However, this has changed in the last few years with the definition and implementation of the Real-Time Specification for Java. In 2002, the specification for the real-time Java (RTSJ) proposed in (Gosling & Bollella, 2000) was finally approved (Microsystems, 2011). The first commercial implementation was issued in the spring of 2003. In 2005, the RTSJ 1.0.1 was released together with the Real-Time Specification (RI). In September 2009 Sun released the Java Real-Time System 2.2 version which is the latest stable one. The use of RTSJ as a development language for real-time systems is not generalized, although there have been many papers on embedded systems implementations based on RTSJ and even several full Java microprocessors on different technologies have been proposed and used (Schoeberl, 2009). However, Java is penetrating into more areas ranging from Internet based products to small embedded mobile products like phones as well as from complex enterprise systems to small components in a sensor network. In order to extend the life of the software, even over a particular device, it becomes necessary to have transparent development platforms to the hardware architecture, as it is the case of RTSJ. This is undoubtedly a new scenario in the development of embedded real time systems. There is a wide range of hardware possibilities in the market (microcontrollers, microprocessors and DSPs); also there are many different programming languages, like C, C++, C#, Java, Ada; and there are more than forty real-time operating systems (RTOS) like RT-Linux, Windows Embedded or FreeRTOS. This chapter offers a road-map for the design of real-time embedded systems evaluating the pros and cons of the different programming languages and operating systems.

Organization: This chapter is organized in the following way. Section 2 describes the main characteristics that a real-time operating system should have. Section 3 discusses the scope of some of the more well known RTOSs. Section 4 introduces the languages used for real-time programming and compares the main characteristics. Section 5 presents and compares different alternatives for the implementation of real-time Java. Finally, Section 6 concludes.

2. Real time operating system

The formal definition of a real-time system was introduced in Section 1. In a nutshell these are systems which have additional non-functional requirements that are as important as the functional ones for the correct operation. It is not enough to produce correct logical-arithmetic results; these results must also be accomplished before a certain deadline (Stankovic, 1988). This timeliness behavior imposes extra constraints that should be carefully considered during the whole design process. If these constraints are not satisfied, the system risks severe consequences. Traditionally, real-time systems are classified as hard, firm and soft. The first class is associated to critical safety systems where no deadlines can be missed. The second class covers some applications where occasional missed deadlines can be tolerated if they follow a certain predefined pattern. The last class is associated to systems where the missed deadlines degrade the performance of the applications but do not cause severe consequences. An embedded system is any computer that is a component of a larger system and relies on its own microprocessor (Wolf, 2002). It is said to work in real-time when it has to comply with time constraints, being hard, firm or soft. In this case, the software is encapsulated in the hardware it controls. There are several examples of real-time embedded systems such as the controller for the power-train in cars, voice processing in digital phones, video codecs for DVD players or Collision Warning Systems in cars and video surveillance cam controllers.

RTOS have special characteristics that make them different to common OS. In the particular case of embedded systems, the OS usually allows direct access to the microprocessor registers, program memory and peripherals. These characteristics are not present in traditional OS as they preserve the kernel areas from the user ones. The kernel is the main part of an operating system. It provides the task dispatching, communication and synchronization functions. For the particular case of embedded systems, the OS is practically reduced to these main functions. Real-time kernels have to provide primitives to handle the time constraints for the tasks and applications (deadlines, periods, worst case execution times (WCET)), a priority discipline to order the execution of the tasks, fast context switching, a small footprint and small overheads.

The kernel provides services to the tasks such as I/O and interrupt handling and memory allocation through *system-calls*. These may be invoked at any instant. The kernel has to be able to preempt tasks when one of higher priority is ready to execute. To do this, it usually has the maximum priority in the system and executes the scheduler and dispatcher periodically based on a timer tick interrupt. At these instants, it has to check a ready task queue structure and if necessary remove the running task from the processor and dispatch a higher priority one. The most accepted priority discipline used in RTOS is fixed priorities (FP) (*eCosCentric,* 2011; *Enea OSE,* 2011; *LynxOS RTOS, The real-time operating system for complex embedded systems,* 2011; *Minimal Real-Time Operating System,* 2011; *RTLinuxFree,* 2011; *The free RTOS Project,* 2011; *VxWorks RTOS,* 2011; *Windows Embedded,* 2011). However, there are some RTOSs that are implementing other disciplines like earliest deadline first (EDF) (*Erika Enterprise: Open Source RTOS for single- and multi-core applications,* 2011; *Service Oriented Operating System,* 2011; *S.Ha.R.K.: Soft Hard Real-Time Kernel,* 2007). Traditionally, real-time systems scheduling theory starts considering independent, preemptive and periodic tasks. However, this simple model is not useful when considering a real application in which tasks synchronize, communicate among each other and share resources. In fact, task synchronization and communication are two central aspects when dealing with real-time applications. The use of semaphores and critical sections should be controlled with a contention policy capable of bounding the unavoidable priority inversion and preventing deadlocks. The most common contention policies implemented at kernel level are the priority ceiling protocol (Sha et al., 1990) and the stack resource policy (Baker, 1990). Usually, embedded systems have a limited memory address space because of size, energy and cost constraints. It is important then to have a small footprint so more memory is available for the implementation of the actual application. Finally, the time overhead of the RTOS should be as small as possible to reduce the interference it produces in the normal execution of the tasks.

The IEEE standard, Portable Operating System Interface for Computer Environments (POSIX 1003.1b) defines a set of rules and services that provide a common base for RTOS (IEEE, 2003). Being POSIX compatible provides a standard interface for the system calls and services that the OS provides to the applications. In this way, an application can be easily ported across different OSs. Even though this is a desirable feature for an embedded RTOS, it is not always possible to comply with the standard and keep a small footprint simultaneously. Among the main services defined in the POSIX standard, the following are probably the most important ones:

- Memory locking and Semaphore implementations to handle shared memory accesses and synchronization for critical sections.

- Execution scheduling based on round robin and fixed priorities disciplines with thread preemption. Thus the threads can be waiting, executing, suspended or blocked.

- Timers are at the core of any RTOS. A real-time clock, usually the system clock should be implemented to keep the time reference for scheduling, dispatching and execution of threads.Memory locking and Semaphore implementations to handle shared memory accesses and synchronization for critical sections.

2.1 Task model and time constraints

A real-time system is temporally described as a set of tasks $S(m) = \{\tau_1, \ldots, \tau_i, \ldots, \tau_m\}$ where each task is described by a tuple $(WCET_i, T_i, D_i)$ where T_i is the period or minimum interarrival time and D_i is the relative deadline that should be greater than or equal to the worst case response time. With this description, the scheduling conditions of the system for different priority disciplines can be evaluated. This model assumes that the designer of the system can measure in a deterministic way the worst case execution time of the tasks. Yet, this assumes knowledge about many hardware dependent aspects like the microprocessor architecture, context switching times and interrupts latencies. It is also necessary to know certain things about the OS implementation such as the timer tick and the priority discipline used to evaluate the kernel interference in task implementation. However, these aspects are not always known beforehand so the designer of a real-time system should be careful while implementing the tasks. Avoiding recursive functions or uncontrolled loops are basic rules that should be followed at the moment of writing an application. Programming real-time applications requires the developer to be specially careful with the nesting of critical sections and the access to shared resources. Most commonly, the kernel does not provide a validation of the time constraints of the tasks, thus these aspects should be checked and validated at the design stage.

2.2 Memory management

RTOS specially designed for small embedded system should have very simple memory management policies. Even if dynamic allocations can provide a better performance and usage, they add an important degree of complexity. If the embedded system is a small one with a small address space, the application is usually compiled together with the OS and the whole thing is burnt into the ROM memory of the device. If the embedded system has a large memory address space, such as the ones used in cell phones or tablets, the OS behaves more like a traditional one and thus, dynamic handling of memory allocations for the different tasks is possible. The use of dynamic allocations of memory also requires the implementation of garbage collector functions for freeing the memory no longer in use.

2.3 Scheduling algorithms

To support multi-task real-time applications, a RTOS must be multi-threaded and preemptible. The scheduler should be able to preempt any thread in the system and dispatch the highest priority active thread. Sometimes, the OS allows external interrupts to be enabled. In that case, it is necessary to provide proper handlers for these. These handlers include a controlled preemption of the executing thread and a safe context switch. Interrupts are usually associated to kernel interrupt service routines (ISR), such as the timer tick or serial port interfaces management. The ISR in charge of handling the devices is seen by the applications like services provided by the OS.

RTOS should provide a predictable behavior and respond in the same way to identical situations. This is perhaps the most important requirement that has to be satisfied. There are two approaches to handle the scheduling of tasks: time triggered or event triggered. The main characteristic of the first approach is that all activities are carried out at certain points in time known a prori. For this, all processes and their time specifications must be known in advance. Otherwise, an efficient implementation is not possible. Furthermore, the communication and the task scheduling on the control units have to be synchronized during operation in order to ensure the strict timing specifications of the system design (Albert, 2004). In this case the task execution schedule is defined off-line and the kernel follows it during run time. Once a feasible schedule is found, it is implemented with a cycle-executive that repeats itself each time. It is difficult to find an optimum schedule but onces it is found the implementation is simple and can be done with a look-up table. This approach does not allow a dynamic system to incorporate new tasks or applications. A modification on the number of executing tasks requires the recomputation of the schedule and this is rather complex to be implemented on line. In the second approach, external or internal events are used to dispatch the different activities. This kind of designs involve creating systems which handle multiple interrupts. For example, interrupts may arise from periodic timer overflows, the arrival of messages on a CAN bus, the pressing of a switch, the completion of an analogue-to-digital conversion and so on. Tasks are ordered following a priority order and the highest priority one is dispatched each time. Usually, the kernel is based on a timer tick that preempts the current executing task and checks the ready queue for higher priority tasks. The priority disciplines most frequently used are round robin and fixed priorities. For example, the Department of Defense of the United States has adopted fixed priorities Rate Monotonic Sheduling (priority is assigned in reverse order to periods, giving the highest priority to the shortest period) and with this has made it a *de facto* standard Obenza (1993). The event triggered scheduling can introduce priority inversions, deadlocks and starvation if the access to shared resources and critical sections is not controlled in a proper manner. These problems are not acceptable in safety critical real-time applications. The main advantage of event-triggered systems is their ability to fastly react to asynchronous external events which are not known in advance (Albert & Gerth, 2003). In addition, event-triggered systems possess a higher flexibility and allow in many cases the adaptation to the actual demand without a redesign of the complete system (Albert, 2004).

2.4 Contention policies for shared resources and critical sections

Contention policies are fundamental in event-triggered schedulers. RTOSs have different approaches to handle this problem. A first solution is to leave the control mechanism in hands of the developers. This is a non-portable, costly and error prone solution. The second one implements a contention protocol based on priority inheritance (Sha et al., 1990). This solution bounds the priority inversions to the longest critical section of each lower priority task. It does not prevent deadlocks but eliminates the possibility of starvation. Finally, the Priority Ceiling Protocol (PCP) (Sha et al., 1990) and the Stack Resource Policy (SRP) (Baker, 1990) bound the priority inversion to the longest critical section of the system, avoid starvation and deadlocks. Both policies require an active kernel controlling semaphores and shared resources. The SRP performs better since it produces an early blocking avoiding some unnecessary preemptions present in the PCP. However, both approaches are efficient.

3. Real time operating system and their scope

This section presents a short review on some RTOS currently available. The list is not exhaustive as there are over forty academic and commercial developments. However, this section introduces the reader to a general view of what can be expected in this area and the kind of OS available for the development of real-time systems.

3.1 RTOS for mobile or small devices

Probably one of the most frequently used RTOS is Windows CE. Windows CE is now known as Windows Embedded and its family includes Windows Mobile and more recently Windows Phone 7 (*Windows Embedded*, 2011). Far from being a simplification of the well known OS from Microsoft, Windows CE is a RTOS with a relatively small footprint and is used in several embedded systems. In its actual version, it works on 32 bit processors and can be installed in 12 different architectures. It works with a timer tick or time quantum and provides 256 priority levels. It has a memory management unit and all processes, threads, mutexes, events and semaphores are allocated in virtual memory. It handles an accuracy of one millisecond for SLEEP and WAIT related operations. The footprint is close to 400 KB and this is the main limitation for its use in devices with small memory address spaces like the ones present in wireless sensor networks microcontrollers.

eCos is an open source real-time operating system intended for embedded applications (*eCosCentric*, 2011). The configurability technology that lies at the heart of the eCos system enables it to scale from extremely small memory constrained SOC type devices to more sophisticated systems that require more complex levels of functionality. It provides a highly optimized kernel that implements preemptive real-time scheduling policies, a rich set of synchronization primitives, and low latency interrupt handling. The eCos kernel can be configured with one of two schedulers: The Bitmap scheduler and the Multi-Level Queue (MLQ) scheduler. Both are preemptible schedulers that use a simple numerical priority to determine which thread should be running. The number of priority levels is configurable up to 32. Therefore thread priorities will be in the range of 0 to 31, with 0 being the highest priority. The bitmap scheduler only allows one thread per priority level, so if the system is configured with 32 priority levels then it is limited to only 32 threads and it is not possible to preempt the current thread in favor of another one with the same priority. Identifying the highest-priority runnable thread involves a simple operation on the bitmap, and an array index operation can then be used to get hold of the thread data structure itself. This makes the bitmap scheduler fast and totally deterministic. The MLQ scheduler allows multiple threads to run at the same priority. This means that there is no limit on the number of threads in the system, other than the amount of memory available. However operations such as finding the highest priority runnable thread are a slightly bit more expensive than for the bitmap scheduler. Optionally the MLQ scheduler supports time slicing, where the scheduler automatically switches from one runnable thread to another when a certain number of clock ticks have occurred.

LynxOS (*LynxOS RTOS, The real-time operating system for complex embedded systems*, 2011) is a POSIX-compatible, multiprocess, multithreaded OS. It has a wide target of hardware architectures as it can work on complex switching systems and also in small embedded products. The last version of the kernel follows a microkernel design and has a minimum footprint of 28KB. This is about 20 times smaller than Windows CE. Besides scheduling, interrupt, dispatch and synchronize, there are additional services that are provided in the form of plug-ins so the designer of the system may choose to add the libraries it needs for a special purposes such as file system administration or TCP/IP support. The addition of these services obviously increases the footprint but they are optional and the designer may choose to have them or not. LynxOS can handle 512 priority levels and can implement several scheduling policies including prioritized FIFO, dynamic deadline monotonic scheduling, prioritized round robin, and time slicing among others.

FreeRTOS is an open source project (*The free RTOS Project*, 2011). It provides porting to 28 different hardware architectures. It is a multi-task operating system where each task has its own stack defined so it can be preempted and dispatched in a simple way. The kernel provides a scheduler that dispatches the tasks based on a timer tick according to a Fixed Priority policy. The scheduler consists of an only-memory-limited queue with threads of different priority. Threads in the queue that share the same priority will share the CPU with the round robin time slicing. It provides primitives for suspending, sleeping and blocking a task if a synchronization process is active. It also provides an interrupt service protocol for handling I/O in an asynchronous way.

MaRTE OS is a Hard Real-Time Operating System for embedded applications that follows the Minimal Real-Time POSIX.13 subset (*Minimal Real-Time Operating System*, 2011). It was developed at University of Cantabria, Spain, and has many external contributions that have provided drivers for different communication interfaces, protocols and I/O devices. MaRTE provides an easy to use and controlled environment to develop multi-thread Real-Time applications. It supports mixed language applications in ADA, C and C++ and there is an experimental support for Java as well. The kernel has been developed with Ada2005 Real-Time Annex (*ISO/IEC 8526:AMD1:2007. Ada 2005 Language Reference Manual (LRM)*, 2005). Ada 2005 Language Reference Manual (LRM), 2005). It offers some of the services defined in the POSIX.13 subset like pthreads and mutexes. All the services have a time bounded response that includes the dynamic memory allocation. Memory is managed as a single address space shared by the kernel and the applications. MaRTE has been released under the GNU General Public License 2.

There are many other RTOS like SHArK (*S.Ha.R.K.: Soft Hard Real-Time Kernel*, 2007), Erika (*Erika Enterprise: Open Source RTOS for single- and multi-core applications*, 2011), SOOS (*Service Oriented Operating System*, 2011), that have been proposed in the academic literature to validate different scheduling and contention policies. Some of them can implement fault-tolerance and energy-aware mechanisms too. Usually written in C or C++ these RTOSs are research oriented projects.

3.2 General purpose RTOS

VxWorks is a proprietary RTOS. It is cross-compiled in a standard PC using both Windows or Linux (*VxWorks RTOS*, 2011). It can be compiled for almost every hardware architecture used in embedded systems including ARM, StrongARM and xScale processors. It provides mechanisms for protecting memory areas for real-time tasks, kernel and general tasks. It implements mutual exclusion semaphores with priority inheritance and local and distributed messages queues. It is able to handle different file systems including high reliability file systems and network file systems. It provides the necessary elements to implement the Ipv6 networking stack. There is also a complete development utility that runs over Eclipse.

RT-Linux was developed at the New Mexico School of Mines as an academic project (*RTLinuxFree*, 2011)(RTLinuxFree, 2011). The idea is simple and consists in turning the base GNU/Linux kernel into a thread of the Real-Time one. In this way, the RTKernel has control over the traditional one and can handle the real-time applications without interference from the applications running within the traditional kernel. Later RT-Linux was commercialized by FMLabs and finally by Wind River that also commercializes VxWorks. GNU/Linux drivers handle almost all I/O. First-In-First-Out pipes (FIFOs) or shared memory can be used to share data between the operating system and RTCore. Several distributions of GNU/Linux include RTLinux as an optional package.

RTAI is another real-time extension for GNU/Linux (*RTAI - the RealTime Application Interface for Linux*, 2010). It stands for Real-Time Application Interface. It was developed for several hardware architectures such as x86, x86_64, PowerPC, ARM and m68k. RTAI consists in a patch that is applied to the traditional GNU/Linux kernel and provides the necessary real-time primitives for programming applications with time constraints. There is also a toolchain provided, RTAI-Lab, that facilitates the implementation of complex tasks. RTAI is not a commercial development but a community effort with base at University of Padova.

QNX is a unix like system that was developed in Canada. Since 2009 it is a proprietary OS (*QNX RTOS v4 System Documentation*, 2011). It is structured in a microkernel fashion with the services provided by the OS in the form of servers. In case an specific server is not required it is not executed and this is achieved by not starting it. In this way, QNX has a small footprint and can run on many different hardware platforms. It is available for different hardware platforms like the PowerPC, x86 family, MIPS, SH-4 and the closely related family of ARM, StrongARM and XScale CPUs. It is the main software component for the Blackberry PlayBook. Also Cisco has derived an OS from QNX.

OSE is a proprietary OS (*Enea OSE*, 2011). It was originally developed in Sweden. Oriented to the embedded mobile systems market, this OS is installed in over 1.5 billion cell phones in the world. It is structured in a microkernel fashion and is developed by telecommunication companies and thus it is specifically oriented to this kind of applications. It follows an event driven paradigm and is capable of handling both periodic and aperiodic tasks. Since 2009, an extension to multicore processors has been available.

4. Real-time programming languages

Real-time software is necessary to comply not only with functional application requirements but also with non functional ones like temporal restrictions. The nature of the applications requires a bottom-up approach in some cases a top-down approach in others. This makes the programming of real-time systems a challenge because different development techniques need to be implemented and coordinated for a successful project.

In a bottom-up approach one programming language that can be very useful is assembler. It is clear that using assembler provides access to the registers and internal operations of the processor. It is also well known that assembler is quite error prone as the programmer has to implement a large number of code lines. The main problem however is that using assembler makes the software platform dependent on the hardware and it is almost impossible to port the software to another hardware platform. Another language that is useful for a bottom-up approach is C. C provides an interesting level of abstraction and still gives access to the details of the hardware, thus allowing for one last optimization pass of the code. There are C compilers developed for almost every hardware platform and this gives an important portability to the code. The characteristics of C limits the software development in some cases and this is why in the last few years the use of C++ has become popular. C++ extends the language to include an object-oriented paradigm. The use of C++ provides a more friendly engineering approach as applications can be developed based on the object- oriented paradigm with a higher degree of abstraction facilitating the modeling aspects of the design. C++ compilers are available for many platforms but not for so many as in the C case. With this degree of abstraction, ADA is another a real-time language that provides resources for many different aspects related to real-time programming as tasks synchronization and semaphores implementations. All the programming languages mentioned up to now require a particular compiler to execute them on a specific hardware platform. Usually the software is customized for that particular platform. There is another approach in which the code is written once and runs anywhere. This approach requires the implementation of a virtual machine that deals with the particularities of the operating system and hardware platform. The virtual machine presents a simple interface for the programmer, who does not have to deal with these details. Java is probably the most well known WORA language and has a real-time extension that facilitates the real-time programming.

In the rest of this section the different languages are discussed highlighting their pros and cons in each case are given so the reader can decide which is the best option for his project.

4.1 Assembler

Assembler gives the lowest possible level access to the microprocessor architecture such as registers, internal memory, I/O ports and interrupts handling. This direct access provides the programmer with full control over the platform. With this kind of programming, the code has very little portability and may produce hazard errors. Usually the memory management, allocation of resources and synchronization become a cumbersome job that results in very complex code structures. The programmer should be specialized on the hardware platform and should also know the details of the architecture to take advantage of such a low level programming. Assembler provides predictability on execution time of the code as it is possible to count the clock states to perform a certain operation.

There is total control over the hardware and so it is possible to predict the instant at which the different activities are going to be done.

Assembler is used in applications that require a high degree of predictability and are specialized on a particular kind of hardware architecture. The verification, validation and maintenance of the code is expensive. The life time of the software generated with this language is limited by the end-of-life of the hardware.

The cost associated to the development of the software, which is high due to the high degree of specialization, the low portability and the short life, make Assembler convenient only for very special applications such as military and space applications.

4.2 C

C is a language that was developed by Denis Ritchie and Brian Kernighan. The language is closely related to the development of the Unix Operating System. In 1978 the authors published a book of reference for programming in C that was used for a 25 years. Later, C was standardized by ANSI and the second edition of the book on included the changes incorporated in the standardization of the language (*ISO/IEC 9899:1999 - Programming languages - C*, 1999). Today, C is taught in all computer science and engineering courses and has a compiler for almost every available hardware platform.

C is a function oriented language. This important characteristic allows the construction of special purpose libraries that implement different functions like Fast Fourier Transforms, Sums of Products, Convolutions, I/O ports handling or Timing. Many of these are available for free and can be easily adapted to the particular requirements of a developer.

C offers a very simple I/O interface. The inclusion of certain libraries facilitates the implementation of I/O related functions. It is also possible to construct a Hardware Adaptation Layer in a simple way and introduce new functionalities in this way . Another important aspect in C is memory management. C has a large variety of variable types that include, among others, char, int, long, float and double. C is also capable of handling pointers to any of the previous types of variables and arrays. The combination of pointers, arrays and types produce such a rich representation of data that almost anything is addressable. Memory management is completed with two very important operations: `calloc` and `malloc` that reserve space memory and the corresponding `free` operation to return the control of the allocated memory to the operating system.

The possibility of writing a code in C and compiling it for almost every possible hardware platform, the use of libraries, the direct access and handling of I/O resources and the memory management functions constitute excellent reasons for choosing this programming language at the time of developing a real-time application for embedded systems.

4.3 C++

The object-oriented extension of C was introduced by Bjarne Stroustrup in 1985. In 1999 the language received the status of standard (*ISO/IEC 14882:2003 - Programming languages C++*, 2003). C++ is backward compatible with C. That means that a function developed in C can be compiled in C++ without errors. The language introduces the concept of Classes, Constructors, Destructors and Containers. All these are included in an additional library that extends the original C one.

In C++ it is possible to do virtual and multiple inheritance. As an object oriented language it has a great versatility for implementing complex data and programming structures. Pointers are extended and can be used to address classes and functions enhancing the rich addressable elements of C. These possibilities require an important degree of expertise for the programmer as the possibility of introducing errors is important.

C++ compilers are not as widespread as the C ones. Although the language is very powerful in the administration of hardware, memory management and modeling, it is quite difficult to master all the aspects it includes. The lack of compilers for different architectures limits its use for embedded systems. Usually, software developers prefer the C language with its limitations to the use of the C++ extensions.

4.4 ADA

Ada is a programming language developed for real-time applications (*ISO/IEC 8526:AMD1:2007. Ada 2005 Language Reference Manual (LRM)*, 2005). Like C++ it supports structured and object-oriented programming but also provides support for distributed and concurrent programming. Ada provides native synchronization primitives for tasks. This is important when dealing with real-time systems as the language provides the tools to solve a key aspect in the programming of this kind of systems. Ada is used in large scale programs. The platforms usually involve powerful processors and large memory spaces. Under these conditions Ada provides a very secure programming environment. On the other hand, Ada is not suitable for small applications running on low end processors like the ones implementing wireless sensors networks with reduced memory spaces and processor capacities.

Ada uses a safe type system that allows the developer to construct powerful abstractions reflecting the real world while the compiler can detect logic errors. The software can be built in modules facilitating development of large systems by teams. It also separates interfaces from implementation providing control over visibility. The strict definition of types and the syntax allow the code to be compiled without changes on different compliant compilers on different hardware platforms. Another important feature is the early standardization of the language. Ada compilers are officially tested and are accepted only after passing the test for military and commercial work. Ada also has support for low level programming features. It allows the programmer to do address arithmetic, directly access to memory address space, perform bit wise operations and manipulations and the insert of machine code. Thus Ada is a good choice for programming embedded systems with real-time or safety-critical applications. These important features have facilitated the maintainability of the code across the life time of the software and this facilitates its use in aerospace, defense, medical, rail-road and nuclear applications.

4.5 C#

Microsoft's integrated development environment (.NET) includes a new programming language C# which targets the .NET Framework. Microsoft does not claim that C# and .NET are intended for real-time systems. In fact, C# and the .NET platform do not support many of the thread management constructs that real-time systems, particularly hard ones, often require. Even Anders Hejlsberg (Microsoft's C# chief architect) states, "I would say that 'hard real-time' kinds of programs wouldn't be a good fit (at least right now)" for the .NET platform (Lutz & Laplante, 2003). For instance, the Framework does not support thread creation at a particular instant in time with the guarantee that it will be completed by a certain in time. C# supports many thread synchronization mechanisms but none with high precision.

Windows CE has significantly improved thread management constructs. If properly leveraged by C# and the .NET Compact Framework, it could potentially provide a reasonably powerful thread management infrastructure. Current enumerations for thread priority in the .NET Framework, however, are largely unsatisfactory for real-time systems. Only five levels exist: AboveNormal, BelowNormal, Highest, Lowest, and Normal. By contrast Windows CE, specifically designed for real time systems has 256 thread priorities. Microsoft's ThreadPriority enumeration documentation also states that "the scheduling algorithm used to determine the order of thread execution varies with each operating system." This inconsistency might cause real-time systems to behave differently on different operating systems.

4.6 Real-time java

Java includes a number of technologies ranging from JavaCard applications running in tens of kilobytes to large server applications running with the Java 2 Enterprise Edition requiring many gigabytes of memory. In this section, the Real-time specification for Java (RTSJ) is described in detail. This specification proposes a complete set of tools to develop real-time applications. None of the other languages used in real-time programming provide classes, templates and structures on which the developer can build the application. When using other languages, the programmer needs to construct classes, templates and structures and then implement the application taking care of the scheduler, periodic and sporadic task handling and the synchronization mechanism.

RTSJ is a platform developed to handle real-time applications on top of a Java Virtual Machine (JVM). The JVM specification describes an abstract stack machine that executes *bytecodes,* the intermediate code of the Java language. Threads are created by the JVM but are eventually scheduled by the operating system scheduler over which it runs. The Real-Time Specification for Java (Gosling & Bollella, 2000; Microsystems, 2011) provides a framework for developing real-time scheduling mostly on uniprocessors systems. Although it is designed to support a variety of schedulers only the `PriorityScheduler` is currently defined and is a preemptive fixed priorities one (FPP). The implementation of this abstraction could be handled either as a middleware application on top of stock hardware and operating systems or by a direct hardware implementation (Borg et al., 2005). RTS Java guarantees backward compatibility so applications developed in traditional Java can be executed together with real-time ones. The specification requires an operating system capable of handling real-time threads like RT-Linux. The indispensable OS capabilities must include a high-resolution timer, program-defined low-level interrupts, and a robust priority-based scheduler with deterministic procedures to solve resource sharing priority inversions. RTSJ models three types of tasks: Periodic, Sporadic and Aperiodic. The specification uses a FPP scheduler (`PriorityScheduler`) with 28 different priority levels. These priority levels are handled under the `Schedulable` interface which is implemented by two classes: `RealtimeThread` and `AsyncEventHandler`. The first ones are tasks that run under the FPP scheduler associated to one of the 28 different priority levels and are implementations of the `javax.realtime.RealtimeThread`, RealtimeThread for short. Sporadic tasks are not in the FPP scheduler and are served as soon as they are released by the `AsyncEventHandler`. The last ones do not have known temporal parameters and are handled as standard `java.lang.Thread` (Microsystems, 2011). There are two classes of parameters that should be attached to a schedulable real-time entity. The first one is specified in the class `SchedulingParameters`. In this class the parameters that are necessary for the scheduling, for example the priority, are defined. The second one, is the class `ReleaseParameters`. In this case, the parameters related to the mode in which the activation of the thread is done such as period, worst case computation time, and offset are defined.

Traditional Java uses a Garbage Collector (GC) to free the region of memory that is not referenced any more. The normal memory space for Java applications is the `HeapMemory`. The GC activity interferes with the execution of the threads in the JVM. This interference is unacceptable in the real-time domain as it imposes blocking times for the currently active threads that are neither bounded nor can they be determined in advance. To solve this, the real-time specification introduces a new memory model to avoid the interference of the GC during runtime. The abstract class `MemoryArea` models the memory by dividing it in regions. There are three types of memory: `HeapMemory`, `ScopedMemory` and `InmortalMemory`. The first one is used by non real time threads and is subject to GC activity. The second one, is used by real time threads and is a memory that is used by the thread while it is active and it is immediately freed when the real-time thread stops. The last one is a very special type of memory that should be used very carefully as even when the JVM finishes it may remain allocated. The RTSJ defines a sub-class `NoHeapRealtimeThread` of `RealtimeThread` in which the code inside the method `run()` should not reference any object within the `HeapMemory` area. With this, a real-time thread will preempt the GC if necessary. Also when specifying an `AsyncEventHandler` it is possible to avoid the use of `HeapMemory` and define instead the use of `ScopedMemory` in its constructor.

4.6.1 Contention policy for shared resources and task synchronization

The RTSJ virtual machine supports priority-ordered queues and performs by default a basic priority inheritance and a ceiling priority inheritance called priority ceiling emulation. The priority inheritance protocol has the problem that it does not prevent deadlocks when a wrong nested blocking occurs. The priority ceiling protocol avoids this by assigning a ceiling priority to a critical section which is equal to the highest priority of any task that may lock it. This is effective but it is more complex to implement. The mix of the two inheritance protocols avoid unbounded priority inversions caused by low priority thread locks.

Each thread has a base and an active priority. The base priority is the priority allocated by the programmer. The active priority is the priority that the scheduler uses to sort the run queue. As mentioned before, the real-time JVM must support priority-ordered queues and perform priority inheritance whenever high priority threads are blocked by low priority ones. The active priority of a thread is, therefore, the maximum of its base priority and the priority it has inherited.

The RTSJ virtual machine supports priority-ordered queues and performs by default a basic priority inheritance and a ceiling priority inheritance called priority ceiling emulation. The priority inheritance protocol has the problem that it does not prevent deadlocks when a wrong nested blocking occurs. The priority ceiling protocol avoids this by assigning a ceiling priority to a critical section which is equal to the highest priority of any task that may lock it. This is effective but it is more complex to implement. The mix of the two inheritance protocols avoid unbounded priority inversions caused by low priority threads locks.

Each thread has a *base* and an *active* priority. The base priority is the priority allocated by the programmer. The active priority is the priority that the scheduler uses to order the run queue. As mentioned before, the real-time JVM must support priority-ordered queues and perform priority inheritance whenever high priority threads are blocked by low priority ones. The active priority of a thread is, therefore, the maximum of its base priority and the priority it has inherited.

4.7 C/C++ or RTJ

In real-time embedded systems development flexibility, predictability and portability are required at the same time. Different aspects such as contention policies implementation and asynchronous handling, are managed naturally in RTSJ. Other languages, on the other hand, require a careful programming by the developer. However, RTSJ has some limitations when it is used in small systems where the footprint of the system should be kept as small as possible. In the last few years, the development of this kind of systems has been dominated by C/C++. One reason for this trend is that C/C++ exposes low-level system facilities more easily and the designer can provide ad-hoc optimized solutions in order to reach embedded-system real time requirements. On the other hand, Java runs on a Virtual Machine, which protects software components from each other. In particular, one of the common errors in a C/C++ program is caused by the memory management mechanism of C/C++ which forces the programmers to allocate and deallocate memory manually. Comparisons between C/C++ and Java in the literature recognize pros and cons for both. Nevertheless, most of the ongoing research on this topic concentrates on modifying and adapting Java. This is because its environment presents some attributes that make it attractive for real-time developers. Another interesting attribute from a software designer point of view is that Java has a powerful, portable and continuously updated standard library that can reduce programming time and costs. In Table 1 the different aspects of the languages discussed are summarized. *VG* stands for very good, *G* for good, *R* for regular and *B* for bad.

Language	Portability	Flexibility	Abstraction	Resource Handling	Predictability
Assembler	B	B	B	VG	VG
C	G	G	G	VG	G
C++	R	VG	VG	VG	G
Ada	R	VG	VG	VG	G
RTSJ	VG	VG	VG	R	R

Table 1. Languages characteristics

5. Java implementations

In this section different approaches to the implementation of Java are presented. As explained, a java application requires a virtual machine. The implementation of the JVM is a fundamental aspect that affects the performance of the system. There are different approaches for this. The simplest one, resolves everything at software level. The jave bytecodes of the application are interpreted by the JVM that passes the execution code to the RTOS and this dispatches the thread. Another option consists in having a Just in Time (JIT) compiler to transform the java code in machine code and directly execute it within the processor. And finally, it is possible to implement the JVM in hardware as a coprocessor or directly as a processor. Each solution has pros and cons that are discussed in what follows for different cases. Figure 1 shows the different possibilities in a schematic way.

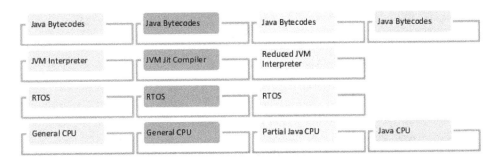

Fig. 1. Java layered implementations

In the domain of small embedded devices, the JVM turns out to be slow and requires an important amount of memory resources and processor capabilities. These are serious drawbacks to the implementation of embedded systems with RTSJ. In order to overcome these problems, advances in JIT compilers promote them as the standard execution mode of the JVM in desktop and server environments. However, this approach introduces uncertainties to the execution time due to runtime compilation. Thus execution times are not predictable and this fact prevents the computation of the WCET forbidding its use in hard real-time applications. Even if the program execution speeds up, it still requires an important amount of memory. The solution is not practical for small embedded systems.

In the embedded domain, where resources are scarce, a Java processors or coprocessors are more promising options. There are two types of hardware JVM implementations:

- A coprocessor works in concert with a general purpose processor translating java byte codes to a sequence of instructions specific to this coupled CPU.

- Java chips entirely replace the general CPU. In the Java Processors the JVM bytecode is the native instruction set, therefore programs are written in Java. This solution can result in quite a small processor with little memory demand.

In the embedded domain, where resources are scarce, a Java processors or coprocessors are more promising options. There are two types of hardware JVM implementations:

- A coprocessor works in concert with a general purpose processor translating java bytecodes to a sequence of instructions specific for this coupled CPU.

- Java chips entirely replace the general CPU. In the Java Processors the JVM bytecode is the native instruction set, therefore programs are written in Java. This solution can result in quite a small processor with little memory demand.

Table 2 shows a short list of Java processors.

Name	Target technology	Size	Speed [MHz]
JOP	Altera, Xilinx FPGA	2050 LCs, 3KB Ram	100
picoJava	No realization	128K gates, 38KB	
picoJava II	Altera Cyclone FPGA	27.5 K LCs; 47.6 KB	
aJile	aJ102 aJ200 ASIC	0.25μ	100
Cjip	ASIC 0.35μ	70K gates, 55MB ROM, RAM	80
Moon	Altera FPGA	3660 LCs, 4KB RAM	
Lightfoot	Xilinx FPGA	3400 LCs	40
LavaCORE	Xilinx FPGA	3800 LCs 30K gates	33
Komodo		2600 LCs	33
FemtoJava	Xilinx FPGA	2710 LCs	56

Table 2. Java Processors List

In 1997 Sun introduced the first version of picoJava and in 1999 it launched the picoJava-II processor. Its core provides an optimized hardware environment for hosting a JVM implementing most of the Java virtual machine instructions directly. Java bytecodes are directly implemented in hardware. The architecture of picoJava is a stack-based CISC processor implementing 341 different instructions (O'Connor & Tremblay, 1997). Simple Java bytecodes are directly implemented in hardware and some performance critical instructions are implemented in microcode. A set of complex instructions are emulated by a sequence of simpler instructions. When the core encounters an instruction that must be emulated, it generates a trap with a trap type corresponding to that instruction and then jumps to an emulation trap handler that emulates the instruction in software. This mechanism has a high variability latency that prevents its use in real-time because of the difficulty to compute the WCET (Borg et al., 2005; Puffitsch & Schoeberl, 2007).

Komodo (Brinkschulte et al., 1999) is a Java microcontroller with an event handling mechanism that allows handling of simultaneous overlapping events with hard real-time requirements. The Komodo microcontroller design adds multithreading to a basic Java design in order to attain predictability of real time threads requirements. The exclusive feature of Komodo is the instruction fetch unit with four independent program counters and status flags for four threads. A priority manager is responsible for hardware real-time scheduling and can select a new thread after each bytecode instruction. The microcontroller holds the contexts of up to four threads. To scale up for larger systems with more than three real-time threads the authors suggest a parallel execution on several microcontrollers connected by a middleware platform.

FemtoJava is a Java microcontroller with a reduced-instruction-set Harvard architecture (Beck & Carro, 2003). It is basically a research project to build an -application specific- Java dedicated microcontroller. Because it is synthesized in an FPGA, the microcontroller can also be adapted to a specific application by adding functions that could includes new Java instructions. The bytecode usage of the embedded application is analyzed and a customized version of FemtoJava is generated (similar to LavaCORE) in order to minimize resource usage: power consumption, small program code size, microarchitecture optimizations (instruction set, data width, register file size) and high integration (memory communications on the same die).

Hardware designs like JOP (Java Optimized Processor) and AONIX PERC processors currently provide a safety certifiable, hard real-time virtual machine that offers throughput comparable to optimized C or C++ solutions (Schoeberl, 2009)

The Java processor JOP (Altera or Xilinx FPGA) is a hardware implementation of the Java virtual machine (JVM). The JVM bytecodes are the native instruction set of JOP. The main advantage of directly executing bytecode instructions is that WCET analysis can be performed at the bytecode level. The WCET tool WCA is part of the JOP distribution. The main characteristics of JOP architecture are presented in (Schoeberl, 2009). They include a dynamic translation of the CISC Java bytecodes to a RISC stack based instruction set that can be executed in a three microcode pipeline stages: microcode fetch, decode and execute. The processor is capable of translating one bytecode per cycle giving a constant execution time for all microcode instructions without any stall in the pipeline. The interrupts are inserted in the translation stage as special bytecodes and are transparent to the microcode pipeline. The four stages pipeline produces short branch delays. There is a simple execution stage with the two top most stack elements (registers A and B). Bytecodes have no time dependencies and the instructions and data caches are time-predictable since ther are no prefetch or store buffers (which could have introduced unbound time dependencies of instructions). There is no direct connection between the core processor and the external world. The memory interface provides a connection between the main memory and the core processor.

JOP is designed to be an easy target for WCET analysis. WCET estimates can be obtained either by measurement or static analysis. (Schoeberl, 2009) presents a number of performance comparisons and finds that JOP has a good average performance relative to other non real-time Java processors, in a small design and preserving the key characteristics that define a RTS platform. A representative ASIC implementation is the aJile aJ102 processor (*Ajile Systems*, 2011). This processor is a low-power SOC that directly executes Java Virtual Machine (JVM) instructions, real-time Java threading primitives, and secured networking. It is designed for a real-time DSP and networking. In addition, the aJ-102 can execute bytecode extensions for custom application accelerations. The core of the aJ102 is the JEMCore-III low-power direct execution Java microprocessor core. The JEMCore-III implements the entire JVM bytecode instructions in silicon.

JOP includes an internal microprogrammed real-time kernel that performs the traditional operating system functions such as scheduling, context switching, interrupt preprocessing, error preprocessing, and object synchronization. As explained above, a low-level analysis of execution times is of primary importance for WCET analysis. Even though the multiprocessors systems are a common solution to general purpose equipments it makes static WCET analysis practically impossible. On the other hand, most real-time systems are multi-threaded applications and performance could be highly improved by using multi core processors on a single chip. (Schoeberl, 2010) presents an approach to a time-predictable chip multiprocessor system that aims to improve system performance while still enabling WCET analysis. The proposed chip uses a shared memory statically scheduled with a time-division multiple access (TDMA) scheme which can be integrated into the WCET analysis. The static schedule guarantees that thread execution times on different cores are independent of each other.

6. Conclusions

In this chapter a critical review of the state of the art in real-time programming languages and real-time operating systems providing support to them has been presented. The programming lan guages are limited mainly to five: C, C++, Ada, RT Java and for very specific applications, Assembler. The world of RTOS is much wider. Virtually every research group has created its own operating system. In the commercial world there is also a range of RTOS. At the top of the preferences appear Vxworks, QNX, Windows CE family, RT Linux, FreeRTOS, eCOS and OSE. However, there are many others providing support in particular areas. In this paper, a short list of the most well known ones has been described.

At this point it is worth asking why while there are so many RTOSs available there are so few programming languages. The answer probably is that while a RTOS is oriented to a particular application area such as communications, low end microprocessors, high end microprocessors, distributed systems, wireless sensors network and communications among others, the requirements are not universal. The programming languages, on the other hand need to be and are indeed universal and useful for every domain.

Although the main programming languages for real-time embedded systems are almost reduced to five the actual trend reduces these to only C/C++ and RT Java. The first option provides the low level access to the processor architecture and provides an object oriented paradigm too. The second option has the great advantage of a WORA language with increasing hardware support to implement the JVM in a more efficient.

In the last few years, there has been an important increase in ad-hoc solutions based on special processors created for specific domains. The introduction of Java processors changes the approach to embedded systems design since the advantages of the WORA programming are added to a simple implementation of the hardware.

The selection of an adequate hardware platform, a RTOS and a programming language will be tightly linked to the kind of embedded system being developed. The designer will choose the combination that best suits the demands of the application but it is really important to select one that has support along the whole design process.

7. References

Ajile Systems (2011). http://www.ajile.com/.

Albert, A. (2004). Comparison of event-triggered and time-triggered concepts with regard to distributed control systems, *Embedded World 2004*, pp. 235–252.

Albert, A. & Gerth, W. (2003). Evaluation and comparison of the real-time performance of can and ttcan, *9th international CAN in Automation Conference*, p. 05/01–05/08.

Baker, T. (1990). A stack-based resource allocation policy for realtime processes, *Real-Time Systems Symposium, 1990. Proceedings., 11th* pp. 191–200.

Beck, A. & Carro, L. (2003). Low power java processor for embedded applications, *12th IFIP International Conference on Very Large Scale Integration*.

Borg, A., Audsley, N. & Wellings, A. (2005). Real-time java for embedded devices: The javamen project, *Perspectives in Pervasive Computing*, pp. 1–10.

Brinkschulte, U., Krakowski, C., Kreuzinger, J. & Ungerer, T. (1999). A multithreaded java microcontroller for thread-oriented real-time event-handling, *Parallel Architectures and Compilation Techniques, 1999. Proceedings. 1999 International Conference on*, pp. 34 –39.

eCosCentric (2011). http://www.ecoscentric.com/index.shtml.

Enea OSE (2011). http://www.enea.com/software/products/rtos/ose/.

Erika Enterprise: Open Source RTOS for single- and multi-core applications (2011). http://www. evidence.eu.com/content/view/27/254/.

Gosling, J. & Bollella, G. (2000). *The Real-Time Specification for Java*, Addison-Wesley Longman Publishing Co., Inc., Boston, MA, USA.

IEEE (2003). *ISO/IEC 9945:2003, Information Technology–Portable Operating System Interface (POSIX)*, IEEE.

ISO/IEC 14882:2003 - Programming languages C++ (2003).

ISO/IEC 8526:AMD1:2007. Ada 2005 Language Reference Manual (LRM) (2005). http://www.adaic.org/standards/05rm/html/RM-TTL.html.

ISO/IEC 9899:1999 - Programming languages - C (1999). http://www.open-std.org/ JTC1/SC22/WG14/ www/docs/n1256.pdf.

Lutz, M. & Laplante, P. (2003). C# and the .net framework: ready for real time?, *Software, IEEE* 20(1): 74–80.

LynxOS RTOS, The real-time operating system for complex embedded systems (2011). http://www.lynuxworks.com/rtos/rtos.php.

Maglyas, A., Nikula, U. & Smolander, K. (2010). Comparison of two models of success prediction in software development projects, *6th Central and Eastern European Software Engineering Conference (CEE-SECR), 2010*, pp. 43–49.

Microsystems, S. (2011). Real-time specification for java documentation, http://www. rtsj.org/.

Minimal Real-Time Operating System (2011). http://marte.unican.es/.

Obenza, R. (1993). Rate monotonic analysis for real-time systems, *Computer* 26: 73–74. URL: *http://portal.acm.org/citation.cfm?id=618978.619872*

O'Connor, J. & Tremblay, M. (1997). picojava-i: the java virtual machine in hardware, *Micro, IEEE* 17(2): 45 –53.

Pleunis, J. (2009). Extending the lifetime of software-intensive systems, *Technical report*, Information Technology for European Advancement, http://www.itea2.org/ innovation_reports.

Puffitsch, W. & Schoeberl, M. (2007). picojava-ii in an fpga, *Proceedings of the 5th international workshop on Java technologies for real-time and embedded systems*, JTRES '07, ACM, New York, NY, USA, pp. 213–221.
URL: *http://doi.acm.org/10.1145/1288940.1288972*

QNX RTOS v4 System Documentation (2011). http://www.qnx.com/developers/qnx4/documentation.html.

Robertz, S. G., Henriksson, R., Nilsson, K., Blomdell, A. & Tarasov, I. (2007). Using real-time java for industrial robot control, *Proceedings of the 5th international workshop on Java technologies for real-time and embedded systems*, JTRES '07, ACM, New York, NY, USA, pp. 104–110.
URL: *http://doi.acm.org/10.1145/1288940.1288955*

RTAI - the RealTime Application Interface for Linux (2010). https://www.rtai.org/.

RTLinuxFree (2011). http://www.rtlinuxfree.com/.

Schoeberl, M. (2009). *JOP Reference Handbook: Building Embedded Systems with a Java Processor*, number ISBN 978-1438239699, CreateSpace. Available at http://www.jopdesign.com/doc/handbook.pdf.
URL: *http://www. jopdesign.com/ doc/handbook.pdf*

Schoeberl, M. (2010). Time-predictable chip-multiprocessor design, *Signals, Systems and Computers (ASILOMAR), 2010 Conference Record of the Forty Fourth Asilomar Conference on*, pp. 2116 –2120.

Service Oriented Operating System (2011). http://www.ingelec.uns.edu.ar/rts/soos.

Sha, L., Rajkumar, R. & Lehoczky, J. P. (1990). Priority inheritance protocols: An approach to real-time synchronization, *IEEE Trans. Comput.* 39(9): 1175–1185.

S.Ha.R.K.: Soft Hard Real-Time Kernel (2007). http://shark.sssup.it/.

Stankovic, J. A. (1988). Misconceptions about real-time computing, *IEEE Computer* 21(17): 10–19.

The Chaos Report (1994). www.standishgroup.com/sample_ research/PDFpages/Chaos1994.pdf.

The free RTOS Project (2011). http://www.freertos.org/.

VxWorks RTOS (2011). http://www.windriver.com/products/vxworks/.

Windows Embedded (2011). http://www.microsoft.com/windowsembedded/en-us/develop/windows-embedded-products-for-developers.aspx.

Wolf, W. (2002). What is Embedded Computing?, *IEEE Computer* 35(1): 136–137.

Zerzelidis, A. & Wellings, A. (2004). Requirements for a real-time .net framework, *Technical Report YCS-2004-377*, Dep. of Computer Science, University of York.

Safely Embedded Software for State Machines in Automotive Applications

Juergen Mottok[1], Frank Schiller[2] and Thomas Zeitler[3]

[1]*Regensburg University of Applied Sciences*
[2]*Beckhoff Automation GmbH*
[3]*Continental Automotive GmbH*
Germany

1. Introduction

Currently, both fail safe and fail operational architectures are based on hardware redundancy in automotive embedded systems. In contrast to this approach, safety is either a result of diverse software channels or of one channel of specifically coded software within the framework of Safely Embedded Software. Product costs are reduced and flexibility is increased. The overall concept is inspired by the well-known Vital Coded Processor approach. There the transformation of variables constitutes an $(AN+B)$-code with prime factor A and offset B, where B contains a static signature for each variable and a dynamic signature for each program cycle. Operations are transformed accordingly.

Mealy state machines are frequently used in embedded automotive systems. The given Safely Embedded Software approach generates the safety of the overall system in the level of the application software, is realized in the high level programming language C, and is evaluated for Mealy state machines with acceptable overhead. An outline of the comprehensive safety architecture is given.

The importance of the non-functional requirement safety is more and more recognized in the automotive industry and therewith in the automotive embedded systems area. There are two safety categories to be distinguished in automotive systems:

- The goal of *active safety* is to prevent accidents. Typical examples are Electronic Stability Control (ESC), Lane Departure Warning System (LDWS), Adaptive Cruise Control (ACC), and Anti-lock Braking System (ABS).

- If an accident cannot be prevented, measures of *passive safety* will react. They act jointly in order to minimize human damage. For instance, the collaboration of safety means such as front, side, curtain, and knee airbags reduce the risk tremendously.

Each safety system is usually controlled by the so called Electronic Control Unit (ECU). In contrast to functions without a relation to safety, the execution of safety-related functions on an ECU-like device necessitates additional considerations and efforts.

The normative regulations of the generic industrial safety standard IEC 61508 (IEC61508, 1998) can be applied to automotive safety functions as well. Independently of its official present and future status in automotive industry, it provides helpful advice for design and development.

In the future, the automotive safety standard ISO/WD 26262 will be available. In general, based on the safety standards, a hazard and risk graph analysis (cf. e. g. (Braband, 2005)) of a given system determines the safety integrity level of the considered system functions. The detailed safety analysis is supported by tools and graphical representations as in the domain of Fault Tree Analysis (FTA) (Meyna, 2003) and Failure Modes, Effects, and Diagnosis Analysis (FMEDA) (Boersoek, 2007; Meyna, 2003).

The required hardware and software architectures depend on the required safety integrity level. At present, safety systems are mainly realized by means of hardware redundant elements in automotive embedded systems (Schaueffele, 2004).

In this chapter, the concept of Safely Embedded Software (SES) is proposed. This concept is capable to reduce redundancy in hardware by adding diverse redundancy in software, i.e. by specific coding of data and instructions. Safely Embedded Software enables the proof of safety properties and fulfills the condition of single fault detection (Douglass, 2011; Ehrenberger, 2002). The specific coding avoids non-detectable common-cause failures in the software components. Safely Embedded Software does not restrict capabilities but can supplement multi-version software fault tolerance techniques (Torres-Pomales, 2000) like N version programming, consensus recovery block techniques, or N self-checking programming. The new contribution of the Safely Embedded Software approaches the constitution of safety in the layer of application software, that it is realized in the high level programming language C and that it is evaluated for Mealy state machines with acceptable overhead.

In a recently published generic safety architecture approach for automotive embedded systems (Mottok, 2006), safety-critical and safety-related software components are encapsulated in the application software layer. There the overall open system architecture consists of an application software, a middleware referred to as Runtime-Environment, a basic software, and an operating system according to e. g. AUTOSAR (AUTOSAR, 2011; Tarabbia, 2005). A safety certification of the safety-critical and the safety-related components based on the Safely Embedded Software approach is possible independently of the type of underlying layers. Therefore, a sufficiently safe fault detection for data and operations is necessary in this layer. It is efficiently realized by means of Safely Embedded Software, developed by the authors.

The chapter is organized as follows: An overview of related work is described in Section 2. In Section 3, the Safely Embedded Software Approach is explained. Coding of data, arithmetic operations and logical operations is derived and presented. Safety code weaving applies these coding techniques in the high level programming language C as described in Section 4. A case study with a *Simplified Sensor Actuator State Machine* is discussed in Section 5. Conclusions and statements about necessary future work are given in Section 6.

2. Related work

In 1989, the Vital Coded Processor (Forin, 1989) was published as an approach to design typically used operators and to process and compute vital data with non-redundant hardware and software. One of the first realizations of this technique has been applied to trains for the metro A line in Paris. The Vital technique proposes a data mapping transformation also referred to in this chapter. The Vital transformation for generating diverse coded data x_c can be roughly described by multiplication of a date x_f with a prime factor A such that $x_c = A * x_f$ holds. The prime A determines the error detection probability, or residual error probability, respectively, of the system. Furthermore, an additive modification by a static signature for each variable B_x and a dynamic signature for each program cycle D lead finally to the code of the type $x_c = A * x_f + B_x + D$. The hardware consists of a single microprocessor, the so called Coded Monoprocessor, an additional dynamic controller, and a logical input/output interface. The dynamic controller includes a clock generator and a comparator function. Further on, a logical output interface is connected to the microprocessor and the dynamic controller. In particular, the Vital Coded Processor approach cannot be handled as standard embedded hardware and the comparator function is separated from the microprocessor in the dynamic controller.

The ED^{4}I approach (Oh, 2002) applies a commercial off-the-shelf processor. Error detection by means of diverse data and duplicated instructions is based on the SIHFT technique that detects both temporary and permanent faults by executing two programs with the same functionality but different data sets and comparing their outputs. An original program is transformed into a new program. The transformation consists of a multiplication of all variables and constants by a diversity factor k. The two programs use different parts of the underlying hardware and propagate faults in different ways. The fault detection probability was examined to determine an adequate multiplier value k. A technique for adding commands to check the correct execution of the logical program flow has been published in (Rebaudengo, 2003). These treated program flow faults occur when a processor fetches and executes an incorrect instruction during the program execution. The effectiveness of the proposed approach is assessed by several fault injection sessions for different example algorithms.

Different classical software fail safe techniques in automotive applications are, amongst others, program flow monitoring methods that are discussed in a survey paper (Leaphart, 2005).

A demonstration of a fail safe electronic accelerator safety concept of electronic control units for automotive engine control can be found in (Schaueffele, 2004). The electronic accelerator concept is a three-level safety architecture with classical fail safe techniques and asymmetric hardware redundancy.

Currently, research is done on the Safely Embedded Software approach. Further results were published in (Mottok, 2007; Steindl, 2009;?; Mottok, 2009; Steindl, 2010; Raab, 2011; Laumer, 2011). Contemporaneous Software Encoded Processing was published (Wappler, 2007). This approach is based on the Vital transformation. In contrast to the Safely Embedded Software approach it provides the execution of arbitrary programs given as binaries on commodity hardware.

3. The safely embedded software approach

3.1 Overview

Safely Embedded Software (SES) can establish safety independently of a specific processing unit or memory. It is possible to detect permanent errors, e. g. errors in the Arithmetic Logical Unit (ALU) as well as temporary errors, e. g. bit-flips and their impact on data and control flow. SES runs on the application software layer as depicted in Fig. 1. Several application tasks have to be safeguarded like e. g. the evaluation of diagnosis data and the check of the data from the sensors. Because of the underlying principles, SES is independent not only of the hardware but also of the operating system.

Fig. 2 shows the method of Safety Code Weaving as a basic principle of SES. Safety Code Weaving is the procedure of adding a second software channel to an existing software channel.

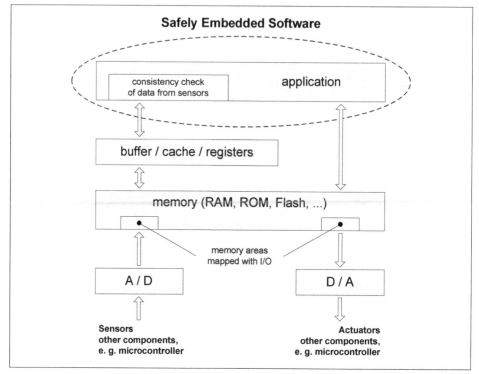

Fig. 1. The Safely Embedded Software approach.

In this way, SES adds a second channel of the transformed domain to the software channel of the original domain. In dedicated nodes of the control flow graph, comparator functionality is added. Though, the second channel comprises diverse data, diverse instructions, comparator and monitoring functionality. The comparator or voter, respectively, on the same ECU has to be safeguarded with voter diversity (Ehrenberger, 2002) or other additional diverse checks.

It is not possible to detect errors of software specification, software design, and software implementation by SES. Normally, this kind of errors has to be detected with software quality assurance methods in the software development process. Alternatively, software fault tolerance techniques (Torres-Pomales, 2000) like N version programming can be used with SES to detect software design errors during system runtime.

As mentioned above, SES is also a programming language independent approach. Its implementation is possible in assembler language as well as in an intermediate or a high programming language like C. When using an intermediate or higher implementation language, the compiler has to be used without code optimization. A code review has to assure, that neither a compiler code optimization nor removal of diverse instructions happened. Basically, the certification process is based on the assembler program or a similar machine language.

Since programming language C is the de facto implementation language in automotive industry, the C programming language is used in this study exclusively. C code quality can be

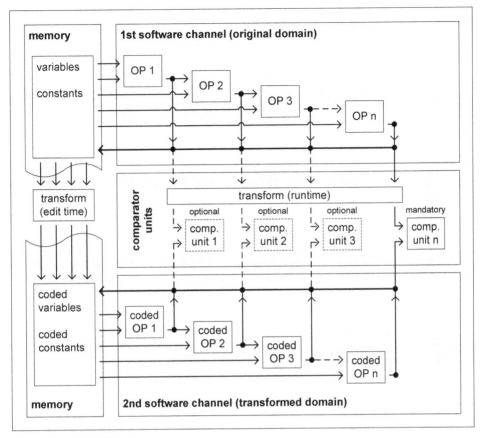

Fig. 2. Safety Code Weaving.

assured by application of e. g. the MISRA-2 (MISRA, 2004). A safety argument for dedicated deviation from MISRA-2 rules can be justified.

3.2 Detectable faults by means of safely embedded software

In this section, the kind of faults detectable by means of Safely Embedded Software is discussed. For this reason, the instruction layer model of a generalized computer architecture is presented in Fig. 3. Bit flips in different memory areas and in the central processing unit can be identified.

Table 1 illustrates the Failure Modes, Effects, and Diagnosis Analysis (FMEDA). Different faults are enumerated and the SES strategy for fault detection is related.

In Fig. 2 and in Table 1, the SES comparator function is introduced. There are two alternatives for the location of the SES comparator. If a local comparator is used on the same ECU, the comparator itself has also to be safeguarded. If an additional comparator on a remote receiving ECU is applied, hardware redundancy is used implicitly, but the inter-ECU communication has to be safeguarded by a safety protocol (Mottok, 2006). In a later system

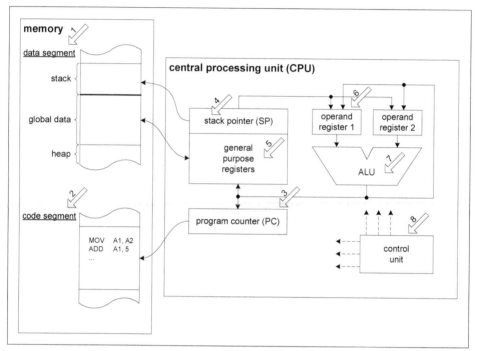

Fig. 3. Model of a generalized computer architecture (instruction layer). The potential occurrence of faults are marked with a label.

FMEDA, the appropriate fault reaction has to be added, regarding that SES is working on the application software layer.

The fault reaction on the application software layer depends on the functional and physical constraints of the considered automotive system. There are various options to select a fault reaction. For instance, fault recovery strategies, achieving degraded modes, shut off paths in the case of fail-safe systems, or the activation of cold redundancy in the case of fail-operational architectures are possible.

3.3 Coding of data

Safely Embedded Software is based on the (AN+B)-code of the Coded Monoprocessor (Forin, 1989) transformation of original integer data x_f into diverse coded data x_c. Coded data are data fulfilling the following relation:

$$x_c = A * x_f + B_x + D \quad where \qquad x_c, x_f \in \mathbb{Z}, \; A \in \mathbb{N}^+, \; B_x, D \in \mathbb{N}_0,$$
$$and \quad B_x + D < A. \tag{1}$$

The duplication of original instructions and data is the simplest approach to achieve a redundant channel. Obviously, common cause failures cannot be detected as they appear in both channels. Data are used in the same way and identical erroneous results could be produced. In this case, fault detection with a comparator is not sufficient.

label	area of action	fault	error	detection
1	stack, global data and heap	bitflip	incorrect data incorrect address	SES comparator SES logical program flow monitoring
2	code segment	bitflip	incorrect operator (but right PC)	SES comparator SES logical program flow monitoring
3	program counter	bitflip	jump to incorrect instruction in the code	SES logical program flow monitoring
4	stack pointer	bitflip	incorrect data incorrect address	SES comparator SES logical program flow monitoring
5	general purpose registers	bitflip	incorrect data incorrect address	SES comparator SES logical program flow monitoring
6	operand register	bitflip	incorrect data	SES comparator
7	ALU	bitflip	incorrect operator	SES comparator
8	control unit		incorrect data incorrect operator	SES comparator SES logical program flow monitoring

Table 1. Faults, errors, and their detection ordered by their area of action. (The labels correspond with the numbers presented in Fig. 3.)

The prime number A (Forin, 1989; Ozello, 1992) determines important safety characteristics like Hamming Distance and residual error probability $P = 1/A$ of the code. Number A has to be prime because in case of a sequence of i faulty operations with constant offset f, the final offset will be $i * f$. This offset is a multiple of a prime number A if and only if i or f is divisible by A. If A is not a prime number then several factors of i and f may cause multiples of A. The same holds for the multiplication of two faulty operands. Additionally, so called deterministic criteria like the above mentioned Hamming distance and the arithmetic distance verify the choice of a prime number.

Other functional characteristics like necessary bit field size etc. and the handling of overflow are also caused by the value of A. The simple transformation $x_c = A * x_f$ is illustrated in Fig. 4.

The static signature B_x ensures the correct memory addresses of variables by using the memory address of the variable or any other variable specific number. The dynamic signature D ensures that the variable is used in the correct task cycle. The determination of the dynamic signature depends on the used scheduling scheme (see Fig. 6). It can be calculated by a clocked counter or it is offered directly by the task scheduler.

The instructions are coded in that way that at the end of each cycle, i. e. before the output starts, either a comparator verifies the diverse channel results $z_c = A * z_f + B_z + D$?, or the coded channel is checked directly by the verification condition $(z_c - B_z - D) \bmod A = 0$? (cf. Equation 1).

In general, there are two alternatives for the representation of original and coded data. The first alternative is to use completely unconnected variables for original data and the coded ones. The second alternative uses a connected but separable code as shown in Fig. 5. In the

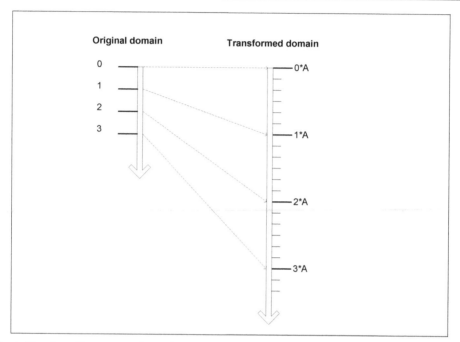

Fig. 4. Simple coding $x_c = A * x_f$ from the original into the transformation domain.

separable code, the transformed value x_c contains the original value x_f. Obviously, x_f can be read out easily from x_c.

The coding operation for separable code is introduced in (Forin, 1989):

Separable coded data are data fulfilling the following relation:

$$x_c = 2^k * x_f + (-2^k * x_f) \text{ modulo } A + B_x + D \qquad (2)$$

The factor 2^k causes a dedicated k-times right shift in the n-bit field. Therefore, one variable can be used for representing original data x_f and coded data x_c.

Without loss of generality, independent variables for original data x_f and coded data x_c are used in this study.

In automotive embedded systems, a hybrid scheduling architecture is commonly used, where interrupts, preemptive tasks, and cooperative tasks coexist, e. g. in engine control units on base of the OSEK operating system. Jitters in the task cycle have to be expected. An inclusion of the dynamic signature into the check will ensure that used data values are those of the current task cycle.

Measures for logical program flow and temporal control flow are added into the SES approach.

One goal is to avoid the relatively high probability that two instruction channels using the original data x_f and produce same output for the same hardware fault. When using the transformation, the corresponding residual error probability is basically given by the

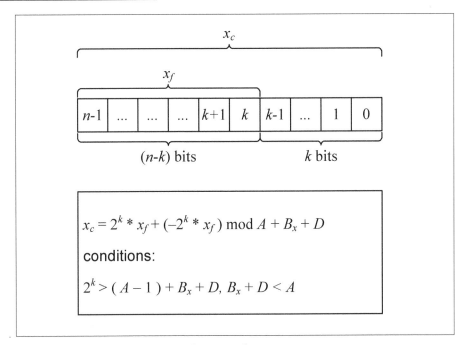

Fig. 5. Separable code and conditions for its application.

reciprocal of the prime multiplier, A^{-1}. The value of A determines the safe failure fraction (SFF) in this way and finally the safety integrity level of the overall safety-related system (IEC61508, 1998).

3.4 Coding of operations

A complete set of arithmetic and logical operators in the transformed domain can be derived. The transformation in Equation (1) is used. The coding of addition follows (Forin, 1989) whereas the coding of the Greater or Equal Zero operator has been developed within the Safely Embedded Software approach.

A coded operator OP_c is an operator in the transformed domain that corresponds to an operator OP in the original domain. Its application to uncoded values provides coded values as results that are equal to those received by transforming the result from the original domain after the application OP for the original values. The formalism is defined, such that the following statement is correct for all x_f, y_f from the original domain and all x_c, y_c from the transformed domain, where $x_c = \sigma(x_f)$ and $y_c = \sigma(y_f)$ is valid:

$$x_f \; \circ\!\!-\!\!\bullet \; x_c$$
$$y_f \; \circ\!\!-\!\!\bullet \; y_c$$
$$z_f \; \circ\!\!-\!\!\bullet \; z_c$$
$$z_f = x_f \, OP \, y_f \quad \circ\!\!-\!\!\bullet \quad x_c \, OP_c \, y_c = z_c \tag{3}$$

Accordingly, the unary operators are noted as:

$$z_f = OP\, y_f \quad \circ\!\!\!-\!\!\!\bullet \quad OP_c\, y_c = z_c \tag{4}$$

In the following, the derivation steps for the addition operation and some logical operations in the transformed domain are explained.

3.4.1 Coding of addition

The addition is the simplest operation of the four basic arithmetic operations. Defining a coded operator (see Equation (3)), the coded operation $\oplus$ is formalized as follows:

$$z_f = x_f + y_f \quad \Rightarrow \quad z_c = x_c \oplus y_c \tag{5}$$

Starting with the addition in the original domain and applying the formula for the inverse transformation, the following equation can be obtained for z_c:

$$z_f = x_f + y_f$$
$$\frac{z_c - B_z - D}{A} = \frac{x_c - B_x - D}{A} + \frac{y_c - B_y - D}{A}$$
$$z_c - B_z - D = x_c - B_x - D + y_c - B_y - D$$
$$z_c = x_c - B_x - D + y_c - B_y + B_z$$
$$z_c = x_c + y_c + \underbrace{(B_z - B_x - B_y)}_{const.} - D \tag{6}$$

The Equations (5) and (6) state two different representations of z_c. A comparison leads immediately to the definition of the coded addition $\oplus$:

$$z_c = x_c \oplus y_c = x_c + y_c + (B_z - B_x - B_y) - D \tag{7}$$

3.4.2 Coding of comparison: Greater or equal zero

The coded (unary) operator geqz$_c$ (greater or equal zero) is applied to a coded value x_c. geqz$_c$ returns TRUE$_c$, if the corresponding original value x_f is greater than or equal to zero. It returns FALSE$_c$, if the corresponding original value x_f is less than zero. (This corresponds to the definition of a coded operator (see Definition 3) and the definition of the ≥ 0 operator of the original domain.)

$$geqz_c(x_c) = \begin{cases} TRUE_c, & if\ x_f \geq 0, \\ FALSE_c, & if\ x_f < 0. \end{cases} \tag{8}$$

Before deriving the transformation steps of the coded operator geqz$_c$, the following theorem has to be introduced and proved.

The original value x_f is greater than or equal to zero, if and only if the coded value x_c is greater than or equal to zero.

$$x_f \geq 0 \Leftrightarrow x_c \geq 0 \text{ with } x_f \in \mathbb{Z} \text{ and } x_c = \sigma(x_f) = A * x_f + B_x + D$$
$$\text{where } A \in \mathbb{N}^+,\ B_x, D \in \mathbb{N}_0,\ B_x + D < A \tag{9}$$

Proof.

$$
\begin{array}{rll}
& x_c & \geq 0 \\
\Leftrightarrow & A * x_f + B_x + D & \geq 0 \\
\Leftrightarrow & A * x_f & \geq -(B_x + D) \\
\Leftrightarrow & x_f & \geq -\underbrace{\dfrac{\overbrace{B_x + D}^{<A}}{A}}_{\in\,]\text{-}1,\,0]} \\
\Leftrightarrow & x_f & \geq 0,\ \text{since } x_f \in \mathbb{Z}
\end{array}
$$

$\square$

The goal is to implement a function returning $TRUE_c$, if and only if the coded value x_c (and thus x_f) is greater or equal to zero. Correspondingly, the function has to return $FALSE_c$, if and only if x_c is less than zero. As an extension to Definition 8, $ERROR_c$ should be returned in case of a fault, e. g. if x_c is not a valid code word.

By applying the $\geq$ operator according to Equation (9), it can be checked whether x_c is negative or non-negative, but it cannot be checked whether x_c is a valid code word. Additionally, this procedure is very similar to the procedure in the original domain. The use of the unsigned modulo function umod is a possible solution to that problem. This function is applied to the coded value x_c. The idea of this approach is based on (Forin, 1989):

$$x_c \text{ umod } A = \text{unsigned}(x_c) \bmod A = \text{unsigned}(A * x_f + B_x + D) \bmod A$$

In order to resolve the unsigned function, two different cases have to be distinguished:

case 1: $x_f \geq 0$

$$x_c \text{ umod } A = \text{unsigned}(\underbrace{A * x_f + B_x + D}_{x_f \geq 0 \;\Rightarrow\; x_c \geq 0 \text{ (cf. Eqn. (9))}}) \bmod A$$

$$= (\underbrace{(A * x_f) \bmod A}_{=0} + \underbrace{B_x + D}_{<A}) \bmod A$$

$$= B_x + D$$

case 2: $x_f < 0$

$$x_c \text{ umod } A = \text{unsigned}(\underbrace{A * x_f + B_x + D}_{x_f < 0 \;\Rightarrow\; x_c < 0 \text{ (cf. Eqn. (9))}}) \bmod A$$

$$= \underbrace{(A * x_f + B_x + D + 2^n)}_{\text{resolved unsigned function}} \bmod A$$

$$= (\underbrace{(A * x_f) \bmod A}_{=0} + B_x + D + 2^n) \bmod A$$

$$= (B_x + D + 2^n) \bmod A$$

$$= (B_x + D + \underbrace{(2^n \bmod A)}_{\text{known constant}}) \bmod A$$

Conclusion of these two cases:

Result of case 1:
$$x_f \geq 0 \quad \Rightarrow \quad x_c \text{ umod } A = B_x + D \tag{10}$$

Result of case 2:
$$x_f < 0 \quad \Rightarrow \quad x_c \text{ umod } A = (B_x + D + (2^n \text{ mod } A)) \text{ mod } A \tag{11}$$

Remark: The index n represents the minimum number of bits necessary for storing x_c. If x_c is stored in an int32 variable, n is equal to 32.

It has to be checked, if in addition to the two implications (10) and (11) the following implications

$$x_c \text{ umod } A = B_x + D \qquad\qquad\qquad \Rightarrow \quad x_f \geq 0$$
$$x_c \text{ umod } A = (B_x + D + (2^n \text{ mod } A)) \text{ mod } A \qquad \Rightarrow \quad x_f < 0$$

hold. These implications are only valid and applicable, if the two terms $B_x + D$ and $(B_x + D + (2^n \text{ mod } A)) \text{ mod } A$ are never equal. In the following, equality is assumed and conditions on A are identified that have to hold for a disproof:

$$B_x + D \;=\; (\underbrace{B_x + D}_{\in\,[0,\,A\text{-}1]} + \underbrace{(2^n \text{ mod } A)}_{\in\,[0,\,A\text{-}1]}) \text{ mod } A$$
$$\underbrace{\qquad\qquad\qquad\qquad\qquad}_{\in\,[0,\,2A\text{-}2]}$$

case 1: $\quad 0 \;\leq\; (B_x + D + (2^n \text{ mod } A)) \;<\; A$

$$B_x + D \;=\; \underbrace{(B_x + D + (2^n \text{ mod } A)) \text{ mod } A}_{\in\,[0,\,A\text{-}1]}$$

$\Leftrightarrow \quad B_x + D \;=\; B_x + D + (2^n \text{ mod } A)$

$\Leftrightarrow \quad 2^n \text{ mod } A \;=\; 0$

$\Leftrightarrow \quad 2^n \;=\; k * A \qquad \forall\, k \in \mathbb{N}^+$

$\Leftrightarrow \quad A \;=\; \dfrac{2^n}{k}$

Since $A \in \mathbb{N}^+$ and 2^n is only divisible by powers of 2, k has to be a power of 2, and, therefore, the same holds for A. That means, if A is not a number to the power of 2, inequality holds in case 1.

case 2: $\quad A \;\leq\; (B_x + D + (2^n \text{ mod } A)) \;\leq\; 2A - 2$

$$B_x + D \;=\; \underbrace{(B_x + D + (2^n \text{ mod } A)) \text{ mod } A}_{\in\,[A,\,2A\text{-}2]}$$

$\Leftrightarrow \quad B_x + D \;=\; B_x + D + (2^n \text{ mod } A) - A$

$\Leftrightarrow \quad A \;=\; \underbrace{2^n \text{ mod } A}_{\in\,[0,\,A\text{-}1]}$

This cannot hold since the result of the modulo-operation is always smaller than A.

The two implications (10) and (11) can be extended to equivalences, if A is chosen not as a number to the power of 2. Thus for implementing the $geqz_c$ operator, the following conclusions can be used:

1. IF x_c umod $A = B_x + D$ THEN $x_f \geq 0$.

2. ELSE IF x_c umod $A = (B_x + D + (2^n \bmod A)) \bmod A$ THEN $x_f < 0$.

3.   ELSE x_c is not a valid code word.

The $geqz_c$ operator is implemented based on this argumentation. Its application is presented in Listing 2, whereas its uncoded form is presented in Listing 1.

4. Safety code weaving for C control structures

In the former sections, a subset of SES transformation was discussed. The complete set of transformations for data, arithmetic operators, and Boolean operators are collected in a C library. In the following, the principle procedure of safety code weaving is motivated for C control structures. An example code is given in Listing 1 that will be safeguarded in a further step.

Listing 1. Original version of the code. It will be safeguarded in further steps.

```
int  af  =  1;
int  xf  =  5;

if  (  xf  >=  0  )
{
    af  =  4;
}
else
{
    af  =  9;
}
```

In general, there are a few preconditions for the original, non-coded, single channel C source code: e.g. operations should be transformable and instructions with short expressions are preferred in order to simplify the coding of operations.

Safety code weaving is realized in compliance with nine rules:

1. *Diverse data.* The declaration of coded variables and coded constants have to follow the underlying code definition.

2. *Diverse operations.* Each original operation follows directly the transformed operation.

3. *Update of dynamic signature.* In each task cycle, the dynamic signature of each variable has to be incremented.

4. *Local (logical) program flow monitoring.* The C control structures are safeguarded against local program flow errors. The branch condition of the control structure is transformed and checked inside the branch.

5. *Global (logical) program flow monitoring.* This technique includes a specific initial key value and a key process within the program function to assure that the program function has completed in the given parts and in the correct order (Leaphart, 2005). An alternative operating system based approach is given in Raab (2011).

6. *Temporal program flow monitoring.* Dedicated checkpoints have to be added for monitoring periodicity and deadlines. The specified execution time is safeguarded.

7. *Comparator function.* Comparator functions have to be added in the specified granularity in the program flow for each task cycle. Either a comparator verifies the diverse channel results $z_c = A * z_f + B_z + D$?, or the coded channel is checked directly by checking the condition $(z_c - B_z - D) \bmod A = 0$?.

8. *Safety protocol.* Safety critical and safety related software modules (in the application software layer) communicate intra or inter ECU via a safety protocol (Mottok, 2006). Therefore a safety interface is added to the functional interface.

9. *Safe communication with a safety supervisor.* Fault status information is communicated to a global safety supervisor. The safety supervisor can initiate the appropriate (global) fault reaction (Mottok, 2006).

The example code of Listing 1 is transformed according to the rules 1, 2, 4, and 5 in Listing 2. The C control structures while-Loop, do-while-Loop, for-Loop, if-statement, and switch-statement are transformed in accordance with the complete set of rules. It can be realized that the geqz$_c$ operator is frequently applied for safeguarding C control structures.

5. The case study: Simplified sensor actuator state machine

In the case study, a simplified sensor actuator state machine is used. The behavior of a sensor actuator chain is managed by control techniques and Mealy state machines.

Acquisition and diagnosis of sensor signals are managed outside of the state machine in the input management whereas the output management is responsible for control techniques and for distributing the actuator signals. For both tasks, a specific basic software above the application software is necessary for communication with D/A- or A/D-converters. As discussed in Fig. 1, a diagnosis of D/A-converter is established, too.

The electronic accelerator concept (Schaueffele, 2004) is used as an example. Here diverse sensor signals of the pedal are compared in the input management. The output management provides diverse shut-off paths, e. g. power stages in the electronic subsystem.

Listing 2. Example code after applying the rule 1, 2, 4 and 5.

```
int  af;                    int  ac;
int  xf;                    int  xc;
int  tmpf;                  int  tmpc;

cf = 152; /* begin basic block 152 */
af   = 1;                   ac   = 1*A + Ba + D;  // coded 1
xf   = 5;                   xc   = 5*A + Bx + D;  // coded 5
tmpf = ( xf >= 0 );         tmpc = geqz_c( xc );
                            // greater/equal zero operator

if ( cf != 152 ) { ERROR } /* end basic block 152 */
```

```
if ( tmpf )
{
    cf = 153; /* begin basic block 153 */
    if ( tmpc - TRUE_C ) { ERROR }
    af = 4;                   ac   = 4*A + Ba + D; //coded 4
    if ( cf != 153 ) { ERROR } /* end basic block 153 */
}
else
{
    cf = 154; /* begin basic block 154 */
    if ( tmpc - FALSE_C ) { ERROR }
    af = 9;                   ac   = 9*A + Ba + D; //coded 9
    if ( cf != 154 ) { ERROR } /* end basic block 154 */
}
```

The input management processes the sensor values (s1 and s2 in Fig. 6), generates an event, and saves them on a blackboard as a managed global variable. This is a widely used implementation architecture for software in embedded systems for optimization performance, memory consumption, and stack usage. A blackboard (Noble, 2001) is realized as a kind of data pool. The state machine reads the current state and the event from the blackboard, if necessary executes a transition and saves the next state and the action on the blackboard. If a fault is detected, the blackboard is saved in a fault storage for diagnosis purposes.

Finally, the output management executes the action (actuator values a1, a2, a3, and a4 in Fig. 6). This is repeated in each cycle of the task.

The Safety Supervisor supervises the correct work of the state machine in the application software. Incorrect data or instruction faults are locally detected by the comparator function inside the state machine implementation whereas the analysis of the fault pattern and the initiation of a dedicated fault reaction are managed globally by a safety supervisor (Mottok, 2006). A similar approach with a software watchdog can be found in (Lauer, 2007).

The simplified state machine was implemented in the Safely Embedded Software approach. The two classical implementation variants given by nested switch statement and table driven design are implemented. The runtime and the file size of the state machine are measured and compared with the non-coded original one for the nested switch statement design.

The measurements of runtime and file size for the original single channel implementation and the transformed one contain a ground load corresponding to a simple task cycle infrastructure of 10,000,000 cycles. Both the NEC Fx3 V850ES 32 bit microcontroller, and the Freescale S12X 16 bit microcontroller were used as references for the Safely Embedded Software approach.

5.1 NEC Fx3 V850ES microcontroller

The NEC Fx3 V850ES is a 32 bit microcontroller, being compared with the Freescale S12X more powerful with respect to calculations. It runs with an 8 MHz quartz and internally with 32 MHz per PLL. The metrics of the Simplified Sensor Actuator State Machine (nested switch implemented) by using the embedded compiler for the NEC are shown in Table 2. The compiler "Green Hills Software, MULTI v4.2.3C v800" and the linker "Green Hills Software, MULTI v4.2.3A V800 SPR5843" were used.

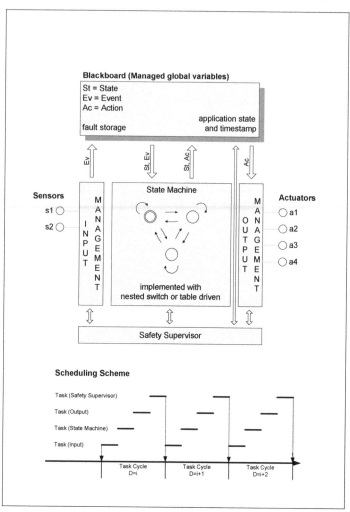

Fig. 6. Simplified sensor actuator state machine and a scheduling schema covering tasks for the input management, the state machine, the output management and the safety supervisor. The task cycle is given by dynamic signature D, which can be realized by a clocked counter.

5.2 Freescale S12X microcontroller

The Freescale S12X is a 16 bit microcontroller and obviously a more efficient control unit compared to the NEC Fx3 V850ES. It runs with an 8 MHz quartz and internally with 32 MHz per PLL. The processor is exactly denominated as "PC9S12X DP512MFV". The metrics of the Simplified Sensor Actuator State Machine (nested switch implemented) by using the compiler for the Freescale S12X are shown in Table 3. The compiler "Metrowerks 5.0.28.5073" and the linker "Metrowerks SmartLinker 5.0.26.5051" were used.

	minimal code	original code	trans-formed code	factor	annotation
CS (init)	2	48	184	3.96	init code, run once
CS (cycle)	2	256	2,402	9.45	state machine, run cyclic
CS (lib)	0	0	252	-	8 functions for the transformed domain used: add_c, div_c, geqz_c, lz_c, ov2cv, sub_c, umod, updD
DS	0	40	84	2.10	global variables
SUM (CS, DS)	4	344	2,922	8.58	sum of CS(init), CS(cycle), CS(lib) and DS
RUN-TIME	0.20	4.80	28.80	6.22	average runtime of the cyclic function in μs
FILE-SIZE	4,264,264	4,267,288	4,284,592	6.72	size (in bytes) of the binary, executable file

Table 2. Metrics of the Simplified Sensor Actuator State Machine (nested switch implemented) using the NEC Fx3 V850ES compiler.

	minimal code	original code	trans-formed code	factor	annotation
CS (init)	1	41	203	5.05	init code, run once
CS (cycle)	1	212	1,758	8.33	state machine, run cyclic
CS (lib)	0	0	234	-	8 functions for the transformed domain used: add_c, div_c, geqz_c, lz_c, ov2cv, sub_c, umod, updD
DS	0	20	42	2.10	global variables
SUM (CS, DS)	2	273	2,237	8.25	sum of CS(init), CS(cycle), CS(lib) and DS
RUN-TIME	0.85	6.80	63.30	10.50	average runtime of the cyclic function in μs
FILE-SIZE	2,079,061	2,080,225	2,088,557	8.16	size (in bytes) of the binary, executable file

Table 3. Metrics of the Simplified Sensor Actuator State Machine (nested switch implemented) using the Freescale S12X compiler.

5.3 Results

The results in this section are based on the nested switch implemented variant of the Simplified Sensor Actuator State Machine of Section 5. The two microcontrollers NEC Fx3 V850ES and Freescale S12X need roundabout nine times memory for the transformed code and data as it is necessary for the original code and data. As expected, there is a duplication of data segement size for both investigated controllers because of the coded data.

There is a clear difference with respect to the raise of runtime compared to the need of memory. The results show that the NEC handles the higher computational efforts as a result of additional transformed code much better than the Freescale does. The runtime of the NEC only increases by factor 6 whereas the runtime of the Freescale increases by factor 10.

5.4 Optimization strategies

There is still a potential for optimizing memory consumption and performance in the SES approach:

- Run time reduction can be achieved by using only the transformed channel.
- Reduction of memory consumption is possible by packed bit fields, but more effort with bit shift operations and masking techniques.
- Using of macros like inline functions.
- Using initializations at compile time.
- Caching of frequently used values.
- Using efficient assembler code for the coded operations from the first beginning.
- First ordering frequently used cases in nested switch(Analogously: entries in the state table).
- Coded constants without dynamic signature.

In the future, the table driven implementation variant will be verified for file size and runtime with cross compilers for embedded platforms and performance measurements on embedded systems.

6. Comprehensive safety architecture and outlook

Safely Embedded Software gives a guideline to diversify application software. A significant but acceptable increase in runtime and code size was measured. The fault detection is realized locally by SES, whereas the fault reaction is globally managed by a Safety Supervisor.

An overall safety architecture comprises diversity of application software realized with the nine rules of Safely Embedded Software in addition to hardware diagnosis and hardware redundancy like e. g. a clock time watchdog. Moreover environmental monitoring (supply voltage, temperature) has to be provided by hardware means.

Temporal control flow monitoring needs control hooks maintained by the operation system or by specialized basic software.

State of the art implementation techniques (IEC61508, 1998; ISO26262, 2011) like actuator activation by complex command sequences or distribution of command sequences (instructions) in different memory areas have been applied. Furthermore, it is recommended to allocate original and coded variables in different memory branches.

Classical RAM test techniques can be replaced by SES since fault propagation techniques ensures the propagation of the detectability up to the check just before the output to the plant.

A system partitioning is possible, the comparator function might be located on another ECU. In this case, a safety protocol is necessary for inter ECU communication. Also a partitioning of different SIL functions on the same ECU is proposed by coding the functions with different prime multipliers A_1, A_2 and A_3 depending on the SIL level. The choice of the prime multiplier is determined by maximizing their pairwise lowest common multiple. In this context, a fault tolerant architecture can be realized by a duplex hardware using in each channel the SES approach with different prime multipliers A_i. In contrast to classical faul-tolerant architectures, here a two channel hardware is sufficient since the correctness of data of each channel are checked individually by determination of their divisibility by A_i.

An application of SES can be motivated by the model driven approach in the automotive industry. State machines are modeled with tools like Matlab or Rhapsody. A dedicated safety code weaving compiler for the given tools has been proposed. The intention is to develop a single channel state chart model in the functional design phase. A preprocessor will add the duplex channel and comparator to the model. Afterwards, the tool based code generation can be performed to produce the required C code.

Either a safety certification (IEC61508, 1998; ISO26262, 2011; Bärwald, 2010) of the used tools will be necessary, or the assembler code will be reviewed. The latter is easier to be executed in the example and seems to be easier in general. Further research in theory as well as in practice will be continued.

7. References

AUTOSAR consortium. (2011). *AUTOSAR*, Official AUTOSAR web site:www.AUTOSAR.org.

Braband, J. (2005). *Risikoanalysen in der Eisenbahn-Automatisierung*,Eurailpress, Hamburg.

Douglass, B. P. (2011). *Safety-Critical Systems Design*, i-Logix, Whitepaper.

Ehrenberger W. (2011). *Software-Verifikation*, Hanser, Munich.

Forin, P. (1989). *Vital Coded Microprocessor Principles and Application for Various Transit Systems*, IFAC Control, Computers, Communications, pp. 79-84, Paris.

Hummel, M., Egen R., Mottok, J., Schiller, F., Mattes, T., Blum, M., Duckstein, F. (2006). *Generische Safety-Architektur für KFZ-Software*, Hanser Automotive, 11, pp. 52-54, Munich.

Mottok, J., Schiller, F., Völkl, T., Zeitler, T. (2007). *Concept for a Safe Realization of a State Machine in Embedded Automotive Applications*, International Conference on Computer Safety, Reliability and Security, SAFECOMP 2007, Springer, LNCS 4680, pp.283-288, Munich.

Wappler, U., Fetzer, C. (2007). *Software Encoded Processing: Building Dependable Systems with Commodity Hardware*, International Conference on Computer Safety, Reliability and Security, SAFECOMP 2007, Springer, LNCS 4680, pp. 356-369, Munich.

IEC (1998). *International Electrotechnical Commission (IEC):Functional Safety of Electrical / Electronic / Programmable Electronic Safety-Related Systems.*

ISO (2011). *ISO26262 International Organization for Standardization Road Vehicles Functional Safety, Final Draft International Standard.*

Leaphart, E.G., Czerny, B.J., D'Ambrosio, J.G., Denlinger, C.L., Littlejohn, D. (2005). *Survey of Software Failsafe Techniques for Safety-Critical Automotive Applications*, SAE World Congress, pp. 1-16, Detroit.

Motor Industry Research Association (2004). *MISRA-C: 2004, Guidelines for the use of the C language in critical systems*, MISRA, Nuneaton.

Börcsök, J. (2007). *Functional Safety, Basic Principles of Safety-related Systems*, Hüthig, Heidelberg.

Meyna, A., Pauli, B. (2003). *Taschenbuch der Zuverlässigkeits- und Sicherheitstechnik*, Hanser, Munich.

Noble, J., Weir, C.(2001). *Small Memory Software, Patterns for Systems with Limited Memory,* Addison Wesley, Edinbourgh.

Oh, N., Mitra, S., McCluskey, E.J. (2002). *4I:Error Detection by Diverse Data and Duplicated Instructions,* IEEE Transactions on Computers, 51, pp. 180-199.

Rebaudengo, M., Reorda, M.S., Torchiano, M., Violante, M. (2003). *Soft-error Detection Using Control Flow Assertions,* 18th IEEE International Symposium on Defect and Fault Tolerance in VLSI Systems, pp. 581-588, Soston.

Ozello, P. (2002). *The Coded Microprocessor Certification,* International Conference on Computer Safety, Reliability and Security, SAFECOMP 1992, Springer, pp. 185-190, Munich.

Schäuffele, J., Zurawka, T. (2004). *Automotive Software Engineering,* Vieweg, Wiesbaden.

Tarabbia, J.-F.(2004), *An Open Platform Strategy in the Context of AUTOSAR,* VDI Berichte Nr. 1907, pp. 439-454.

Torres-Pomales, W.(2000). *Software Fault Tolerance: A Tutorial,* NASA, Langley Research Center, Hampton, Virginia.

Chen, X., Feng, J., Hiller, M., Lauer, V. (2007). *Application of Software Watchdog as Dependability Software Service for Automotive Safety Relevant Systems,* The 37th Annual IEEE/IFIP International Conference on Dependable Systems and Networks, DSN 2007, Edinburgh.

Steindl, M., Mottok, J., Meier,H., Schiller, F., and Fruechtl, M. (2009). *Diskussion des Einsatzes von Safely Embedded Software in FPGA-Architekturen,* In Proceedings of the 2nd Embedded Software Engineering Congress, ISBN 978-3-8343-2402-3, pp. 655-661, Sindelfingen.

Steindl, M. (200). *Safely Embedded Software (SES) im Umfeld der Normen für funktionale Sicherheit,* Jahresrückblick 2009 des Bayerischen IT-Sicherheitsclusters, pp. 22-23, Regensburg.

Mottok, J. (2009) *Safely Embedded Software,*In Proceedings of the 2nd Embedded Software Engineering Congress, pp. 10-12, Sindelfingen.

Steindl, M., Mottok, J. and Meier, H. (2010) *SES-based Framework for Fault-tolerant Systems,* in Proceedings of the 8th IEEE Workshop on Intelligent Solutions in Embedded Systems, Heraklion.

Raab, P., Kraemer, S., Mottok, J., Meier, H., Racek, S. (2011). *Safe Software Processing by Concurrent Execution in a Real-Time Operating System,* in Proceedings, International Conference on Applied Electronics, Pilsen.

Laumer, M., Felis, S., Mottok, J., Kinalzyk, D., Scharfenberg, G. (2011). *Safely Embedded Software and the ISO 26262,* Electromobility Conference, Prague.

Bärwald, A., Hauff, H., Mottok, J. (2010). *Certification of safety relevant systems - Benefits of using pre-certified components,* In Automotive Safety and Security, Stuttgart.

4

Vulnerability Analysis and Risk Assessment for SoCs Used in Safety-Critical Embedded Systems

Yung-Yuan Chen and Tong-Ying Juang
National Taipei University
Taiwan

1. Introduction

Intelligent systems, such as intelligent automotive systems or intelligent robots, require a rigorous reliability/safety while the systems are in operation. As system-on-chip (*SoC*) becomes more and more complicated, the *SoC* could encounter the reliability problem due to the increased likelihood of faults or radiation-induced soft errors especially when the chip fabrication enters the very deep submicron technology [Baumann, 2005; Constantinescu, 2002; Karnik et al., 2004; Zorian et al., 2005]. *SoC* becomes prevalent in the intelligent safety-related applications, and therefore, fault-robust design with the safety validation is required to guarantee that the developed *SoC* is able to comply with the safety requirements defined by the international norms, such as IEC 61508 [Brown, 2000; International Electrotechnical Commission [IEC], 1998-2000]. Therefore, safety attribute plays a key metric in the design of *SoC* systems. It is essential to perform the safety validation and risk reduction process to guarantee the safety metric of *SoC* before it is being put to use.

If the system safety level is not adequate, the risk reduction process, which consists of the vulnerability analysis and fault-robust design, is activated to raise the safety to the required level. For the complicated IP-based *SoCs* or embedded systems, it is unpractical and not cost-effective to protect the entire *SoC* or system. Analyzing the vulnerability of microprocessors or *SoCs* can help designers not only invest limited resources on the most crucial regions but also understand the gain derived from the investments [Hosseinabady et al., 2007; Kim & Somani, 2002; Mariani et al., 2007; Mukherjee et al., 2003; Ruiz et al., 2004; Tony et al., 2007; Wang et al., 2004].

The previous literature in estimating the vulnerability and failure rate of systems is based on either the analytical methodology or the fault injection approach at various system modeling levels. The fault injection approach was used to assess the vulnerability of high-performance microprocessors described in Verilog hardware description language at RTL design level [Kim & Somani, 2002; Wang et al., 2004]. The authors of [Mukherjee et al., 2003] proposed a systematic methodology based on the concept of architecturally correct execution to compute the architectural vulnerability factor. [Hosseinabady et al., 2007] and [Tony et al., 2007] proposed the analytical methods, which adopted the concept of timing vulnerability factor and architectural vulnerability factor [Mukherjee et al., 2003] respectively to estimate the vulnerability and failure rate of *SoCs*, where a UML-based real time description was employed to model the systems.

The authors of [Mariani et al., 2007] presented an innovative failure mode and effects analysis (FMEA) method at *SoC*-level design in RTL description to design in compliance with IEC61508. The methodology presented in [Mariani et al., 2007] was based on the concept of sensible zone to analyze the vulnerability and to validate the robustness of the target system. A memory sub-system embedded in fault-robust microcontrollers for automotive applications was used to demonstrate the feasibility of their FMEA method. However, the design level in the scheme presented in [Mariani et al., 2007] is RTL level, which may still require considerable time and efforts to implement a *SoC* using RTL description due to the complexity of oncoming *SoC* increasing rapidly. A dependability benchmark for automotive engine control applications was proposed in paper [Ruiz et al., 2004]. The work showed the feasibility of the proposed dependability benchmark using a prototype of diesel electronic control unit (ECU) control engine system. The fault injection campaigns were conducted to measure the dependability of benchmark prototype. The domain of application for dependability benchmark specification presented in paper [Ruiz et al., 2004] confines to the automotive engine control systems which were built by commercial off-the-shelf (COTS) components. While dependability evaluation is performed after physical systems have been built, the difficulty of performing fault injection campaign is high and the costs of re-designing systems due to inadequate dependability can be prohibitively expensive.

It is well known that FMEA [Mikulak et al., 2008] and fault tree analysis (FTA) [Stamatelatos et al., 2002] are two effective approaches for the vulnerability analysis of the *SoC*. However, due to the high complexity of the *SoC*, the incorporation of the FMEA/FTA and fault-tolerant demand into the *SoC* will further raise the design complexity. Therefore, we need to adopt the behavioral level or higher level of abstraction to describe/model the *SoC*, such as using SystemC, to tackle the complexity of the *SoC* design and verification. An important issue in the design of *SoC* is how to validate the system dependability as early in the development phase to reduce the re-design cost and time-to-market. As a result, a *SoC*-level safety process is required to facilitate the designers in assessing and enhancing the safety/robustness of a *SoC* with an efficient manner.

Previously, the issue of *SoC*-level vulnerability analysis and risk assessment is seldom addressed especially in SystemC transaction-level modeling (TLM) design level [Thorsten et al., 2002; Open SystemC Initiative [OSCI], 2003]. At TLM design level, we can more effectively deal with the issues of design complexity, simulation performance, development cost, fault injection, and dependability for safety-critical *SoC* applications. In this study, we investigate the effect of soft errors on the *SoCs* for safety-critical systems. An IP-based *SoC*-level safety validation and risk reduction (SVRR) process combining FMEA with fault injection scheme is proposed to identify the potential failure modes in a *SoC* modeled at SystemC TLM design level, to measure the risk scales of consequences resulting from various failure modes, and to locate the vulnerability of the system. A *SoC* system safety verification platform was built on the SystemC *CoWare Platform Architect* design environment to demonstrate the core idea of SVRR process. The verification platform comprises a system-level fault injection tool and a vulnerability analysis and risk assessment tool, which were created to assist us in understanding the effect of faults on system behavior, in measuring the robustness of the system, and in identifying the critical parts of the system during the *SoC* design process under the environment of *CoWare Platform Architect*.

Since the modeling of *SoCs* is raised to the level of TLM abstraction, the safety-oriented analysis can be carried out efficiently in early design phase to validate the safety/robustness of the *SoC* and identify the critical components and failure modes to be protected if necessary. The proposed SVRR process and verification platform is valuable in that it provides the capability to quickly assess the *SoC* safety, and if the measured safety cannot meet the system requirement, the results of vulnerability analysis and risk assessment will be used to help us develop a feasible and cost-effective risk reduction process. We use an ARM-based *SoC* to demonstrate the robustness/safety validation process, where the soft errors were injected into the register file of ARM CPU, memory system, and AMBA AHB.

The remaining paper is organized as follows. In Section 2, the SVRR process is presented. A risk model for vulnerability analysis and risk assessment is proposed in the following section. In Section 4, based on the SVRR process, we develop a *SoC*-level system safety verification platform under the environment of *CoWare Platform Architect*. A case study with the experimental results and a thorough vulnerability and risk analysis are given in Section 5. The conclusion appears in Section 6.

2. Safety validation and risk reduction process

We propose a SVRR process as shown in Fig. 1 to develop the safety-critical electronic systems. The process consists of three phases described as follows:

Phase 1 (fault hypothesis): this phase is to identify the potential interferences and develop the fault injection strategy to emulate the interference-induced errors that could possibly occur during the system operation.

Phase 2 (vulnerability analysis and risk assessment): this phase is to perform the fault injection campaigns based on the Phase 1 fault hypothesis. Throughout the fault injection campaigns, we can identify the failure modes of the system, which are caused by the faults/errors injected into the system while the system is in operation. The probability distribution of failure modes can be derived from the fault injection campaigns. The risk-priority number (RPN) [Mollah, 2005] is then calculated for the components inside the electronic system. A component's *RPN* aims to rate the risk of the consequence caused by component's failure. RPN can be used to locate the critical components to be protected. The robustness of the system is computed based on the adopted robustness criterion, such as safety integrity level (SIL) defined in the IEC 61508 [IEC, 1998-2000]. If the robustness of the system meets the safety requirement, the system passes the validation; else the robustness/safety is not adequate, so Phase 3 is activated to enhance the system robustness/safety.

Phase 3 (fault-tolerant design and risk reduction): This phase is to develop a feasible risk-reduction approach by fault-tolerant design, such as the schemes presented in [Austin, 1999; Mitra et al., 2005; Rotenberg, 1999; Slegel et al., 1999;], to improve the robustness of the critical components identified in Phase 2. The enhanced version then goes to Phase 2 to recheck whether the adopted risk-reduction approach can satisfy the safety/robustness requirement or not.

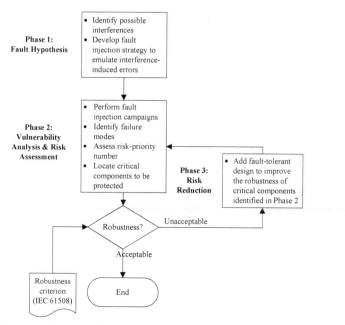

Fig. 1. Safety validation and risk reduction process.

3. Vulnerability analysis and risk assessment

Analyzing the vulnerability of *SoCs* or systems can help designers not only invest limited resources on the most crucial region but also understand the gain derived from the investment. In this section, we propose a *SoC*-level risk model to quickly assess the *SoC's* vulnerability at SystemC TLM level. Conceptually, our risk model is based on the FMEA method with the fault injection approach to measure the robustness of *SoCs*. From the assessment results, the rank of component vulnerability related to the risk scale of causing the system failure can be acquired. The notations used in the risk model are developed below.

- n: number of components to be investigated in the *SoC*;
- z: number of possible failure modes of the *SoC*;
- $C(i)$: the ith component, where $1 \leq i \leq n$;
- $ER\_C(i)$: raw error rate of the ith component;
- $SFR\_C(i)$: the part of *SoC* failure rate contributed from the error rate of the ith component;
- SFR: *SoC* failure rate;
- $FM(k)$: the kth failure mode of the *SoC*, where $1 \leq k \leq z$;
- NE: no effect which means that a fault/error happening in a component has no impact on the *SoC* operation at all;
- $P(i, FM(K))$: probability of $FM(K)$ if an error occurs in the ith component;
- $P(i, NE)$: probability of no effect for an error occurring in the ith component;
- $P(i, SF)$: probability of *SoC* failure for an error occurring in the ith component;

- $SR\_FM(k)$: severity rate of the effect of kth failure mode, where $1 \leq k \leq z$;
- $RPN\_C(i)$: risk priority number of the ith component;
- $RPN\_FM(k)$: risk priority number of the kth failure mode.

3.1 Fault hypothesis

It is well known that the rate of soft errors caused by single event upset (SEU) increases rapidly while the chip fabrication enters the very deep submicron technology [Baumann, 2005; Constantinescu, 2002; Karnik et al., 2004; Zorian et al., 2005]. Radiation-induced soft errors could cause a serious dependability problem for *SoCs*, electronic control units, and nodes used in the safety-critical applications. The soft errors may happen in the flip-flop, register file, memory system, system bus and combinational logic. In this work, single soft error is considered in the derivation of risk model.

3.2 Risk model

The potential effects of faults on *SoC* can be identified from the fault injection campaigns. We can inject the faults into a specific component, and then investigate the effect of component's errors on the *SoC* behaviors. Throughout the injection campaigns for each component, we can identify the failure modes of the *SoC*, which are caused by the errors of components in the *SoC*. The parameter $P(i, FM(k))$ defined before can be derived from the fault injection campaigns.

In general, the following failure behaviors: fatal failure (FF), such as system crash or process hang, silent data corruption (SDC), correct data/incorrect time (CD/IT), and infinite loop (IL) (note that we declare the failure as IL if the execution of benchmark exceeds the 1.5 times of normal execution time), which were observed from our previous work, represent the possible *SoC* failure modes caused by the faults occurring in the components. Therefore, we adopt those four *SoC* failure modes in this study to demonstrate our risk assessment approach. We note that a fault may not cause any trouble at all, and this phenomenon is called no effect of the fault.

One thing should be pointed out that to obtain the highly reliable experimental results to analyze the robustness/safety and vulnerability of the target system we need to perform the adequate number of fault injection campaigns to guarantee the validity of the statistical data obtained. In addition, the features of benchmarks could also affect the system response to the faults. Therefore, several representative benchmarks are required in the injection campaigns to enhance the confidence level of the statistical data.

In the derivation of $P(i, FM(K))$, we need to perform the fault injection campaigns to collect the fault simulation data. Each fault injection campaign represents an experiment by injecting a fault into the ith component, and records the fault simulation data, which will be used in the failure mode classification procedure to identify which failure mode or no effect the *SoC* encountered in this fault injection campaign. The failure mode classification procedure inputs the fault-free simulation data, and fault simulation data derived from the fault injection campaigns to analyze the effect of faults occurring in the ith component on the *SoC* behavior based on the classification rules for potential failure modes.

The derivation process of $P(i, FM(K))$ by fault injection process is described below. Several notations are developed first:

- *SoC_FM*: a set of *SoC* failure modes used to record the possible *SoC* failure modes happened in the fault injection campaigns.
- *counter(i, k)*: an array which is used to count the number of the kth *SoC* failure mode occurring in the fault injection experiments for the ith component, where $1 \leq i \leq n$, and $1 \leq k \leq z$. *counter(i, z+1)* is used to count the number of no effect in the fault injection campaigns.
- *no_fi(i)*: the number of fault injection campaigns performed in the ith component, where $1 \leq i \leq n$.

Fault injection process:

$z = 4$; *SoC_FM* = {FF, SDC, CD/IT, IL};
for i = 1 to n   //fault injection experiments for the ith component;//
{for j = 1 to no_fi(i)
    {//injecting a fault into the ith component, and investigating the effect of component's fault on the *SoC* behavior by *failure mode classification procedure*; the result of classification is recorded in the parameter *'classification'.*//
    switch (*classification*)
    { case 'FF': *counter(i, 1) = counter(i, 1) + 1;*
    case 'SDC': *counter(i, 2) = counter(i, 2) + 1;*
    case 'CD/IT': *counter(i, 3) = counter(i, 3) + 1;*
    case 'IL': *counter(i, 4) = counter(i, 4) + 1;*
    case 'NE': *counter(i, 5) = counter(i, 5) + 1;*}

}}

The failure mode classification procedure is used to classify the *SoC* failure modes caused by the component's faults. For a specific benchmark program, we need to perform a fault-free simulation to acquire the golden results that are used to assist the failure mode classification procedure in identifying which failure mode or no effect the *SoC* encountered in this fault injection campaign.

Failure mode classification procedure:

Inputs: fault-free simulation golden data and fault simulation data for an injection campaign;

Output: *SoC* failure mode caused by the component's fault or no effect of the fault in this injection campaign.

{if (execution of fault simulation is complete)

   then if (execution time of fault simulation is the same as execution time of fault-free simulation)

      then if (execution results of fault simulation are the same as execution results of fault-free simulation)

         then *classification* := 'NE';

         else *classification* := 'SDC';

      else if (execution results of fault simulation are the same as execution results of fault-free simulation) then *classification* := 'CD/IT';

      else *classification* := 'SDC';

  else if (execution of benchmark exceeds the 1.5 times of normal execution time)

    then *classification* := 'IL';

    else //execution of fault simulation was hung or crash due to the injected fault;//

      classification := 'FF';

}

After carrying out the above injection experiments, the parameter of $P(i, FM(K))$ can be computed by

$$P(i, FM(K)) = \frac{counter(i,k)}{no\_fi(i)}$$

Where $1 \leq i \leq n$ and $1 \leq k \leq z$. The following expressions are exploited to evaluate the terms of $P(i, SF)$ and $P(i, NE)$.

$$P(i, SF) = \sum_{k=1}^{z} P(i, FM(k))$$

$$P(i, NE) = 1 - P(i, SF)$$

The derivation of the component's raw error rate is out of the scope of this paper, so we here assume the data of $ER\_C(i)$, for $1 \leq i \leq n$, are given. The part of *SoC* failure rate contributed from error rate of the ith component can be calculated by

$$SFR\_C(i) = ER\_C(i) \times P(i, SF)$$

If each component $C(i)$, $1 \leq i \leq n$, must operate correctly for the *SoC* to operate correctly and also assume that other components not shown in $C(i)$ list are fault-free, the *SoC* failure rate can be written as

$$SFR = \sum_{i=1}^{n} SFR\_C(i)$$

The meaning of the parameter $SR\_FM(k)$ and the role it playing can be explained from the aspect of FMEA process [Mollah, 2005]. The method of FMEA is to identify all possible failure modes of a *SoC* and analyze the effects or consequences of the identified failure modes. In general, an FMEA records each potential failure mode, its effect in the next level, and the cause of failure. We note that the faults occurring in different components could cause the same *SoC* failure mode, whereas the severity degree of the consequences resulting from various *SoC* failure modes could not be identical. The parameter $SR\_FM(k)$ is exploited to express the severity rate of the consequence resulting from the kth failure mode, where $1 \leq k \leq z$.

We illustrate the risk evaluation with FMEA idea using the following example. An ECU running engine control software is employed for automotive engine control. Its outputs are used to control the engine operation. The ECU could encounter several types of output failures due to hardware or software faults in ECU. The various types of failure mode of ECU outputs would result in different levels of risk/criticality on the controlled engine. A risk assessment is performed to identify the potential failure modes of ECU outputs as well as the likelihood of failure occurrence, and estimate the resulting risks of the ECU-controlled engine.

In the following, we propose an effective SoC-level FMEA method to assess the risk-priority number (RPN) for the components inside the SoC and for the potential SoC failure modes. A component's RPN aims to rate the risk of the consequences caused by component's faults. In other words, a component's RPN represents how serious is the impact of component's errors on the system safety. A risk assessment should be carried out to identify the critical components within a SoC and try to mitigate the risks caused by those critical components. Once the critical components and their risk scales have been identified, the risk-reduction process, for example fault-tolerant design, should be activated to improve the system dependability. RPN can also give the protection priority among the analyzed components. As a result, a feasible risk-reduction approach can be developed to effectively protect the vulnerable components and enhance the system robustness and safety.

The parameter $RPN\_C(i)$, i.e. risk scale of failures occurring in the ith component, can be computed by

$$RPN\_C(i) = ER\_C(i) \times \sum_{k=1}^{z} P(i, FM(k)) \times SR\_FM(k)$$

where $1 \leq i \leq n$. The expression of $RPN\_C(i)$ contains three terms which are, from left to right, error rate of the ith component, probability of $FM(K)$ if a fault occurs in the ith component, and severity rate of the kth failure mode. As stated previously, a component's fault could result in several different system failure modes, and each identified failure mode has its potential impact on the system safety. So, $RPN\_C(i)$ is the summation of the following expression $ER\_C(i) \times P(i, FM(K)) \times SR\_FM(k)$, for k from one to z. The term of $ER\_C(i) \times P(i, FM(K))$ represents the occurrence rate of the kth failure mode, which is caused by the ith component failing to perform its intended function.

The $RPN\_FM(k)$ represents the risk scale of the kth failure mode, which can be calculated by

$$RPN\_FM(k) = SR\_FM(k) \times \sum_{i=1}^{n} ER\_C(i) \times P(i, FM(k))$$

where $1 \leq k \leq z$. $\sum_{i=1}^{n} ER\_C(i) \times P(i, FM(k))$ expresses the occurrence rate of the kth failure mode in a SoC. This sort of assessment can reveal the risk levels of the failure modes to its system and identify the major failure modes for protection so as to reduce the impact of failures to the system safety.

4. System safety verification platform

We have created an effective safety verification platform to provide the capability to quickly handle the operation of fault injection campaigns and dependability analysis for the system design with SystemC. The core of the verification platform is the fault injection tool [Chang & Chen, 2007; Chen et al., 2008] under the environment of *CoWare Platform Architect* [CoWare, 2006], and the vulnerability analysis and risk assessment tool. The tool is able to deal with the fault injection at the following levels of abstraction [Chang & Chen, 2007; Chen et al., 2008]: bus-cycle accurate level, untimed functional TLM with primitive channel sc_fifo, and timed functional TLM with hierarchical channel. An interesting feature of our fault injection tool is to offer not only the time-triggered but also the event-triggered methodologies to decide when to inject a fault. Consequently, our injection tool can significantly reduce the effort and time for performing the fault injection campaigns. Combining the fault injection tool with vulnerability analysis and risk assessment tool, the verification platform can dramatically increase the efficiency of carrying out the system robustness validation and vulnerability analysis and risk assessment. For the details of our fault injection tool, please refer to [Chang & Chen, 2007; Chen et al., 2008].

However, the IP-based *SoCs* designed by *CoWare Platform Architect* in SystemC design environment encounter the injection controllability problem. The simulation-based fault injection scheme cannot access the fault targets inside the IP components imported from other sources. As a result, the injection tool developed in SystemC abstraction level may lack the capability to inject the faults into the inside of the imported IP components, such as CPU or DSP. To fulfill this need, we exploit the software-implemented fault injection scheme [Sieh, 1993; Kanawati et al., 1995] to supplement the injection ability. The software-implemented fault injection scheme, which uses the system calls of Unix-type operating system to implement the injection of faults, allows us to inject the faults into the targets of storage elements in processors, like register file in CPU, and memory systems. As discussed, a complete IP-based *SoC* system-level fault injection tool should consist of the software-implemented and simulation-based fault injection schemes.

Due to the lack of the support of Unix-type operating system in *CoWare Platform Architect*, the current version of safety verification platform cannot provide the software-implemented fault injection function in the tool. Instead, we employed a physical system platform built by ARM-embedded *SoC* running Linux operating system to validate the developed software-implemented fault injection mechanism. We note that if the *CoWare Platform Architect* can support the UNIX-type operating system in the SystemC design environment, our software-implemented fault injection concept should be brought in the SystemC design platform. Under the circumstances, we can implement the so called hybrid fault injection approach, which comprises the software-implemented and simulation-based fault injection methodologies, in the SystemC design environment to provide more variety of injection functions.

5. Case study

An ARM926EJ-based *SoC* platform provided by *CoWare Platform Architect* [CoWare, 2006] was used to demonstrate the feasibility of our risk model. The illustrated *SoC* platform was modeled at the timed functional TLM abstraction level. This case study is to investigate three important components, which are register file in ARM926EJ, AMBA Advanced High-performance Bus (AHB), and the memory sub-system, to assess their risk scales to the *SoC*-controlled system. We exploited the safety verification platform to perform the fault injection process associated with the risk model presented in Section 3 to obtain the risk-related parameters for the components mentioned above. The potential *SoC* failure modes classified from the fault injection process are fatal failure (FF), silent data corruption (SDC), correct data/incorrect time (CD/IT), and infinite loop (IL). In the following, we summarize the data used in this case study.

- $n = 3$, $\{C(1),\ C(2),\ C(3)\}$ = {AMBA AHB, memory sub-system, register file in ARM926EJ}.
- $z = 4$, $\{FM(1), FM(2), FM(3), FM(4)\}$ = {FF, SDC, CD/IT, IL}.
- The benchmarks employed in the fault injection process are: JPEG (pixels: 255 × 154), matrix multiplication (M-M: 50 × 50), quicksort (QS: 3000 elements) and FFT (256 points).

5.1 AMBA AHB experimental results

The system bus, such as AMBA AHB, provides an interconnected platform for IP-based *SoC*. Apparently, the robustness of system bus plays an important role in the *SoC* reliability. It is evident that the faults happening in the bus signals will lead to the data transaction errors and finally cause the system failures. In this experiment, we choose three bus signals HADDR[31:0], HSIZE[2:0], and HDATA[31:0] to investigate the effect of bus errors on the system. The results of fault injection process for AHB system bus under various benchmarks are shown in Table 1 and 2. The results of a particular benchmark in Table 1 and 2 were derived from the six thousand fault injection campaigns, where each injection campaign injected 1-bit flip fault to bus signals. The fault duration lasts for the length of one-time data transaction. The statistics derived from six thousand times of fault injection campaigns have been verified to guarantee the validity of the analysis.

From Table 1, it is evident that the susceptibility of the *SoC* to bus faults is benchmark-dependent and the rank of system bus vulnerability over different benchmarks is JPEG > M-M > FFT > QS. However, all benchmarks exhibit the same trend in that the probabilities of FF show no substantial difference, and while a fault arises in the bus signals, the occurring probabilities of SDC and FF occupy the top two ranks. The results of the last row offer the average statistics over four benchmarks employed in the fault injection process. Since the probabilities of *SoC* failure modes are benchmark-variant, the average results illustrated in Table 1 give us the expected probabilities for the system bus vulnerability of the developing *SoC*, which are very valuable for us to gain the robustness of the system bus and the probability distribution of failure modes. The robustness measure of the system bus is only 26.78% as shown in Table 1, which means that a fault occurring in the system bus, the *SoC* has the probability of 26.78% to survive for that fault.

The experimental results shown in Table 2 are probability distribution of failure modes with respect to the various bus signal errors for the used benchmarks. From the data illustrated in the NE column, we observed that the most vulnerable part is the address bus HADDR[31:0]. Also from the data displayed in the FF column, the faults occurring in address bus will have the probability between 38.9% and 42.3% to cause a serious fatal failure for the used benchmarks. The HSIZE and HDATA signal errors mainly cause the SDC failure. In summary, our results reveal that the address bus HADDR should be protected first in the design of system bus, and the SDC is the most popular failure mode for the demonstrated *SoC* responding to the bus faults or errors.

	FF (%)	SDC (%)	CD/IT (%)	IL(%)	SF (%)	NE (%)
JPEG	18.57	45.90	0.16	15.88	80.51	19.49
M-M	18.95	55.06	2.15	3.57	79.73	20.27
FFT	20.18	21.09	15.74	6.38	63.39	36.61
QS	20.06	17.52	12.24	5.67	55.50	44.50
Avg.	19.41	38.16	7.59	8.06	73.22	26.78

Table 1. P (1, FM(K)), P (1, SF) and P (1, NE) for the used benchmarks.

	FF (%)				SDC (%)				CD/IT (%)			
	1	2	3	4	1	2	3	4	1	2	3	4
HADDR	38.9	39.7	42.3	42	42.9	43.6	18.2	15.2	0.08	1.94	14.4	11.4
HSIZE	0.16	0.0	0.0	0	68.2	67.6	25.6	22.6	0.25	9.64	37.4	38.5
HDATA	0.0	0.0	0.0	0	46.8	65.4	23.6	19.4	0.24	1.66	15.0	10.6

	IL (%)				NE (%)			
	1	2	3	4	1	2	3	4
HADDR	11.5	2.02	3.41	2.02	6.62	12.7	21.7	29.4
HSIZE	11.6	2.38	6.97	7.53	19.8	20.4	30.0	31.4
HDATA	20.7	5.23	9.29	9.15	32.3	27.7	52.1	60.9

Table 2. Probability distribution of failure modes with respect to various bus signal errors for the used benchmarks (1, 2, 3 and 4 represent the jpeg, m-m, fft and qs benchmark, respectively).

5.2 Memory sub-system experimental results

The memory sub-system could be affected by the radiation articles, which may cause the bit-flipped soft errors. However, the bit errors won't cause damage to the system operation if one of the following situations occurs:

- Situation 1: The benchmark program never reads the affected words after the bit errors happen.
- Situation 2: The first access to the affected words after the occurrence of bit errors is the 'write' action.

Otherwise, the bit errors could cause damage to the system operation. Clearly, if the first access to the affected words after the occurrence of bit errors is the 'read' action, the bit errors will be propagated and could finally lead to the failures of SoC operation. So, whether the bit errors will become fatal or not, it all depends on the occurring time of bit errors, the locations of affected words, and the benchmark's memory access patterns after the occurrence of bit errors.

According to the above discussion, two interesting issues arise; one is the propagation probability of bit errors and another is the failure probability of propagated bit errors. We define the propagation probability of bit errors as the probability of bit errors which will be read out and propagated to influence the execution of the benchmarks. The failure probability of propagated bit errors represents the probability of propagated bit errors which will finally result in the failures of SoC operation.

Initially, we tried performing the fault injection campaigns in the *CoWare Platform Architect* to collect the simulation data. After a number of fault injection and simulation campaigns, we realized that the length of experimental time will be a problem because a huge amount of fault injection and simulation campaigns should be conducted for each benchmark and several benchmarks are required for the experiments. From the analysis of the campaigns, we observed that a lot of bit-flip errors injected to the memory sub-system fell into the Situation 1 or 2, and therefore, we must carry out an adequate number of fault injection campaigns to obtain the validity of the statistical data.

To solve this dilemma, we decide to perform two types of experiments termed as Type 1 experiment and Type 2 experiment, or called hybrid experiment, to assess the propagation probability and failure probability of bit errors, respectively. As explained below, Type 1 experiment uses a software tool to emulate the fault injection and simulation campaigns to quickly gain the propagation probability of bit errors, and the set of propagated bit errors. The set of propagated bit errors will be used in the Type 2 experiment to measure the failure probability of propagated bit errors.

Type 1 experiment: we develop the experimental process as described below to measure the propagation probability of bit errors. The following notations are used in the experimental process.

- N_{bench}: the number of benchmarks used in the experiments.
- $N_{inj}(j)$: the number of fault injection campaigns performed in the jth benchmark's experiment.
- $C_{p-b-err}$: counter of propagated bit errors.
- $N_{p-b-err}$: the expected number of propagated bit errors.
- S_m: address space of memory sub-system.
- N_{d-t}: the number of read/write data transactions occurring in the memory sub-system during the benchmark execution.
- T_{error}: the occurring time of bit error.
- A_{error}: the address of affected memory word.
- $S_{p-b-err}(j)$: set of propagated bit errors conducted in the jth benchmark's experiment.
- $P_{p-b-err}$: propagation probability of bit errors.

Experimental Process: We injected a bit-flipped error into a randomly chosen memory address at random read/write transaction time for each injection campaign. As stated earlier, this bit error could either be propagated to the system or not. If yes, then we add one to the parameter $C_{p-b-err}$. The parameter $N_{p-b-err}$ is set by users and employed as the terminated condition for the current benchmark's experiment. When the value of $C_{p-b-err}$ reaches to $N_{p-b-err}$, the process of current benchmark's experiment is terminated. The $P_{p-b-err}$ can then be derived from $N_{p-b-err}$ divided by N_{inj}. The values of N_{bench}, S_m and $N_{p-b-err}$ are given before performing the experimental process.

for j = 1 to N_{bench}
{
Step 1: Run the jth benchmark in the experimental *SoC* platform under *CoWare Platform Architect* to collect the desired bus read/write transaction information that include address, data and control signals of each data transaction into an operational profile during the program execution. The value of N_{d-t} can be obtained from this step.

Step 2: $C_{p-b-err} = 0$; $N_{inj}(j) = 0$;

While $C_{p-b-err} < N_{p-b-err}$ do

{T_{error} can be decided by randomly choosing a number x between one and N_{d-t}. It means that T_{error} is equivalent to the time of the xth data transaction occurring in the memory sub-system. Similarly, A_{error} is determined by randomly choosing an address between one and S_m. A bit is randomly picked up from the word pointed by A_{error}, and the bit selected is flipped. Here, we assume that the probability of fault occurrence of each word in memory sub-system is the same.

If ((Situation 1 occurs) or (Situation 2 occurs))

then {the injected bit error won't cause damage to the system operation;}

else {$C_{p-b-err} = C_{p-b-err} + 1$;

record the related information of this propagated bit error to $S_{p-b-err}(j)$ including T_{error}, A_{error} and bit location.}

//Situation 1 and 2 are described in the beginning of this Section. The operational profile generated in Step 1 is exploited to help us investigate the resulting situation caused by the current bit error. From the operational profile, we check the memory access patterns beginning from the time of occurrence of bit error to identify which situation the injected bit error will lead to. //

$N_{inj}(j) = N_{inj}(j) + 1$;}

}

For each benchmark, we need to perform the Step 1 of Type 1 experimental process once to obtain the operational profile, which will be used in the execution of Step 2. We then created a software tool to implement the Step 2 of Type 1 experimental process. We note that the created software tool emulates the fault injection campaigns required in Step 2 and checks the consequences of the injected bit errors with the support of operational profile derived from Step 1. It is clear to see that the Type 1 experimental process does not utilize the simulation-based fault injection tool implemented in safety verification platform as described in Section 4. The reason why we did not exploit the safety verification platform in this experiment is the consideration of time efficiency. The comparison of required simulation time between the methodologies of hybrid experiment and the pure simulation-based fault injection approach implemented in *CoWare Platform Architect* will be given later.

The Type 1 experimental process was carried out to estimate $P_{p-b-err}$, where N_{bench}, S_m and $N_{p-b-err}$ were set as the values of 4, 524288, and 500 respectively. Table 3 shows the propagation probability of bit errors for four benchmarks, which were derived from a huge amount of fault injection campaigns to guarantee their statistical validity. It is evident that the propagation probability is benchmark-variant and a bit error in memory would have the probability between 0.866% and 3.551% to propagate the bit error from memory to system. The results imply that most of the bit errors won't cause damage to the system. We should emphasize that the size of memory space and characteristics of the used benchmarks (such as amount of memory space use and amount of memory read/write) will affect the result of $P_{p-b-err}$. Therefore, the data in Table 3 reflect the results for the selected memory space and benchmarks.

Type 2 experiment: From Type 1 experimental process, we collect $N_{p-b-err}$ bit errors for each benchmark to the set $S_{p-b-err}(j)$. Those propagated bit errors were used to assess the failure probability of propagated bit errors. Therefore, $N_{p-b-err}$ simulation-based fault injection

Benchmark	N_{inj}	$N_{p\text{-}b\text{-}err}$	$P_{p\text{-}b\text{-}err}$
M-M	14079	500	3.551%
QS	23309	500	2.145%
JPEG	27410	500	1.824%
FFT	57716	500	0.866%

Table 3. Propagation probability of bit errors.

campaigns were conducted under CoWare Platform Architect, and each injection campaign injects a bit error into the memory according to the error scenarios recorded in the set $S_{p\text{-}b\text{-}err}(j)$. Therefore, we can examine the SoC behavior for each injected bit error.

As can be seen from Table 3, we need to conduct an enormous amount of fault injection campaigns to reach the expected number of propagated bit errors. Without the use of Type 1 experiment, we need to utilize the simulation-based fault injection approach to assess the propagation probability and failure probability of bit errors as illustrated in Table 3, 5, and 6, which require a huge number of simulation-based fault injection campaigns to be conducted. As a result, an enormous amount of simulation time is required to complete the injection and simulation campaigns. Instead, we developed a software tool to implement the experimental process described in Type 1 experiment to quickly identify which situation the injected bit error will lead to. Using this approach, the number of simulation-based fault injection campaigns performed in Type 2 experiment decreases dramatically. The performance of software tool adopted in Type 1 experiment is higher than that of simulation-based fault injection campaign employed in Type 2 experiment. Therefore, we can save a considerable amount of simulation time.

The data of Table 3 indicate that without the help of Type 1 experiment, we need to carry out a few ten thousand simulation-based fault injection campaigns in Type 2 experiment. As opposite to that, with the assistance of Type 1 experiment, only five hundred injection campaigns are required in Type 2 experiment. Table 4 gives the experimental time of the Type 1 plus Type 2 approach and pure simulation-based fault injection approach, where the data in the column of ratio are calculated by the experimental time of Type 1 plus Type 2 approach divided by the experimental time of pure simulation-based approach. The experimental environment consists of four machines to speed up the validation, where each machine is equipped with Intel® Core™2 Quad Processor Q8400 CPU, 2G RAM, and CentOS 4.6. In the experiments of Type 1 plus Type 2 approach and pure simulation-based approach, each machine is responsible for performing the simulation task for one benchmark. According to the simulation results, the average execution time for one simulation-based fault injection experiment is 14.5 seconds. It is evident that the performance of Type 1 plus Type 2 approach is quite efficient compared to the pure simulation-based approach because Type 1 plus Type 2 approach employed a software tool to effectively reduce the number of simulation-based fault injection experiments to five hundred times compared to a few ten thousand simulation-based fault injection experiments for pure simulation-based approach.

Given $N_{p\text{-}b\text{-}err}$ and $S_{p\text{-}b\text{-}err}(j)$, i.e. five hundred simulation-based fault injection campaigns, the Type 2 experimental results are illustrated in Table 5. From Table 5, we can identify the potential failure modes and the distribution of failure modes for each benchmark. It is clear that the susceptibility of a system to the memory bit errors is benchmark-variant, and the M-

M is the most critical benchmark among the four adopted benchmarks, according to the results of Table 5.

We then manipulated the data of Table 3 and 5 to acquire the results of Table 6. Table 6 shows the probability distribution of failure modes if a bit error occurs in the memory sub-system. Each datum in the row of 'Avg.' was obtained by mathematical average of the benchmarks' data in the corresponding column. This table offers the following valuable information: the robustness of memory sub-system, the probability distribution of failure modes and the impact of benchmark on the *SoC* dependability. Probability of *SoC* failure for a bit error occurring in the memory is between 0.738% and 3.438%. We also found that the *SoC* has the highest probability to encounter the SDC failure mode for a memory bit error. In addition, the vulnerability rank of benchmarks for memory bit errors is M-M > QS > JPEG > FFT.

Table 7 illustrates the statistics of memory read/write for the adopted benchmarks. The results of Table 7 confirm the vulnerability rank of benchmarks as observed in Table 6. Situation 2 as mentioned in the beginning of this section indicates that the occurring probability of Situation 2 increases as the probability of performing the memory write operation increases. Consequently, the robustness of a benchmark rises with an increase in the probability of Situation 2.

Benchmark	Type 1 + 2 (minute)	Pure approach (minute)	Ratio
M-M	312	1525	20.46%
QS	835	2719	30.71%
JPEG	7596	15760	48.20%
FFT	3257	9619	33.86%

Table 4. Comparison of experimental time between type 1 + 2 & pure simulation-based approach.

Benchmark	FF	SDC	CD/IT	IL	NE
M-M	0	484	0	0	16
QS	0	138	103	99	160
JPEG	0	241	1	126	132
FFT	0	177	93	156	74

Table 5. Type 2 experimental results.

	FF (%)	SDC (%)	CD/IT (%)	IL (%)	SF (%)	NE (%)
M-M	0.0	3.438	0.0	0.0	3.438	96.562
QS	0.0	0.592	0.442	0.425	1.459	98.541
JPEG	0.0	0.879	0.004	0.460	1.343	98.657
FFT	0.0	0.307	0.161	0.270	0.738	99.262
Avg.	0.0	1.304	0.152	0.289	1.745	98.255

Table 6. P (2, FM(K)), P (2, SF) and P (2, NE) for the used benchmarks.

	#R/W	#R	R(%)	#W	W(%)
M-M	265135	255026	96.187%	10110	3.813%
QS	226580	196554	86.748%	30027	13.252%
JPEG	1862291	1436535	77.138%	425758	22.862%
FFT	467582	240752	50.495%	236030	49.505%

Table 7. The statistics of memory read/write for the used benchmarks.

5.3 Register file experimental results

The ARM926EJ CPU used in the experimental *SoC* platform is an IP provided from *CoWare Platform Architect*. Therefore, the proposed simulation-based fault injection approach has a limitation to inject the faults into the register file inside the CPU. This problem can be solved by software-implemented fault injection methodology as described in Section 4. Currently, we cannot perform the fault injection campaigns in register file under *CoWare Platform Architect* due to lack of the operating system support. We note that the literature [Leveugle et al., 2009; Bergaoui et al., 2010] have pointed out that the register file is vulnerable to the radiation-induced soft errors. Therefore, we think the register file should be taken into account in the vulnerability analysis and risk assessment. Once the critical registers are located, the SEU-resilient flip-flop and register design can be exploited to harden the register file. In this experiment, we employed a similar physical system platform built by ARM926EJ-embedded *SoC* running Linux operating system 2.6.19 to derive the experimental results for register file.

The register set in ARM926EJ CPU used in this experiment is R0 ~ R12, R13 (SP), R14 (LR), R15 (PC), R16 (CPSR), and R17 (ORIG_R0). A fault injection campaign injects a single bit-flip fault to the target register to investigate its effect on the system behavior. For each benchmark, we performed one thousand fault injection campaigns for each target register by randomly choosing the time instant of fault injection within the benchmark simulation duration, and randomly choosing the target bit to inject 1-bit flip fault. So, eighteen thousand fault injection campaigns were carried out for each benchmark to obtain the data shown in Table 8. From Table 8, it is evident that the susceptibility of the system to register faults is benchmark-dependent and the rank of system vulnerability over different benchmarks is QS > FFT > M-M. However, all benchmarks exhibit the same trend in that while a fault arises in the register set, the occurring probabilities of CD/IT and FF occupy the top two ranks. The robustness measure of the register file is around 74% as shown in Table 8, which means that a fault occurring in the register file, the *SoC* has the probability of 74% to survive for that fault.

	FF (%)	SDC (%)	CD/IT (%)	IL (%)	SF (%)	NE (%)
M-M	6.94	1.71	10.41	0.05	19.11	80.89
FFT	8.63	1.93	15.25	0.04	25.86	74.14
QS	5.68	0.97	23.44	0.51	30.59	69.41
Avg.	7.08	1.54	16.36	0.2	25.19	74.81

Table 8. *P* (3, *FM(K)*), *P* (3, *SF*) and *P* (3, *NE*) for the used benchmarks.

REG #	SoC failure probability			REG #	SoC failure probability		
	M-M (%)	FFT (%)	QS (%)		M-M (%)	FFT (%)	QS (%)
R0	7.9	13.0	5.6	R9	12.4	7.3	20.6
R1	31.1	18.3	19.8	R10	23.2	32.5	19.9
R2	19.7	14.6	19.2	R11	37.5	25.3	19.2
R3	18.6	17.0	15.4	R12	22.6	13.1	25.3
R4	4.3	12.8	21.3	R13	34.0	39.0	20.3
R5	4.0	15.2	20.4	R14	5.1	100.0	100.0
R6	7.4	8.8	21.6	R15	100.0	100.0	100.0
R7	5.0	14.6	23.9	R16	3.6	8.3	49.4
R8	4.0	9.7	24.7	R17	3.6	15.9	24.0

Table 9. Statistics of *SoC* failure probability for each target register with various benchmarks.

Table 9 illustrates the statistics of *SoC* failure probability for each target register under the used benchmarks. Throughout this table, we can observe the vulnerability of each register for different benchmarks. It is evident that the vulnerability of registers quite depends on the characteristics of the benchmarks, which could affect the read/write frequency and read/write syndrome of the target registers. The bit errors won't cause damage to the system operation if one of the following situations occurs:

- Situation 1: The benchmark never uses the affected registers after the bit errors happen.
- Situation 2: The first access to the affected registers after the occurrence of bit errors is the 'write' action.

It is apparent to see that the utilization and read frequency of R4 ~ R8 and R14 for benchmark M-M is quite lower than FFT and QS, so the *SoC* failure probability caused by the errors happening in R4 ~ R8 and R14 for M-M is significantly lower than FFT and QS as illustrated in Table 9. We observe that the usage and write frequency of registers, which reflects the features and the programming styles of benchmark, dominates the soft error sensitivity of the registers. Without a doubt, the susceptibility of register R15 (program counter) to the faults is 100%. It indicates that the R15 is the most vulnerable register to be protected in the register set. Fig. 2 illustrates the average SoC failure probabilities for the registers R0 ~ R17, which are derived from the data of the used benchmarks as exhibited in Table 9. According to Fig. 2, the top three vulnerable registers are R15 (100%), R14 (68.4%), as well as R13 (31.1%), and the SoC failure probabilities for other registers are all below 30%.

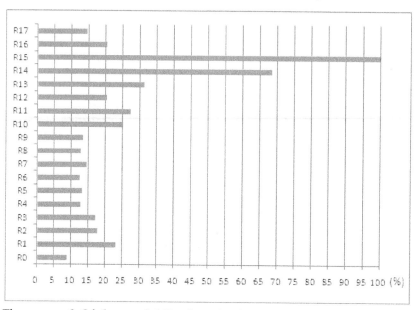

Fig. 2. The average SoC failure probability from the data of the used benchmarks.

5.4 SoC-level vulnerability analysis and risk assessment

According to IEC 61508, if a failure will result in a *critical effect* on system and lead human's life to be in danger, then such a failure is identified as a *dangerous failure or hazard*. IEC 61508 defines a system's safety integrity level (SIL) to be the Probability of the occurrence of a dangerous Failure per Hour (PFH) in the system. For continuous mode of operation (high demand rate), the four levels of SIL are given in Table 10 [IEC, 1998-2000].

SIL	PFH
4	$\geq 10^{-9}$ to $< 10^{-8}$
3	$\geq 10^{-8}$ to $< 10^{-7}$
2	$\geq 10^{-7}$ to $< 10^{-6}$
1	$\geq 10^{-6}$ to $< 10^{-5}$

Table 10. Safety integrity levels.

In this case study, three components, ARM926EJ CPU, AMBA AHB system bus and memory sub-system, were utilized to demonstrate the proposed risk model to assess the scales of failure-induced risks in a system. The following data are used to show the vulnerability analysis and risk assessment for the selected components $\{C(1), C(2), C(3)\}$ = {AMBA AHB, memory sub-system, register file in ARM926EJ}: $\{ER\_C(1), ER\_C(2), ER\_C(3)\}$ = $\{10^{-6} \sim 10^{-8}/hour\ \}$; $\{SR\_FM(1), SR\_FM(2), SR\_FM(3), SR\_FM(4)\}$ = {10, 8, 4, 6}. According to the expressions presented in Section 3 and the results shown in Section 5.1 to 5.3, the SoC failure rate, SIL and RPN are obtained and illustrated in Table 11, 12 and 13.

ER_C/hour	1×10^{-6}	0.5×10^{-6}	1×10^{-7}	0.5×10^{-7}	1×10^{-8}
SFR_C(1)	7.32×10^{-7}	3.66×10^{-7}	7.32×10^{-8}	3.66×10^{-8}	7.32×10^{-9}
SFR_C(2)	1.75×10^{-8}	8.73×10^{-9}	1.75×10^{-9}	8.73×10^{-10}	1.75×10^{-10}
SFR_C(3)	2.52×10^{-7}	1.26×10^{-7}	2.52×10^{-8}	1.26×10^{-8}	2.52×10^{-9}
SFR	1.0×10^{-6}	5.0×10^{-7}	1.0×10^{-7}	5.0×10^{-8}	1.0×10^{-8}
SIL	1	2	2	3	3

Table 11. SoC failure rate and SIL.

ER_C/hour	1×10^{-6}	0.5×10^{-6}	1×10^{-7}	0.5×10^{-7}	1×10^{-8}
RPN_C(1)	5.68×10^{-6}	2.84×10^{-6}	5.68×10^{-7}	2.84×10^{-7}	5.68×10^{-8}
RPN_C(2)	1.28×10^{-7}	6.38×10^{-8}	1.28×10^{-8}	6.38×10^{-9}	1.28×10^{-9}
RPN_C(3)	1.5×10^{-6}	7.49×10^{-7}	1.5×10^{-7}	7.49×10^{-8}	1.5×10^{-8}

Table 12. Risk priority number for the target components.

ER_C/hour	1×10^{-6}	0.5×10^{-6}	1×10^{-7}	0.5×10^{-7}	1×10^{-8}
RPN_FM(1)	2.65×10^{-6}	1.32×10^{-6}	2.65×10^{-7}	1.32×10^{-7}	2.65×10^{-8}
RPN_FM(2)	3.28×10^{-6}	1.64×10^{-6}	3.28×10^{-7}	1.64×10^{-7}	3.28×10^{-8}
RPN_FM(3)	9.64×10^{-7}	4.82×10^{-7}	9.64×10^{-8}	4.82×10^{-8}	9.64×10^{-9}
RPN_FM(4)	5.13×10^{-7}	2.56×10^{-7}	5.13×10^{-8}	2.56×10^{-8}	5.13×10^{-9}

Table 13. Risk priority number for the potential failure modes.

We should note that the components' error rates used in this case study are only for the demonstration of the proposed robustness/safety validation process, and the more realistic components' error rates for the considered components should be determined by process and circuit technology [Mukherjee et al., 2003]. According to the given components' error rates, the data of SFR in Table 11 can be used to assess the safety integrity level of the system. One thing should be pointed out that a SoC failure may or may not cause the dangerous effect on the system and human life. Consequently, a SoC failure could be classified into safe failure or dangerous failure. To simplify the demonstration, we make an assumption in this assessment that the SoC failures caused by the faults occurring in the components are always the dangerous failures or hazards. Therefore, the SFR in Table 11 is used to approximate the PFH, and so the SIL can be derived from Table 10.

With respect to safety design process, if the current design does not meet the SIL requirement, we need to perform the risk reduction procedure to lower the PFH, and in the meantime to reach the SIL requirement. The vulnerability analysis and risk assessment can be exploited to identify the most critical components and failure modes to be protected. In such approach, the system safety can be improved efficiently and economically.

Based on the results of $RPN\_C(i)$ as exhibited in Table 12, for $i = 1, 2, 3$, it is evident that the error of AMBA AHB is more critical than the errors of register set and memory sub-system. So, the results suggest that the AHB system bus is more urgent to be protected than the register set and memory. Moreover, the data of $RPN\_FM(k)$ in Table 13, k from one to four, infer that SDC is the most crucial failure mode in this illustrated example. Throughout the above vulnerability and risk analyses, we can identify the critical components and failure modes, which are the major targets for design enhancement. In this demonstration, the top priority of the design enhancement is to raise the robustness of the AHB HADDR bus signals to significantly reduce the rate of SDC and the scale of system risk if the system reliability/safety is not adequate.

6. Conclusion

Validating the functional safety of system-on-chip (*SoC*) in compliance with international standard, such as IEC 61508, is imperative to guarantee the dependability of the systems before they are being put to use. It is beneficial to assess the *SoC* robustness in early design phase in order to significantly reduce the cost and time of re-design. To fulfill such needs, in this study, we have presented a valuable *SoC*-level safety validation and risk reduction process to perform the hazard analysis and risk assessment, and exploited an ARM-based *SoC* platform to demonstrate its feasibility and usefulness. The main contributions of this study are first to develop a useful SVRR process and risk model to assess the scales of robustness and failure-induced risks in a system; second to raise the level of dependability validation to the untimed/timed functional TLM, and to construct a *SoC*-level system safety verification platform including an automatic fault injection and failure mode classification tool on the SystemC *CoWare Platform Architect* design environment to demonstrate the core idea of SVRR process. So the efficiency of the validation process is dramatically increased; third to conduct a thorough vulnerability analysis and risk assessment of the register set, AMBA bus and memory sub-system based on a real ARM-embedded *SoC*.

The analyses help us measure the robustness of the target components and system safety, and locate the critical components and failure modes to be guarded. Such results can be used to examine whether the safety of investigated system meets the safety requirement or not, and if not, the most critical components and failure modes are protected by some effective risk reduction approaches to enhance the safety of the investigated system. The vulnerability analysis gives a guideline for prioritized use of robust components. Therefore, the resources can be invested in the right place, and the fault-robust design can quickly achieve the safety goal with less cost, die area, performance and power impact.

7. Acknowledgment

The author acknowledges the support of the National Science Council, R.O.C., under Contract No. NSC 97-2221-E-216-018 and NSC 98-2221-E-305-010. Thanks are also due to the

National Chip Implementation Center, R.O.C., for the support of SystemC design tool – *CoWare Platform Architect.*

8. References

Austin, T. (1999). DIVA: A Reliable Substrate for Deep Submicron Microarchitecture Design, *Proceedings of 32nd Annual IEEE/ACM International Symposium on Microarchitecture*, pp. 196-207, ISBN 076950437X, Haifa, Israel, Nov. 1999

Baumann, R. (2005). Soft Errors in Advanced Computer Systems. *IEEE Design & Test of Computers*, Vol. 22, No. 3, (May-June 2005), pp. (258 – 266), ISSN 0740-7475

Bergaoui, S.; Vanhauwaert, P. & Leveugle, R. (2010) A New Critical Variable Analysis in Processor-Based Systems. *IEEE Transactions on Nuclear Science*, Vol. 57, No. 4, (August 2010), pp. (1992-1999), ISSN 0018-9499

Brown, S. (2000). Overview of IEC 61508 Design of electrical/electronic/programmable electronic safety-related systems. *Computing & Control Engineering Journal*, Vol. 11, No. 1, (February 2000), pp. (6-12), ISSN 0956-3385

International Electrotechnical Commission [IEC], (1998-2000). CEI International Standard IEC 61508, 1998-2000

Chang, K. & Chen, Y. (2007). System-Level Fault Injection in SystemC Design Platform, *Proceedings of 8th International Symposium on Advanced Intelligent Systems*, pp. 354-359, Sokcho-City, Korea, Sept. 05-08, 2007

Chen, Y.; Wang, Y. & Peng, J. (2008). SoC-Level Fault Injection Methodology in SystemC Design Platform, *Proceedings of 7th International Conference on System Simulation and Scientific Computing*, pp. 680-687, Beijing, China, Oct. 10-12, 2008

Constantinescu, C. (2002). Impact of Deep Submicron Technology on Dependability of VLSI Circuits, *Proceedings of IEEE International Conference on Dependable Systems and Networks*, pp. 205-209, ISBN 0-7695-1597-5, Bethesda, MD, USA, June 23-26, 2002

CoWare, (2006). Platform Creator User's Guide, IN: CoWare Model Library Product Version V2006.1.2

Grotker, T.; Liao, S.; martin, G. & Swan, S. (2002). *System Design with SystemC*, Kluwer Academic Publishers, ISBN 978-1-4419-5285-1, Boston, Massachusetts, USA

Hosseinabady, M.; Neishaburi, M.; Lotfi-Kamran P. & Navabi, Z. (2007). A UML Based System Level Failure Rate Assessment Technique for SoC Designs, *Proceedings of 25th IEEE VLSI Test Symposium*, pp. 243 – 248, ISBN 0-7695-2812-0, Berkeley, California, USA, May 6-10, 2007

Kanawati, G.; Kanawati, N. & Abraham, J. (1995). FERRARI: A Flexible Software-Based Fault and Error Injection System. *IEEE Transactions on Computers*, Vol. 44, No. 2, (Feb. 1995), pp. (248-260), ISSN 0018-9340

Karnik, T.; Hazucha, P. & Patel, J. (2004). Characterization of Soft Errors Caused by Single Event Upsets in CMOS Processes. *IEEE Transactions on Dependable and Secure Computing*, Vol. 1, No. 2, (April-June 2004), pp. (128-143), ISSN 1545-5971

Kim, S. & Somani, A. (2002). Soft Error Sensitivity Characterization for Microprocessor Dependability Enhancement Strategy, *Proceedings of IEEE International Conference on Dependable Systems and Networks*, pp. 416-425, ISBN 0-7695-1597-5, Bethesda, MD, USA, June 23-26, 2002

Leveugle, R.; Pierre, L.; Maistri, P. & Clavel, R. (2009). Soft Error Effect and Register Criticality Evaluations: Past, Present and Future, *Proceedings of IEEE Workshop on*

Silicon Errors in Logic - System Effects, pp. 1-6, Stanford University, California, USA, March 24-25, 2009

Mariani, R.; Boschi, G. & Colucci, F. (2007). Using an innovative SoC-level FMEA methodology to design in compliance with IEC61508, *Proceedings of 2007 Design, Automation & Test in Europe Conference & Exhibition*, pp. 492-497, ISBN 9783981080124, Nice, France, April 16-20, 2007

Mikulak, R.; McDermott, R. & Beauregard, M. (2008). *The Basics of FMEA* (Second Edition), CRC Press, ISBN 1563273772, New York, NY, USA

Mitra, S.; Seifert, N.; Zhang, M.; Shi, Q. & Kim, K. (2005). Robust System Design with Built-in Soft-Error Resilience. *IEEE Computer*, Vol. 38, No. 2, (Feb. 2005), pp. 43-52, ISSN 0018-9162

Mollah, A. (2005). Application of Failure Mode and Effect Analysis (FMEA) for Process Risk Assessment. *BioProcess International*, Vol. 3, No. 10, (November 2005), pp. (12–20)

Mukherjee, S.; Weaver, C.; Emer, J.; Reinhardt, S. & Austin, T. (2003). A Systematic Methodology to Compute the Architectural Vulnerability Factors for a High Performance Microprocessor, *Proceedings of 36th Annual IEEE/ACM International Symposium on Microarchitecture*, pp. 29-40, ISBN 0-7695-2043-X, San Diego, California, USA, Dec. 03-05, 2003

Open SystemC Initiative (OSCI), (2003). SystemC 2.0.1 Language Reference Manual (Revision 1.0), IN: *Open SystemC Initiative*, Available from: < homes.dsi.unimi.it/~pedersin/AD/SystemC_v201_LRM.pdf>

Rotenberg, E. (1999). AR-SMT: A Microarchitectural Approach to Fault Tolerance in Microprocessor, *Proceedings of 29th Annual IEEE International Symposium on Fault-Tolerant Computing*, pp. 84-91, ISBN 076950213X, Madison , WI, USA, 1999

Ruiz, J.; Yuste, P.; Gil, P. & Lemus, L. (2004). On Benchmarking the Dependability of Automotive Engine Control Applications, *Proceedings of IEEE International Conference on Dependable Systems and Networks*, pp. 857 – 866, ISBN 0-7695-2052-9, Palazzo dei Congressi, Florence, Italy, June 28 – July 01, 2004

Sieh, V. (1993). Fault-Injector using UNIX ptrace Interface, IN: *Internal Report No.: 11/93, IMMD3, Universität Erlangen-Nürnberg*, Available from: < http://www3.informatik.uni-erlangen.de/Publications/Reports/ir_11_93.pdf>

Slegel, T. et al. (1999). IBM's S/390 G5 Microprocessor Design. *IEEE Micro*, Vol. 19, No. 2, (March/April, 1999), pp. (12-23), ISSN 0272-1732

Stamatelatos, M.; Vesely, W.; Dugan, J.; Fragola, J.; Minarick III, J. & Railsback, J. (2002). Fault Tree Handbook with Aerospace Applications (version 1.1), IN: *NASA*, Available from: <www.hq.nasa.gov/office/codeq/doctree/fthb.pdf>

Tony, S.; Mohammad, H.; Mathew, J. & Pradhan, D. (2007). Soft-Error induced System-Failure Rate Analysis in an SoC, *Proceedings of 25th Norchip Conf.*, pp. 1-4, Aalborg, DK, Nov. 19-20, 2007

Wang, N.; Quek, J.; Rafacz, T. & Patel, S. (2004). Characterizing the Effects of Transient Faults on a High-Performance Processor Pipeline, *Proceedings of IEEE International Conference on Dependable Systems and Networks*, pp. 61-70, ISBN 0-7695-2052-9, Palazzo dei Congressi, Florence, Italy, June 28 – July 01, 2004

Zorian, Y.; Vardanian, V.; Aleksanyan, K. & Amirkhanyan, K. (2005). Impact of Soft Error Challenge on SoC Design, *Proceedings of 11th IEEE International On-Line Testing Symposium*, pp. 63 – 68, ISBN 0-7695-2406-0, Saint Raphael, French Riviera, France, July 06-08, 2005

5

Simulation and Synthesis Techniques for Soft Error-Resilient Microprocessors

Makoto Sugihara
Kyushu University
Japan

1. Introduction

A single event upset (SEU) is a change of state which is caused by a high-energy particle striking to a sensitive node in semiconductor devices. An SEU in an integrated circuit (IC) component often causes a false behavior of a computer system, or a soft error. A soft error rate (SER) is the rate at which a device or system encounters or is predicted to encounter soft errors during a certain time. An SER is often utilized as a metric for vulnerability of an IC component.

May first discovered that particles emitted from radioactive substances caused SEUs in DRAM modules (May & Wood, 1979). Occurrence of SEUs in SRAM memories is increasing and becoming more critical as technology continues to shrink (Karnik et al., 2001; Seifert et al., 2001a, 2001b). The feature size of integrated circuits has reached nanoscale and the nano-scale transistors have become more soft-error sensitive (Baumann, 2005). Soft error estimation and highly-reliable design have become of utmost concern in mission-critical systems as well as consumer products. Shivakumar et al. predicted that the SER of combinational logic would increase to be comparable to the SER of memory components in the future (Shivakumar et al., 2002). Embedding vulnerable IC components into a computer system deteriorates its reliability and should be carefully taken into account under several constraints such as performance, chip area, and power consumption. From the viewpoint of system design, accurate reliability estimation and design for reliability (DFR) are becoming critical in order that one applies reasonable DFR to vulnerable part of the computer system at an early design stage. Evaluating reliability of an entire computer system is essential rather than separately evaluating that of each component because of the following reasons.

1. A computer system consists of miscellaneous IC components such as a CPU, an SRAM module, a DRAM module, an ASIC, and so on. Each IC component has its own SER which may be entirely different from one another.
2. Depending on DFR techniques such as parity coding, the SER, access latency and chip area may be completely different among SRAM modules. A DFR technique should be chosen to satisfy the design requirement of the computer system so that one can avoid a superfluous cost rise, performance degradation, and power rise.
3. The behavior of a computer system is determined by hardware, software, and input to the system. Largely depending on a program, the behavior of the computer system varies from program to program. Some programs use large memory space and the

others do not. Furthermore, some programs efficiently use as many CPU cores of a multiprocessor system as possible and the others do not. The behavior of a computer system determines temporal and spatial usage of vulnerable components.

This chapter reviews a simulation technique for soft error vulnerability of a microprocessor system (Sugihara et al., 2006, 2007b) and a synthesis technique for a reliable microprocessor system (Sugihara et al., 2009b, 2010b).

2. Simulation technique for soft error vulnerability of microprocessors

2.1 Introduction

Recently, several techniques for estimating reliability were proposed. Fault injection techniques were discussed for microprocessors (Degalahal et al., 2004; Rebaudengo et al., 2003; Wang et al., 2004). Soft error simulation in logic circuits was also studied and developed (Tosaka, 1997, 1999, 2004a, 2004b). In contrast, the structure of memory modules is so regular and monotonous that it is comparatively easy to estimate their vulnerability because that can be calculated with the SERs obtained by field or accelerated tests. Mukherjee et al. proposed a vulnerability estimation method for microprocessors (Mukherjee et al., 2003). Their methodology estimates only vulnerability of a microprocessor whereas a computer system consists of various components such as CPUs, SRAM modules and DRAM modules. Their approach would be effective in case the vulnerability of a CPU is most dominant in a computer system. Asadi et al. proposed a vulnerability estimation method for computer systems that had L1 caches (Asadi et al., 2005). They pointed out that SRAM-based L1 caches were most vulnerable in most of current designs and gave a reliability model for computing critical SEUs in L1 caches. Their assumption is true in most of current designs and false in some designs. Vulnerability of DRAM modules would be dominant in entire vulnerability of a computer system if plain DRAM modules and ECC SRAM ones are utilized. As technology proceeds, a latch becomes more vulnerable than an SRAM memory cell (Baumann, 2005). It is important to obtain a vulnerability estimate of an entire system by considering which part of a computer system is vulnerable.

An SER for a memory module is a vulnerability measurement characterizing it rather than one reflecting its actual behavior. SERs of memory modules become pessimistic when they are embedded into computer systems. More specifically, every SEU occurring in memory modules is regarded as a critical error when memory modules are under field or accelerated tests. This implicitly assumes that every SEU on memory cells of a memory module makes a computer system faulty. Since memory modules are used spatially and temporally in computer systems, some of SEUs on the memory modules make the computer system faulty and the others not. Therefore, the soft errors in an entire computer system should be estimated in a different way from the way used for memory modules.

Accurate soft error estimation of an entire computer system is one of the themes of urgent concern. The SER is the rate at which a device or system encounters or is predicted to encounter soft errors. The SER is quite effective measurement for evaluating memory modules but not for computer systems. Accumulating SERs of all memories in a computer system causes pessimistic soft error estimation because memory cells are used spatially and temporally during program execution and some of SEUs make the computer system faulty. This chapter models soft errors at the architectural level for a computer system, which has several memory hierarchies with it, in order that one can accurately estimate the reliability of the computer system within reasonable computation time. We define a *critical SEU* as one which is a possible cause of faulty behavior of a computer system. We also define an *SEU vulnerability factor* for a job to run on a computer system as the expected number of critical SEUs which occur during executing the job on the computer system, unlike a classical vulnerability factor such as the SER one. The architectural-level soft-error model identifies which part of memory modules is utilized temporally and spatially and which SEUs are critical to the program execution of the computer system at the cycle-accurate ISS (instruction set simulation) level. Our architectural-level soft-error model is capable of estimating the reliability of a computer system that has several memory hierarchies with it and finding which memory module is vulnerable in the computer system. Reliability estimation helps one apply reliable design techniques to vulnerable part of their design.

2.2 SEUs on a word item

Unlike memory components, the SER of a computer system varies every moment because the computer system uses memory modules spatially and temporally. Since only active part of the memory modules affects reliability of the computer system, it is essential to identify the active part of memory modules for accurately estimating the number of soft errors occurring in the computer system. A universal soft error metric other than an SER is necessary to estimate reliability of computer systems because an SER is a reliability metric suitable for components of regular and monotonous structure like memory modules but not for computer systems. In this chapter, the number of soft errors which occur during execution of a program is adopted as a soft error metric for computer systems. In computer systems, a word item is a basic element for computation in CPUs. A word item is an instruction item in an instruction memory while that is a data item in a data memory. A collective of word items is required to be processed in order to run a program. We consider the reliability to process all word items as the reliability of a computer system. The total number of SEUs which are expected to occur on all the word items is regarded as the number of SEUs of the computer system. This section discusses an estimation model for the number of soft errors on a word item. A CPU-centric computer system typically has the hierarchical structure of memory modules which includes a register file, cache memory modules, and main memory modules. The computer system at which we target has N_{mem} levels of memory modules, $M_1, M_2, \cdots, M_{N_{mem}}$ in order of accessibility from/to the CPU. In the hierarchical memory system, instruction items are generally processed as follows.

1. Instruction items are generated by a compiler and loaded into a main memory. The birth time of an instruction item is the time when the instruction item is loaded into the main memory, from the viewpoint of program execution.
2. When the CPU requires an instruction item, it fetches the instruction item from the memory module closest to it. The instruction item is duplicated into all levels of memory modules which reside between the CPU and the source memory module.

Note that instruction items are basically read-only. Duplication of instruction items are unidirectionally made from a low level to a high level of a memory module. Data items in data memory are processed as follows.

1.   Some data items are given as initial values of a program when the program is generated with a compiler. The birth time of such a data item is the time when the program is loaded into a main memory. The other data items are generated during execution of the program by the CPU. The birth time of the data item which is made on-line is the time when the data item is made and saved to the register file.
2.   When a data item is required by a CPU, the CPU fetches it from the memory module closest to the CPU. If the write allocate policy is adopted, the data item is duplicated at all levels of memory modules which reside between the CPU and the master memory module, and otherwise it is not duplicated at the interjacent memory modules.

Note that data items are writable as well as readable. This means that data items can be copied from a high level to a low level of a memory module, and vice versa. In CPU centric computer systems, data items are utilized as constituent elements. The data items vary in lifetime and the numbers of soft errors on the data items vary from data item to data item.

Let an SER of a word item in Memory Module M_i be SER_{M_i}. When a word item w is retained during Time $time(w)$ in Memory Module M_i, the number of soft errors, $error_{M_i}(w)$, which is expected to occur on the word item, is described as follows:

$$error_{M_i}(w) = SER_{M_i} \cdot time(w). \tag{1}$$

Word item w is required to be retained during Time $retain\_time_{M_i}(w)$ in Memory Module M_i to transfer to the CPU. The number of soft errors, $error_{all\_mems}(w)$, which occur from the birth time to the time when the CPU fetches is given as

$$error_{all\_mems}(w) = \sum_i SER_{M_i} \cdot retain\_time_{M_i}(w) \tag{2}$$

where $retain\_time_{M_i}(w)$ is necessary and minimal time to transfer the word item from the master memory module to the CPU, and depends on the memory architecture. This kind of retention time is exactly obtained with cycle-accurate simulation of the computer system.

2.3 SEUs in instruction memory

Each instruction item has its own lifetime while a program runs. The lifetime of each instruction item is different from that of one another and is not necessarily equal to the execution time of a program. Generally speaking, the birth time of instruction items is the time when they are loaded into main memory, from the viewpoint of program execution. It is necessary to identify which part of retention time of an instruction item in a memory module affects reliability of the computer system. Now let us break down into the number of soft errors in an instruction item before we discuss the total number of soft errors in instruction memory. The time when a CPU fetches an instruction item of Address a for the i-th time is shown by $if(a, i)$. $if(a, 0)$ denotes the time when the instruction is loaded into the main memory. An example of several instruction fetches is shown in Fig. 1. In this figure, the boxes show that the copies of the instruction item reside in the corresponding memory modules. The labels on the boxes show when the copies of the instruction items are born. In this example, the instruction item is fetched three times by the CPU.

On the first instruction fetch for the instruction item, a copy of the instruction item exists in neither the L1 nor L2 cache memories. The instruction item resides only in the main

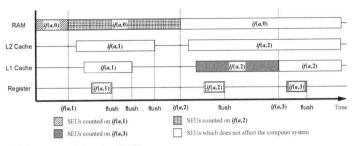

Fig. 1. SEUs which are read by the CPU.

memory. The instruction item is required to be transferred from the main memory to the CPU. On transferring the instruction item to the CPU, its copies are made in the L1 and L2 cache memory modules. In this example, we assume that some latency is necessary to transfer the instruction item between memory modules. When the instruction item in a source memory module is fetched by the CPU, any SEUs which occur after completing transferring the instruction item have no influence on the instruction fetch. In the figure, the boxes with slanting lines are the retention times whose SEUs make the instruction fetch at $if(a,1)$ faulty. The SEUs during any other retention times are unknown to make the computer system faulty.

On the second instruction fetch for the instruction item, the instruction item resides only in the main memory, same as on the first instruction fetch. The instruction item is fetched from the main memory to the CPU, same as on the first instruction fetch. The dotted boxes are found to be the retention times whose SEUs make the instruction fetch at $if(a,2)$ faulty. Note that the SEUs on the box with slanting lines in the main memory are already treated on the instruction fetch at $if(a,1)$ and are not treated on the one at $if(a,2)$ in order to avoid counting SEUs duplicately.

On the third instruction fetch for the instruction item, the highest level of memory module that retains the instruction item is the L1 cache memory. SEUs on the gray boxes are treated as the ones which make Instruction Fetch $if(a,3)$ faulty. The SEUs on any other boxes are not counted for the instruction fetch at $if(a,3)$. Now assume that a program is executed in a computer system. Given an input data to a program, let an instruction fetch sequence be $i_1, i_2, \cdots, i_{N_{inst}}$ to run the program. And let the necessary and minimal retention time for Instruction Fetch i_i to be on Memory Module M_j be $retain\_time_{M_j}(i_i)$. The number of soft errors on Instruction Fetch i_i, $error(i_i)$, is given as follows.

$$error_{single\_inst}(i_i) = \sum_j SER_{M_j} \cdot retain\_time_{M_j}(i_i). \tag{3}$$

The total number of soft errors in the computer system is shown as follows:

$$\begin{aligned} error_{all\_insts}(i) &= \sum_i error_{single\_inst}(i_i) \\ &= \sum_{i,j} SER_{M_j} \cdot retain\_time_{M_j}(i_i) \end{aligned} \tag{4}$$

where i={ i_1,i_2,...,i_N_inst}. Given the program of the computer system, $retain\_time_{M_j}(i_i)$ can be exactly obtained by performing cycle-accurate simulation for the computer system.

2.4 SEUs in data memory

Data memory is writable as well as readable. It is more complex than instruction memory because word items are bidirectionally transferred between a high level of memory and a low level of memory. Some data items are given as an input to a program and the others are born during the program execution. Some data items are used and the others are unused even if they reside in memory modules. The SEUs which occur during some retention time of a data item are influential in a computer system. The SEUs which occur during the other retention time are not influential even if the data item is used by the CPU. A data item has valid or invalid part of time with regard to soft errors of the computer system. It is quite important to identify valid or invalid part of retention time of a data item in order to accurately estimate the number of soft errors of a computer system. In this chapter, valid retention time is sought out by using the following rules.

- A data item which is generated on compilation is born when it is loaded into main memory.
- A data item as input to a computer system is born when it is inputted to the computer system.
- A data item is born when the CPU issues a store instruction for the data item.
- A data item is valid at least until the time when the CPU loads the data item and uses it in its operation.
- A data item which a user explicitly specifies as a valid one is valid even if the CPU does not issue a load instruction for the data item.

The bidirectional copies between high-level and low-level memory modules must be taken into account in data memory because data memory is writable as well as readable. There are two basic options on cache hit when writing to the cache as follows (Hennessy & Patterson, 2002).

- *Write through*: the information is written to both the block in the cache and to the block in the lower-level memory.
- *Write back*: the information is written only to the block in the cache. The modified cache block is written to main memory only when it is replaced.

The write policies affect the estimation for the number of soft errors and should be taken into account.

2.4.1 Soft error model in a write-back system

A soft-error estimation model in write-back systems is discussed in this section. Let the time when the i-th store operation of a CPU at Address a is issued be $s(a, i)$ and the time when the j-th load operation at Address a is issued be $l(a, j)$. Fig. 2 shows an example of the behavior of a write-back system. Each box in the figure shows the existence of the data item in the corresponding memory module. The labels on the boxes show when the data items are born. In the example, two store operations and two load operations are executed. First, a store operation is executed and only the L1 cache is updated with the data item. The L2 cache or main memory is not updated with the store operation. A load operation on the data item which resides at Address a follows. The data item resides in the L1 cache memory and is transferred from the L1 cache to the CPU. The SEUs on the boxes with slanting lines are influential in reliability of the computer system by the issue of a load at $l(a,1)$. The other boxes with Label $s(a,1)$ are unknown to be influential in the reliability. Next, the data item in the L1 cache goes out to the L2 cache by the other data item. The L2 cache memory becomes the highest level of memory which retains the data item. Next, a load operation at $l(a,2)$ is issued and the data item is transferred from the L2 cache memory to the CPU. With the load operation at $l(a,2)$, the SEUs on the dotted boxes are found to be influential in reliability of the computer system. SEUs on the white boxes labeled as $s(a,2)$ are not counted on the load at $l(a,2)$.

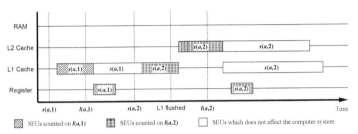

Fig. 2. Critical time in the write-back system.

2.4.2 Soft error model in a write-through system

A soft-error estimation model in write-through systems is discussed in this section. An example of the behavior of a write-through system is shown in Fig. 3. First, a store operation at Address a is issued. The write-through policy makes multiple copies of the data item in the cache memories and the main memory. Next, a load operation follows. The CPU fetches the data item from the L1 cache and SEUs on the boxes with slanting lines are found to be influential in reliability of the computer system. Next, a store operation at $s(a,2)$ comes. The previous data item at Address a is overridden and the white boxes labeled as $s(a,1)$ are no longer influential in reliability of the computer system. Next, the data item in the L1 cache is replaced with the other data item. The L2 cache becomes the highest level of memory which has the data item of Address a. Next, a load operation at $l(a,2)$ follows and the data item is transferred from the L2 cache to the CPU. With the load operation at $l(a,2)$, SEUs on the dotted boxes are found to be influential in reliability of the computer system.

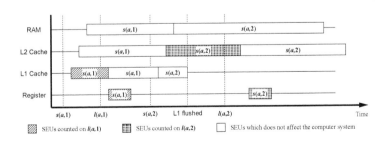

Fig. 3. Critical time in the write-through system.

2.5 Simulation-based soft error estimation

As discussed in the previous sections, the retention time of every word item in memory modules needs to be obtained so that the number of soft errors in a computer system can be estimated. We adopted a cycle-accurate ISS which can obtain the retention time of every word item. A simplified algorithm to estimate the number of soft errors for a computer system to finish a program is shown in Fig. 4. The input to the algorithm is an instruction sequence, and the output from the algorithm is the accurate number of soft errors, $error_{system}$, which occur during program execution.

First, several variables are initialized. Variable $error_{system}$ is initialized with 0. The birth times of all data items are initialized with the time when the program starts. A for-loop sentence follows. A cycle-accurate ISS is executed in the for-loop. An iteration loop corresponds to an execution of an instruction. The number of soft errors is counted for every instruction item and is accumulated to variable $error_{system}$. When variable $error_{system}$ is updated, the birth time of the corresponding word item is also updated with the present time. Some computation is additionally done when the present instruction is a store or a load operation. If the instruction is a load operation, the number of SEUs on the data item which is found to be critical in the reliability of the computer system is added to variable $error_{system}$. A load operation updates the birth time of the data item with the present time. If the instruction is a store operation, the birth time of all changed word items is updated with the present time. After the above procedure is applied to all instructions, $error_{system}$ is outputted as the number of soft errors which occur during the program execution.

Procedure EstimateSoftError

Input: Instruction sequence given by a trace.

Output: the number of soft errors for the system, $error_{system}$

begin

　　$error_{system}$ is initialized with 0.

　　Birth time of every word iterm is initialized with the beginning time.

　　for all instructions **do**

　　　　// Computation for soft errors in instruction memory

　　　　Add the number of critical soft errors of the instruction item to $error_{system}$.

　　　　Update the birth time on the instruction item with the present time.

　　　　// Computation for soft errors in data memory

　　　　if the current instruction is a load **then**

Fig. 4. A soft error estimation algorithm.

2.6 Experiments

Using several programs, we examined the number of soft errors during executing each of them.

2.6.1 Experimental setup

We targeted a microprocessor-based system consisting of an ARM processor (ARMv4T, 200MHz), an instruction cache module, and a data cache module, and a main memory module as shown in Fig. 5. The cache line size and the number of cache-sets are 32-byte and 32, respectively. We adopted the least recently used (LRU) policy as the cache replacement policy. We evaluated reliability of computer systems with the two write policies, write-through and write-back ones. The cell-upset rates of both SRAM and DRAM modules are shown in Table 1. We used the cell-upset rates shown in (Slayman, 2005) as the cell-upset rates of plain SRAMs and DRAMs. According to Baumann, error detection and correction (EDAC) or error correction codes (ECC) protection will provide a significant reduction in failure rates (typically 10k or more times reduction in effective error rates) (Baumann, 2005). We assumed that introducing an ECC circuit makes reliability of memory modules 10k times higher.

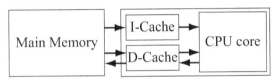

Fig. 5. The target system.

	Cell Upset Rate			
	[FIT/bit]		[errors/word/cycle]	
	w/o ECC	w. ECC	w/o ECC	w. ECC
SRAM	1.0×10^{-4}	1.0×10^{-8}	4.4×10^{-24}	4.4×10^{-28}
DRAM	1.0×10^{-8}	1.0×10^{-12}	4.4×10^{-24}	4.4×10^{-32}

Table 1. Cell upset rates for experiments.

We used three benchmark programs: Compress version 4.0 (Compress), JPEG encoder version 6b (JPEG), and MPEG2 encoder version 1.2 (MPEG2). We used the GNU C compiler and debugger to generate address traces. We chose to execute 100 million instructions in each benchmark program. This allowed the simulations to finish in a reasonable amount of time. All programs were compiled with "-O3" option. Table 2 shows the code size, activated code size, and activated data size in words for each benchmark program. The activated code and data sizes represent the number of instruction and data addresses which were accessed during the execution of 100 million instructions, respectively.

	Code size S_{code} [words]	Activated code size AS_{code} [words]	Activated data size AS_{data} [words]
Compress	10,716	1,874	140,198
JPEG	30,867	6,129	33,105
MPEG2	33,850	7,853	258,072

Table 2. Specification for benchmark programs.

2.6.2 Experimental results

Figures 6, 7, and 8 show the results of our soft error estimation method. Four different memory configurations were considered as follows:

1. non-ECC L1 cache memory and non-ECC main memory,
2. non-ECC L1 cache memory and ECC main memory,
3. ECC L1 cache memory and non-ECC main memory,
4. and ECC L1 cache memory and ECC main memory.

Note that Asadi's vulnerability estimation methodology (Asadi et al., 2005) does not cover vulnerability estimation for the second configuration above because their approach is dedicated to estimating vulnerability of L1 caches. The vertical axis presents the number of soft errors occurring during the execution of 100 million instructions. The horizontal axis presents the number of cache ways in a data cache. The other cache parameters, i.e., the line size and the number of lines in a cache way, are unchanged. The size of the data cache is, therefore, linear to the number of cache ways in this experiment. The cache sizes corresponding to the values shown on the horizontal axis are 1 KB, 2 KB, 4 KB, 8 KB, 16 KB, 32 KB, and 64 KB, respectively.

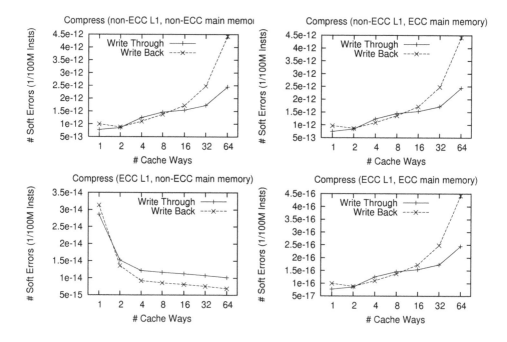

Fig. 6. Experimental results for Compress.

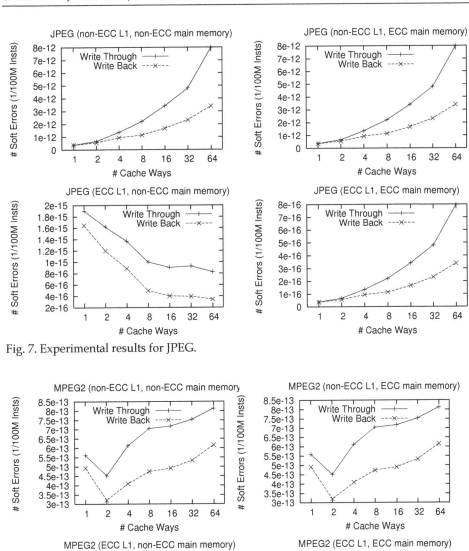

Fig. 7. Experimental results for JPEG.

Fig. 8. Experimental results for MPEG2.

According to the experimental results shown in Figures 6, 7, and 8, the number of soft errors which occurred during a program execution depends on the reliability design of the memory hierarchy. When the cell-upset rate of SRAMs was higher than that of DRAMs, the soft errors on cache memories became dominant in the whole soft errors of the computer systems. The number of soft errors in a computer system, therefore, increased as the size of cache memories increased. When the cell-upset rate of SRAM modules was equal to that of DRAM ones, the soft errors on main memories became dominant in the system soft errors in contrast. The number of soft errors in a computer system, therefore, decreased as the size of cache memories increased because the larger size of cache memories reduced runtime of a program as well as usage of the main memory. Table 3 shows the number of CPU cycles to finish executing the 100 million instructions of each program.

		The number of cache ways in a cache memory (1 way = 1 KB)						
		1	2	4	8	16	32	64
Compress	WT	968	523	422	405	390	371	348
	WB	1,058	471	325	303	286	267	243
JPEG	WT	548	455	364	260	247	245	244
	WB	474	336	237	129	110	104	101
MPEG2	WT	497	179	168	168	167	167	167
	WB	446	124	110	110	110	110	110

Table 3. The number of CPU cycles for 100 million instructions.

Table 4 shows the results of more naive approaches and our approach. The two naive approaches, M1 and M2, calculated the number of soft errors using the following equations.

$$SE_1 = \{S_{cache} \cdot SER_S + (S_{code} + AS_{data}) \cdot SER_D\} \cdot N_{cycle} \tag{5}$$
$$SE_2 = \{S_{cache} \cdot SER_S + (AS_{code} + AS_{data}) \cdot SER_D\} \cdot N_{cycle} \tag{6}$$

where S_{cache}, S_{code}, AS_{code}, AS_{data}, N_{cycle}, SER_S, SER_D denote the cache size, the code size, the activated code size, the activated data size, the number of CPU cycles, the SER per word per cycle for SRAM, and the SER per word per cycle for DRAM, respectively. M1 and M2 appearing in Table 4 correspond to the calculations using Equations (5) and (6), respectively. Our method corresponds to M3. It is obvious that the simple summation of SERs resulted in large overestimation of soft errors. This indicates that accumulating SERs of all memory modules in a system resulted in pessimistic estimation. The universal soft error metric other than the SER is necessary to estimate reliability of computer systems which behave dynamically. The number of soft errors which occur during execution of a program would be the universal soft error metric of computer systems.

			The number of cache ways						
			1	2	4	8	16	32	64
Compress	WT	M1	2267	2417	3869	7394	14216	27068	50755
		M2	2263	2415	3867	7393	14214	27067	50754
		M3	776	852	1248	1458	1541	1724	2446
	WB	M1	2478	2175	2976	5530	10423	19461	35410
		M2	2474	2173	2975	5529	10439	19460	35410
		M3	999	881	1101	1372	1722	2484	4426
JPEG	WT	M1	1262	2083	3324	4735	9013	17867	35556
		M2	1255	2078	3320	4732	9010	17864	35553
		M3	384	670	1355	2209	3417	4801	7977
	WB	M1	1092	1540	2160	2355	4024	7593	14759
		M2	1087	1536	2157	2354	4023	7592	14758
		M3	369	558	941	1147	1664	2323	3407
MPEG2	WT	M1	1197	838	1550	3167	6310	12217	24411
		M2	1191	836	1548	3069	6118	12215	24410
		M3	561	453	613	705	718	754	813
	WB	M1	1073	578	1019	2016	4016	8017	16016
		M2	1067	577	1018	2015	4015	8016	16015
		M3	494	321	410	474	492	534	616

Table 4. The number of soft errors which occur during execution [10^{-17} errors/instruction].

2.7 Conclusion

This section discussed the simulation-based soft error estimation technique which sought the accurate number of soft errors for a computer system to finish running a program. Depending on application programs which are executed on a computer system, its reliability changes. The important point to emphasize is that seeking for the number of soft errors to run a program is essential for accurate soft-error estimation of computer systems. We estimated the accurate number of soft errors of the computer systems which were based on ARM V4T architecture. The experimental results clearly showed the following facts.

- It was found that there was a great difference between the number of soft errors derived with our technique and that derived from the simple summations of the static SERs of memory modules. The dynamic behavior of computer systems must be taken into account for accurate reliability estimation.
- The SER of a computer system virtually increases with a larger cache memory adopted because the SER is calculated by summing up the SERs of memory modules utilized in the system. It was, however, found that the number of soft errors to finish a program was reduced with larger cache memories in the computer system that had an ECC L1 cache and a non-ECC main memory. This is because the soft errors in cache memories were negligible and the retention time of data items in the main memory was reduced by the performance improvement.

3. Reliable microprocessor synthesis for embedded systems

DFR is one of the themes of urgent concern. Coding and parity techniques are popular design techniques for detecting or correcting SEUs in memory modules. Exploiting triple modular redundancy (TMR) is also a popular design technique which decides a correct value by voting on a correct value among three identical modules. These techniques have been well studied and developed. Elakkumanan et al. proposed a DFR technique for logic circuits, which exploits time redundancy by using scan flip-flops (Elakkumanan, 2006). Their approach updates a pair of flip-flops at different moments for an output signal to duplicate for higher reliability. Their approach is effective in ICs which have scan paths. We reported that there exists a trade-off between performance and reliability in a computer system and proposed a DFR technique by adjusting the size of vulnerable cache memory online (Sugihara et al., 2007a, 2008b). The work presented a reliable cache architecture which offered performance and reliability modes. More cache memory is used in the performance mode while less cache memory is used in the reliability mode to avoid SEUs. All tasks are statically scheduled under real-time and reliability constraints. The demerit of the approach is that switching operation modes causes performance and area overheads and might be unacceptable to high-performance or general-purpose microprocessors. We also proposed a task scheduling scheme which minimized SEU vulnerability of a heterogeneous multiprocessor under real-time constraints (Sugihara, 2008a, 2009a). Architectural heterogeneity among CPU cores offers a variety of reliability for a task. We presented a task scheduling problem which minimized SEU vulnerability of an entire system under a real-time constraint. The demerit of the approach is that the fixed heterogeneous architecture loses general-purpose programmability. We also presented a dynamic continuous signature monitoring technique which detects a soft error on a control signal (Sugihara, 2010a, 2011).

This section reviews a system synthesis approach for a heterogeneous multiprocessor system under performance and reliability constraints (Sugihara, 2009b, 2010b). To our best knowledge, this is the first study to synthesize a heterogeneous multiprocessor system with a soft error issue taken into account. In this section we use the SEU vulnerability factor as a vulnerability factor. The other vulnerability factors, however, are applicable to our system synthesis methodology as far as they are capable to estimating task-wise vulnerability on a processor. If a single event transient (SET) is a dominant factor to fail a system, a vulnerability factor which can treat SETs should be used in our heterogeneous multiprocessor synthesis methodology. Our methodology assumes that a set of tasks are given and that several variants of processors are given as building blocks. It also assumes that real-time and vulnerability constraints are given by system designers. Simulation with every combination of a processor model and a task characterizes performance and reliability. Our system synthesis methodology uses the values of the chip area of every building block, the characterized runtime and vulnerability, and the given real-time and vulnerability constraints in order to synthesize a heterogeneous multiprocessor system whose chip area is minimal under the constraints.

3.1 Performance and reliability in various processor configurations

A processor configuration, which specifies instruction set architecture, the number of pipeline stages, the size of cache memory, cache architecture, coding redundancy, structural redundancy, temporal redundancy, and so on, is a major factor to determine chip area, performance and reliability of a computer system. One must carefully select a processor configuration for each processor core of their products so that they can make the price of their products competitive. From the viewpoint of reliability, processor configurations are mainly characterized by the following design parameters.

- Coding techniques, i.e. parity and Hamming codes.
- Modular redundancy techniques i.e. double modular redundancy (DMR) and triple modular redundancy (TMR).
- Temporal redundancy techniques, i.e. multiple executions of a task and multi-timing sampling of outputs of a combinational circuit.
- The size of cache memory. We reported that SRAM is a vulnerable component and the size of cache memory would be one of the factors which characterize processor reliability (Sugihara et al., 2006, 2007b).

Design parameters are required to offer various alternatives which cover a wide range of chip area, performance, and reliability for building a reliable and small multiprocessor. This chapter mainly focuses on the size of cache memory as an example of variable design parameters in explanation of our design methodology. The other design parameters as mentioned above, however, are applicable to our heterogeneous multiprocessor synthesis paradigm.

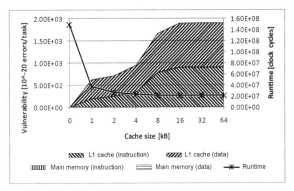

Fig. 9. Cache size vs SEU vulnerability and performance for susan (input_small, smooth).

Fig. 9 is an example that the cache size, which is one of design parameters, changes runtime and reliability of a computer system. We assumed that the cache line size is 32 bytes and that the number of cache-sets is 32. Changing the number of cache ways from 0 to 64 ranges from 0 to 64 KB of cache memory. For plotting the graph, we utilized an ARM CPU core (ARMv4T instruction set, 200 MHz) and a benchmark program susan, which is a program from the MiBench benchmark suite (Guthaus et al., 2001), with an input file input small and an option "-s". We utilized the vulnerability estimation approach we had formerly proposed (Sugihara, 2006, 2007b). For the processor configuration, we assumed that SRAM and DRAM modules have their own SEC-DED (single error correction and double error detection) circuits. We regarded SETs in logic circuitry as negligible ones because of its infrequency. Note that vulnerability of SRAM in the L1 cache is dominant in the entire vulnerability of the system and that of DRAM in main memory is too small to see in the figure. The figure shows that, as the cache size increases, runtime decreases and SEU vulnerability increases. The figure shows that the SEU vulnerability converged at 16 KB of a cache memory. This is because using more cache ways than 16 ones did not contribute to reducing conflict misses and did not increase temporal and spatial usage of the cache memory, which determined the SEU vulnerability factor. The cache size at which SEU vulnerability converges depends on a program, input to the program, and cache parameters such as the size of a cache line, the number of cache sets, the number of cache ways, and its replacement policy. The figure shows that most of SEU vulnerability of a system is caused by SRAM circuitry. It clearly shows that there is a trade-off between performance and reliability. A design paradigm in which chip area, performance and reliability can be taken into account is of critical importance in the multi-CPU core era.

3.2 Heterogeneous multiprocessor synthesis

It is quite important to consider the trade-off among chip area, performance, and reliability of a system which one develops. As we discussed in the previous section, chip area, performance and reliability vary among processor configurations. This section discusses a heterogeneous multiprocessor synthesis methodology in which an optimal set of processor configurations are sought under real-time and reliability constraints so that the chip area of a multiprocessor system is minimized.

3.2.1 Overview of heterogeneous multiprocessor synthesis

We show an overview of a heterogeneous multiprocessor synthesis methodology, that is a design paradigm in which a heterogeneous multiprocessor is synthesized and its chip area is minimized under real-time and SEU vulnerability constraints. Figure 10 shows the design flow based on our design paradigm. In the design flow, designers begin with specifying their system. Once they fix their specification, they begin to develop their hardware and software. They may use IP (intellectual property) of processor cores which they designed or purchased before. They may also develop a new processor core if they do not have one appropriate to their system. Various processor configurations are to be prepared by changing design parameters such as their cache size, structural redundancy, temporal redundancy, coding redundancy, and anything else which strongly affects vulnerability, performance, and chip area. Increasing design parameters expands the number of processor configurations, enlarges design space to explore, and causes a long synthesis time. Design parameters should be chosen to offer design alternatives among chip area, performance, and reliability. Even if any design parameter can be treated in a general optimization procedure, design parameters should be carefully chosen in order to avoid large design space exploration. A design parameter which offers slight difference regarding chip area, performance, and reliability would result in a long synthesis time and should be possibly excluded from our multiprocessor synthesis. Software is mainly developed at a granularity level of tasks. ISS is performed with the object codes for obtaining accurate runtime and SEU vulnerability on every processor configuration. SEU vulnerability can be easily obtained with the vulnerability estimation techniques previously mentioned. We used the reliability estimation technique (Sugihara et al., 2006, 2007b) throughout this chapter but any other technique can be used as far as it is capable of estimating task-wise reliability on a processor configuration. When SETs become dominant in reliability of a computer system, one should use a reliability estimation technique which treats SETs. Our heterogeneous multiprocessor synthesis paradigm is basically independent of a reliability estimation technique as far as it characterizes task-wise runtime and vulnerability. One should specify reliability and performance constraints from which one obtains the upper bound of the SEU vulnerability factor for every task, the upper bound of the SEU vulnerability for total tasks, and arrival and deadline times of all tasks. From the specification and the hardware and software components which one has given, a mixed integer linear programming (MILP) model to synthesize a heterogeneous multiprocessor system is automatically generated. By solving the MILP model with the generic solving procedure, an optimal configuration of the heterogeneous multiprocessor is sought. This chapter mainly focuses on defining the heterogeneous multiprocessor synthesis problem and building an MILP model to synthesize a heterogeneous multiprocessor system. Subsection 3.2.2 formally defines the heterogeneous multiprocessor synthesis problem and Subsection 3.2.3 gives an MILP model for the problem.

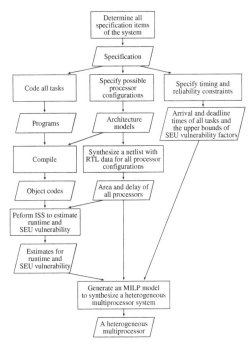

Fig. 10. Our design paradigm.

3.2.2 Problem definition

We now address a mathematical problem in which we synthesize a heterogeneous multiprocessor system and minimize its chip area under real-time and SEU vulnerability constraints. We synthesize a heterogeneous multiprocessor on which N_{task} tasks are executed. N_{CPU} processor configurations are given as building blocks for the heterogeneous multiprocessor system. The chip area of Processor Configuration k, $1 \leq k \leq N_{CPU}$, is given with A_k. We assume that all the tasks are non-preemptive on the heterogeneous multiprocessor system. Preemption causes large deviations between the worst-case execution times (WCET) of tasks that can be statically guaranteed and average-case behavior. Non-preemptivity gives a better predictability on runtime since the worst-case is closer to the average case behavior. Task i, $1 \leq i \leq N_{\text{task}}$, becomes available to start at its arrival time T_{arrival_i} and must finish by its deadline time T_{deadline_i}. Task i runs for Duration $D_{\text{runtime}_{i,k}}$ on Processor Configuration k. The SEU vulnerability factor for Task i to run on Processor Configuration k, $V_{i,k}$, is the number of critical SEUs which occur during the task execution. We assume that one specifies the upper bound of the SEU vulnerability factor of Task i, V_{const_i}, and the upper bound of the SEU vulnerability factor of the total tasks, $V_{\text{const}_{\text{all}}}$.

The heterogeneous multiprocessor synthesis problem that we address in this subsection is to minimize the chip area of a heterogeneous multiprocessor system by optimally determining a set of processor cores constituting a heterogeneous multiprocessor system, the start times $s_1, s_2, \cdots, s_{N_{\text{task}}}$ for all tasks, and assignments of a task to a processor core. The heterogeneous multiprocessor synthesis problem P_{HMS} is formally stated as follows.

- P_{HMS}: For given N_{task} tasks, N_{CPU} processor configurations, the chip area A_k of Processor Configuration k, arrival and deadline times of Task i, T_{arrival_i} and T_{deadline_i}, duration $D_{\text{runtime}_{i,k}}$ for which Task i runs on Processor Configuration k, the SEU vulnerability factor $V_{i,k}$ for Task i to run on Processor Configuration k, the upper bound of the SEU vulnerability factor for Task i, V_{const_i}, and the upper bound of the SEU vulnerability factor for total tasks, $V_{\text{const}_{\text{all}}}$, determine an optimal set of processor cores, assign every task to an optimal processor core, and determine the optimal start time of every task such that (1) every task is executed on a single processor core, (2) every task starts at or after its arrival time and completes by its deadline, (3) the SEU vulnerability of every task is less than or equal to that given by system designers, (4) the total SEU vulnerability of the system is less than or equal to that given by system designers and (5) the chip area is minimized.

3.2.3 Problem definition

We now build an MILP model for Problem P_{HMS}. From the assumption of non-preemptivity, the upper bound of the number of processors of the multiprocessor system is given by the number of tasks, N_{task}. Let $x_{i,j}$, $1 \leq i \leq N_{\text{task}}$, $1 \leq j \leq N_{\text{task}}$ be a binary variable defined as follows:

$$x_{i,j} = \begin{cases} 1 & \text{if Task } i \text{ is assigned to Processor } j, \\ 0 & \text{otherwise.} \end{cases} \tag{7}$$

Let $y_{j,k}$, $1 \leq j \leq N_{\text{task}}$, $1 \leq k \leq N_{\text{CPU}}$ be a binary variable defined as follows:

$$y_{j,k} = \begin{cases} 1 & \text{if one takes Processor Configuration } k \text{ as the one of Processor } j, \\ 0 & \text{otherwise.} \end{cases} \tag{8}$$

The chip area of the heterogeneous multiprocessor is the sum of the total chip areas of all processor cores used in the system. The total chip area A_{chip}, which is the objective function, is, therefore, stated as follows:

$$A_{\text{chip}} = \sum_{j,k} A_k y_{j,k}. \tag{9}$$

The assumption of non-preemptivity causes a task to run on only a single processor. The following constraint is, therefore, introduced.

$$\sum_j x_{i,j} = 1, 1 \leq \forall i \leq N_{\text{task}}.\tag{10}$$

If a task is assigned to a single processor, the processor must have its entity. The following constraint, therefore, is introduced.

$$x_{i,j} = 1 \rightarrow \sum_k y_{j,k} = 1, 1 \leq \forall i \leq N_{\text{task}}, 1 \leq \forall j \leq N_{\text{task}}.\tag{11}$$

The reliability requirement varies among tasks, depending on the disprofit of a failure event of a task. We assume that one specifies the upper bound of the SEU vulnerability factor for each task. The SEU vulnerability factor of Task i must be less than or equal to V_{const_i}. The SEU vulnerability factor of a task is determined by assignment of the task to a processor. The following constraint, therefore, is introduced.

$$\sum_{j,k} V_{i,k} x_{i,j} y_{j,k} \leq V_{\text{const}_i}, 1 \leq \forall i \leq N_{\text{task}}.\tag{12}$$

The SEU vulnerability factor of the heterogeneous multiprocessor system is the sum of the SEU vulnerability factors of all tasks. The SEU vulnerability of the computer system V_{chip}, therefore, is stated as follows.

$$V_{\text{chip}} = \sum_{i,j,k} V_{i,k} x_{i,j} y_{j,k}.\tag{13}$$

We assume that one specifies an SEU vulnerability constraint, which is the upper bound of the SEU vulnerability of the system, and so the following constraint is introduced.

$$V_{\text{chip}} \leq V_{\text{const}_{\text{all}}}.\tag{14}$$

Task i starts between its arrival time T_{arrival_i} and its deadline time T_{deadline_i}. A variable for start time s_i is, therefore, bounded as follows.

$$T_{\text{arrival}_i} \leq s_i \leq T_{\text{deadline}_i}, 1 \leq \forall i \leq N_{\text{task}}\tag{15}$$

Task i must finish by its deadline time T_{deadline_i}. A constraint on the deadline time of the task is introduced as follows.

$$s_i + \sum_{j,k} D_{\text{runtime}_{i,k}} x_{i,j} y_{j,k} \leq T_{\text{deadline}_i}, 1 \leq \forall i \leq N_{\text{task}}\tag{16}$$

Now assume that two tasks $i1$ and $i2$ are assigned to Processor j and that its processor configuration is Processor Configuration k. Formal expressions for these assumptions are shown as follows:

$$x_{i1,j} = x_{i2,j} = y_{j,k} = 1.\tag{17}$$

Two tasks are simultaneously inexecutable on the single processor. The two tasks must be sequentially executed on the single processor. Two tasks i1 and i2 are inexecutable on the single processor if $s_{i1} < s_{i2} + D_{\text{runtime}_{i2,k}}$ and $s_{i1} + D_{\text{runtime}_{i1,k}} > s_{i2}$. The two tasks, inversely, are executable on the processor under the following constraints.

$$x_{i1,j} = x_{i2,j} = y_{j,k} = 1 \rightarrow \left\{\left(s_{i1} + D_{\text{runtime}_{i1,k}} \leq s_{i2}\right) \vee \left(s_{i2} + D_{\text{runtime}_{i2,k}} \leq s_{i1}\right)\right\},$$

$$1 \leq \forall i1 < \forall i2 \leq N_{\text{task}}, 1 \leq \forall j \leq N_{\text{task}}, \text{ and } 1 \leq \forall k \leq N_{\text{CPU}}. \tag{18}$$

The heterogeneous multiprocessor synthesis problem is now stated as follows.

Minimize the cost function $A_{\text{chip}} = \sum_{j,k} A_k \, y_{j,k}$

subject to

1. $\sum_j x_{i,j} = 1, 1 \leq \forall i \leq N_{\text{task}}.$

2. $x_{i,j} = 1 \rightarrow \sum_k y_{j,k} = 1, 1 \leq \forall i \leq N_{\text{task}}, 1 \leq \forall j \leq N_{\text{task}}.$

3. $\sum_{j,k} V_{i,k} \, x_{i,j} y_{j,k} \leq V_{\text{const}_i}, 1 \leq \forall i \leq N_{\text{task}}.$

4. $\sum_{j,k} V_{i,k} \, x_{i,j} y_{j,k} \leq V_{\text{const}_{\text{all}}}.$

5. $s_i + \sum_{j,k} D_{\text{runtime}_{i,k}} x_{i,j} y_{j,k} \leq T_{\text{deadline}_i}, 1 \leq \forall i \leq N_{\text{task}}.$

6. $x_{i1,j} = x_{i2,j} = y_{j,k} = 1 \rightarrow \{(s_{i1} + D_{\text{runtime}_{i1,k}} \leq s_{i2}) \vee (s_{i2} + D_{\text{runtime}_{i2,k}} \leq s_{i1})\}, \quad 1 \leq \forall i1 <$
$\forall i2 \leq N_{\text{task}}, 1 \leq \forall j \leq N_{\text{task}}, \text{ and } 1 \leq \forall k \leq N_{\text{CPU}}.$

Variables

- $x_{i,j}$ is a binary variable, $1 \leq \forall i \leq N_{\text{task}}, 1 \leq \forall j \leq N_{\text{task}}.$

- $y_{j,k}$ is a binary variable, $1 \leq \forall j \leq N_{\text{task}}, 1 \leq \forall k \leq N_{\text{CPU}}.$

- s_i is a real variable, $1 \leq \forall i \leq N_{\text{task}}.$

Bounds

- $T_{\text{arrival}_i} \leq s_i \leq T_{\text{deadline}_i}, \ 1 \leq \forall i \leq N_{\text{task}}.$

The above nonlinear mathematical model can be transformed into a linear one using standard techniques (Williams, 1999) and can be solved with an LP solver. Seeking optimal values for the above variables determines hardware and software for the heterogeneous system. Variables $x_{i,j}$ and s_i determine the optimal software and Variable $y_{j,k}$ determines the optimal hardware. The other variables are the intermediate ones in the problem. As we showed in Subsection 3.2.2, the values $N_{\text{task}}, N_{\text{CPU}}, A_k, T_{\text{arrival}_i}, D_{\text{runtime}_{i,k}}, V_{i,k}, V_{\text{const}_i}$, and $V_{\text{const}_{\text{all}}}$ are given. Once these values are given, the above MILP model can be generated automatically. Solving the generated MILP model optimally determines a set of processors, assignment of every task to a processor core, and start time of every task. The set of processors constitutes a heterogeneous multiprocessor system which satisfies the minimal chip area under real-time and SEU vulnerability constraints.

3.3 Experiments and results

3.3.1 Experimental setup

We experimentally synthesized heterogeneous multiprocessor systems under real-time and SEU vulnerability constraints. We prepared several processor configurations in which the system consists of multiple ARM CPU cores (ARMv4T, 200 MHz). Table 5 shows all the processor configurations we hypothetically made. They are different from one another regarding their cache sizes. For the processor configurations, we adopted write-through policy (Hennessy & Patterson, 2002) as write policy on hit for the cache memory. We also adopted the LRU policy (Hennessy & Patterson, 2002) for cache line replacement. For experiment, we assumed that each of ARM cores has its own memory space and does not interfere the execution of the others. The cache line size and the number of cache-sets are 32 bytes and 32, respectively. We did not adopt error check and correct (ECC) circuitry for all memory modules. Note that the processor configurations given in Table 5 are just examples and the other design parameters such as coding redundancy, structural redundancy, temporal redundancy, and anything else which one wants, are available. The units for runtime and vulnerability in the table are M cycles/execution and 10^{-18} errors/execution respectively.

	L1 cache size [KB]	Hypothetical chip area [a.u.]
Conf. 1	0	64
Conf. 2	1	80
Conf. 3	2	96
Conf. 4	4	128
Conf. 5	8	192
Conf. 6	16	320

Table 5. Hypothetical processor configurations for experiment.

We used 11 benchmark programs from MiBench, the embedded benchmark suite (Guthaus et al., 2001). We assumed that there were 25 tasks with the 11 benchmark programs. Table 6 shows the runtime, the SEU vulnerability, and the SER of a task on every processor configuration.

As the size of input to a program affects its execution time, we regarded execution instances of a program, which are executed for distinct input sizes, as distinct jobs. We also assumed that there was no inter-task dependency. The table shows runtime and SEU vulnerability for every task to run on all processor configurations. These kinds of vulnerabilities can be obtained by using the estimation techniques formerly mentioned. In our experiments, we assumed that the SER of SRAM modules is 1.0×10^{-4} [FIT/bit], for which we referred to Slayman's paper (Slayman, 2005), and utilized the SEU vulnerability estimation technique which mainly estimated the SEU vulnerability of the memory hierarchy of systems (Sugihara et al., 2006, 2007b). Note that our synthesis methodology does not restrict designers to a certain estimation technique. Our synthesis technique is effective as far as the trade-off between performance and reliability exists among several processor configurations.

We utilized an ILOG CPLEX 11.2 optimization engine (ILOG, 2008) for solving MILP problem instances shown in Section 3.2 so that optimal heterogeneous multiprocessor systems whose chip area was minimal were synthesized. We solved all heterogeneous multiprocessor synthesis problem instances on a PC which has two Intel Xeon X5365 processors with 2 GB memory. We gave 18000 seconds to each problem instance for computation. We took a temporal schedule for unfinished optimization processes.

	Task 1	Task 2	Task 3	Task 4	Task 5	Task 6	Task 7
Program name	bscmth	bitcnts	bf	bf	bf	crc	dijkstra
Input	bscmth_sml	bitcnts_sml	bf_sml1	bf_sml2	bf_sml3	crc_sml	dijkstra_sml
Runtime on Conf. 1	1980.42	239.91	328.69	1.37	2.46	188.22	442.41
Runtime on Conf. 2	1011.63	53.32	185.52	1.05	1.66	43.72	187.67
Runtime on Conf. 3	834.11	53.25	93.68	0.32	0.63	42.97	134.31
Runtime on Conf. 4	684.62	53.15	75.03	0.26	0.51	42.97	93.31
Runtime on Conf. 5	448.90	53.15	74.86	0.26	0.51	42.97	86.51
Runtime on Conf. 6	205.25	53.15	74.86	0.26	0.51	42.97	83.05
Vulnerability on Conf. 1	4171.4	315.1	376.1	1.7	3.1	171.2	2370.3
Vulnerability on Conf. 2	965179.8	41038.1	334963.9	1708.0	2705.0	132178.3	277271.4
Vulnerability on Conf. 3	1459772.8	94799.9	546614.4	1540.6	3154.7	152849.7	385777.1
Vulnerability on Conf. 4	2388614.3	222481.6	709463.0	1301.9	3210.0	186194.8	591639.0
Vulnerability on Conf. 5	5602028.0	424776.5	740064.1	1354.9	3367.6	191300.9	846289.5
Vulnerability on Conf. 6	6530436.1	426503.9	740064.1	1354.9	3367.6	193001.8	1724177.3

Task 8	Task 9	Task 10	Task 11	Task 12	Task 13	Task 14	Task 15	Task 16
dijkstra	fft	fft	jpeg	jpeg	jpeg	jpeg	qsort	sha
dijkstra_lrg	fft_sml1	fft_sml2	jpeg_sml1	jpeg_sml2	jpeg_lrg1	jpeg_lrg2	qsort_sml	sha_sml
2057.38	850.96	1923.92	238.82	66.30	896.22	229.97	153.59	95.28
832.04	412.71	935.99	86.04	32.56	319.03	111.72	75.57	20.04
626.39	286.91	641.06	58.85	18.51	270.63	59.29	46.12	17.23
434.72	224.98	479.29	52.79	14.62	198.36	51.36	45.00	17.06
400.41	183.04	417.04	51.17	14.12	192.59	50.00	44.05	16.74
382.88	182.60	417.02	50.89	14.12	191.62	49.23	43.04	16.74
11417.5	3562.3	12765.0	4160.3	169.2	56258.2	755.9	10589.2	140.6
1252086.8	463504.7	1091299.2	140259.8	53306.2	11540509.4	161705.0	118478.2	30428.2
1811976.1	667661.5	1598447.8	1844171.5	70113.3	11850739.6	206141.0	130503.2	46806.2
2880579.7	1133958.1	2651166.5	316602.2	118874.8	1151005.5	415712.0	174905.9	88481.7
4148898.8	1476214.0	3038682.2	501870.4	197558.2	1855734.6	620950.8	223119.3	153368.5
8638330.6	4042453.5	3223703.4	655647.4	283364.1	2480431.9	1181311.0	323458.3	153589.2

Task 17	Task 18	Task 19	Task 20	Task 21	Task 22	Task 23	Task 24	Task 25
sha	strsrch	strsrch	ssn	ssn	ssn	ssn	ssn	ssn
sha_lrg	strgsrch_sml	strsrch_lrg	ssn_sml1	ssn_sml2	ssn_sml3	ssn_lrg1	ssn_lrg2	ssn_lrg3
991.69	1.75	43.02	143.30	28.42	12.13	2043.75	849.21	226.69
208.21	1.04	23.63	30.08	11.71	5.10	390.87	379.17	105.44
177.25	0.62	14.33	20.96	7.45	2.82	282.18	245.82	58.83
173.88	0.45	10.49	20.25	5.09	2.42	279.57	148.28	43.05
173.88	0.45	10.48	20.24	5.07	2.42	279.48	147.57	43.02
173.88	0.45	10.48	20.24	5.05	2.42	279.45	147.57	43.01
1465.8	1.2	68.7	222.9	121.9	44.3	16179.7	38144.7	11476.0
317100.1	1106.5	27954.0	52800.4	12776.3	7369.5	515954.7	467280.9	267585.5
487613.4	1611.7	51986.9	55307.3	21487.3	8247.0	665690.1	930325.9	309314.3
929878.2	1732.8	80046.3	79470.4	24835.8	10183.9	2215638.8	1152520.6	315312.6
1618482.9	1773.3	87641.1	168981.9	31464.6	13495.2	2748450.9	1373224.1	377518.1
1620777.6	1773.3	89015.0	196048.8	46562.1	16895.8	2896506.3	1662613.3	439999.9

Table 6. Benchmark programs.

3.3.2 Experimental results

We synthesized heterogeneous multiprocessor systems under various real-time and SEU vulnerability constraints so that we could examine their chip areas. We assumed that the arrival time of every task was zero and that the deadline time of every task was same as the others. We also assumed that there was no SEU vulnerability constraint on each task, that is $V_{constraint_i} = \infty$. Generally speaking, the existence of loosely-bounded variables causes long computation time. It is quite easy to guess that the assumptions make exploration space huge and result in long computation time. The assumption, however, is helpful to obtaining the lower bound on chip area for given SEU vulnerability constraints. The deadline time of all tasks ranged from 3500 to 9500 million cycles and SEU vulnerability constraints of an entire system ranged from 500 to 50000 [10^{-15} errors/system]. Fig. 11 shows the results of heterogeneous multiprocessor synthesis. Chip area ranged from 80 to 320 in arbitrary unit. When we tightened the SEU vulnerability constraints under fixed real-time constraints, more processor cores which have no cache memory were utilized. Similarly, when we tightened the real-time constraints under fixed SEU vulnerability constraints, more processor cores which had a sufficient and minimal size of cache memory were utilized. Tighter SEU vulnerability constraints worked for selecting a smaller size of a cache memory while tighter real-time constraints worked for selecting a larger size of a cache memory. The figure clearly shows that relaxing constraints reduced the chip area of a multiprocessor system.

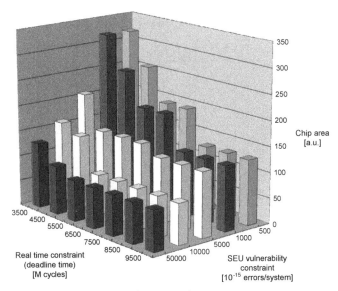

Fig. 11. Heterogeneous multiprocessor synthesis result.

We show four synthesis examples in Tables 7, 8, 9, and 10. We name them HS_1, HS_2, HS_3, and HS_4 respectively. For Synthesis HS_1, we gave the constraints that $T_{deadline_i} = 3500$ [M cycles] and $V_{const_{all}} = 5000$ [10^{-15} errors/system]. In this synthesis, a heterogeneous multiprocessor was synthesized which had two Conf. 1 processor cores and a Conf. 2 processor core as shown in Table 7.

For Synthesis HS_2, we gave the constraints that $T_{\text{deadline}_i} = 3500$ [M cycles] and $V_{\text{const}_{\text{all}}} = 500$ [10^{-15} errs/syst]. Only the constraint on $V_{\text{const}_{\text{all}}}$ became tighter in Synthesis HS_2 than in Synthesis HS_1. Table 8 shows that more reliable processor cores were utilized for achieving the tighter vulnerability constraint.

For Synthesis HS_3, we gave the constraints that $T_{\text{deadline}_i} = 3500$ [M cycles] and $V_{\text{const}_{\text{all}}} = 50000$ [10^{-15} errs/syst]. Only the constraint on $V_{\text{const}_{\text{all}}}$ became looser than in Synthesis HS_1. In this synthesis, a single Conf. 4 processor core was utilized as shown in Table 9. The looser constraint caused that a more vulnerable and greater processor core was utilized. The chip area was reduced in total.

For Synthesis HS_4, we gave the constraints that $T_{\text{deadline}_i} = 4500$ and $V_{\text{const}_{\text{all}}} = 5000$ [10^{-15} errs/syst]. Only the constraint on T_{deadline_i} became looser than in Synthesis HS_1. In this synthesis, a Conf. 1 processor core and a Conf. 2 processor core were utilized as shown in Table 10. The looser constraint on deadline time caused that a subset of the processor cores in Synthesis HS_1 were utilized to reduce chip area.

	Tasks
CPU 1 (Conf. 1)	{10, 13, 20, 25}
CPU 2 (Conf. 1)	{17, 23}
CPU 3 (Conf. 2)	{1, 2, 3, 4, 5, 6, 7, 8, 9, 11, 12, 14, 15, 16, 18, 19, 21, 22, 24}

Table 7. Result for HS_1 ($T_{\text{deadline}_i} = 3.5 \times 10^9$ cycles, $V_{\text{const}_{\text{all}}} = 5 \times 10^{-12}$ errs/syst).

	Tasks
CPU 1 (Conf. 1)	{1, 2, 3, 4, 5, 6, 7, 11, 18, 22}
CPU 2 (Conf. 1)	{8, 9, 14, 15, 16, 21}
CPU 3 (Conf. 1)	{10, 12, 13, 19, 25}
CPU 4 (Conf. 1)	{17, 20, 23}
CPU 5 (Conf. 1)	{24}

Table 8. Result for HS_2 ($T_{\text{deadline}_i} = 3.5 \times 10^9$ cycles, $V_{\text{const}_{\text{all}}} = 5 \times 10^{-13}$ errs/syst).

	Tasks
CPU 1 (Conf. 4)	{1, 2, 3, 4, 5, 6, 7, 8, 9, 10, 11, 12, 13, 14, 15, 16, 17, 18, 19, 20, 21, 22, 23, 24, 25}

Table 9. Result for HS_3 ($T_{\text{deadline}_i} = 3.5 \times 10^9$ cycles, $V_{\text{const}_{\text{all}}} = 5 \times 10^{-11}$ errs/syst).

	Tasks
CPU 1 (Conf. 1)	{1, 6, 10, 14, 16, 19, 21, 25}
CPU 2 (Conf. 2)	{2, 3, 4, 5, 7, 8, 9, 11, 12, 13, 14, 15, 17, 18, 20, 22, 23, 24}

Table 10. Result for HS_4 ($T_{\text{deadline}_i} = 4.5 \times 10^9$ cycles, $V_{\text{const}_{\text{all}}} = 5 \times 10^{-12}$ errs/syst).

3.3.3 Conclusion

We reviewed a heterogeneous multiprocessor synthesis paradigm in which we took real-time and SEU vulnerability constraints into account. We formally defined a heterogeneous multiprocessor synthesis problem in the form of an MILP model. By solving the problem instances, we synthesized heterogeneous multiprocessor systems. Our experiment showed that relaxing constraints reduced chip area of heterogeneous multiprocessor systems. There exists a trade-off between chip area and another constraint (performance or reliability) in synthesizing heterogeneous multiprocessor systems.

In the problem formulation we mainly focused on heterogeneous "multi-core" processor synthesis and ignored inter-task communication overhead time under two assumptions: (i) computation is the most dominant factor in execution time, (ii) sharing main memory and communication circuitry among several processor cores does not affect execution time. From a practical point of view, runtime of a task changes, depending on the other tasks which run simultaneously because memory accesses from multiple processor cores may collide on a shared hardware resource such as a communication bus. If task collisions on a shared communication mechanism cause large deviation on runtime, system designers may generate a customized on-chip network design with both a template processor configuration and the Drinic's technique (Drinic et al., 2006) before heterogeneous system synthesis so that such collisions are reduced.

From the viewpoint of commodification of ICs, we think that a heterogeneous multiprocessor consisting of a reliable but slow processor core and a vulnerable but fast one would be sufficient for many situations in which reliability and performance requirements differ among tasks. General-purpose processor architecture should be studied further for achieving both reliability and performance in commodity processors.

4. Concluding remarks

This chapter presented simulation and synthesis technique for a computer system. We presented an accurate vulnerability estimation technique which estimates the vulnerability of a computer system at the ISS level. Our vulnerability estimation technique is based on cycle-accurate ISS level simulation which is much faster than logic, transistor, and device simulations. Our technique, however, is slow for simulating large-scale programs. From the viewpoint of practicality fast vulnerability estimation techniques should be studied.

We also presented a multiprocessor synthesis technique for an embedded system. The multiprocessor synthesis technique is powerful to develop a reliable embedded system. Our synthesis technique offers system designers a way to a trade-off between chip area, reliability, and real-time execution. Our synthesis technique is mainly specific to "multi-core" processor synthesis because we simplified overhead time for bus arbitration. Our synthesis technique should be extended to "many-core" considering overhead time for arbitration of communication mechanisms.

5. References

Asadi, G. H.; Sridharan, V.; Tahoori, M. B. & Kaeli, D. (2005). Balancing performance and reliability in the memory hierarchy, *Proc. IEEE Int'l Symp. on Performance Analysis of Systems and Software*, pp. 269-279, ISBN 0-7803-8965-4, Austin, Texas, USA, March 2005

Asadi, H.; Sridharan, V.; Tahoori, M. B. & Kaeli, D. (2006). Vulnerability analysis of L2 cache elements to single event upsets, *Proc. Design, Automation and Test in Europe Conf.*, pp. 1276–1281, ISBN 3-9810801-0-6, Leuven, Belgium, March 2006

Baumann, R. B. Radiation-induced soft errors in advanced semiconductor technologies, *IEEE Trans. on device and materials reliability*, Vol. 5, No. 3, (September 2005), pp. 305-316, ISSN 1530-4388

Biswas, A.; Racunas, P.; Cheveresan, R.; Emer, J.; Mukherjee, S. S. & Rangan, R. (2005). Computing architectural vulnerability factors for address-based structures, *Proc. IEEE Int'l Symp. on Computer Architecture*, pp. 532–543, ISBN 0-7695-2270-X, Madison, WI, USA, June 2005

Degalahal, V.; Vijaykrishnan, N.; Irwin, M. J.; Cetiner, S.; Alim, F. & Unlu, K. (2004). SESEE: soft error simulation and estimation engine, *Proc. MAPLD Int'l Conf.*, Submission 192, Washington, D.C., USA, September 2004

Drinic, M.; Krovski, D.; Megerian, S. & Potkonjak, M. (2006). Latency guided on-chip bus-network design, *IEEE Trans. on Computer-Aided Design of Integrated Circuits and Systems*, Vol. 25, No. 12, (December 2006), pp. 2663-2673, ISSN 0278-0070

Elakkumanan, P.; Prasad, K. & Sridhar, R. (2006). Time redundancy based scan flip-flop reuse to reduce SER of combinational logic, *Proc. IEEE Int'l Symp. on Quality Electronic Design*, pp. 617-622, ISBN 978-1-4244-6455-5, San Jose, CA, USA, March 2006

Guthaus, M. R.; Ringenberg, J. S.; Ernst, D.; Austin, T. M.; Mudge, T. & Brown, R. B. (2001). MiBench: A Free, commercially representative embedded benchmark suite, *Proc. IEEE Workshop on Workload Characterization*, ISBN 0-7803-7315-4, Austin, TX, USA, December 2001

Hennessy, J. L. & Patterson, D. A. (2002). *Computer architecture: a quantitative approach*, Morgan Kaufmann Publishers Inc., ISBN 978-1558605961, San Francisco, CA, USA

Karnik, T.; Bloechel, B.; Soumyanath, K.; De, V. & Borkar, S. (2001). Scaling trends of cosmic ray induced soft errors in static latches beyond 0.18 μm, *Proc. Symp. on VLSI Circuits*, pp. 61–62, ISBN 4-89114-014-3, Tokyo, Japan, June 2001

Li, X.; Adve, S. V.; Bose, P. & Rivers, J. A. (2005). SoftArch: An architecture level tool for modeling and analyzing soft errors, *Proc. IEEE Int'l Conf. on Dependable Systems and Networks*, pp. 496–505, ISBN 0-7695-2282-3, Yokohama, Japan, June 2005

May, T. C. & Woods, M. H. (1979). Alpha-particle-induced soft errors in dynamic memories, *IEEE Trans. on Electron Devices*, vol. 26, Issue 1, (January 1979), pp. 2–7, ISSN 0018-9383

Mukherjee, S. S.; Weaver, C.; Emer, J.; Reinhardt, S. K. & Austin, T. (2003). A systematic methodology to compute the architectural vulnerability factors for a high-performance microprocessor, *Proc. IEEE/ACM Int'l Symp. on Microarchitecture*, pp. 29-40, ISBN 0-7695-2043-X, San Diego, CA, USA, December 2003.

Mukherjee, S. S.; Emer, J. & Reinhardt, S. K. (2005). The soft error problem: an architectural perspective, *Proc. IEEE Int'l Symp. on HPCA*, pp.243-247, ISBN 0-7695-2275-0, San Francisco, CA, USA, February 2005

Rebaudengo, M.; Reorda, M. S. & Violante, M. (2003). An accurate analysis of the effects of soft errors in the instruction and data caches of a pipelined microprocessor, *Proc. Design, Automation and Test in Europe*, pp.10602-10607, ISBN 0-7695-1870-2, Munich, Germany, 2003

Seifert, N.; Moyer, D.; Leland, N. & Hokinson, R. (2001a). Historical trend in alpha-particle induced soft error rates of the Alpha(tm) microprocessor," *Proc. IEEE Int'l Reliability Physics Symp.*, pp. 259–265, ISBN 0-7803-6587-9, Orlando, FL, USA, April 2001.

Seifert, N.; Zhu, X.; Moyer, D.; Mueller, R.; Hokinson, R.; Leland, N.; Shade, M. & Massengill, L. (2001b). Frequency dependence of soft error rates for sub-micron CMOS technologies, *Technical Digest of Int'l Electron Devices Meeting*, pp. 14.4.1–14.4.4, ISBN 0-7803-7050-3, Washington, DC, USA, December 2001

Shivakumar, P.; Kistler, M.; Keckler, S. W.; Burger, D. & Alvisi, L. (2002). Modeling the effect of technology trends of the soft error rate of combinational logic, *Proc. Int'l Conf. on Dependable Systems and Networks*, pp. 389-398, ISBN 0-7695-1597-5, Bethesda, MD, June 2002

Slayman, C. W. (2005) Cache and memory error detection, correction and reduction techniques for terrestrial servers and workstations, *IEEE Trans. on Device and Materials Reliability*, vol. 5, no. 3, (September 2005), pp. 397-404, ISSN 1530-4388

Sugihara, M.; Ishihara, T.; Hashimoto, K. & Muroyama, M. (2006). A simulation-based soft error estimation methodology for computer systems, *Proc. IEEE Int'l Symp. on Quality Electronic Design*, pp. 196-203, ISBN 0-7695-2523-7, San Jose, CA, USA, March 2006

Sugihara, M.; Ishihara, T. & Murakami, K. (2007a). Task scheduling for reliable cache architectures of multiprocessor systems, *Proc. Design, Automation and Test in Europe Conf.*, pp. 1490-1495, ISBN 978-3-98108010-2-4, Nice, France, April 2007

Sugihara, M.; Ishihara, T. & Murakami, K. (2007b). Architectural-level soft-error modeling for estimating reliability of computer systems, *IEICE Trans. Electron.*, Vol. E90-C, No. 10, (October 2007), pp. 1983-1991, ISSN 0916-8524

Sugihara, M. (2008a). SEU vulnerability of multiprocessor systems and task scheduling for heterogeneous multiprocessor systems, *Proc. Int'l Symp. on Quality Electronic Design*, ISBN 978-0-7695-3117-5, pp. 757-762, San Jose, CA, USA, March 2008

Sugihara, M.; Ishihara, T. & Murakami, K. (2008b). Reliable cache architectures and task scheduling for multiprocessor systems, *IEICE Trans. Electron.*, Vol. E91-C, No. 4, (April 2008), pp. 410-417, ISSN 0916-8516

Sugihara, M. (2009a). Reliability inherent in heterogeneous multiprocessor systems and task scheduling for ameliorating their reliability, *IEICE Trans. Fundamentals*, Vol. E92-A, No. 4, (April 2009), pp. 1121-1128, ISSN 0916-8508

Sugihara, M. (2009b). Heterogeneous multiprocessor synthesis under performance and reliability constraints, *Proc. EUROMICRO Conf. on Digital System Design*, pp. 333-340, ISBN 978-0-7695-3782-5, Patras, Greece, August 2009.

Sugihara, M. (2010a). Dynamic control flow checking technique for reliable microprocessors, *Proc. EUCROMICRO Conf. on Digital System Design*, pp. 232-239, ISBN 978-1-4244-7839-2, Lille, France, September 2010

Sugihara, M. (2010b). On synthesizing a reliable multiprocessor for embedded systems, *IEICE Trans. Fundamentals*, Vol. E93-A, No. 12, (December 2010), pp. 2560-2569, ISSN 0916-8508

Sugihara, M. (2011). A dynamic continuous signature monitoring technique for reliable microprocessors, *IEICE Trans. Electron.*, Vol. E94-C, No. 4, (April 2011), pp. 477-486, ISSN 0916-8524

Tosaka, Y.; Satoh, S. & Itakura, T. (1997). Neutron-induced soft error simulator and its accurate predictions, *Proc. IEEE Int'l Conf. on SISPAD*, pp. 253–256, ISBN 0-7803-3775-1, Cambridge, MA , USA, September 1997

Tosaka, Y.; Kanata, H.; Itakura, T. & Satoh, S. (1999). Simulation technologies for cosmic ray neutron-induced soft errors: models and simulation systems, *IEEE Trans. on Nuclear Science*, vol. 46, (June, 1999), pp. 774-780, ISSN 0018-9499

Tosaka, Y.; Ehara, H.; Igeta, M.; Uemura, T & Oka, H. (2004a). Comprehensive study of soft errors in advanced CMOS circuits with 90/130 nm technology, *Technical Digest of IEEE Int'l Electron Devices, pp. 941–948*, ISBN 0-7803-8684-1, San Francisco, CA, USA, December 2004

Tosaka, Y.; Satoh, S. & Oka, H. (2004b). Comprehensive soft error simulator NISES II, *Proc. IEEE Int'l Conf. on SISPAD*, pp. 219–226, ISBN 978-3211224687, Munich, Germany, September 2004

Wang, N. J.; Quek, J.; Rafacz, T. M. & Patel, S. J. (2004). Characterizing the effects of transient faults on a high-performance processor pipeline, *Proc. IEEE Int'l Conf. on Dependable Systems and Networks*, pp.61-70, ISBN 0-7695-2052-9, Florence, Italy, June 2004

Williams, H. P. (1999). Model Building in Mathematical Programming, John Wiley & Sons, 1999

ILOG Inc., CPLEX 11.2 User's Manual, 2008

Part 2

Design/Evaluation Methodology, Verification, and Development Environment

6

Architecting Embedded Software for Context-Aware Systems

Susanna Pantsar-Syväniemi
VTT Technical Research Centre of Finland
Finland

1. Introduction

During the last three decades the architecting of embedded software has changed by i) the ever-enhancing processing performance of processors and their parallel usage, ii) design methods and languages, and iii) tools. The role of software has also changed as it has become a more dominant part of the embedded system. The progress of hardware development regarding size, cost and energy consumption is currently speeding up the appearance of smart environments. This necessitates the information to be distributed to our daily environment along with smart, but separate, items like sensors. The cooperation of the smart items, by themselves and with human beings, demands new kinds of embedded software.

The architecting of embedded software is facing new challenges as it moves toward smart environments where physical and digital environments will be integrated and interoperable. The need for human beings to interact is decreasing dramatically because digital and physical environments are able to decide and plan behavior by themselves in areas where functionality currently requires intervention from human beings, such as showing a barcode to a reader in the grocery store. The smart environment, in our mind, is not exactly an Internet of Things (IoT) environment, but it can be. The difference is that the smart environment that we are thinking of does not assume that all tiny equipment is able to communicate via the Internet. Thus, the smart environment is an antecedent for the IoT environment.

At the start of the 1990s, hardware and software co-design in real time and embedded systems were seen as complicated matters because of integration of different modeling techniques in the co-design process (Kronlöf, 1993). In the smart environment, the co-design is radically changing, at least from the software perspective. This is due to the software needing to be more and more intelligent by, e.g., predicting future situations to offer relevant services for human beings. The software needs to be interoperable, as well as scattered around the environment, with devices that were previously isolated because of different communication mechanisms or standards.

Research into pervasive and ubiquitous computing has been ongoing for over a decade, providing many context-aware systems and a multitude of related surveys. One of those surveys is a literature review of 237 journal articles that were published between 2000 and

2007 (Hong et al., 2009). The review presents that context-aware systems i) are still developing in order to improve, and ii) are not fully implemented in real life. It also emphasizes that context-awareness is a key factor for new applications in the area of ubiquitous computing, i.e., pervasive computing. The context-aware system is based on pervasive or ubiquitous computing. To manage the complexity of pervasive computing, the context-aware system needs to be designed in new way — from the bottom up — while understanding the eligible ecosystem, and from small functionalities to bigger ones. The small functionalities are formed up to the small architectures, micro-architectures. Another key issue is to reuse the existing, e.g., communication technologies and devices, as much as possible, at least at the start of development, to minimize the amount of new things.

To get new perspective on the architecting of context-aware systems, Section two introduces the major factors that have influenced the architecting of embedded and real-time software for digital base stations, as needed in the ecosystem of the mobile network. This introduction also highlights the evolution of the digital base station in the revolution of the Internet. The major factors are standards and design and modeling approaches, and their usefulness is compared for architecting embedded software for context-aware systems. The context of pervasive computing calms down when compared to the context of digital signal processing software as a part of baseband computing which is a part of the digital base station. It seems that the current challenges have similarities in both pervasive and baseband computing. Section two is based on the experiences gathered during software development at Nokia Networks from 1993 to 2008 and subsequently in research at the VTT Technical Research Centre of Finland. This software development included many kinds of things, e.g., managing the feature development of subsystems, specifying the requirements for the system and subsystem levels, and architecting software subsystems. The research is related to enable context-awareness with the help of ontologies and unique micro-architecture.

Section three goes through the main research results related to designing context-aware applications for smart environments. The results relate to context modeling, storing, and processing. The latter includes a new solution, a context-aware micro-architecture (CAMA), for managing context when architecting embedded software for context-aware systems. Section four concludes this chapter.

2. Architecting real-time and embedded software in the 1990s and 2000s

2.1 The industrial evolution of the digital base station

Figure 1 shows the evolution of the Internet compared with a digital base station (the base station used from now on) for mobile networks. It also shows the change from proprietary interfaces toward open and Internet-based interfaces. In the 1990s, the base station was not built for communicating via the Internet. The base station was isolated in the sense that it was bound to a base station controller that controlled a group of base stations. That meant that a customer was forced to buy both the base stations and the base station controller from the same manufacturer.

In the 2000s, the industrial evolution brought the Internet to the base station and it opened the base station for module business by defining interfaces between modules. It also dissolved the "engagement" between the base stations and their controllers as it moved from the second generation mobile network (2G) to third one (3G). Later, the baseband module of the base station was also reachable via the Internet. In the 2010s, the baseband module will go to the cloud to be able to meet the constantly changing capacity and coverage demands on the mobile network. The baseband modules will form a centralized baseband pool. These demands arise as smartphone, tablet and other smart device users switch applications and devices at different times and places (Nokia Siemens Networks, 2011).

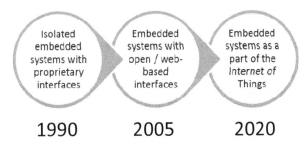

Fig. 1. The evolution of the base station.

The evolution of base-band computing in the base station changes from distributed to centralized as a result of dynamicity. The estimation of needed capacity per mobile user was easier when mobiles were used mainly for phone calls and text messaging. The more fancy features that mobiles offer and users demand, the harder it is to estimate the needed base-band capacity.

The evolution of the base station goes hand-in-hand with mobile phones and other network elements, and that is the strength of the system architecture. The mobile network ecosystem has benefited a lot from the system architecture of, for example, the Global System for Mobile Communications (GSM). The context-aware system is lacking system architecture and that is hindering its breakthrough.

2.2 The standardization of mobile communication

During the 1980s, European telecommunication organizations and companies reached a common understanding on the development of a Pan-European mobile communication standard, the Global System for Mobile Communications (GSM), by establishing a dedicated organization, the European Telecommunications Standards Institute (ETSI, www.etsi.org), for the further evolvement of the GSM air-interface standard. This organization has produced the GSM900 and 1800 standard specifications (Hillebrand, 1999). The development of the GSM standard included more and more challenging features of standard mobile technology as defined by ETSI, such as High Speed Circuit Switched Data (HSCSD), General Packet Radio Service (GPRS), Adaptive Multirate Codec (AMR), and Enhanced Data rates for GSM Evolution (EDGE) (Hillebrand, 1999).

The Universal Mobile Telecommunication System (UMTS) should be interpreted as a continuation of the regulatory regime and technological path set in motion through GSM, rather than a radical break from this regime. In effect, GSM standardization defined a path of progress through GPRS and EDGE toward UMTS as the major standard of 3G under the 3GPP standardization organization (Palmberg & Martikainen, 2003). The technological path from GSM to UMTS up to LTE is illustrated in Table 1. High-Speed Downlink Packet Access (HSDPA) and High-Speed Uplink Packet Access (HSUPA) are enhancements of the UMTS to offer a more interactive service for mobile (smartphone) users.

GSM -> HSCD, GPRS, AMR, EDGE	UMTS -> HSDPA, HSUPA	LTE
2G =>	3G =>	4G

Table 1. The technological path of the mobile communication system

It is remarkable that standards have such a major role in the telecommunication industry. They define many facts via specifications, like communication between different parties. The European Telecommunications Standards Institute (ETSI) is a body that serves many players such as network suppliers and network operators. Added to that, the network suppliers have created industry forums: OBSAI (Open Base Station Architecture Initiative) and CPRI (Common Public Radio Interface). The forums were set up to define and agree on open standards for base station internal architecture and key interfaces. This, the opening of the internals, enabled new business opportunities with base station modules. Thus, module vendors were able to develop and sell modules that fulfilled the open, but specified, interface and sell them to base station manufacturers. In the beginning the OBSAI was heavily driven by Nokia Networks and the CPRI respectively by Ericsson. Nokia Siemens Networks joined CPRI when it was merged by Nokia and Siemens.

The IoT ecosystem is lacking a standardization body, such as ETSI has been for the mobile networking ecosystem, to create the needed base for the business. However, there is the Internet of Things initiative (IoT-i), which is working and attempting to build a unified IoT community in Europe, www.iot-i.eu.

2.3 Design methods

The object-oriented approach became popular more than twenty years ago. It changed the way of thinking. Rumbaugh et al. defined object-oriented development as follows, i) it is a conceptual process independent of a programming language until the final stage, and ii) it is fundamentally a new way of thinking and not a programming technique (Rumbaugh et al., 1991). At the same time, the focus was changing from software implementation issues to software design. In those times, many methods for software design were introduced under the Object-Oriented Analysis (OOA) method (Shlaer & Mellor, 1992), the Object-Oriented Software Engineering (OOSE) method (Jacobson et al., 1992), and the Fusion method (Coleman et al., 1993). The Fusion method highlighted the role of entity-relationship graphs in the analysis phase and the behavior-centered view in the design phase.

The Object Modeling Technique (OMT) was introduced for object-oriented software development. It covers the analysis, design, and implementation stages but not integration and maintenance. The OMT views a system via a model that has two dimensions (Rumbaugh et al., 1991). The first dimension is viewing a system: the object, dynamic, or functional model. The second dimension represents a stage of the development: analysis, design, or implementation. The object model represents the static, structural, "data" aspects of a system. The dynamic model represents the temporal, behavioral, "control" aspects of a system. The functional model illustrates the transformational, "function" aspects of a system. Each of these models evolves during a stage of development, i.e. analysis, design, and implementation.

The OCTOPUS method is based on the OMT and Fusion methods and it aims to provide a systematic approach for developing object-oriented software for embedded real-time systems. OCTOPUS provides solutions for many important problems such as concurrency, synchronization, communication, interrupt handling, ASICs (application-specific integrated circuit), hardware interfaces and end-to-end response time through the system (Awad et al., 1996). It isolates the hardware behind a software layer called the hardware wrapper. The idea for the isolation is to be able to postpone the analysis and design of the hardware wrapper (or parts of it) until the requirements set by the proper software are realized or known (Awad et al., 1996).

The OCTOPUS method has many advantages related to the system division of the subsystems, but without any previous knowledge of the system under development the architect was able to end up with the wrong division in a system between the controlling and the other functionalities. Thus, the method was dedicated to developing single and solid software systems separately. The OCTOPUS, like the OMT, was a laborious method because of the analysis and design phases. These phases were too similar for there to be any value in carrying them out separately. The OCTOPUS is a top-down method and, because of that, is not suitable to guide bottom-up design as is needed in context-aware systems.

Software architecture started to become defined in the late 1980s and in the early 1990s. Mary Shaw defined that i) architecture is design at the level of abstraction that focuses on the patterns of system organization which describe how functionality is partitioned and the parts are interconnected and ii) architecture serves as an important communication, reasoning, analysis, and growth tool for systems (Shaw, 1990). Rumbaugh et al. defined software architecture as the overall structure of a system, including its partitioning into subsystems and their allocation to tasks and processors (Rumbaugh et al., 1991). Figure 2 represents several methods, approaches, and tools with which we have experimented and which have their roots in object-oriented programming.

For describing software architecture, the 4+1 approach was introduced by Philippe Krüchten. The 4+1 approach has four views: logical, process, development and physical. The last view, the +1 view, is for checking that the four views work together. The checking is done using important use cases (Krüchten, 1995). The 4+1 approach was part of the foundation for the Rational Unified Process, RUP. Since the introduction of the 4+1 approach software architecture has had more emphasis in the development of software systems. The most referred definition for the software architecture is the following one:

> The structure or structures of the system, which comprises software elements, the externally visible properties of those elements, and the relationships among them, (Bass et al., 1998)

Views are important when documenting software architecture. Clements et al. give a definition for the view: "A view is a representation of a set of system elements and the relationships associated with them". Different views illustrate different uses of the software system. As an example, a layered view is relevant for telling about the portability of the software system under development (Clements, 2003). The views are presented using, for example, UML model elements as they are more descriptive than pure text.

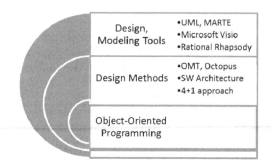

Fig. 2. From object-oriented to design methods and supporting tools.

Software architecture has always has a role in base station development. In the beginning it represented the main separation of the functionalities, e.g. operation and maintenance, digital signal processing, and the user interface. Later on, software architecture was formulated via architectural views and it has been the window to each of these main functionalities, called software subsystems. Hence, software architecture is an efficient media for sharing information about the software and sharing the development work, as well.

2.4 Modeling

In the model-driven development (MDD) vision, models are the primary artifacts of software development and developers rely on computer-based technologies to transform models into running systems (France & Rumpe, 2007). The Model-Driven Architecture (MDA), standardized by the Object Management Group (OMG, www.omg.org), is an approach to using models in software development. MDA is a known technique of MDD. It is meant for specifying a system independently of the platform that supports it, specifying platforms, choosing a particular platform for the system, and transforming the system specification into a particular platform. The three primary goals of MDA are portability, interoperability and reusability through the architectural separation of concerns (Miller & Mukerji, 2003).

MDA advocates modeling systems from three viewpoints: computational-independent, platform-independent, and platform-specific viewpoints. The computational-independent viewpoint focuses on the environment in which the system of interest will operate in and on the required features of the system. This results in a computation-independent model (CIM). The platform-independent viewpoint focuses on the aspects of system features that are not likely to change from one platform to another. A platform-independent model (PIM) is used to present this viewpoint. The platform-specific viewpoint provides a view of a system in which platform-specific details are integrated with the elements in a PIM. This view of a system is described by a platform-specific model (PSM), (France & Rumpe, 2007).

The MDA approach is good for separating hardware-related software development from the application (standard-based software) development. Before the separation, the maintenance of hardware-related software was done invisibly under the guise of application development. By separating both application- and hardware-related software development, the development and maintenance of previously invisible parts, i.e., hardware-related software, becomes visible and measurable, and costs are easier to explicitly separate for the pure application and the hardware-related software.

Two schools exist in MDA for modeling languages: the Extensible General-Purpose Modeling Language and the Domain Specific Modeling Language. The former means Unified Modeling Language (UML) with the possibility to define domain-specific extensions via profiles. The latter is for defining a domain-specific language by using meta-modeling mechanisms and tools. The UML has grown to be a de facto industry standard and it is also managed by the OMG. The UML has been created to visualize object-oriented software but also used to clarify the software architecture of a subsystem that is not object-oriented.

The UML is formed based on the three object-oriented methods: the OOSE, the OMT, and Gary Booch's Booch method. A UML profile describes how UML model elements are extended using stereotypes and tagged values that define additional properties for the elements (France & Rumpe, 2007). A Modeling and Analysis of Real-Time Embedded Systems (MARTE) profile is a domain-specific extension for UML to model and analyze real time and embedded systems. One of the main guiding principles for the MARTE profile (www.omgmarte.org) has been that it should support independent modeling of both software or hardware parts of real-time and embedded systems and the relationship between them. OMG's Systems Modeling Language (SysML, www.omgsysml.org) is a general-purpose graphical modeling language. The SysML includes a graphical construct to represent text-based requirements and relate them to other model elements.

Microsoft Visio is usually used for drawing UML–figures for, for example, software architecture specifications. The UML–figures present, for example, the context of the software subsystem and the deployment of that software subsystem. The MARTE and SysML profiles are supported by the Papyrus tool. Without good tool support the MARTE profile will provide only minimal value for embedded software systems.

Based on our earlier experience and the MARTE experiment, as introduced in (Pantsar-Syväniemi & Ovaska, 2010), we claim that MARTE is not as applicable to embedded systems as base station products. The reason is that base station products are dependent on long-term maintenance and they have a huge amount of software. With the MARTE, it is not possible to i) model a greater amount of software and ii) maintain the design over the years. We can conclude that the MARTE profile has been developed from a hardware design point of view because software reuse seems to have been neglected.

Many tools exist, but we picked up on Rational Rhapsody because we have seen it used for the design and code generation of real-time and embedded software. However, we found that the generated code took up too much of the available memory, due to which Rational Rhapsody was considered not able to meet its performance targets. The hard real-time and embedded software denotes digital signal processing (DSP) software. DSP is a central part of the physical layer baseband solutions of telecommunications (or mobile wireless) systems, such as mobile phones and base stations. In general, the functions of the physical layer have been implemented in hardware, for example, ASIC (application-specific integrated circuits), and FPGA (field programmable gate arrays), or near to hardware (Paulin et al., 1997), (Goossens et al., 1997).

Due to the fact that Unified Modeling Language (UML) is the most widely accepted modeling language, several model-driven approaches have emerged (Kapitsaki et al., 2009), (Achillelos et al., 2010). Typically, these approaches introduce a meta-model enriched with context-related artifacts, in order to support context-aware service engineering. We have also used UML for designing the collaboration between software agents and context storage during our research related to the designing of smart spaces based on the ontological approach (Pantsar-Syväniemi et al., 2011a, 2012).

2.5 Reuse and software product lines

The use of C language is one of the enabling factors of making reusable DSP software (Purhonen, 2002). Another enabling factor is more advanced tools, making it possible to separate DSP software development from the underlying platform. Standards and underlying hardware are the main constraints for DSP software. It is essential to note that hardware and standards have different lifetimes. Hardware evolves according to 'Moore's Law' (Enders, 2003), according to which progress is much more rapid than the evolution of standards. From 3G base stations onward, DSP software has been reusable because of the possibility to use C language instead of processor-specific assembly language. The reusability only has to do with code reuse, which can be regarded as a stage toward overall reuse in software development, as shown in Figure 3.

Regarding the reuse of design outputs and knowledge, it was the normal method of operation at the beginning of 2G base station software developments and was not too tightly driven by development processes or business programs. We have presented the characteristics of base station DSP software development in our previous work (Pantsar-Syväniemi et al., 2006) that is based on experiences when working at Nokia Networks. That work introduces the establishment of reuse actives in the early 2000s. Those activities were development 'for reuse' and development 'with reuse'. 'For reuse' means development of reusable assets and 'with reuse' means using the assets in product development or maintenance (Karlsson, 1995).

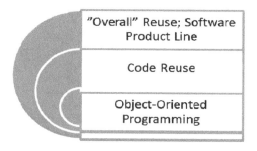

Fig. 3. Toward the overall reuse in the software development.

The main problem within this process-centric, 'for reuse' and 'with reuse', development was that it produced an architecture that was too abstract. The reason was that the domain was too wide, i.e., the domain was base station software in its entirety. In addition to that, the software reuse was "sacrificed" to fulfill the demand to get a certain base station product market-ready. This is paradoxical because software reuse was created to shorten products' time-to-market and to expand the product portfolio. The software reuse was due to business demands.

In addition to Karlsson's 'for and with reuse' book, we highlight two process-centric reuse books among many others. To design and use software architectures is written by Bosch (Bosch, 2000). This book has reality aspects when guiding toward the selection of a suitable organizational model for the software development work that was meant to be built around software architecture. In his paper, (Bosch, 1999), Bosch presents the main influencing factors for selecting the organization model: geographical distribution, maturity of project management, organizational culture, and the type of products. In that paper, he stated that a software product built in accordance with the software architecture is much more likely to fulfill its quality requirements in addition to its functional requirements.

Bosch emphasized the importance of software architecture. His software product line (SPL) approach is introduced according to these phases: development of the architecture and component set, deployment through product development and evolution of the assets (Bosch, 2000). He presented that not all development results are sharable within the SPL but there are also product-specific results, called artifacts.

The third interesting book introduces the software product line as compared to the development of a single software system at a time. This book shortly presents several ways for starting software development according to the software product line. It is written by Pohl et al. (Pohl et al., 2005) and describes a framework for product-line engineering. The book stresses the key differences of software product-line engineering in comparison with single-software system development:

- The need for two distinct development processes: domain engineering and application engineering. The aim of the domain-engineering process is to define and realize the commonality and the variability of the software product line. The aim of the application-engineering process is to derive specific applications by exploiting the variability of the software product line.
- The need to explicitly define and manage variability: During domain engineering, variability is introduced in all domain engineering artifacts (requirements, architecture, components, test cases, etc.). It is exploited during application engineering to derive applications tailored to the specific needs of different customers.

A transition from single-system development to software product-line engineering is not easy. It requires investments that have to be determined carefully to get the desired benefits (Pohl et al., 2005). The transition can be introduced via all of its aspects: process, development methods, technology, and organization. For a successful transition, we have to change all the relevant aspects, not just some of them (Pohl et al., 2005). With the base station products, we have seen that a single-system development has been powerful when products were more hardware- than software-oriented and with less functionality and complexity. The management aspect, besides the development, is taken into account in the product line but how does it support long-life products needing maintenance over ten years? So far, there is no proposal for the maintenance of long-life products within the software product line. Maintenance is definitely an issue to consider when building up the software product line.

The strength of the software product line is that it clarifies responsibility issues in creating, modifying and maintaining the software needed for the company's products. In software product-line engineering, the emphasis is to find the commonalities and variabilities and that is the huge difference between the software product-line approach and the OCTOPUS method. We believe that the software product-line approach will benefit if enhanced with a model-driven approach because the latter strengthens the work with the commonalities and variabilities.

Based on our experience, we can identify that the software product-line (SPL) and model-driven approach (MDA) alike are used for base station products. Thus, a combination of SPL and MDA is good approach when architecting huge software systems in which hundreds of persons are involved for the architecting, developing and maintaining of the software. A good requirement tool is needed to keep track of the commonalities and variabilities. The more requirements, the more sophisticated tool should be with the possibility to tag on the requirements based on the reuse targets and not based on a single business program.

The SPL approach needs to be revised for context-aware systems. This is needed to guide the architecting via the understanding of an eligible ecosystem toward small functionalities or subsystems. Each of these subsystems is a micro-architecture with a unique role. Run-time security management is one micro-architecture (Evesti & Pantsar-Syväniemi, 2010) that reuses context monitoring from the context-awareness micro-architecture, CAMA (Pantsar-Syväniemi et al., 2011a). The revision needs a new mindset to form reusable micro-architectures for the whole context-aware ecosystem. It is good to note that micro-architectures can differ in the granularity of the reuse.

2.6 Summary of section 2

The object-oriented methods, like Fusion, OMT, and OCTOPUS, were dedicated for single-system development. The OCTOPUS was the first object-oriented method that we used for an embedded system with an interface to the hardware. Both the OCTOPUS and the OMT were burdening the development work with three phases: object-oriented analysis (OOA) object-oriented design (OOD), and implementation. The OOD was similar to the implementation. In those days there was a lack of modeling tools. The message sequence charts (MSC) were done with the help of text editor.

When it comes to base station development, the software has become larger and more complicated with the new features needed for the mobile network along with the UML, the modeling tools supporting UML, and the architectural views. Thus, software development is more and more challenging although the methods and tools have become more helpful. The methods and tools can also hinder when moving inside the software system from one subsystem to another if the subsystems are developed using different methods and tools.

Related to DSP software, the tight timing requirements have been reached with optimized C-code, and not by generating code from design models. Thus, the code generators are too ineffective for hard real time and embedded software. One of the challenges in DSP software is the memory consumption because of the growing dynamicity in the amount of data that flows through mobile networks. This is due to the evolution of mobile network features like HSDPA and HSUPA that enable more features for mobile users. The increasing dynamicity demands simplification in the architecture of the software system. One of these simplifications is the movement from distributed baseband computing to centralized computing.

Simplification has a key role in context-aware computing. Therefore, we recall that by breaking the overall embedded software architecture into smaller pieces with specialized functionality, the dynamicity and complexity can be dealt with more easily. The smaller pieces will be dedicated micro-architectures, for example, run-time performance or security management. We can see that in smart environments the existing wireless networks are working more or less as they currently work. Thus, we are not assuming that they will converge together or form only one network. By taking care of and concentrating the data that those networks provide or transmit, we can enable the networks to work seamlessly together. Thus, the networks and the data they carry will form the basis for interoperability within smart environments. The data is the context for which it has been provided. Therefore, the data is in a key position in context-aware computing.

The MSC is the most important design output because it visualizes the collaboration between the context storage, context producers and context consumers. The OCTOPUS method is not applicable but SPL is when revised with micro-architectures, as presented earlier. The architecting context-aware systems need a new mindset to be able to i) handle dynamically changing context by filtering to recognize the meaningful context, ii) be designed bottom-up, while keeping in mind the whole system, and iii) reuse the legacy systems with adapters when and where it is relevant and feasible.

3. Architecting real-time and embedded software in the smart environment

Context has always been an issue but had not been used as a term as widely with regard to embedded and real-time systems as it has been used in pervasive and ubiquitous computing. Context was part of the architectural design while we created architectures for the subsystem of the base station software. It was related to the co-operation between the subsystem under creation and the other subsystems. It was visualized with UML figures showing the offered and used interfaces. The exact data was described in the separate interface specifications. This can be known as external context. Internal context existed and it was used inside the subsystems.

Context, both internal and external, has been distributed between subsystems but it has been used inside the base station. It is important to note that external context can be context that is dedicated either for the mobile phone user or for internal usage. The meaning of context that is going to, or coming from, the mobile phone user is meaningless for the base station but it needs memory to be processed. In pervasive computing, external context is always meaningful and dynamic. The difference is in the nature of context and the commonality is in the dynamicity of the context.

Recent research results into the pervasive computing state that:

- due to the inherent complexity of context-aware applications, development should be supported by adequate context-information modeling and reasoning techniques (Bettini et al., 2010)
- distributed context management, context-aware service modeling and engineering, context reasoning and quality of context, security and privacy, have not been well addressed in the Context-Aware Web Service Systems (Truong & Dustdar, 2009)
- development of context-aware applications is complex as there are many software engineering challenges stemming from the heterogeneity of context information sources, the imperfection of context information, and the necessity for reasoning on contextual situations that require application adaptations (Indulska & Nicklas, 2010)
- proper understanding of context and its relationship with adaptability is crucial in order to construct a new understanding for context-aware software development for pervasive computing environments (Soylu et al., 2009)
- ontology will play a crucial role in enabling the processing and sharing of information and knowledge of middleware (Hong et al., 2009)

3.1 Definitions

Many definitions for context as well for context-awareness are given in written research. The generic definition by Dey and Abowd for context and context-awareness are widely cited (Dey & Abowd, 1999):

'**Context** is any information that can be used to characterize the situation of an entity. An entity is a person, place, or object that is considered relevant to the interaction between a user and an application, including the user and the application themselves. '

'**Context-awareness** is a property of a system that uses context to provide relevant information and/or services to the user, where relevancy depends on the user's task. '

Context-awareness is also defined to mean that one is able to use context-information (Hong et al., 2009). Being context-aware will improve how software adapts to dynamic changes influenced by various factors during the operation of the software. Context-aware techniques have been widely applied in different types of applications, but still are limited to small-scale or single-organizational environments due to the lack of well-agreed interfaces, protocols, and models for exchanging context data (Truong & Dustdar, 2009).

In large embedded-software systems the user is not always the human being but can also be the other subsystem. Hence, the user has a wider meaning than in pervasive computing where the user, the human being, is in the center. We claim that pervasive computing will come closer to the user definition of embedded-software systems in the near future. Therefore, we propose that '*A context defines the limit of information usage of a smart space application*' (Toninelli et al., 2009). That is based on the assumption that any piece of data, at a given time, can be context for a given smart space application.

3.2 Designing the context

Concentrating on the context and changing the design from top-down to bottom-up while keeping the overall system in the mind is the solution to the challenges in the context-aware computing. Many approaches have been introduced for context modeling but we introduce one of the most cited classifications in (Strang & Linnhoff-Popien, 2004):

1. Key-Value Models

    The model of key-value pairs is the most simple data structure for modeling contextual information. The key-value pairs are easy to manage, but lack capabilities for sophisticated structuring for enabling efficient context retrieval algorithms.

2. Markup Scheme Models

    Common to all markup scheme modeling approaches is a hierarchical data structure consisting of markup tags with attributes and content. The content of the markup tags is usually recursively defined by other markup tags. Typical representatives of this kind of context modeling approach are profiles.

3. Graphical Model

    A very well-known general purpose modeling instrument is the UML which has a strong graphical component: UML diagrams. Due to its generic structure, UML is also appropriate to model the context.

4. Object-Oriented Models

    Common to object-oriented context modeling approaches is the intention to employ the main benefits of any object-oriented approach – namely encapsulation and reusability – to cover parts of the problems arising from the dynamics of the context in ubiquitous environments. The details of context processing are encapsulated on an object level and hence hidden to other components. Access to contextual information is provided through specified interfaces only.

5. Logic-Based Models

    A logic defines the conditions on which a concluding expression or fact may be derived (a process known as reasoning or inferencing) from a set of other expressions or facts. To describe these conditions in a set of rules a formal system is applied. In a logic-based context model, the context is consequently defined as facts, expressions and rules. Usually contextual information is added to, updated in and deleted from a logic based system in terms of facts or inferred from the rules in the system respectively. Common to all logic-based models is a high degree of formality.

6. Ontology-Based Models

    Ontologies are particularly suitable to project parts of the information describing and being used in our daily life onto a data structure utilizable by computers. Three ontology-based models are presented in this survey: i) Context Ontology Language (CoOL), (Strang et al., 2003); ii) the CONON context modeling approach (Wang et al., 2004); and iii) the CoBrA system (Chen et al., 2003a).

The survey of context modeling for pervasive cooperative learning covers the above-mentioned context modeling approaches and introduces a Machine Learning Modeling (MLM) approach that uses machine learning (ML) techniques. It concludes that to achieve the system design objectives, the use of ML approaches in combination with semantic context reasoning ontologies offers promising research directions to enable the effective implementation of context (Moore et al., 2007).

The role of ontologies has been emphasized in multitude of the surveys, e.g., (Baldauf et al., 2007), (Soylu et al., 2009), (Hong et al., 2009), (Truong & Dustdar, 2009). The survey related to context modeling and reasoning techniques (Bettini et al., 2010) highlights that ontological models of context provide clear advantages both in terms of heterogeneity and interoperability. Web Ontology Language, OWL, (OWL, 2004) is a de facto standard for describing context ontology. OWL is one of W3C recommendations (www.w3.org) for a Semantic Web. Graphical tools, such as Protégé and NeOnToolkit, exist for describing ontologies.

3.3 Context platform and storage

Eugster et al. present the middleware classification that they performed for 22 middleware platforms from the viewpoint of a developer of context-aware applications (Eugster et al., 2009). That is one of the many surveys done on the context-aware systems but it is interesting because of the developer viewpoint. They classified the platforms according to i) the type of context, ii) the given programming support, and iii) architectural dimensions such as decentralization, portability, and interoperability. The most relevant classification criteria of those are currently the high-level programming support and the three architectural dimensions.

High-level programming support means that the middleware platform adds a context storage and management. The three architectural dimensions are: (1) decentralization, (2) portability, and (3) interoperability. Decentralization measures a platform's dependence on specific components. Portability classifies platforms into two groups: portable platforms can run on many different operating systems, and operating system-dependent platforms, which can only run on few operating systems (usually one). Interoperability then measures the ease with which a platform can communicate with heterogeneous software components.

Ideal interoperable platforms can communicate with many different applications, regardless of the operating system on which they are built or of the programming language in which they are written. This kind of InterOperabilility Platform (IOP) is developed in the SOFIA-project (www.sofia-project.eu). The IOP's context storage is a Semantic Information Broker (SIB), which is a Resource Description Framework, RDF, (RDF, 2004) database. Software agents which are called Knowledge Processors (KP) can connect to the SIB and exchange information through an XML-based interaction protocol called Smart Space Access Protocol (SSAP). KPs use a Knowledge Processor Interface (KPI) to communicate with the SIB. KPs consume and produce RDF triples into the SIB according to the used ontology.

The IOP is proposed to be extended, where and when needed, with context-aware functionalities following 'the separation of concern' principle to keep application free of the context (Toninelli et al., 2009).

Kuusijärvi and Stenius illustrate how reusable KPs can be designed and implemented, i.e., how to apply 'for reuse' and 'with reuse' practices in the development of smart environments (Kuusijärvi & Stenius, 2011). Thus, they cover the need for programming level reusability.

3.4 Context-aware micro-architecture

When context information is described by OWL and ontologies, typically reasoning techniques will be based on a semantic approach, such as SPARQL Query Language for RDF (SPARQL), (Truong & Dustdar, 2009).

The context-awareness micro-architecture, CAMA, is the solution for managing adaptation based on context in smart environments. Context-awareness micro-architecture consists of three types of agents: context monitoring, context reasoning and context-based adaptation agents (Pantsar-Syväniemi et al., 2011a). These agents share information via the semantic database. Figure 4 illustrates the structural viewpoint of the logical context-awareness micro-architecture.

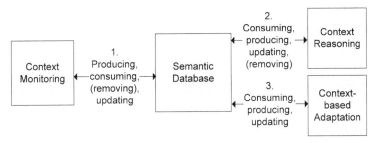

Fig. 4. The logical structure of the CAMA.

The context-monitoring agent is configured via configuration parameters which are defined by the architect of the intelligent application. The configuration parameters can be updated at run-time because the parameters follow the used context. The configuration parameters can be given by the ontology, i.e., a set of triples to match, or by a SPARQL query, if the monitored data is more complicated. The idea is that the context monitoring recognizes the current status of the context information and reports this to the semantic database. Later on, the reported information can be used in decision making.

The rule-based reasoning agent is based on a set of rules and a set of activation conditions for these rules. In practice, the rules are elaborated 'if-then-else' statements that drive activation of behaviors, i.e., activation patterns. The architect describes behavior by MSC diagrams with annotated behavior descriptions attached to the agents. Then, the behavior is transformed into SPARQL rules by the developer who exploits the MSC diagrams and the defined ontologies to create SPARQL queries. The developer also handles the dynamicity of the space by providing the means to change the rules at run-time. The context reasoning is a fully dynamic agent, whose actions are controlled by the dynamically changing rules (at run-time).

If the amount of agents producing and consuming inferred information is small, the rules can be checked by hand during the development phase of testing. If an unknown amount of agents are executing an unknown amount of rules, it may lead to a situation where one rule affects another rule in an unwanted way. A usual case is that two agents try to change the state of an intelligent object at the same time resulting in an unwanted situation. Therefore, there should be an automated way of checking all the rules and determining possible problems prior to executing them. Some of these problems can be solved by bringing priorities into the rules, so that a single agent can determine what rules to execute at a given time. This, of course, implies that only one agent has rules affecting certain intelligent objects.

CAMA has been used:

- to activate required functionality according to the rules and existing situation(s) (Pantsar-Syväniemi et al., 2011a)
- to map context and domain-specific ontologies in a smart maintenance scenario for a context-aware supervision feature (Pantsar-Syväniemi et al., 2011b)
- in run-time security management for monitoring situations (Evesti & Pantsar-Syväniemi, 2010)

The Context Ontology for Smart Spaces, (CO4SS), is meant to be used together with the CAMA. It has been developed because the existing context ontologies were already few years old and not generic enough (Pantsar-Syväniemi et al, 2012). The objective of the CO4SS is to support the evolution management of the smart space: all smart spaces and their applications 'understand' the common language defined by it. Thus, the context ontology is used as a foundational ontology to which application-specific or run-time quality management concepts are mapped.

4. Conclusion

The role of software in large embedded systems, like in base stations, has changed remarkably in the last three decades; software has become more dominant compared to the role of hardware. The progression of processors and compilers has prepared the way for reuse and software product lines by means of C language, especially in the area of DSP software. Context-aware systems have been researched for many years and the maturity of the results has been growing. A similar evolution has happened with the object-oriented engineering that comes to DSP software. Although the methods were mature, it took many years to gain proper processors and compilers that support coding with C language. This shows that without hardware support there is no room to start to use the new methods.

The current progress of hardware development regarding size, cost and energy consumption is speeding up the appearance of context-aware systems. This necessitates that the information be distributed to our daily environment along with smart but separated things like sensors. The cooperation of the smart things by themselves and with human beings demands new kinds of embedded software. The new software is to be designed by the ontological approach and instead of the process being top-down, it should use the bottom-up way. The bottom-up way means that the smart space applications are formed from the small functionalities, micro-architecture, which can be configured at design time, on instantiation time and during run-time.

The new solution to designing the context management of context-aware systems from the bottom-up is context-aware micro-architecture, CAMA, which is meant to be used with CO4SS ontology. The CO4SS provides generic concepts of the smart spaces and is a common 'language'. The ontologies can be compared to the message-based interface specifications in the base stations. This solution can be the grounds for new initiatives or a body to start forming the 'borders', i.e., the system architecture, for the context-aware ecosystem.

5. Acknowledgment

The author thanks Eila Ovaska from the VTT Technical Research Centre and Olli Silvén from the University of Oulu for their valuable feedback.

6. References

Achillelos, A.; Yang, K. & Georgalas, N. (2009). Context modelling and a context-aware framework for pervasive service creation: A model-driven approach, *Pervasive and Mobile Computing*, Vol.6, No.2, (April, 2010), pp. 281-296, ISSN 1574-1192

Awad, M.; Kuusela, J. & Ziegler, J. (1996). *Object-Oriented Technology for Real-Time Systems. A Practical Approach Using OMT and Fusion*, Prentice-Hall Inc., ISBN 0-13-227943-6, Upper Saddle River, NJ, USA

Baldauf, M.; Dustdar, S. & Rosenberg, F. (2007). A survey on context-aware systems, *International Journal of Ad Hoc and Ubiquitous Computing*, Vol.2, No.4., (June, 2007), pp. 263-277, ISSN 1743-8225

Bass, L.; Clements, P. & Kazman, R. (1998*). Software Architecture in Practice*, first ed., Addison-Wesley, ISBN 0-201-19930-0, Boston, MA, USA

Bettini, C.; Brdiczka, O.; Henricksen, K.; Indulska, J.; Nicklas, D.; Ranganathan, A. & Riboni D. (2010). A survey of context modelling and reasoning techniques. *Pervasive and Mobile Computing*, Vol.6, No.2, (April, 2010), pp.161 – 180, ISSN 1574-1192

Bosch, J. (1999). Product-line architectures in industry: A case study, *Proceedings of ICSE 1999 21st International Conference on Software Engineering*, pp. 544-554, ISBN 1-58113-074-0, Los Angeles, CA, USA, May 16-22, 1999

Bosch, J. (2000). *Design and Use of Software Architectures. Adopting and evolving a product-line approach*, Addison-Wesley, ISBN 0-201-67484-7, Boston, MA, USA

Chen, H.; Finin, T. & Joshi, A. (2003a). Using OWL in a Pervasive Computing Broker, *Proceedings of AAMAS 2003 Workshop on Ontologies in Open Agent Systems*, pp.9-16, ISBN 1-58113-683-8, ACM, July, 2003

Clements, P.C.; Bachmann, F.; Bass L.; Garlan, D.; Ivers, J.; Little, R.; Nord, R. & Stafford, J. (2003). *Documenting Software Architectures, Views and Beyond*, Addison-Wesley, ISBN 0-201-70372-6, Boston, MA, USA

Coleman, D.; Arnold, P.; Bodoff, S.; Dollin, C.; Gilchrist, H.; Hayes, F. & Jeremaes, P. (1993). *Object-Oriented Development – The Fusion Method*, Prentice Hall, ISBN 0-13-338823-9, Englewood Cliffs, NJ, USA

CPRI. (2003). Common Public Radio Interface, 9.10.2011, Available from http://www.cpri.info/

Dey, A. K. & Abowd, G. D. (1999). *Towards a Better Understanding of Context and Context-Awareness*. Technical Report GIT-GVU-99-22, Georgia Institute of Technology, College of Computing, USA

Enders, A. & Rombach, D. (2003). *A Handbook of Software and Systems Engineering, Empirical Observations, Laws and Theories*, Pearson Education, ISBN 0-32-115420-7, Harlow, Essex, England, UK

Eugster, P. Th.; Garbinato, B. & Holzer, A. (2009) Middleware Support for Context-aware Applications. In: *Middleware for Network Eccentric and Mobile Applications* Garbinato, B.; Miranda, H. & Rodrigues, L. (eds.), pp. 305-322, Springer-Verlag, ISBN 978-3-642-10053-6, Berlin Heidelberg, Germany

Evesti, A. & Pantsar-Syväniemi, S. (2010). Towards micro architecture for security adaption, Proceedings of ECSA 2010 4th European Conference on Software Architecture Doctoral Symposium, Industrial Track and Workshops, pp. 181-188, Copenhagen, Denmark, August 23-26, 2010

France, R. & Rumpe, B. (2007). Model-driven Development of Complex Software: A Research Roadmap. Proceedings of FOSE'07 International Conference on Future of Software Engineering, pp. 37-54, ISBN 0-7695-2829-5, IEEE Computer Society, Washington DC, USA, March, 2007

Goossens, G.; Van Praet, J.; Lanneer, D.; Geurts, W.; Kifli, A.; Liem, C. & Paulin, P. (1997) Embedded Software in Real-Time Signal Processing Systems: Design Technologies. *Proceedings of the IEEE*, Vol. 85, No.3, (March, 1997), pp.436–454, ISSN 0018-9219

Hillebrand, F. (1999). The Status and Development of the GSM Specifications, In: *GSM Evolutions Towards 3rd Generation Systems*, Zvonar, Z.; Jung, P. & Kammerlander, K., pp. 1-14, Kluwer Academic Publishers, ISBN 0-792-38351-6, Boston, USA

Hong, J.; Suh, E. & Kim, S. (2009). Context-aware systems: A literature review and classification. *Expert System with Applications*, Vol.36, No.4, (May 2009), pp. 8509-8522, ISSN 0957-4174

Indulska, J. & Nicklas, D. (2010). Introduction to the special issue on context modelling, reasoning and management, *Pervasive and Mobile Computing*, Vol.6, No.2, (April 2010), pp. 159-160, ISSN 1574-1192

Jacobson, I., et al. (1992). *Object-Oriented Software Engineering – A Use Case Driven Approach*, Addison-Wesley, ISBN 0-201-54435-0, Reading, MA, USA

Karlsson, E-A. (1995). *Software Reuse. A Holistic Approach*, Wiley, ISBN 0-471-95819-0, Chichester, UK

Kapitsaki, G. M.; Prezerakos, G. N.; Tselikas, N. D. & Venieris, I. S. (2009). Context-aware service engineering: A survey, *The Journal of Systems and Software*, Vol.82, No.8, (August, 2009), pp.1285-1297, ISSN 0164-1212

Kronlöf, K. (1993). *Method Integration: Concepts and Case Studies*, John Wiley & Sons, ISBN 0-471-93555-7, New York, USA

Krüchten, P. (1995). Architectural Blueprints—The "4+1" View Model of Software Architecture, *IEEE Software*, Vol.12, No.6, (November, 1995), pp.42-50, ISSN 0740-7459

Kuusijärvi, J. & Stenudd, S. (2011). Developing Reusable Knowledge Processors for Smart Environments, *Proceedings of SISS 2011 The Second International Workshop on "Semantic Interoperability for Smart Spaces" on 11th IEEE/IPSJ International Symposium on Applications and the Internet (SAINT 2011)*, pp. 286-291, Munich, Germany, July 20, 2011

Miller J. & Mukerji, J. (2003). MDA Guide Version 1.0.1. http://www.omg.org/docs/omg/03-06-01.pdf

Moore, P.; Hu, B.; Zhu, X.; Campbell, W. & Ratcliffe, M. (2007). A Survey of Context Modeling for Pervasive Cooperative Learning, *Proceedings of the ISITAE'07 1st IEEE International Symposium on Information Technologies and Applications in Education*, pp.K51-K56, ISBN 978-1-4244-1385-0, Nov 23-25, 2007

Nokia Siemens Networks. (2011). Liquid Radio - Let traffic waves flow most efficiently. White paper. 17.11.2011, Available from http://www.nokiasiemensnetworks.com/portfolio/liquidnet

OBSAI. (2002). Open Base Station Architecture Initiative, 10.10.2011, Available from http://www.obsai.org/

OWL. (2004). Web Ontology Language Overview, W3C Recommendation, 29.11.2011, Available from http://www.w3.org/TR/owl-features/

Palmberg, C. & Martikainen, O. (2003) *Overcoming a Technological Discontinuity - The case of the Finnish telecom industry and the GSM*, Discussion Papers No.855, The Research Institute of the Finnish Economy, ETLA, Helsinki, Finland, ISSN 0781-6847

Pantsar-Syväniemi, S.; Taramaa, J. & Niemelä, E. (2006). Organizational evolution of digital signal processing software development, *Journal of Software Maintenance and Evolution: Research and Practice*, Vol.18, No.4, (July/August, 2006), pp. 293-305, ISSN 1532-0618

Pantsar-Syväniemi, S. & Ovaska, E. (2010). Model based architecting with MARTE and SysML profiles. *Proceedings of SE 2010 IASTED International Conference on Software Engineering*, 677-013, Innsbruck, Austria, Feb 16-18, 2010

Pantsar-Syväniemi, S.; Kuusijärvi, J. & Ovaska, E. (2011a) Context-Awareness Micro-Architecture for Smart Spaces, *Proceedings of GPC 2011 6th International Conference on Grid and Pervasive Computing*, pp. 148–157, ISBN 978-3-642-20753-2, LNCS 6646, Oulu, Finland, May 11-13, 2011

Pantsar-Syväniemi, S.; Ovaska, E.; Ferrari, S.; Salmon Cinotti, T.; Zamagni, G.; Roffia, L.; Mattarozzi, S. & Nannini, V. (2011b) Case study: Context-aware supervision of a smart maintenance process, *Proceedings of SISS 2011 The Second International Workshop on "Semantic Interoperability for Smart Spaces", on 11th IEEE/IPSJ International Symposium on Applications and the Internet (SAINT 2011)*, pp.309-314, Munich, Germany, July 20, 2011

Pantsar-Syväniemi, S.; Kuusijärvi, J. & Ovaska, E. (2012) Supporting Situation-Awareness in Smart Spaces, *Proceedings of GPC 2011 6th International Conference on Grid and Pervasive Computing Workshops*, pp. 14–23, ISBN 978-3-642-27915-7, LNCS 7096, Oulu, Finland, May 11, 2011

Paulin, P.G.; Liem, C.; Cornero, M.; Nacabal, F. & Goossens, G. (1997). Embedded Software in Real-Time Signal Processing Systems: Application and Architecture Trends, *Proceedings of the IEEE,* Vol.85, No.3, (March, 2007), pp.419-435, ISSN 0018-9219

Pohl, K.; Böckle, G. & van der Linden, F. (2005). *Software Product Line Engineering*, Springer-Verlag, ISBN 3-540-24372-0, Berlin Heidelberg

Purhonen, A. (2002). *Quality Driven Multimode DSP Software Architecture Development*, VTT Electronics, ISBN 951-38-6005-1, Espoo, Finland

RDF. Resource Description Framework, 29.11.2011, Available from http://www.w3.org/RDF/

Rumbaugh, J.; Blaha, M.; Premerlani, W.; Eddy, F. & Lorensen, W. (1991) *Object-Oriented Modeling and Design*, Prentice-Hall Inc., ISBN 0-13-629841-9, Upper Saddle River, NJ, USA

Shaw, M. (1990). Toward High-Level Abstraction for Software Systems, *Data and Knowledge Engineering,* Vol. 5, No.2, (July 1990), pp. 119-128, ISSN 0169-023X

Shlaer, S. & Mellor, S.J. (1992) *Object Lifecycles: Modeling the World in States*, Prentice-Hall, ISBN 0-13-629940-7, Upper Saddle River, NJ, USA

Soylu, A.; De Causmaecker1, P. & Desmet, P. (2009). Context and Adaptivity in Pervasive Computing Environments: Links with Software Engineering and Ontological

Engineering, *Journal of Software*, Vol.4, No.9, (November, 2009), pp.992-1013, ISSN 1796-217X

SPARQL. SPARQL Query Language for RDF, W3C Recommendation, 29.11.2011, Available from http://www.w3.org/TR/rdf-sparql-query/

Strang, T.; Linnhoff-Popien, C. & Frank, K. (2003). CoOL: A Context Ontology Language to enable Contextual Interoperability, *Proceedings of DAIS2003 4th IFIP WG 6.1 International Conference on Distributed Applications and Interoperable Systems*, pp.236-247, LNCS 2893, Springer-Verlag, ISBN 978-3-540-20529-6, Paris, France, November 18-21, 2003

Strang, T. & Linnhoff-Popien, C. (2004). A context modelling survey, *Proceedings of UbiComp 2004 1st International Workshop on Advanced Context Modelling, Reasoning and Management*, pp.31-41, Nottingham, England, September, 2004

Toninelli, A.; Pantsar-Syväniemi, S.; Bellavista, P. & Ovaska, E. (2009) Supporting Context Awareness in Smart Environments: a Scalable Approach to Information Interoperability, *Proceedings of M-PAC'09 International Workshop on Middleware for Pervasive Mobile and Embedded Computing*, session: short papers, Article No: 5, ISBN 978-1-60558-849-0, Urbana Champaign, Illinois, USA, November 30, 2009

Truong, H. & Dustdar, S. (2009). A Survey on Context-aware Web Service Systems. *International Journal of Web Information Systems*, Vol.5, No.1, pp. 5-31, ISSN 1744-0084

Wang, X. H.; Zhang, D. Q.; Gu, T. & Pung, H. K. (2004). Ontology Based Context Modeling and Reasoning using OWL, *Proceedings of PerComW '04 2nd IEEE Annual Conference on Pervasive Computing and Communications Workshops*, pp. 18–22, ISBN 0-7695-2106-1, Orlando, Florida, USA, March 14-17, 2004

7

FSMD-Based Hardware Accelerators for FPGAs

Nikolaos Kavvadias, Vasiliki Giannakopoulou and Kostas Masselos
Department of Computer Science and Technology,
University of Peloponnese, Tripoli
Greece

1. Introduction

Current VLSI technology allows the design of sophisticated digital systems with escalated demands in performance and power/energy consumption. The annual increase of chip complexity is 58%, while human designers productivity increase is limited to 21% per annum (ITRS, 2011). The growing technology-productivity gap is probably the most important problem in the industrial development of innovative products. A dramatic increase in designer productivity is only possible through the adoption of methodologies/tools that raise the design abstraction level, ingeniously hiding low-level, time-consuming, error-prone details. New EDA methodologies aim to generate digital designs from high-level descriptions, a process called High-Level Synthesis (HLS) (Coussy & Morawiec, 2008) or else hardware compilation (Wirth, 1998). The input to this process is an algorithmic description (for example in C/C++/SystemC) generating synthesizable and verifiable Verilog/VHDL designs (IEEE, 2006; 2009).

Our aim is to highlight aspects regarding the organization and design of the targeted hardware of such process. In this chapter, it is argued that a proper Model of Computation (MoC) for the targeted hardware is an adapted and extended form of the FSMD (Finite-State Machine with Datapath) model which is universal, well-defined and suitable for either data- or control-dominated applications. Several design examples will be presented throughout the chapter that illustrate our approach.

2. Higher-level representations of FSMDs

This section discusses issues related to higher-level representations of FSMDs (Gajski & Ramachandran, 1994) focusing on textual intermediate representations (IRs). It first provides a short overview of existing approaches focusing on the well-known GCC GIMPLE and LLVM IRs. Then the BASIL (Bit-Accurate Symbolic Intermediate Language) is introduced as a more appropriate lightweight IR for self-contained representation of FSMD-based hardware architectures. Lower-level graph-based forms are presented focusing on the CDFG (Control-Data Flow Graph) procedure-level representation using Graphviz (*Graphviz*, 2011) files. This section also illustrates a linear CDFG construction algorithm from BASIL. In addition, an end-to-end example is given illustrating algorithmic specifications in ANSI

C, BASIL, Graphviz CDFGs and their visualizations utilizing a 2D Euclidean distance approximation function.

2.1 Overview of compiler intermediate representations

Recent compilation frameworks provide linear IRs for applying analyses, optimizations and as input for backend code generation. GCC (GCC, 2011) supports the GIMPLE IR. Many GCC optimizations have been rewritten for GIMPLE, but it is still undergoing grammar and interface changes. The current GCC distribution incorporates backends for contemporary processors such as the Cell SPU and the baseline Xtensa application processor (Gonzalez, 2000) but it is not suitable for rapid retargeting to non-trivial and/or custom architectures. LLVM (LLVM, 2011) is a compiler framework that draws growing interest within the compilation community. The LLVM compiler uses the homonymous LLVM bitcode, a register-based IR, targeted by a C/C++ companion frontend named clang (clang homepage, 2011). It is written in a more pleasant coding style than GCC, but similarly the IR infrastructure and semantics are excessive.

Other academic infrastructures include COINS (COINS, 2011), LANCE (LANCE, 2011) and Machine-SUIF (Machine-SUIF, 2002). COINS is written entirely in Java, and supports two IRs: the HIR (high level) and the LIR (low-level) which is based on S-expressions. COINS features a powerful SSA-based optimizer, however its LISP-like IR is unsuitable for directly expressing control and data dependencies and to fully automate the construction of a machine backend. LANCE (Leupers et al., 2003) introduces an executable IR form (IR-C), which combines the simplicity of three-address code with the executability of ANSI C code. LANCE compilation passes accept and emit IR-C, which eases the integration of LANCE into third-party environments. However, ANSI C semantics are neither general nor neutral enough in order to express vastly different IR forms. Machine-SUIF is a research compiler infrastructure built around the SUIFvm IR which has both a CFG (control-flow graph) and SSA form. Past experience with this compiler has proved that it is overly difficult both to alter or extend its semantics. It appears that the Phoenix (Microsoft, 2008) compiler is a rewrite and extension of Machine-SUIF in C#. As an IR, the CIL (Common Intermediate Language) is used which is entirely stack-based, a feature that hinders the application of modern optimization techniques. Finally, CoSy (CoSy, 2011) is the prevalent commercial retargetable compiler infrastructure. It uses the CCMIR intermediate language whose specification is confidential. Most of these frameworks fall short in providing a minimal, multi-purpose compilation infrastructure that is easy to maintain and extend.

The careful design of the compiler intermediate language is a necessity, due to its dual purpose as both the program representation and an abstract target machine. Its design affects the complexity, efficiency and ease of maintenance of all compilation phases; frontend, optimizer and effortlessly retargetable backend.

The following subsection introduces the BASIL intermediate representation. BASIL supports semantic-free n-input/m-output mappings, user-defined data types, and specifies a virtual machine architecture. BASIL's strength is its simplicity: it is inherently easy to develop a CDFG (control/data flow graph) extraction API, apply graph-based IR transformations for

Data type	Regular expression	Example
UNSIGNED_INT	[Uu][1-9][0-9]*	u32
SIGNED_INT	[Ss][1-9][0-9]*	s11
UNSIGNED/ SIGNED_FXP	[Qq][0-9]+.[0-9]+[S|U]	q4.4u, q2.14s
FLP	[Ff][0|1].[0-9]+.[0-9]+	F1.8.23 fields: sign, exponent, mantissa

Table 1. Data type specifications in BASIL.

domain specialization, investigate SSA (Static Single Assignment) construction algorithms and perform other compilation tasks.

2.2 Representing programs in BASIL

BASIL provides arbitrary n-to-m mappings allowing the elimination of implicit side-effects, a single construct for all operations, and bit-accurate data types. It supports scalar, single-dimensional array and streamed I/O procedure arguments. BASIL statements are labels, n-address instructions or procedure calls.

BASIL is similar in concept to the GIMPLE and LLVM intermediate languages but with certain unique features. For example, while BASIL supports SSA form, it provides very light operation semantics. A single construct is required for supporting any given operation as an m-to-n mapping between source and destination sites. An n-address operation is actually the specification of a mapping from a set of n ordered inputs to a set of m ordered outputs. An n-address instruction (or else termed as an n, m-operation) is formatted as follows:

outp1, ..., outpm <= operation inp1, ..., inpn; where:

- operation is a mnemonic referring to an IR-level instruction
- outp1, ..., outpm are the m outputs of the operation
- inp1, ..., inpn are the n inputs of the operation

In BASIL all declared objects (global variables, local variables, input and output procedure arguments) have an explicit static type specification. BASIL uses the notions of "globalvar" (a global scalar or single-dimensional array variable), "localvar" (a local scalar or single-dimensional array variable), "in" (an input argument to the given procedure), and "out" (an output argument to the given procedure).

BASIL supports bit-accurate data types for integer, fixed-point and floating-point arithmetic. Data type specifications are essentially strings that can be easily decoded by a regular expression scanner; examples are given in Table 1.

The EBNF grammar for BASIL is shown in Fig. 1 where it can be seen that rules "nac" and "pcall" provide the means for the n-to-m generic mapping for operations and procedure calls, respectively. It is important to note that BASIL has no predefined operator set; operators are defined through a textual mnemonic.

For instance, an addition of two scalar operands is written: a <= add b, c;. Control-transfer operations include conditional and unconditional jumps explicitly visible in

```
basil_top = {gvar_def} {proc_def}.
gvar_def = "globalvar" anum decl_item_list ";".
proc_def = "procedure" [anum] "(" [arg_list] ")"
                "{" [{lvar_decl}] [{stmt}] "}".
stmt = nac | pcall | id ":".
nac = [id_list "<="] anum [id_list] ";".
pcall = ["(" id_list ")" "<="] anum ["(" id_list ")"] ";".
id_list = id {"," id}.
decl_item_list = decl_item {"," decl_item}.
decl_item = (anum | uninitarr | initarr).
arg_list = arg_decl {"," arg_decl}.
arg_decl = ("in" | "out") anum (anum | uninitarr).
lvar_decl = "localvar" anum decl_item_list ";".
initarr = anum "[" id "]" "=" "{" numer {"," numer} "}".
uninitarr = anum "[" [id] "]".
anum = (letter | "_") {letter | digit}.
id = anum | (["-"] (integer | fxpnum)).
```

Fig. 1. EBNF grammar for BASIL.

the IR. An example of an unconditional jump would be: BB5 <= jmpun; while conditional jumps always declare both targets: BB1, BB2 <= jmpeq i, 10;. This statement enables a control transfer to the entry of basic block BB1 when i equals to 10, otherwise to BB2. Multi-way branches corresponding to compound decoding clauses can be easily added.

An interesting aspect of BASIL is the support of procedures as non-atomic operations by using a similar form to operations. In (y) <= sqrt(x); the square root of an operand x is computed; procedure argument lists are indicated as enclosed in parentheses.

2.3 BASIL program structure and encoding

A specification written in BASIL incorporates the complete information of a translation unit of the original program comprising of a list of "globalvar" definitions and a list of procedures (equivalently: control-flow graphs). A single BASIL procedure is captured by the following information:

- procedure name
- ordered input (output) arguments
- "localvar" definitions
- BASIL statements.
- basic block labels.

Label items point to basic block (BB) entry points and are defined as name, bb, addr 3-tuples, where *name* is the corresponding identifier, *bb* the basic block enumeration, and *addr* the absolute address of the statement succeeding the label.

Statements are organized in the form of a C struct or equivalently a record (in other programming languages) as shown in Fig. 2.

The Statement ADT therefore can be used to model an (n, m)-operation. The input and output operand lists collect operand items, as defined in the OperandItem data structure definition shown in Fig. 3.

```
typedef struct {
  char *mnemonic;      /* Designates the statement type. */
  NodeType ntype;      /* OPERATION or PROCEDURE_CALL. */
  List opnds_in;       /* Collects all input operands. */
  List opnds_out;      /* Collects all output operands. */
  int  bb;             /* Basic block number. */
  int  addr;           /* Absolute statement address. */
} _Statement;
typedef _Statement *Statement;
```

Fig. 2. C-style record for encoding a BASIL statement.

```
typedef struct {
  char *name;          /* Identifier name. */
  char *dataspec;      /* Data type string spec. */
  OperandType otype;   /* Operand type representation. */
  int  ix;             /* Absolute operand item index. */
} _OperandItem;
typedef _OperandItem *OperandItem;
```

Fig. 3. C-style record for encoding an OperandItem.

The OperandItem data structure is used for representing input arguments (INVAR), output arguments (OUTVAR), local (LOCALVAR) and global (GLOBALVAR) variables and constants (CONSTANT). If using a graph-based intermediate representation, arguments and constants could use node and incoming or outgoing edge representations, while it is meaningful to represent variables as edges as long as their storage sites are not considered.

The typical BASIL program is structured as follows:

```
<Global variable declarations>

procedure name_1 (
  <comma-separated input arguments>,
  <comma-separated output arguments>
) {
  <Local variable declarations>
  <BASIL labels, instructions, procedure calls>
}
...
procedure name_n (
  <comma-separated input arguments>,
  <comma-separated output arguments>
) {
  <Local variable declarations>
  <BASIL labels, instructions, procedure calls>
}
```

Fig. 4. Translation unit structure for BASIL.

Mnemonic	Description	(N_i, N_o)
ldc	Load constant	(1,1)
neg, mov	Unary arithmetic op.	(1,1)
add, sub, abs, min, max, mul, div, mod, shl, shr	Binary arithmetic op.	(2,1)
not, and, ior, xor	Logical	(2,1)
szz	Comparison for zz: (eq,ne,lt,le,gt,ge)	(2,1)
muxzz	Conditional selection	(3,1)
load, store	Load/Store register from/to memory	(2,1)
sxt, zxt, trunc	Type conversion	(1,1)
jmpun	Unconditional jump	(0,1)
jmpzz	Conditional jump	(2,2)
print	Diagnostic output	(1,0)

Table 2. A set of basic operations for a BASIL-based IR.

2.4 A basic BASIL implementation

A basic operation set for RISC-like compilation is summarized in Table 2. N_i (N_o) denotes the number of input (output) operands for each operation.

The memory access model defines dedicated address spaces per array, so that both loads and stores require the array identifier as an explicit operand. For an indexed load in C (b = a[i];), a frontend would generate the following BASIL: b <= load a, i;, while for an indexed store (a[i] = b;) it is a <= store b, i;.

Pointer accesses can be handled in a similar way, although dependence extraction requires careful data flow analysis for non-trivial cases. Multi-dimensional arrays are handled through matrix flattening transformations.

2.5 CDFG construction

A novel, fast CDFG construction algorithm has been devised for both SSA and non-SSA BASIL forms producing flat CDFGs as Graphviz files (Fig. 5). A CDFG symbol table item is a node (operation, procedure call, globalvar, or constant) or edge (localvar) with user-defined attributes: the unique name, label and data type specification; node and edge type enumeration; respective order of incoming or outgoing edges; input/output argument order of a node and basic block index. Further attributes can be defined, e.g. for scheduling bookkeeping.

This approach is unique since it focuses on building the CDFG symbol table (st) from which the associated graph (cdfg) is constructed as one possible of many facets. It naturally supports loop-carried dependencies and array accesses.

2.6 Fixed-point arithmetic

The use of fixed-point arithmetic (Yates, 2009) provides an inexpensive means for improved numerical dynamic range, when artifacts due to quantization and overflow effects can be tolerated. Rounding operators are used for controlling the numerical precision involved in a series of computations; they are defined for inexact arithmetic representations such as fixed-

```
BASILtoCDFG()
  input List BASILs, List variables, List labels, Graph cfg;
  output SymbolTable st, Graph cdfg;
begin
  Insert constant, input/output arguments and global
  variable operand nodes to st;
  Insert operation nodes;
  Insert incoming {global/constant/input, operation} and
  outgoing {operation, global/output} edges;
  Add control-dependence edges among operation nodes;
  Add data-dependence edges among operation nodes,
  extract loop-carried dependencies via cfg-reachability;
  Generate cdfg from st;
end
```

Fig. 5. CDFG construction algorithm accepting BASIL input.

and floating-point. Proposed and in-use specifications for fixed-point arithmetic of related practice include:

- the C99 standard (ISO/IEC JTC1/SC22, 2007)

- lightweight custom implementations such as (Edwards, 2006)

- explicit data types with open source implementations (Mentor Graphics, 2011; SystemC, 2006)

Fixed-point arithmetic is a variant of the typical integral representation (2's-complement signed or unsigned) where a binary point is defined, purely as a notational artifact to signify integer powers of 2 with a negative exponent. Assuming an integer part of width $IW > 0$ and a fractional part with $-FW < 0$, the VHDL-2008 sfixed data type has a range of $2^{IW-1} - 2^{|FW|}$ to -2^{IW-1} with a representable quantum of $2^{|FW|}$ (Bishop, 2010a;b). The corresponding ufixed type has the following range: $2^{IW} - 2^{|FW|}$ to 0. Both are defined properly given a IW-1:-FW vector range.

BASIL currently supports a proposed list of extension operators for handling fixed-point arithmetic:

- conversion from integer to fixed-point format: i2ufx, i2sfx

- conversion from fixed-point to integer format: ufx2i, sfx2i

- operand resizing: resize, using three input operands; source operand src1 and src2, src3 as numerical values that denote the new size (high-to-low range) of the resulting fixed-point operand

- rounding primitives: ceil, fix, floor, round, nearest, convergent for rounding towards plus infinity, zero, minus infinity, and nearest (ties to greatest absolute value, plus infinity and closest even, respectively).

2.7 Scan-based SSA construction algorithms for BASIL

In our experiments with BASIL we have investigated minimal SSA construction schemes – the Appel (Appel, 1998) and Aycock-Horspool (Aycock & Horspool, 2000) algorithms – that don't require the computation of the iterated dominance frontier (Cytron et al., 1991).

App.	LOC (BASIL)	LOC (dot)	$P/V/E$	#ϕs	#Instr.
atsort	155	484	2/136/336	10	6907
coins	105	509	2/121/376	10	405726
cordic	56	178	1/57/115	7	256335
easter	47	111	1/46/59	2	3082
fixsqrt	32	87	1/29/52	6	833900
perfect	31	65	1/23/36	4	6590739
sieve	82	199	2/64/123	12	515687
xorshift	26	80	1/29/45	0	2000

Table 3. Application profiling with a BASIL framework.

In traditional compilation infrastructures (GCC, LLVM) (GCC, 2011; LLVM, 2011), Cytron's approach (Cytron et al., 1991) is preferred since it enables bit-vector dataflow frameworks and optimizations that require elaborate data structures and manipulations. It can be argued that rapid prototyping compilers, integral parts of heterogeneous design flows, would benefit from straightforward SSA construction schemes which don't require the use of sophisticated concepts and data structures (Appel, 1998; Aycock & Horspool, 2000).

The general scheme for these methods consists of series of passes for variable numbering, ϕ-insertion, ϕ-minimization, and dead code elimination. The lists of BASIL statements, localvars and labels are all affected by the transformations.

The first algorithm presents a "really-crude" approach for variable renaming and ϕ-function insertion in two separate phases (Appel, 1998). In the first phase, every variable is split at BB boundaries, while in the second phase ϕ-functions are placed for each variable in each BB. Variable versions are actually preassigned in constant time and reflect a specific BB ordering (e.g. DFS). Thus, variable versioning starts from a positive integer n, equal to the number of BBs in the given CFG.

The second algorithm does not predetermine variable versions at control-flow joins but accounts ϕs the same way as actual computations visible in the original CFG. Due to this fact, ϕ-insertion also presents dissimilarities. Both methods share common ϕ-minimization and dead code elimination phases.

2.8 Application profiling with BASILVM

BASIL programs can be translated to low-level C for the easy evaluation of nominal performance on an abstract machine, called BASILVM. To show the applicability of BASILVM profiling, a set of small realistic integer/fixed-point kernels has been selected: *atsort* (an all topological sorts algorithm (Knuth, 2011)), *coins* (compute change with minimum amount of coins), *easter* (Easter date calculations), *fixsqrt* (fixed-point square root (Turkowski, 1995)), *perfect* (perfect number detection), *sieve* (prime sieve of Eratosthenes) and *xorshift* (100 calls to George Marsaglia's PRNG (Marsaglia, 2003) with a $2^{128} - 1$ period, which passes Diehard tests).

Static and dynamic metrics have been collected in Table 3. For each application (App.), the lines of BASIL and resulting CDFGs are given in columns 2-3, number of CDFGs (P:

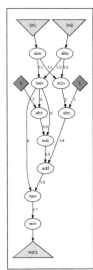

```
void eda(int in1, int in2,
  int *out1)
{
  int t1, t2, t3,
      t4, t5, t6, t7;
  int x, y;

  t1 = ABS(in1);
  t2 = ABS(in2);
  x = MAX(t1, t2);
  y = MIN(t1, t2);
  t3 = x >> 3;
  t4 = y >> 1;
  t5 = x - t3;
  t6 = t4 + t5;
  t7 = MAX(t6, x);
  *out1 = t7;
}
```

(a) ANSI C code.

```
procedure eda (in s16 in1, in s16 in2,
  out u16 out1)
{
  localvar u16 x, y,
             t1, t2, t3,
             t4, t5, t6, t7;
S_1:
  t1 <= abs in1;
  t2 <= abs in2;
  x <= max t1, t2;
  y <= min t1, t2;
  t3 <= shr x, 3;
  t4 <= shr y, 1;
  t5 <= sub x, t3;
  t6 <= add t4, t5;
  t7 <= max t6, x;
  out1 <= mov t7;
}
```

(b) BASIL code.

(c) CDFG code.

Fig. 6. Different facets of an euclidean distance approximation computation.

procedures), vertices and edges (for each procedure) in columns 4-5, amount of ϕ statements (column 6) and the number of dynamic instructions for the non-SSA case. The latter is measured using *gcc*-3.4.4 on Cygwin/XP by means of the executed code lines with the *gcov* code coverage tool.

2.9 Representative example: 2D Euclidean distance approximation

A fast linear algorithm for approximating the euclidean distance of a point (x, y) from the origin is given in (Gajski et al., 2009) by the equation: $eda = MAX((0.875 * x + 0.5 * y), x)$ where $x = MAX(|a|, |b|)$ and $y = MIN(|a|, |b|)$. The average error of this approximation against the integer-rounded exact value ($dist = \sqrt{a^2 + b^2}$) is 4.7% when compared to the rounded-down $\lfloor dist \rfloor$ and 3.85% to the rounded-up $\lceil dist \rceil$ value.

Fig. 6 shows the three relevant facets of *eda*: ANSI C code (Fig. 6(a)), a manually derived BASIL implementation (Fig. 6(b)) and the corresponding CDFG (Fig. 6(c)). Constant multiplications have been reduced to adds, subtracts and shifts. The latter subfigure naturally also shows the ASAP schedule of the data flow graph, which is evidently of length 7.

3. Architecture and organization of extended FSMDs

This section deals with aspects of specification and design of FSMDs, especially their interface, architecture and organization, as well as communication and integration issues. The section is wrapped-up with realistic examples of CDFG mappings to FSMDs, alongside their performance investigation with the help of HDL simulations.

3.1 FSMD overview

A Finite State Machine with Data (FSMD) specification (Gajski & Ramachandran, 1994) is an upgraded version of the well-known Finite State Machine representation providing the same information as the equivalent CDFG (Gajski et al., 2009). The main difference is the introduction of embedded actions within the next state generation logic. An FSMD specification is timing-aware since it must be decided that each state is executed within a certain amount of machine cycles. Also the precise RTL semantics of operations taking place within these cycles must be determined. In this way, an FSMD can provide an accurate model of an RTL design's performance as well as serve as a synthesizable manifestation of the designer's intent. Depending on the RT-level specification (usually VHDL or Verilog) it can convey sufficient details for hardware synthesis to a specific target platform, e.g. Xilinx FPGA devices (Xilinx, 2011b).

3.2 Extended FSMDs

The FSMDs of our approach follow the established scheme of a Mealy FSM with computational actions embedded within state logic (Chu, 2006). In this work, the extended FSMD MoC describing the hardware architectures supports the following features, the most relevant of which will be sufficiently described and supported by short examples:

- Support of scalar and array input and output ports.
- Support of streaming inputs and outputs and allowing mixed types of input and output ports in the same design block.
- Communication with embedded block and distributed LUT memories.
- Design of a latency-insensitive local interface of the FSMD units to master FSMDs, assuming the FSMD is a locally-interfaced slave.
- Design of memory interconnects for the FSMD units.

Advanced issues in the design of FSMDs that are not covered include the following:

- Mapping of SSA-form (Cytron et al., 1991) low-level IR (BASIL) directly to hardware, by the hardware implementation of variable-argument ϕ functions.
- External interrupts.
- Communication to global aggregate type storage (global arrays) from within the context of both root and non-root procedures using a multiplexer-based bus controlled by a scalable arbiter.

3.2.1 Interface

The FSMDs of our approach use fully-synchronous conventions and register all their outputs (Chu, 2006; Keating & Bricaud, 2002). The control interface is rather simple, yet can service all possible designs:

- *clk*: signal from external clocking source
- *reset* (*rst* or *arst*): synchronous or asynchronous reset, depending on target specification

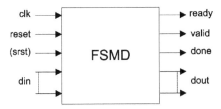

Fig. 7. FSMD I/O interface.

- *ready*: the block is ready to accept new input
- *valid*: asserted when a certain data output port is streamed-out from the block (generally it is a vector)
- *done*: end of computation for the block

ready signifies only the ability to accept new input (non-streamed) and does not address the status of an output (streaming or not).

Multi-dimensional data ports are feasible based on their equivalent single-dimensional flattened array type definition. Then, port selection is a matter of bitfield extraction. For instance, data input din is defined as din: in std_logic_vector(M*N-1 downto 0);, where M, N are generics. The flattened vector defines M input ports of width N. A selection of the form din((i+1)*N-1 downto i*N) is typical for a for-generate loop in order to synthesize iterative structures.

The following example (Fig. 8) illustrates an element-wise copy of array b to c without the use of a local array resource. Each interface array consists of 10 elements. It should be assumed that the physical content of both arrays lies in distributed LUT RAM, from which custom connections can be implemented.

Fig. 8(a) illustrates the corresponding function func1. The VHDL interface of func1 is shown in Fig. 8(b), where the derived array types b_type and c_type are used for b, c, respectively. The definitions of these types can be easily devised as aliases to a basic type denoted as: type cdt_type is array (9 downto 0) of std_logic_vector(31 downto 0);. Then, the alias for b is: alias b_type is cdt_type;

3.2.2 Architecture and organization

The FSMDs are organized as computations allocated into $n + 2$ states, where n is the number of required control steps as derived by an operation scheduler. The two overhead states are the entry (S_ENTRY) and the exit (S_EXIT) states which correspond to the source and sink nodes of the control-data flow graph of the given procedure, respectively.

Fig. 9 shows the absolute minimal example of a compliant FSMD written in VHDL. The FSMD is described in a two-process style using one process for the current state logic and another process for a combined description of the next state and output logic. This code will serve as a running example for better explaining the basic concepts of the FSMD paradigm.

```
procedure func1 (in s32 b[10],
                out s32 c[10]) {
  localvar s32 i, t;
S_1:
  i <= ldc 0;
  S_2 <= jmpun;
S_2:
  S_3, S_EXIT <= jmplt i, 10;
S_3:
  t <= load b, i;
  c <= store t, i;
  i <= add i, 1;
  S_2 <= jmpun;
S_EXIT:
  nop;
}
```

(a) BASIL code.

```
entity func1 is
  port (
    clk    : in   std_logic;
    reset  : in   std_logic;
    start  : in   std_logic;
    b      : in   b_type;
    c      : out  c_type;
    done   : out  std_logic;
    ready  : out  std_logic
  );
end func1;
```

(b) VHDL interface.

Fig. 8. Array-to-array copy without intermediate storage.

The example of Fig. 9(a), 9(b) implements the computation of assigning a constant value to the output port of the FSMD: `outp <= ldc 42;`. Thus, lines 5–14 declare the interface (entity) for the hardware block, assuming that `outp` is a 16-bit quantity. The FSMD requires three states. In line 17, a state type enumeration is defined consisting of types `S_ENTRY`, `S_EXIT` and `S_1`. Line 18 defines the signal 2-tuple for maintaining the state register, while in lines 19–20 the output register is defined. The current state logic (lines 25–34) performs asynchonous reset to all storage resources and assigns new contents to both the state and output registers. Next state and output logic (lines 37–57) decode `current_state` in order to determine the necessary actions for the computational states of the FSMD. State `S_ENTRY` is the idle state of the FSMD. When the FSMD is driven to this state, it is assumed ready to accept new input, thus the corresponding status output is raised. When a start prompt is given externally, the FSMD is activated and in the next cycle, state `S_1` is reached. In `S_1` the action of assigning `CNST_42` to `outp` is performed. Finally, when state `S_EXIT` is reached, the FSMD declares the end of all computations via `done` and returns to its idle state.

It should be noted that this design approach is a rather conservative one. One possible optimization that can occur in certain cases is the merging of computational states that immediately prediate the sink state (`S_EXIT`) with it.

Fig. 9(c) shows the timing diagram for the "minimal" design. As expected, the overall latency for computing a sample is three machine cycles.

In certain cases, input registering might be desired. This intent can be made explicit by copying input port data to an internal register. For the case of the *eda* algorithm, a new localvar, a would be introduced to perform the copy as `a <= mov in1;`. The VHDL counterpart is given as `a_1_next <= in1;`, making this data available through register `a_1_reg` in the following cycle. For register `r`, signal `r_next` represents the value that is available at the register input, and `r_reg` the stored data in the register.

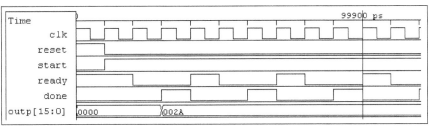

```
1   library IEEE;
2   use IEEE.std_logic_1164.all;
3   use IEEE.numeric_std.all;
4
5   entity minimal is
6     port (
7       clk    : in  std_logic;
8       reset  : in  std_logic;
9       start  : in  std_logic;
10      outp   : out std_logic_vector(15 downto 0);
11      done   : out std_logic;
12      ready  : out std_logic
13      );
14  end minimal;
15
16  architecture fsmd of minimal is
17    type state_type is (S_ENTRY, S_EXIT, S_1);
18    signal current_state, next_state: state_type;
19    signal outp_next: std_logic_vector(15 downto 0);
20    signal outp_reg: std_logic_vector(15 downto 0);
21    constant CNST_42: std_logic_vector(15 downto 0)
22                      := "0000000000101010";
23  begin
24    -- current state logic
25    process (clk, reset)
26    begin
27      if (reset = '1') then
28        current_state <= S_ENTRY;
29        outp_reg <= (others => '0');
30      elsif (clk = '1' and clk'EVENT) then
```

(a) VHDL code.

```
31        current_state <= next_state;
32        outp_reg <= outp_next;
33      end if;
34    end process;
35
36    -- next state and output logic
37    process (current_state, start, outp_reg)
38    begin
39      done <= '0';
40      ready <= '0';
41      outp_next <= outp_reg;
42      case current_state is
43      when S_ENTRY =>
44        ready <= '1';
45        if (start = '1') then
46          next_state <= S_1;
47        else
48          next_state <= S_ENTRY;
49        end if;
50      when S_1 =>
51        outp_next <= CNST_42;
52        next_state <= S_EXIT;
53      when S_EXIT =>
54        done <= '1';
55        next_state <= S_ENTRY;
56      end case;
57    end process;
58    outp <= outp_reg;
59  end fsmd;
```

(b) VHDL code (cont.)

(c) Timing diagram.

Fig. 9. Minimal FSMD implementation in VHDL.

3.2.3 Communication with embedded memories

Array objects can be synthesized to block RAMs in contemporary FPGAs. These embedded memories support fully synchronous read and write operations (Xilinx, 2005). A requirement for asynchronous read mandates the use of memory residing in distributed LUT storage.

In BASIL, the load and store primitives are used for describing read and write memory access. We will assume a RAM memory model with write enable, and separate data input (din) and output (dout) sharing a common address port (rwaddr). To control access to such block, a set of four non-trivial signals is needed: mem_we, a write enable signal, and the corresponding signals for addressing, data input and output.

store is the simpler operation of the two. It requires raising mem_we in a given single-cycle state so that data are stored in memory and made available in the subsequent state/machine cycle.

```
when STATE_1 =>
  mem_addr <= index;
  waitstate_next <= not (waitstate_reg);
  if (waitstate_reg = '1') then
    mysignal_next <= mem_dout;
    next_state <= STATE_2;
  else
    next_state <= STATE_1;
  end if;
when STATE_2 =>
  . . .
```

Fig. 10. Wait-state-based communication for loading data from a block RAM.

Synchronous `load` requires the introduction of a `waitstate` register. This register assists in devising a dual-cycle state for performing the load. Fig. 10 illustrates the implementation of a load operation. During the first cycle of STATE_1 the memory block is addressed. In the second cycle, the requested data are made available through mem_dout and are assigned to register `mysignal`. This data can be read from `mysignal_reg` during STATE_2.

3.2.4 Hierarchical FSMDs

Our extended FSMD concept allows for hierarchical FSMDs defining entire systems with calling and callee CDFGs. A two-state protocol can be used to describe a proper communication between such FSMDs. The first state is considered as the "preparation" state for the communication, while the latter state actually comprises an "evaluation" superstate where the entire computation applied by the callee FSMD is effectively hidden.

The calling FSMD performs computations where new values are assigned to *_next signals and registered values are read from *_reg signals. To avoid the problem of multiple signal drivers, callee procedure instances produce *_eval data outputs that can then be connected to register inputs by hardwiring to the *_next signal.

Fig. 11 illustrates a procedure call to an integer square root evaluation procedure. This procedure uses one input and one output std_logic_vector operands, both considered to represent integer values. Thus, a procedure call of the form (m) <= isqrt(x); is implemented by the given code segment in Fig. 11.

STATE_1 sets up the callee instance. The following state is a superstate where control is transferred to the component instance of the callee. When the callee instance terminates its computation, the ready signal is raised. Since the start signal of the callee is kept low, the generated output data can be transferred to the *m* register via its m_next input port. Control then is handed over to state STATE_3.

The callee instance follows the established FSMD interface, reading x_reg data and producing an exact integer square root in m_eval. Multiple copies of a given callee are supported by versioning of the component instances.

```
when STATE_1 =>
  isqrt_start <= '1';
  next_state <= SUPERSTATE_2;
when SUPERSTATE_2 =>
  if ((isqrt_ready = '1') and (isqrt_start = '0')) then
    m_next <= m_eval;
    next_state <= STATE_3;
  else
    next_state <= SUPERSTATE_2;
  end if;
when STATE_3 =>
...
isqrt_0 : entity WORK.isqrt(fsmd)
  port map (
    clk, reset,
    isqrt_start, x_reg, m_eval,
    isqrt_done, isqrt_ready
  );
```

Fig. 11. State-superstate-based communication of a caller and callee procedure instance in VHDL.

```
(B) <= func1 (A);
(C) <= func2 (B);
(D) <= func3 (C);
...
```

Fig. 12. Example of a functional pipeline in BASIL.

3.2.5 Steaming ports

ANSI C is the archetypical example of a general-purpose imperative language that does not support streaming primitives, i.e. it is not possible for someone to express and process streams solely based on the semantics of such language. Streaming (e.g. through queues) suits applications with near-complete absence of control flow. Such example would be the functional pipeline of the form of Fig. 12 with A, B, C, D either compound types (arrays/vectors). Control flow in general applications is complex and it is not easy to intermix streamed and non-streamed inputs/outputs for each FSMD, either calling or callee.

3.2.6 Other issues

3.2.6.1 VHDL packages for implicit fixed-point arithmetic support

The latest approved IEEE 1076 standard (termed VHDL-2008) (IEEE, 2009) adds signed and unsigned (sfixed, ufixed) fixed-point data types and a set of primitives for their manipulation. The VHDL fixed-point package provides synthesizable implementations of fixed-point primitives for arithmetic, scaling and operand resizing (Ashenden & Lewis, 2008).

3.2.6.2 Design organization of an FSMD hardware IP

A proper FSMD hardware IP should seamlessly integrate to a hypothetical system. FSMD IPs would be viewed as black boxes adhering to certain principles such as registered outputs.

```
globalvar B [...]=...;
...
() <= func1 (A);
() <= func2 ();
() <= func3 ();
```

Fig. 13. The functional pipeline of Fig. 12 after argument globalization.

Unconstrained vectors help in maintaining generic blocks without the need of explicit generics, and it is an interesting idea, however not easily applicable when derived types are involved.

The outer product of two vectors A and B could be a theoretical case for a hardware block. The outer (or "cross") product is given by $C = A \times B$ or $C = cross(A, B)$ for reading two matrices A, B to calculate C. Matrices A, B, C will have appropriate derived types that are declared in the cross_pkg.vhd package; a prerequisite for using the cross.vhd design file.

Regarding the block internals, the cross product of A, B is calculated and stored in a localvar array called *Clocal*. *Clocal* is then copied (possibly in parallel) to the C interface array with the help of a for-generate construct.

3.2.6.3 High-level optimizations relevant to hardware block development

Very important optimizations for increasing the efficiency of system-level communication are matrix flattening and argument globalization. The latter optimization is related to choices at the hardware interconnect level.

Matrix flattening deals with reducing the dimensions of an array from N to one. This optimization creates multiple benefits:

• addressing simplification

• direct mapping to physical memory (where addressing is naturally single-dimensional)

• interface and communication simplifications

Argument globalization is useful for replacing multiple copies of a given array by a single-access "globalvar" array. One important benefit is the prevention of exhausting interconnect resources. This optimization is feasible for single-threaded applications. For the example in Fig. 12 we assume that all changes can be applied sequentially on the B array, and that all original data are stored in A.

The aforementioned optimization would rapidly increase the number of "globalvar" arrays. A "safe" but conservative approach would apply a restriction on "globalvar" access, allowing access to globals only by the root procedure of the call graph. This can be overcome by the development of a bus-based hardware interface for "globalvar" arrays making globals accessible by any procedure.

3.2.6.4 Low-level optimizations relevant to hardware block development

A significant low-level optimization that can boost performance while operating locally at the basic block level is operation chaining. A scheduler supporting this optimization would assign to a single control step, multiple operations that are associated through data dependencies. Operation chaining is popular for deriving custom instructions or superinstructions that can be added to processor cores as instruction-set extensions (Pozzi et al., 2006). Most techniques require a form of graph partitioning based on certain criteria such as the maximum acceptable path delay.

A hardware developer could resort in a simpler means for selective operation chaining by merging ASAP states to compound states. This optimization is only possible when a single definition site is used per variable (thus SSA form is mandatory). Then, an intermediate register is eliminated by assigning to a *_next signal and reusing this value in the subsequent chained computation, instead of reading from the stored *_reg value.

3.3 Hardware design of the 2D Euclidean distance approximation

The *eda* algorithm shows good potential for speedup via operation chaining. Without this optimization, 7 cycles are required for computing the approximation, while chaining allows to squeeze all computational states into one; thus three cycles are needed to complete the operation. Fig. 14 depicts VHDL code segments for an ASAP schedule with chaining disabled (Fig. 14(a)) and enabled (Fig. 14(b)). Figures 14(c) and 14(d) show cycle timings for the relevant I/O signals for both cases.

4. Non-trivial examples

4.1 Integer factorization

The prime factorization algorithm (*pfactor*) is a paramount example of the use of streaming outputs. Output outp is streaming and the data stemming from this port should be accessed based on the valid status. The reader can observe that outp is accessed periodically in context of basic block BB3 as shown in Fig. 15(b).

Fig. 15 shows the four relevant facets of *pfactor*: ANSI C code (Fig. 15(a)), a manually derived BASIL implementation (Fig. 15(b)) and the corresponding CFG (Fig. 15(c)) and CDFG (Fig. 15(d)) views.

Fig. 16 shows the interface signals for factoring values 6 (a composite), 7 (a prime), and 8 (a composite which is also a power-of-2).

4.2 Multi-function CORDIC

This example illustrates a universal CORDIC IP core supporting all directions (ROTATION, VECTORING) and modes (CIRCULAR, LINEAR, HYPERBOLIC) (Andraka, 1998; Volder, 1959). The input/ouput interface is similar to e.g. the CORDIC IP generated by Xilinx Core Generator (Xilinx, 2011a). It provides three data inputs (x_{in}, y_{in}, z_{in}) and three data outputs $(x_{out}, y_{out}, z_{out})$ as well as the direction and mode control inputs. The testbench will test the core for computing $\cos(x_{in})$, $\sin(y_{in})$, $arctan(y_{in}/x_{in})$, y_{in}/x_{in}, $\sqrt{w}$, $1/\sqrt{w}$, with $x_{in} = w + 1/4$, $y_{in} = w - 1/4$, but it can be used for anything computable by CORDIC iterations. The computation of $1/\sqrt{w}$ is performed in two stages: a) $y = 1/w$, b) $z = \sqrt{y}$. The

```
type state_type is (S_ENTRY, S_EXIT, S_1_1, S_1_2,
  S_1_3, S_1_4, S_1_5, S_1_6, S_1_7);
signal current_state, next_state: state_type;
...
  case current_state is
    when S_ENTRY =>
      ready <= '1';
      if (start = '1') then
        next_state <= S_1_1;
      else
        next_state <= S_ENTRY;
      end if;
    ...
    when S_1_3 =>
      t3_next <= "000" & x_reg(15 downto 3);
      t4_next <= "0" & y_reg(15 downto 1);
      next_state <= S_1_4;
    when S_1_4 =>
      t5_next <= std_logic_vector(unsigned(x_reg)
                   - unsigned(t3_reg));
      next_state <= S_1_5;
    when S_1_5 =>
      t6_next <= std_logic_vector(unsigned(t4_reg)
                   + unsigned(t5_reg));
      next_state <= S_1_6;
    ...
    when S_1_7 =>
      out1_next <= t7_reg;
      next_state <= S_EXIT;
    when S_EXIT =>
      done <= '1';
      next_state <= S_ENTRY;
```

(a) VHDL code without chaining.

```
type state_type is (S_ENTRY, S_EXIT, S_1_1);
signal current_state, next_state: state_type;
...
case current_state is
  ...
  when S_ENTRY =>
    ready <= '1';
    if (start = '1') then
      next_state <= S_1_1;
    else
      next_state <= S_ENTRY;
    end if;
  when S_1_1 =>
    ...
    t3_next <= "000" & x_next(15 downto 3);
    t4_next <= "0" & y_next(15 downto 1);
    t5_next <= std_logic_vector(unsigned(x_next)
                 - unsigned(t3_next));
    t6_next <= std_logic_vector(unsigned(t4_next)
                 + unsigned(t5_next));
    ...
    out1_next <= t7_next;
    ...
```

(b) VHDL code with chaining.

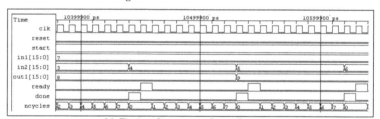

(c) Timing diagram without chaining.

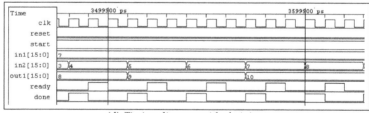

(d) Timing diagram with chaining.

Fig. 14. FSMD implementation in VHDL and timing for the *eda* algorithm.

```
void pfactor(unsigned int x,
  unsigned int *outp)
{

  unsigned int i, n;
  i = 2;
  n = x;
  while (i <= n)
  {
    while ((n % i) == 0)
    {
      n = n / i;
      *outp = i;
      // emitting to file stream
      PRINT(i);
    }
    i = i + 1;
  }
}
```

(a) ANSI C code.

```
procedure pfactor (in u16 x, out u16 outp)
{
  localvar u16 i, n, t0;
BB1:
  n <= mov x;
  i <= ldc 2;
  BB2 <= jmpun;
BB2:
  BB3, BB_EXIT <= jmple i, n;
BB3:
  t0 <= rem n, i;
  BB4, BB5 <= jmpeq t0, 0;
BB4:
  n <= div n, i;
  outp <= mov i;
  BB3 <= jmpun;
BB5:
  i <= add i, 1;
  BB2 <= jmpun;
BB_EXIT:
  nop;
}
```

(b) BASIL code.

(c) CFG.

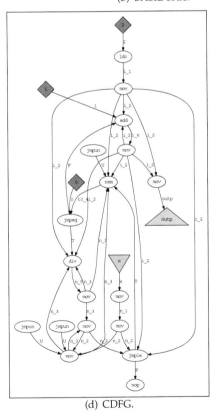

(d) CDFG.

Fig. 15. Different facets of a prime factorization algorithm.

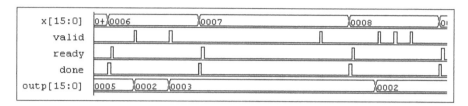

Fig. 16. Non-trivial interface signals for the *pfactor* FSMD design.

Design	Description	Max. frequency	Area (LUTs)
cordic1cyc	1-cycle/iteration; uses asynchronous read LUT RAM	204.5	741
cordic5cyc	5-cycles/iteration; uses synchronous read (Block) RAM	271.5	571, 1 BRAM

Table 4. Logic synthesis results for multi-function CORDIC.

design is a monolithic FSMD that does not include post-processing needed such as the scaling operation for the square root.

The FSMD for the CORDIC uses Q2.14 fixed-point arithmetic. While the required lines of ANSI C code are 29, the hand-coded BASIL representation uses 56 lines; the CDFG representation and the VHDL design, 178 and 436, respectively, showing a clear tendency among the different abstraction levels used for design representation.

The core achieves 18 (CIRCULAR, LINEAR) and 19 cycles (HYPERBOLIC) per sample or $n + 4$ and $n + 5$ cycles, respectively, where n is the fractional bitwidth. When the operation chaining optimization is not applied, 5 cycles per iteration are required instead of a single cycle where all operations all collapsed. A single-cycle per iteration constraint imposes the use of distributed LUT RAM, otherwise 3 cycles are required per sample.

Fig.17(a) shows a C-like implementation of the multi-function CORDIC inspired by recent work (Arndt, 2010; Williamson, 2011). CNTAB is equivalent to fractional width n, HYPER, LIN and CIRC are shortened names for CORDIC modes and ROTN for the rotation direction, cordic_tab is the array of CORDIC coefficients and cordic_hyp_steps an auxiliary table handling repeated iterations for hyperbolic functions. cordic_tab is used to access coefficients for all modes with different offsets (0, 14 or 28 for our case).

Table 4 illustrates synthesis statistics for two CORDIC designs. The logic synthesis results with Xilinx ISE 12.3i reveal a 217MHz (estimated) design when branching is entirely eliminated in the CORDIC loop, otherwise a faster design can be achieved (271.5 MHz). Both cycles and MHz could be improved by source optimization, loop unrolling for pipelining, and the use of embedded multipliers (pseudo-CORDIC) that would eliminate some of the branching needed in the CORDIC loop.

```
void cordic(dir, mode, xin, yin, zin, *xout, *yout, *zout) {
  ...
  x = xin; y = yin; z = zin;
  offset = ((mode == HYPER) ? 0 : ((mode == LIN) ? 14 : 28));
  kfinal = ((mode != HYPER) ? CNTAB : CNTAB+1);
  for (k = 0; k < kfinal; k++) {
    d = ((dir == ROTN) ? ((z>=0) ? 0 : 1) : ((y<0) ? 0 : 1));
    kk = ((mode != HYPER) ? k :
          cordic_hyp_steps[k]);
    xbyk = (x>>kk);
    ybyk = ((mode == HYPER) ? -(y>>kk) : ((mode == LIN) ? 0 :
           (y>>kk)));
    tabval = cordic_tab[kk+offset];
    x1 = x - ybyk; x2 = x + ybyk;
    y1 = y + xbyk; y2 = y - xbyk;
    z1 = z - tabval; z2 = z + tabval;
    x = ((d == 0) ? x1 : x2);
    y = ((d == 0) ? y1 : y2);
    z = ((d == 0) ? z1 : z2);}
  *xout = x; *yout = y; *zout = z;
}
```

(a) C-like code.

```
process (*)
begin
  ...
  case current_state is ...
    when S_3 =>
      t1_next <= cordic_hyp_steps(
                 to_integer(unsigned(k_reg(3 downto 0))));
      if (mode /= CNST_2) then
        kk_next <= k_reg;
      else
        kk_next <= t1_next;
      end if;
      t2_next <= shr(y_reg, kk_next, '1');
      ...
      x1_next <= x_reg - ybyk_next;
      y1_next <= y_reg + xbyk_next;
      z1_next <= z_reg - tabval_next;
      ...
    when S_4 =>
      xout_next <= x_5_reg;
      yout_next <= y_5_reg;
      zout_next <= z_5_reg;
      next_state <= S_EXIT;
  ...
end process;
zout <= zout_reg;
yout <= yout_reg;
xout <= xout_reg;
```

(b) Partial VHDL code.

Fig. 17. Multi-function CORDIC listings.

5. Conclusion

In this chapter, a straightforward FSMD-style model of computation was introduced that augments existing approaches. Our FSMD concept supports inter-FSMD communication, embedded memories, streaming outputs, and seamless integration of user IPs/black boxes. To raise the level of design abstraction, the BASIL typed assembly language is introduced which can be used for capturing the user's intend. We show that it is possible to convert this intermediate representation to self-contained CDFGs and finally to provide an easier path for designing a synthesizable VHDL implementation.

Along the course of this chapter, representative examples were used to illustrate the key concepts of our approach such as a prime factorization algorithm and an improved FSMD design of a multi-function CORDIC.

6. References

Andraka, R. (1998). A survey of CORDIC algorithms for FPGA based computers, *1998 ACM/SIGDA sixth international symposium on Field programmable gate arrays*, Monterey, CA, USA, pp. 191–200.

Appel, A. W. (1998). SSA is functional programming, *ACM SIGPLAN Notices* 33(4): 17–20.
URL: *http://doi.acm.org/10.1145/278283.278285*

Arndt, J. (2010). *Matters Computational: Ideas, Algorithms, Source Code*, Springer.
URL: *http://www.jjj.de/fxt/*

Ashenden, P. J. & Lewis, J. (2008). *VHDL-2008: Just the New Stuff*, Elsevier/Morgan Kaufmann Publishers.

Aycock, J. & Horspool, N. (2000). Simple generation of static single assignment form, *Proceedings of the 9th International Conference in Compiler Construction*, Vol. 1781 of *Lecture Notes in Computer Science*, Springer, pp. 110–125.
URL: *http://citeseer.ist.psu.edu/aycock00simple.html*

Bishop, D. (2010a). *Fixed point package user's guide*.
URL: *http://www.eda.org/fphdl/fixed_ug.pdf*

Bishop, D. (2010b). VHDL-2008 support library.
URL: *http://www.eda.org/fphdl/*

Chu, P. P. (2006). *RTL Hardware Design Using VHDL: Coding for Efficiency, Portability, and Scalability*, Wiley-IEEE Press.

clang homepage (2011).
URL: *http://clang.llvm.org*

COINS (2011).
URL: *http://www.coins-project.org*

CoSy, A. (2011). ACE homepage.
URL: *http://www.ace.nl*

Coussy, P. & Morawiec, A. (eds) (2008). *High-Level Synthesis: From Algorithm to Digital Circuits*, Springer.

Cytron, R., Ferrante, J., Rosen, B. K., Wegman, M. N. & Zadeck, F. K. (1991). Efficiently computing static single assignment form and the control dependence graph, *ACM Transactions on Programming Languages and Systems* 13(4): 451–490.
URL: *http://doi.acm.org/10.1145/115372.115320*

Edwards, S. A. (2006). Using program specialization to speed SystemC fixed-point simulation, *Proceedings of the Workshop on Partial Evaluation and Progra Manipulation (PEPM)*, Charleston, South Carolina, USA, pp. 21–28.

Gajski, D. D., Abdi, S., Gerstlauer, A. & Schirner, G. (2009). *Embedded System Design: Modeling, Synthesis and Verification*, Springer.

Gajski, D. D. & Ramachandran, L. (1994). Introduction to high-level synthesis, *IEEE Design & Test of Computers* 11(1): 44–54.

GCC (2011). The GNU compiler collection homepage.
URL: *http://gcc.gnu.org*

Gonzalez, R. (2000). Xtensa: A configurable and extensible processor, *IEEE Micro* 20(2): 60–70.

Graphviz (2011).
URL: *http://www.graphviz.org*

IEEE (2006). *IEEE 1364-2005, IEEE Standard for Verilog Hardware Description Language*.

IEEE (2009). *IEEE 1076-2008 Standard VHDL Language Reference Manual*.

ISO/IEC JTC1/SC22 (2007). *ISO/IEC 9899:TC3 International Standard (Programming Language: C), Committee Draft*.
URL: *http://www.open-std.org/jtc1/sc22/WG14/www/docs/n1256.pdf*

ITRS (2011). International technology roadmap for semiconductors.
URL: *http://www.itrs.net/reports.html*

Keating, M. & Bricaud, P. (2002). *Reuse Methodology Manual for System-on-a-Chip Designs*, third edition edn, Springer-Verlag. 2nd printing.

Knuth, D. E. (2011). *Art of Computer Programming: Combinatorial Algorithms*, number pt. 1 in *Addison-Wesley Series in Computer Science*, Addison Wesley Professional.

LANCE (2011). LANCE retargetable C compiler.
URL: *http://www.lancecompiler.com*

Leupers, R., Wahlen, O., Hohenauer, M., Kogel, T. & Marwedel, P. (2003). An Executable Intermediate Representation for Retargetable Compilation and High-Level Code Optimization, *Int. Conf. on Inf. Comm. Tech. in Education*.

LLVM (2011).
URL: *http://llvm.org*

Machine-SUIF (2002).
URL: *http://www.eecs.harvard.edu/hube/software/*

Marsaglia, G. (2003). Xorshift RNGs, *Journal of Statistical Software* 8(14).

Mentor Graphics (2011). Algorithmic C data types.
URL: *http://www.mentor.com/esl/catapult/algorithmic*

Microsoft (2008). Phoenix compiler framework.
URL: *http://connect.microsoft.com/Phoenix*

Pozzi, L., Atasu, K. & Ienne, P. (2006). Exact and approximate algorithms for the extension of embedded processor instruction sets, *IEEE Transactions on CAD of Integrated Circuits and Systems* 25(7): 1209–1229.

SystemC (2006). *IEEE 1666™-2005: Open SystemC Language Reference Manual*.

Turkowski, K. (1995). Graphics gems v, Academic Press Professional, Inc., San Diego, CA, USA, chapter Fixed-point square root, pp. 22–24.

Volder, J. E. (1959). The CORDIC Trigonometric Computing Technique, *IRE Transactions on Electronic Computers* EC-8: 330–334.

Williamson, J. (2011). Simple C code for fixed-point CORDIC.
    URL: *http://www.dcs.gla.ac.uk/ jhw/cordic/*

Wirth, N. (1998). Hardware compilation: Translating programs into circuits, *IEEE Computer* 31(6): 25–31.

Xilinx (2005). *Spartan-3 FPGA Family Using Block Spartan-3 Generation FPGAs (v2.0).*

Xilinx (2011a). *CORDIC v4.0 - Product Specifications, XILINX LogiCORE, DS249 (vl.5).*

Xilinx (2011b). Xilinx.
    URL: *http://www.xilinx.com*

Yates, R. (2009). Fixed-point arithmetic: An introduction, *Technical reference*, Digital Signal Labs.

8

A Visual Software Development Environment that Considers Tests of Physical Units [*]

Takaaki Goto[1], Yasunori Shiono[2], Tomoo Sumida[2], Tetsuro Nishino[1],
Takeo Yaku[3] and Kensei Tsuchida[2]
[1]*The University of Electro-Communications*
[2]*Toyo University*
[3]*Nihon University*
Japan

1. Introduction

Embedded systems are extensively used in various small devices, such as mobile phones, in transportation systems, such as those in cars or aircraft, and in large-scale distributed systems, such as cloud computing environments. We need a technology that can be used to develop low-cost, high-performance embedded systems. This technology would be useful for designing, testing, implementing, and evaluating embedded prototype systems by using a software simulator.

So far, embedded systems are typically used only in machine controls, but it seems that they will soon also have an information processing function. Recent embedded systems target not only industrial products but also consumer products, and this appears to be spreading across various fields. In the United States and Europe, there are large national projects related to the development of embedded systems. Embedded systems are increasing in size and becoming more complicated, so the development of methodologies and efficient testing for them is highly desirable.

The authors have been engaged in the development of a software development environment based on graph theory, which includes graph drawing theory and graph grammars [2–4]. In our research, we use Hichart, which is a program diagram methodology originally introduced by Yaku and Futatsugi [5].

There has been a substantial amount of research devoted to Hichart. A prototype formulation of attribute graph grammar for Hichart was reported in [6]. This grammar consists of Hichart syntax rules, which use a context-free graph grammar [7], and semantic rules for layout. The authors have been developing a software development environment based on graph theory that includes graph drawing theory and various graph grammars [2, 8]. So far, we have developed bidirectional translators that can translate a Pascal, C, or DXL source into Hichart and can alternatively translate Hichart into Pascal, C, or DXL [2, 8]. For example, HiChart Graph Grammar (HCGG) [9] is an attribute graph grammar with an underlying

---

[*]Part of the results have previously been reported by [1]

graph grammar based on edNCE graph grammar [10] and intended for use with DXL. It is problematic, however, in that it cannot parse very efficiently. Hichart Precedence Graph Grammar (HCPGG) was introduced in [11].

In recent years, model checking methodologies have been applied to embedded systems. In our current work, we constructed a visual software development environment to support a developed embedded system. The target of this research is NQC, which is the program language for LEGO MINDSTORM. Our visual software development system for embedded systems can

1. generate Promela codes for given Hichart diagrams, and
2. detect problems by using visual feedback features.

Our previously developed environment was not sufficiently functional, so we created an effective testing environment for the visual environment.

In this chapter, we describe our visual software development environment that supports the development of embedded systems.

2. Preliminaries

2.1 Embedded systems

An embedded system is a system that controls various components and specific functions of the industrial equipment or consumer electronic device it is built into [12, 13]. Product life cycles are currently being shortened, and the period from development to verification has now been trimmed down to about three months. Four requirements are needed to implement modern embedded systems.

- Concurrency
  Multi-core and/or multi processors are becoming dominant in the architecture of processors as a solution to the limits in circuit line width (manufacturing process), increased generation of heat, and clock speed limits. Therefore, it is necessary to implement applications by using methods with parallelism descriptions.
- Hierarchy
  System modules are arranged in a hierarchal fashion in main systems, subsystems, and sub-subsystems. Diversity and recycling must be improved, and the number of development processes should be reduced as much as possible.
- Resource Constraints
  It is necessary to comply with the constraints of built-in factors like memory and power consumption.
- Safety and Reliability
  System failure is a serious problem that can cause severe damage and potentially fatal accidents. It is extremely important to guarantee the safety of a system.

LEGO MINDSTORMS [14] is a robotics environment that was jointly developed by the REGO and MIT. MINDSTORMS consists of a block with an RCX or NXT micro processor. Robots that are constructed with RCX or NXT and sensors can work autonomously, so a block with RCX or NXT can control a robot's behavior. RCX or NXT detects environment information through attached sensors and then activates motors in accordance with the programs. RCX and NXT are micro processors with a touch sensor, humidity sensor, photodetector, motor, and lamp.

ROBOLAB is a programming environment developed by National Instruments, the REGO, and Tufts University. It is based on LABVIEW (developed by National Instruments) and provides a graphical programming environment that uses icons.

It is easy for users to develop programs in a short amount of time because ROBOLAB uses templates. These templates include various icons that correspond to different functions which then appear in the developed program in pilot level. ROBOLAB has fewer options than LABVIEW, but it does have some additional commands that have been customized for RCX.

Two programming levels, pilot level and inventor level, can be used in ROBOLAB. The steps then taken to construct a program are as follows.

1. Choose icons from palette.
2. Put icons in a program window.
3. Set orders of icons and then connect them.
4. Transfer obtained program to the RCX.

Not Quite C (NQC) [15] is a language that can be used in LEGO MINDSTORM RCX. Its specification is similar to that of C language, but differs in that it does not provide a pointer but instead has functions specialized for LEGO MINDSTORMS, including "turn on motors," "check touch sensors value," and so on.

A typical NQC program starts from a "main" task and can handle a maximum of ten tasks. When we write NQC source codes, the below description is required.

Listing 1. Example1

```
task main()
{
}
```

Here, we investigate functions and constants. The below program shows MINDSTORMS going forward for four seconds, then backward for four seconds, and then stopping.

Listing 2. Example2

```
task main()
{
  OnFwd(OUT_A+OUT_C);
  Wait(400);
  OnRev(OUT_A+OUT_C);
  Wait(400);
  Off(OUT_A+OUT_C);
}
```

Here, the functions "OnFwd," "OnRev," etc. control RCX. Table 1 shows an example of functions customized for NQC.

Functions	Explanation	Example of description
SetSensor(<sensor name>, <configuration>)	set type and mode of sensors	SetSensor(SENSOR_1, SENSOR_TOUCH)
SetSensorMode(<sensor name>, <mode>)	set a sensor's mode	SetSensorMode(SENSOR_2, SENSOR_MODE_PERCENT)
OnFwd(<outputs>)	set direction and turn on	OnFwd(OUT_A)

Table 1. Functions of RCX

As for the constants, they are constants with names and work to improve programmers' understanding of NQC programs.

Table 2 shows an example of constants.

Constants category	Constants
Setting for SetSensor()	SENSOR_MODE_RAW, SENSOR_MODE_BOOL, SENSOR_MODE_EDGE, SENSOR_MODE_PULSE, SENSOR_MODE_PERCENT, SENSOR_MODE_CELCIUS, SENSOR_MODE_FAHRENHEIT, SENSOR_MODE_ROTATION
Mode for SetSensorMode	SENSOR_MODE_RAW, SENSOR_MODE_BOOL, SENSOR_MODE_EDGE, SENSOR_MODE_PULSE, SENSOR_MODE_PERCENT, SENSOR_MODE_CELCIUS, SENSOR_MODE_FAHRENHEIT, SENSOR_MODE_ROTATION

Table 2. Constants of RCX

We adopt REGO MINDSTORMS as an example of embedded systems with sensors.

2.2 Program diagrams

In software design and development, program diagrams are often used for software visualization. Many kinds of program diagrams, such as the previously mentioned hierarchical flowchart language (Hichart), problem analysis diagram (PAD), hierarchical and compact description chart (HCP), and structured programming diagram (SPD), have been used in software development [2, 16]. Moreover, software development using these program diagrams is steadily on the increase.

In our research, we used the Hichart program diagram [17], which was first introduced by Yaku and Futatsugi [5]. Figure 1 shows a program called "Tower of Hanoi" that was written in Hichart.

Hichart has three key features:

1. A tree-flowchart diagram that has the flow control lines of a Neumann program flowchart,

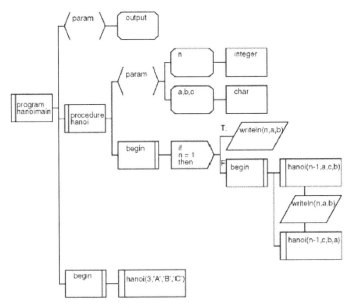

Fig. 1. Example of Hichart: "Tower of Hanoi".

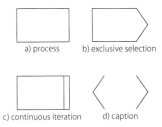

a) process   b) exclusive selection

c) continuous iteration   d) caption

Fig. 2. Example of Hichart symbols.

2. Nodes of the different functions in a diagram that are represented by differently shaped cells, and

3. A data structure hierarchy (represented by a diagram) and a control flow that are simultaneously displayed on a plane, which distinguishes it from other program diagram methodologies.

Hichart is described by cell and line. There are various type of cells, such as "process," "exclusive selection," "continuous iteration," "caption," and so on. Figure 2 shows an example of some of the Hichart symbols.

3. Program diagrams for embedded systems

In this section, we describe program diagrams for embedded systems, specifically, a detailed procedure for constructing program diagrams for an embedded system using Hichart for NQC.

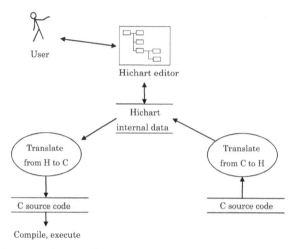

Fig. 3. Overview of our previous study.

Figure 3 shows an overview of our previous study on a Hichart-C translation system.

In our previous system, it is possible to obtain internal Hichart data from C source code via a C-to-H translator implemented using JavaCC. Users can edit a Hichart diagram on a Hichart editor that visualizes the internal Hichart data as a Hichart diagram. The H-to-C translator can generate C source codes from the internal Hichart data, and then we can obtain the C source code corresponding to the Hichart diagrams. Our system can illustrate programs as diagrams, which leads to an improved understanding of programs.

We expanded the above framework to treat embedded system programming. Specifically we extended H-to-C and C-to-H specialized for NQC. Some of the alterations we made are as follows.

1. task
   The "task" is a unique keyword of NQC, and we therefore added it to the C-to-H function.
2. start, stop
   We added "start" and "stop" statements in Hichart (as shown in List 3) to control tasks.

Listing 3. Example3

```
task main()
{
    SetSensor (SENSOR_1, SENSOR_TOUCH);
    start check_sensors;
    start move_square;
}
task move_square()
{
    while(true)
    {
        OnFwd(OUT_A+OUT_C);  Wait(100);
```

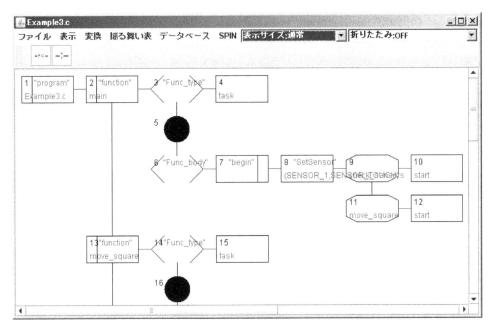

Fig. 4. Screenshot of Hichart for NQC that correspond to List 3.

```
    OnRev(OUT_C);  Wait(68);
  }
}

task check_sensors()
{
  while(true)
  {
    if (SENSOR_1 == 1)
    {
      stop move_square;
      OnRev(OUT_A+OUT_C);  Wait(50);
      OnFwd(OUT_A);  Wait(85);
      start move_square;
    }
  }
}
```

There are some differences between C syntax and NQC syntax; therefore, we modified JavaCC, which defines syntax, to cover them. Thus, we obtained program diagrams for embedded systems.

Figure 4 shows a screenshot of Hichart for NQC that correspond to List 3.

4. A visual software development environment

We propose a visual software development environment based on Hichart for NQC. We visualize NQC code by the abovementioned Hichart diagrams through a Hichart visual software development environment called Hichart editor. Hichart diagrams or NQC source codes are inputted into the editor, and the editor outputs NQC source codes after editing code such as parameter values in diagrams.

In the Hichart editor, the program code is shown as a diagram. List 4 shows a sample program of NQC, and Figure 5 shows the Hichart diagram corresponding to List 4.

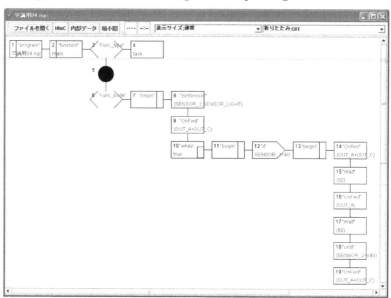

Fig. 5. Screen of Hichart editor.

Listing 4. anti-drop program

```
task main()
{
  SetSensor (SENSOR_2,SENSOR_LIGHT);
  OnFwd(OUT_A+OUT_C);
  while(true)
  {
    if (SENSOR_2 < 40)
    {
      OnRev(OUT_A+OUT_C);
      Wait(50);
      OnFwd(OUT_A);
      Wait(68);
      until(SENSOR_2 >= 40);
      OnFwd(OUT_A+OUT_C);
```

```
    }
  }
}
```

This Hichart editor for NQC has the following characteristics.

1. Generation of Hichart diagram corresponding to NQC
2. Editing of Hichart diagrams
3. Generation of NQC source codes from Hichart diagrams
4. Layout modification of Hichart diagrams

Users can edit each diagram directly on the editor. For example, cells can be added by double-clicking on the editor screen, after which cell information, such as type and label, is embedded into the new cell.

Figure 6 shows the Hichart screen after diagram editing. In this case, some of the parameter's values have been changed.

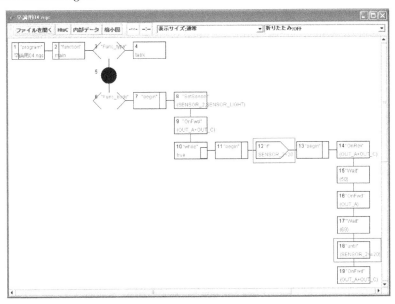

Fig. 6. Hichart editor screen after editing.

The Hichart editor can read NQC source codes and convert them into Hichart codes using the N-to-H function, and it can generate NQC source codes from Hichart codes by using the H-to-N function. The Hichart codes consist of tree data structure. Each node of the structure has four pointers (to parent node, to child cell, to previous cell, and to next cell) and node information such as node type, node label, node label, and so on. To generate NQC codes by the H-to-N function, tree structures can be traversed in preorder.

The obtained NQC source code can be transferred to the LEGO MINDSTORM RCX via BricxCC. Figure 7 shows a screenshot of NQC source code generated by the Hichart editor.

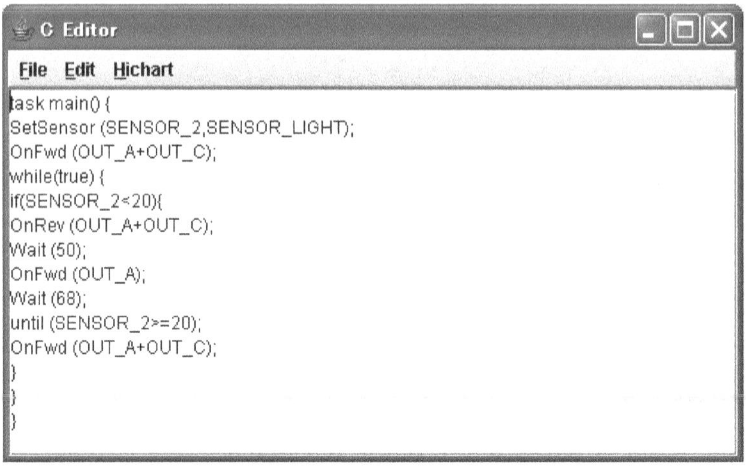

```
C Editor                                              [_][□][×]

File  Edit  Hichart

task main() {
SetSensor (SENSOR_2,SENSOR_LIGHT);
OnFwd (OUT_A+OUT_C);
while(true) {
if(SENSOR_2<20){
OnRev (OUT_A+OUT_C);
Wait (50);
OnFwd (OUT_A);
Wait (68);
until (SENSOR_2>=20);
OnFwd (OUT_A+OUT_C);
}
}
}
```

Fig. 7. Screenshot of NQC source code generated by Hichart editor.

Sensitivity s	0-32	33-49	50-100
Recognize a table edge	×	◯	◯
Turn in its tracks	◯	◯	×

Table 3. Behavioral specifications table.

5. Testing environment based on behavioral specification and logical checking

To test embedded system behaviors, especially for those that have physical devices such as sensors, two areas must be checked: the value of the sensors and the logical correctness of the embedded system. Embedded systems with sensors are affected by the environment around the machine, so it is important that developers are able to set the appropriate sensor value. Of course, even if the physical parameters are appropriate, if there are logical errors in a machine's program, the embedded systems will not always work as we expect.

In this section, we propose two testing methods to check the behaviors of embedded systems.

5.1 Behavioral specifications table

A behavioral specifications table is used when users set the physical parameters of RCX. An example of such a table is shown in Table 3. The leftmost column lists the behavioral specifications and the three columns on the right show the parameter values. A circle indicates an expected performance; a cross indicates an unexpected one. The numerical values indicate the range of sensitivity parameters s.

For example, when the sensitivity parameter s was between 0 and 32, the moving object did not recognize a table edge (the specifications for "recognizes a table edge" were not met) and did not spin around on that spot. When the sensitivity parameter s was between 33 and 49, the specifications for "recognizes a table edge" and "does not spin around on that spot" were both met.

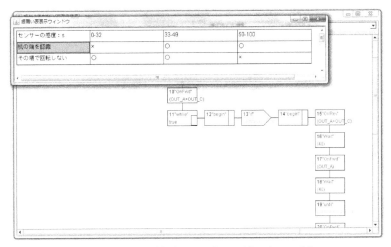

Fig. 8. Screenshot of Hichart editor and behavioral specifications table.

The results in the table show that the RCX with a sensor value from 0 to 32 cannot distinguish the edge of the table and so falls off. Therefore, users need to change the sensor value to the optimum value by referencing the table and choosing the appropriate value. In this case, if users only choose the column with the values from 33 to 49, the chosen value is reflected in the Hichart diagram. This modified Hichart diagram can then generate an NQC source code. This is an example of how developers can easily set appropriate physical parameters by using behavioral specifications tables.

The behavioral specifications function has the following characteristics.

1. The editor changes the colors of Hichart cells that are associated with the parameters in the behavioral specifications table.
2. The editor sets the parameter value of Hichart cells that are associated with the parameters in the behavioral specifications table.

Here, we show an example in which an RCX runs without falling off a desk. In this example, when a photodetector on the RCX recognizes the edge of the desk, the RCX reverses and turns. Figure 8 shows a screenshot of the Hichart editor and the related behavioral specifications table.

In the Hichart editor, the input-output cells related to a behavioral specifications table are redrawn in green when the user chooses a menu that displays the behavioral specifications table.

Figure 9 shows the behavior of an RCX after setting the appropriate physical parameters. The RCX can distinguish the table edge and turn after reversing.

We also constructed a function that enables a behavioral specification table to be stored in a database that was made using MySQL. After we test a given device, we can input the results via the database function in the Hichart editor. Using stored information, we can construct a behavioral specification table with an optimized parameter's value.

Fig. 9. Screenshot of RCX that recognizes table edge.

5.2 Model checking

We propose a method for checking behavior in the Hichart development environment by using the model checking tool SPIN [18, 19] to logically check whether a given behavior specification is fulfilled before applying the program to a real machine. As described previously, the behavioral specifications table can check the physical parameters of a real machine. However, it cannot check logical behavior. We therefore built a model checking function into our editor that can translate internal Hichart data into Promela code.

The major characteristics of the behavior specification verification function are listed below.

- Generation of Promela codes
  Generating Promela codes from Hichart diagrams displayed on the Hichart editor.
- Execution of SPIN
  Generating pan.c or LTL-formulas.
- Compilation
  Compiling obtained pan.c to generate .exe file for model checking.
- Analyzing

- Analysis
  We found that programs do not bear the behavior specification by model checking and so generated trail files. The function then analyzes the trail files and feeds them back to the Hichart diagrams.

The Promela code is used to check whether a given behavior specification is fulfilled. Feedback from the checks is then sent to a Hichart graphical editor. If a given behavioral specification is not fulfilled, the result of the checking is reflected in the implicated location of the Hichart.

To give an actual example, we consider the specifications that make the RCX repeat forward movements and turn left. If it is touch sensitive, the RCX changes course. This specification means that RCX definitely swerves when touched. In this study, we checked whether the created program met the behavior specification by using SPIN before applying the program to real machines.

Listing 5. Source code of NQC

```
task move_square(){
  while(true){
    OnFwd(OUT_A + OUT_C);
    Wait(1000);
    OnRev(OUT_C);
    Wait(85);
  }
}
```

Listing 6. Promela code

```
proctype move_square(){
  do
  ::
    state = OnFwd;
    state = Wait;
    state = OnRev;
    state = Wait;
  od
}
```

Lists 5 and 6 show part of the NQC source code corresponding to the above specification and the automatically generated Promela source code.

We explain the feedback procedure, which is shown in Fig. 10.

An assertion statement of "state == OnFwd" is an example. If a moving object (RCX) is moving forward at the point where the assertion is set, the statement is true. Otherwise, it is false. For example, we can verify by steps (3)-(7) in Fig. 10 whether the moving object is always moving forward or not.

Here, we show an example of manipulating our Hichart editor. We can embed an assertion description through the Hichart editor, as shown in Fig. 11, and then obtain a Promela code from the Hichart code. When we obtain this code, we have to specify the behaviors that we want to check. Figure 12 shows a result obtained through this process.

Next, we execute SPIN. If we embed assertions in the Hichart code, we execute SPIN as it currently stands, while if we use LTL-formulas, we execute SPIN with an "-f" option and then obtain pan.c. The model is checked by compiling the obtained pan.c. Figure 13 is a screenshot of the model checking result using the Hichart editor.

If there are any factors that do not meet the behavioral specifications, trail files are generated. Figure 14 shows some of the result of analyzing the trail file.

The trail files contain information on how frequently the processing calls and execution paths were made. We use this information to narrow the search area of the entire program by using the visual feedback. Users can detect a problematic area interactively by using the Hichart editor with the help of this visual feedback.

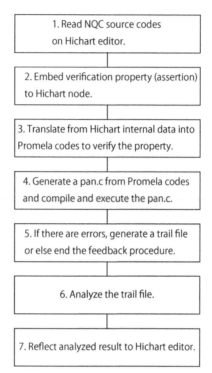

Fig. 10. Feedback procedure.

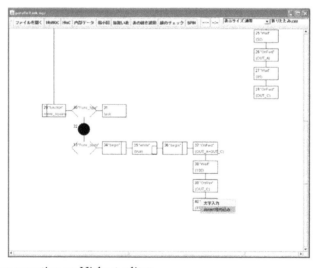

Fig. 11. Embed an assertion on Hichart editor.

Fig. 12. Result of generating a Promela code.

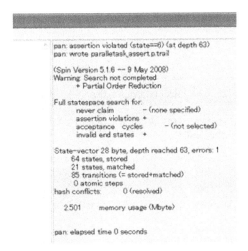

Fig. 13. Result of model checking.

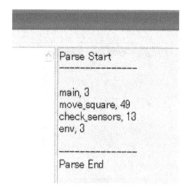

Fig. 14. Result of analyzing trail file.

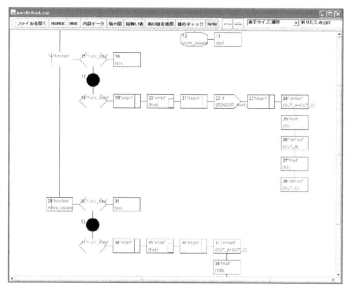

Fig. 15. Part of Hichart editor feedback screen.

After analyzing the trail files, we can obtain feedback from the Hichart editor. Figure 15 shows part of a Hichart editor feedback screen.

If the result is that programs did not meet the behavior specification by using SPIN, the tasks indicated as the causes are highlighted. The locations that do not meet the behavior specifications can be seen by using the Hichart feedback feature. This is an example of efficient assistance for embedded software.

6. Conclusion

We described our application of a behavioral specification table and model-checking methodologies to a visual software development environment we developed for embedded software.

A key element of our study was the separation of logical and physical behavioral specifications. It is difficult to verify behaviors such as those of robot sensors without access to the behaviors of real machines, and it is also difficult to simulate behaviors accurately. Therefore, we developed behavioral specification tables, a model-checking function, and a method of giving visual feedback.

It is rather difficult to set exact values for physical parameters under development circumstances using a tool such as MATLAB/simulink because the physical parameters vary depending on external conditions (e.g., weather), and therefore, there were certain limitations to the simulations. We obtained a couple of examples demonstrating the validity of our approach in both the behavioral specification table and the logical specification check by using SPIN.

In our previous work, some visual software development environments were developed based on graph grammar; however, the environment for embedded systems described in this article is not yet based on graph grammars. A graph grammar for Hichart that supports NQC is currently under development.

In our future work, we will construct a Hichart development environment with additional functions that further support the development of embedded systems.

7. References

[1] T. Goto, Y. Shiono, T. Nishino, T. Yaku, and K. Tsuchida. Behavioral verification in hichart development environment for embedded software. In *Computer and Information Science (ICIS), 2010 IEEE/ACIS 9th International Conference on*, pages 337 –340, aug. 2010.
[2] K. Sugita, A. Adachi, Y. Miyadera, K. Tsuchida, and T. Yaku. A visual programming environment based on graph grammars and tidy graph drawing. In *Proceedings of The 20th International Conference on Software Engineering (ICSE '98)*, volume 2, pages 74–79, 1998.
[3] T. Goto, T. Kirishima, N. Motousu, K. Tsuchida, and T. Yaku. A visual software development environment based on graph grammars. In *Proc. IASTED Software Engineering 2004*, pages 620–625, 2004.
[4] Takaaki Goto, Kenji Ruise, Takeo Yaku, and Kensei Tsuchida. Visual software development environment based on graph grammars. *IEICE transactions on information and systems*, 92(3):401–412, 2009.
[5] Takeo Yaku and Kokichi Futatsugi. Tree structured flow-chart. In *Memoir of IEICE*, pages AL–78, 1978.
[6] T. Nishino. Attribute graph grammars with applications to hichart program chart editors. In *Advances in Software Science and Technology*, volume 1, pages 89–104, 1989.
[7] C. Ghezzi P. D. Vigna. Context-free graph grammars. In *Information Control*, volume 37, pages 207–233, 1978.
[8] Y. Adachi, K. Anzai, K. Tsuchida, and T. Yaku. Hierarchical program diagram editor based on attribute graph grammar. In *Proc. COMPSAC*, volume 20, pages 205–213, 1996.
[9] Masahiro Miyazaki, Kenji Ruise, Kensei Tsuchida, and Takeo Yaku. An NCE Attribute Graph Grammar for Program Diagrams with Respect to Drawing Problems. *IEICE Technical Report*, 100(52):1–8, 2000.

[10] Grzegorz Rozenberg. *Handbook of Graph Grammar and Computing by Graph Transformation Volume 1*. World Scientific Publishing, 1997.

[11] K. Ruise, K. Tsuchida, and T. Yaku. Parsing of program diagrams with attribute precedence graph grammar. In *Technical Report of IPSJ*, number 27, pages 17–20, 2001.

[12] R. Zurawski. *Embedded systems design and verification*. CRC Press, 2009.

[13] S. Narayan. Requirements for specification of embedded systems. In *ASIC Conference and Exhibit, 1996. Proceedings., Ninth Annual IEEE International*, pages 133 –137, sep 1996.

[14] LEGO. LEGO mindstorms. http://mindstorms.lego.com/en-us/Default.aspx.

[15] Not Quite C. http://bricxcc.sourceforge.net/nqc/.

[16] Kenichi Harada. *Structure Editor*. Kyoritsu Shuppan, 1987. (in Japanese).

[17] T. Yaku, K. Futatsugi, A. Adachi, and E. Moriya. HICHART -A hierarchical flowchart description language-. In *Proc. IEEE COMPSAC*, volume 11, pages 157–163, 1987.

[18] G.J. Holzmann. The model checker spin. *Software Engineering, IEEE Transactions on*, 23(5):279 –295, may 1997.

[19] M. Ben-Ari. *Principles of the SPIN Model Checker*. Springer, 2008.

9

A Methodology for Scheduling Analysis Based on UML Development Models

Matthias Hagner and Ursula Goltz
Institute for Programming and Reactive Systems
TU Braunschweig
Germany

1. Introduction

The complexity of embedded systems and their safety requirements have risen significantly in the last years. The model based development approach helps to handle the complexity. However, the support for analysis of non-functional properties based on development models, and consequently the integration of these analyses in a development process exist only sporadically, in particular concerning scheduling analysis. There is no methodology that covers all aspects of doing a scheduling analysis, including process steps concerning the questions, how to add necessary parameters to the UML model, how to separate between experimental decisions and design decisions, or how to handle different variants of a system. In this chapter, we describe a methodology that covers these aspects for an integration of scheduling analyses into a UML based development process. The methodology describes process steps that define how to create a UML model containing the timing aspects, how to parameterise it (e.g., by using external specialised tools), how to do an analysis, how to handle different variants of a model, and how to carry design decision based on analysis results over to the design model. The methodology specifies guidelines on how to integrate a scheduling analysis for systems using static priority scheduling policies in a development process. We present this methodology on a case study on a robotic control system.

To handle the complexity and fulfil the sometimes safety critical requirements, the model based development approach has been widely appreciated. The UML (Object Management Group (2003)) has been established as one of the most popular modelling languages. Using extension, e.g., SysML (Object Management Group (2007)), or UML profiles, e.g., MARTE (Modelling and Analysis of Real-Time and Embedded Systems) (Object Management Group (2009)), UML can be better adapted to the needs of embedded systems, e.g., the non functional requirement scheduling. Especially MARTE contains a large number of possibilities to add timing and scheduling aspects to a UML model. However, because of the size and complexity of the profile it is hard for common developers to handle it. Hence, it requires guidance in terms of a methodology for a successful application of the MARTE profile.

Besides specification and tracing of timing requirements through different design stages, the major goal of enriching models with timing information is to enable early validation and verification of design decisions. As designs for an embedded or safety critical systems may have to be discarded if deadlines are missed or resources are overloaded, early timing analysis has become an issue and is supported by a number of specialised analysis tools, e.g., SymTA/S (Henia et al. (2005)), MAST (Harbour et al. (2001)), and TIMES (Fersman & Yi

(2004)). However, the meta models used by these tools differ from each other and in particular from UML models used for design. Thus, to make an analysis possible and to integrate it into a development process, the developer has to remodel the system in the analysis tool. This leads to more work and possibly errors made by the remodelling. Additionally, the developer has to learn how to use the chosen analysis tool. To avoid this major effort, an automatic model transformation is needed to build an interface that enables automated analysis of a MARTE extended UML model using existing real-time analysis technology.

There has been some work done developing support for the application of the MARTE profile or to enable scheduling analysis based on UML models. The Scheduling Analysis View (SAV) (Hagner & Huhn (2007), Hagner & Huhn (2008)) is one example for guidelines to handle the complexity of the UML and the MARTE profile. A transformation from the SAV to an analysis tool SymTA/S is already realised (Hagner & Goltz (2010)). Additional tool support was created (Hagner & Huhn (2008)) to help the developer to adapt to guidelines of the SAV. Espinoza et al. (2008) described how to use design decisions based on analysis results and showed the limitations of the UML concerning these aspects. There are also methodical steps identified, how the developer can make such a design decision. However, there are still important steps missing to integrate the scheduling analysis into a UML based development process. In Hagner et al. (2008), we observed the possibilities MARTE offers for the development in the rail automation domain. However, no concrete methodology is described. In this chapter, we want to address open questions like: Where do the scheduling parameters come from (e.g., priorities, execution patterns, execution times), considering the development stages (early development stage: estimated values or measured values from components-off-the-shelf, later development stages: parameters from specialised tools, e.g., aiT (Ferdinand et al. (2001))? How to bring back design decision based on scheduling analysis results into a design model? How to handle different criticality levels or different variants of the same system (e.g., by using different task distributions on the hardware resources)? In this chapter, we want to present a methodology to integrate the scheduling analysis into a UML based development process for embedded real-time systems by covering these aspects. All implementations presented in this chapter are realised for the case tool Papyrus for UML[1].

This chapter is structured as follows: Section 2 describes our methodology, Section 3 gives a case study of a robotic control system on which we applied our methodology, Section 4 shows how this approach could be adopted to other non-functional properties, and Section 5 concludes the chapter.

2. A methodology for the integration of scheduling analysis into a UML based development process

The integration of scheduling analysis demands specified methodologies, because the UML based development models cannot be used as an input for analysis tools. One reason is that these tools use their own input format/meta model, which is not compatible with UML. Another reason is that there is important scheduling information missing in the development model. UML profiles and model transformation help to bridge the gap between development models and analysis tools. However, these tools have to be adapted well to the needs of the development. Moreover, the developer needs guidelines to do an analysis as this cannot be fully automated.

---

[1] http://www.papyrusuml.org

Figure 1 depicts our methodology for integrating the scheduling analysis into a UML based development process. On the left side, the Design Model is the starting point of our methodology. It contains the common system description by using UML and SysML diagrams. We assume that it is already part of the development process before we add our methodology. Everything else depicted in Figure 1 describes the methodology.

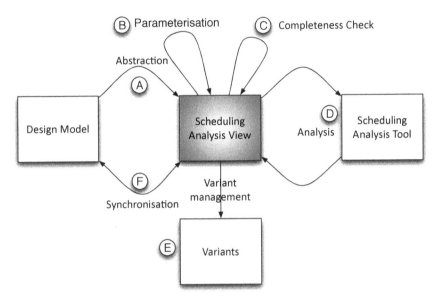

Fig. 1. Methodology for the integration of scheduling analysis in a UML based development process

The centre of the methodology is the Scheduling Analysis View (SAV). It is a special view on the system under a scheduling analysis perspective. It leaves out not relevant information for a scheduling analysis, but offers possibilities to add important scheduling information that are usually difficult to specify in a common UML model and are often left out of the normal Design Model. The SAV consists of UML diagrams and MARTE elements. It is an intermediate step between the Design Model and the scheduling analysis tools. The rest of the methodology is based on the SAV. It connects the different views and the external analysis tools. It consists of:

- an abstraction, to create a SAV based on the Design Model using as much information from the Design Model as possible,

- a parameterisation, to add the missing information relevant for the analysis (e.g., priorities, execution times),

- a completeness check, to make sure the SAV is properly defined,

- the analysis, to perform the scheduling analysis,

- variant management, to handle different variants of the same system (e.g., using different distribution, other priorities), and

- a synchronisation, to keep the consistency between the Design Model and the SAV.

The developer does not need to see or learn how to use the analysis tools, as a scheduling analysis can be performed automatically from the SAV as an input.

The following subsections describe these steps in more detail. Figure 1 gives an order in which the steps should be executed (using the letters A, B, . . .). A (the abstraction) is performed only once and F (the synchronisation) only if required. Concerning the other steps, B, C, D, E can be executed repeatedly until the developer is satisfied. Then, F can be performed.

2.1 The scheduling analysis view

Independent, non-functional properties should be handled separately to allow the developer to concentrate on the particular aspect he/she is working on and masking those parts of a model that do not contribute to it. This is drawn upon the cognitive load theory (Sweller (2003)), which states that human cognitive productivity dramatically decreases when more different dimensions have to be considered at the same time. As a consequence in software engineering a number of clearly differentiated views for architecture and design have been proposed (Kruchten (1995)).

As a centre of this methodology, we use the Scheduling Analysis View (SAV) (Hagner & Huhn (2008)) as a special view on the system. The SAV is based on UML diagrams and the MARTE profile (stereotypes and tagged values). MARTE is proposed by the "ProMarte" consortium with the goal of extending UML modelling facilities with concepts needed for real-time embedded systems design like timing, resource allocation, and other non-functional runtime properties. The MARTE profile is a successor of the profile for Schedulability, Performance, and Time (SPT profile) (Object Management Group (2002)) and the profile for Modelling Quality of Service and Fault Tolerance Characteristics and Mechanisms (QoS profile) (Object Management Group (2004)).

The profile consists of three main packages. The MARTE Foundations package defines the basic concepts to design and analyse an embedded, real-time system. The MARTE Design Model offers elements for requirements capturing, the specification, the design, and the implementation phase. Therefore, it provides a concept for high-level modelling and a concept for detailed hard- and software description. The MARTE Analysis Model defines specific model abstractions and annotations that could be used by external tools to analyse the described system. Thus, the analysis package is divided into three parts, according to the kind of analysis. The first part defines a general concept for quantitative analysis techniques; the second and third parts are focused on schedulability and performance analysis.

Because runtime properties and in particular timing are important in each development phase, the MARTE profile is applicable during the development process, e.g., to define and refine requirements, to model the partitioning of software and hardware in detail, or to prepare and complete UML models for transformation to automated scheduling or performance analysis. One application of the MARTE profile is shown in Figure 2. MARTE is widespread in the field of developing of embedded systems (e.g., Argyris et al. (2010); Arpinen et al. (2011); Faugere et al. (2007)).

We only use a small amount of the stereotypes and tagged values for the SAV, as the MARTE profile offers much more applications. One goal of the SAV is to keep it as simple as possible. Therefore, only elements are used that are necessary to describe all the information that is needed for an analysis. In Table 1 all used stereotypes and tagged values are presented. Additionally, we offer guidelines and rules, how to define certain aspects of the systems in the SAV. The SAV was designed regarding the information required by a number of scheduling

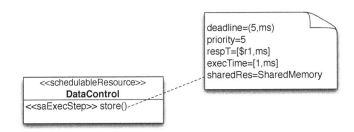

Fig. 2. Example of a UML profile

analysis tools. It concentrates on and highlights timing and scheduling aspects. It is based on
the Design Model, but abstracts/leaves out all information that is not needed for a scheduling
analysis (e.g., data structure). On the other side, it includes elements that are usually not
part of the Design Model, but necessary for scheduling analysis (e.g., priorities, deadlines,
scheduling algorithms, execution times of tasks).

Stereotype	used on	Tagged Values
«saExecHost»	Classes, Objects	Utilization, mainScheduler, isSched
«saCommHost»	Classes, Objects	Utilization, mainScheduler, isSched
«scheduler»	Classes, Objects	schedPolicy, otherSchedPolicy
«schedulableResource»	Classes, Objects	
«saSharedResources»	Classes, Objects	
«saExecStep»	Methods	deadline, priority, execTime, usedResource, respT
«saCommStep»	Methods	deadline, priority, execTime, msgSize, respT
«saEndToEndFlow»	Activities	end2endT, end2endD, isSched
«gaWorkloadEvent»	Initial-Node	pattern
«allocated»	Associations	

Table 1. The MARTE stereotypes and tagged values used for the SAV

Another advantage of the SAV is the fact, that it is separate from the normal Design Model.
Besides the possibility to focus just on scheduling, it also gives the developer the possibility to
test variants/design decisions in the SAV without changing anything in the Design Model. As
there is no automatic and instant synchronisation (see Section 2.6), it does not automatically
change the Design Model if the developer wants to experiment or e.g., has to add provisional
priorities to the system to analyse it, although at an early stage these priorities are not a design
decision.

Moreover, an advantage of using the SAV is that the tagged values help the developer to keep
track of timing requirements during the development, as these parameters are part of the
development model. This especially helps to keep considering them during refinement.

Class diagrams are used to describe the architectural view/the structure of the modelled system. The diagrams show resources, tasks, and associations between these elements. Furthermore, schedulers and other resources, like shared memory, can be defined. Figure 3 shows a class diagram of the SAV that describes the architecture of a sample system. The functionalities/the tasks and communication tasks are represented by methods. The tasks are described using the «saExecStep» stereotype. The methods that represent the communication tasks (transmitting of data over a bus) are extended with the «saCommStep» stereotype. The tasks or communication tasks, represented as methods, are part of schedulable resource classes (marked with the «schedulabeResource» stereotype), which combine tasks or communications that belong together, e.g., since they are part of the same use case or all of them are service routines. Processor resources are represented as classes with the «saExecHost» stereotype and bus resources are classes with the «saCommHost» stereotype. The tasks and communications are mapped on processors or busses by using associations between the schedulable resources and the corresponding bus or processor resource. The associations are extended with the «allocated» stereotype. Scheduling relevant parameters (deadlines, execution times, priorities, etc.) are added to the model using tagged values (see an example in Figure 2).

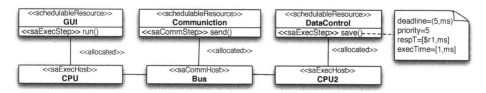

Fig. 3. Architectural Part of the SAV

The object diagram or runtime view is based on the class diagram/architectural view of the SAV. It defines how many instances are parts of the runtime system respectively and what parts are considered for the scheduling analysis. It is possible that only some elements defined in the class diagram are instantiated. Furthermore, some elements can be instantiated twice or more (e.g., if elements are redundant). Only instantiated objects will later be taken into account for the scheduling analysis.

Activity diagrams are used to describe the behaviour of the system. Therefore, workload situations are defined that outline the flow of tasks that are executed during a certain mode of the system. The dependencies of tasks and the execution order are illustrated. The «gaWorkloadEvent» and the «saEnd2EndFlow» stereotypes and their corresponding tagged values are used to describe the workload behaviour parameters like the arrival pattern of the event that triggers the flow or the deadline of the outlined task chain. For example, in Figure 4 it is well defined that at first *cpu.run()* has to be completely executed, before *communication.send()* is scheduled etc.. As activity diagrams are more complex concerning their behaviour than most analysis tools, there are restrictions for the modelling of runtime situations, e.g., no hierarchy is allowed.

The SAV can be easily extended, if necessary. If a scheduling analysis tool offers more possibilities to describe or to analyse a system (e.g., a different scheduling algorithm) and needs more system parameters for it, these parameters have to be part of the SAV. Therefore, the view can be extended with new tagged values that offer the possibility to add the necessary parameters to the system description (added to Table 1).

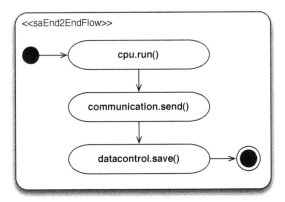

Fig. 4. Workload situation in a SAV

2.2 Abstraction of the design model

The first step of the methodology is the abstraction of the Design Model to the SAV. The Design Model is used as a basis for the scheduling analysis. The basic idea is to find the relevant parts from the Design Model and abstract them in the format of the SAV. Hence, all relevant information for the analysis is identified and transformed into the format of the SAV.

The UML offers many possibilities to describe things. Consequently, most UML Design Models do look different. Even similar things can be described using different expressions (e.g., behaviour could be described using activity diagrams, sequence diagrams, or state charts; deployment can be described using deployment diagrams, but it is also possible to describe it using class diagrams). As a result, an automatic abstraction of the parts necessary for a scheduling analysis is not possible.

As the integration of the scheduling analysis in a UML based development process should be an adaption to the already defined and established development process and not the other way around, our approach offers a flexibility to abstract different Design Models. Our approach uses a rule-based abstraction. The developer creates rules, e.g., "all elements of type device represent a CPU". Based on these rules, the automatic abstraction creates a SAV with the elements of the Design Model. This automatic transformation is implemented for Papyrus for UML[2].

There are two types of rules for the abstraction. The first type describes the element in the Design Model and its representation in the SAV:

ID(element_type, diagram_name, limit1, ...) −> sav_element_type

The rule begins with a unique ID, afterwards the element type is specified (element_type). The following element types can be abstracted: method, class, device, artifact. Then, the diagram can be named on which the abstraction should be done (diagram_name). Finally, it is possible to define limitations, all separated by commas. Limitations can be string filtering or stereotypes. After the arrow, the corresponding element in the SAV can be named. All elements that have a stereotype in the SAV are possible (see Table 1).

---

[2] http://www.papyrusuml.org

The second type of rules abstracts references:

```
(element_type, diagram_name, ID_ref1, ID_ref2)-> Allocation
```

The rule specifies mappings in the SAV. It begins with the element type. Here, only deploys or associations are allowed. After the name of the diagram, the developer has to give two IDs of the basic rules. The abstraction searches for all elements that are affected by the first given rule (ID_ref1) and the second given rule (ID_ref2) and checks, if there is a connection between them, specified through the given element_type. If this is the case, an allocation between the abstracted elements in the SAV is created.

Additionally, it is possible to use the ID_ref as a starting point to use different model elements that are connected to the affected element (e.g., ID_ref1 affects methods, then ID_ref1.class affects the corresponding classes that contain the methods).

Figure 5 gives a simple example of an abstraction. On the left side the Design Model is represented and on the right side, the abstracted SAV. At the beginning, only the left side exists. In this example, one modelling convention for the Design Model was to add the string "_task" to all method names that represent tasks. Another convention was to add "_res" to all class names that represent a CPU.

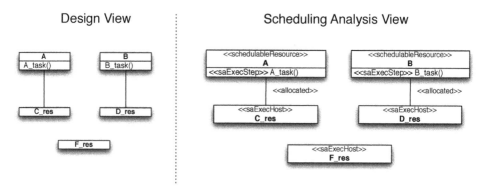

Fig. 5. Simple example of an abstraction from the Design Model to the SAV

The following rules define the abstraction of tasks and CPUs:

```
A1(Class,  ''*'',  ''*_res'')->CPU
A2(Method, ''*'',  ''*_task'')->Task
```

The mapping is described using the following rule:

```
(Association, ''*'', A2.class, A1)->Allocation
```

This rule is used on associations in all diagrams (Association, "*"). All methods that are part of classes (A2.class), which are affected by rule A2, that do have an association with a class that is affected by rule A1, are abstracted to allocations.

It is also possible to define, that model elements in one diagram are directly connected to a model element in another diagram using "<=>" (e.g., a package in one diagram represents a device in another diagram by using the construct "package<=>device", for more information see our case study in Section 3 and Bruechert (2011).

The automatic abstraction of the behaviour using activity diagrams for scheduling analysis is as follows: Using the defined rules, it will be determined which methods are to be considered in the SAV. The corresponding activity diagrams are analysed (all actions that represent a task). All other actions will be deleted and skipped. All activities that do not contain a method representing a task will be removed. In a similar way this is done with sequence diagrams and state machines.

Besides the creating of the SAV during the process of abstraction, there is also a synchronisation table created that documents the abstraction. The table describes the elements in the Design Model and their representation in the SAV. This table is later used for the synchronisation (see Section 2.6). More details about the abstraction and the synchronisation (including a formal description) can be found in Bruechert (2011).

As it is possible that there is still architectural or behaviour information missing after the abstraction, we created additional tool support for the UML case tool Papyrus to help the developer add elements to the SAV (Hagner & Huhn (2008)). We implemented a palette for simpler adding of SAV elements to the system model. Using this extension, the developer does not need to know the relevant stereotypes of how to apply them.

2.3 Parameterisation

After the abstraction, there is still important information missing, e.g., priorities, execution times. The MARTE profile elements are already attached to the corresponding UML element but the values to the parameters are missing. Depending on the stage of the development, these parameters must be added by experts or specialised tools. In early development phases, an expert might be able to give information or, if COTS[3] are used, measured values from earlier developments can be used. In later phases, tools, like aiT (Ferdinand et al. (2001)), T1[4], or Traceanalyzer[5] can be used for automatic parameterisation of the SAV. These tools use static analysis or simple measurement for finding the execution times or the execution patterns of tasks. aiT observes the binary and finds the worst-case execution cycles. As the tool also knows the processor the binary will be executed on, it can calculate the worst-case execution times of the tasks. T1 orchestrates the binary and logs parameters while the tasks are executed on the real platform. Traceanalyzer uses measured values and visualises them (e.g., examines patterns, execution times).

In other development approaches, the parameters are classified with an additional parameter depending on its examination. For example, AUTOSAR[6] separates between worst-case execution time, measured execution time, simulated execution time, and rough estimation of execution time. There are possibilities to add these parameters to the SAV, too. This helps the developer understanding the meaningfulness of the analysis results (e.g., results based on worst-case execution times are more meaningful than results based on rough estimated values).

---

[3] Components-off-the-shelf
[4] http://www.gliwa.com/e/products-T1.html
[5] http://www.symtavision.com/traceanalyzer.html
[6] The AUTOSAR Development Partnership. Automotive Open System Architecture. http://www.autosar.org

Additionally, depending on the chosen scheduling algorithm, one important aspect in this step is the definition of the task priorities. Especially in early phases of a development this can be difficult. There are approaches to find automatically parameters like priorities based on scheduling analysis results. In our method, we suggest to define the priorities manually, do the analysis, and create new variants of the system (see Section 2.5). If, at an early stage, priorities are not known and (more or less) unimportant, the priorities can be set arbitrary, as analysis tools demand these parameters to be set.

2.4 Completeness check and analysis

After the parameterisation is finished and the system is completely described, with respect to the scheduling parameters, an analysis is possible. Before the analysis is done, the system is checked if all parameters are set correctly (e.g., every tasks has to have an execution time; if round robin is set as a scheduling algorithm, tasks need to have a parameter that defines the slot size).

For the analysis, specialised tools are necessary. There are e.g., SymTA/S (Henia et al. (2005)), MAST (Harbour et al. (2001)), and TIMES (Fersman & Yi (2004)). All of these tools are using different meta models. Additionally, these tools have different advantages and abilities.

We created an automatic transformation of the SAV to the scheduling analysis tool SymTA/S (Hagner & Goltz (2010)) and to TIMES (Werner (2006)) by using transformation languages (e.g., ATLAS Group (INRIA & LINA) (2003)). As all information necessary for an analysis is already included in the SAV, a transformation puts all information of the SAV into the format of the analysis tool, triggers the analysis, and brings back the analysis results into the SAV. The developer does not need to see SymTA/S or TIMES, remodel the system in the format of the analysis tool, and does not need to know how the analysis tool works.

SymTA/S links established analysis algorithms with event streams and realises a global analysis of distributed systems. At first, the analysis considers each resource on its own and identifies the response time of the mapped tasks. From these response times and the given input event model it calculates the output event model and propagates it by the event stream. If there are cyclic dependencies, the system is analysed from a starting point iteratively until reaching convergence.

SymTA/S is able to analyse distributed systems using different bus architectures and different scheduling strategies for processors. However, SymTA/S is limited concerning behavioural description, as it is not possible to describe different workload situations. The user has to define the worst-case workload situation or has to analyse different situation independently. Anyhow, as every analysis tool has its advantages it is useful not to use only one analysis tool.

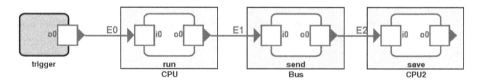

Fig. 6. Representation in SymTA/S

The example depicted in Figure 6 is the SymTA/S representation of the system described in Section 2.1 and illustrated in Figure 3 and Figure 4. There is one source (trigger), two

CPUs (CPU and CPU2), which execute two tasks (run and save), and a bus (Bus) with one communication task (send). All tasks are connected using event streams, representing task chains.

As already mentioned, it is also possible to use other tools for scheduling analysis, e.g., TIMES (Fersman & Yi (2004)). TIMES is based on UPPAAL (Behrmann et al. (2004)) and uses timed automata (Alur & Dill (1994)) for an analysis. Consequently, the results are more precise compared to the over approximated results from SymTA/S. Besides this feature, it also offers code generator for automatic synthesis of C-code on LegoOS platform from the model and a simulator, in which the user can validate the dynamic behaviour of the system and see how the tasks execute according to the task parameters and a given scheduling policy. The simulator shows a graphical representation of the generated trace showing the time points when the tasks are released, invoked, suspended, resumed, and completed. On the other side, as UPPAAL is a model checker, the analysis time could be very long for complex systems due to state space explosion. TIMES is only able to analyse one processor systems. Consequently, for an analysis of distributed systems other tools are necessary.

Figure 7 gives a TIMES representation of the system we described in Section 2.1, with the limitation that all tasks are executed on the same processor. The graph describes the dependencies of the tasks.

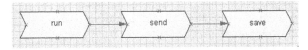

Fig. 7. Representation in TIMES

In TIMES it is also possible to specify a more complex task behaviour/dependency description by using timed automata. Figure 8 gives the example from Section 2.1 using timed automata to describe the system. Timed automata contain locations (in Figure 8 Location_1, Location_2, and Location_3) and switches, which connect the locations. Additionally, the system can contain clocks and other variables. A state of a system is described using the location, the value of the clocks, and the value of other variables. The locations describe the task triggering. By entering a location, the task connected to the location is triggered. Additionally, invariants in locations or guards on the switches are allowed. The guards and the invariants can refer on clocks or other variables.

After the analysis is finished, the analysis results are published in the SAV. In the SAV, the developer can see if there are tasks or task chains that miss their deadlines or if there are resources with a utilisation higher than 100%. The SAV provides tagged values that are used to give the developer a feedback about the analysis results. One example is given in Figure 2, where the respT tagged value is set with a variable ($r1), which means that the response time of the corresponding task is entered at this point after the analysis (this is done automatically by our implemented transformations). There are also other parameters, which give a feedback to the developer (see also Table 1, all are set automatically by the transformations):

- The **respT** tagged values gives a feedback about the worst-case response time of the (communication) tasks and is offered by the «saExecStep» and the «saCommHost» stereotype.

- As the respT, the **end2endT** tagged values offers the worst case response time, in this case for task paths/task chains and is offered by the «saEnd2EndFlow» stereotype. It is not

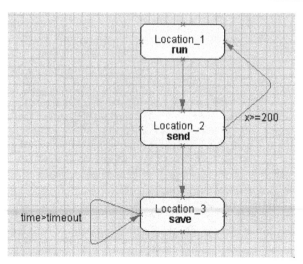

Fig. 8. More advanced representation in TIMES

a summation of all worst-case response times of the tasks that are part of the path, but a worst-case calculated response time of the whole path examined by the scheduling analysis tool (for more details see Henia et al. (2005)).

- The «saExecHost» and the «saCommHost» stereotypes offer a **Utilization** tagged value that gives a feedback about the load of CPUs or busses. If the value is higher than 100% this resource is not schedulable (and the isShed tagged value is false, too). If this value is under 100%, the system might be schedulable (depending on the other analysis results). A high value for this variable always indicates a warning that the resource could be overloaded.

- The tagged value **isShed** gives a feedback if the tasks mapped on this resource are schedulable or not and is offered by the «saExecHost» and the «saCommHost» stereotypes. The tagged values are connected to the Utilization tagged value (e.g., if the utilisation is higher than 100%, the isShed tagged value is false). The **isShed** is also offered by the «saEnd2EndFlow» stereotype. As the «saEnd2EndFlow» stereotype defines parameters for task paths/task chains, the **isShed** tagged value gives a feedback whether the deadline for the path is missed or not.

Using these tagged values, the developer can find out if the system is schedulable by checking the isShed tagged value of the «seEnd2EndFlow» stereotype. If the value is false, the developer has to find the reason why the scheduling failed using the other tagged values. The end2EndT tagged value shows to what extent the deadline is missed, as it gives the response time of the task paths/task chains. The response times of the tasks and the utilisation of the resources give also a feedback where the bottleneck might be (e.g., a resource with a high utilisation and tasks scheduled on it with long response times are more likely a bottleneck compared to resources with low utilisation).

If this information is not sufficient, the developer has to use the scheduling analysis tools for more detailed information. TIMES offers a trace to show the developer where deadlines are missed. SymTA/S offers Gantt charts for more detailed information.

2.5 Variant management

Variant management helps the developer to handle different versions of a SAV. In case of an unsuccessful analysis result (e.g., system is not schedulable) the developer might want to change parameters or distributions directly in the SAV without having to synchronise with the Design Model first, but wants to keep the old version as a backup. Even when the system is schedulable, the developer might want to change parameters to see if it is possible to save resources by using lower CPU frequencies, slower CPUs, or slower bus systems.

It is also possible to add external tools that find good distributions of tasks on resources. Steiner et al. (2008) explored the problem to determine an optimised mapping of tasks to processors, one that minimises bus communication and still, to a certain degree, balances the algorithmic load. The number of possibilities for the distribution of N tasks to M resources is M^N. A search that evaluates all possible patterns for their suitability can be extremely costly and will be limited to small systems. However, not all patterns represent a legal distribution. Data dependencies between tasks may cause additional bus communication if they are assigned to different resources and communication over a bus is much slower than a direct communication via shared memory or message passing on a single processor. Thus, minimising bus communication is an important aspect when a distribution pattern is generated. To use additionally provided CPU resources and create potential for optimisations also the balance of the algorithmic load has to be considered.

In Steiner et al. (2008) the distribution pattern generation is transformed into a graph partitioning problem. The system is represented as an undirected graph, its node weights represent the worst-case execution time of a task and an edge weight corresponds to the amount of data that is transferred between two connected tasks. The algorithm presented searches for a small cut that splits the graph into a number of similar sized partitions. The result is a good candidate for a distribution pattern, where bus communication is minimised and the utilisation of CPU resources is balanced.

Another need for variant management is different criticality levels, necessary e.g., in the ISO 26262 (Road Vehicles Functional Safety (2008)). Many safety-critical embedded systems are subject to certification requirements; some systems are required to meet multiple sets of certification requirements from different certification authorities. For every Safety Integrity Level (SIL) a different variant of the system can be used. In every different variant, the mapping of the tasks and the priorities will be the same. However, the values for the scheduling parameters can be different, e.g., the execution times, as they have to be examined using different methods for each different SIL and consequently for each variant representing a different SIL (see Section 2.3 for different possibilities to parameterise the SAV).

2.6 Synchronisation

If the developer changes something in the SAV (due to analysis results) later and wants to synchronise it with the Design Model, it is possible to use the rule-based approach. During the abstraction (Section 2.2), a matching table/synchronisation table is created and can be used for synchronisation. This approach also works the other way around (changes in the Design Model are transferred to the SAV). During a synchronisation, our implementation is updating the synchronisation table automatically.

One entry in the synchronisation table has two columns. The first specifies the item in the Design Model and the second the corresponding element in the SAV. According to the two rule types (basic rule or reference rule), two types of entries are distinguished in the

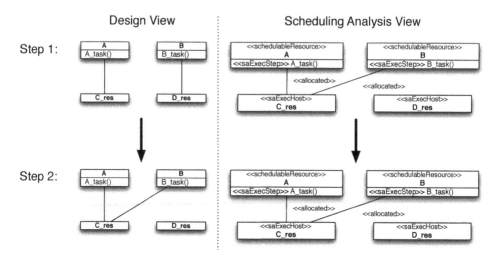

Fig. 9. Synchronisation of the Design Model and the SAV

synchronisation table. The basic entry corresponds to the abstraction of an item that is described by a basic rule. The single entry is described in a Design Model column and a SAV column. The Design Model column contains the element type in the Design Model, the XMI[7] ID in the Design Model, and the name in the Design Model. The SAV column contains the element type, the XMI ID, and the name in the SAV. Regarding a reference entry, based on the reference rules, the Design Model column contains the element type, the XMI ID, the XMI IDs of the two elements with the connection from the Design Model. The SAV column contains the element type, the XMI ID, and, again the XMI IDs from the elements that are connected.

Design Model	SAV
Class, ID_C_res, C_res	CPU, ID_C_res, C_res
Class, ID_D_res, D_res	CPU, ID_D_res, D_res
Method, ID_A_task, A_task	Task, ID_A_task, A_task
Method, ID_B_task, B_task	Task, ID_B_task, B_task
Association, ID, ID_A_task, ID_C_res	Allocation, ID, ID_A_task, ID_C_res
Association, ID, ID_B_task, ID_D_res	Allocation, ID, ID_B_task, ID_D_res

Table 2. The synchronisation table before the synchronisation

Figure 9 gives a simple example, where synchronisation is done. It is based on the example given in Section 2.2 and illustrated in Figure 5. Table 2 gives the corresponding synchronisation table before the synchronisation (for simplification we use a variable name for the XMI IDs).

Because of analysis results, the mapping has been changed and B_task() will now be executed on CPU C_res. Consequently, the mapping has changed in the SAV column in the synchronisation table (see last row in Table 3). Additionally, this is happening in the Design

---

[7] XML Interchange Language (Object Management Group (1998))

Design Model	SAV
Class, ID_C_res, C_res	CPU, ID_C_res, C_res
Class, ID_D_res, D_res	CPU, ID_D_res, D_res
Method, ID_A_task, A_task	Task, ID_A_task, A_task
Method, ID_B_task, B_task	Task, ID_B_task, B_task
Association, ID, ID_A_task, ID_C_res	Allocation, ID, ID_A_task, ID_C_res
Association, ID, ID_B_task, ID_C_res	Allocation, ID, ID_B_task, ID_C_res

Table 3. The synchronisation table after the synchronisation

Model column and finally in the Design Model, too (see Figure 9). More details can be found in Bruechert (2011)

3. Case study

In this Section we want to apply the above introduced methodology to the development of a robotic control system of a parallel robot developed in the Collaborative Research Centre 562 (CRC 562)[8]. The aim of the Collaborative Research Centre 562 is the development of methodological and component-related fundamentals for the construction of robotic systems based on closed kinematic chains (parallel kinematic chains - PKMs), to improve the promising potential of these robots, particularly with regard to high operating speeds, accelerations, and accuracy (Merlet (2000)). This kind of robots features closed kinematic chains and has a high stiffness and accuracy. Due to low moved masses, PKMs have a high weight-to-load-ratio compared to serial robots. The demonstrators which have been developed in the research centre 562 move very fast (up to 10 m/s) and achieve high accelerations (up to 100 m/s^2). The high velocities induced several hard real-time constraints on the software architecture *PROSA-X* (Steiner et al. (2009)) that controls the robots. *PROSA-X* (**P**arallel **R**obots **S**oftware **A**rchitecture - e**X**tended) can use multiple control PCs to distribute its algorithmic load. A middleware (*MiRPA-X*) and a bus protocol that operates on top of a FireWire bus (IEEE 1394, Anderson (1999)) (*IAP*) realise communication satisfying the hard real-time constraints (Kohn et al. (2004)). The architecture is based on a layered design with multiple real-time layers within QNX[9] to realise e.g., a deterministic execution order for critical tasks (Maass et al. (2006)). The robots are controlled using cyclic frequencies between 1 and 8 kHz. If these hard deadlines are missed, this could cause damage to the robot and its environment. To avoid such problems, a scheduling analysis based on models ensures the fulfilment of real-time requirements.

Figure 10 and Figure 11 present the Design Model of the robotic control architecture. Figure 10 shows a component diagram of the robotic control architecture containing the hardware resources. In this variant, there is a "Control_PC1" that performs various computations. The "Control_PC1" is connected via a FireWire data bus with a number of digital signal processors ("DSP_1-7"), which are supervising and controlling the machine. Additionally, there are artefacts («artifact») that are deployed (using the associations marked with the «deploy» stereotype) to the resources. These artefacts represent software that is executed on the corresponding resources.

The software is depicted in Figure 10. This diagram contains packages where every package represents an artefact depicted in Figure 11 (the packages IAP_Nodes_2-7 have been omitted

---

[8] http://www.tu-braunschweig.de/sfb562
[9] QNX Neutrino is a micro kernel real-time operating system.

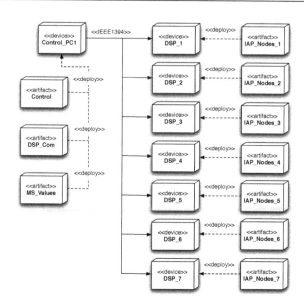

Fig. 10. Component diagram of the robotic control architecture

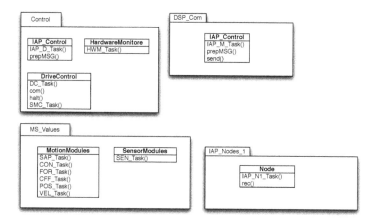

Fig. 11. Package diagram of the robotic control architecture

due to space and are only represented by IAP_Nodes_1). The packages are containing the software that is executed on the corresponding resource. The packages are containing classes and the classes are containing methods. Some methods represent tasks. These methods are marked using the addition of "_Task" to their name (e.g., the package "Control" contains the class "DriveControl" and this class contains three methods, where method DC_Task() represents a task). The tasks that are represented using methods have the following functionality:

- *IAP_D*: This instance of the *IAP* bus protocol receives the *DDTs (Device Data Telegram)* that contain the instantaneous values of the DSP nodes over the FireWire bus.

- *HWM*: The *Hardware Monitoring* takes the instantaneous values received by the *IAP_D* and prepares them for the control.

- *DC*: The *Drive Controller* operates the actuators of the parallel kinematic machine.

- *SMC*: The *Smart Material Controller* operates the active vibration suppression of the machine.

- *IAP_M*: This instance of the bus protocol *IAP* sends the setpoint values, calculated by DC and SMC, to the DSP node.

- *CC*: The *Central Control* activates the currently required sensor and motion modules (see below) and collects their results.

- *CON*: *Contact Planner*. Combination of power and speed control. For the end effector of the robot to make contact with a surface.

- *FOR*: *Force Control*, sets the force for the end effector of the robot.

- *CFF*: Another *Contact Planner*, similar to CON.

- *VEL*: *Velocity Control*, sets the speed for the end effector of the robot.

- *POS*: The *Position Controller* sets the position of the end effector.

- *SAP*: The *Singularity Avoidance Planner* plans paths through the work area to avoid singularities.

- *SEN*: An exemplary *Sensor Module*.

There are three task paths/task chains with real-time requirements. The first task chain receives the instantaneous values and calculates the new setpoint values (using the tasks IAP_D, HWM, DC, SMC). The deadline for this is 250 microseconds. The second task chain contains the sending of the setpoint values to the DSPs and their processing (using tasks IAP_M, MDT, IAP_N1, ..., IAP_N7, DDT1, ..., DDT7). This must be finished within 750 microseconds. The third chain comprises the control of the sensor and motion modules (using tasks CC, CON, FOR, CFF, POS, VEL, SEN, SAP) and has to be completed within 1945 microseconds. The tasks chains including their dependencies were described using activity diagrams.

To verify these real-time requirements we adapted out methodology to the Design Model of the robotic control architecture. The first step was the abstraction of the scheduling relevant information and the creation of the corresponding SAV. As described in Section 2.2, we had to define rules for the abstraction. The following rules were used:

```
A1(Device, ''ComponentDiagram'', ''*'')->CPU
A2(Method, ''PackageDiagram'', ''*_Task'')->Task
```

Rule A1 creates all CPUs in the SAV (classes containing the «saExecHost» stereotype). Rule A2 creates schedulable resources containing the tasks (methods with the «saExecStep» stereotype). Here, we were using the option to sum all tasks that are scheduled on one resource into one schedulable resource representing class (see Figure 12). The corresponding rule to abstract the mapping is:

```
(Deploy, ''*'', A2.class.package<=>Artifact, A1)->Allocation
```

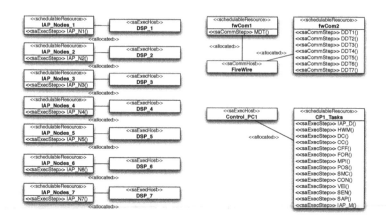

Fig. 12. The architectural view of the PROSA-X system

The packages that contain classes that contain methods that are effected by rule A2, under the assumption that there is an artefact that represents the package in another diagram, are taken into account. It is observed if there is a deploy element between the corresponding artefact and a device element that is effected by rule A1. If this is the case, there is an allocation between these elements. As not all necessary elements are described in the Design Model, e.g., the FireWire bus was not abstracted; it has to be modelled manually in the SAV, as it is important for the scheduling analysis. The result (the architectural view of the SAV) is presented in Figure 3

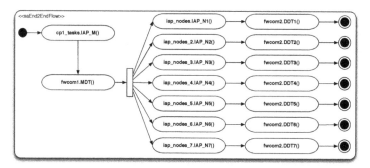

Fig. 13. Sending of the setpoint values to the DSPs

Additionally, a runtime view is created and the behaviour (the workload situations) are created. Figure 13 represents the task chain that sends the setpoint values to the DSPs and describes their processing (IAP_M, MDT, IAP_N1, …, IAP_N7, DDT1, …, DDT7). The deadline is 750 microseconds.

Besides the SAV, a synchronisation table is created. Exemplarily, it is presented in Table 4.

After the SAV is created, it can be parameterised. We have done this by expert knowledge, measuring, and monitoring prototypes. Using these methods, we were able to set the necessary parameters (e.g., execution times, activation pattern, priorities).

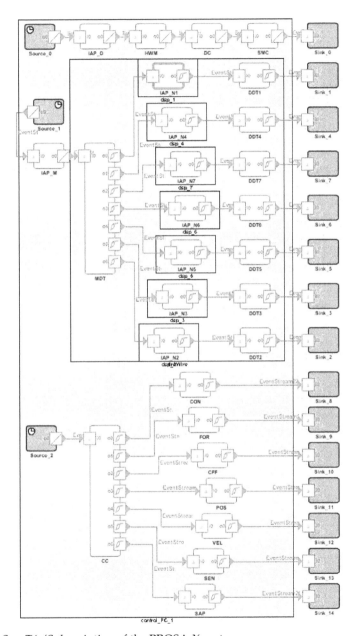

Fig. 14. The SymTA/S description of the PROSA-X system

Design View	SAV
Method, ID, IAP_D_Task	Task, ID, IAP_D_Task
Device, ID, Control_PC1	CPU, ID, Control_PC1
Deploy, ID, IAP_D_Task.IAP_Control.Control <=>Control, Control_PC1	Association, ID, IAP_D_Task, Control_PC1
...	...

Table 4. The synchronisation table of the robotic control system

As we have created automatic transformation to the scheduling analysis tool SymTA/S, the transformation creates a corresponding SymTA/S model and makes it possible to analyse the system. The completeness check is included in the transformation. Afterwards, the output model was analysed by SymTA/S and the expectations were confirmed: The analysis was successful, all paths keep their real-time requirements, and the resources are not overloaded. The SymTA/S model is depicted in Figure 14.

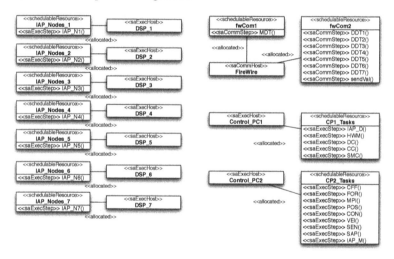

Fig. 15. The new architectural view of the PROSA-X system containing a second control pc

After the successful analysis, the results are automatically published back into the SAV (see Section 2.4). However, we created a new variant of the same system to observe if a faster distribution is possible by adding a new control pc ("Control_PC2"). Consequently, we changed the distribution and added tasks to the second control pc that were originally executed on "Control_PC1") (see Figure 15). As the tasks are more distributed now, we had to add an additional communication task (*sendVal()*) to transfer the results of the calculations. We went through the parameterisation and the analysis again and found out, that this distribution is also valid in terms of scheduling.

As a next step, we can synchronise our results with the Design Model. During the synchronisation, the relevant entries in the synchronisation table were examined. New entries (e.g., for the new control pc) are created and, consequently, the mapping of the artefact "Control" is created corresponding to the SAV. The result is depicted in Figure 16.

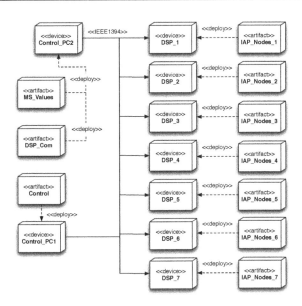

Fig. 16. Component diagram after the synchronisation containing the new device

4. Adapting the approach to other non-functional properties

The presented approach can be adapted to other non-functional requirements (e.g., power consumption or reliability). For every non-functional requirement, there can be an individual view to help the developer concentrate on the aspect he/she is working on. This is drawn upon the cognitive load theory (Sweller (2003)). Consequently, besides the view, a methodology (like the one in this paper) is necessary. Depending on which requirements are considered, the methodologies differ from each other; other steps are necessary and the analysis is different. Additionally, there can be dependencies between the different views (e.g., between the SAV and a view for power consumption as we will explain later).

Power is one of the important metrics for optimisation in the design and operation of embedded systems. One way to reduce power consumption in embedded computing systems is processor slowdown using frequency or voltage. Scaling the frequency and voltage of a processor leads to an increase in the execution time of a task. In real-time systems, we want to minimise energy while adhering to the deadlines of the tasks. Dynamic voltage scaling (DVS) techniques exploit the idle time of the processor to reduce the energy consumption of a system (Aydin et al. (2004); Ishihara & Yasuura (1998); Shin & Kim (2005); Walsh et al. (2003); Yao et al. (1995)).

We defined a Power Consumption Analysis View (PCAV), according to the SAV (Hagner et al. (2011)), to give the developer the possibility to add energy and power consumption relevant parameters to the UML model. Therefore, we created the PCAV profile as an extension of the MARTE profile and an automatic analysis algorithm. The PCAV supports DVS systems. In Figure 17 an example for a PCAV is given. It uses different stereotypes than the SAV as there are different parameters to describe. However, the implementation is similar to the SAV. Additionally, we developed and implemented an algorithm to find a most power aware, but still real-time schedulable system configuration for a DVS system.

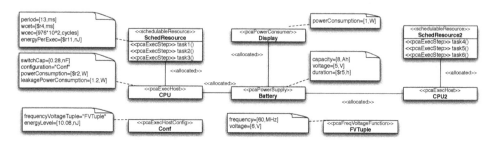

Fig. 17. Power Consumption Analysis View (PCAV)

The power consumption and the scheduling depend on each other (Tavares et al. (2008)). If slower hardware is used to decrease the power consumption, the scheduling analysis could fail due to deadlines that are missed because tasks are executed slower. If faster hardware is used, the power consumption increases. The solution is to find a system configuration that is most power aware but still real-time with respect to their deadline. For our algorithm, we were using both, the SAV and the PCAV. Based on the Design Model we created both views, used the PCAV to do the power consumption analysis and to calculate the execution times and then used the SAV to check the real-time capabilities (Aniculaesei (2011)).

5. Conclusion

In this chapter we have presented a methodology to integrate the scheduling analysis in a UML based development. The methodology is based on the Scheduling Analysis View and contains steps, how to create this view, independently how the UML Design Model looks like, how to process with this view, analyse it, handle variants, and synchronise it with the Design Model. We have presented this methodology in a case study of a robotic control system. Additionally, we have given an outlook on the possibility to create new views for other non-functional requirements.

Future work can be to add additional support concerning the variant management to comply with standards (e.g., Road Vehicles Functional Safety (2008)). Other work can be done by creating different views for other requirements and observe the dependencies between the views.

6. Acknowledgment

The authors would like to thank Symtavision for the grant of free licenses.

7. References

Alur, R. & Dill, D. L. (1994). A theory of timed automata, *Theoretical Computer Science* 126(2): 183 – 235.
    URL: *http://www.sciencedirect.com/science/article/pii/0304397594900108*
Anderson, D. (1999). *FireWire system architecture (2nd ed.): IEEE 1394a*, Addison-Wesley Longman Publishing Co., Inc., Boston, MA, USA.
Aniculaesei, A. (2011). *Uml based analysis of power consumption in real-time embedded systems*, Master's thesis, TU Braunschweig.

Argyris, I., Mura, M. & Prevostini, M. (2010). Using marte for designing power supply section of wsns, *M-BED 2010: Proc. of the 1st Workshop on Model Based Engineering for Embedded Systems Design (a DATE 2010 Workshop)*, Germany.

Arpinen, T., Salminen, E., Hännikäinen, T. D. & Hännikäinen, M. (2011). Marte profile extension for modeling dynamic power management of embedded systems, *Journal of Systems Architecture, In Press, Corrected Proof*.

ATLAS Group (INRIA & LINA) (2003). Atlas transformation language, http://www.eclipse.org/m2m/atl/.

Aydin, H., Melhem, R., Mossé, D. & Mejía-Alvarez, P. (2004). Power-aware scheduling for periodic real-time tasks, *IEEE Trans. Comput.* pp. 584–600.

Behrmann, G., David, R. & Larsen, K. G. (2004). A tutorial on uppaal, *A tutorial on UPPAAL*, Springer, pp. 200–236.

Bruechert, A. (2011). *Abstraktion und synchronisation von uml-modellen fÃijr die scheduling-analyse*, Master's thesis, TU Braunschweig.

Espinoza, H., Servat, D. & Gérard, S. (2008). Leveraging analysis-aided design decision knowledge in uml-based development of embedded systems, *Proceedings of the 3rd international workshop on Sharing and reusing architectural knowledge*, SHARK '08, ACM, New York, NY, USA, pp. 55–62.
URL: *http://doi.acm.org/10.1145/1370062.1370078*

Faugere, M., Bourbeau, T., Simone, R. & Gerard, S. (2007). MARTE: Also an UML profile for modeling AADL applications, *Engineering Complex Computer Systems, 2007. 12th IEEE International Conference on*, pp. 359–364.

Ferdinand, C., Heckmann, R., Langenbach, M., Martin, F., Schmidt, M., Theiling, H., Thesing, S. & Wilhelm, R. (2001). Reliable and precise wcet determination for a real-life processor, *EMSOFT '01: Proc. of the First International Workshop on Embedded Software*, Springer-Verlag, London, UK, pp. 469–485.

Fersman, E. & Yi, W. (2004). A generic approach to schedulability analysis of real-time tasks, *Nordic J. of Computing* 11(2): 129–147.

Hagner, M., Aniculaesei, A. & Goltz, U. (2011). Uml-based analysis of power consumption for real-time embedded systems, *8th IEEE International Conference on Embedded Software and Systems (IEEE ICESS-11), Changsha, China*, Changsha, China.

Hagner, M. & Goltz, U. (2010). Integration of scheduling analysis into uml based development processes through model transformation, *5th International Workshop on Real Time Software (RTS'10) at IMCSIT'10*.

Hagner, M. & Huhn, M. (2007). Modellierung und analyse von zeitanforderungen basierend auf der uml, *in* H. Koschke (ed.), *Workshop*, Vol. 110 of *LNI*, pp. 531–535.

Hagner, M. & Huhn, M. (2008). Tool support for a scheduling analysis view, *Design, Automation and Test in Europe (DATE 08)*.

Hagner, M., Huhn, M. & Zechner, A. (2008). Timing analysis using the MARTE profile in the design of rail automation systems, *4th European Congress on Embedded Realtime Software (ERTS 08)*.

Harbour, M. G., García, J. J. G., Gutiérrez, J. C. P. & Moyano, J. M. D. (2001). Mast: Modeling and analysis suite for real time applications, *ECRTS '01: Proc. of the 13th Euromicro Conference on Real-Time Systems*, IEEE Computer Society, Washington, DC, USA, p. 125.

Henia, R., Hamann, A., Jersak, M., Racu, R., Richter, K. & Ernst, R. (2005). System level performance analysis - the SymTA/S approach, *IEEE Proc. Computers and Digital Techniques* 152(2): 148–166.

Ishihara, T. & Yasuura, H. (1998). Voltage scheduling problem for dynamically variable voltage processors, *Proc. of the 1998 International Symposium on Low Power Electronics and Design (ISLPED '98)* pp. 197–202.

Kohn, N., Varchmin, J.-U., Steiner, J. & Goltz, U. (2004). Universal communication architecture for high-dynamic robot systems using QNX, *Proc. of International Conference on Control, Automation, Robotics and Vision (ICARCV 8th)*, Vol. 1, IEEE Computer Society, Kunming, China, pp. 205–210. ISBN: 0-7803-8653-1.

Kruchten, P. (1995). The 4+1 view model of architecture, *IEEE Softw.* 12(6): 42–50.

Maass, J., Kohn, N. & Hesselbach, J. (2006). Open modular robot control architecture for assembly using the task frame formalism, *International Journal of Advanced Robotic Systems* 3(1): 1–10. ISSN: 1729-8806.

Merlet, J.-P. (2000). *Parallel Robots*, Kluwer Academic Publishers.

Object Management Group (1998). XML model interchange(XMI).

Object Management Group (2002). UML profile for schedulability, performance and time.

Object Management Group (2003). Unified modeling language specification.

Object Management Group (2004). UML profile for modeling quality of service and fault tolerance characteristics and mechanisms.

Object Management Group (2007). Systems Modeling Language (SysML).

Object Management Group (2009). UML profile for modeling and analysis of real-time and embedded systems (MARTE).

Road Vehicles Functional Safety, i. O. f. S. (2008). Iso 26262.

Shin, D. & Kim, J. (2005). Intra-task voltage scheduling on dvs-enabled hard real-time systems, *IEEE Transactions on Computer-aided Design of Integrated Circuits and Systems* .

Steiner, J., Amado, A., Goltz, U., Hagner, M. & Huhn, M. (2008). Engineering self-management into a robot control system, *Proceedings of 3rd International Colloquium of the Collaborative Research Center 562*, pp. 285–297.

Steiner, J., Goltz, U. & Maass, J. (2009). Dynamische verteilung von steuerungskomponenten unter erhalt von echtzeiteigenschaften, *6. Paderborner Workshop Entwurf mechatronischer Systeme*.

Sweller, J. (2003). Evolution of human cognitive architecture, *The Psychology of Learning and Motivation*, Vol. 43, pp. 215–266.

Tavares, E., Maciel, P., Silva, B. & Oliveira, M. (2008). Hard real-time tasks' scheduling considering voltage scaling, precedence and . . . , *Information Processing Letters* .
URL: *http://linkinghub.elsevier.com/retrieve/pii/S0020019008000951*

Walsh, B., Van Engelen, R., Gallivan, K., Birch, J. & Shou, Y. (2003). Parametric intra-task dynamic voltage scheduling, *Proc. of COLP 2003* .

Werner, T. (2006). *Automatische transformation von uml-modellen fuer die schedulability analyse*, Master's thesis, Technische UniversitâĂŕt Braunschweig.

Yao, F., Demers, A. & Shenker, S. (1995). A scheduling model for reduced cpu energy, *Proc. of the 36th Annual Symposium on Foundations of Computer Science* .

10

Context Aware Model-Checking
for Embedded Software

Philippe Dhaussy[1], Jean-Charles Roger[1]
and Frédéric Boniol[2]
[1]*Ensta-Bretagne*
[2]*ONERA*
France

1. Introduction

Reactive systems are becoming extremely complex with the huge increase in high technologies. Despite technical improvements, the increasing size of the systems makes the introduction of a wide range of potential errors easier. Among reactive systems, the asynchronous systems communicating by exchanging messages via buffer queues are often characterized by a vast number of possible behaviors. To cope with this difficulty, manufacturers of industrial systems make significant efforts in testing and simulation to successfully pass the certification process. Nevertheless revealing errors and bugs in this huge number of behaviors remains a very difficult activity. An alternative method is to adopt formal methods, and to use exhaustive and automatic verification tools such as model-checkers.

Model-checking algorithms can be used to verify requirements of a model formally and automatically. Several model checkers as (Berthomieu et al., 2004; Holzmann, 1997; Larsen et al., 1997), have been developed to help the verification of concurrent asynchronous systems. It is well known that an important issue that limits the application of model checking techniques in industrial software projects is the combinatorial explosion problem (Clarke et al., 1986; Holzmann & Peled, 1994; Park & Kwon, 2006). Because of the internal complexity of developed software, model checking of requirements over the system behavioral models could lead to an unmanageable state space.

The approach described in this chapter presents an exploratory work to provide solutions to the problems mentioned above. It is based on two joint ideas: first, to reduce behaviors system to be validated during model-checking and secondly, help the user to specify the formal properties to check. For this, we propose to specify the behavior of the entities that compose the system environment. These entities interact with the system. Their behaviors are described by use cases (scenarios) called here *contexts*. They describe how the environment interacts with the system. Each context corresponds to an operational phase identified as system initialization, reconfiguration, graceful degradation, etc.. In addition, each context is associated with a set of properties to check. The aim is to guide the model-checker to focus on a restriction of the system behavior for verification of specific properties instead on exploring the global system automaton.

In this chapter, we describe the formalism called CDL (Context Description Language), such as DSL[1]. This language serves to support our approach to reduce the state space. We report a feedback on several case studies industrial field of aeronautics, which was conducted in close collaboration with engineers in the field.

This chapter is organized as follows: Section 2 presents related work on the techniques to improve model checking by state reduction and property specification. Section 3 presents the principles of our approach for context aware formal verification. Section 4 describes the CDL language for context specification. Our toolset used for the experiments is presented section 5. In Section 6, we give results of industrial case studies. Section 7 discusses our approach and presents future work.

2. Related works

Several model checkers such as SPIN (Holzmann, 1997), Uppaal (Larsen et al., 1997), TINA-SELT (Berthomieu et al., 2004), have been developed to assist in the verification of concurrent asynchronous systems. For example, the SPIN model-checker based on the formal language Promela allows the verification of LTL (Pnueli, 1977) properties encoded in "never claim" formalism and further converted into Buchi automata. Several techniques have been investigated in order to improve the performance of SPIN. For instance the state compression method or partial-order reduction contributed to the further alleviation of combinatorial explosion (Godefroid, 1995). In (Bosnacki & Holzmann, 2005) the partial-order algorithm based on a depth-first search (DFS) has been adapted to the breadth first search (BFS) algorithm in the SPIN model-checker to exploit interesting properties inherent to the BFS. Partial-order methods (Godefroid, 1995; Peled, 1994; Valmari, 1991) aim at eliminating equivalent sequences of transitions in the global state space without modifying the falsity of the property under verification. These methods, exploiting the symmetries of the systems, seemed to be interesting and were integrated into many verification tools (for instance SPIN).

Compositional (modular) specification and analysis techniques have been researched for a long time and resulted in, e.g., assume/guarantee reasoning or design-by-contract techniques. A lot of work exists in applying these techniques to model checking including, e.g. (Alfaro & Henzinger, 2001; Clarke et al., 1999; Flanagan & Qadeer, 2003; Tkachuk & Dwyer, 2003) These works deal with model checking/analyzing individual components (rather than whole systems) by specifying, considering or even automatically determining the interactions that a component has or could have with its environment so that the analysis can be restricted to these interactions. Design by contract proposes to verify a system by verifying all its components one by one. Using a specific composition operator preserving properties, it allows assuming that the system is verified.

Our approach is different from compositional or modular analysis. We propose to formally specify the context behavior of components in a way that allows a fully automatic divide-and-conquer algorithm. We choose to explicit contexts separately from the model to be validated. However, our approach can be used in conjunction with design by contract process. It is about using the knowledge of the environment of a whole system (or model) to conduct a verification to the end.

Another difficulty is about requirement specification. Embedded software systems integrate more and more advanced features, such as complex data structures, recursion,

---

[1] Domain Specific Language

multithreading. Despite the increased level of automation, users of finite-state verification tools are still constrained to specify the system requirements in their specification language which is often informal. While temporal logic based languages (example LTL or CTL (Clarke et al., 1986)) allow a great expressivity for the properties, these languages are not adapted to practically describe most of the requirements expressed in industrial analysis documents. Modal and temporal logics are rather rudimentary formalisms for expressing requirements, i.e., they are designed having in mind the straightforwardness of its processing by a tool such as a model-checker rather than the user-friendliness. Their concrete syntax is often simplistic, tailored for easing its processing by particular tools such as model checkers. Their efficient use in practice is hampered by the difficulty to write logic formula correctly without extensive expertise in the idioms of the specification languages.

It is thus necessary to facilitate the requirement expression with adequate languages by abstracting some details in the property description, at a price of reducing the expressivity. This conclusion was drawn a long time ago and several researchers (Dwyer et al., 1999; Konrad & Cheng, 2005; Smith et al., 2002) proposed to formulate the properties using definition patterns in order to assist engineers in expressing system requirements. Patterns are textual templates that capture common logical and temporal properties and that can be instantiated in a specific context. They represent commonly occurring types of real-time properties found in several requirement documents for embedded systems.

3. Context aware verification

To illustrate the explosion problem, let us consider the example in Figure 1. We are trying to verify some requirements by model checking using the TINA-SELT model checker. We present the results for a part of the S_CP model. Then, we introduce our approach based on context specifications.

3.1 An illustration

We present one part of an industrial case study: the software part of an anti-aircraft system (S_CP). This controller controls the internal modes, the system physical devices (sensors, actuators) and their actions in response to incoming signals from the environment. The S_CP system interacts with devices (Dev) that are considered to be *actors* included in the S_CP environment called here *context*.

The sequence diagrams of Figure 2 illustrate interactions between context actors and the S_CP system during an initialization phase. This context describes the environment we want to consider for the verification of the S_CP controller. This context is composed of several actors Dev running in parallel or in sequence. All these actors interleave their behavior. After the initializing phase, all actors Dev_i ($i \in [1\ldots n]$) wait for orders $goInitDev$ from the system. Then, actors Dev_i send $login_i$ and receive either $ackLog(id)$ (Figure 2.a and 2.c) or $nackLog(err)$ (Figure 2.b) as responses from the system. The logged devices can send $operate(op)$ (Figure 2.a and 2.c) and receive either $ackOper(role)$ (Figure 2.a) or $nackOper(err)$ (Figure 2.c). The messages $goInitDev$ can be received in parallel in any order. However, the delay between messages $login_i$ and $ackLog(id)$ (Figure 1) is constrained by $maxD\_log$. The delay between messages $operate(op)$ and $ackOper(role)$ (Figure 1) is constrained by $maxD\_oper$. And finally all Dev_i send $logout_i$ to end the interaction with the S_CP controller.

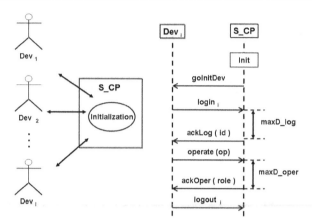

Fig. 1. S_CP system: partial description during the initialization phase.

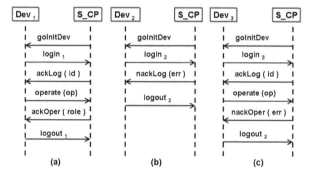

Fig. 2. An example of S_CP context scenario with 3 devices.

3.2 Model-checking results

To verify requirements on the system model[2], we used the TINA-SELT model checker. To do so, the system model is translated into FIACRE format (Farail et al., 2008) to explore all the S_CP model behaviors by simulation, S_CP interacting with its environment (devices). Model exploration generates a labeled transition system (LTS) which represents all the behaviors of the controller in its environment. Table 1 shows[3] the exploration time and the amount of configurations and transitions in the LTS for different complexities (n indicates the number of considered actors). Over four devices, we see a state explosion because of the limited memory of our computer.

3.3 Combinatorial explosion reduction

When checking the properties of a model, a model-checker explores all the model behaviors and checks whether the properties are true or not. Most of the time, as shown by previous

---

[2] Here by system or system model, we refer to the model to be validated.
[3] Tests were executed on Linux 32 bits - 3 Go RAM computer, with TINA vers.2.9.8 and Frac parser vers.1.4.2.

N.of devices	Exploration time (sec)	N.of LTS configurations	N.of LTS transitions
1	10	16 766	82 541
2	25	66 137	320 388
3	91	269 977	1 297 987
4	118	939 689	4 506 637
5	Explosion	–	–

Table 1. Table highlighting the verification complexity for an industrial case study (S_CP).

results, the number of reachable configurations is too large to be contained in memory (Figure 3.a). We propose to restrict model behavior by composing it with an environment that interacts with the model. The environment enables a subset of the behavior of the model. This technique can reduce the complexity of the exploration by limiting the scope of the verification to precise system behaviors related to some specific environmental conditions.

This reduction is computed in two stages: Contexts are first identified by the user (context$_i$, $i \in$ [1..n] in Figure 3.b). They correspond to patterns of use of the component being modeled. The aim is to circumvent the combinatorial explosion by restricting the behavior system with an environment describing different configurations in which one wishes to check requirements. Then each context is automatically partitioned into a set of sub-contexts. Here we precisely define these two aspects implemented in our approach.

The context identification focuses on a subset of behavior and a subset of properties. In the context of reactive embedded systems, the environment of each component of a system is often well known. It is therefore more effective to identify this environment than trying reduce the configuration space of the model system to explore.

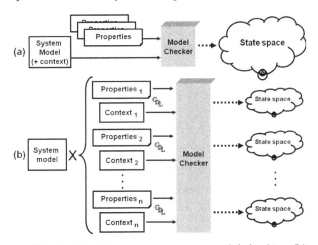

Fig. 3. Traditional model checking (a) vs. context-aware model checking (b).

In this approach, we suppose that the designer is able to identify all possible interactions between the system and its environment. We also consider that each context expressed initially is finite, (i.e., there is a non infinite loop in the context). We justify this strong hypothesis, particularly in the field of embedded systems, by the fact that the designer of a software component needs to know precisely and completely the perimeter (constraints, conditions) of its system for properly developing it. It would be necessary to study formally the validity of this working hypothesis based on the targeted applications. In this chapter, we do not address this aspect that gives rise to a methodological work to be undertaken.

Moreover, properties are often related to specific use cases (such as initialization, reconfiguration, degraded modes). Therefore, it is not necessary for a given property to take into account all possible behaviors of the environment, but only the subpart concerned by the verification. The context description thus allows a first limitation of the explored space search, and hence a first reduction in the combinatorial explosion.

The second idea is to automatically split each identified context into a set of smaller sub-contexts (Figure 4). The following verification process is then equivalent: (i) compose the context and the system, and then verify the resulting global system, (ii) partition the environment into k sub-contexts (scenarios), and successively deal each scenario with the model and check the properties on the outcome of each composition. Actually, we transform the global verification problem into k smaller verification sub problems. In our approach, the complete context model can be split into pieces that have to be composed separately with the system model. To reach that goal, we implemented a recursive splitting algorithm in our OBP tool. Figure 4 illustrates the function $explore\_mc()$ for exploration of a *model*, with a *context* and model-checking of a set of properties *pty*. The context is represented by acyclic graph. This graph is composed with the model for exploration. In case of explosion, this context is automatically split into several parts (taking into account a parameter d for the depth in the graph for splitting) until the exploration succeeds.

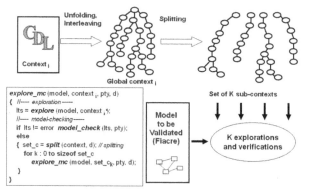

Fig. 4. Context splitting and verification for each partition (sub-context).

In summary, the context aware method provides three reduction axes: the context behavior is constrained, the properties are focused and the state space is split into pieces. The reduction in the model behavior is particularly interesting while dealing with complex embedded systems, such as in avionic systems, since it is relevant to check properties over specific system modes (or use cases) which is less complex because we are dealing with a subset of the system automata. Unfortunately, only few existing approaches propose operational ways to precisely capture these contexts in order to reduce formal verification complexity and thus improve the scalability of existing model checking approaches. The necessity of a clear methodology has also to be identified, since the context partitioning is not trivial, i.e., it requires the formalization of the context of the subset of functions under study. An associated methodology must be defined to help users for modeling contexts (out of scope of this chapter).

4. CDL language for context and property specification

We propose a formal tool-supported framework that combines context description and model transformations to assist in the definition of requirements and of the environmental conditions in which they should be satisfied. Thus, we proposed (Dhaussy et al., 2009) a context-aware verification process that makes use of the CDL language. CDL was proposed to fill the gap between user models and formal models required to perform formal verifications. CDL is a Domain Specific Language presented either in the form of UML like graphical diagrams (a subset of activity and sequence diagrams) or in a textual form to capture environment interactions.

4.1 Context hierarchical description

CDL is based on Use Case Charts of (Whittle, 2006) using activity and sequence diagrams. We extended this language to allow several entities (actors) to be described in a context (Figure 5). These entities run in parallel. A CDL[4] model describes, on the one hand, the context using activity and sequence diagrams and, on the other hand, the properties to be checked using property patterns. Figure 5 illustrates a CDL model for the partial use cases of Figures 1 and 2. Initial use cases and sequence diagrams are transformed and completed to create the context model. All context scenarios are represented, combined with parallel and alternative operators, in terms of CDL.

A diagrammatical and textual concrete syntax is created for the context description and a textual syntax for the property expression. CDL is hierarchically constructed in three levels: Level-1 is a set of use case diagrams which describes hierarchical activity diagrams. Either alternative between several executions (alternative/merge) or a parallelization of several executions (fork/join) is available. Level-2 is a set of scenario diagrams organized in alternatives. Each scenario is fully described at Level-3 by sequence diagrams. These diagrams are composed of lifelines, some for the context actors and others for processes composing the system model. Counters limit the iterations of diagram executions. This ensures the generation of finite context automata.

From a semantic point of view, we can consider that the model is structured in a set of sequence diagrams (MSCs) connected together with three operators: sequence (*seq*), parallel (*par*) and alternative (*alt*). The interleaving of context actors described by a set of MSCs generates a graph representing all executions of the actors of the environment. This graph is then partitioned in such a way as to generate a set of subgraphs corresponding to the sub-contexts as mentioned in 3.3.

The originality of CDL is its ability to link each expressed property to a context diagram, i.e. a limited scope of the system behavior. The properties can be specified with property pattern definitions that we do not describe here but can be found in (Dhaussy & Roger, 2011). Properties can be linked to the context description at Level 1 or Level 2 (such as $P1$ and $P3$ in Figure 5) by the stereotyped links property/scope. A property can have several scopes and several properties can refer to a single diagram. CDL is designed so that formal artifacts

---

[4] For the detailed syntax, see (Dhaussy & Roger, 2011) available (currently in french) on http://www.obpcdl.org.

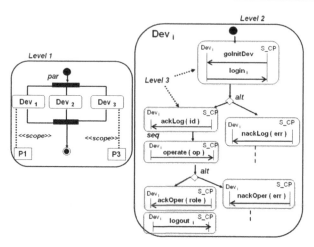

Fig. 5. *S_CP* case study: partial representation of the context.

required by existing model checkers could be automatically generated from it. This generation is currently implemented in our prototype tool called OBP (*Observer Based Prover*) described briefly in Section 5. We will now present the CDL formal syntax and semantics.

4.2 Formal syntax

A CDL model (also called "context") is a finite generalized MSC C, following the formal grammar:

$$C ::= M \mid C_1; C_2 \mid C_1 + C_2 \mid C_1 \| C_2$$
$$M ::= 0 \mid a!; M \mid a?; M$$

In other words, a context is either (1) a single MSC M composed as a sequence of event emissions $a!$ and event receptions $a?$ terminated by the empty MSC (**0**) which does nothing, or (2) a sequential composition (*seq* denoted ;) of two contexts ($C_1; C_2$), or (3) a non deterministic choice (*alt* denoted +) between two contexts ($C_1 + C_2$), or (4) a parallel composition (*par* denoted $\|$) between two contexts ($C_1 \| C_2$).

For instance, let us consider the context Figure 5 graphically described. This context describes the environment we want to consider for the validation of the system model. We consider that the environment is composed of 3 actors Dev_1, Dev_2 and Dev_3. All these actors run in parallel and interleave their behavior. The model can be formalized, with the above textual grammar as follows[5].

$$C \quad = Dev_1 \| Dev_2 \| Dev_2$$
$$Dev_i = Log_i; (Oper + (nackLog\ (err)?; \ldots .0))$$
$$Log_i = (goInitDev\ ?\ ; \ login_i\ !)$$
$$Oper = (\ ackLog\ (id)\ ?\ ; \ operate\ (op)\ !\ (\ Ack_i\ + \ (\ nackOper\ (err)\ ?\ ; \ldots ;0)))$$
$$Ack_i = (\ ackOper\ (role)\ ?\ ; \ logout_i\ !\ ; \ldots ; \mathbf{0})$$
$$Dev_1, Dev_2, Dev_3 \ = \ Dev_i\ with\ i\ = \ 1,\ 2,\ 3$$

---

[5] In this chapter, as an illustration, we consider that the behavior of actors extends, noted by the "…".

4.3 Semantics

The semantics is based on the semantics of the scenarios and expressed by construction rules of sets of traces built using *seq*, *alt* and *par* operators. A scenario trace is an ordered events sequence which describes a history of the interactions between the context and the model.

To describe the formal semantics, let us define a function $wait(C)$ associating the context C with the set of events awaited in its initial state:

$$Wait\,(0) \stackrel{\text{def}}{=} \varnothing \qquad Wait\,(a!;M) \stackrel{\text{def}}{=} \varnothing \qquad Wait\,(a?;M) \stackrel{\text{def}}{=} \{a\}$$

$$Wait\,(C_1 + C_2) \stackrel{\text{def}}{=} Wait\,(C_1) \cup Wait\,(C_2) \qquad Wait\,(C_1;C_2) \stackrel{\text{def}}{=} Wait\,(C_1)\ if\ C_1 \neq 0$$

$$Wait\,(0;C_2) \stackrel{\text{def}}{=} Wait\,(C_2) \qquad Wait\,(C_1\|C_2) \stackrel{\text{def}}{=} Wait\,(C_1) \cup Wait\,(C_2)$$

We consider that a context is a process communicating in an asynchronous way with the system, memorizing its input events (from the system) in a *buffer*. The semantics of CDL is defined by the relation $(C,B) \stackrel{a}{\rightarrow} (C',B')$ to express that the context C with the buffer B "produces" a (which can be a sending or a receiving signal, or the $null_\sigma$ signal if C does not evolve) and then becomes the new context C' with the new buffer B'. This relation is defined by the 8 rules in Figure 6 (In these rules, a represents an event which is different from $null_\sigma$).

The *pref1* rule (without any preconditions) specifies that an MSC beginning with a sending event a! emits this event and continues with the remaining MSC. The *pref2* rule expresses that if an MSC begins by a reception a? and faces an input buffer containing this event at the head of the buffer, the MSC consumes this event and continues with the remaining MSC. The *seq1* rule establishes that a sequence of contexts $C_1;C_2$ behaves as C_1 until it has terminated. The *seq2* rule says that if the first context C_1 terminates (i.e., becomes 0), then the sequence becomes C_2. The *par1* and *par2* rules say that the semantics of the parallel operation is based on an asynchronous interleaving semantics. The *alt* rule expresses that the alternative context $C_1 + C_2$ behaves either as C_1 or as C_2. Finally, the *discard* rule says that if an event a at the head of the input buffer is not expected, then this event is lost (removed from the head of the buffer).

4.4 Context and system composition

We can now formally define the "closure" composition $< (C,B_1)\ |\ (s,\mathcal{S},B_2) >$ of a system $\mathcal{S}$ in a state $s \in \Sigma$ (Σ is the set of system states), with its input buffer B_2, with its context C, with its input buffer B_1 (note that each component, system and context, has its own buffer). The evolution of $\mathcal{S}$ closed by C is given by two relations: the relation (1):

$$< (C,B_1)|(s,\mathcal{S},B_2) > \stackrel{a}{\rightarrow} < (C',B'_1)|(s',\mathcal{S},B'_2) > (1)$$

to express that $\mathcal{S}$ in the state s evolves to state s' receiving event a, potentially empty ($null_e$), (sent by the context) and producing the sequence of events σ, potentially empty ($null_\sigma$) (to the context). and the relation (2):

$$< (C,B_1)|(s,\mathcal{S},B_2) > \stackrel{t}{\rightarrow} < (C,B_1)|(s',\mathcal{S},B'_2) > (2)$$

to express that $\mathcal{S}$ in state s evolves to the state s' by progressing time t, and producing the sequence of events σ potentially empty ($null_\sigma$) (to the context). Note that in the case of timed

$$\frac{}{(a!; M, B) \xrightarrow{a!} (M, B)} \text{[pref1]} \qquad \frac{}{(a?; M, a.B) \xrightarrow{a?} (M, B)} \text{[pref2]}$$

$$\frac{\begin{array}{c} C_1' \neq \mathbf{0} \\ (C_1, B) \xrightarrow{a} (C_1', B') \end{array}}{(C_1.C_2, B) \xrightarrow{a} (C_1'.C_2, B')} \text{[seq1]} \qquad \frac{(C_1, B) \xrightarrow{a} (\mathbf{0}, B')}{(C_1.C_2, B) \xrightarrow{a} (C_2, B')} \text{[seq2]}$$

$$\frac{\begin{array}{c} C_1' \neq \mathbf{0} \\ (C_1, B) \xrightarrow{a} (C_1', B') \end{array}}{\begin{array}{c} (C_1 \| C_2, B) \xrightarrow{a} (C_1' \| C_2, B') \\ (C_2 \| C_1, B) \xrightarrow{a} (C_2 \| C_1', B') \end{array}} \text{[par1]} \qquad \frac{(C_1, B) \xrightarrow{a} (\mathbf{0}, B')}{\begin{array}{c} (C_1 \| C_2, B) \xrightarrow{a} (C_2, B') \\ (C_2 \| C_1, B) \xrightarrow{a} (C_2, B') \end{array}} \text{[par2]}$$

$$\frac{(C_1, B) \xrightarrow{a} (C_1', B')}{\begin{array}{c} (C_1 + C_2, B) \xrightarrow{a} (C_1', B') \\ (C_2 + C_1, B) \xrightarrow{a} (C_1', B') \end{array}} \text{[alt]} \qquad \frac{a \notin wait(C)}{(C, a.B) \xrightarrow{null_\sigma} (C, B)} \text{[discard}_C\text{]}$$

Fig. 6. Context semantics.

evolution, only the system evolves, the context is not timed. The semantics of this composition is defined by the four following rules (Figure 7).

Rule $cp1$: If $\mathcal{S}$ can produce σ, then $\mathcal{S}$ evolves and σ is put at the end of the buffer of C. Rule $cp2$: If C can emit a, C evolves and a is queued in the buffer of $\mathcal{S}$. Rule $cp3$: If C can consume a, then it evolves whereas $\mathcal{S}$ remains the same. Rule $cp4$: If the time can progress in $\mathcal{S}$, then the time progress in the composition $\mathcal{S}$ and C.

Note that the "closure" composition between a system and its context can be compared with an asynchronous parallel composition: the behavior of C and of $\mathcal{S}$ are interleaved, and they communicate through asynchronous buffers. We will denote $< (C, B) | (s, \mathcal{S}, B') > \nrightarrow$ to express that the system and its context cannot evolve (the system is blocked or the context terminated). We then define the set of traces (called *runs*) of the system closed by its context from a state s, by:

$$\llbracket C \mid (s, \mathcal{S}) \rrbracket \overset{\text{def}}{=} \{a_1 \cdot \sigma_1 \cdot \ldots \cdot a_n \cdot \sigma_n \cdot end_C \mid$$
$$< (C, null_\sigma) \mid (s, null_\sigma) > \xrightarrow[\sigma_1]{a_1} < (C_1, B_1) \mid (s_1, \mathcal{S}, B_1') >$$
$$\xrightarrow[\sigma_2]{a_2} \ldots \xrightarrow[\sigma_n]{a_n} < (C_n, B_n) \mid (s_n, \mathcal{S}, B_n') > \nrightarrow \}$$

$\llbracket C | (s, \mathcal{S}) \rrbracket$ is the set runs of $\mathcal{S}$ closed by C from the state s. Note that a context is built as sequential or parallel compositions of finite loop-free MSCs. Consequently the *runs* of a system model closed by a CDL context are necessarily finite. We then extend each *run* of $\llbracket C | (s, \mathcal{S}) \rrbracket$ by a specific terminal event end_C allowing the observer to catch the ending of a scenario and accessibility properties to be checked.

$$\frac{(s, \mathcal{S}, B_2) \xrightarrow[\sigma]{} (s', \mathcal{S}, B_2')}{< (C, B_1)|(s, \mathcal{S}, B_2) > \xrightarrow[\sigma]{null_\triangle} < (C, B_1.\sigma)|(s', \mathcal{S}, B_2') >} \quad [\text{cp1}]$$

$$\frac{(C, B_1) \xrightarrow{a!} (C', B_1')}{< (C, B_1)|(s, \mathcal{S}, B_2) > \xrightarrow[null_\sigma]{a} < (C', B_1')|(s, \mathcal{S}, B_2.a) >} \quad [\text{cp2}]$$

$$\frac{(C, B_1) \xrightarrow{a?} (C', B_1')}{< (C, B_1)|(s, \mathcal{S}, B_2) > \xrightarrow[null_\sigma]{null_\triangle} < (C', B_1')|(s, \mathcal{S}, B_2) >} \quad [\text{cp3}]$$

$$\frac{(s, \mathcal{S}, B_2) \xrightarrow[\sigma]{t} (s', \mathcal{S}, B_2')}{< (C, B_1)|(s, \mathcal{S}, B_2) > \xrightarrow[\sigma]{t} < (C, B_1)|(s', \mathcal{S}, B_2') >} \quad [\text{cp4}]$$

Fig. 7. CDL context and system composition semantics.

4.5 Property specification patterns

Property specifying needs to use powerful yet easy mechanisms for expressing temporal requirements of software source code. As example, let's see a requirement of the S_CP system described in section 3.1. This requirement was found in a document of our partner and is shown in Listing 1. It refers to many events related to the execution of the model or environment. It also depends on an execution history that has to be taken into account as a constraint or pre-condition.

Requirement R: *During initialization procedure, S_CP shall associate an identifier to each device (Dev), after login request and before $maxD\_log$ time units.*

Listing 1. Initialization requirement for the S_CP system described in section 3.

If we want to express this requirement with a temporal logic based language as LTL or CTL, the logical formulas are of great complexity and become difficult to read and to handle by engineers. So, for the property specification, we propose to reuse the categories of Dwyer patterns (Dwyer et al., 1999) and extend them to deal with more specific temporal properties which appear when high-level specifications are refined. Additionally, a textual syntax is proposed to formalize properties to be checked using property description patterns (Konrad & Cheng, 2005). To improve the expressiveness of these patterns, we enriched them with options (*Pre-arity, Post-arity, Immediacy, Precedence, Nullity, Repeatability*) using annotations as (Smith et al., 2002). Choosing among these options should help the user to consider the relevant alternatives and subtleties associated with the intended behavior. These annotations allow these details to be explicitly captured. During a future work, we will adapt these patterns taking into account the taxonomy of relevant properties, if this appears necessary.

We integrate property patterns description in the CDL language. Patterns are classified in families, which take into account the timed aspects of the properties to be specified. The identified patterns support properties of answer (*Response*), the necessity one (*Precedence*), of absence (*Absence*), of existence (*Existence*) to be expressed. The properties refer to detectable events like transmissions or receptions of signals, actions, and model state changes. The property must be taken into account either during the entire model execution, before, after or between occurrences of events. Another extension of the patterns is the possibility of handling sets of events, ordered or not ordered similar to the proposal of (Janssen et al., 1999). The operators AN and ALL respectively specify if an event or all the events, ordered (Ordered) or not (Combined), of an event set are concerned with the property.

We illustrate these patterns with our case study. The given requirement R (Listing 1) must be interpreted and can be written with CDL in a property $P1$ as follow (cf. Listing 2). $P1$ is linked to the communication sequence between the S_CP and device (Dev_1). According to the sequence diagram of figure 5, the association to other devices has no effect on $P1$.

> *Property P1;*
> *ALL Ordered*
>   *exactly one occurence of S_CP_hasReachState_Init*
>   *exactly one occurence of login1*
> *end*
> *eventually leads − to* [0..maxD_log]
> *AN*
>   *one or more occurence of ackLog(id)*
> *end*
> *S_CP_hasReachState_Init may never occurs*
> *login1 may never occurs*
> *one of ackLog(id) cannot occur before login1*
> *repeatibility : true*

Listing 2. S_CP case study: A response pattern from R requirement.

$P1$ specifies an observation of event occurrences in accordance with figure 5. $login1$ refers to $login_1$ reception event in the model, $ackLog$ refers to ackLog reception event by Dev_1. $S\_CP\_hasReachState\_Init$ refers a state change in the model under study.

For the sake of simplicity, we consider in this chapter that properties are modeled as observers. Our OBP toolset transforms each property into an observer automaton including a reject node. An observer is an automaton which *observes* the set of events exchanged by the system S and its context C (and thus events occurring in the *runs* of $[\![C|(init, S)]\!]$) and which produces an event *reject* whenever the property becomes false. With observers, the properties we can handle are of safety and bounded liveness type. The accessibility analysis consists of checking if there is a reject state reached by a property observer. In our example, this reject node is reached after detecting the event sequence of $S\_CP\_hasReachState\_Init$ and $login_1$, in that order, if the sequence of one or more of $ackLog$ is not produced before $maxD\_log$ time units. Conversely, the reject node is not reached either if $S\_CP\_hasReachState\_Init$ or $login_1$ are never received, or if $ackLog$ event above is correctly produced with the right delay. Consequently, such a property can be verified by using reachability analysis implemented in our OBP Explorer. For that purpose, OBP translates the property into an observer automaton, depicted in figure 8.

4.6 Formalization of observers

The third part of the formalization relies on the expression of the properties to be fulfilled. We consider in the following that an observer is an automaton $\mathcal{O} = \langle \Sigma_o, init_o, T_o, Sig, \{reject\}, Sv_o \rangle$

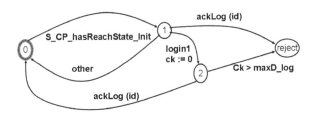

Fig. 8. Observer automaton for the property P1 of Listing 2.

(a) emitting a single output event: *reject*, (b) where *Sig* is the set of matched events by the observer; events produced and received by the system and its context and (c) such that all transitions labelled *reject* arrive in a specific state called "unhappy".

Semantics. We say that S in the state $s \in \Sigma$. S closed by C satisfies $\mathcal{O}$, denoted $C|(s,S) \models \mathcal{O}$, if and only if no execution of $\mathcal{O}$ faced to the runs r of $[\![C|(s,S)]\!]$ produces a *reject* event. This means:

$$C \mid (s,S) \models \mathcal{O} \iff \forall r \in [\![C \mid (s,S)]\!],$$

$$(init_o, \mathcal{O}, r) \xrightarrow[null_\sigma]{} (s_1, \mathcal{O}, r_1) \xrightarrow[null_\sigma]{} \cdots \xrightarrow[null_\sigma]{} (s_n, \mathcal{O}, r_n) \not\rightarrow$$

Remark: executing $\mathcal{O}$ on a run r of $[\![C|(s,S)]\!]$ is equivalent to put r in the input buffer of $\mathcal{O}$ and to execute $\mathcal{O}$ with this buffer. This property is satisfied if and only if only the empty event ($null_\sigma$) is produced (i.e., the *reject* event is never emitted).

5. OBP toolset

To carry out our experiments, we used our OBP[6] tool (Figure 9). OBP is an implementation of a CDL language translation in terms of formal languages, i.e. currently FIACRE (Farail et al., 2008). As depicted in Figure 9, OBP leverages existing academic model checkers such as TINA or simulators such as our explorer called OBP Explorer. From CDL context diagrams, the OBP tool generates a set of context graphs which represent the sets of the environment runs. Currently, each generated graph is transformed into a FIACRE automaton. Each graph represents a set of possible interactions between model and context. To validate the model under study, it is necessary to compose each graph with the model. Each property on each graph must be verified. To do so, OBP generates either an observer automaton (Halbwachs et al., 1993) from each property for OBP Explorer, or SELT logic formula (Berthomieu et al., 2004) for the TINA model checker. With OBP Explorer, the accessibility analysis is carried out on the result of the composition between a graph, a set of observers and the system model as described in (Dhaussy et al., 2009). If, for a given context, we face state explosion, the accessibility analysis or model-checking is not possible. In this case, the context is split into a subset of contexts and the composition is executed again as mentioned in 3.3.

To import models with standard format such as UML, SysML, AADL, SDL, we necessarily need to implement adequate translators such as those studied in TopCased[7] or Omega[8] projects to generate FIACRE programs.

---

[6] OBP$_t$ (OBP for TINA) is available on http://www.obpcdl.org.
[7] http://www.topcased.org
[8] http://www-Omega.imag.fr

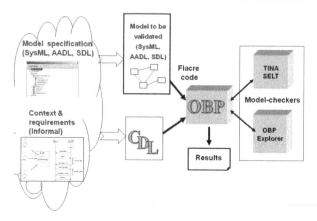

Fig. 9. CDL model transformation with OBP.

6. Experiments and results

Our approach was applied to several embedded systems applications in the avionic or electronic industrial domain. These experiments were carried out with our French industrial partners. We reported here the results of these experiments.

6.1 Requirement specification

This section reports on six case studies (CS_1 to CS_6). Four of the software components come from an industrial A and two from a B[9]. For each industrial component, the industrial partner provided requirement documents (use cases, requirements in natural language) and the component executable model. Component executable models are described with UML, completed by ADA or JAVA programs, or with SDL language. The number of requirements in Table 2 evaluates the complexity of the component. To validate these models, we specify properties and contexts.

	CS_1	CS_2	CS_3	CS_4	CS_5	CS_6
Modeling language	SDL	SDL	SDL	SDL	UML2	UML2
Number of code lines	4 000	15 000	30 000	15 000	38 000	25 000
Number of requirements	49	94	136	85	188	151

Table 2. Industrial case study classification.

6.1.1 Property specification

Requirements are inputs of our approach. Here, the work consists in transforming natural language requirements into temporal properties. To create the CDL models with patterns-based properties, we analyzed the software engineering documents of the proposed case studies. We transformed textual requirements. We focused on requirements which

---

[9] CS_5 corresponds to the case study partially described in section 3.1.

can be translated into observer automata. Firstly, we note that most of requirements had to be rewritten into a set of several properties. Secondly, model requirements of different abstraction levels are mixed. We extracted requirement sets corresponding to the model abstraction level. Finally, we observe that most of the textual requirements are ambiguous. We had to rewrite them consequently to discussion with industrial partners. Table 3 shows the number of properties which are translated from requirements. We consider three categories of requirements. *Provable* requirements correspond to requirements which can be captured with our approach and can be translated into observers. The proof technique can be applied on a given context without combinatorial explosion. *Non-Computable* requirements are requirements which can be interpreted by a pattern but cannot be translated into an observer. For example, liveness properties cannot be translated because they are unbounded. Observers capture only bounded liveness properties. From the interpretation, we could generate another temporal logic formula, which could feed a model checker as TINA. *Non-Provable* requirements are requirements which cannot be interpreted at all with our patterns. It is the case when a property refers to undetectable events for the observer, such as the absence of a signal.

	CS_1	CS_2	CS_3	CS_4	CS_5	CS_6	Average
Provable properties	38/49 (78%)	73/94 (78%)	72/136 (53%)	49/85 (58%)	155/188 (82%)	41/151 (27%)	428/703 (61%)
Non-computable properties	0/49 (0%)	2/94 (2%)	24/136 (18%)	2/85 (2%)	18/188 (10%)	48/151 (32%)	94/703 (13%)
Non-Provable properties	11/49 (22%)	19/94 (20%)	40/136 (29%)	34/85 (40%)	15/188 (8%)	62/151 (41%)	181/703 (26%)

Table 3. Table highlighting the number of expressible properties in 6 industrial case studies.

For the CS_5, we note that the percentage (82%) of provable properties is very high. One reason is that the most of 188 requirements was written with a good property pattern matching. For the CS_6, we note that the percentage (27%) is very low. It was very difficult to re-write the requirements from specification documentation. We should have spent much time to interpret requirements with our industrial partner to formalize them with our patterns.

6.2 Context specification

For the S_CP case study, we constructed several CDL models with different complexities depending on the number of devices. The tests are performed on each CDL model composed with S_CP system.

N.of devices	Exploration time (sec)	N.of sub-contexts	N.of LTS config.	N.of LTS trans.
1	11	3	16 884	82 855
2	26	3	66 255	320 802
3	92	3	270 095	1 298 401
4	121	3	939 807	4 507 051
5	240	3	2 616 502	12 698 620
6	2161	40	32 064 058	157 361 783
7	4 518	55	64 746 500	322 838 592

Table 4. Exploration with TINA explorer with context splitting using OBP_t (S_CP case study).

Table 4 shows the amount of TINA exploration[10] for CDL examples with the use of context splitting. The first column depicts the number n of Dev asking for login to the S_CP. The other columns depict the exploration time and the cumulative amount of configurations and transitions of all LTS generated during exploration by TINA with context splitting. Table 4 also shows the number of contexts split by OBP. For example, with 7 devices, we needed to split the CDL context in 55 parts for successful exploration. Without splitting, the exploration is limited to 4 devices by state explosion as shown Table 1. It is clear that device number limit depends on the memory size of used computer.

7. Discussion and future work

CDL is a prototype language to formalize contexts and properties. However, CDL concepts can be implemented in another language. For example, context diagrams are easily described using full UML2. CDL permits us to study our methodology. In future work, CDL can be viewed as an intermediate language. Today, the results obtained using the currently implemented CDL language and OBP are very encouraging. For each case study, it was possible to build CDL models and to generate sets of context graphs with OBP.

CDL contributes to overcoming the combinatorial explosion by allowing partial verification on restricted scenarios specified by the context automata. CDL permits contexts and non ambiguous properties to be formalized. Property can be linked to whole or specific contexts. During experiments, we noted that some contexts and requirements were often described in the available documentation in an incomplete way. With the collaboration between engineers responsible for developing this documentation and ourselves, these engineers were motivated to consider a more formal approach to express their requirements, which is certainly a positive improvement.

In some case study, 70% textual requirements can be rewritten more easily with pattern property. So, CDL permits a better formal verification appropriation by industrial partners. Contexts and properties are verification data useful to perform proof activities and to validate models. These data have to be capitalized if the implementation evolves over the development life cycle.

In case studies, context diagrams were built, on the one hand, from scenarios described in the design documents and, on the other hand, from the sentences of requirement documents. Two major difficulties have arisen. The first is the lack of complete and coherent description of the environment behavior. Use cases describing interactions between the system (S_CP for instance) and its environment are often incomplete. For instance, data concerning interaction modes may be implicit. CDL diagram development thus requires discussions with experts who have designed the models under study in order to make explicit all context assumptions. The problem comes from the difficulty in formalizing system requirements into formal properties. These requirements are expressed in several documents of different (possibly low) levels. Furthermore, they are written in a textual form and many of them can have several interpretations. Others implicitly refer to an applicable configuration, operational phase or history without defining it. Such information, necessary for verification, can only be deduced by manually analyzing design and requirement documents and by interviewing expert engineers.

---

[10] Tests with same computer as for Table 1.

The use of CDL as a framework for formal and explicit context and requirement definition can overcome these two difficulties: it uses a specification style very close to UML and thus readable by engineers. In all case studies, the feedback from industrial collaborators indicates that CDL models enhance communication between developers with different levels of experience and backgrounds. Additionally, CDL models enable developers, guided by behavior CDL diagrams, to structure and formalize the environment description of their systems and their requirements. Furthermore, constraints from CDL can guide developers to construct formal properties to check against their models. Using CDL, they have a means of rigorously checking whether requirements are captured appropriately in the models using simulation and model checking techniques.

One element highlighted when working on embedded software case studies with industrial partners, is the need for formal verification expertise capitalization. Given our experience in formal checking for validation activities, it seems important to structure the approach and the data handled during the verifications. That can lead to a better methodological framework, and afterwards a better integration of validation techniques in model development processes. Consequently, the development process must include a step of environment specification making it possible to identify sets of bounded behaviors in a complete way.

Although the CDL approach has been shown scalable in several industrial case studies, the approach suffers from a lack of methodology. The handling of contexts, and then the formalization of CDL diagrams, must be done carefully in order to avoid combinatorial explosion when generating context graphs to be composed with the model to be validated. The definition of such a methodology will be addressed by the next step of this work.

8. References

Alfaro, L. D. & Henzinger, T. A. (2001). Interface automata, *Proceedings of the Ninth Annual Symposium on Foundations of Software Engineering (FSE), ACM*, Press, pp. 109–120.

Berthomieu, B., Ribet, P.-O. & Verdanat, F. (2004). The tool TINA - Construction of Abstract State Spaces for Petri Nets and Time Petri Nets, *International Journal of Production Research* 42.

Bosnacki, D. & Holzmann, G. J. (2005). Improving spin's partial-order reduction for breadth-first search, *SPIN*, pp. 91–105.

Clarke, E., Emerson, E. & Sistla, A. (1986). Automatic verification of finite-state concurrent systems using temporal logic specifications, *ACM Trans. Program. Lang. Syst.* 8(2): 244–263.

Clarke, E. M., Long, D. E. & Mcmillan, K. L. (1999). Compositional model checking, MIT Press.

Dhaussy, P., Pillain, P.-Y., Creff, S., Raji, A., Traon, Y. L. & Baudry, B. (2009). Evaluating context descriptions and property definition patterns for software formal validation, in B. S. Andy Schuerr (ed.), *12th IEEE/ACM conf. Model Driven Engineering Languages and Systems (Models'09)*, Vol. LNCS 5795, Springer-Verlag, pp. 438–452.

Dhaussy, P. & Roger, J.-C. (2011). Cdl (context description language) : Syntax and semantics, *Technical report*, ENSTA-Bretagne.

Dwyer, M. B., Avrunin, G. S. & Corbett, J. C. (1999). Patterns in property specifications for finite-state verification, *21st Int. Conf. on Software Engineering*, IEEE Computer Society Press, pp. 411–420.

Farail, P., Gaufillet, P., Peres, F., Bodeveix, J.-P., Filali, M., Berthomieu, B., Rodrigo, S., Vernadat, F., Garavel, H. & Lang, F. (2008). FIACRE: an intermediate language for model verification in the TOPCASED environment, *European Congress on Embedded Real-Time Software (ERTS), Toulouse, 29/01/2008-01/02/2008*, SEE.

Flanagan, C. & Qadeer, S. (2003). Thread-modular model checking, *SPIN'03*.

Godefroid, P. (1995). The Ulg partial-order package for SPIN, *SPIN Workshop* .

Halbwachs, N., Lagnier, F. & Raymond, P. (1993). Synchronous observers and the verification of reactive systems, *in* M. Nivat, C. Rattray, T. Rus & G. Scollo (eds), *Third Int. Conf. on Algebraic Methodology and Software Technology, AMAST'93*, Workshops in Computing, Springer Verlag, Twente.

Holzmann, G. (1997). The model checker SPIN, *Software Engineering* 23(5): 279–295.

Holzmann, G. & Peled, D. (1994). An improvement in formal verification, *Proc. Formal Description Techniques, FORTE94*, Chapman & Hall, Berne, Switzerland, pp. 197–211.

Janssen, W., Mateescu, R., Mauw, S., Fennema, P. & Stappen, P. V. D. (1999). Model checking for managers, *SPIN*, pp. 92–107.

Konrad, S. & Cheng, B. (2005). Real-time specification patterns, *27th Int. Conf. on Software Engineering (ICSE05), St Louis, MO, USA*.

Larsen, K. G., Pettersson, P. & Yi, W. (1997). UPPAAL in a nutshell, *International Journal on Software Tools for Technology Transfer* 1(1-2): 134–152.
URL: *citeseer.nj.nec.com/larsen97uppaal.html*

Park, S. & Kwon, G. (2006). Avoidance of state explosion using dependency analysis in model checking control flow model, *ICCSA (5)*, pp. 905–911.

Peled, D. (1994). Combining Partial-Order Reductions with On-the-fly Model-Checking, *CAV '94: Proceedings of the 6th International Conference on Computer Aided Verification*, Springer-Verlag, London, UK, pp. 377–390.

Pnueli, A. (1977). The temporal logic of programs, *SFCS '77: Proceedings of the 18th Annual Symposium on Foundations of Computer Science*, IEEE Computer Society, Washington, DC, USA, pp. 46–57.

Smith, R., Avrunin, G., Clarke, L. & Osterweil, L. (2002). Propel: An approach supporting property elucidation, *24st Int. Conf. on Software Engineering(ICSE02), St Louis, MO, USA*, ACM Press, pp. 11–21.

Tkachuk, O. & Dwyer, M. B. (2003). Automated environment generation for software model checking, *In Proceedings of the 18th International Conference on Automated Software Engineering*, pp. 116–129.

Valmari, A. (1991). Stubborn sets for reduced state space generation, *Proceedings of the 10th International Conference on Applications and Theory of Petri Nets*, Springer-Verlag, London, UK, pp. 491–515.

Whittle, J. (2006). Specifying precise use cases with use case charts, *MoDELS'06, Satellite Events*, pp. 290–301.

11

Formal Foundations for the Generation of Heterogeneous Executable Specifications in SystemC from UML/MARTE Models

Pablo Peñil, Fernando Herrera and Eugenio Villar
Microelectronics Engineering Group of the University of Cantabria
Spain

1. Introduction

Technological evolution is provoking an increase in the complexity of embedded systems derived from the capacity to implement a growing number of elements in a single, multi-processing, system-on-chip (MPSoC).

Embedded system heterogeneity leads to the need to understand the system as an aggregation of components in which different behavioural semantics should cohabit. Heterogeneity has two dimensions. On the one hand, during the design process, different execution semantics, specifically in terms of time (untimed, synchronous, timed) can be required in order to provide specific behaviour characteristics for the concurrent system elements. On the other hand, different system components may require different models of computation (MoCs) in order to better capture their functionality, such as Kahn Process Networks (KPN), Synchronous Reactive (SR), Communicating Sequential Processes (CSP), TLM, Discrete Event (DE), etc.

Another aspect affecting the complexity of current embedded systems derives from their structural concurrency. The system should be conceived as an understandable architecture of cooperating, concurrent processes. The cooperation among these concurrent processes is implemented through information exchange and synchronization mechanisms. Therefore, it is essential to deal with the massive concurrency and parallelism found in current embedded systems and provide adequate mechanisms to specify and verify the system functionality, taking into account the effects of the different architectural mappings to the platform resources.

In this context, the challenge of designing embedded systems is being dealt with by application of methodologies based on Model Driven Architecture (MDA) (MDA guide, 2003). MDA is a developing framework that enables the description of systems by means of models at different abstraction levels. MDA separates the specification of the system's generic characteristics from the details of the platform where the system will be implemented. Specifically, in Platform Independent Models (PIMs), designers capture the relevant properties that characterize the system; the internal structure, the communication mechanisms, the behavior of the different components, etc. Therefore, PIMs provide a general, synthetic representation that is independent and, thus, decoupled from the final system implementation. High-level PIM models are the starting point of ESL methodologies, and they are crucial for fast validation and Design Space Exploration (DSE). PIMs can be implemented on different platforms leading to different Platform Specific Models (PSMs). PSMs enable the analysis of performance characteristics of the system implementation.

The most widely accepted and used language for MDA is the Unified Modelling Language (UML) (UML, 2010). UML is a standard graphical language to visualize, specify and document the system. From the first application as object-oriented software system modelling, the application domain of UML has been extended. Nowadays, UML is used to deal with electronic system design (Lavagno et al. 2003). Nevertheless, UML lacks the specific semantics required to support embedded system specification, modelling and design. This lack of expressivity is dealt with by means of specific profiles that provide the UML elements with the necessary, precise semantics to apply the UML modelling capabilities to the corresponding domain.

Specifically in the embedded system domain, UML should be able to deal with design aspects such as specification, analysis, architectural mapping and implementation of complex, HW/SW embedded systems. The MARTE UML profile (UML Profile for MARTE, 2009), which was created recently, was developed in order to model and analyze real-time embedded systems, providing the concepts needed to describe real-time features that specify the semantics of this kind of systems at different abstraction levels. The MARTE profile has the necessary concepts to create models of embedded systems and provide the capabilities that enable the analysis of different aspects of the behaviour of such systems in the same framework. By using this UML profile, designers will be able to specify the system both as a generic entity, capturing the high-level system characteristics and, after a refinement process, as a detailed architecture of heterogeneous components. In this way, designers will be assisted by design flows with a generic system model as an initial stage. Then, by means of a refinement process supported by modelling and analysis tools, they will be able to decide on the most appropriate architectural mapping.

As with any UML profile, MARTE is not associated with any explicit execution semantics. As a consequence, no executable model can be directly extracted for simulation, functional verification and performance estimation purposes. In order to address this need, SystemC (Open SystemC) has been proposed as the specification and simulation framework for MARTE models. From the MARTE model, an executable model in SystemC can be inferred establishing a MARTE/SystemC relationship.

The MARTE/SystemC relationship is established in a formal way. The corresponding formalism should be as general as possible in order to enable the integration of heterogeneous components interacting in a predictable and well-understood way (horizontal heterogeneity) and to support the vertical heterogeneity, that is, refinement of the model from one abstraction level to another. Finally, this formalism should remove the ambiguity in the execution semantics of the models in order to provide a basis for supporting methodologies that tackle embedded system design.

For this purpose, the ForSyDe (Formal System Design) meta-model (Jantsch, 2004) was introduced. ForSyDe was developed to support the design of heterogeneous embedded systems by means of a formal notation. ForSyDe enables the production of a formal specification that captures the functionality of the system as a high abstraction-level model.

From these initial formal specifications, a set of transformations can be applied to refine the model into the final system model. This refinement process generally involves MoC transformation.

A system-level modelling and specification methodology based on UML/MARTE is proposed. A subset of UML and MARTE elements is selected in order to provide a generic model of the system. This subset of UML/MARTE elements is focused on capturing the generic concurrency and the communication aspects among concurrent elements. Here, system-level refers to a PIM able to capture the system structure and functionality independently of its final implementation on the different platform resources. The internal system structure is modelled by means of Composite Structure diagrams. MARTE *concurrency resources* are used to model the concurrent processes composing the concurrent structure of the system. The communication elements among the concurrent processes are modelled using the CommunicationMedia stereotype. The concurrent processes and the communication media compose the Concurrent&Communication (C&C) structure of the system. The explicit identification of the concurrent elements facilitates the allocation of the system application to platforms with multiple processing elements in later design phases.

In order to avoid any restrictions on the designer, the methodology does not impose any specific functionality modelling of concurrent processes. Nevertheless, with no loss of generality, UML activity diagrams are used as a meta-model of functionality. The activity diagram will provide formal support to the C&C structure of the system, explaining when each concurrent process takes input values, how it computes them and when the corresponding outputs are delivered.

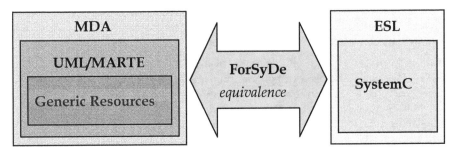

Fig. 1. ForSyDe formal link between MDA and ESL.

Based on the MARTE/SystemC formal link supported by ForSyDe, the methodology enables untimed SystemC executable specifications to be obtained from UML/MARTE models. The untimed SystemC executable specification allows the simulation, validation and analysis of the corresponding UML/MARTE model based on a clear simulation semantics provided by the underlying formal model. Although the formal model could be kept transparent to the user, the model defines clear simulation semantics associated with the MARTE model and its implementation in the SystemC model, which can be fully understood by any designer. Therefore, the ForSyDe meta-model formally supports interoperability between MARTE and SystemC.

In this way, the gap between MDA and ESL is formally bridged by means of a conceptual mapping. The mapping established among UML/MARTE and SystemC will provide consistency in order to ensure that the SystemC executable specification obtained is equivalent to the original UML/MARTE model. The formal link provided by ForSyDe enables the abstract executive semantics of both the UML/MARTE model and its corresponding SystemC executable specification to be reflected (Figure 4.). This demonstrates the equivalence among the two design flow stages, provides the required consistency to the mapping established between the two languages and ensures that the transformation process is correct-by-construction.

2. Related work

Several works have shown the advantages of using the MARTE profile for embedded system design. For instance, in (Taha et al, 2007) a methodology for modelling hardware by using the MARTE profile is proposed. In (Vidal et al, 2009), a co-design methodology for high-quality real-time embedded system design from MARTE is presented.

Several research lines have tackled the problem of providing an executive semantics for UML. In this context, two main approaches for generating SystemC executable specifications from UML can be distinguished. One research line is to create a SystemC profile in order to capture the semantics of SystemC facilities in UML diagrams (Bocchio et al., 2008). In this case, SystemC is used both as modelling and action language, while UML enables a graphical capture. A second research line for relating UML and SystemC consists in establishing mapping rules between the UML metamodel and the SystemC constructs. In this case, pure UML is used for system modelling, while the SystemC model generated is used as the action language. Mapping rules enable automatic generation of the executable SystemC code (Andersson & Höst, 2008). In (Kreku et al., 2007) a mapping between UML application models and the SystemC platform models is proposed in order to define transformation rules to enable semi-automatic code generation.

A few works have focused on obtaining SystemC executable models from MARTE. Gaspard2 (Piel et al. 2008) is a design environment for data-intensive applications which enables MARTE description of both the application and the hardware platform, including MPSoC and regular structures. Through model transformations, Gaspard2 is able to generate an executable TLM SystemC platform at the timed programmers view (PVT) level. Therefore, Gaspard2 enables flows starting from the MARTE post-partitioning models, and the generation of their corresponding post-partitioning SystemC executables.

Several works have confronted the challenge of providing a formal basis for UML and SystemC-based methodologies. Regarding UML formalization, most of the effort has been focused on providing an understanding of the different UML diagrams under a particular formalism. In (Störrle & Hausmann, 2005) activity diagrams are understood through the Petri net formalism. In (Eshuis & Wieringa, 2001) formal execution semantics for the activity diagrams is defined to support the execution workflow. In the context of MARTE, the Clock Constraint Specification Language (CCSL) (Mallet, 2008) is a formalism developed for capturing timing information from MARTE models. However, further formalization effort is still required.

A significant formalization effort has also been made in the SystemC context. The need to conceive the whole system in a model has brought about the formalization of abstract and heterogeneous specifications in SystemC. In (Kroening & Sharygna, 2005) SystemC specifications including software and hardware domains are formalized to support verification. In (Maraninchi et al., 2005) TLM descriptions are related to synchronous systems are formalized. In (Traulsem et al., 2007) TLM descriptions related to asynchronous systems are formalized. Comprehensive untimed SystemC specification frameworks have been proposed, such as SysteMoC (Falk et al., 2006) and HetSC (Herrera & Villar 2006). These methodologies take advantage of the formal properties of the specific MoCs they support but do not provide formal support for untimed SystemC specifications in general. Previous work on the formalization of SystemC was focused on simulation semantics. These approaches were inspired by previous formalization work carried out for hardware design languages such as VHDL and Verilog. In (Mueller et al., 2001), SystemC processes were seen as distributed abstract state machines which consume and produce data in each delta cycle. In this way the corresponding model is strongly related to the simulation semantics. In (Salem, 2003), denotation semantics was provided for the synchronous domain. Efforts towards more abstract levels address the formalization of TLM specifications. In (Ecker et al., 2006), SystemC specifications including software and hardware functions are formalized. In (Moy et al., 2008) TLM descriptions are related to synchronous and asynchronous formalisms.

Nevertheless, a formal framework for UML/MARTE-SystemC mapping based on common formal models of both languages is required. A good candidate to provide this formal framework is the ForSyDe metamodel (Janstch, 2004). The Formal System Design (ForSyDe) formalism is able to provide a synthetic notation and understanding of concurrent and heterogeneous specifications. ForSyDe covers modelling of time at different abstraction levels, such as untimed, synchronous and timed. Moreover, ForSyDe supports verification and transformational design (Raudvere et al. 2008).

3. ForSyDe

ForSyDe provides the mechanism to enable a formal description of a system. ForSyDe is mainly focused on understanding concurrency and time in a formal way representing a system as a concurrent model, where processes communicate through signals. In this way, ForSyDe provides the foundations for the formalization of the C&C structure of the system. Furthermore, ForSyDe formally supports the functionality descriptions associated with each concurrent process.

Processes and signals are metamodelling concepts with a precise and unambiguous mathematical definition. A ForSyDe signal is a sequence of events where each event has a tag and a value. The tag is often given implicitly as the position in the signal and it is used to denote the partial order of events. In ForSyDe, processes have to be seen as mathematical relations among signals. The processes are concurrent elements with an internal state machine. The relation among processes and signals is shown in Figure 2.

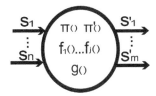

Fig. 2. ForSyDe metamodel representation.

From a general point of view; a ForSyDe process p is characterized by the expression:

$$p(s_1...s_n) = s'_1...s'_m \tag{1}$$

The process p takes a set of signals $(s_1...s_n)$ as inputs and produces a set of outputs $(s'_1...s'_m)$, where $\forall\ 1 \le i \le n \wedge 1 \le j \le m$ with $n, m \in \mathbb{N}$; $s_i, s_j \in S$ where s_k are individual signals and S is the set of all ForSyDe signals.

ForSyDe distinguishes three kinds of signals namely untimed signals, synchronous signals and timed signals. Each kind of MoC is determined by a set of characteristics which define it. Based on these generic characteristics, it is possible to define a particular MoC's specific semantics.

Expressions (2) and (4) denote an important, relevant aspect that characterizes the ForSyDe processes, the data consumed/produced.

$$\pi(v_1, s_1) = \left\langle a_{s_1}(z) \right\rangle$$
$$... \tag{2}$$
$$\pi(v_n, s_n) = \left\langle a_{s_n}(z) \right\rangle$$

with
$$v_n(z) = \gamma(\omega_q) \tag{3}$$

$$\pi(v_1', \hat{s}_1') = \left\langle a'_{s_1}(z) \right\rangle$$
$$... \tag{4}$$
$$\pi(v_m', \hat{s}_m') = \left\langle a'_{s_m}(z) \right\rangle$$

with
$$v_m'(z) = length(a'_{s_m}(z)) \tag{5}$$

A partition $\pi(v,s)$ of a signal s defines an ordered set of signals $\langle a_n \rangle$ that "almost" forms the original signal s. The brackets $\langle ... \rangle$ denote a set of ordered elements (events or signals). The function $v(z)$ defines the length of the subsignal $a_n(z)$; the semantics associated with the $v(z)$ function is: $v_n(0) = length(a_n(0))$; $v_n(1) = length(a_n(1))$ ... where z denotes the number of the data partition.

For the input signals, the length of these subsignals depends on which state the process is, denoted by the expression (3), where γ is the function that determines the number of events consumed in this state. The internal state of the process is denoted by ω_q with $q \in \mathbb{N}_0$. In some cases, $v_n(z)$ does not depend on the process state and thus $v_n(z)$ is a constant, denoted by the expression $v(z) = c$ with $c \in \mathbb{N}$.

For the output signals, the length is denoted by expression (5). The output subsignals $a'_1...a'_m$ are determined by the corresponding output function f_a that depends on the input subsignals $a_1...a_n$ and the internal state of the process ω_q, expression (6).

$$f_\alpha\big((a_1 ... a_n), \omega_q\big) = \big(a'_1 \ ... \ a'_m\big) \tag{6}$$

where $\forall \ 1 \le \alpha \le j \land j \in \mathbb{N}$

The next internal state of the process is calculated using the function g:

$$g\big((a_1 ... a_n), \omega_q\big) = \omega_{q+1} \tag{7}$$

where $\forall \ 1 \le i \le n \land n \in \mathbb{N}_0$, $a_i \in S$, $\forall \ q \in \mathbb{N}_0$, $\omega_q \in E$. E is the set of all events, that is, untimed events, synchronous events and timed events respectively.

ForSyDe processes can be characterized by the four tuple TYPEs $\langle$TI, TO, NI, NO$\rangle$. TI and TO are the sets of signal types for the input and output signals respectively. The signal type is specified by the value type of its corresponding events that made up the signal. NI = $\{v_1(i)...v_n(i)\}$ is the set of partitioning functions for the n input signals; NO=$\{v_1'(i)...v_n'(i)\}$ is the set of partitioning functions of the m output signals.

The advance of time in ForSyDe processes is understood as a totally ordered sequence of evaluation cycles. In each evaluation cycle (ec) "a process consumes inputs, computes its new internal state, and emits outputs" (Jantsch, 2004). After receiving the inputs, the process reacts and then, it computes the outputs depending on its inputs and the process's internal state.

4. AVD system

In order to illustrate the formal foundations between UML/MARTE and SystemC a video decoder is used, specifically an Adaptive Video decoder (AVD) system. Adaptive software is a new paradigm in software programming which addresses the need to make the software more effective and thus reusable for new purposes or situations it was not originally designed for. Moreover, adaptive software has to deal with a changing environment and changing goals without the chance of rewriting and recompiling the program. Therefore, dynamic adaptation is required for these systems. Adaptive software requires the representation of the set of alternative actions that can be taken, the goals that the program is trying to achieve and the way in which the program automatically manages change, including the way the information from the environment and from the system itself is taken.

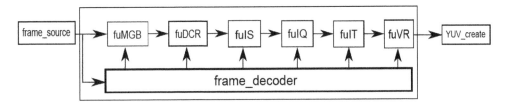

Fig. 3. Block diagram of the Adaptive Video decoder.

Specifically, the AVD specification is based on the RVC decoder architecture (Jang et al., 2008). Figure 3 illustrates a simplified scheme of the AVD architecture. The RVC architecture divides the decoder functionality into a set of functional units (fu). Each of these functional units is in charge of a specific video decoding functionality. The *frame_decoder* functional unit is in charge of parsing and decoding the incoming MPEG frame. This functional unit is enabled to parse and extract the forward coding information associated with every frame of the input video stream. The coding information is provided to the functional units *fuIS* and *fuIQ*. The macroblock generator (*fuMGB*) is in charge of structuring the frame information into macroblocks (where a macroblock is a basic video information unit, composed of a group of blocks). The inverse scan functional unit (*fuIS*) implements the Inverse zig-zag scan. The normal process converts a matrix of any size into a one-dimensional array by implementing the zig-zag scan procedure. The inverse function takes in a one-dimensional array and by specifying the desired number of rows and columns, it returns a matrix having the specified dimensions. The inverse scan constructs an array of 8x8 DCT coefficients from a one-dimensional sequence. The *fuIQ* functional unit performs the Inverse Quantization. This functional unit implements a parameter-based adaptive process. The *fuIT* functional unit can perform the Inverse Transformation by applying an inverse DCT algorithm (IDCT), or an inverse Haar algorithm (IHAAR). Finally, the *fuVR* functional unit is in charge of video reconstruction.

The *frame_source* and the *YUV_create* blocks make up the environment of the AVD system. The *frame_source* block provides the frames of a video file that the AVD system decodes later. The *YUV_create* block rebuilds the video (in a .YUV video file) and checks the results obtained.

4.1 UML/MARTE model from the AVD system

The system is designed as a concurrent entity; the functionality of each functional unit is implemented by concurrent elements. Each one of these concurrent elements is allocated to an UML component and identified by the MARTE stereotype <<ConcurrencyResource>>. This MARTE generic resource models the elements that are capable of performing its associated execution flow concurrently with others. *Concurrency resources* enable the functional specification of the system as a set of concurrent processes. The information is transmitted among the *concurrent resources* by means of communicating elements identified by the MARTE stereotype <<CommunicationMedia>>. Both *ConcurrencyResource* and *CommunicationMedia* are included in MARTE subprofile Generic Resource Modelling (GRM). This gives the designer complete freedom in deciding on the most appropriate mapping of the different functional components of the system specification to the available executing resources. These MARTE elements are generic in the sense that they do not assume a specific platform mapping to HW or to SW. Thus, they are suitable for system-level pre-partition modelling.

Depending on the parameters defining the *communication media*, several types of channels can be identified. Based on the type of channels used, several MoCs can be identified (Peñil et al, 2009). When a specific MoC is found, the design methodologies associated with it can be used taking advantage of the properties that that MoC provides. Additional kinds of channels can be identified, the border channels. A border channel is a communication media that enables the connections of different MoC domains, which have their own properties and characteristics. The basic principle of the border channel semantics is that from each MoC side, the border channel is seen as the channel associated with the MoC. In the case of

channel_4 of Figure 4, this communication media establishes the connection among the KPN
MoC domains (Kanh,1974) and the CSP MoC domains (Hoare, 1978). This border channel is
inferred from a *communication media* with a storage capacity provided by the stereotype
<<StorageResource>>. In order to capture the unlimited storage capacity that characterizes
the KPN channels, the tag *resMult* should not be defined. The communication is carried by
the calls to a set of methods that a *communication media* provides. These methods are MARTE
<<RtService>>. The *RtService* associated with the KPN side should be *asynchronous* and
writer. In the CSP side, the *RtService* should be d*elayedSynchronous*. This attribute value
expresses synchronization with the invoked service when the invoked service returns a
value. In this *RtService* the value of *concPolicy* should be *writer* so that the data received from
the *communication media* in the synchronization is consumed and, thus, producing side
effects in the *communication media*. The *RtServices* are the methods that should be called by
the *concurrency resources* in order to obtain/transmit the information.

Another communication (and interaction) mechanisms used for communicating threads is
performed through protected shared objects. The most simple is the shared variable. A
shared variable is inferred from a *communication media* that requires storage capacity
provided by the MARTE stereotype <<StorageResource>>. Shared variables use the same
memory block to store the value of a variable. In order to model this memory block, the tag
resMult of the *StorageResource* stereotype should be one. The communication media accesses
that enable the writings are performed using *Flowport* typed as *in*. A *RtService* is provided by
this *FlowPort* and this *RtService* is specified as *asynchronous* and as *writer* in the tags
synchKind and *concPolicy* respectively. The tag value *writer* expresses that a call to this
method produces side effects in the *communication media*, that is, the stored data is modified
in each writing access. Regarding the reading accesses, they are performed through *out* flow
ports. The value of the *synchKind* should be *synchronous* to denote that the corresponding
concurrency resource waits until receiving the data that should be delivered by the
communication media. The value of *concPolicy* should be *reader* to denote that the stored data
is not modified and, thus, several readings of the same data are enabled.

Figure 4 shows a sketch of a complete UML/MARTE PIM that describes the AVD system.
Figure 4 is focused on the *MGB* component showing the components that are connected to
the *MGB* component and the channels used for the exchange of information between this
component and its specific environment. Based on this AVD component, a complete
example of the ForSyDe interrelation between UML/MARTE and SystemC will be
presented. However, before introducing this example, it is necessary to describe the
ForSyDe formalization of the subset of UML/MARTE elements selected. For that purpose,
the *IS* component is used.

4.2 Computation & communication structure

The formalization is done by providing a semantically equivalent ForSyDe model of the
UML/MARTE PIM. Such a model guarantees the determinism of the specification and
enables the application of the formal verification and refinement methodologies associated
with ForSyDe. As was mentioned before, the ForSyDe metamodel is focused on the formal
understanding of the communication and processing structure of a system and the timing
semantics associated with each processing element's behaviour. Therefore, in order to obtain
a ForSyDe model, all the system information associated with an UML/MARTE model

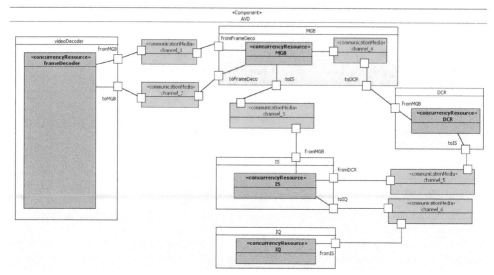

Fig. 4. Sketch of the UML/MARTE model that describes the AVD system.

related to the system structure has to be ignored. All the model elements that determine the hierarchy system structure such as UML components, UML ports, etc. have to be removed. In this way, the resulting abstraction is a model composed of the processing elements (*concurrency resources*) and the communicating elements (*communication media*). This C&C model determines the abstract semantics associated with the model and, by extension, determines the system execution semantics. Figure 5 shows the C&C abstraction of Figure 4 where only the *concurrency resources* and the *communication media* are presented.

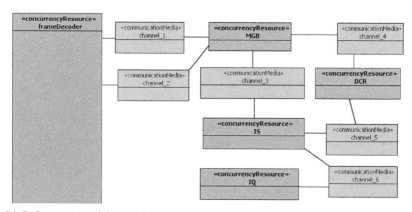

Fig. 5. C&C abstraction of the model in Figure 4.

4.3 ForSyDe representation of C&C structure

While the extraction of the C&C model is maintained in the UML/MARTE domain, the second step of the formalization consists in the abstraction of this UML/MARTE C&C

model as the semantically equivalent ForSyDe model. More specifically, the ForSyDe
abstraction means the specification from the UML/MARTE C&C model of the
corresponding processes and signals; the timing abstraction (untimed, synchronous, etc); the
input and output partitions; and the specific type of process constructors, which establish
the relationships between the input partitions and the output partitions. The first step of the
ForSyDe abstraction is to obtain a ForSyDe model in which the different processes and
signals are identified. In order to obtain this abstract model, a direct mapping between
ConcurrencyResource-processes and *CommunicationMedia*-signals is established. Figure 6
shows the C&C abstract model of Figure 5 using ForSyDe processes and signals. Therefore,
with this first abstraction, the ForSyDe C&C system structure is obtained.

There is a particular case related to the ForSyDe abstraction of the *CommunicationMedia*-
signal. Assume that in *channel_6* of the example in Figure 4 another MARTE stereotype has
been applied, specifically the <<ConcurrencyResource>> stereotype. In this way, the
communicating element has the characteristic of performing a specific functionality. This
combination of *concurrency resource* and *communication media* semantics can be used in order
to model system elements that transmit data and, moreover, perform a transformation of
this data. The ForSyDe representation of this kind of channels consists in a process that
represents the functionality associated with the channel and a signal that represents the
output data generated by the channel after the input data is computed.

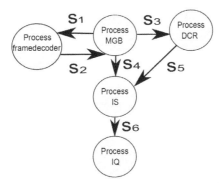

Fig. 6. ForSyDe representation of the C&C model of the Figure 5.

4.4 Concurrency resource's behaviour description

A concurrent element can be described by a finite state machine where in each state the
concurrent element receives inputs, computes these inputs and calculates their new state
and the corresponding outputs. The structure of the behaviour of each concurrency resource
is modelled by means of an Activity Diagram. The activity diagram can model the complete
resource behaviour. In this case, there is no clear identification of the class states; the states
executed by the class during its execution are implicit. Activity diagrams represent activity
executions that are composed of single steps to be performed in order to model the complete
behaviour of a particular class. These activities can be composed of single actions that
represent different behaviours, related to method calls or algorithm descriptions. In this
case, the complete behaviour captured in an activity diagram can be structured as a
sequence of states fulfilling the following definition: each state is identified as a stage where

the concurrency resource receives the data from its environment; these data are computed by an atomic function, producing the corresponding output data. Therefore, in the most general approach, an implicit state in an activity diagram is determined between two waiting stages, that is, between two stages that represent input data. In this kind of stages, the concurrency resource has to wait until the required data are available in all the inputs associated with the corresponding function. In the same way, if code were directly written, an equivalent activity diagram could be derived. Additionally, the behavioural modelling of the concurrent resources can be modelled by an explicit UML finite state machine. This UML diagram is focused on which states the object covers throughout its execution and the well-defined conditions that trigger the transitions among these states (the states are explicitly identified). Each UML state can have an associated behaviour denoted by the label *do*. This label identifies the specific behaviour that is performed as long as the concurrent element is in the particular state. Therefore, in order to describe the functionality in each state, UML activity diagrams is used.

Figure 7 shows the activity diagram that captures the functionality performed by the *concurrency resource* of the *IS* component. According to the aforementioned internal state definition, this diagram identifies two states; one state where the *concurrency resource* is only initialized and another state where the tuple data-consumption/computation/data generation is modelled. The data consumption is modelled by a set of *AcceptEventAction*. In the general case, this UML action represents a service call owned by a *communication media* from which the data are required. Then, these data are computed by the atomic function *Scan*. The data generated from this computation (in this case, *data3*) are sent to another system component; the sending of data is modelled by *SendObjectAction* that represents the corresponding service call for the computing data transmissions.

Apart from the UML elements related to the data transmission and the data computation, another set of UML elements are used in order completely specify the functionality to be modelled. The fork node (▭▭▭▭▭) establishes concurrent flows in order to enable the modelling of data inputs required from different channels in the same state. The UML pins (the white squares) associated to the *AcceptEventAction*, function *Scan* and *SendObjectAction* represent the data received from the communication, the data required/generated by the atomic function execution and the data sending, respectively. An important characteristic needed to define the *concurrency resource* functionality behaviour is the number of data required/generated by a specific atomic function. This characteristic is denoted by the multiplicity value. Multiplicity expresses the minimum and the maximum number of data that can be accepted by or generated from each invocation of a specific atomic function. Additionally, the minimum multiplicity value means that some atomic functions cannot be executed until the receipt of the minimum number of data in all atomic function incoming edges. In Figure 7, the multiplicity values are annotated in blue UML comments.

As was mentioned, *concurrent resource* behaviour is composed of pure functionality represented by atomic functions and *communication media* accesses; the structure of the behaviour of a *concurrency resource* specifies how pure functionality and communication accesses are interlaced. This structure is as relevant as the C&C structure, since both are involved in the executive semantics of the process network.

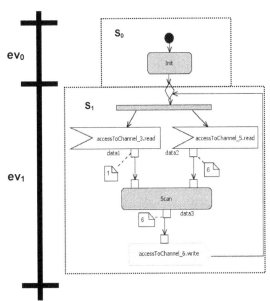

Fig. 7. Activity diagram that describes the functionality implemented by the IS component.

4.5 ForSyDe representation of concurrency resource functionality modelling

In the behavioural model in Figure 7 two implicit states (S_0 and S_1) can be indentified. The activity diagram implicit states are represented as ω_j in ForSyDe. A state ω_j is understood to be a state composed of two different states, P_j and D_j. In the general case, P_j denotes segments of the behavioural description that are between two consecutive waiting stages. In this case, such waiting stages are identified by two consecutive sets of *AcceptEventActions*. Therefore, P_j corresponds to the basic structure described in the previous section. D_j expresses all internal values that characterize the state. The change in the internal state of a *concurrency resource* is denoted by the *next state function* $g((a_1...a_n), \omega_j) = \omega_{j+1}$ where ω_j represents the current state and $a_1...a_n$ the input data consumed in this state. The function $g()$ calculates both D_{j+1} and P_{j+1}.

The atomic function implemented in a state ω_j (for instance, in the example in Figure 7 the function *Scan*) is represented by the ForSyDe *output function* $f_i()$. This function generates the outputs (represented as the subsignals $a'_1...a'_m$) as a result of computing the data inputs.

The multiplicity values of the input and output data sequences are abstracted by a *partition function* ν :

$$\nu_1(z) = \gamma(\omega_i) = p$$

Input partition functions ...

$$\nu_n(z) = \gamma(\omega_i) = q$$

(8)

$$\forall z, i \in \mathbb{N}_0 \wedge \{p, q\} \in \mathbb{N}$$

$$\text{Output partition functions } length(f_i(a_1 \dots a_n), \omega_i) = \begin{cases} v'_1(z) = length(a'_1) = a \\ \dots \\ v'_M(z) = length(a'_M) = b \end{cases} \tag{9}$$

$$\forall z, i \in \mathbb{N}_0 \wedge \{a, b\} \in \mathbb{N}$$

A *partition function* enables a signal partition $\pi(v,s)$, that is, the division of a signal s into a sequence of sub-signals a_i. The partition function denotes the amount of data consumed/produced in each input/output in each ForSyDe process computation, referred to as evaluation cycle.

The data received by the concurrency resource through the *AcceptEventActions* are represented by the ForSyDe signal $a_1 \dots a_n$. Regarding the data transmitted through *SendObjectActions*, they are represented by $a'_1 \dots a'_m$.

In addition, the behavioural description has a ForSyDe time interpretation; Figure 7 corresponds to two evaluation cycles (ev_0 and ev_1) in ForSyDe. The corresponding time interpretation can be different depending on the specific time domain. These evaluation cycles will have different meanings depending on which MoC the designer desires to capture in the models. In this case, the timing semantics of interest is the untimed semantics.

5. UML/MARTE-SystemC mapping

The UML/MARTE-SystemC mapping enables the generation of SystemC executable code from UML/MARTE models.

This mapping enables the association of a corresponding SystemC executable code which reflects the same concurrency and communication structure through processes and channels. Similarly, the SystemC code can reflect the same hierarchical structure as the MARTE model by means of modules, ports, and the different types of SystemC binding schemes (port-port, channel-port, etc). However, other mapping alternatives maintaining the semantic correspondence, using port- export connections, are feasible thanks to the ForSyDe formal link. Figure 8 shows the first approach to the UML/MARTE-SystemC mapping regarding the C&C structure and the system hierarchy. The correspondence among the system hierarchy elements, component-module and port-port, is straightforward. In the same way, the correspondence *concurrency resource*-process is straightforward. A different case is the communicating elements. As a general approach, a communication media corresponds to a SystemC channel. However, the type of SystemC channel depends on the communication semantics captured in the corresponding *communication media*. As can be seen in (Peñil et al., 2009), depending on the characteristics allocated to the *communication media*, different communication semantics can be identified in UML/MARTE models which implies that the SystemC channel to be mapped should implement the same communication semantics.

Regarding the functional description, the *AcceptEventActions* and *SendObjectActions* are mapped to channel accesses. If channel instances are beyond the scope of the module, the accesses to them become port accesses. The multiplicity value of each data transmission in

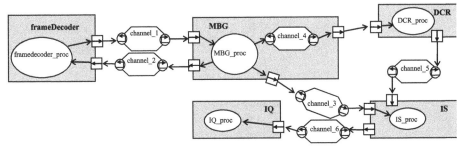

Fig. 8. SystemC representation of the UML/MARTE model in Figure 4.

the activity diagram corresponds to multiple channel accesses (of a single data value) in the
SystemC code. Execution of pure functionality captured as atomic functions represents the
individual functions that compose the complete *concurrency resource* functionality. The
functions can correspond to a representation of functions to be implemented in a later
design step according to a description attached to this function or pure C/C++ code
allocated to the model. Additionally, loops and conditional structures are considered in
order to complement the behaviour specification of the *concurrency resource*. Figure 9 shows
the SystemC code structure that corresponds to the functional description of Figure 7. Lines
(2-3-4) are the declarations of the variables typed as T_i used for communication and
computation. Then, an atomic function for initializing some internal aspects of the
concurrency resource is executed. Line 5 denotes the statement that defines the infinite loop.
Line 6 is the data access to the *communication media channel_3*. In this case, the channel access
is done through the port *fromMGB*. In the same way, line 7 is the statement for reading the
six data from *channel_5* through the port *fromDCR*. The atomic functions *Scan* is represented
as a function call, specifying the function parameters (line 9). Finally, the output data
resulting from the *Scan* computation (*data3*) are sent through the port *toIQ* by using the
communication media channel_6.

```
(1) void IS::IS_proc(){
(2) T1  data1;
(3) T2  data2[ ];
(4) T3  data3[ ];
(5) Init();
(6) while (true) {
(7)   data1 = fromMGB.read();
(8)   for(int i=0;i<6;i++) data2[i]= fromDCR.read();
(9)   Scan (dat1, data2, data3);
(10)  for(int i=0;i<6;i++) toIQ.write(data3[i]);
(11) }}
```

Fig. 9. SystemC code corresponding to the model in Figure 7.

5.1 UML/MARTE-SystemC mapping: ForSyDe formal foundations

As was described, there are similarities which lead to the conclusion that the link of these
MARTE and SystemC methodologies is feasible. However, there are obvious differences in

terms of UML and SystemC primitives. Moreover, there is no exact a one to one correspondence, e.g., in the elements for hierarchical structure. Even when correspondence seems to be straightforward (e.g. *ConcurrencyResource* = SystemC Process), doubts can arise about whether every type of SystemC process can be considered in this relationship. A more subtle, but important consideration in the relationship is that the SystemC code is executable over a Discrete Event (DE) timed simulation kernel, which provides the code with low level execution semantics. SystemC channel implementation internally relies on event synchronizations, shared variables, etc, which map the abstract communication mechanism of the channel onto the DE time axis. In contrast, the execution semantics of the MARTE model relies on the attributes of the *communication media* (Peñil et al, 2009) and on CCSL (Mallet, 2008). A common representation of the abstract semantics of the SystemC channel and of the *communication media* is required. All these reasons make the proposed formal link necessary.

The UML/MARTE-SystemC mapping enables the generation of SystemC executable code from UML/MARTE models. The transformation process should maintain the C&C structure, the behaviour semantics, and the timing information captured in the UML/MARTE models in the corresponding SystemC executable model. This information preservation is supported by ForSyDe, which provides the required semantic consistency. This consistency is provided by a common formal annotation that captures the previous relevant information that characterizes the behaviour of a *concurrency resource* and additional relevant information such as the internal states of the process, the atomic functionality performed in each state, the inputs and the number of inputs required for this atomic functionality to be performed and the resulting data generated outputs from this atomic function execution.

An important characteristic is the timing domain. This article is focused on high-level (untimed) UML/MARTE PIMs. In the untimed models, the time modelling is abstracted as a causality relation; the events communicated by the concurrent elements do not contain any timing information. An order relation is denoted; the event sent first by a producer is received first by a consumer, but there is no relation among events that form different signals. Additionally, the computation and the communication take an arbitrary and unknown amount of time.

Figure 10 shows the ForSyDe abstract, formal annotation of the *IS concurrency resource* behaviour description and the functional specification of the SystemC process *IS_proc*. Line 1 specifies the type of *processor constructor*; in this case the *processor constructor* is a *mealyU*. The U suffix denotes untimed execution semantics. The *mealyU* process constructor defines a process with internal states that take the output function $f()$, the next state functions $g()$, the function $\gamma()$ for defining the signal partitions, and the initial state ω_0 as arguments. In general $\gamma()$, $f()$ and $g()$ are state-dependent functions. In this case, the abstraction splits $f()$, $g()$ and $\gamma()$ into state-independent functions. The function $\gamma()$ is the function used to calculate the new partition functions v_{sk} of the inputs signals. Specifically, output function $f()$ of the *IS* process is divided into 2 functions corresponding to the two internal state that the concurrency resource has. The first output function $f_0()$ models the *Init()* function; the output function $f_1()$ models the function *Scan()*. In this function, the partition functions v_{sk} of each input data required for the computing of the *Scan()* (line [7]) are annotated. Line [9] represents the partition function of the resulting output signal s'_1. In the same way as in the case of the function $f()$, next state of the function $g()$ is divided into 2 functions, in order to specify the state transitions (lines [5] and [10]) identified in the activity diagram. The data communicated by the *IS concurrent resource data1, data2, data3* are represented by the signals S_1 and S_2 for the inputs (data1, data2) and S'_1 for the output signal *data3*. The implicit states identified in the activity diagram St_0 and St_1 are abstracted as the states ω_0 and ω_1, respectively.

[1] $IS = \text{mealyU}(\gamma, g, f, \omega_0)$
[2] $IS\ (s_1, s_2) = <s'_1>$

[3] if $(state_i = \omega_0)$ then
[4] $f_0()_i = Init()$
[5] $state_{i+1} = g(\omega_0) = \omega_1$
[6] elseif $(state_i = \omega_1)$

[7] $\begin{cases} v_{s1}(i) = 6 , \pi(v_{s1}, s_1) = <a1_i> \\ v_{s2}(i) = 1 , \pi(v_{s1}, s_1) = <a2_i> \end{cases}$

[8] $a1'_i = f_1(a1_i, a2_i) = Scan(a1_i, a2_i)$
[9] $v_{s'1}(i) = 6.\ \pi(v_{s'1}, s'_1) = < a1'_i>$
[10] $state_{i+1} = g(\omega_1) = \omega_1$

Fig. 10. ForSyDe annotation of the UML/MARTE model in Figure 7 and the SystemC code in Figure 9.

According to the definition of evaluation cycle presented in section 3, both implicit states that can be identified in the activity diagram shown in Figure 7 correspond to a specific ForSyDe evaluation cycle (*ev0* and *ev1*).

Therefore, the abstract, formal notation shown in Figure 10 captures the same, common behaviour semantics modelled in Figure 7 and specified in Figure 9, and, thus, provides consistency in the mapping between UML/MARTE and SystemC in order to enable the later code generation (Figure 11).

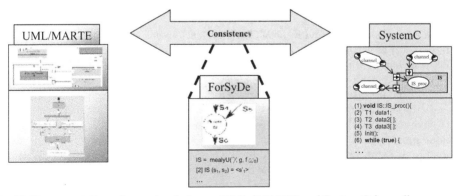

Fig. 11. Representation of mapping between UML/MARTE and SystemC formally supported by ForSyDe.

5.2 Formal support for untimed UML/MARTE-SystemC models

The main problem when trying to define a formal mapping between MARTE and SystemC is to define the untimed semantics of a DE simulation language such as SystemC. Under this untimed semantics, the strict ordering of events imposed by the DE simulation mechanism of SystemC's simulation kernel has to be relaxed. In principle, the consecutive events in a particular SystemC object (a channel, accesses to a shared variable, etc.) should be considered as totally ordered as they originate from the execution of a sequential algorithm. Any change in this order in any implementation of the algorithm should be based on a sound optimization methodology or should be clearly explained by the designer. Events in objects corresponding to different concurrent processes related by causal dependencies are also ordered and, again, any change should be fully justified. However, events in objects corresponding to different concurrent processes without any causal dependency can be implemented in any order. This is the flexibility required by the design process in order to ensure optimal implementations under the imposed design constraints.

As was commented previously, SystemC processes and MARTE *concurrency resources* can be directly abstracted as ForSyDe processes. Nevertheless, and in the most general case, the abstraction of a SystemC communication mechanism and the *communication media* relating two processes is more complex. The type of communication in this article is addressed through channels and shared variables. When the communication mechanism fulfils the required conditions, then, it can be straightforwardly abstracted as a ForSyDe signal.

The *MGB* component shown in figure 4 is connected to its particular environment through four *communication media*. Assuming that in these *communication media* four different communication semantics can be identified. The *communication media channel_1* represents an infinite FIFO that implements the semantics associated to the KPN MoC. The *channel_3* establishes a rendezvous communication with data transmission. The way to identify the properties that characterize these communication mechanisms in UML/MARTE models was presented in (Peñil et al, 2009). The *channel_2* represents a shared variable and the *channel_4* is a border channel between the domains KPN-CSP. Therefore, the MGB *concurrency resource* is a border process. A border process is a sort of process which channel accesses are connections to different *communication media* that captured different communication semantics. In this way, the AVD system is a heterogeneous entity where different behaviour semantics can exist.

The data transmission dealt with the MGB *concurrency resource* is carried out by means of a different sort of *communication media*: unlimited FIFO, shared memory, rendezvous and a KPN-CSP border channel. Those communication media accesses are denoted by the corresponding *AcceptEventActions* and *SendObjectActions* identified by the port or channel used by the data transmission and the service called for that data transmission (see Figure 1a)). All these communication semantics captured in the UML/MARTE *communication media* have to be mapped to specific SystemC communication mechanism ensuring the semantic preservation. The *communication media channel_1, channel_2* and *channel_4* can be mapped to SystemC channels provided by the HetSC methodology (HetSC, 2007). HetSC is a system methodology based on the ForSyDe foundations for the creation of formal execution specifications for heterogeneous systems. Additionally, HetSC provides a set of communications mechanisms required to implement the semantics of several MoCs. Therefore, the mapping process from the previous *communication media* to the SystemC channels ensures the semantic equivalence since HetSC provides the required SystemC channels that implement the same communication semantics captured in the corresponding *communication media*. Additionally, these *communication media* fulfil, by construction, the condition that the data obtained by the consumer process are the same and in the same order as the data generated by the producer process. In this way, they can be abstracted as a ForSyDe signal which implies that the *communication media*-SystemC channel mapping is correct-by-construction. As an example of SystemC channel accesses, in Figure 12 b), line (5) denotes a channel access through a port and line (7) specifies a direct channel access.

An additional application of the extracted ForSyDe model is the generation of some properties that the SystemC specification should satisfy under any dynamic condition in any feasible testbench. Note that the ForSyDe model is static in nature and does not include the synchronization and firing mechanism used by the SystemC model. In the example of *MGB* component, a mechanism for communication among processes can be implemented through a shared variable, specifically the *channel_2*. Nevertheless, the communication of concurrent processes through shared variables is a well-known problem in system engineering. As the SystemC simulation semantics is non-preemptive, protecting the access to the shared variables does not make any difference. However, this is an implementation issue when mapping SystemC processes to SW or HW. A variable shared between two SystemC processes correctly implements a ForSyDe signal when the following conditions apply:

1. Every data token written by the producer process is read by the consumer process.
2. Every data token written by the producer process is read only once by the consumer process.

In some cases, in order to simplify the design, the designer may decide to use the shared variable as local memory. As commented above, this problem can be avoided by renaming. A new condition can be applied:

1. If a consumer uses a shared variable as local memory, no new data can be written by the producer until after the last access to local memory by the consumer, that is, during the local memory lifetime of the shared variable.

Additionally, other conditions have to be considered in order to enable a ForSyDe abstraction to be obtained which provides properties to be satisfied in the system design. Another condition to be considered in the *concurrent resource* behaviour description is the use of fork nodes and thus, the modelling of the internal concurrency in a concurrent element. As a design condition, the specification of internal concurrency is not permitted in the *concurrency resource* behaviour (except for the previously mentioned modelling of the data requirements from different inputs). The behaviour description consists of a sequence of internal states to create a complete activity diagram that models the *concurrent resource* behaviour. As a general first approach, it is possible to use the fork node to describe internal concurrent behaviour of a concurrent element if and only if the corresponding inputs and outputs of each concurrent flow are univocal. Among several concurrent flows, it is essential to know from which inputs the data are being taken and to which the outputs are being sent; in a particular state, only one concurrent flow can access specific communication media.

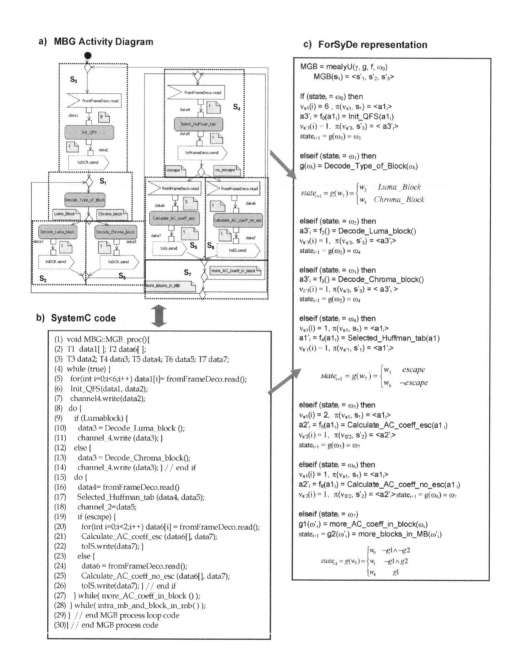

Fig. 12. ForSyDe abstraction (c) of the MBG *concurrency resource* functionality model (a) and its corresponding SystemC code (b).

Another modelling condition that can be considered in the *concurrency resource* behaviour description is the specification of the multiplicity values of the data inputs and outputs. This multiplicity specification has to be explicit and unequivocal, that is, expressions such as [1...3] are not allowed. A previous multiplicity specification is not consistent with the ForSyDe formalization since ForSyDe defines that in each process state, each input and output partition is well defined. The multiplicity specification [a...b] presents indeterminacy in order to define the process behaviour; it is not possible to know univocally the number of data required/produced by a computation. This fact can yield an inconsistent functionality and, thus, can present risks of incorrect performance.

As was mentioned before, not only the communication semantics defined in the communication media is necessary to specify the behaviour semantics of the system, but the way that each communication access is interlaced with pure functionality is also required in order to specify the execution semantics of the processes network. The *communication media channel_3* implements a rendezvous communication among the *MGB concurrency resource* and the *IS concurrency resource* which involves a synchronization and, thus, a partial order in the execution of functions of the two processes. The atomic function *Scan* shown in Figure 7 requires a datum provided by the communication media *channel_3*. This data is provided when either the function *Calculate_AC_coeff_esc* has finished or when the function *Calculate_AC_coeff_no_esc* has finished, depending on which internal state the *MGB concurrency resource* is in. In the same way, the *MGB concurrency resource* needs the *IS concurrency resource* to finish the atomic function *Scan()* in order to go on with the block computation. In this way, the two processes synchronize their independent execution flows, waiting for each other at this point for data exchange. Therefore, besides the semantics captured in the *communication media*, the way the calls to this *communication media* and the computation stages are established in order to model the *concurrency resource*'s behaviour defines its execution semantics, affecting the behaviour of others *concurrency resources*.

The ForSyDe model is a formal representation that enables the capture of the relevant properties that characterize the behaviour of a system. Figure 12 c) shows the ForSyDe formal annotation of the functional model of the MGB *concurrency resource*'s behaviour shown in Figure 12 a) and the SystemC code in Figure 12 b), which is the execution specification of the previous UML/MARTE model. This ForSyDe model specifies the different internal states that can be identified in the activity diagram in Figure 12 a) (all of them identified by a rectangle and the annotation S_i). Additionally, ForSyDe formally describes all data requirements for the computations, the functions executed in each state, the data generated in each of these computations and the conditions for the state transitions. This relevant information defines the *concurrency resource*'s behaviour. Therefore, the ForSyDe model provides an abstract untimed semantics associated with the UML/MARTE model which could be used as a reference model for any specification generated from it, specifically, a SystemC specification, in order to guarantee the equivalence between the two system representations.

6. Conclusions

This chapter proposes ForSyDe as a formal link between MARTE and SystemC. This link is necessary to maintain the coherence between MARTE models and their corresponding

SystemC executable specifications, in order to provide safe and productive methodologies integrating MDA and ESL design methodologies. Moreover, the chapter provides the formal foundations for enabling this ForSyDe-based link between PIM UML/MARTE models and their corresponding SystemC executable code. The most immediate application of the results of this work will be in the automation of the generation of heterogeneous executable SystemC specifications from untimed UML/MARTE models which specify the system concurrency and communication structure and the behaviour of concurrency resources.

7. Acknowledgments

This work was financed by the ICT SATURN (FP7-216807) and COMPLEX (FP7-247999) European projects and by the Spanish MICyT project TEC 2008-04107.

8. References

[1] Andersson, P. & M.Höst. (2008). "UML and SystemC a Comparison and Mapping Rules for Automatic Code

[2] Generation", in E. Villar (ed.): "Embedded Systems Specification and Design Languages", Springer, 2008.

[3] Bocchio, S.; Riccobene, E.; Rosti, A. & Scandurra, P. (2008). "An Enhanced SystemC UML Profile for Modeling at

[4] Transaction-Level", in E. Villar (ed.): "Embedded Systems Specification and Design Languages", Springer, 2008.

[5] Ecker, W.; Esen, V. &, Hull, M. (2006). Execution Semantics and Formalisms for Multi-Abstraction TLM Assertions. In Proc. of MEMOCODES'06. Napa, California. July, 2006.

[6] Eshuis, R. & Wieringa, R. (2001). "A Formal Semantics for UML Activity Diagrams–Formalizing Workflow Models",

[7] CTIT Technical Reports Series (01-04).

[8] Falk, J.; Haubelt, C. & Teich, J. (2006). "Efficient Representation and Simulation of Model-Based Designs in SystemC", in proc. of FDL'2006, ECSI, 2006.

[9] Herrera, F & Villar, E. (2006). "A framework for Embedded System Specification under Different Models of Computation in SystemC", in proc. of the Design Automation Conference, DAC'2006, ACM, 2006.

[10] Hoare, C. A. R. (1978). Communicating sequential processes. Commun. ACM 21, 8. 1978.

[11] Jang, E. S.; Ohm, J. & Mattavelli, M. (January 2008). Whitepaper on Reconfigurable Video Coding (RVC). ISO/IEC JTC1/SC29/WG11 N9586. Antalya, Turkey. Available in http://www.chiariglione.org/mpeg/technologies/mpb-rvc/index.htm.

[12] Jantsch, A. (2004). Modeling Embedded Systems and SoCs. Morgan Kaufmann Elsevier Science. ISBN 1558609253.

[13] Kahn, G. (1974). The semantics of a simple language for parallel programming. In Proceedings of the International Federation for Information Processing Working Conference on Data Semantics.

[14] Kreku, J. ; Hoppari, M. & Kestilä, T. (2007). "SystemC workload model generation from UML for performance simulation", in proc. of FDL'2007, ECSI, 2007.

[15] Kroening, D. & Sharygna, N. (2005). "Formal Verification of SystemC by Automatic Hardware/Software Partitioning", in

[16] proc. of MEMOCODES'05.

[17] Lavagno, L.; Martin, G. & Selic, B. (2003). UML for real: design of embedded real-time systems. ISBN 1-4020-7501-4.

[18] Mallet, F. (2008). "Clock constraint specification language: specifying clock constraints with UML/MARTE", Innovations in Systems and Software Engineering, V.4, N.3, October, 2008.

[19] Maraninchi, F.; Moy, M. & L. Maillet-Contoz. (2005). "Lussy: An Open Tool for the Analysis of Systems-on-a-Chip at the Transaction Level", Design Automation of Embedded Systems, V.10, N.2-3, 2005.

[20] Moy, M.; Maraninchin, F. & Maillet-Contoz, L. (2008). "SystemC/TLM Semantics for Heterogeneous System-on-Chip Validation", in proc. of NEWCAS and TAISA Conference, IEEE, 2008.

[21] Mueller, W.; Ruf, J.; Hoffmann, D.; Gerlach, J.; Kropf, T. & W. Rosenstiel. (2001). "The Simulation Semantics of SystemC", in proc. of Design, Automation and Test in Europe, DATE'2001, IEEE, 2001.

[22] Peñil, P; Medina, J. & Posadas, H. & Villar, E. (2009). "Generating Heterogeneous Executable Specifications in SystemC from UML/MARTE Models", in proc. of the 11th Int. Conference on Formal Engineering Methods, IEEE, 2009.

[23] Piel, E.; Attitalah, R. B.; Marquet, P.; Meftali, S. ; Niar, S.; Etien, A.; Dekeyser, J.L. & P. Boulet. (2008). "Gaspard2: from MARTE to SystemC Simulation", in proc. of Design, Automation and Test in Europe, DATE'2008, IEEE, 2008.

[24] UML Specification v2.3. (2010).

[25] UML Profile for MARTE, v1.0. (2009).

[26] MDA guide, Version 1.1, June 2003.

[27] Open SystemC Initiative. www.systemc.org.

[28] Raudvere, T.; Sander, I. & Jantsch, A. (2008). "Application and Verification of Local Non Semantic-Preserving Transformations in System Design", IEEE Trans. on CAD of ICs and Systems, V.27, N.6, 2008.

[29] Salem, A. (2003). "Formal Semantics of Synchronous SystemC", in proc. of Design, Automation and Test in Europe, DATE'2003, IEEE, 2003.

[30] Störrle, H. & Hausmann, J.H. (2005). "Towards a Formal Semantics of UML 2.0 Activities", Software Engineering Vol. 64.

[31] Taha, S.; Radermacher, A.; Gerard, S. & Dekeyser, J. L. (2007). "MARTE: UML-based Hardware Design from Modeling to Simulation", in proc. of FDL'2007, ECSI 2007.

[32] Traulsem, C.; Cornet, J.; Moy, M. & Maraninchi, F. (2007). "A SystemC/TLM semantics in PROMELA and its possible Applications", in proc. of the Workshop on Model Checking Software, SPIN'2007, 2007.

[33] Vidal, J.; de Lamotte, F.; Gogniat, G.; Soulard, P. & Diguet, J.P. (2009). "A Code-Design Approach for Embedded System Modeling and Code Generation with UML and MARTE", proc. of the Design, Automation & Test in Europe Conference, DATE'09, IEEE 2009.

12

Concurrent Specification of Embedded Systems: An Insight into the Flexibility vs Correctness Trade-Off

F. Herrera and I. Ugarte
University of Cantabria
Spain

1. Introduction

In 2002, (Kish, 2002) warned about the danger of the abrupt break in Moore's law. Fortunately, nowadays integration capabilities are still growing and 20nm and 14nm technologies are envisaged, (Chiang, 2011). However, the frequency of integrated circuits cannot grow anymore. Therefore, in order to achieve a continuous improvement of performance, computer architectures are evolving towards the integration of more and more parallel computing resources. Examples of this include modern Graphical Processing Units (GPUs), such as the new CUDA architecture, named Fermi, which will use 512 cores, (Halfhill, 2012). Embedded system architectures show a similar trend with General Purpose Processors (GPPs), and some mobile phones already included between 2 and 8 RISC processors a few years ago, (Martin, 2006). Moreover, many embedded architectures are heterogeneous, and enclose different types of truly parallel computing resources such as (GPPs), Co-Processors, Digital Signal Processors, GPUs, custom-hardware accelerators, etc.

The evolution of HW architectures is driving the change in the programming paradigm. Several languages, such as (OpenMP, 2008), and (MPI, 2009), are defining the de facto programming paradigm for multi-core platforms. Embedded MPSoC platforms, with a growing number of general purpose RISC processors, are necessitating the adoption of a task-level centric approach in order to enable applications which efficiently use the computational resources provided by the underlying hardware platform.

Parallelism can be exploited at different levels of granularity. GPU-related languages enable the handling of a finer level of granularity, in order to exploit the inherent data parallelism of graphical applications. These languages also enable some explicit handling of the underlying architecture. MPSoC homogenous architectures require and enable a task-level approach, which provides a larger granularity in the handling of concurrency, and a higher level of abstraction to hide architectural details. A task-level approach enables the acceleration problem to be seen as a partition of functionality into *tasks* or high-level processes. A standard language which enables a task-level specification of concurrent functionality, and its communication and synchronization is convenient. In this scenario, SystemC (IEEE, 2005) standard has become the most widespread language for the specification of embedded systems. The main reason is that SystemC extends C/C++ with a set of features for a rich, standard modelling of concurrency, time, data types and modular hierarchical.

Summing up, concurrency is becoming a must in embedded system specification as it has become necessary for exploiting the underlying concurrency of MPSoC platforms. However, it brings a higher degree of complexity which introduces new challenges in embedded system specification, (Lee, 2006). In this chapter, the challenges and solutions for producing concurrent and correct specifications through simulation-based verification techniques are reviewed, and an alternative based on correct-by-construction specification methodologies is introduced. The chapter mainly addresses abstract concurrent specifications formed by asynchronous processes (formally speaking, untimed models of computation, MoCs, (Jansch, 2004). This type of modelling is required for speeding up the simulation of complex systems in new design activities, such as Design Space Exploration (DSE). This chapter does not assume a single definition of "correct" specification. For instance, functional determinism can be required or not, depending on the application and on the intention of the specification. However, to check whether such a property is fulfilled for every case requires the provision of the means for considering the different execution paths enabled by the control statements of an initially sequential algorithm, and, moreover, for considering the additional paths raised by a concurrent partition of such an algorithm.

The chapter will review different approaches and techniques for ensuring the correctness of concurrent specifications, to finally establish the trade-off between the flexibility in the usage of a specification language and the correctness of the coded specification. The rest of the chapter is structured as follows. Section 2 introduces an apparently simple specification problem in order to show how a rich specification language such as SystemC enables many different correct solutions, but also similar incorrect ones. Then, section 3 explores the possibilities and limitations of checking a SystemC specification through the application of simulation-based verification techniques. Finally, section 4 introduces an alternative, based on methodologies for correct-by-construction specifications and/or specification for verification. Section 5 gives conclusions about the trade-off between specification flexibility and verification cost and feasibility.

2. A "simple" specification problem

Some users may identify the knowledge of a specification language with the specification methodology itself. These users will take for granted that knowing the syntax, semantics and grammatical rules of the language is enough to build a "correct", or suitable, specification for a given design flow. Later on, in section 3, the benefits of this will be discussed. For now, let's see how a specification problem can be tackled in different ways.

A rich language provides great flexibility to tackle a similar specification problem in different ways, which in many cases is seen as a benefit by designers. In this sense, a simple experiment enabled the authors to deduce that this richness is actually employed when different users tackle the same specification problem. Let's assume we want to build a specification able to solve the functionality sketched in Fig.1.

This functionality is summarized by the following equations:

$$y= f_Y(a,b)= f_{12} (f_{11}(a), f_{21}(b)) \tag{1}$$

$$z = f_Z(a,b) = f_{22}(f_{11}(a), f_{21}(b)) \tag{2}$$

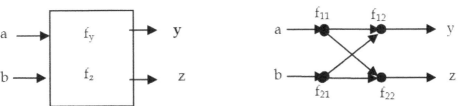

Fig. 1. Specification Intent.

In principle, the specification problem posed in Fig.1 is sufficiently general and simple to enable reasoning about it. The simple set of instances of f_{ij} functionalities, given by equation (3) will be used later on for facilitating the explanation of examples. However, the same reasoning and conclusions can be extrapolated to heavier and more complex functionalities.

$$f_{11}(x) = x+1 \quad f_{21}(x) = x+2 \tag{3}$$

$$f_{12}(x_1, x_2) = x_1 + x_2 \quad f_{22}(x_1, x_2) = (x_1 = 25,713)? \ 2x_1 - x_2 + 5 : x_2 - x_1$$

Initially, this is a straightforward specification problem which can be solved with a sequential specification, e.g., written in C/C++. The only condition to be fulfilled is to obey the dependency graph among f_{ij} functionalities shown on the right hand side of Fig.1. Thus, for instance, if the program executes the sequence $\{f_{11}, f_{21}, f_{12}, f_{22}\}$, it will be considered a correct model, and the model will produce its corresponding output as expected. For example, for $(a,b)=(1,2)$, an output $(y,z) = (6,2)$, where $f_{11}(1)=2$, $f_{21}(2)=4$, $f_{12}=2+4=6$ and $f_{22}=4-2=2$ (since $x_1=2 \neq 25,713$). Here, a user will already find some flexibility, once the order of f_{ij} executions can be permuted without impact on the intended functionality. Things start to get more complex when concurrency enters the stage. Once a pair of functionalities f_{ij} and f_{mn} can run concurrently no assumption about their execution order can be made. Assuming an atomic execution (non-preemptive) of f_{ij} functions, the basic principle for getting a solution fulfilling the specification intent of Fig. 1 is to guarantee the fulfilment of the following conditions:

$$T(f_{12}) > T(f_{11}) \tag{4}$$

$$T(f_{12}) > T(f_{21}) \tag{5}$$

$$T(f_{22}) > T(f_{21}) \tag{6}$$

$$T(f_{22}) > T(f_{11}) \tag{7}$$

Where $T(f_{ij})$ stands for the time tag associated with the computation of functionality f_{ij}. Equations (4-7) are conditions which define a partial order (PO) in the execution of f_{ij} functionalities. It is a partial order because it defines an execution order relationship only for a subset of the whole set of pairs of f_{ij} functionalities. In other words, there are pairs of functionalities, f_{ij} and f_{mn}, with $i \neq m \ \lor j \neq n$, which do not have any order relationship. This no order relationship is denoted $f_{ij} >< f_{mn}$. Some specification methodologies, such as HetSC, help the designer capture untimed specifications, which implicitly capture a PO. Untimed specifications reflect conditions only in terms of execution order, without assuming specific physical time conditions, thus they are the most abstract ones in terms of time handling. The PO is sufficient for ensuring the same specific global system functionality, while it reflects the available flexibility for further design steps. Indeed, no-order relationships spot functionalities which can be run in natural parallelism (that is, they are functionalities which do not require pipelining for running in actual parallelism) or which can be freely scheduled.

SystemC has a discrete event (DE) semantics, which means that the time tag is twofold, that is, $T=(t, \Delta)$. Any computation or event happens in a specific delta cycle (Δ_i). Additionally, each delta has an associated physical time stamp (t_i), in such a way that a set of consecutive deltas can share the same time stamp (this way, instantaneous reactions can be modelled as reactions in terms of delta advance, but no physical time advance). Complementarily, it is possible that two consecutive delta cycles present a jump in physical time ranging from the minimum to the maximum physical time which can be represented.

Since SystemC provides different types of processes, communication and synchronization mechanisms for ensuring the PO expressed by equations (4-7), it is easy to imagine that there are different ways to solve the specification intent in Fig.1 as a SystemC concurrent specification, even if only untimed specifications are considered. In order to check how such a specification would be solved by users knowing SystemC, but without knowledge of particular specification methodologies or experience in specification, six master students were asked to provide a concurrent solution. No conditions on the use of SystemC were set.

Five students managed to provide a correct solution. By "correct" solution it is understood that for any value of 'a' and 'b', and for any valid execution (that is, fulfilling SystemC execution semantics) the output results were the expected ones, that is $y=f_Y(a,b)$ and $z=f_Z(a,b)$. In other words, we were looking for solutions with functional determinism, (Jantsch, 2004). A first interesting observation was that, from the five correct solutions, four different solutions were provided. These solutions were considered different in terms of the concurrency structure (number of processes used, which functionality is associated to each process), communication and synchronization structure (how many channels, events and shared variables are used, and how they are used for process communication), and the order of computation, communication and synchronization within a process.

Fig. 2, 3 and 4 sketch some possible solutions where functionality is divided into 2 or 4 processes. These solutions are based on the most primitive synchronization facilities provided by SystemC ('wait' statements and SystemC events), using shared variables for data transfer among functionalities. Therefore, the solutions in Fig. 2, 3 and 4 reflect only a subset of the many coding possibilities. For instance, SystemC provides additional specification facilities, e.g. standard channels, which can be used for providing alternative solutions.

Fig.2, Fig.3a and Fig.3b show two-process-based solutions. In Fig. 2, the two processes P1 and P2 execute f_{i1} functionalities before issuing a wait(d) statement, with d of 'sc_time' type and where 'd' can be either a single delta cycle delay (d=SC_ZERO_TIME) or a timed delay (s>SC_ZERO_TIME), that is, an advance of one or more deltas (Δ) with an associated physical time advance (t). Notice that this actually means two different solutions in SystemC, under the SystemC semantics. In the former case, f_{11} and f_{21} are executed in Δ_0,

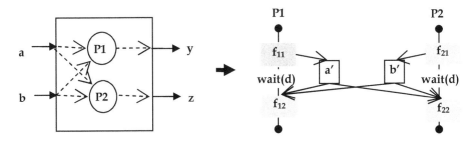

Fig. 2. Solution based on two processes and on wait statements.

while f_{21} and f_{22} are executed in Δ_1, without t advance, while in the latter case, f_{21} and f_{22} are executed in a T with a different t coordinate. Anyhow, in both cases the same untimed and abstract semantics is fulfilled, in the sense that both fulfil the same PO, that is, equations (4-7) are fulfilled. Notice that there are more solutions derived from the sketch in Fig. 2. For instance, several 'wait(d)' statements can be used on each side.

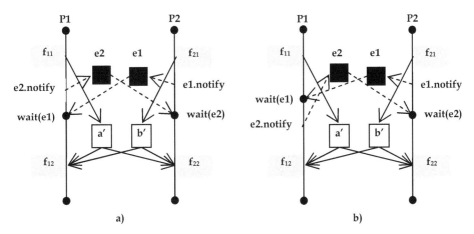

Fig. 3. Solutions based on two processes and on SystemC events.

Fig.3a and Fig.3b show two solutions based on SystemC events. In the Fig.3a solution, both processes compute f_{11} and f_{21} in Δ_0 and schedule a notification to a SystemC event which will resume the other process in the next delta. Then, both processes get blocked. The crossed notification sketch ensures the fulfilment of equations (5) and (7). Equations (4) and (6) are fulfilled since f_{11} and f_{12} are sequentially executed within the same process (P_1), and similarly, f_{21} and f_{22} are sequentially executed by process P_2. Notice that several variants based on the Fig.3a sketch can be coded without impact on the fulfilment of equations (4-7). For instance, it is possible to use notifications after a given amount of delta cycles, or after physical time and still fulfil (4-7). It is also possible to swap the execution of f_{11} and e_2 notification, and/or to swap the execution of f_{11} and e_1 notification.

Fig.3b represents another variant of the Fig.3a solution where one of the processes (specifically P_1 in Fig.3b) makes the notification after the wait statement. It adds an order condition, described by the equation $T(f_{22}) > T(f_{12})$, and which obliges the execution to require one delta cycle more (f_{22} will be executed in a delta cycle after f_{12}). Anyhow, this additional constraint on the execution order still preserves the partial order described by equations (4-7) and guarantees the functional determinism of the specification represented by Fig. 3b.

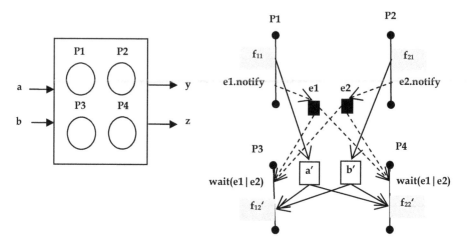

Fig. 4. Solution based on four finite and non-blocking processes.

Finally, Fig.4 shows a solution with a higher degree of concurrency, since it is based on four finite non-blocking processes. In this solution, each process computes f_{ij} functionality without blocking. P3 and P4 processes compute f_{12} and f_{22} respectively only after two events, e_1 and e_2, have been notified. These events denote that the inputs for f_{12} and for f_{22} functionalities, $a' = f_{11}(a)$ and $b' = f_{21}(b)$, are ready. In general, P3 and P4 have to handle a local status variable (not-represented in Fig.4) for registering the arrival of each event since e_1 and e_2 notifications could arrive in different deltas. Such handling is an additional functionality wrapping the original f_{i2} functionality, which results in a functionality f_{i2}', as shown in Fig.4.

The sketch in Fig. 4 enables several equivalent codes based on the fact that processes P_3 and P_4 can be written either as SC_METHOD processes with a static sensitivity list, or as SC_THREAD processes with an initial and unique wait statement (coded as a SystemC dynamic sensitivity list, but used as a static one), before the function computation. Moreover, as with the Fig. 3 cases, both in P_1 and in P_2, the execution of f_{i1} functionalities and event notifications can be swapped without repercussion on the fulfilment of equations (4-7).

Summarizing, the solutions shown are samples of the wide range of coding solutions for a simple specification problem. The richness of specification facilities and flexibility of SystemC enable each student to find at least one solution, and furthermore, to provide some different alternatives. However, such an open use of the language also leads to a variety of possible incorrect solutions. Fig. 5 illustrates only two of them.

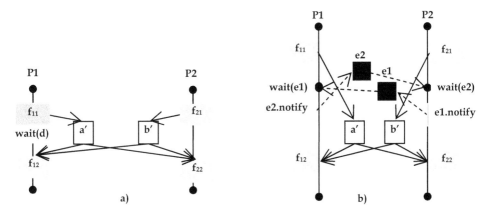

Fig. 5. Solution based on four finite and non-blocking processes.

In the Fig.5a example, the order condition (7) might be broken, and thus the specification intent in Fig.5a is not fulfilled. Under SystemC execution semantics, f_{22} may happen either before or after f_{11}. The former case can happen if P2 starts its execution first. SystemC is non-pre-emptive, thus f_{22} will execute immediately after f_{21}, and thus before the start of P1, which violates condition (7). Moreover, the example in Fig. 5a does not provide functional determinism because condition (7) might be fulfilled or not, which means that output z can present different output values for the same inputs. Therefore, it is not possible to make a deterministic prediction of what output z will be for the same set of inputs, since sometimes it can be $z=f_{22}(a,f_{21}(b))$, while others it can be $z=f_{22}(f_{11}(a),f_{21}(b))$. In many specification contexts functional determinism is required or at least desirable.

The Fig. 5b example shows another typical issue related to concurrency: deadlock. In Fig. 5b, a SystemC execution will always reach a point where both processes P_1 and P_2 get blocked forever, since the condition for them to reach the resumption can never be fulfilled. This is due to a circular dependency between their unblocking conditions. After reaching the wait statement, unblocking P_1 requires a notification on event e1. This notification will never come since P_2 is in turn waiting for a notification on event e2.

Even for the small parallel specification used in our experiment, al least one student was not able to find a correct solution. However, even for experienced designers it is not easy to validate and deal with concurrent specifications just by inspecting the code, relying and reasoning based on the execution semantics, even if they are supported by a graphical representation of the concurrency, synchronization and communication structure. Relatively small concurrent examples can present many alternatives for analysis. Things get worse with complex examples, where the user might need to compose blocks whose code is not known or even visible. Moreover, even simple concurrent codes, can present subtle bug conditions, which are hard to detect, but risky and likely to happen in the final implementation.

For example, let's consider a new solution of the 'simple' specification example based on the Fig.3a structure. It was already explained that this structure works well when considering either delta notification or timed notification. A user could be tempted to use immediate notification for speeding up the simulation with the Fig.3a structure. However, this specification would be non-deterministic. In effect, at the beginning of the simulation, both P1 and P2 are ready to execute in the first delta cycle. SystemC simulation semantics do not state which process should start in a valid simulation. If P1 starts, it will mean that the e2 immediate notification will get lost. This is because SystemC does not register immediate notification and requires the process receiving it (in this case P2) to be waiting for it already. Thus, there will be a partial deadlock in the specification. P2 will get blocked in the 'wait(e2)' statement forever and the output of P2 will be the null sequence $z=\{\}$, while $y=\{f_{21}(f_{11}(a),f_{21}(b))\}$. Assuming the functions of equations (3), for $(a,b)=(\{1\},\{2\})$, $(y,z) = (\{6\},\{\})$. Symmetrically, if P2 starts the execution first, then P1 will get blocked forever at its wait statement, and the output will be $y=\{\}$, $z=\{f_{22}(f_{11}(a),f_{21}(b))\}$. Assuming the functions of equations (3), for $(a,b)=(\{1\},\{2\})$, $(y,z) = (\{\},\{2\})$. Thus, in this case, no outputs correspond to the initial intention. There is functional non-determinism, and partial deadlock.

It is not recommended here that some properties should always be present (e.g., not every application requires functional determinism). Nor is the prohibition of some mechanisms for concurrent specification recommended. For instance, immediate notification was introduced in SystemC for SW modelling and can speed up simulation. Indeed, the Fig.3a example can deterministically use immediate notification with some modifications in the code for explicit registering of immediate events. However, such modification shows that the solution was not as straightforward as designers could initially think. Therefore, the definition of when and how to use such a construct is convenient in order to save wastage of time in debugging, or what it would be worse, a late detection of unexpected results.

Actually, what it is being stated is that concurrent specification becomes far from straightforward when the user wants to ensure that the specification avoids the plethora of issues which may easily appear in concurrent specifications (non-determinism, deadlock, starvation, etc), especially when the number of processes and their interrelations grow. Therefore, a first challenge which needs to be tackled is to provide methods or tools to detect that a specification can present any of the aforementioned issues. The following sections will introduce this problem in the context of SystemC simulation. The difficulty in being exhaustive with simulation-based techniques will be shown. Then the possibility to rely on correct by construction specification approaches will be discussed.

In order to simplify the discussion, the following sections will focus on functional determinism. In general, other issues, e.g. deadlock, are orthogonal to functional determinism. For instance, the Fig. 5b case presents deadlock while still being deterministic (whatever the input, each output is always the same, a null sequence). However, non-determinism is usually a source of other problems, since it usually leads to unexpected process states, for which the code was not prepared to avoid deadlock or other problems. Fig. 4a example with immediate notification was an example of this.

3. Simulation-based verification for flexible coding

Simulation-based verification requires the development of a verification environment. Fig. 6 represents a conventional SystemC verification environment. It includes a test bench, that is, a SystemC model of the actual environment where the system will be encrusted. The test bench is connected and compiled together with the SystemC description of the system as a single executable specification. When the OSCI SystemC library is used, the simulation kernel is also included in the executable specification. In order to simulate the model, the executable specification is launched. Then, the test bench provides the input stimuli to the system model, which produces the corresponding outputs. Those outputs are in turn collected and validated by the test bench.

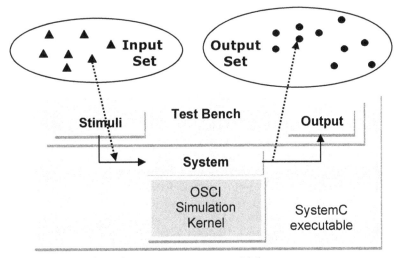

Fig. 6. Simulation-based verification environment with low coverage.

The Fig. 6 framework has a significant problem. A single execution of the executable specification provides very low verification coverage. This is due to two main factors:

- The test bench only reflects a subset of the whole set of possible inputs which can be fed by the actual environment (Input Set).
- Concurrency implies that, for each fixed input (triangle in Fig. 6), there are in general more than one feasible execution order or *scheduling*, thus potentially, more than one feasible output. However, a single simulation shows only one *scheduling*.

The first point will be addressed in section 3.1. The following sections will focus on dealing with how to tackle verification when concurrency appears in the specification.

3.1 Stimuli generation

Assuming a fully sequential system specification, the first problem consists in finding a sufficient number of stimuli for a 'satisfactory' verification of the specification code. Satisfactory can mean 100% or a sufficiently high percentage of a specific coverage metric.

Therefore, an important question is which coverage metrics to use. A typical coverage metric is branch coverage, but there are more code coverage metrics, such as lines, blocks, branches, expressions, paths, and boundary-path. Other techniques (Fallah, 1998); (Gupta, 2002); (Ugarte, 2011) are based on functional coverage metrics. Functional coverage metrics are defined by the engineer, and thus rely on engineer experience. They can provide better performance in bug detection than code coverage metrics. However, code coverage metrics do not depend on the engineer, thus they can be more easily automated. They are also simpler, and provide a first quality metric of the input set.

In complex cases, an exhaustive generation of input vectors is not feasible. Then, the question is which vectors to generate and how to generate them. A basic solution is random generation of input vectors, (Kuo, 2007). The advantages are simplicity, fast execution speed and many uncovered bugs with the first stimulus. However, the main disadvantages are twofold: first, many sets of input values might lead to the same observable behaviour and are thus redundant, and second, the probability of selecting particular inputs corresponding to corner cases causing buggy behaviour may be very small.

An alternative to random generation is, constrained random vector generation, (Yuan, 2004). Environments enabling constrained random generation enable a random, but controlled generation of input vectors by imposing some bounds (constraints) on the input data. This enables a generation of input vectors that are more representative of the expected environment. For instance, one can generate values for an address bus in a certain range of the memory map. Constrained randomization also enables a more efficient generation of input vectors, once they can be better directed to reach parts of code that a simple random generation will either be unlikely to reach or will reach at the cost of a huge number of input stimuli. In the SystemC context, the SystemC Verification library (SCV) (OSCI, 2003), is an open source freely available library which provides facilities for constrained randomization of input vectors. Moreover, the SCV library provides facilities for controlling the statistical profile in the vector generation. That is, the user can apply typical distribution functions, and even define customized distribution functions, for the stimuli generated. There are also commercial versions such as Incisive Specman Cadence (Kuhn, 2001), VCS of Synopsys, and Questa Advanced Simulator of Mentor Graphics. The inconvenience of constrained random generation of input vectors is the effort required to generate the constraints. It already requires extracting information from the specification, and relies on the experience of the engineer. Moreover, there is a significant increase in the computational effort required for the generation of vectors, which needs solvers.

More recently, techniques for automatic generation of input vectors have been proposed (Godefroid, 2005); (Sen, 2005); (Cadar, 2008). These techniques use a coverage metric to guide (or direct) the generation of vectors, and bound the amount of vectors generated as a function of a certain target coverage. However, these techniques for automatic vector generation require constrained usage of the specification language, which limits the complexity of the description that they can handle.

In order to explain these strategies, we will use an example consisting in a sequential specification which executes the f_{ij} functionalities in Fig. 1 in the following order $\{f_{11}, f_{21}, f_{12}, f_{22}\}$. Therefore, this is an execution sequence fulfilling the specification intent, provided the dependency graph in Fig. 1b. Let's assume that the specific functions of this sequential system are given by equations (3), and that the metric to guide the vector generation is branch coverage. It will also be assumed that the inputs ('a' and 'b') are of integer type with range [-2,147,483,648 to 2,147,483,647]. A first observation to make is that our example will have two execution paths, defined by the control statements, specifically, the conditional function f_{22}. Entering one or another path depends on the value of the 'x_1' input of f_{22}, which in turn depends on the input to f_{11}, that is, on the input 'a'.

By following the first strategy, namely, running the executable specification with random vectors of 'a' and 'b', it will be unlikely to reach the true branch of the control sentence within f_{22}, since the probability of reaching it is less than 2.5E-10 for each input vector. Even if we provide means to avoid repeating an input vector, we could need 2.5E10 simulations to reach the true path.

Under the second strategy, the verification engineer has to define a constraint to increase the probability of reaching the true branch. In this simple example, the constraint could be the creation of a weighted distribution for the x input, so that some values are chosen more often than others. For instance, the following sentence: *dist {[min_value:25713]:= 33, 25714:= 34, [25715:max_value]:=33}*, states that the value that reaches the true branch of f_{22}, that is, 25,714, has a 33.3% probability to be produced by the random generator. The likelihood of generation of values below 25.714 would be 33.3%, and similarly 33.3% for values over 25,714. Thus, the average number of vectors required for covering the two paths would be 3. Then, the user could prepare the environment for producing three input vectors (or a slightly bigger number of them for safety). One possible vector set generated could be: (a,b) = {(12390, -2344), (-3949, 1234), (25714, -34959)}. The efficiency of this method relies on the user experience. Specifically, the user has to know or guess which values can lead to different execution paths, and thus which groups of input values will likely involve different behaviours.

The latter strategy would be directed vector generation. This strategy analyses the code in order to generate the minimum set of vectors for covering all branches. Directing the generation in order to cover all execution paths would be the ideal goal. However, this makes the problem explode. In the simple case in Fig. 1, branch and path coverage is the same since there is only one control statement. In this case, only one vector is required per branch. For example, the first value generated could be random, e.g., (a = 39349, b= -1024). As a result, the system executes the false path of the control statement. The constraint of the executed path is detected and the constraint of the other branch generated. In this case, the constraint is a=25714. The generator solves the constraint and produces the next vector (a, b) = (25714, 203405). With this vector, the branch coverage reaches 100% of coverage and vector generation finishes. Therefore, the stimulus set is (a,b) = { (39349, 1024), (25714, 203405)}.

3.2 Introducing concurrency: scheduling coverage

In the previous section, the generation of input vectors for reaching certain coverage (usually of branches or of execution paths) has been discussed. For this, we assumed a sequential specification, which means that for a fixed input vector, a fixed output vector is expected. Thus, the work focuses on finding vectors for exercising the different paths which can be executed by the real code, since these paths reflect the different behaviours that the code can exhibit for each input. Each type of behaviour is a relationship between the input and the output. Functional behaviour will imply a single output for given input.

As was mentioned at the beginning of section 3, the injection of concurrency in the specification raises a second issue. Concurrency makes it necessary to consider the possibility of several schedulings for the execution of the system functionality for a fixed input vector. This can potentially lead to different behaviours for the same input. At specification level, there are no design decisions imposing timing and thus no strict ordering

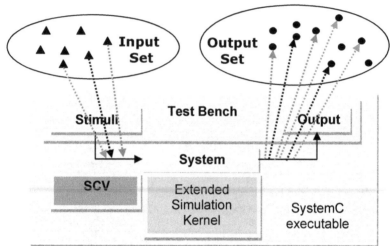

Fig. 7. Higher coverage by checking several inputs and several schedulings per input.

to the computation of the concurrent functionality, thus all feasible order must be taken into account. The only exception is the timing of the environment, which can be neglected for generality. In other words, inputs can be considered as arriving in any order.

In order to tackle this issue, Fig. 7 shows the verification environment based on multiple simulations proposed by (Herrera, 2006). Using multiple simulations, that is, multiple executions (ME) in a SystemC-based framework, enables the possibility of feeding different input combinations. SystemC LRM comprises the possibility of launching several simulations from the same executable specification through several calls to the *sc_elab_and_sim* function. (Herrera, 2006), and (Herrera, 2009), explain how this could be done in SystemC. However, SystemC LRM also states that such support depends on the implementation of the SystemC simulator. Currently, the OSCI simulator does not support this feature. Thus, it can be assumed that running N_E simulations currently means running the SystemC executable specification N_E times. In (Herrera, 2006), and (Herrera, 2009), the launch of several simulations is automated through an independent launcher application.

The problem is how to simulate different scheduling, and thus potentially different behaviour, for each single input. Initially, one can try to perform several simulations for a fixed input test bench (one triangle in the Fig. 7 schema,). However, by using the OSCI SystemC simulator, and most of the available SystemC simulators, only one scheduling is simulated. In order to demonstrate the problem, we define a scheduling as a sequence of segments (s_{ij}). A scheduling reflects a possible execution order of segments under SystemC semantics. A segment is a piece of code executed without any pre-emption between calls to the SystemC scheduler, which can then make a scheduling decision (SDi). A segment is usually delimited by blocking statements. A scheduling can be characterized by a specific sequence of scheduling decisions. In turn, the set of feasible schedulings of a specification can be represented in a compact way through a scheduling decision tree (SDT). For instance, Fig. 8 shows the SDT of the Fig. 2 (and Fig. 3) specification. This SDT shows that there are 4 possible schedulings (S_i in Fig. 8). Each segment is represented as a line ended with a black

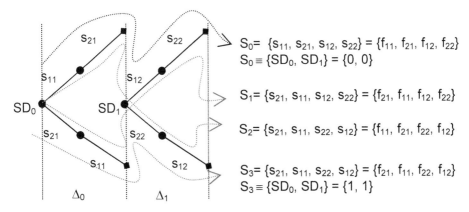

$$S_0 = \{s_{11}, s_{21}, s_{12}, s_{22}\} = \{f_{11}, f_{21}, f_{12}, f_{22}\}$$
$$S_0 \equiv \{SD_0, SD_1\} = \{0, 0\}$$

$$S_1 = \{s_{21}, s_{11}, s_{12}, s_{22}\} = \{f_{21}, f_{11}, f_{12}, f_{22}\}$$

$$S_2 = \{s_{21}, s_{11}, s_{22}, s_{12}\} = \{f_{11}, f_{21}, f_{22}, f_{12}\}$$

$$S_3 = \{s_{21}, s_{11}, s_{22}, s_{12}\} = \{f_{21}, f_{11}, f_{22}, f_{12}\}$$
$$S_3 \equiv \{SD_0, SD_1\} = \{1, 1\}$$

Fig. 8. Scheduling Decision Tree for the examples in Fig. 2 and Fig. 3.

dot. Moreover, in the Fig. 8 example, each s_{ij} segment corresponds to a f_{ij} functionality, computed in this execution segment. Each dot in Fig. 8 reflects a call to the SystemC scheduler. Therefore, each simulation of the Fig. 2, and Fig. 3 examples, either with delta or timed notification, always involves 4 calls to the SystemC scheduler after simulation starts. However, only two of them require an actual selection among two or more processes ready to execute, that is, a scheduling decision (SD_i). As was mentioned, multiple executions of the executable simulation compiled against the existing simulators would exhibit only a single scheduling, for instance S_0 in the Fig. 8 example. Therefore, the remaining schedulings, S_1, S_2 and S_3 would never be checked, no matter how many times the simulation is launched.

As was explained in section 2, the Fig. 2 and Fig. 3 examples fulfil the partial order defined by equations (4-7), so the unchecked schedulings will produce the same result. This is easy to deduce by considering that each segment corresponds to a f_{ij} functionality of the example.

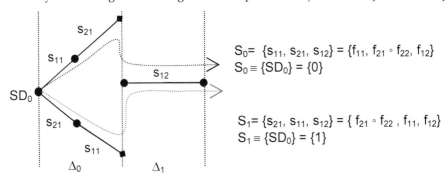

$$S_0 = \{s_{11}, s_{21}, s_{12}\} = \{f_{11}, f_{21} \circ f_{22}, f_{12}\}$$
$$S_0 \equiv \{SD_0\} = \{0\}$$

$$S_1 = \{s_{21}, s_{11}, s_{12}\} = \{f_{21} \circ f_{22}, f_{11}, f_{12}\}$$
$$S_1 \equiv \{SD_0\} = \{1\}$$

Fig. 9. Scheduling Decision Tree for the Fig.2 and Fig. 3 examples.

However, let's consider the Scheduling Decision Tree (SDT) in the Fig. 5a example, shown in Fig. 9. The lack of a wait statement between f_{21} and f_{22} in P2 in the Fig. 5a example implies that P_2 executes all its functionality (f_{21} and f_{22}) in a single segment (s_{21}). Notice that a segment can comprise different functionalities, or, as in this case, one functionality as a result of composition of f_{21} and f_{22} (denoted $f_{21} \circ f_{22}$). Therefore, for the Fig. 5a example, the SystemC kernel executes three segments, instead of four as in the case of Fig. 4 example. Notice also that several scheduler calls can appear within the boundaries of a delta cycle.

The SDT of the Fig. 5 example has only a single scheduling decision. Therefore, two schedulings are feasible, denoted S_0 and S_1. However, only one of them, S_0, fulfils the partial order defined by equations (4-7). As was mentioned, the OSCI simulator will execute only one, either S_0 or S_1, even if we run the simulation several times. This is due to practical reasons, since OSCI and other SystemC simulators implement a fast and straightforward scheduling based on a first-in first-out (FIFO) policy. If we are lucky, S_1 will be executed, and we will establish that there is a bug in our concurrent specification. However, if we are not lucky, and S_0 is always executed, then the bug will never be apparent. Thus, we can get the false impression of facing a deterministic concurrent specification.

Therefore, a simulation-based environment requires some capability for observing the different schedulings, ideally 100% coverage of schedulings, which are feasible for a fixed input. Current OSCI implementation of the SystemC simulation kernel fulfils the SystemC semantics and enables fast scheduling decisions. However, it produces a deterministic sequence of scheduling decisions, which is not changed from simulation to simulation for a fixed input. This has leveraged several techniques for enabling an improvement of the scheduling coverage. Before introducing them, a set of metrics for comparing different techniques for improving scheduling coverage of simulation-based verification techniques, proposed in (Herrera, 2006), will be introduced. They can be used for a more formal comparison of the techniques discussed here. These metrics are dependent on each input vector, calculated by means of any of the techniques explained in section 3.1.

Let's denote the whole set of schedulings S, where $S = \{S_0, S_1, ..., S_{size(s)}\}$, and size(S) is the total number of feasible schedulings for a fixed input. Then, the Scheduling Coverage, C_S, is the number of checked schedulings with regard to the total number of possible schedulings.

$$C_S = \frac{N_S}{size(S)} \tag{8}$$

The Multiple Execution Efficiency η_{ME} is the actual number of (non-repeated) schedulings N_S covered after N_E simulations (executions in SystemC).

$$\eta_{ME} = \frac{N_S}{N_E} = \frac{N_S}{N_S + N_R} = \frac{1}{1 + R_E} \tag{9}$$

N_R stands for the amount of repeated schedulings, which are not useful. As can be seen, η_{ME} can be expressed in terms of R_S. R_S is a factor which accounts for the number of repeated schedulings out of the total number of simulations N_E.

The total number of simulations to be performed to reach a specific scheduling coverage, $N_T(C_S)$ can be expressed as a function of the desired coverage, the number of possible schedulings, and the multiple execution efficiency.

$$N_T(C_S) = \frac{C_S \cdot size(S)}{\eta_{ME}} \tag{10}$$

Finally, the Time Cost for achieving a coverage C_S is approximated by the following equation:

$$T_E \approx \frac{C(TE) \cdot size(TE)}{\eta_{ME}} \cdot \bar{t} \tag{11}$$

Where $\bar{t}$ is the average simulation time of each scheduling. It is actually a rough approximation, since each scheduling can derive in shorter or longer schedulings. It also depends on the actual scheduling technique. However, equations (8-11) will be sufficiently useful for comparing the techniques introduced in the following sections, and the yield of conventional SystemC simulators, including the OSCI SystemC library in the simulation-based verification environments shown in Fig. 7. Conventional SystemC simulators provide a very limited scheduling coverage, $C_S = \frac{1}{size(S)}$, since $N_S=1$. Moreover, the scheduling coverage is fixed and cannot grow with further simulations. Since size(S) exponentially grows when adding tasks and synchronization mechanisms, the scheduling coverage quickly becomes low even with small examples. For instance, in (Herrera, 2006), a simple extension of the Fig. 2 example to three processes, each of three segments, leads to size(S)=216, thus C_S=0.46%.

3.2.1 Random and pseudo-random scheduling

The user of an OSCI simulator can try a trick to check different schedulings in a SystemC specification. It consists in changing the order of declaration of SystemC processes in the module constructor. Thus, the result of the first dispatching of the OSCI simulator at the beginning of the simulation can be changed. However, this trick gives no control over further scheduling decisions. Moreover, checking a different scheduling requires the modification of the specification code.

A simple alternative for getting multiple executions to exhibit different schedulings is changing the simulation kernel to enable a random selection among the processes ready to execute in each scheduling decision. Random scheduling enables $\frac{1}{size(S)} \leq C_S \leq 1$, and a monotonic growth of C_S with the number of simulations N_E. The dispatching is still fast, since it only requires the random generation of an index suitable for the number of processes ready to execute in each scheduling decision. The implementation can range from more complex ones guaranteeing the equal likelihood in the selection of each process in the ready-to-execute list, to simpler ones, such as the one proposed in (Herrera, 2006), which is faster and has low impact in the equal likelihood of the selection.

There are still better alternatives to pure random scheduling. In (Herrera, 2006), pseudorandom (PR) scheduling is proposed. Pseudorandom scheduling consists in enabling a pseudo-random, but deterministic, sequence of scheduling decisions from an initial seed. This provides the advantage of making each scheduling reproducible in a further execution. This reproducibility is important since it enables to debug the system with the scheduling which showed an issue (unexpected result, deadlock, etc) as many times as desired. Without this reproducibility, the simulation-based verification framework would be able to detect there is an issue, but would not be practically applicable for debugging it. Therefore, Pseudorandom scheduling presents the same coverage, $\dfrac{1}{size(S)} \leq C_S \leq 1$, and monotonic growth as C_S with the number of simulations of pure random scheduling. A freely available extension of the OSCI kernel, which implements and makes available Pseudorandom scheduling (for SC_THREAD processes) is provided in (UCSCKext, 2011).

Pseudorandom scheduling still presents issues. One issue is that, despite the monotonic growth of C_S with N_E, this growth is approximately logarithmic, due to the probability of finding a new scheduling with the number of simulations performed. Each new scheduling found reduces the number of new schedulings to be found, and Pseudorandom schedulings have no mechanisms to direct the search of new schedulings. Thus, in pseudorandom scheduling, $\eta_{ME} \leq 1$ in general, and it quickly tends to 0 when N_E grows. Another issue is that it does not provide specification-independent criteria to know when a specific C_S or a size(S) has been reached. C_S or size(S) can be guessed for some concurrency structures.

3.2.2 Exhaustive scheduling

In (Herrera, 2009), a technique for directing scheduling decisions for an efficient and exhaustive coverage of schedulings, called DEC scheduling, was proposed. The basic idea, was to direct scheduling decisions in such a way that the sequence of simulations perform a depth-first search (DFS) of the SDT. For an efficient implementation, (Herrera, 2009), proposes to use a scheduling decision register (SDR), which stores the sequence of decisions taken in the last simulation.

For instance, for the Fig. 8 SDT, corresponding to examples in Fig.2 and 3, the first simulation will produce the S_0 scheduling. This means that the SDR will be $SDR_0=\{0,0\}$, matching the FIFO scheduling semantics of conventional SystemC simulators, where the first process in the ready-to-execute queue is always selected. Then, a second simulation under the DEC scheduling, will use the SDR to reproduce the scheduling sequence until the penultimate decision (also included). Then, the last decision is changed. Remember that a scheduling decision SD_i is taken whenever a selection among at least two ready-to-execute processes is required. Since in the previous simulation the last scheduling decision was to select the 0-th process (denoted in the example as $SD_1=0$), in the current simulation the next process available in the ready-to-execute queue is selected (that is, $SD_1=1$). Therefore, the second execution in the example simulates the next scheduling of the SDT, $S_1=\{0,1\}$.

In a general case, the change in the selection of the last decision can mean an extension of the SDT (which means that the simulation must go on, and so go deeper into the SDT). Another possibility is what happens in the example shown, where the branch at the current depth level has been fully explored and a back trace is required. In our example, the third simulation will go back to SD_0 decision and will look for a different scheduling decision ($SD_0=1$). What will occur in this case is that the simulation can go on and new scheduling decisions, will be required, thus requiring the extension of the SDR again, and thus leading to the $S_2=\{1,0\}$ scheduling. Following the same reasoning, it is straightforward to deduce that the next simulation will produce the scheduling $S_3=\{1,0\}$.

Therefore, the main advantage of DEC scheduling with regard to PR scheduling is that $\eta_{ME} = 1$. That is, each new simulation guarantees the exploration of a new scheduling. This provides a more efficient search since the scheduling coverage grows linearly with the number of simulations. That is, for DEC scheduling:

$$\frac{1}{size(S)} \leq C_S = \frac{N_E}{size(S)} \leq 1 \tag{12}$$

Another advantage of DEC scheduling is that it provides criteria for finishing the exploration of schedulings which does not require an analysis of the specification. It is possible thanks to the ordered exploration of the SDT, (Herrera, 2009). The condition for finishing the exploration is fulfilled once a simulation (indeed the N_E=size(S)-th simulation) has selected the last available process for each scheduling decision of the SDR, and no SDT extension (that is, no further events and longer simulation) is required. In the example in Fig. 8, this corresponds to the scheduling S_3={1,1}. When this condition is fulfilled, 100% scheduling coverage (C_S) has been reached. Notice that, in order to check the fulfilment of the condition, no estimation of size(S) is necessary, thus no analysis of the concurrency and synchronization structure of the specification is required. In the case that size(S) can be calculated, e.g. because the concurrency and synchronization structure of the specification is regular or sufficiently simple, then C_S, can be calculated through equation (12). For instance, in the Fig. 8 example size(S)=4, then, applying equation (8), C_S=0.25N_S.

The main limitation of DEC scheduling is that size(S) has an exponentially growth for a linear growth of concurrency. Thus, although $\eta_{ME} = 1$ is fulfilled, the specification will exhibit a state explosion problem. The state explosion problem is exemplified in (Godefroid, 1995), which shows how a simple philosopher's example can pass from 10 states to almost 10^6 states when the number of philosophers grows from two up to twelve. Another related downside is that a long SDR has to be stored in hard disk, thus the reproduction of scheduling decisions will include the time penalties for accessing the file system. This means a growth of $\bar{t}$ in equation (11) for the calculation of the simulation-based verification time, which has to be taken into account when comparing DEC scheduling with Pseudo-random or pure random techniques, where scheduling decisions are lighter.

3.3 Partial Order Reduction techniques

A set of simulation-based techniques, based on Partial Order Reduction (POR) has been proposed for tackling the state explosion problem. POR is a partition-based testing technique, based on the execution of a single representative scheduling for each class of equivalent schedulings. This reduces the number of schedulings to be explored, from size(S) feasible schedulings, to M, with M<size(S). M is the number of sets of non-equivalent scheduling classes, each one enclosing a set of equivalent schedulings. The equivalence is understood in functional terms. That is, the simulation of two schedulings of an equivalent scheduling class will lead to the same state, and therefore to the same effect on the system behaviour. When applying POR techniques, the objective is not to achieve C_S=100%, but C_M=100%, where C_M stands for the coverage of representative (non-equivalent) schedulings. Expressed in other terms, a single simulation serves to check on average a set of $\bar{L}$ equivalent simulations. Thus POR techniques enable a scheduling

coverage of $\dfrac{N_E \cdot \overline{L}}{size(S)}$ and efficiencies greater than 1, that is, $\eta_{ME} = \dfrac{N_S}{N_E} \geq 1$. Obviously, the efficiency in the exploration of non-equivalent schedulings will always remain below or equal to 1.

In order to deduce which schedulings are equivalent, POR methods require the extraction and analysis of information from the specification, in order to study when the possible interactions and dependencies between processes may lead or not to functionally equivalent paths. For instance, the detection of shared variables, and the analysis of write-after-write, read-after-write, and write-after-read situations in them, enable the extraction of non-equivalent paths which can lead to race conditions. Similarly, event synchronization has to be analyzed (notification after wait, wait after notification, etc) since non-persistence of events can lead to misses and to unexpected deadlock situations, non-determinism or other undesirable effects. (Helmstetter, 2006) and (Helmstetter, 2007) propose dynamic POR (DPOR) of SystemC models, by adapting dynamic POR techniques initially developed for software (Flanagan, 2005). Dynamic POR selects the paths to be checked during the simulation, in each scheduling decision, performing the analysis among ready-to-execute processes. Later works, such as the 'Satya' framework (Kundu, 2008), have proposed the combination of static POR techniques with dynamic POR techniques. The basic idea is that the runtime overhead is reduced by computing the dependency information statically; to later use it during runtime.

As an example, let's consider the first scheduling decision (SD_0) in the SDT in Fig. 8 for any of the specifications represented by Fig. 2 and 3. Depending on SD_0, the scheduling executed can start either by $\{s_{11}, s_{21}, \ldots\}$ or by $\{s_{21}, s_{11}, \ldots\}$, each one representing two different classes of schedulings, $\{S_0, S_1\}$ and $\{S_2, S_3\}$ respectively. A POR analysis focused on the impact on functionality, will establish that those scheduling classes actually account for the following two possible starting sequences in functional terms, either $\{f_{11}, f_{21}, \ldots\}$ or $\{f_{21}, f_{11}, \ldots\}$. A POR technique will establish that f_{11} and f_{21} have impact on some intermediate and shared variables, 'a' and 'b', which reflect the state of the concurrent system and which imply dependencies between P_1 and P_2, thus requiring a specific analysis. Specifically, the POR technique will establish that those two possible initializations of the schedulings lead to the same state (in the next delta, Δ_1), described by $a'=f_{11}(a)$ and $b'=f_{11}(b)$. In other words, since there are no dependencies, any starting sequence leads to the same intermediate state, and schedulings starting with $SD_0=0$, that is, starting by $\{s_{11}, s_{21}, \ldots\}$, and schedulings starting with $SD_0=1$, that is, starting by $\{s_{21}, s_{11}, \ldots\}$ will be equivalent if they keep the same sequence of decisions in the rest of the sequence of scheduling decisions (SD_0). Therefore only one of the alternatives in SD_0 has to be explored. This idea can be iteratively applied generally leading to a drastic reduction in the number of paths which have to be explored, thus fulfilling M<<size(s). Such a drastic reduction can be observed in our simple example if we continue with it. Let's take, for instance, $SD_0=0$ in the example, and let's continue the application of a dynamic POR. At this stage, in the worst case, we will need to execute S_0 and S_1, thus M=2 simulations for a complete coverage of functional equivalent schedulings. Furthermore, DPOR is again applied for the second delta, Δ_1. Considering y and z as state variables directly forwarded to the outputs, there is no read after write, write after read or write after write dependency among them. Therefore, it can be concluded that the decision on SD_1 will be irrelevant in reaching the same (y, z) state after the Δ_1 delta. Therefore, M=1,

and $\eta_{ME} = 4$ in this case, since any of the four schedulings exposed by a single simulation will be representative of a single class of schedulings, equivalent in functional terms.

The method described in (Helmstetter, 2006) is complete, but not minimal, since it is feasible to think about specifications where M non-equivalent schedulings lead to different states, but where those different states are not translated into different outputs. This means that M would still admit a further reduction. This reduction would require an additional analysis of the actual relationship between state variables and the outputs. As an example, let's consider that in our examples in Fig. 2, z was not considered as a system output, but as informative or debugging data, resulting from post-processing, through f_{22}, an internal state variable), and that the only output is y. Thus, it would demonstrate the irrelevance of the SD_1 scheduling decision, which would save the last DPOR analysis in Δ_1.

The approach of (Helmstetter, 2006) is also fork-based. Whenever a scheduling decision finds non-equivalent or potentially non-equivalent paths, the simulation is spawned in order to enable a concurrent check. Thus, several non-equivalent groups of schedulings can be explored by launching a single simulation. This makes η_{ME} even bigger, and $\eta_{ME} = N_S \geq 1$, up to the point where a single simulation could cover all the scheduling classes. However, this optimization should be carefully considered. In order to give an actual speed up to the verification, it is necessary that the simulation engine can take advantage of a multi-core host machine. In (Helmstetter, 2006), the first advances for a parallel SystemC simulator are given. If the simulation is sequential, then a fork-based approach can easily be counter-productive in terms of time cost even if SystemC simulators with actual parallel simulation capabilities are available.

In general, the main limitation of POR-based approaches is their need for extracting information from the specification. The limitations of the front-end tools used for extracting the information used for static dependency analysis, and the need to make the analysis feasible limit the supported input code. Specifically, the approach of (Helmstetter, 2006) is restricted to the SystemC subset admitted by the open-source and freely available Pinapa front-end (Moy, 2005). Satya is based in the commercial EDG C++ front-end, which provides wider support than Pinapa. However, it still presents limitations for supporting features such as dynamic casting and process creation. The work of (Sen, 2008) claims its independency from any external parser, while being able to detect potential errors in an observed execution, even if the error does not take place in the actual simulation. However, its goal is temporal assertion-based verification, rather than improving test coverage.

3.4 Merging scheduling techniques

In (Herrera, 09), the local application and cooperation of different scheduling techniques (PR, DEC and POR) is proposed. Two types of localities are distinguished:

- Spatial Locality: in order to improve scheduling coverage for a specific group of processes of the system specification.
- Temporal Locality: in order to improve scheduling coverage in a specific interval of the simulation time.

For instance, in some parts of the specification where SystemC is used in a flexible manner, e.g., a high-level concurrent model of an intellectual property (IP) block, DEC scheduling could be applied. Then POR could be applied to other parts, e.g., an in-house TLM platform, where the IP block is connected, and whose code can be bound to the specification rules stated by the POR technique. Table 1 summarizes the main characteristics of the different scheduling techniques reviewed.

Scheduling Technique	C_S	η_{ME}	Reproducibility	Linear growth of Cs with N_E	Specification Independent Detection of $C_S=1$	Specification Analysis Required
FIFO (OSCI simulator)	$\dfrac{1}{size(S)}$	$\dfrac{1}{N_E}$	yes	no	no	no
Random	$\geq \dfrac{1}{size(S)}$ ≤ 1	$\geq \dfrac{1}{N_E}$ ≤ 1	no	no	no	no
Pseudo Random	$\geq \dfrac{1}{size(S)}$ ≤ 1	$\geq \dfrac{1}{N_E}$ ≤ 1	yes	no	no	no
DEC	$\dfrac{N_E}{size(S)}$	1	yes	yes	yes	no
POR	$\dfrac{N_E \cdot \overline{L}}{size(S)}$	≥ 1 $\leq L$	yes	yes	yes	yes

Table 1. Comparison of scheduling techniques for simulation-based verification.

4. Methodologies for early correct specification

As shown in the previous sections, the success of a simulation-based verification methodology greatly depends on the ability to explore the effects of all the feasible execution alternatives, or at least, the "equivalent ones". The problem is already challenging for sequential specifications, especially for control-oriented algorithms, and becomes practically intractable when concurrency appears in the specification, since the number of execution paths grows exponentially.

As has been shown, a way to tackle the explosion problem, for finding both a more reduced and efficient set of input vector generation, and an efficient set of schedulings, is the usage of information from the specification. Automated test generation techniques direct vector generation by detecting control statements and looking for vectors which exercise their different branches. Similarly, partial order reduction techniques need to analyze, either statically or dynamically, which variables or events produce dependencies among processes in order to extract the representative schedulings which need to be simulated.

This means that some conditions for making the specification wrong and hard to verify are already known. Thus, a different perspective is possible. Why not build specification methodologies which oblige, or at least help, the user to avoid such source problems, instead of letting them appear in the specification, with the consequential requirement of a costly verification.

An alternative consists in building specification methodologies which selectively adopt certain specification rules. Such rules will enable enough expressivity to solve the specification problem, but at the same time they rely on formal conditions for building correct specifications. By assuming the fulfilment of such specification rules, the formal support ensures the fulfilment of the properties pursued, or at least enables the application of analysis techniques for assessing such fulfilment. This idea is generally applicable. For instance, a methodology could forbid the usage of control statements. Then the specification would have just one data path, and the generation of test input vectors would be drastically simplified. However, this type of coding constraint would be very restrictive in many application domains, where user needs control sentences. Each specification methodology has its expressivity requirements, which puts bounds on the specification rules.

Embedded system specification requires expressing concurrency and abstraction. It might easily lead to the SystemC user to run into the plethora of issues associated to concurrency (non-determinism, deadlock, starvation, etc), as was illustrated in section 3. However, if certain smart rules are imposed on how concurrency is expressed in SystemC, it can highly facilitate to build early correct concurrent specifications. This principle has inspired several works, such as SystemC-H (Patel, 2004), SysteMoC (Haubelt, 2007), HetSC (Herrera, 2007), and HetMoC (Zhu, 2010), which have proposed SystemC specification methodologies to ensure, or facilitate the verification, of certain properties. These methodologies state a set of SystemC facilities (and provide additional ones when they are not provided by the standard core of SystemC) and state a set of specification rules. Methodologies such as HetSC, SystemC-H and SysteMoC rely on well-known formalisms, related to specific Models of Computation (MoC), such as Khan Process Networks (KPN) (Kahn, 1974), Synchronous Data Flows (SDF) (Lee, 1987), Concurrent Sequential Processes (CSP), Synchronous Reactive (SR) systems, and Dynamic Data Flows (DDF). HetMoC, relies on the ForSyDe formalism (Jantsch, 2004), which targets the unification of several MoCs. Finally, a standard extension of the SystemC language, such as SystemC-AMS, adopts a variation of the SDF MoC, called T-SDF, which annotates a time advance after each cluster execution.

Two important factors which characterize these types of specification methodologies are the properties targeted and the way these are achieved, that is, which specification facilities, specification rules, and assumptions configure the methodology. Two typical properties pursued are functional determinism and deadlock protection. A relatively flexible way to ensure functional determinisms is to build the specification methodology according to the KPN formalism. The adoption of a more constrained specification style, through a specification methodology which fulfils the SDF formalism, enables the application of an analysis for ensuring deadlock protection, as well as functional determinism. This is illustrated through the Fig. 10 example.

Fig. 10a shows the structure of a HetSC specification for solving the Fig.1 specification problem. HetSC states the rules to be followed in the SystemC coding for building the concurrent solution as a Khan Process Network. There are rules regarding the facilities to use (SC_THREADS for P1 and P2, and blocking fifo channels with infinite buffering capability, that is, channels of uc_inf_fifo type, provided by the HetSC library). There are rules regarding how to write the processes, e.g., only one channel instance can be accessed (either for reading or for writing) at a time. Finally, there are rules regarding communication and computation, e.g., no more than one process can access a channel instance either as a reader or as a writer. More details on the rules can be found at the (HetSC website, 2012). All these SystemC coding rules are designed to fulfil the rules and assumptions stated in Kahn, 1974. Provided they are fulfilled, as happens in the Fig. 10a case, it can be said that the Fig.10a specification is functionally deterministic. Notice that read accesses to the uc_inf_fifo instances are blocking, thus they ensure the partial order stated by equations (4-7).

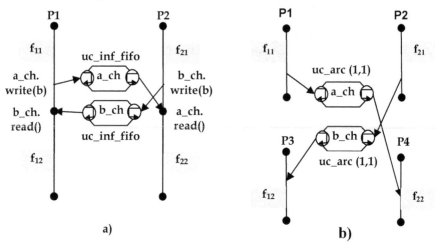

Fig. 10. Specification of Fig.1 solved as a) a Kahn process network and b) as a static dataflow.

Fig.10b shows a second possibility, where the specification is built fulfilling the SDF MoC rules, by using the HetSC methodology and facilities. To fulfil the SDF MoC, the specification style has to be more restrictive than in KPN in several ways. First of all, the KPN specification rules as in the Fig. 10a case, still apply. For instance, only one reader and one writer process can access each channel instance. Furthermore, there are additional rules. For example, each of the specification processes has to be coded without any blocking statement in the middle. Due to this, a single process has been used for each f_{ij} function, enabling a correspondence between a process firing and the execution of function f_{ij}. Moreover, the specific amount of data consumed and produced for each f_{ij} firing has to be known in advance. In HetSC, that information is associated to uc_arc channel instances. The advantage provided by the Fig. 10b solution is that not only does it ensure functional determinism by construction, but it also enables a static analysis based on the extraction of the SDF graph. The Fig. 10b direct SDFG easily leads to the conclusion that the specification is protected against deadlock, and moreover, that a static scheduling is also possible.

5. Conclusions

There is a trade off (shown in qualitative terms in Fig. 11) between the flexibility in the usage of a language and the verification cost for ensuring certain degree of correctness in a specification. In practice, simulation-based methodologies are in the best position for the verification of complex specifications, since formal and semiformal verification techniques easily explode. However, concurrency has become a necessary feature in specification methodologies. Therefore, the capability of simulation based techniques for verification of complex embedded systems has to be reconsidered. A reasonable alternative seems to be the development of cooperative techniques which combine simulation-based methods and specification methodologies which constrain the usage of the language under some formal rules, oriented to fulfilling the desired properties. Specifically, while SystemC is a language with a rich expressivity, it is still necessary to build abstract specification methodologies using SystemC as host language, by constraining the specification facilities and the way they can be used. This way, certain key properties can be guaranteed by construction, and the fulfilment of others can be analyzed. The set of properties to be guaranteed depend on the application domain. Moreover, a formally supported specification methodology can help to validate additional properties through simulation-based verification techniques with a drastic improvement in the detection capabilities and time spent on simulation.

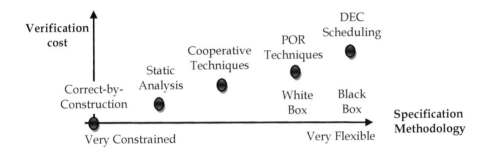

Fig. 11. Trade off between flexibility and verification time after considering concurrency.

6. Acknowledgement

This work has been partially funded by the EU FP7-247999 COMPLEX project and by the Spanish government through the MICINN TEC2008-04107 project.

7. References

Burton, M. et al. (2007). *ESL Design and Verification,* Morgan Kaufman, ISBN 0-12-373551-3

Bergeron, J. (2003) *Writing Testbenches. Functional Verification of HDL Models.* Springer, ISBN 1-40-207401-8.

Cadar, C., Ganesh, V., Pawlowski, P.M., Dill, D.L., & Engler, D.R., (2008). EXE: Automatically Generating Inputs of Death. *ACM Transactions on Information and System Security (TISSEC).* V12, Issue 2, Article 10. December, 2008.

Chiang, S. Y. (2011). Keynote Speech. *Proceedings of ARM Techcom Conference.* October, 25th, 2011. Santa Clara, USA.

EDG website, (2012). EDG Website. http://www.edg.com/. Checked in November, 2011.

Fallah, F., Devadas, S. & Keutzer, K. (1998) Functional vector generation for HDL models using linear programming and 3-satisfiability. *Proceedings of the 35th annual Design Automation Conference (DAC '98)*. ACM, New York, NY, USA, pp. 528-533.

Flanagan, C. & Godefroid, P. (2005) Dynamic Partial Order Reduction for Model Checking Software. *Proceedings of ACM SIGPLAN-SIGACT Symposium on Principles of Programming Languages*. 2005.

Godefroid, P. (1995) Partial-Order Methods for the Verification of Concurrent Systems; An approach to the State-Explosion Problem. *PhD thesis*. University of Liege. 1995.

Godefroid, P., Klarlund, N. & Sen, K. (2005) DART: Directed Automated Random Testing. *Proceedings of the 2005 ACM SIGPLAN conference on Programming language design and implementation (PLDI '05)*. ACM, New York, NY, USA, pp. 213-223.

Grant, M. (2006). Overview of the MPSoC Design Challenge. *Proceedings of Design Automation Conference 2006, DAC'06.*, ISBN 1-59593-381-6 San Francisco, USA.

Gupta, A., Casavant, A.E., Ashar, P., Mukaiyama, A., Wakabayashi, K. & Liu, X. G. (2002). Property-Specific Testbench Generation for Guided Simulation. *Proceedings of the 2002 Asia and South Pacific Design Automation Conference (ASP-DAC '02)*. IEEE Computer Society, Washington, DC, USA. 2002.

Halfhill, T. (2012). Looking beyond Graphics. 2012. Whipe paper, Available in http://www.nvidia.com/object/fermi_architecture.html.

Haubelt, C., Falk , J., Keinert, J. , Schlichter, T., Streubühr, M. , Deyhle, A. , Hadert, A., Teich, J. (2007). A SystemC-Based Design Methodology for Digital Signal Processing Systems. EURASIP Journal on Embedded Systems. V. 2007, Article ID 47580, 22 pages. January, 2007.

Helmstetter, C. & Maraninchi, F., Maillet-Contoz & Moy, M. (2006) Automatic Generation of Schedulings for Improving the Test Coverage of Systems-on-a-Chip. *Proceedings of Formal Methods in Computer Aided Design, FMCAD'06*. November, 2006.

Helmstetter, C. (2007). Validation de Modèles de Systèmes sur Puce en présence d'ordonnancements Indétermnistes et de Temps Imprecis. *PhD thesis*. March. 2007.

Herrera, F., & Villar, E. (2006). Extension of the SystemC kernel for Simulation Coverage Improvement of System-Level Concurrent Specifications. *Proceedings of the Forum on Specification and Design Languages, FDL'06*. Darmstad. Germany. Sept., 2006.

Herrera, F. & Villar, E. (2007). A Framework for Heterogeneous Specification and Design of Electronic Embedded Systems in SystemC. ACM Transactions on Design Automation of Electronic Systems, Special Issue on Demonstrable Software Systems and Hardware Platforms, V.12, Issue 3, N.22. August, 2007.

Herrera, F., & Villar, E. (2009). Local Application of Simulation Directed for Exhaustive Coverage of Schedulings of SystemC Specifications. *Proc. of the Forum on Specification and Design Languages, FDL'09*. Sophia Antipolis. France. September, 2009. ISBN 1636-9874.

HetSC website, (2012). HetSC website. www.teisa.unican.es/HetSC. 2012.

IEEE, (2005). SystemC Language Reference Manual. Available in http://standards.ieee.org/getieee/1666/download/1666-2005.pdf.

Incisive, (2009). Incisive Enterprise Simulator Datasheet. Available in http://www.cadence.com/rl/Resources/datasheets/incisive_enterprise_specman.pdf. March, 2009

Jantsch, A. (2004). *Modelling Embedded Systems and SoCs. Concurrency and Time in Models of Computation*. Elsevier Science (USA), 2004. ISBN 1-55860-925-3.

Kahn, G. 1974. The Semantics of a simple Language for Parallel Programming. *Proceedings of the IFIP Conference 1974*, North-Holland, 1974.

Kish, L. B. (2002). End of Moore's Law: thermal (noise) death of integration in micro and nano electronics. Physics Letters A 305. pp. 144-149. Elselvier.

Kuhn,T., Oppold, T., Winterholer, M., Rosenstiel, W., Edwards, M. and Kashai, Y. (2001). A Framework for Object Oriented Hardware Specification Verification, and Synthesis. *Proceedings of the Design Automation Conference*, 2001.DAC'01. 2001.

Kundu, S., Ganai, M., Gupta, R. (2008) Partial Order Reduction for Scalable Testing of SystemC TLM Designs. Proceedings of the Design Automation Conference, DAC'08. Anaheim, CA, USA. June, 2008.

Kuo,Y.M., Lin, C.H., Wang, C.Y., Chang, S.H. & Ho, P.H. (2007). Intelligent Random Vector Generator Based on Probability Analysis of Circuit Structure. *Proceedings of the 8th International Symposium on Quality Electronic Design (ISQED '07)*. IEEE Computer Society, Washington, DC, USA, pp. 344-349.

Lee, E. A. & Messerschmitt, D.G. (1987). Static Scheduling of Synchronous Data Flow Programs for Digital Signal Processing. *IEEE Transactions on Computers. V. C-36. N.1.* pp. 24-35, January, 1987.

Lee, E.A. (2006). What's the Problem with Threads. *IEEE Computer*, Vol. 36, No. 5, pp. 33-42, May, 2006.

Moy, M., Maraninchi, F., Maillet-Contoz, L. (2005) PINAPA: An Extraction Tool for SystemC Descriptions of Systems on a Chip. *Proceedings of EMSOFT*, September, 2005.

MPI: A Message-Passing Interface Standard. Version 2.2. September, 2009. Available from http://www.mcs.anl.gov/research/projects/mpi/

OSCI Verification WG (2003). SystemC Verification Standard. Version 1.0e. May 16, 2003. Available at www.systemc.org.

OpenMP. (2008). Application Program Interface. 4 Version 3.0 May 2008. Available from http://openmp.org/wp/.

Patel, H.D. & Shukla, S.K. (2004). SystemC kernel extensions for Heterogeneous System Modelling: A Framework for Multi-MoC Modelling and Simulation. Kluwer. 2004.

Sen, K., Marinov, D. & Agha, G.. (2005). CUTE: a Concolic Unit Testing Engine for C. *Proceedings of the 10th European Software Engineering Conference (ESEC/FSE-13)*. ACM, New York, NY, USA, 263-272.

Sen, A., Ogale, V., Abadir, M. S. (2008). Predictive Runtime Verification of multi-processor SoCs in SystemC. Proceedings of Design Automation Conference, DAC'08. Anaheim, CA, USA June, 2008.

UCSCKext, (2011). Website for SystemC kernel extensions provided by University of Cantabria. http://www.teisa.unican.es/HetSC/kernel_ext.html. November, 2011.

Ugarte, I. & Sanchez, P. (2011) Automatic vector generation guided by a functional metric. *Proceedings of SPIE*. 8067, 80670U (2011)

Yuan, J., Aziz, A., Pixley, C., Albin, K. (2004). Simplifying Boolean constraint solving for random simulation-vector generation. *IEEE Transactions on Computer-Aided Design of Integrated Circuits and Systems*. V. 23, N. 3, pp. 412-20, March, 2004.

Zhu, J., Sander, I., & Jantsch, A. (2010). HetMoC: heterogeneous modelling in SystemC. Proceedings of Forum for Design Languages (FDL '10). Southampton, UK, 2010.

13

The Innovative Design of Low Cost Embedded Controller for Complex Control Systems

Meng Shao[1], Zhe Peng[2] and Longhua Ma[2]
[1]Computer Centre, Hangzhou First People's Hospital, Hangzhou,
[2]School of Aeronautics and Astronautics, Zhejiang University, Hangzhou,
China

1. Introduction

With the availability of ever more powerful and cheaper products, the number of embedded devices deployed in the real world has been far greater than that of the various general-purpose computers such as desktop PCs. An embedded system is an application-specific computer system that is physically encapsulated by the device it controls. It is generally a part of a larger system and is hidden from end users. There are a few different architectures for embedded processors, such as ARM, PowerPC, x86, MIPS, etc. Some embedded systems have no operating system, while many more run real-time operating systems and complex multithreaded programs. Nowadays embedded systems are used in numerous application areas, for example, aerospace, instrument, industrial control, transportation, military, consumer electronics, and sensor networks. In particular, embedded controllers that implement control functions of various physical processes have become unprecedentedly popular in computer-controlled systems (Wittenmark et al., 2002 ; Xia, F. & Sun, Y.X., 2008). The use of embedded processors has the potential of reducing the size and cost, increasing the reliability, and improving the performance of control systems.

The majority of embedded control systems in use today are implemented on microcontrollers or programmable logic controllers (PLC). Although microcontrollers and programmable logic controllers provide most of the essential features to implement basic control systems, the programming languages for embedded control software have not evolved as in other software technologies (Albertos, P. 2005). A large number of embedded control systems are programmed using special programming languages such as sequential function charts (SFC), function block languages, or ladder diagram languages, which generally provide poor programming structures. On the other hand, the complexity of control software is growing rapidly due to expanding requirements on the system functionalities. As this trend continues, the old way of developing embedded control software is becoming less and less efficient.

There are quite a lot of efforts in both industry and academia to address the above-mentioned problem. One example is the ARTIST2 network of excellence on embedded systems design (http://www.artist-embedded.org). Another example is the CEMACS project (http://www.hamilton.ie/cemacs/) that aims to devise a systematic, modular, model-based approach for designing complex automotive control systems. From a technical point of view, a classical solution for developing complex embedded control software is to use the Matlab/Simulink platform that has been commercially available for many years. For instance, Bucher and Balemi (Bucher, R.; Balemi, S., 2006) developed a rapid controller prototyping system based on Matlab, Simulink and the Real-Time Workshop toolbox; Chindris and Muresan (Chindris, G.; Muresan, M., 2006) presented a method for using Simulink along with code generation software to build control applications on programmable system-on-chip devices. However, these solutions are often complicated and expensive. Automatic generation of executable codes directly from Matlab/Simulink models may not always be supported. It is also possible that the generated codes do not perform satisfactorily on embedded platforms, even if the corresponding Matlab/Simulink models are able to achieve very good performance in simulations on PC. Consequently, the developers often have to spend significant time dealing with such situations. As computer hardware is becoming cheaper and cheaper, embedded software dominates the development cost in most cases. In this context, more affordable solutions that use low-cost, even free, software tools rather than expensive proprietary counterparts are preferable.

The main contributions of this book are multifold. First, a design methodology that features the integration of controller design and its implementation is introduced for embedded control systems. Secondly, a low-cost, reusable, reconfigurable platform is developed for designing and implementing embedded control systems based on Scilab and Linux, which are freely available along with source code. Finally, a case study is conducted to test the performance of the developed platform, with preliminary results presented.

The platform is built on the Cirrus Logic EP9315 (ARM9) development board running a Linux operating system. Since Scilab was originally designed for general-purpose computers such as PCs, we port Scilab to the embedded ARM-Linux platform. To enable data acquisition from sensors and control of physical processes, the drivers for interfacing Scilab with several communication protocols including serial, Ethernet, and Modbus are implemented, respectively. The developed platform has the following main features:

- It enables developers to perform all phases of the development cycle of control systems within a unified environment, thus facilitating rapid development of embedded control software. This has the potential of improving the performance of the resulting system.
- It makes possible to implement complex control strategies on embedded platforms, for example, robust control, model predictive control, optimal control, and online system optimization. With this capability, the embedded platform can be used to control complex physical processes.
- It significantly reduces system development cost thanks to the use of free and open source software packages. Both Scilab and Linux can be freely downloaded from the Internet, thus minimizing the cost of software.

While Scilab has attracted significant attention around the world, limited work has been conducted in applying it to the development/implementation of practically applicable control applications. Bucher et al. presented a rapid control prototyping environment based on Scilab/Scicos, where the executable code is automatically generated for Linux RTAI(Bucher, R.; Balemi, S, 2005). The generated code runs as a hard real-time user space application on a standard PC. The changes in the Scilab/Scicos environment needed to interface the generated code to the RTAI Linux OS are described. Hladowski et al. (Hladowski et al., 2006) developed a Scilab-compatible software package for the analysis and control of repetitive processes. The main features of the implemented toolkit include visualization of the process dynamics, system stability analysis, control law design, and a user-friendly interface. Considering a control law designed with Scicos and implemented on a distributed architecture with the SynDEx tool, Ben Gaid et al. proposed a design methodology for improving the software development cycle of embedded control systems(Ben Gaid et al., 2008). Mannori et al. presented a complete development chain, from the design tools to the automatic code generation of standalone embedded control and user interface program, for industrial control systems based on Scilab/Scicos (Mannori et al., 2008).

2. Embedded control systems design

In this paper, we develop an embedded controller for complex control applications. The key software used is the Scilab/Scicos package, a free and open source alternative to commercial packages for dynamical system modeling and simulation such as Matlab/Simulink. Since hardware devices are becoming cheaper by the day, software development cost has dominated the cost of most embedded systems. As a consequence, the use of the free and open source software minimizes the cost of the embedded controller. On the other hand, Scilab is a software package providing a powerful open computing environment for engineering and scientific applications. It features a variety of powerful primitives for numerical computations. There exist a number of mature Scilab toolboxes, such as Scicos, fuzzy logic control, genetic algorithm, artificial neural network, model predictive control, etc. All these features of Scilab make it possible, and quite easy, to implement complex control algorithms on the embedded platform we develop in this work.

To satisfy the ever-increasing requirement of complex control systems with respect to computational capability, we use the Cirrus Logic EP9315 ARM chip in this project. The platform runs on an ARM-Linux system. Since Scilab and Scicos were originally developed for general-purpose computers such as desktop PCs, we port Scilab/Scicos to the ARM-Linux platform (Longhua Ma, et al., 2008 ; Feng Xia, et al., 2008). Several interfaces and toolboxes are implemented to facilitate embedded control.

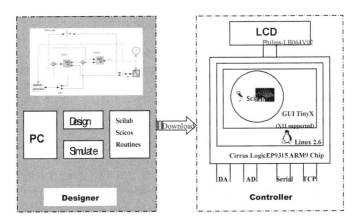

Fig. 1. Design of an embedded controller.

With the developed platform, the design and implementation of a complex control system will become relatively simple, as shown in Figure 1. The main procedures involved in this process are as follows: model, design, and simulate the control system with Scilab/Scicos on a host PC, then download the well designed control algorithm(s) to the target embedded system. The Scilab code on the embedded platform is completely compatible with that on the PC. Consequently, the development time can be significantly reduced.

2.1 Architecture

As control systems increase in complexity and functionality, it becomes impossible in many cases to use analog controllers. At present almost all controllers are digitally implemented on computers. The introduction of computers in the control loop has many advantages. For instance, it makes possible to execute advanced algorithms with complicated computations, and to build user-friendly GUI. The general structure of an embedded control system with one single control loop is shown in Figure 2. The main components consist of the physical process being controlled, a sensor that contains an A/D (Analog-to-Digital) converter, an embedded computer/controller, an actuator that contains a D/A (Digital-to-Analog) converter, and, in some cases, a network.

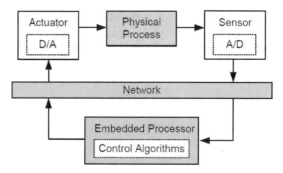

Fig. 2. General structure of embedded control systems.

The most basic operations within the control loop are sensing, control, and actuation. The controlled system is usually a continuous-time physical process, e.g. DC motor, inverted pendulum, etc. The inputs and outputs of the process are continuous-time signals. The A/D converter transforms the outputs of the process into digital signals at sampling instants. It can be either a separated unit, or embedded into the sensor. The controller takes charge of executing software programs that process the sequence of sampled data according to specific control algorithms and then produce the sequence of control commands. To make these digital signals applicable to the physical process, the D/A converter transforms them into continuous-time signals with the help of a hold circuit that determines the input to the process until a new control command is available from the controller. The most common method is the zero-order-hold that holds the input constant over the sampling period. In a networked environment, the sequences of sampled data and the control commands need to be transmitted from the sensor to the controller and from the controller to the actuator, respectively, over the communication network. The network could either be wire line (e.g. field bus, Ethernet, and Internet) or be wireless (e.g. WLAN, ZigBee, and Bluetooth). In a multitasking/multi-loop environment, as illustrated in Figure 3, different tasks will have to compete for the use of the same embedded processor on which they run concurrently.

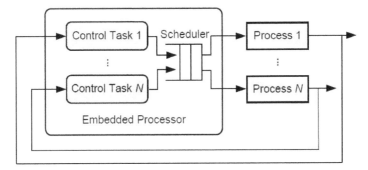

Fig. 3. A multitasking embedded control system.

2.2 Design methodology

There is no doubt that embedded control systems constitute an important subclass of real-time systems in which the value of the task depends not only on the correctness of the computation but also on the time at which the results are available. From a real-time systems point of view, the temporal behavior of a system highly relies on the availability of resources. Therefore, it is compulsory for the system to gain sufficient resources within a certain time interval in order that the execution of individual tasks can be completed in time. Unfortunately, most embedded platforms are suffering from resource limitations, which is in contrast to general-purpose computer systems. There are many reasons behind. For instance, embedded devices are often subject to various limitations on physical factors such as size and weight due to the stringent application requirements. In this context, care must be taken when developing embedded control systems such that the timing requirements of the target application can be satisfied.

Traditionally, the development cycle of a control system consists of two main steps: controller design and its implementation. These two steps are often separated, as shown in Figure 4, where the so-called V-model is given. While the controller design is usually done by control engineers, the implementation is the responsibility of system (software) engineers. In the first step, the control engineers model the physical processes using mathematical equations. According to the requirements specification, the control engineers then design the control algorithms. The parameters of the control algorithms are often determined through extensive simulations to achieve the best possible performance. A widely used tool in this step is Matlab/Simulink that supports modeling, synthesis, and simulation of control systems. In this environment the physical processes are usually modeled in continuous time while the control algorithms are to facilitate digital implementation. In the second step, the software engineers produce the programs executing the control algorithms with the parameters designed in the first step. There are a number of mature programming languages available for the implementation. The system will be tested, possibly many times before the satisfactory performance is achieved.

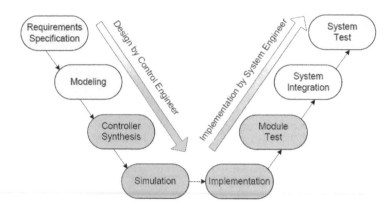

Fig. 4. Traditional development process of control software.

The traditional development process features separation of control and scheduling. The control engineers pay no attention to how the designed control algorithms will be implemented, while the software engineers have no idea about the requirements of the control applications with respect to temporal attributes. In resource-constrained embedded environments, the traditional design methodology cannot guarantee that the desired temporal behavior is achieved, which may lead to much worse-than-possible control performance. Furthermore, the development cycle of a system that can deliver good performance may potentially take a long time, making it difficult to support rapid development that is increasingly important for commercial embedded products.

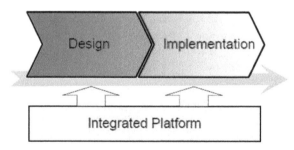

Fig. 5. Integrated design and implementation on a unified platform.

In this paper we adopt a design methodology that bridges the gaps between the traditionally separated two steps of the development process. As shown in Figure 5, we develop an integrated platform that provides support for all phases of the whole development cycle of embedded control systems. With this platform, the modeling, synthesis, simulation, implementation, and test of control software can be performed in a unified environment. Thanks to the seamless integration of the controller design and its implementation, this design methodology enables rapid development of high quality embedded controllers that can be used in real-world systems.

3. Hardware platform

3.1 SoC system

SoC is believed to be more cost effective than a system in package, particularly in large volumes. One of the most typical application areas of SoC is embedded systems. In this work, the processor of SoC is chosen to be the Cirrus Logic EP9315 ARM9 chip, which contains a Maverick Crunch coprocessor. A snapshot of the hardware board is shown in Figure 6.

Fig. 6. Hardware platform.

Using this SoC board, it is easy to communicate with other components of the system, for example, to sample data from sensors and to send control commands to actuators, thanks to its support for A/D, D/A, Serial and Ethernet interfaces, etc. To keep the system user-friendly, the embedded controller also includes a LCD with touch screen.

3.2 Maverick crunch coprocessor

The Maverick Crunch coprocessor accelerates IEEE-754 floating point arithmetic and 32-bit fixed point arithmetic operations such as addition, subtraction, multiplication, etc. It provides an integer multiply-accumulate (MAC) that is considerably faster than the native MAC implementation in the ARM920T. The single-cycle integer multiply-accumulate instruction in the Maverick Crunch coprocessor allows the EP9315 to offer unique speed and performance while dealing with math-intensive computing and data processing functions in industrial electronics. The computational speed of the system becomes 10 to 100 times faster when the Maverick Crunch coprocessor is used.

In Table 1 we list the time needed to execute every test function 360,000 times, both with the Maverick Crunch coprocessor and without it. Compared with the case without the Maverick Crunch coprocessor, the computational speed of the system becomes 10 to 100 times faster when the Maverick Crunch coprocessor is used.

Functions	ADD	SUB	MUL	SIN	LOG	EXP
HPF (ms) With Maverick Crunch	1	1	25	950	950	902
SFP (ms) Without Maverick Crunch	187	190	310	7155	7468	6879
Ratio	1:187	1:190	1:12.8	1:7.6	1:7.8	1:7.6

Table 1. Comparison of computational capability of PC and ARM.

The reason of this coprocessor selection is due to its high computation performance compared to normal embedded coprocessor.

4. Software design

There are a number of considerations in implementing control algorithms on embedded platforms including the ARM9 board we use. One of the most important is that embedded platforms are usually limited in resource such as processor speed and memory. Therefore, control software must be designed in a resource-efficient fashion, in a sense that the limited resources are efficiently used.

The key software packages used in this paper includes Linux, TinyX, JWM, Scilab/Scicos, the Scilab SCADA (Supervisory Control and Data Acquisition) toolbox we develop, and other related Scilab toolboxes. The system software architecture is shown in Figure 7. In the following, we detail the software design of the embedded controller.

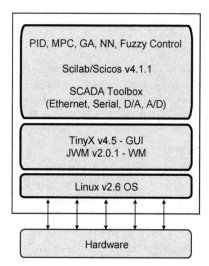

Fig. 7. Software architecture.

4.1 The Scilab/Scicos environment

Scilab is a free and open source scientific software package for numerical computations, which provides a powerful open computing environment for engineering and scientific applications.

It has been developed by researchers from INRIA and ENPC, France, since 1990 and distributed freely and in open source via the Internet since 1994. It is currently the responsibility of the Scilab Consortium, which was launched in 2003. Scilab is becoming increasingly popular in both educational/academic and industrial environments worldwide. Scilab provides hundreds of built-in powerful primitives in the form of mathematical functions. It supports all basic operations on matrices such as addition, multiplication, concatenation, extraction, and transpose, etc. It has an open programming environment in which the user can define new data types and operations on these data types. In particular, it supports a character string type that allows the online creation of functions. It is easy to interface Scilab with FORTRAN, C, C++, Java, Tcl/Tk, LabView, and Maple, for example, to add interactively FORTRAN or C programs. Scilab has sophisticated and transparent data structures including matrices, lists, polynomials, rational functions, linear systems, among others. It includes a high-level programming language, an interpreter, and a number of toolboxes for linear algebra, signal processing, classic and robust control, optimization, graphs and networks, etc. In addition, a large (and increasing) number of contributions can be downloaded from the Scilab website. The latest stable release of Scilab (version 4.1.2) can work on GNU/Linux, Windows 2000/XP/VISTA, HP-UX, and Mac OS.

Scilab includes a graphical system modeler and simulator toolbox called Scicos (http://www.scicos.org), which corresponds to Simulink in Matlab. Scicos is particularly useful in signal processing, systems control, and study of queuing, physical, and biological systems. It enables the user to model and simulate the dynamics of hybrid dynamical systems through creating block diagrams using a GUI-based editor and to compile models into executable codes. There are a large number of standard blocks available in the palettes. It is possible for the user to program new blocks in C, FORTRAN, or Scilab Language and constructs a library of reusable blocks that can be used in different systems. Scicos allows running simulations in real time and generating C code from Scicos model using a code generator. Scilab/Scicos is the open source alternative to commercial software packages for system modeling and simulation such as Matlab/Simulink. Figure 8 gives a screen shot of the Scilab/Scicos package.

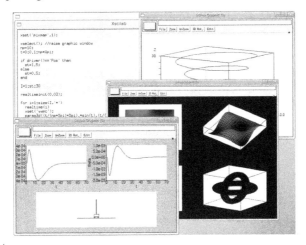

Fig. 8. Scilab environment.

4.2 Software packages

Underneath is the list of the software packages.

- Linux. The developed embedded controller is built on the Linux kernel (www.linux.org). The Linux kernel provides a level of flexibility and reliability simply impossible to achieve with any other operating system such as Windows, UNIX and Mac OS. Mainly for this reason, many embedded systems choose Linux OS.
- TinyX. TinyX is an X server written by Keith Packard. It was designed for low memory environments. On Linux/x86, a TinyX server with RENDER support but without support for scalable fonts compiles into less than 700 KB of text. TinyX tends to avoid large memory allocations at runtime, and tries to perform operations on-the-fly whenever possible. Unlike the usual XFree86 server, a TinyX server is completely self-contained: it does not require any configuration files, and will function even if no on disk fonts are available. All configurations are done at compile time and through command-line flags. It is easy to build user-specified GUI applications with TinyX. More information about TinyX can be found at http://www.xfree86.org.
- Scilab. Scilab/Scicos is utilized in this work to build the development environment for control software executing control algorithms. Developed initially by researchers from INRIA and ENPC, France, since 1990, Scilab is currently a free and open source scientific software package for numerical computations. Scilab has many toolboxes for modelling, designing, simulating, implementing, and evaluating hybrid control systems. It is now used in academic, educational, and industrial environments around the world. Scilab includes hundreds of mathematical functions with the possibility to add interactively programs from various languages, e.g., FORTRAN, C, C++, and Java. It has sophisticated data structures including, among others, lists, polynomials, rational functions, and linear systems, an interpreter, and a high level programming language, i.e., the Scilab language.
- Scicos. Although it is possible to model and design a hybrid dynamical system through writing scripts using the primitives of the Scilab language, this is often time consuming and the developers are prone to insert bugs during the manual coding. To simplify this task, Scilab includes a graphical dynamical system modeller and simulator toolbox called Scicos. Scicos can be used for applications in control, communication, signal processing, queuing systems, and study of physical and biological systems, etc. Using the Scicos graphical editor, it is possible to model and simulate hybrid dynamical systems by simply placing, configuring, and connect blocks. To achieve complete integration with Scilab, easy customization, and the maximum flexibility, most of the Scicos GUIs are written in the Scilab language.
- Scilab SCADA toolbox. To facilitate data acquisition and control operations, we develop the Scilab SCADA toolbox that interfaces Scilab with several kinds of I/O ports including serial port, Ethernet, and Modbus on the embedded Linux system. These communication interfaces make it possible to connect the embedded controller with other entities in the system, e.g., sensors, actuators, and the controlled physical process, using various communication mechanisms/networks. In a complex, possibly large-scale, control system in industry, a huge amount of data, e.g. system output samples and control commands, will be produced during run time. These data usually has to be stored in order to provide support for, e.g., historical data query and higher-layer system optimization. To meet this requirement, we develop the interface to MySQL

database in the Scilab SCADA toolbox. In addition, to provide a standard-compatible solution for the industrial control field, the Scilab SCADA toolbox conforms to the OPC (OLE for Process Control) standard. OPC is a widely accepted industrial communication standard that enables the exchange of data between multi-vendor devices and control applications. It helps provide solutions that are truly open, which in turn gives users more choices in their control applications. The interoperability between heterogeneous entities is assured through the support for non-proprietary specifications. A GUI of the OPC toolbox we develop is shown in Figure 9. With this OPC interface, it is possible to use Scilab as the core control software, and the communications with other (third-party) hardware devices and software tools will be effortless. These help to fully exploit the powerful functionalities of Scilab in complex control applications.

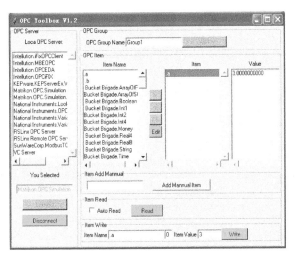

Fig. 9. OPC interface.

4.3 Building cross-compilation tool chain

A cross compiler is a compiler that is able to create executable code for a platform other than the one on which it is run. The basic role of a cross compiler is to separate the build environment from the target environment. This will be particularly useful for the development of the embedded controller based on Scilab/Scicos, which typically works in a general purpose computing environment other than the embedded platform. To port related software packages from PC to the ARM-Linux system, it is essential to build the cross compilation tool chain environment first. There exist several approaches to setting up a cross-compilation tool chain. In this work, we build the cross compiler for the ARM-Linux system using the build root toolkits. Build root is a set of Makefiles and patches that allow to easily generating both a cross-compilation tool chain and a root file for the target system. The cross compilation tool chain makes use of uClibc, a tiny C standard library. Several tools, such as bison, flex, and build-essential, are also exploited. It is worth mentioning that the g77 compiler option should be enabled during this process. Since most of the Scilab code is written in FORTRAN, the g77 compiler is necessary when compiling Scilab.

4.4 Porting Scilab/Scicos to ARM-Linux

Scilab/Scicos was originally designed for PC-based systems but not embedded ARM-Linux systems. Therefore, it is necessary to port Scilab/Scicos onto the embedded platform. Since the majority of core codes of Scilab are written in FORTRAN, we first build a cross-compiler for g77 in order to support cross-compilation of GUI, for example. The GUI system of Scilab/Scicos is based on X11, and therefore the X11 server TinyX is included. To reduce runtime overheads, we optimize/modify some programs in Scilab/Scicos. We have successfully ported Scilab/Scicos to the ARM-Linux system (see Figure 14). To achieve this goal, a number of files in Scilab and Linux have been modified. The main tasks involved in this process are as follows:

- Port Linux to the ARM platform;
- Port TinyX to ARM-Linux;
- Port JWM to ARM-Linux;
- Port Scilab/Scicos to ARM-Linux;
- Configure and optimize the embedded Scilab/Scicos.

The more details of how to porting Scilab/Scicos can be found at Book The embedded ARM-Linux computation develop based Scilab(Ma Longhua, Peng Zhe, 2008).

4.5 Software programming

Once all the necessary software packages are ported to ARM Linux, programming with Scilab in the embedded ARM Linux environment will be the same as on a PC. In this section we address some key issues closely related to embed software programming using Scilab in the ARM Linux platform. Scilab supports numerous data types, such as list, matrix, polynomial, scalar, string, and vector, among others. The syntax is designed to be natural and easy to use. The basic data type is a matrix. All basic operations on matrices, e.g., addition, multiplication, concatenation, and extraction, are provided by means of built-in functions. Scilab can also handle more complex objects such as polynomial matrices and transfer matrices. The syntax for manipulating these matrices is identical with that for constant matrices. This powerful capability of Scilab to handle matrices makes it particularly useful for systems control and signal processing. For instance, it is easy to obtain a natural symbolic representation of complicated mathematical objects such as transfer functions, dynamic systems, and graphs.

In addition, the Scicos toolbox allows users to model and simulate the dynamics of complex hybrid systems using a block-diagram graphical editor. Scilab is composed of three main parts: an interpreter, libraries of functions and libraries of FORTRAN and C routines. It provides an open programming environment in which users can easily create new functions and libraries of functions. In Scilab, functions are treated as data objects. As a consequence, they can be created and manipulated as other data objects. For instance, it is possible to define and/or treat a Scilab function as an input or output argument of other functions. In particular, Scilab supports a character string data type allowing for on-line creation of functions. Scilab has a high level programming language, i.e., the Scilab language. It can be easily interfaced with external FORTRAN or C programs by using dynamic links, or by building an interface program. Dynamic links can be realized using the link primitive. The linked routine can then be interactively called by the call primitive, which transmits Scilab variables to the linked program and transforms back the output parameters into Scilab variables. In the next section, we will use this technique in developing the interfaces to hardware devices. The interface program can be produced by intersci, which is a built-in Scilab program for building an interface file between Scilab and external functions. It describes the routine called and the associated Scilab function. In addition, the interface program can also be written by the user using mexfiles. With an appropriate interface, it is possible to add a permanent new primitive to Scilab through making a new executable code for Scilab. In addition to the Scilab language and the interface program, Scilab includes hundreds of powerful primitives in the form of mathematical functions. A large number of toolboxes for simulation, control, optimization, signal processing, graphics and networks, etc., are also available. These built-in functions and toolboxes allow users to program software with ease. Figure 10 gives an example of Scilab scripts in which a PID controller is implemented. In this program, GetSample() and UpdateState() are user-defined functions, which may be built by exploiting the I/O port drivers to be presented in the next section. The former obtains the sampled data from sensors, while the latter sends the new control command to actuators.

```
Digital PID Controller
//SP: Setpoint; y: System output; u: Control input
//Ts: Sampling period
//Kc, Td, Ti: Controller parameters
mode(-1)
Ts=2; Kc=1; Td=1; Ti=1; SP=1; u=0;
e(1)=0; e(2)=0; i=3;
Ki=Kc*Td/Ti;
Kd=Kc*Td/Ts;
realtimeinit(Ts);
realtime(0);
while 1
y=GetSample();
e(i)=SP-y;
du=Kc*(e(i)-e(i-1))+Ki*e(i)+Kd*(e(i)-2*e(i-1)+e(i-2));
u=du+u;
UpdateState(u);
e(i-2)=e(i-1);
e(i-1)=e(i);
i=i+1;
realtime(i-3);
end
```

Fig. 10. Example of Scilab scripts in which a PID controller.

5. Platform performance & interface

5.1 Rapid prototyping of control algorithms

The use of Scilab makes it easy to model, design, and implement complex control algorithms in the embedded controller developed in this work. Scilab has a variety of powerful primitives for programming control applications. Additionally, there are several different ways to realize a control algorithm in the Scilab/Scicos environment. For instance, it can be programmed as a Scilab .sci file using the Scilab language, or visualized as a Scicos block linked to a specific function written in FORTRAN or C. In addition, there are an increasing number of contributions that provide support for implementing advanced control strategies in Scilab using, e.g., fuzzy logic, genetic algorithm, neural networks, and online optimization. As a simple example for system modeling and simulation in Scicos, Figure 11 shows a control system for a water tank. The models of the controller and the water tank are highlighted by the dashed and solid rectangles, respectively. The step response of the control system is depicted in Figure 12.

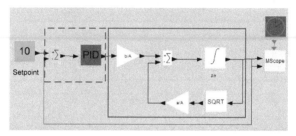

Fig. 11. An example control system in Scicos.

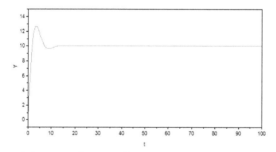

Fig. 12. Step response of the example control system.

5.2 Hardware drivers

Almost all embedded systems in practice need to interact with other related components (i.e. hardware devices) via I/O ports. In order for the developers to build practically useful embedded software with communication ability, it is necessary to provide hardware drivers in the embedded Scilab environment. To address this issue, we have developed the drivers for several types of communication interfaces including serial port, Ethernet, and Modbus. Illustrated below is how to program these drivers using Scilab in ARM Linux, while taking the serial port interface as an example. In the process of communication via a serial port, there are several basic operations, including open connection, set communication parameters, read data, write data, and close connection. Each basic operation is implemented as a separate C function. To facilitate dynamic links with Scilab, all arguments of the C functions are defined as pointers, as shown in the following example figure 13 where the function for reading and writing data from a serial port is implemented.

```
int serialread(int *handle, char *readbuff)
{
int nread;
readbuff[0]='\0';
while((nread=read(*handle,buff,512))>0)
{
printf('\nLen %d\n',nread);
buff[nread]='\0';
strcat(readbuff, buff);
}
}

int serialwrite(int *handle, char *writebuff)
{
int nwrite;
nwrite = write(*handle, writebuff,
strlen(writebuff));
printf('serialwrite%d\n %d\n %d\n', *handle,
nwrite, strlen(writebuff));
if (nwrite==strlen(writebuff))
printf('%d successfully written!\n',
nwrite);
else printf('write error!\n');
}
```

Fig. 13. Example of serial port reading and writing script.

As such, the hardware drivers are implemented as Scilab functions. These functions can be used by Scilab software programs in the same way as using other built-in Scilab functions. The developed hardware drivers, in the form of functions, serve as the gateway linking the different entities. Figure 14 gives a snapshot of the Scilab-based embedded ARM Linux system we develop using the programming techniques described in this Book(Peng, Z, 2008).

Fig. 14. The embedded control developed.

5.3 Computational capability analysis

Computational capability is a critical attribute of the embedded controller since the execution of the control program affects the temporal behavior of the control system, especially when complex control algorithms are employed. Therefore, we assess the computational capability of the developed embedded controller in comparison with that of a PC (Intel Pentium M CPU @1.60 GHz, with 760 MB of RAM) running Linux. The time for executing different algorithms is summarized in Table 2.

	Rand(800, 800)	DeJoy Algorithm
PC (s)	0.029	3.486
ARM (s)	1.176	92.3
Ratio	1:40	1:30

Table 2. Comparison of computational capability of PC and ARM.

6. Experimental test

In this section, we will test the performance of the developed embedded controller via experiments. For a research laboratory, however, it is very costly, if not impossible, to build the real controlled physical processes for experiments on complex control applications. For this reason, we construct a virtual control laboratory to facilitate the experiments on the embedded controller.

6.1 Virtual control platform

The schematic diagram of the structure of the experimental system is shown in Figure 15. The basic idea behind the virtual control laboratory is to use a PC running a dynamical system modeling software to simulate the physical process to be controlled. The control algorithms are implemented on the embedded controller, which exchanges data with the PC via a certain communication protocol, e.g., serial, Ethernet, or Modbus.

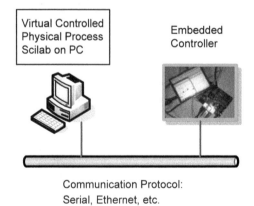

Fig. 15. Experimental system.

Both of the PC and the embedded controller use Scilab/Scicos as core software. Using this virtual control platform, experiments on various (virtual) physical processes are possible given that they can be modeled using Scilab/Scicos.

6.2 Case study

In the following, the control of a water tank is taken as an example for the experimental study. The water tank is modeled as shown in Figure 15 and implemented on the PC (Figure 16). The controller implemented on the embedded controller is shown in Figure 17. The control objective is to keep the water level (denoted y) in the tank to 10. The PC and the embedded controller are connected using Ethernet, and they communicate based on the UDP protocol. The PID algorithm is used for control

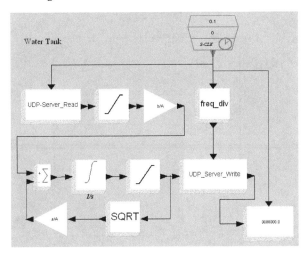

Fig. 16. Controlled process.

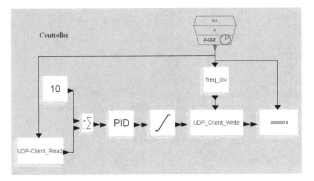

Fig. 17. Controller.

Figure 18 depicts the water level in the tank when different sampling periods are used, i.e., h = 0.1s, 0.2s and 0.5s, respectively. It can be seen that the control system achieve satisfactory performance. The water level is successfully controlled at the desired value in all cases.

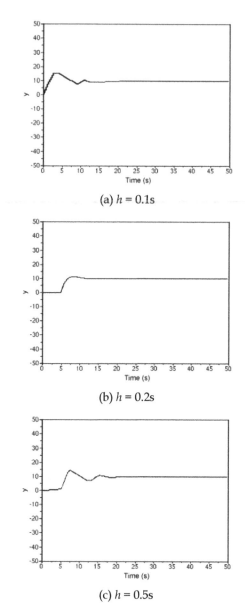

(a) $h = 0.1s$

(b) $h = 0.2s$

(c) $h = 0.5s$

Fig. 18. Control performance.

7. Conclusion

We have developed an embedded platform that can be used to design and implement embedded control systems in a rapid and cost-efficient fashion. This platform is built on free and open source software such as Scilab and Linux. Therefore, the system development cost can be minimized. Since the platform provides a unified environment in which the users are able to perform all phases of the development cycle of control systems, the development time can be reduced while the resulting performance may potentially be improved. In addition to industrial control, the platform can also be applied to many other areas such as optimization, image processing, instrument, and education. Our future work includes test and application of the developed platform in real-world systems where real sensors and actuators are deployed.

8. Acknowledgment

This work is supported in part by Natural Science Foundation of China under Grant No. 61070003, and Zhejiang Provincial Natural Science Foundation of China under Grant No. R1090052 and Grant No. Y108685.

9. References

Albertos, P.; Crespo, A.; Vallés, M. & Ripoll, I. Embedded control systems: some issues and solutions, Proc. of the 16th IFAC World Congress, pp. 257-262, Prague, 2005

Ben Gaid, M.; Kocik, R.; Sorel, Y.; Hamouche, R. A methodology for improving software design lifecycle in embedded control systems, Proc. of Design, Automation and Test in Europe (DATE), Munich, Germany, March 2008

Bucher, R.; Balemi, S. Rapid controller prototyping with Matlab/Simulink and Linux, Control Eng. Pract. , pp. 185-192, 2006

Bucher, R.; Balemi, S. Scilab/Scicos and Linux RTAI - a unified approach, Proc. of the IEEE Conf. on Control Applications, pp. 1121-1126, Toronto, Canada, August 2005

Chindris, G.; Muresan, M. Deploying Simulink Models into System-On-Chip Structures, Proc. of 29th Int. Spring Seminar on Electronics Technology, 2006

Feng Xia, Longhua Ma, Zhe Peng, Programming Scilab in ARM Linux, ACM SIGSOFT Software Engineering Notes, Volume 33 number 5, 2008

Hladowski, L.; Cichy, B.; Galkowski, K.; Sulikowski B.; Rogers, E. SCILAB compatible software for analysis and control of repetitive processes, Proc. of the IEEE Conf. on Computer Aided Control Systems Design, pp. 3024-3029, Munich, Germany, October 2006

Longhua Ma, Feng Xia, and Zhe Peng, Integrated Design and Implementation of Embedded Control Systems with Scilab, Sensors, vol.8, no.9, pp. 5501- 5515, 2008.

Mannori, S.; Nikoukhah, R.; Steer, S. Free and Open Source Software for Industrial Process Control Systems, 2008, Available from http://www.scicos.org/ScicosHIL/angers2006eng.pdf

Ma Longhua, Peng Zhe, Embedded ARM-Linux computation develop based Scilab, China Science publication, Beijing, China, 2008

Peng, Z. Research and Development of the Embedded Computing Platform Scilab-EMB Based on ARM-Linux, Master Thesis, Zhejiang University, Hangzhou, 2008.

Wittenmark, B.; Åström, K.J.; Årzén, K.-E. Computer control: An Overview, IFAC Professional Brief, 2002

Xia, F. & Sun, Y.X. Control and Scheduling Codesign: Flexible Resource Management in Real Time Control Systems, Springer, Heidelberg, Germany, 2008

14

SW Annotation Techniques and RTOS Modelling for Native Simulation of Heterogeneous Embedded Systems

Héctor Posadas, Álvaro Díaz and Eugenio Villar
Microelectronics Engineering Group of the University of Cantabria
Spain

1. Introduction

The growing complexity of electronic systems has resulted in the development of large multiprocessor architectures. Many advanced consumer products such as mobile phones, PDAs and media players are based on System on Chip (SoC) solutions. These solutions consist of a highly integrated chip and associated software. SoCs combine hardware IP cores (function specific cores and accelerators) with one or several programmable computing cores (CPUs, DSPs, ASIPs). On top of those HW resources large functionalities are supported.

These functionalities can present different characteristics that result in non homogeneous solutions. For example, different infrastructure to support both hard and soft real time application can be needed.. Additionally, large designs rely on SW reuse and thus on legacy codes developed for different platforms and operating systems. As a consequence, design flows require managing not only large functionalities but also heterogeneous architectures, with different computing cores and different operating systems.

The increasing complexity, heterogeneity and flexibility of the SoCs result in large design efforts, especially for multi-processor SoCs (MpSoC). The high interaction among all the SoC components results in large number of cross-effects to be considered during the development process. Additionally, the huge number of design possibilities of complex SoCs makes very difficult to find optimal solutions. As a consequence, most design decisions can no longer depend only on designers' experience. New solutions for early modeling and evaluating all the possible system configurations are required. These solutions require very high simulation speeds, in order to allow analyzing the different configurations in acceptable amounts of time. Nevertheless, sufficient accuracy must be ensured, which requires considering the performance and interactions of all the design components (e.g. processors, busses, memories, peripherals, etc.).

Static solutions have been proposed to estimate the performance of electronic designs. However, these solutions usually result too pessimistically and are difficult to scale to very complex designs. Instead, performance of complex designs can be more easily evaluated with simulation based approaches. Thus, virtual platforms have been proposed as one of the main ways to solve one of the resulting biggest challenges in these electronic designs:

perform software development and system performance optimization before the hardware board is available. As a result, engineers can start developing and testing the software from the beginning of the design process, at the same time they obtain system performance estimations of the resulting designs.

However, with the increase of system complexity, traditional virtual platform solutions require extremely large times to model these multiprocessor systems and evaluate the results. To overcome this limitation, new tools capable of modeling such complex systems in more efficient ways are required. First, it is required to reduce simulation times. Second, it is required to have tools capable of modeling and evaluating initial, partial designs with a low effort. For example, it is not acceptable to require complete operating system ports to initially evaluate different platform possibilities. Only when the platform is decided OS ports must be done, due to the large design effort required.

Virtual platform technologies based on simulations at different abstraction levels have been proposed, providing different tradeoffs between accuracy and speed. As early evaluation of complex designs requires very high simulation speeds, only the use of faster simulation techniques can be considered. Among them, simulations based on instruction set simulators (ISSs) and binary translation are the most important ones. However, none of them really provides the required trade-off for early evaluation.

ISSs are usually very accurate but too slow to execute the thousands of simulations required to evaluate complete SoC design spaces. ISS-based simulations usually can take hours, which means that the execution of thousand of simulation can require years, something not acceptable in any design process.

Simulations based on binary translation are commonly faster than ISSs. However, these solutions are more oriented to functional execution than to performance estimation. Effects as cache modeling are usually not considered when applying binary translation. Furthermore, this simulations also result too slow to explore large design spaces.

Additionally, in both cases, the simulation requires a completely developed SW and HW platform. Completely operational peripheral models, operating systems, libraries, compilers and device drivers are needed to enable system modeling. However, all these elements are usually not available early in the design process. Then, these simulation techniques are not only too slow but also difficult to perform. The dependence on such kind of platforms also results in low flexibility. Evaluating different allocations in heterogeneous platforms, different kind of processors and different operating systems is limited by the refining effort required to simulate all the options. Similarly, the evaluation of the effect of reusing legacy code in those infrastructures is not an easy task. As a consequence, faster and more flexible simulation techniques, capable of modeling the effect of all the components that impact on system performance, are required for initial system development and performance evaluation.

The solution described in this chapter is to increase the abstraction level, moving the SW simulation and evaluation from binary-based virtual platforms to native-based infrastructures. Using cross-compiled codes to simulate a platform in a host computer requires compulsory using some kind of processor models and a developed target SW platform. Thus, the simulation overhead provided by the processor model, and the development effort to develop the SW platform are items that cannot be avoided. On the contrary, simulations based on native or host-compiled executions avoid requiring a functional processor model, since no binary interpretation is done. Furthermore, a complete SW platform is not required, since the native SW platform can be partially used.

Nevertheless, in order to accurately modeling the system behavior and its performance modelling, a set of additional elements have been included in the native simulation infrastructures. Capabilities for modeling the delay of the SW execution in the target processor, the operation of the different level of caches, the target operating system and the other components in the HW platform, have been added. In the literature, some partial solutions have been proposed to support some of the elements of this list. However, some other features have not been solved in previous approaches, such as the support of different operating systems. Additionally as most of the proposed works are partial proofs of concept, there is a lack of complete integrated solutions.

The modeling of the application SW and its execution time in the target platform is a key element in native simulation, since it is the part of the infrastructure with more impact both in the simulation speed and in the modelling accuracy. Thus, in order to enable the designers to adjust the speed/accuracy ratio according to their needs, different solutions for SW annotation are presented and analyzed in the chapter. All solutions enable very easily exploring the effect of using different processors in the system. Only a generic compiler for the target processor is used. No specific OS ports, linker scripts or libraries are required.

With respect to the operating system, a basic OS modeling infrastructure has been developed, providing the user the possibility of simulating code based on Linux (POSIX), uC/os-II and Windows. The model has been developed starting from an OS modelling infrastructure providing a POSIX API. This infrastructure has been extended to support at the same time the other two APIs. This is an important step ahead to the state of the art, since very few proposed infrastructures support real operating systems, and to the best of our knowledge none of them considers these different APIs.

The resulting virtual platforms are about two-three times slower that functional execution when caches are not considered, and about one order of magnitude slower when using cache models. Processor modeling accuracy in terms of execution times is lower than 5% of error and the number of cache misses has an error of about 10%.

2. Related work

The modelling and performance evaluation of common MpSoC systems focuses in the modelling of the SW components. Since most of the functionality is located in SW this part is the one requiring more simulation times. Additionally the evaluation accuracy of the SW is also critical in the entire infrastructure accuracy. SW components are usually simulated and evaluated using two different approaches: approaches based on the execution of cross-compiled binary code and solutions based on native simulation.

Simulations based on cross-compiled binary code are based on the execution of code compiled for a target different from the host computer. As a consequence, it is required to use an additional tool capable or reading and executing the code. Furthermore, this tool is in charge of obtaining performance estimations. To do so, the tool requires information about the cycles and other effects each instruction of the target machine will have in the system. Three different types of cross-compiled binary code can be performed depending on the type of this tool: simulations with processor models, compiled simulation and binary translation.

Instruction set simulators (ISSs) are commonly used as processor models capable of executing the cross-compiled code. These simulators can model the processor internals in detail (pipeline, register banks, etc.). As a consequence, they achieve very accurate results. However, the resulting simulation speed is very slow. This kind of simulators has been the most commonly used in industrial environments. CoWare Processor Designer (Cowar), CoMET de VaST Systems Technology (CoMET), Synopsys Virtual Platforms (Synopsys), MPARM (Benini et al, 2003) provide examples of these tools. However, due to the slow simulation speeds obtained with those tools, new faster simulation techniques are obtaining increasing interest.

Compiled simulation improves the performance of the ISSs while maintaining a very high accuracy. This solution relies on the possibility of moving part of the computational cost of the model from the simulation to the compilation time. Some of the operations of the processor model are performed during the compilation. For example, decoding stage of the pipeline can be performed in compilation time. Then, depending on the result of this stage, the simulation compiler selects the native operations required to simulate the application (Nohl et al, 2002). Compiled simulations based on architectural description languages have been developed in different projects, such as Sim-nML (Hartoog et al, 1997), ISDL (XSSIM) (Hadjiyiannis et al, 1997) y MIMOLA (Leupers et al, 2099). However, the resulting simulation is still slow and complex and difficult to port.

The third approach is to simulate the cross-compiled code using binary translation (Gligor et al, 2009). In this technique assembler instructions of the target processor are dynamically translated into native assembler instructions. Then, it is not necessary to have a virtual model describing the processor internals. As a result, the SW code is simulated much faster than in the two previous techniques. However, as there is no model of the processor, it is a bit more difficult to obtain accurate performance estimations, especially for specific elements as caches. Some examples of binary translation simulators are IBM PowerVM (PowerVM), QEMU (Qemu) or UQBT (UQBT).

Although these techniques result in quite fast simulators, the need of modelling very complex system early in the design process requires searching for much faster solution. For example, the exploration of wide design spaces can require thousands of simulations, so simulation speed have to be as close to functional execution speed as possible. The previous simulation techniques require a completely developed SW and HW platform, which are usually not available early in the design process. Then, these simulation techniques are not only too slow but also difficult to perform. Additionally, the simulation of heterogeneous platforms, with different kind of processors and different operating systems is limited by the refining effort required to evaluate all the options.

In order to overcome all these limitations, native simulation techniques have been proposed (Gerslauer et al, 2010).

2.1 Native simulation

In native simulation, the SW code is directly executed in the host computer. Thus, it is not required any kind of interpreter. As a consequence, very high simulation speeds can be achieved. However, in order to model not only the functionality but also the performance expected in the target platform additional information has to be added to the original code.

Furthermore, a model of the SW platform is also required. If the target operating system API is different than the native one, an API model is required to enable the execution of the SW code. A scheduler only controlling the tasks of the system model, not the entire host computer processes, specific time controller, or different drivers and peripheral communications are elements the SW infrastructure must provide.

Several solutions have been proposed for both issues in the last years.

2.2 SW performance estimation

Native simulation (Hwang et al, 2008; Schnerr et al, 2008; Bouchima et al, 2009) obtains target performance information from an analysis of the source code of the application SW to be executed. The common technique used to perform native simulations is to divide the code in fragments, estimate the time for each one of the fragments before the compilation process and annotate this information in the code. Usually basic blocks are used as code fragments because the entire block is always completely executed in the same way. Thus, basic blocks can be annotated as a single unit without introducing estimation errors. Such annotated code is then compiled and executed in the host computer, together with an infrastructure capable of capturing the timing estimations generated, and applying the corresponding delays to the simulation. As a consequence a timed model of the SW is obtained; a model which is ready to interact with other timed SW and HW components, to model the entire system.

Several techniques have been proposed to obtain the time information for each code fragment. These techniques can be divided in three main groups: pure source code estimations, estimations of intermediate code and cross-compiled code analysis.

Performance estimations based on source code analysis consider directly the C/C++ instructions of the basic block. They associate a number of cycles per instruction to each C operator. Using these values the total number of cycles required to execute each block is estimated. The associated time per instruction is obtained depending on the compiler and the target platform. Using simple mathematical operations, the number of cycles required to execute large sections of code is obtained (Brandolese et al, 2001; Posadas et al, 2004). Compared with the other two solution types described below, this solution is the most platform-independent one. No operational SW infrastructure for the target platform is required: no compiler, no operating system or libraries, etc. However, the other two solutions are more accurate, especially because no compiler optimizations can be considered in this one.

Estimations obtained from analysis of the intermediate code enable considering compiler optimizations, at least the optimizations that do not depend on the target instruction set. The basic idea is to identify the instructions of the basic blocks of the source code in the intermediate code. Analyzing the blocks in the intermediate code it is possible to obtain more accurate information than that obtained with the source level analysis. The main benefit obtained from using intermediate code is that the task of extracting the relationships among the basic blocks of the source code and the intermediate code is much simpler than with final cross-compiled code (Kempf et al, 2006; Hwang et al, 2008; Bouchima et al, 2009).

However, this technique presents several limitations. First, not all compiler optimizations can be analyzed. Second, the intermediate code is completely dependent on the compiler, so the portability of the solutions is limited. To solve those limitations, a few proposals for analyzing the cross-compiled binary code have been also presented.

Estimations based on binary code are based in the relationships between the basic blocks of the source code and the cross-compiled code (Schnerr et al, 2008). Since the code analyzed is the real binary that is executed in the target platform, no estimation errors are added for wrong consideration of the compiler effects. The problem with these estimations is how to associate the basic blocks of the source code to the binary code (Castillo et al, 2010). Compiler optimizations can provoke important changes in the code structure. As a consequence, techniques capable of making correct associations in a portable way are required.

Moreover, different efforts for modelling the effect of the processor caches in the SW execution have been proposed. In (Schnerr et al, 2008) a first dynamic solution for instruction cache modelling has been proposed. Another interesting proposal was presented in (Castillo et al, 2010). Additionally, also solutions for data cache modelling have been proposed (Gerslauer et al, 2010; Posadas et al, 2011).

This chapter proposes some solutions for making the basic block estimations, providing different ratios between speed and accuracy, always maintaining complete portability for its application to different platforms. Cache solutions provided in (Castillo et al, 2010) and (Posadas et al, 2011) have been applied to optimize the final accuracy and speed.

2.3 Operating system modeling

The second element required to perform a correct native simulation is the modeling of the SW platform. That is, it is required to model the operating system (Zabel et al, 2009; Becker et al, 2010). Concurrency support, scheduling, management of priorities and policies and services for communication and synchronization are critical issues in SW execution. Several solutions have been proposed to simulate SW codes on specific OSs. Some operating system providers include OS simulators in their SW development kits (ENEA; AXLOG). These simulators enable the development and verification of SW functionality without requiring the HW platform. However, these simulators only model the processor execution, without considering other elements of the final system. This limitation has two different drawbacks. First the simulators are not adequate for evaluating the system performance. Additionally, the simulation of the SW with application-specific HW components is not possible. As a result they are not adequate for its integration in co-design flows.

In order to obtain optimal HW/SW co-simulation environments with good relations between accuracy and speed for the early stages of the design process, it is necessary to develop models of RTOS based on high-level modeling languages. Several models based on SpecC (Tomiyama et al, 2001; Gerstlauer et al, 2003) and SystemC (Hassan et al, 2005; He et al, 2005; Schirner et al, 2007) have been proposed. However, most of these solutions have limited functionality and proprietary interfaces, which greatly complicate the modeling of real application SW codes (Gerstlauer et al, 2003; He et al, 2005; Yoo et al, 2002). Most of these models are limited to providing scheduling capabilities. Later a few models of specific operating systems have been proposed (Honda et al, 2004; Hassan et al, 2005). However, these RTOS models were very light and with reduced functionality.

Given the need of providing more complete models for simulating MPSoC operating systems, the infrastructure presented in this chapter starts from a very complete operating system model based on the POSIX interface and the implementation of the Linux operating system (Posadas et al, 2006). This chapter proposes an extension of this work to support different operating Systems. The models of the common operating systems uC/OS and Windows APIs are provided. As a result, the increasing complexity and heterogeneity of the MpSoCs can be managed in a flexible way.

3. Previous technology

As stated above, one of the main elements in a system modelling environment based in native simulation is the operating system model. It is in charge of controlling the execution of the different tasks, providing services to the application SW and controlling the interconnection of the SW and the HW. For that purpose, a model based on the POSIX API is used. The model uses the facilities for thread control of the high-level language SystemC to implement a complete OS model (Figure 1). Threads, mutexes, semaphores, message queues, signals, timers, policies, priorities, I/O and other common POSIX services are provided by the model. This work has been presented in (Posadas et al, 2006).

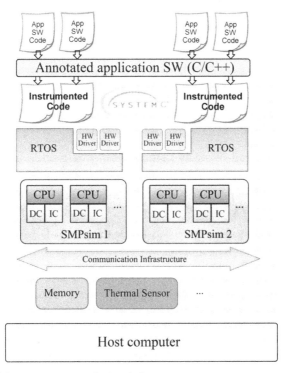

Fig. 1. Structure of the previous simulation infrastructure.

Special interest in the operating system model has the modeling of separated memory spaces in the simulation. As SystemC is a single host process, the integration of SW components containing functions or global variables with the same names in a single executable, or the execution of multiples copies of components that use global variables result in name collisions. To solve that, an approach based on the use of protected dynamic variables has been developed (Posadas et al, 2010).

However, the OS model is not only in charge of managing the application SW tasks. The interconnection between the native SW execution and the HW platform model is also performed by this component. For that goal, the model provides functions for handling interrupts and including device drivers following the Linux kernel 2.6 interfaces.

Additionally, a solution capable of detecting and redirecting accesses to the peripherals directly through the memory map addresses has been implemented. Most embedded systems access the peripherals by accessing their registers directly through pointers. However, in a native simulation, pointer accesses do not interact with the target HW platform model, but with the host peripherals. In fact, accesses to peripherals result in segmentation faults, since the user code has no permission to perform this kind of accesses. To solve that, these accesses are automatically detected and redirected using memory mappings ("mmap()"), interruption handlers, and code injection, in order to work properly (Posadas et al, 2009).

Furthermore, a TCP/IP stack has been integrated in the model. For that purpose, the open-source, stand-alone lwIP stack has been used. The stack has been adapted for its integration into the proposed environment both for connecting different nodes in the simulation through network models, and for connecting the simulation with the IP stack of the host computer, in order to communicate the simulation with other applications.

As a consequence, the infrastructure has demonstrated to be powerful enough to support the development of complete virtual platform models. However, improvements in the API support and performance modelling of the application SW are required. This work proposes solutions to improve them.

4. Virtual platform based on native simulation: goals and benefits

The goal of the native infrastructure is to provide a tool capable of assisting the designer during the initial design steps. More specifically, the infrastructure has been developed to provide the following services to the designers:

- Simulate the initial system models to check the complete functionality, before the platform is available, including timing effects.
- Provide performance estimations of the system models to evaluate the design decisions taken.
- Provide an infrastructure to start the refinement of the HW and SW components and their interconnections from the initial functional specification
- Work as a simulation tool integrated in design space exploration flows together with other tools required in the process

The first goal is to provide the designer with information about the system performance in terms of execution time and power consumption to make possible the verification of the fulfilment of the design constraints. This verification can be performed in two ways. First, the infrastructure reports metrics of the whole system performance at the end of the simulation, in order to enable the verification of global constraints. This solution allows "black box" analysis, where designers can execute several system simulations running different use cases, to easily verify the correct operation in all the working environments expected for the system.

A second option enabled by the infrastructure is to perform the verification of the system functionality and the checking of internal constraints. These internal constraints must be inserted in the application code using assertions. For that purpose, the use of the standard POSIX function "assert" is highly recommended. The infrastructure offers to the designer functions that provide punctual information about execution time and power consumption during simulation. Using that functions, internal assertions can check the accomplishment of parameters as delays, latencies, throughputs, etc.

A second goal of the infrastructure is to provide useful information to guide the designers during the development process. The co-design process of any system starts by making decisions about system architecture, HW/SW partitioning and resource allocation. To take the optimal decisions the infrastructure provides a fast solution to easily evaluate the performance of the different solutions considered by the designer. Task execution times, CPU utilization, cache miss rates, traffic in the communication channel, and power consumption in some HW components are some of the metrics the designer can obtain to analyze the effects of the different decisions in the system.

Another goal of the infrastructure is to provide the designers with a virtual platform where the development of all the components of the system can start very early in the design process. In traditional development flows, some components, such as SW components, cannot start their development process until a prototype of the target platform is built. However, it increases the overall design time since HW and SW components cannot be developed in parallel.

To reduce the design time, it is provided a solution for HW/SW modeling where the design of the SW components can be started. To enable that, the infrastructure provides a fast simulation of the SW components considering the effects of the operating system, the execution time of the SW in the target platform and enabling the interaction of the SW with a complete HW platform model. Even, the use of interruptions and drivers can be modelled in the simulation. The execution of the SW is then transformed in a timed simulation, where the use of services such as alarms, timeouts or timers can be explored in order to ensure certain real-time characteristics in the system.

Furthermore, the simulation of the SW using a native execution improves the debugging possibilities. Designers can directly use the debuggers of the host system, which has a double advantage: first, it is not necessary to learn how to use new debugging tools; second, the correct operation of the debuggers are completely guaranteed, and does not depend on possible errors in the porting of the tool-set to the target platform. Additionally, designers can easily access to all the internal values of both the SW and HW components, since all are modelled using a C++ simulation.

In order to achieve all these goals, the infrastructure implements a modeling infrastructure capable of supporting complete native co-simulation. The infrastructure provides novel solutions to enable automatic annotation of the application SW, a complete RTOS model, models of most common HW platform components and an infrastructure for native execution of the SW and its interconnection with the HW platform. Additionally, it is possible to describe configurable systems obtaining system metrics.

5. SW estimation and modeling

As a stated before, SW modeling solutions have become one of the most important areas of native simulation technology. The fastest possible execution of the system functionality is the direct compilation and execution of the code in the host computer. Thus, the goal is to provide a modeling solution capable of evaluating system performance, but maintaining a similar execution speed, as long as possible. Specially, the modeling solution has to overcome the three main limitations of functional execution with a minimum simulation overhead. First, functional executions do not consider any timing effect resulting of executing the code in the target platform. As a consequence, no performance information and no constraint checkings are available. Second, these executions cannot interact with the functionality implemented as HW components in the target platform. Thus, the simulation of the entire system functionality and the verification of the HW/SW integration are not possible. Finally, there is a problem when trying to execute a SW code developed for other OS APIs different from the native API.

To solve the first limitation, the solution proposed is to automatically modify the application SW in order to model performance effects. These performance effects include the execution of the code in the target processor core and the operation of the processor caches. The general solution applied for that modeling is based on estimating the effects during SW execution and apply them to the simulation, just before the points where the SW tasks start communications with the rest of the system, usually system calls. Four main solutions have been explored for obtaining the estimations: modified host times, the use of operator overloading and static annotation of basic-blocks at source and binary level. As a consequence, designers can modify the simulation speed and accuracy according to their needs on each moment.

The general annotation infrastructure enables using any of the estimation techniques with a virtual platform. Even, they can be combined in the same simulation. It depends on the method selected how to apply the estimated times for each SW component to increase the simulation time. The basic idea is to apply the estimated times when a system call is performed. This is caused because system calls are the points where communications and synchronizations are executed, that is, when SW tasks interacts with the rest of the system.

5.1 SW estimation based on modified host times

The first technique implemented is based on the use of the execution times of the host computer. As the time required for a processor to execute a code depends on the size of the functionality, there is a relationship between the time a SW execution takes in the host computer and in the target platform. Thus the idea is to run the simulation on the native PC getting the time required to execute each code segment. The estimated time costs of the components in the target platform are estimated by multiplying the time required to execute in the host computer by an adjustment factor. This factor is based on the characteristics of the native PC and the target platform.

Unlike the other techniques presented below, this solution does not require the generation of annotated SW code. The original code is executed as it is, without additional sentences. Estimation and time modeling is done automatically when the system calls of the OS model are executed. The execution time of each segment is obtained by calling the function "clock_gettime ()" of the native operating system (Figure 2). To minimize the error produced by the other PC tasks, the simulation must be launched with the highest possible priority.

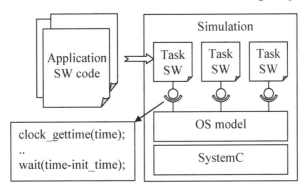

Fig. 2. Modeled by native time setting.

This solution has the advantage of being very fast, because no annotations increasing the execution time are needed. Nevertheless, a number of disadvantages hinder their use in most cases. First, we must be able to ensure that the simulation times obtained are really due to the execution of system code, and no caused by other parasite processes that were running on the computer. Second, the solution is not able to model cache behaviour adequately. Moreover, as only the execution time information can be obtained from the simulation, the transformations applied to obtain times of the target platform are reduced to a linear transformation. However, there is no guarantee that the cost of the native PC and the platform fits a linear relationship. On the contrary, the existence of different hardware structures, such as different caches, memory architectures or mathematical co-processors can produce significant errors in the estimation.

Summarizing, this solution is recommended only for very large simulations or codes where the accuracy obtained in performance estimations is not critical. Additionally, it is a good solution to estimate time of SW components that cannot be annotated. For example, some libraries are provided only in binary format. Thus, annotations are not possible since source code is not present. As a result, this solution is the only applicable of the four proposed.

5.2 SW estimation based on operator overloading

The estimation technique using operator overloading calculates the cost of SW as it progresses. Each operation executed must be accompanied by a consideration of the time cost it requires in the target platform. The temporal estimation of an entire SW code segment is obtained accumulating the times required to perform all the operations of a segment. This solution will avoid costly algorithms and static calculations, avoiding getting oversized times, as in the case of techniques for estimating worst case (WCET), or the consideration of false paths. That way, the estimated time depends on exactly the code that is executed.

The solution relies on the capability of C++ to automatically overload the operators of the user-defined classes. Using that ability, the real functional code can be extended with performance information without requiring any code modification. New C++ classes (generic_int, generic_char, generic_float, ...) have been developed to replace the basic C data types (int, char, float, ...) . These classes replicate the behavior of the basic data type operators, but adding to all the operator functions the expected cost of the operator in the target platform, in terms of binary instructions, cycles and power consumption. The replacement of the basic data types by the new classes is done by the compiler by including an additional header with macros of the type:

"#define int generic_int"

A similar solution is applied to consider the cost of the control statements.

To apply that technique, a table with the cost of all the operators and control statements in the target platform must be provided by the user.

The operating mechanism of this estimation technique can be seen in Figure 3. First, the original code is modified by replacing the original data types of the SW by new classes overloaded. This is done automatically using compiler preprocessor C. The new classes are provided by the simulation infrastructure. There is a class for each basic data type, which stores the value of the data type and the cost of each operation for this operator. The resulting code is executed using the overloaded operators.

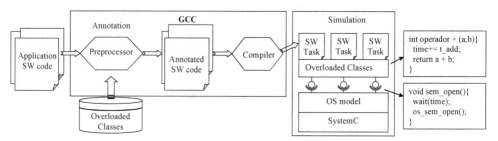

Fig. 3. Temporal model with operator overloading.

The original application code is compiled without any prior analysis or modification. Therefore, the operator overloading modeling technique is completely dynamic. All operations performed in the code are monitored by the annotation technique. This implies that the technique has enormous potential as a technique for code analysis. Studies on the number of operations, or monitoring data types of variables can be easily performed minimally modifying the overloading of operators.

This solution has demonstrated to be easy to implement, and very flexible to support additional evaluations, since all the information is managed dynamically, including the data values. Nevertheless, this solution has several limitations if the solely objective of the simulation is the estimation of execution times. Compiler optimizations are not accurately considered. Only, a mean optimization factor can be applied. Furthermore, the use of operator overloading for all the data types implies a certain overhead, which slows down the simulation speed.

5.3 Annotation from source-code analysis

To obtain simulations with really low overhead, it is needed to move analysis effort from simulation to compilation time. Solutions based on static annotation divides the performance modeling in two steps. First, the source code is statically analyzed, obtaining performance information for each basic block of the source code. After that, this information is annotated in the code, and the cost of each basic block executed is accumulated during the simulation and applied at system calls.

As in the technique of operator overloading, this estimation technique is based on assigning a time cost to each C operator. The total cost of each segment of SW code is estimated by adding the time of the operators executed in the segment. The cost of each operator is calculated in the same manner as shown in the previous technique. As a consequence, the effects of compiler optimizations are difficult to estimate from the analysis of source code. For this reason, an adjustment factor can be provided to the simulation to consider improvements introduced by compiler optimizations. This factor is obtained comparing the sizes of SW code segments both optimized and not optimized.

For the static analysis, a parser based on an open-source C++ grammar has been implemented. The parser analyzes the source code, obtaining the number and type of operators used on each basic block, as long as the control statements at the beginning of each block. Using that information and the table with the cost of each operator used for the previous technique it is possible to obtain the cost for the entire basic block. Then, this cost is applied in the source code in the following way:

"segment_cycles += 120; segment_instructions += 20;"

As a result, the variables segment_cycles and segment_instructions accumulate the total cycles and instructions required to execute the entire code in the target platform. The complete sequence of tasks necessary to perform the estimation based on source code analysis is shown in the next figure.

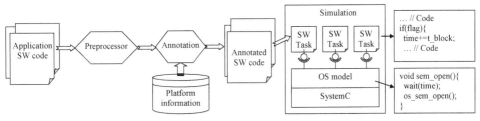

Fig. 4. temporal modeling with source-code analysis.

This solution requires more development effort than the operator overloading technique, especially for the implementation of the parser using the yacc/lex grammar. However, the simulation speed is really improved, achieving simulation times very close to the functional execution times (only two or three times slower). The main limitation of the technique is, again, the impossibility of accurately considering the compiler optimizations, since no analysis of the compiler output is performed.

5.4 Source annotations based on binary analysis

The last solution proposed is capable of maintaining the qualities of the previous annotation technique, but providing more accurate results, including compiler optimizations. In this solution, the analysis of the source code is replaced by an analysis of the cross-compiled binary code. The use of compiled code instead of source code enables accurately considering all the effects of cross compiler optimizations. Once identified the assembler instructions corresponding to each basic block of the SW code, the number of instructions of the blocks and the cycles required to execute them are annotated in the source code.

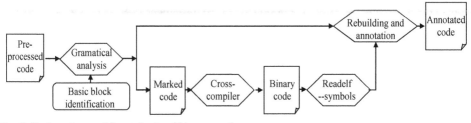

Fig. 5. Estimations with analysis of binary code.

However, estimations based on binary code usually present two limitations: first, it is difficult to identify the basic blocks of the source code in the binary code, and second, these solutions are usually very dependent on the processor. In order to build a simulation infrastructure fast and capable of modelling complex heterogeneous embedded systems, both issues have to be solved.

The correlation between source code and compiled code is sometimes very complex (Cifuentes) This is mainly due to results of the compiler optimizations as the reordering of instructions and dead code elimination. Furthermore, the technique should be easily portable to allow evaluation of different processors with minimal effort. To easily extract the correlation between source code and binary code, the proposed solution is to mark the code using labels. Both the annotation and identification of the positions of the labels can be done in a manner completely independent of the instruction set of the target processor. The annotation of labels in the code is a standard C feature, so it is extremely portable. Additionally, there are several standard ways to know the address of the labels in the target code, such as using the bin-utils or reading the resulting assembler code. Thus, the technique is extremely portable, and well suited to handle heterogeneous systems.

However, including compiler optimizations implies another problem. Compilation without optimizations enables easily identifying points in the binary code by inserting labels in the source code. However, the optimizations have the ability to move or even remove those labels. For example, if we insert a label in a loop, and apply an optimization of loop unrolling, the label loses its meaning. In order to avoid the compiler to eliminate the labels, they are added to the code of the form:

asm volatile("etiqueta_xx:");

The use of volatile labels forces the compiler to keep the labels in the right place. Thus, inserting labels at the beginning and end of each basic block we can easily obtain the number of assembly instructions of each basic block. The identification of basic blocks in the source code is made by a grammatical analysis. This grammatical analysis is done by a pre-compiler developed using "lex" and "yacc" tools, as in the estimation technique of source code analysis. This will locate the positions where the labels first and add annotations later.

Getting the value of the labels can easily be done using the command:

readelf –s binary_code.o | grep label_

The estimated time required to execute each basic block in the target platform is obtained by multiplying the number of instructions by the number of cycles per instruction (CPI) provided by the manufacturer. Although this solution carries a small error, such as not considering stops by data dependencies, it has the advantage of being fast and generic. To evaluate the behavior of a program on one processor, only a cross compiler for that processor is need. Libraries, operating systems or simulators as ISSs adapted specifically for the target platform are not required, resulting in a very portable and flexible approach.

However, with the introduction of volatile labels the compiler behaviour is still partially changed. Most of the optimizations, such as the elimination of memory accesses by reusing registers are correctly applied. But a few optimizations, with minor effects cannot be performed. Loop unrolling is not possible, although its use for processors with cache is unusual because it increases cache misses. The reordering of instructions to avoid data dependencies is also altered, but since the processor's internal effects are not modeled, this optimization has small effect on the estimation technique.

5.5 Cache modelling and pre-emption modeling

Nevertheless, the performance of the SW in the target platform does not only depend on the binary instructions executed. Processor caches also have an important impact on it. Common cache models are based on memory access traces. However, in native co-simulation no traces about the accesses in the target platform are obtained. As a consequence, new solutions for modeling both instruction and data caches have been explored and included in the infrastructure.

The modeling of instruction caches is based on the fact that instructions are placed sequentially in memory, in a place known at compilation time. Knowing the amount of assembler instruction for each basic block it is possible to obtain a relative address for the instructions with respect to the beginning of the "text" section of the "elf" file. This information is used as variables' address to access the cache model, instead of the real access trace. Additionally, the use of static structs has been applied in order to speed-up the simulation speed, achieving a similar error and overhead for instruction cache modeling than for the static time annotation (Castillo et al, 2010).

For data caches, the solution proposed uses corrected host addresses for each data variable used in the code. Additionally, global arrays handling information about the status of all the possible memory cache lines are used to improve the simulation speed maintaining the balance of the two previous techniques. The technique is described more in detail in (Posadas et al, 2011).

A final issue related to modeling the performance of the application SW is how to consider pre-emption. With the proposed modeling solutions, the segments of code between function calls are executed in "0" time, and after that, the time estimated for the segment is applied using "wait" statements. As a consequence, pre-emption events are always received in the "wait" statements. Thus, the segment has been completely executed before the information about the pre-emption arrives. As a consequence, the task execution order and the values of global variables can be wrong. In order to solve these problems, several solutions have been proposed in "Real-time Operating System modeling in SystemC for HW/SW co-simulation" (Posadas et al, 2005). The final solution applied is to use interruptible "wait" statements. This approach solves the problems in the task execution order. Additionally, it is considered that possible modifications in the values of global variables are not a simulation error but an effect of the indeterminism resulting of using unprotected global variables. In other words, it is not really an error but only a possible solution.

6. Operating system modeling

6.1 Support of multiple APIs

On of the main advantages of the underlying infrastructure selected to create the virtual platform infrastructure is the use of a real API. Since an implementation of a complete POSIX infrastructure is provided, most of the platforms based on Linux-like operating systems or other operating systems providing this API can be modelled. Then the infrastructure is able to support real software for a certain amount of platforms. However, other operating systems are used in embedded systems. As a really useful infrastructure has the goal of providing wide support in order to decide at the beginning of the design process the most adequate platforms for an application, support of other operating systems is recommended. Thus, in this work the extension of the infrastructure in that way has been evaluated. To do so, two different operating systems of wide use in embedded systems have been considered: a simple operating system and a complex one. As simple OS, uC/os-II has been selected. As complex OS, the integration of a win32 API has been performed.

6.1.1 Support of uC/os-II

μC/OS-II is a portable, small operating system developed by the Micrium company to be integrated in small devices. It is configurable and scalable, requiring footprints between 5 Kbytes to 24 Kbytes. This operating system provides a preemptive, real-time deterministic multitasking kernel for microprocessors, microcontrollers and DSPs. As a real-time kernel, the execution time for most services provided by μC/OS-II is both constant and deterministic; execution times do not depend on the number of tasks running in the application.

In order to easily implement the μC/OS-II API support the adopted approach has been to generate a layer on top of the existing POSIX API. Then, the implementation of the services only requires in most of the cases to adapt the interface of the μC/OS-II API to call a similar function in the POSIX infrastructure. Following that way, a list of 81 functions of the μC/OS-II API has been implemented. The following services have been implemented:

- Functions for OS management, such as starting the kernel, controlling the scheduler, or managing interrupts.

- Functions for task management, such as starting, stopping and resuming a task or modifying the priority
- Services for task synchronization: mutexes, semaphores and event flag groups.
- Services for task communication: message queues and mailboxes
- Memory management
- Time management and timers

As the POSIX infrastructure is quite complete, the task of generating this layer has resulted relatively easy. This demonstrates the validity of the infrastructure proposed to support other small operating systems.

6.1.2 Support of Win32

Although in the embedded system market Microsoft does not have the dominant position than in the PC (Laptop, Desktop and Server) market, the company through their Windows CE and Windows Mobile, now Windows Phone, holds an important market share which can even increase in the near future once Windows CE is offered under 'shared source' license and after the Nokia-Microsoft partnership. Thus, solutions to support of win32 API in a virtual platform modeling infrastructure results of great interest.

The proposed approach is to integrate virtualization of Win32 on the POSIX API of the performance analysis framework. As it is shown below, the overload of this approach is small. The virtualization framework is provided by the open-source code WINE. WINE is a free software application that aims to allow Unix-like computer operating systems to execute programs written for Microsoft Windows. WINE implements a Windows Application Programming Interface (Win32 API) library, acting as a bridge between the Windows application and Linux.

One of the reasons to use WINE is that, in accordance with the "Wine Developer's Guide", its architecture and kernel are based on the architecture and kernel of Windows NT, so that its behavior will be the same as most of the Windows operating systems, particularly those mostly used in embedded applications like Windows CE and Windows Phone.

Figure 6 shows in grey color the Windows NT architecture allowing the execution of Win32 application by the NT kernel. The white part of the Figure 6 represents the modules added for the construction of the Wine architecture.

Using the complete WINE architecture, the complete Windows NT architecture of Dynamic Link Libraries (DLL) is encapsulated by the WINE server and the WINE executable. The WINE executable virtualizes the underlying Unix kernel. For that purpose, additional DLLs and Unix-shared libraries are used.

The "WINE Server" acts as a Windows kernel emulator, executing the Win32 calls for thread creation, synchronization and destruction. It provides Inter-Process Communication (IPC). When a thread needs to synchronize or communicate with any other thread or process, is the Wine Server the handler of these actions making as an intermediary. The Wine server itself is a single and separated Unix process and does not have its own threading. Instead, it alerts whenever anything happens, such as a client having send a command, or a wait condition having been satisfied.

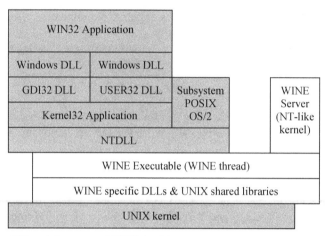

Fig. 6. Windows NT architecture + WINE Architecture.

The architecture of the integration of WINE on top of the POSIX model is shown in Figure 7. The most significant change from the WINE architecture of Figure 6 is the substitution of the POSIX subsystem, responsible for implementing the POSIX API functionality. In this way, the Win32 application is executed and its performance estimated by the native simulation infrastructure after the Win32 to POSIX translation.

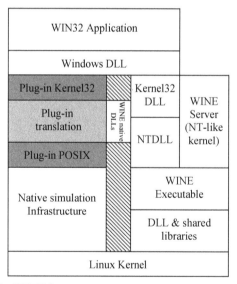

Fig. 7. Architecture of the WINE/native integration.

The WINE use is justified for the integration of WIN32 API in the native simulation framework. WINE allows us to abstract from the redeployment of Win32 functions for the execution in a POSIX system. Ideally, through this we can handle Win32's functions automatically by adding to our architecture the necessary libraries (DLLs).

However, when a simulation is being run, the user code can carry out calls to the API WIN32 functions. However, depending on which functions are being called, they are treated in two different ways. On the one hand, we have all those functions that are completely managed by WINE and that just need to be taken into account by native co-simulation in order to estimate the system performance in terms of execution times, bus loads and power consumption. On the other hand, there are other functions that are internally managed by the abstract POSIX native simulation kernel under the supervision of the WINE functions as they directly affect its kernel. The plug-in translation is responsible for these functions of thread creation, synchronization and destruction. When an API Win32 function is called, the plug-in analyzes and manages the handlers that have been generated by WINE. By default, the native WINE function is run, but in case the handle makes reference to a thread or object based on the synchronization of threads, it runs the translation to an equivalent POSIX function. In this way, the execution of these objects is completely transparent to the user.

As we said, part of the plug-in translation code is aimed at the internal management of the object's handles that are created and destructed in Wine as the user code requires. In the process of creating threads and synchronization objects, the code stores the resulting handle and the information that may be necessary for that regard. Thus, when any operation is performed on such handle, the plug-in can analyze and perform the necessary steps to carry out such operation.

The kind of services affected by such analysis are:

- Concurrency services (e.g. threads)
- Synchronization services (as semaphores, mutexes, events)
- Timing services (e.g. waitable timers)

In case that the handle belongs to any of the previous objects, it would be necessary to run the translation into an equivalent POSIX of the operation to be performed on this object so that it be performed by SCoPE correctly. Nonetheless, there are also other objects that are directly managed by the plug-in translation and do not require a previous analysis like Critical sections or Asynchronous Procedure Calls.

As shown in Figure 7, it is the "WINE Server" which acts as Windows kernel emulation, so that the thread creation, synchronization and destruction are performed through calls to this kernel. That is the reason why there is no literal translation for the behavior of these functions from the Win32 standard into the POSIX standard. An important contribution to this work and, therefore, an innovative solution to this problem, is the creation of a new code that is in charge of performing this task, maintaining the semantic and syntactic behavior of the functions of the affected Win32 standard. This is important in order to perform a translation by using only the calls to the POSIX standard functions, so that through the supervision of "WINE Server" our application is able to run those functions by respecting the Win32 standard at all times.

Finally, Graphics (GDI32) and User (USER32) libraries have been removed because they are not necessary in the functions currently implemented. As commented above, graphic interfaces are not supported yet as their modeling requires additional effort that is out of the scope of the current chapter. The user interface is not necessary when modeling usual embedded applications. Nevertheless, the proposed methodology for abstract modeling of complex OSs opens the way to solve this particular problem.

All the collection of functions of the API Win32 has been faithfully respected in accordance with the on-line standard of MSDN. To check it, a battery of simple tests has been developed to verify the correctness of some critical functions closely related with the integration of WINE with the simulation infrastructure. The tests generated include management of threads, synchronization means, file system functions and timers. The results have been compared with the same tests compiled and executed on a Windows platform (XP SP2 winver 0x0502) and in an embedded Windows CE platform, obtaining the same results in all the cases.

In the compilation process of a Win32 application in WINE, this one generated the scripts that are necessary to create a dynamic library from the application's source code, which is later loaded and run after the initialization process of WINE.

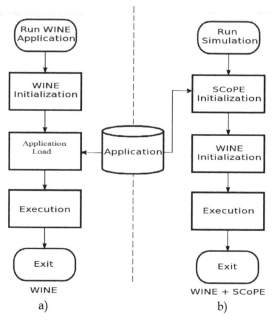

Fig. 8. WINE integration in the native simulation.

The process to generate a POSIX WINE executable from a Win32 application is shown in Figure 8-a. After WINE initialization, the scripts that are necessary to create a dynamic library from the application's source code are generated. Then, using these scripts, the application is loaded and executed. This application initialization and loading process is not compatible with the native co-simulation methodology.

The alternative process implemented is shown in Figure 8-b. The default initialization process of WINE is performed after the native co-simulation initialization process. The application is instrumented and loaded into the native simulation environment in this step. In order to support the parsing and back-annotation required by native co-simulation, it is necessary to integrate in the native co-simulation compiler the options required by WINE in order to recognize the application.

7. Results

Several experiments have been set-up in order to assess the proposed methodology. Firstly, simulation performance has been measured and compared with different execution environments of Win32 applications through small examples. Furthermore, a complete co-simulation case study has been developed showing the full potential of the proposed technology on a realistic embedded system design. After that some experiments have been performed to check the accuracy of the performance estimations.

7.1 Win32 simulation

In order to measure the simulation overhead of the proposed infrastructure, several tests focused on the use of OS services have been developed and instrumented. The tests have been carried out in four different scenarios, all on the same host computer:

- Proposed Win32 native simulation running on a native Linux platform (Fedora 11).
- WINE running on the same Linux platform.
- Windows XP SP2 running in a virtual machine (VirtualMachine 2.2.4) on the same Linux platform.
- Windows XP SP2 installed directly in the host.

The resulting execution times of the tests on the different scenarios are shown in Figure 9. As expected, the execution of Windows on a virtual machine is always slower than the OS directly installed in the host. Nevertheless, this is not the case when virtualising Windows with WINE. Results show that WINE can be faster than XP installed directly on the same host. This is not a surprising result and it has been already reported.

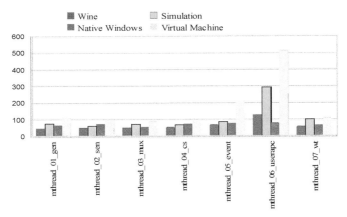

Fig. 9. Execution times.

As shown in Figure 9, native simulation is only 46% slower in average than WINE although the simulation is modeling execution times, data and instructions cache, memory and peripheral accesses, power consumption, etc. This result is coherent with the comparison figures between native simulation including performance estimations and functional execution. This explains why native simulation can be faster in some cases than functional execution on a Windows platform. This result shows the advantage of using WINE; we can integrate native simulation on a virtualization of Windows, implementing most of its functionality and taking advantage of its fast implementation.

In order to assess the Win32 simulation technology in its final application of performance analysis of complex embedded systems including processing nodes using Windows, a heterogeneous system has been modeled, simulated and the performance figures obtained. The system is a low cost surveillance system taking low quality images from a camera at low speed (1 image per second) and coding and sending them through a serial link.

Apart from those simple examples, a complex example, a H.264 coder has been used for global correctness. This example makes an exhaustive use of calls to memory dynamic management functions, and there is also a writing of all the logs resulting from the codification when running. This part of the reference model has been modified so that the calls to the equivalent functions of the API Win32 are carried out in order to verify the correct operation of the plug-in this sort of operations. Dynamic memory management has been carried out through calls to the Global, Local and Heap memory management functions, and the file management through calls to the respective data input and output functions (e.g. CreateFile and WriteFile).

The system architecture is shown in Figure 10. It is composed of a Windows ARM node executing the H.264 coder, the camera taking the images, a memory where the input data are stored and the serial link taking the images and sending them out. The architectural exploration affects the selection of the most appropriate voltage-frequency and data and instruction cache sizes ensuring a CPU usage lower than 90% and a power consumption less than 1W.

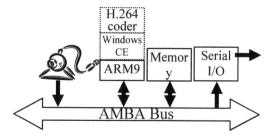

Fig. 10. Case study architecture.

Results of CPU usage and power consumption are shown in Figure 11. As can be seen, in this example, the size of the data and instruction caches do not affect too much the power consumption but the CPU usage.

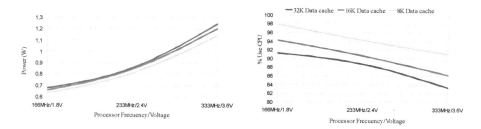

Fig. 11. CPU usage and Power consumption.

7.2 Win32 simulation performance

The proposed approach has been also applied to an ARM9 platform, in order to evaluate the accuracy of each on the techniques presented above. The ARM9 platform has been used to compare the estimation results of the different modeling solutions, in order to obtain the error when applied to one of the most popular processors in the embedded world.

As a summary of the final results achieved, the following tables show the estimation accuracy of the SW modelling, and the simulation times for a list of examples:

	Modified Host Time		Operator Overloading		Source Code analysis		Binary Code analysis	
	Error	Time	Error	Time	Error	Time	Error	Time
Bubble 1000	24.4	0.012s	14.8	0.75s	14.5	0.032s	12.5	0.030s
Bubble 10000	13.5	1.281s	3.5	81.6s	3.2	3.501s	0.01	3.486s
Vocoder	54.2	0.003s	24.2	0.41s	26.4	0.015s	18.3	0.014s
Factorial	34.5	0.013s	4.5	0.85s	4.1	0.042s	0.01	0.043s
Hanoi	47.9	0.082s	17.9	0.82s	16.9	0.271s	14.9	0.262s

Table 1. Comparison of estimation error (%) and simulation time for an ARM9 platform

As can be seen, the most accurate annotation technique is the solution based on the analysis of the binary cross-compiled code. After that, the technique based on source code analysis and the operator overloading are similar, since both rely on the same information (cycles of each C operator) and the same main source of error (optimizations). Finally, the modified host time is the less accurate one.

However, the technique of modified host tome is about 3 times faster than the annotation techniques based on code analysis, and more than 60 times than the operator overloading solution.

Finally, the results for cache modelling are shown in the next tables:

	Instruction Cache Misses					
	Without optimizations (-o0)			With optimizations (-o2)		
	Skyeye	Proposal	Error (%)	Skyeye	Proposal	Error (%)
Bubble 1000	15	16	6.66	6	5	16.67
Bubble 10000	25	27	8	7	7	0
Vocoder	8	7	12,5	5	4	20
Factorial	20	18	10	12	10	16.67
Hanoi	46074	46761	1.49	25842	28607	10.70

Table 2. Comparison of instruction cache misses ARM926t platform.

	Data Cache Misses					
	Without optimizations (-o0)			With optimizations (-o2)		
	Skyeye	Proposal	Error (%)	Skyeye	Proposal	Error (%)
Bubble 1000	126	127	0.80	126	126	0
Bubble 10000	5199772	5209087	0.18	5199310	5211595	0.24
Vocoder	375	500	33.33	375	500	33.33
Factorial	38	45	18.42	41	45	9.76
Hanoi	6018	5908	1.82	6026	5915	1.84

Table 3. Comparison of data cache misses ARM926t platform.

Summarizing, simulation speed-ups of two or more orders of magnitude can be achieved by assuming an acceptable error, below 20%.

8. Conclusions

In this chapter, several solutions have been developed in order to cover all the features required to create an infrastructure capable of obtaining sufficiently accurate performance estimation with very fast simulation speeds. These solutions are based on the idea of native co-simulation, which consists in the combination of native simulation of annotated SW codes with time-approximate HW platform models. All these techniques have been integrated in a simulation tool which can be used as an independent simulator or can be used integrated in different design space exploration flows.

The modeling solutions can be divided in two main groups: solutions for modeling in the native execution the operation of the application SW in the target platform, and a complete operative system modelling infrastructure. These solutions have been implemented as SystemC extensions, using the features of the language to provide multiple execution flows, events and time management.

The modeling of the application SW considers the execution times and power consumption of the code in the target platform, as long as the operation of the processor caches. Four different solutions for modeling the processor performance have been explored in the chapter (modified host times, operator overloading, annotation based on source code analysis and annotation based on binary code analysis), in order to find an approach capable of obtaining accurate solutions with minimal simulation overheads and as flexible as possible, to minimize the effort required to evaluate different target processors and platforms. As a result of the study, the annotation based on binary code analysis has demonstrated to obtain the best results with minimal simulation overhead. Additionally, the technique is very flexible, since only requires a cross-compiler for the target platform capable of generating object files from the source code. No additional libraries, ported operating systems, or linkage scripts are required. Additionally, it has been demonstrated that cache analysis for both instruction and data caches can be performed obtaining accurate results with adequate simulation times.

A POSIX-based operating system model has been also extended to support other APIs. Two different operating system APIs of wide use in embedded systems have been considered: a simple operating system and a complex one. Support for a simple OS, uC/os-II, has been integrated. As complex OS, the integration of a win32 API has been performed.

Summarizing, this chapter demonstrates that the SystemC language can be extended to enable the early modeling and evaluation of electronic systems, and providing important information to help the designers during the first steps of the design process. These extensions allow using a SystemC-based infrastructure for functional simulation, performance evaluation, constraint checking and HW/SW refinement.

9. Acknowledgments

This work has been supported by the FP7-ICT-2009- 4 (247999) Complex and Spanish MICyT TEC2008-04107 projects.

10. References

AXLOG, http://www.axlog.fr.

M.Becker, T.Xie, W.Mueller, G. Di Guglielmo, G. Pravadelli and F.Fummi, "RTOS-Aware Refinement for TLM2.0-Based HW/SW Designs", in DATE, 2010.

Benini et al, "MPARM: Exploring the Multi-Processor SoC Design Space with SystemC", Journal of VLSI Signal Processing n 41, 2005.

A. Bouchima, P. Gerin & F. Pétrot: "Automatic Instrumentation of Embedded Software for High-level HS/SW Co-simulation. ASP-DAC, 2009.

C. Brandolese, W. Fornaciari, F. Salice, and D. Sciuto, "Source-level execution time estimation of c programs," CODES 2001.

J. Castillo, H. Posadas, E. Villar, M. Martínez, "Fast Instruction Cache Modeling for Approximate Timed HW/SW Co-Simulation", 20th Great Lakes Symposium on VLSI (GLSVLSI'10), Providence, USA. 2010

C. Cifuentes. "Reverse Compilation Techniques". PhD thesis, Queensland University of Technilogy, 1994.

VaST Systems Technology. CoMET R.
    http://www.vastsystems.com/docs/CoMET_mar2007.pdf

CoWare Processor Designer, http://www.coware.com/products/processordesigner.php

ENEA: "OSE Soft Kernel Environment", in http://www.ose.com/products.

Gerstlauer, A. Yu, H. & Gajski, D.D.: "RTOS Modeling for System Level Design", Proc. of DATE, IEEE, 2003.

A. Gerslauer, "Host-Compiled Simulation of Multi-Core Platforms", Rapid System Prototyping, 2010

M. Gligor, N. Fournel, and F. Petrot, "Using binary translation in event driven simulation for fast and flexible MPSoC simulation", in CODES+ISSS, France, Oct. 2009.

G. Hadjiyiannis, S. Hanono & S. Devadas. ISDL: An Instruction Set Description Language for Retargetability. Design Automation Conference, 1997.

M. Hartoog J.A. Rowson, P.D. Reddy, S. Desai, D.D. Dunlop, E.A. Harcourt & N. Khullar. "Generation of Software Tools from Processor Descriptions for Hardware/Software Codesign". Design Automation Conference, 1997.

M.A. Hassan, K. Sakanushi, Y. Takeuchi and M. Imai: "RTK-Spec TRON: A simulation model of an ITRON based RTOS kernel in SystemC", Proceedings of the Design, Automation and Test Conference, IEEE, 2005.

Z. He, A. Mok and C. Peng: "Timed RTOS modeling for embedded System Design", Proceedings of the Real Time and Embedded Technology and Applications Symposium, IEEE, 2005.

S. Honda, T. Wakabayashi, H. Tomiyama and H. Takada: "RTOS-centric HW/SW co-simulator for embedded system design", Proceedings of CoDes-ISSS'04, ACM, 2004.

Y. Hwang, S. Abdi, D. Gajski. Cycle-approximate Retargetable Performance Estimation at the Transaction Level. DATE, 2008

T. Kempf, K. Karuri, S. Wallentowitz, G. Ascheid, R. Leupers, H. Meyr. "A SW Performance Estimation Framework for Early System-Level-Design Using Fine-Grained Instrumentation". DATE, 2006

R. Leupers, J. Elste, and B. Landwehr. "Generation of interpretive and compiled instruction set simulators". Asia and South Pacific Design Automation Conference, 1999.

A. Nohl, G. Braun, O. Schliebusch, R. Leupers, H. Meyr & Andreas Hoffmann, "A Universal Technique for Fast and Flexible Instruction-Set Architecture Simulation", DAC, 2002

H. Posadas, F. Herrera, P. Sánchez, E. Villar, F. Blasco: "System-Level Performance Analysis in SystemC", Proc. of DATE, IEEE CS Press. 2004

H. Posadas, E. Villar, F. Blasco: "Real-time Operating System modeling in SystemC for HW/SW co-simulation", XX Conference on Design of Circuits and Integrated Systems, DCIS. 2005

H. Posadas, J. Adámez, P. Sánchez, E. Villar, F. Blasco: "POSIX modeling in SystemC", 11th Asia and South Pacific Design Automation Conference, ASP-DAC, 2006

H. Posadas, E. Villar: "Automatic HW/SW interface modeling for scratch-pad & memory mapped HW components in native source-code co-simulation", A. Rettberg, M. Zanella, M. Amann, M. Keckeiser & F. Rammig (Eds.): "Analysis, Architectures and Modelling of Embedded Systems", Springer, 2009

H. Posadas, E. Villar, Dominique Ragot, M. Martínez: "Early Modeling of Linux-based RTOS Platforms in a SystemC Time-Approximate Co-Simulation Environment", IEEE International Symposium on Object/Component/Service-Oriented Real-Time Distributed Computing (ISORC), 2010

H. Posadas, L. Diaz, E. Villar: "Fast Data-Cache Modeling for Native Co-Simulation", Asia and South-Pacific Design Automation Conference, ASP-DAC, 2011

IBM PowerVM, http://www-03.ibm.com/systems/power/software/virtualization/

Qemu, http://www.qemu.org/

G. Schirner, A. Gerstlauer, and R. Dömer. "Abstract, Multifaceted Modeling of Embedded Processors for System Level Design". Asia and South Pacific Design Automation Conference (ASP-DAC), 2007.

J. Schnerr, O. Bringmann, A. Viehl, W. Rosenstiel. High-Performance Timing Simulation of Embedded Software. DAC, 2008

SkyEye web page, http://www.skyeye.org/index.shtml

Synopsys, Platform Architect tool, http://www.synopsys.com/Systems/ArchitectureDesign/pages/PlatformArchitect.aspx

H. Tomiyama, Y. Cao and K. Murakami: "Modeling fixed-priority preemptive multi-task systems in SpecC", Proceedings of the 10th Workshop on System And System Integration of Mixed Technologies (SASIMI'01), IEEE, 2001.

UQBT, http://www.itee.uq.edu.au/~cristina/uqbt.html

S. Yoo, G. Nicolescu, L. Gauthier, A. Jerraya, "Automatic generation of fast timed simulation models for operating systems in SoC design", Proc. of DATE, IEEE, 2002.

H. Zabel, W. Müller, and A. Gerstlauer, "Accurate RTOS modeling and analysis with SystemC", in "Hardware-dependent Software: Principles and Practice", W. Ecker, W. Mü ller, and R. Dömer, Eds. Springer, 2009.

15

Choosing Appropriate Programming Language to Implement Software for Real-Time Resource-Constrained Embedded Systems

Mouaaz Nahas[1] and Adi Maaita[2]

[1]*Department of Electrical Engineering, College of Engineering and Islamic Architecture,*
Umm Al-Qura University, Makkah,
[2]*Software Engineering Department, Faculty of Information Technology,*
Isra University, Amman,
[1]*Saudi Arabia*
[2]*Jordan*

1. Introduction

In embedded systems development, engineers are concerned with both software and hardware aspects of the system. Once the design specifications of a system are clearly defined and converted into appropriate design elements, the system implementation process can take place by translating those designs into software and hardware components. People working on the development of embedded systems are often concerned with the software implementation of the system in which the system specifications are converted into an executable system (Sommerville, 2007; Koch, 1999). For example, Koch interpreted the implementation of a system as the way in which the software program is arranged to meet the system specifications.

Having decided on the software architecture of the embedded design, the first key decision to be made in the implementation stage is the choice of programming language to implement the embedded software (including the scheduler code, for example). The choice of programming language is an important design consideration as it plays a significant role in reducing the total development time (Grogono, 1999) (as well as the complexity and thus maintainability and expandability of the software).

This chapter is intended to be a useful reference on "computer programming languages" in general and on "embedded programming languages" in particular. The chapter provides a review of (almost) all common programming languages used in computer science and real-time embedded systems. The chapter then discusses the key challenges faced by an embedded systems developer to select a suitable programming language for their design and provides a detailed comparison between the available languages. A detailed literature review of the work done in this area is also provided. The chapter also provides real data which shows that – among the wide range of available choices – "C" remains the most popular language for use in the programming of real-time, resource-constrained embedded systems. The key features of "C" which made it so popular are provided in a great detail.

The chapter is organized as follows. Section 2 provides various definitions of the term "programming language" from a wide range of well-known references. Section 3 and Section 4 provide classification and history of programming languages (respectively). Section 5 provides a review of programming languages used in the fields of real-time embedded systems. Section 6 discusses the choice of programming languages for embedded designs. Section 7 and Section 8 provide the main advantages of "C" which made it the most popular language to use in real-time, resource-constrained embedded systems and a detailed comparison with alternative languages (respectively). Real data which shows the prevalence of "C" against other available languages is also provided in Section 8. Section 9 presents a brief literature review of using "C" to implement software for real-time embedded systems. The overall chapter conclusions are drawn in Section 10.

2. What is a programming language?

Simply, programming as a problem has only arisen since computer machines were first created. The magnitude of the problem is however relative to the size (and complexity) of the computer machine used (Cook, 1999). To program a computer system, a programming language is required. The latter is seen as the major way of communication (interface) between a person who has a problem and the computer system used to solve the problem.

Programming language has been defined in several ways. For example, American Standard Vocabulary for Information Processing (ANSVIP, 1970) defined a programming language as "A language used to prepare computer programs". The IFIP-ICC Vocabulary of Information Processing (IFIP-ICC, 1966) defined it as "A general term for a defined set of symbolic and rules or conventions governing the manner and sequence in which the symbols may be combined into a meaningful communication". The IFIP-ICC glossary also noted that "An unambiguous language, intended for expressing programs, is called a PROGRAMMING LANGUAGE". Other definitions for a programming language include:

- "A computer tool that allows a programmer to write commands in a format that is more easily understood or remembered by a person, and in such a way that they can be translated into codes that the computer can understand and execute." (Budlong, 1999).
- "An artificial language for expressing programs." (ISO, 2001).
- "A self-consistent notation for the precise description of computer programs" (Wizitt, 2001).
- "A standard which specifies how (sort of) human readable text is run on a computer." (Sanders, 2007).
- "A precise artificial language for writing programs which can be automatically translated into machine language." (Holyer, 2008).

However, it was noted elsewhere (e.g. Sammet, 1969) that standard definitions are usually too general as they do not reflect the language usage. A more specific definition for a programming language was given by Sammet as a set of characters and rules (used to combine the characters) that have the following characteristics:

- A programming language requires no knowledge of the machine code by the programmer, thus the programmer can write a program without much knowledge about the physical characteristics of the machine on which the program is to be run.

- A programming language should be machine independent.
- When a program written in a programming language is translated into the machine code, each statement should explode to generate a large set of machine instructions.
- A programming language must have problem-oriented notations which are closer to the specific problem intended to be solved.

It is worth mentioning that a vast number of different programming languages have already been created, and new languages are still being created.

3. Classification of programming languages

This section provides a classification of programming languages. Sources for this section include (Sammet, 1969; Booch, 1991; Grogono, 1999; Lambert & Osborne, 2000; Mitchell, 2003; Calgary, 2005; Davidgould, 2008; Network Dictionary, 2008).

In general, programming languages can be divided into programming paradigms and classified by their intended domain of use. Paradigms include procedural programming, object-oriented (O-O) programming, functional programming, and logic programming. Note that some languages combine multiple paradigms. Each of these paradigms is briefly introduced here.

Procedural programming (or imperative programming) is based on the concept of decomposing the program into a set of procedures (i.e. series of computational steps). Examples of procedural languages are: FORTRAN (**FOR**mula **TRAN**slator), Algol (**ALGO**rithmic Language), COBOL (**CO**mmon **B**usiness **O**riented **L**anguage), PL/I (Programming Language I), Pascal, BASIC (**B**eginner's **A**ll-purpose **S**ymbolic **I**nstruction **C**ode), Modula-2, "C" and Ada. Object-Oriented (O-O) programming is a method where the program is organized as cooperative collections of "objects". This style of programming was not commonly used in software application development until the early 1990s, but nowadays most of the modern programming languages support this type of programming paradigm. Examples of object-oriented languages are: Simula, Smalltalk, C++, Eiffel and Java. Functional programming treats computation as the evaluation of mathematical functions. In functional programming, a high order function can take another function as a parameter or returns a function. An example of functional languages is LISP (**LIS**t **P**rocessor). Finally, logic programming uses mathematical logic in which the program enables the computer to reason logically. An example of logic languages is Prolog (**PRO**gramming in **LOG**ic). It is often argued that languages with support for an O-O programming style have advantages over those from earlier generations (Pont, 2003). For example, Jalote (1997) noted that using O-O helps to represent the problem domain, which makes it easier to produce and understand designs.

In addition to programming paradigm, the purpose of use is an important characteristic of a language: it is unlikely to see one language fitting all needs for all purposes (Sammet, 1969). Programming languages can be divided, according to their purpose, into general-purpose languages, system programming languages, scripting languages, domain-specific languages, and concurrent / distributed languages (or a combination of these). A general-purpose language is a type of programming language that is capable of creating various types of programs for various applications, e.g. "C" language. There has been an argument that some of the general-purpose languages were designed mainly for educational purposes

(Wirth, 1993). A system programming language is a language used to produce software which services the computer hardware rather than the user, e.g. Assembly and Embedded C. Scripting language is a language in which programs are a series of commands that are interpreted and then executed sequentially at run-time without compilation, e.g. JavaScript (used for web page design). Domain-specific programming languages are, in contrast to general-purpose languages, designed for a specific kind of tasks, e.g. Csound (used to create audio files), and GraphViz (used to create visual representations of directed graphs). Concurrent languages are programming languages that have abstractions for writing concurrent programs. A concurrent program is the program that can execute multiple tasks simultaneously, where these tasks can be in the form of separate programs or a set of processes or threads created by a single program. Concurrent programming can support distributed computing, message passing or shared resources. Examples of concurrent programming languages include Java, Eiffel and Ada.

In his famous book (i.e. "Programming Languages: History and Fundamentals", 1969), Jean E. Sammet used the following set of defining categories as a way of classifying programming languages: 1) procedural and non-procedural languages; 2) problem-oriented, application-oriented and special purpose languages; 3) problem-defining, problem describing and problem solving languages; 4) hardware, publication and reference languages. Sammet however underlined that any programming language can fall into more than one of these categories simultaneously: for further details see Sammet (1969).

4. History of programming languages

It has been argued that studying the history of programming languages is essential as it helps developers avoid previously-committed mistakes in the development of new languages (Wilson & Clark, 2000). It was also pointed out that an unfortunate trend in Computer Science is creating new language features without carefully studying previous work in this field (Grogono, 1999). Most books and articles on the history of programming languages tend to discuss languages in terms of generations where languages are classified by age (Cook, 1999). Many articles and books have discussed the generations of programming languages (e.g. Wexelblat, 1981; Martin & Leben, 1986; Watson, 1989; Zuse, 1995; Flynn, 2001). Pont (2003) provides a list of widely-used programming languages classified according to their generations (see Table 1).

Language generation	Example languages
-	Machine code
First generation linguage (1GL)	Assembly
Second generation languages (2GL)	COBOL, FORTRAN
Third generation languages (3GL) "process-oriented'	C, Pascal, Ada 83
Fourth generation languages (4GL) 'object-oriented'	C++, Java, Ada 95

Table 1. Classification of programming languages by generations (Pont, 2003).

A brief history of the most popular programming languages (including the ones presented in Table 1) is provided in this section. Sources for the following material mainly include (Wexelblat, 1981; Martin & Leben, 1986; Watson, 1989; Halang & Stoyenko, 1990; Grogono, 1999; Flynn, 2001).

In the 1940s, the first electrically powered digital computers were created. The computers of the early 1950s used machine language which was quickly superseded by a second generation of programming languages known as Assembly languages. The limitations in resources (e.g. computer speed and memory space) enforced programmers to write their hand-tuned assembly programs. However, it was shortly realized that programming in assembly required a great deal of intellectual effort and was prone to error. It is important to note that although many people consider Assembly as a standard programming language, some others believe it is too low-level to bring satisfactory of communication for user, hence was excluded from the programming languages list (Sammet, 1969).

1950s saw the development of a range of high-level programming languages (some of which are still in widespread use), e.g. FORTRAN, LISP, and COBOL, and other languages such as Algol 60 that had a substantial influence on most of the lately developed programming languages. In 1960s, languages such as APL (**A** **P**rogramming **L**anguage), Simula, BASIC and PL/I were developed. PL/I incorporated the best ideas from FORTRAN and COBOL. Simula is considered to be the first language designed to support O-O programming.

The period between late 1960s and late 1970s brought a great prosperity to programming languages most of which are used nowadays. In the mid-1970s, Smalltalk was introduced with a complete design of an O-O language. The programming language "C" was developed between 1969 and 1973 as a systems programming language, and remained popular. In 1972, Prolog was designed as the first logic programming language. In 1978, ML (**M**eta-**L**anguage) was developed to found statically-typed functional programming languages in which type checking is performed during compile-time allowing more efficient program execution. It is important to highlight that each of these languages originated an entire family of descendants. Some other key languages which were developed in this period include: Pascal, Forth and SQL (**S**tructured **Q**uery **L**anguage).

In 1980s, C++ was developed as a combined O-O and systems programming language. Around the same time, Ada was developed and standardized by the United States government as a systems programming language intended for use in defense systems. One noticeable tendency of language design during the 1980s was the increased focus on programming large-scale systems through the use of modules, or large-scale organizational units of code. Therefore, languages such as Modula-2, Ada, and ML were all extended to support such modular programming in 1980s. Some other languages that were developed in this period include: Eiffel, PEARL (**P**ractical **E**xtraction and **R**eport **L**anguage) and FL (**F**unction **L**evel).

In mid-1990s, the rapid growth of the Internet created opportunities for new languages to emerge. For example, PEARL (which is originally a Unix scripting tool first released in 1987) became widely adopted in dynamic web sites design. Another example is Java which was commonly used in server-side programming. These language developments provided no fundamental novelty: instead, they were modified versions of existing languages and paradigms and largely based on the "C" family of programming languages.

It is difficult to determine which programming languages are most widely used, as there have been various ways to measure language popularity (see O'Reilly, 2006; Bieman & Murdock, 2001). Mostly, languages tend to be popular in particular types of applications. For example, COBOL is a leading language in business applications (Carr & Kizior, 2000),

FORTRAN is widely used in engineering and science applications (Chapman, 2004), and "C" is a genuine language for programming embedded applications and operating systems (Barr, 1999; Pont, 2002; Liberty & Jones, 2004).

5. Programming languages for real-time embedded systems

To develop a real-time embedded system, a number of tools and techniques would be required: the key one is the programming language used to develop the application code (Burns, 2006). Assembly was the first programming language used to implement the software for embedded applications. However, it was argued that the development environments that used the first generation languages such as Assembly lacked the basic support for debugging and testing (Halang & Stoyenko, 1990). Therefore, in 1960s, the need for high-level programming languages to program real-time systems, instead of continuing to use Assembly language, was agreed among many real-time system designers; due to advantages such as ease of learning, programming, understanding, debugging, maintaining and documenting and also code portability (see Boulton & Reid, 1969; Sammet, 1969).

The work in this area began by identifying the essential requirements for a high-level language to fulfill the objectives of real-time applications (Opler, 1966). Such requirements were summarized by Boulton & Reid (1969) as methods of handling real-time signals and interrupts, and methods of scheduling real-time tasks. Opler (1966) argued that to achieve such requirements, one can make extensions / modifications to an existing programming language, where an alternative solution is to develop new languages dedicated specifically for real-time software. Some success, in extending existing languages to real-time computing, was achieved using languages such as FORTRAN (e.g. Jarvis, 1968; Roberts, 1968; Hohmeyer, 1968; Mensh & Diehl, 1968; Kircher & Turner, 1968) and PL/I (e.g. Boulton & Reid, 1969). Some other studies, however, attempted to develop new real-time languages but with some similarity to existing languages, e.g. PROSPRO (Bates, 1968), SPL (Oerter, 1968) and RTL (Schoeffler & Temple, 1970).

In 1970s, a major concern of many researchers became the programming of real-time applications which involve concurrent processing. Useful work in this area demonstrated that, same as before, concurrent programming can be achieved by either extending available general-purpose languages (e.g. Hansen, 1975; Wirth, 1977) or developing entirely new concurrent-processing languages (e.g. Schutz, 1979). However, it was noticed that extended general-purpose languages still lacked genuine concurrency and real-time concepts (Steusloff, 1984). This led to the development of more efficient concurrent real-time languages such as PEARL (DIN, 1979), ILIAD (Schutz, 1979) and Ada (Ada, 1980).

Ada is a well-designed and widely used language for implementing real-time systems (Burns, 2006). Therefore, it is worth discussing it in greater detail. As previously noted, Ada is an object-oriented, high-level programming language which was first developed and adopted by the U.S. Department of Defense (DoD) to implement various defense mission-critical software applications (Ada, 1980; Baker & Shaw, 1989). Ada appeared as a standard language in 1983 – when Ada83 was released – and was later reviewed and improved in 1995 by producing Ada95. Since developed, Ada has gained a great deal of interest by many real-time and embedded systems developers (e.g. see Real-Time Systems (RTS) Group webpage, The University of York, UK). It was declared that Ada embodies features which facilitate the achievement of safety, reliability and predictability in the system behavior (Halang & Stoyenko, 1990). Halang & Stoyenko (1990) carried out a detailed survey on a number of representative real-time programming languages including Ada, FORTRAN, HALL/S, LTR, PEARL, PL/I and Euclid, and concluded that Ada and PEARL were the most widely available and used languages among the others which had been surveyed.

In addition to the previous sets of modified and specialized real-time languages, it was accepted that universal, procedural programming languages (such as C) can also be used for real-time programming although they contain just rudimentary real-time features: this is mainly because such languages are more popular and widely available than genuine real-time languages (Halang & Stoyenko, 1990). Later generations of O-O languages such as C++ and Java also have popularity in embedded programming (Fisher et al., 2004). Embedded versions of famous ".Net" languages are gaining more popularity in the field of embedded systems development. However, they are not a favorite choice when it comes to resource constrained embedded systems as they are O-O languages, hence, they require a lot of resources as compared to the requirements of "C".

6. Choosing a suitable programming language for embedded design

In real-time embedded systems development, the choice of programming language is an important design consideration since it plays a significant role in reducing the total development time (Grogono, 1999).

Overall, it has been widely accepted that the low-level Assembly language suffers high development costs and lack of code portability, and only very few highly-skilled Assembly programmers can be found today (see Barr, 1999; Walls, 2005). If the decision is therefore made not to use the Assembly language due to its inevitable drawbacks, there is no scientific way to select the most optimal high-level programming language for a particular application (Sammet, 1969; Pont, 2002). Instead, researchers tend to discuss the important factors which should be considered in the choice of a language. For example, Sammet (1969) indicated that a major factor in selecting a language is the language suitability to solve the particular classes of problems for which it is intended, and the type of the actual user (i.e. user level of professionalism). It has also been noted by Sammet that factors such as availability on the desired computer hardware, history and previous evaluation, implementation consequences of the language are also key factors to take into account during the language selection process. However, Sammet stressed that a successful choice can only be made if the language includes the required technical features.

Specifically, when choosing a language for embedded systems development, the following factors must be considered (Pont, 2003):

- Embedded processors normally have limited speed and memory, therefore the language used must be efficient to meet the system resource constraints.
- Programming embedded systems requires a low-level access to the hardware. For example, there might be a need to read from / write to particular memory locations. Such actions require appropriate accessing mechanisms, e.g. pointers.
- The language must support the creation of flexible libraries, making it easy to re-use code components in various projects. It is also important that the developed software

should be easily ported and adapted to work on different processors with minimal changes.

- The language must be widely used in order to ensure that the developer can continue to recruit experienced professional programmers, and to guarantee that the existing programmers can have access to information sources (such as books, manuals, websites) for examples of good design and programming practices.

Of course, there is no perfect choice of programming language. However, the chosen language is required to be well-defined, efficient, supports low-level access to hardware, and available for the platform on which it is intended to be used. Against all of these factors, "C" language scores well, hence it turns out to be the most appropriate language to implement software for low-cost resource-constrained embedded systems. Pont (2003) stated that *"C's strengths for embedded system greatly outweigh its weaknesses. It may not be an ideal language for developing embedded systems, but it is unlikely that a 'perfect' language will be created"*.

7. The "C" programming language

In his famous book "Programming Embedded Systems in "C" and C++", Michael Barr (1999) emphasized that "C" language has been a constant factor across all embedded software development due to the following advantages:

- It is small and easy to learn.
- Its compilers are available for almost every processor in use today.
- There are so many experienced "C" programmers around the world.
- It is a hardware-independent programming language, a feature which allows the programmer to concentrate only on the algorithm rather than on the architecture of the processor on which the program will be running.

Despite this, Barr highlighted that the key advantage of "C" which made it the favorite choice for many embedded programmers is its low-level nature that provides the programmer with the ability to interact easily with the underlying hardware without sacrificing the benefits of using high-level programming.

In (Grogono, 1999), it was declared that "C" is based on a small number of primitive concepts, therefore it is an easy language to learn and program by both skilled and unskilled programmers. Moreover, Grogono stated that "C" can be easily compiled to produce efficient object code.

In a more recent publication, Pont (2002) stated that *"C's strengths for embedded system greatly outweigh its weaknesses. It may not be an ideal language for developing embedded systems, but it is unlikely that a 'perfect' language will be created"*. According to (Pont, 2002, 2003), the key features of the "C" language can be summarized as follows.

- It is a mid-level language with both high-level features (such as support for functions and modules) and low-level features (such as access to hardware via pointers).
- It is very efficient, popular and well understood even by desktop developers who programmed on C++ or Java.
- It has well-proven compilers available nowadays for every embedded processor (e.g. 8-, 16-, 32-bit or more).

- Books, training courses, code examples and websites that discuss the use of the language are all widely available.

In (Jones, 2002), it was noted that features such as easy access to hardware, low memory requirements, and efficient run-time performance make the "C" language popular and foremost among other languages. In (Brosgol, 2003), it was made clear that "C" is the typical choice for programming embedded applications as it is processor-independent, has low-level features, can be implemented on any architecture, has reasonable run-time performance, is an international standard, and is familiar to almost all embedded systems programmers. Fisher et al. (2004) emphasized that, in addition to portability and low-level features of the language, C structured programming drives embedded programmers to choose "C" language for their designs. Moreover, it has been clearly noted that "C" cannot be competed in producing a compact, efficient code for almost all processors used today (Ciocarlie & Simon, 2007).

Furthermore, since "C" was recognized as the de facto language for coding embedded systems including those which are safety-related (Jones, 2002; Pont, 2002; Walls, 2005), there have been attempts to make "C" a standard language for such applications by improving its safety characteristics rather than promoting the use of safer languages that are less popular (such as Ada). For example, The UK-based Motor Industry Software Reliability Association (MISRA) has produced a set of guidelines (and rules) for the use of "C" language in safety-critical software: such guidelines are well known as "MISRA C". For more details, see (Jones, 2002).

8. Why does "C" outperform other languages?

When comparing "C" to other alternative languages such as C++ or Ada, the following observations have been made. C++ is a good alternative to "C" as it provides better data abstraction and offers a better O-O programming style, but some of its features may cause degradation in program efficiency (Barr, 1999). Also, such a new generation O-O language is not readily available for the small embedded systems, primarily because of the overheads inherent in the O-O approach, e.g. CPU-time overhead (Pont, 2003).

Despite that Ada was a leading language that provided full support for concurrent and real-time programming, it has not gained much popularity (Brosgol, 2003) and has rarely been used outside the areas related to defense and aerospace applications (Barr, 1999; Ciocarlie & Simon, 2007). Unlike C, not many programmers nowadays are experienced in Ada, therefore only a small number of embedded systems are currently developed using this language (Ciocarlie & Simon, 2007). In addition, despite their approved efficiency, Ada compilers are not widely available for small embedded microcontrollers and usually need hard work to accept the program; especially by new programmers (Dewar, 2006). Indeed, both Ada and C++ have too large demand on low-cost embedded systems resources (e.g. memory requirements) and therefore cannot be suitable languages for such applications[1] (Walls, 2005).

---

[1] However, despite the indicated limitations of Ada, there has been a great deal of work on assessing a new version of Ada language (i.e. Ada-2005) to widen its application domain (see Burns, 2006; Taft et al., 2007). It has been noted that Ada-2005 can have the potential to overwhelm the use of "C" and its descendants in embedded systems programming (Brosgol and Ruiz, 2007).

In a survey carried out by Embedded Systems Design (ESD) in 2006, it was shown that the majority of existing and future embedded projects to which the survey applied were programmed (and likely to be programmed) in C. In particular, the results show that for 2006 projects, 51% were programmed in C, 30% in C++, and less than 5% were programmed in Ada. The survey shows that 47% of the embedded programmers were likely to continue to use "C" in their next projects. See Fig. 1 for further details.

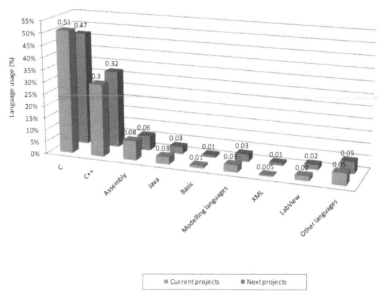

Fig. 1. Programming languages used in embedded system projects surveyed by ESD in 2006. The figure is derived from the data provided in (Nahas, 2008).

9. Using "C" to implement software for real-time embedded systems

Since "C" remains the most popular means for developing software in real-time embedded systems, it has been extensively used in the implementation of real-time schedulers and operating systems for embedded applications. In general, "C" was adopted in the software development of almost all operating systems (including RTOSs) in which schedulers are the core components (Laplante, 2004). In Michael Barr's book on embedded systems programming (i.e. Barr, 1999), it was noted that "C" is the main focus of any book about embedded programming. Therefore, most of the sample codes presented in Barr's book – for both schedulers and operating systems – were written in "C" and the key focus of the discussion was on how to use "C" language for 'in-house' embedded software development. However, some of the example code presented later in the book was written in C++ while Assembly language was avoided as much as possible. In (Barr & Massa, 2006), possible ways for implementing the eCos and the Embedded Linux, as a small and a large open-source operating systems (respectively), in "C" language were discussed. Other books which discuss the use of "C" language in the software implementation of real-time embedded systems include (Ganssle, 1992; Brown, 1994; Sickle, 1997; Zurell, 2000; Labrosse , 2000; Samek, 2002; Barnett et al., 2003; Laplante, 2004).

More specifically, using "C" language to implement the software code for particular scheduling algorithms is quite common. For example, Mooney et al. (1997) described a strategy for implementing a dynamic run-time scheduler using both hardware and software components: the software part was implemented using "C" language. Kravetz & Franke (2001) described an alternative implementation of the Linux operating system scheduler using "C" programming. It was emphasized that the new implementation can maintain the existing scheduler behavior / semantics with very little changes in the existing code.

Rao et al. (2008) discussed the implementation of a new pre-emptive scheduler framework using "C" language. The study basically reviewed and extracted the positive characteristics of existing pre-emptive algorithms (e.g. rate monotonic, EDF and LLF) to implement a new robust, fully pre-emptive real-time scheduler aimed at providing better performance in terms of timing and resource utilization.

Researchers of the Embedded Systems Laboratory (ESL), University of Leicester, UK have been greatly concerned with developing techniques and tools to support the design and implementation of reliable embedded systems, mainly using "C" programming language. An early work in this area was carried out by Pont (2001) which described techniques for implementing Time-Triggered Co-operative (TTC) architectures using a comprehensive set of "software design patterns" written in "C" language. The resulting "pattern language" was referred to as "PTTES[2] Collection" which contained more than seventy different patterns. As experience in this area has grown, this pattern collection has expanded and subsequently been revised in a series of ESL publications (e.g. Pont & Ong, 2003; Pont & Mwelwa, 2003; Mwelwa et al., 2003; Mwelwa & Pont, 2003; Pont et al., 2003; Pont & Banner, 2004; Mwelwa et al., 2004; Kurian & Pont, 2005; Kurian & Pont, 2006b; Pont et al., 2006; Wang et al., 2007, Kurian & Pont, 2007).

In (Nahas et al., 2004), a low-jitter TTC scheduler framework was described using "C" language. Phatrapornnant and Pont (2004a, 2004b) looked at ways for implementing low-power TTC schedulers by applying "dynamic voltage scaling" (DVS) algorithm programmed in "C" language. Moreover, Hughes & Pont (2008) described an implementation of TTC schedulers – in "C" language – with a wide range of "task guardian" mechanisms that aimed to reduce the impact of a task-overrun problem on the real-time performance of a TTC system. On the other hand, various ways in which Time-Triggered Hybrid (TTH) scheduler can be implemented in practice using "C" have been described in (Pont, 2001; Maaita & Pont, 2005; Hughes & Pont, 2008; Phatrapornnant, 2007). The ESL group has also been involved in creating software platforms for distributed embedded systems in which Shared-Clock (S-C) scheduling protocols are employed to achieve time-triggered operation over standard network protocols. All different S-C schedulers were implemented using "C" (for further details, see Pont, 2001; Ayavoo et al., 2007).

10. Conclusions

Selecting a suitable programming language is a key aspect in the success of the software development process. It has been shown that there is no specific method for selecting an appropriate programming language for the development of a specific project. However, the

---

[2] PTTES stands for Patterns for Time-Triggered Embedded Systems.

accumulation of experience along with subjective judgment enables software developers to make intelligent choices of programming languages for different application types.

Embedded software developers utilize different programming languages such as: Assembly, Ada, C, and C++. We have demonstrated that C is the most dominant programming language for embedded systems development. Although other languages may be winning ground when it comes to usage, C remains the de facto language for developing resource-constrained embedded systems which comprise a large portion of today's embedded applications.

11. Acknowledgement

The research summarized in this paper was partially carried out in the Embedded Systems Laboratory (ESL) at University of Leicester, UK, under the supervision of Professor Michael Pont, to whom the authors are thankful.

12. References

Ada (1980) "Reference Manual for the Ada Programming Language", proposed standard document, U.S. Department of Defense.

ANSVIP (1970) "American National Standard Vocabulary for Information Processing", American National Standards Institute, Inc., 1430 Broadway, New York, N.Y.

Ayavoo, D., Pont, M.J., Short, M. and Parker, S. (2007) "Two novel shared-clock scheduling algorithms for use with CAN-based distributed systems", Microprocessors and Microsystems, Vol. 31(5), pp. 326-334.

Baker, T.P. and Shaw, A. (1989) "The cyclic executive model and Ada. Real-Time Systems", Vol. 1 (1), pp. 7-25.

Barnett, R.H., O'Cull, L. and Cox, S. (2003) "Embedded C Programming and the Atmel Avr", Thomson Delmar Learning.

Barr, M. (1999) "Programming Embedded Systems in C and C++", O'Reilly Media.

Bates, D.G. (1968) "PROSPRO/1800", IEEE Transactions on Industrial Electronics and Control Instrumentation, Vol. 15, pp. 70-75.

Bieman, J.M., and Murdock, V. (2001) "Finding code on the World Wide Web: a preliminary investigation", Proceedings First IEEE International Workshop on Source Code Analysis and Manipulation, pp. 73-78.

Booch, G. (1991) "Object Oriented Design with Applications", Benjamin / Cummings.

Boulton, P.I.P. and Reid, P.A. (1969) "A Process-Control Language", IEEE Transactions on Computers, Vol. 18 (11), pp. 1049-1053.

Brosgol, B. and Ruiz, J. (2007) "Ada enhances embedded-systems development", Embedded.com, WWW website (Last accessed: November 2010) http://www.embedded.com/columns/technicalinsights/196800175?_requestid=1 67577

Broster, I. (2003) "Flexibility in dependable real-time communication", PhD thesis, University of York, York, U.K.

Brown, J.F. (1994) "Embedded Systems Programming in C and Assembly", Kluwer Academic Publishers.

Budlong, M. (1999) "Teach Yourself COBOL in 21 days", Sams.

Burns, A. (2006) "Real-Time Languages", Network of Excellence on Embedded Systems Design, WWW website (Last accessed: November 2010) http://www.artist-embedded.org/artist/Real-Time-Languages.html

Calgary (2005) "Calgary Ecommerce Services – Glossary", WWW website (Last accessed: November 2010) http://www.calgary-ecommerce-services.com/glossary.html

Carr, D. and Kizior, R.J. (2000) "The case for continued Cobol education", IEEE Software, Vol. 17 (2), pp. 33-36.

Chapman, S.J (2004) "Fortran 90/95 for Scientists and Engineers", McGraw-Hill Science Engineering.

Ciocarlie, H. and Simon, L. (2007) "Definition of a High Level Language for Real-Time Distributed Systems Programming", EUROCON 2007 The International Conference on "Computer as a Tool", Warsaw, September 9-12.

Cook, D. (1999) "Evolution of Programming Languages and Why a Language is Not Enough to Solve Our Problems", Software Technology Support Center, available online (Last accessed: November 2010)
http://www.stsc.hill.af.mil/crosstalk/1999/12/cook.asp

Davidgould (2008) "Davidgould – Glossary", WWW website (Last accessed: November 2010) http://www.davidgould.com/Glossary/Glossary.htm

Dewar, R.B.K. (2006) "Safety-critical design for secure systems: The languages, tools and methods needed to build error-free-software", WWW website (Last accessed: November 2010)
http://www.embedded.com/columns/technicalinsights/190400498?_requestid=1 77701

DIN (1979) "Programming language PEARL", Part 1. Basic PEARL, Part 2: Full PEARL, Deutsches Institut für Normung (DIN) German Standards Institute, Berlin, DIN 66253, 1979 (in English).

Fisher, J.A., Faraboschi, P. and Young, C. (2004) "Embedded Computing: A VLIW Approach to Architecture, Compilers and Tools", Morgan Kaufmann.

Flynn, I.M. (2001) "Generations, Languages", Macmillan Science Library: Computer Sciences, WWW website (Last accessed: November 2010)
http://www.bookrags.com/research/generations-languages-csci-01/

Ganssle, J. (1992) "The art of programming embedded systems", Academic Press, San Diego, USA.

Grogono, P. (1999) "The Evolution of Programming Languages", Course Notes, Department of Computer Science, Concordia University, Montreal, Quebec, Canada.

Halang, W.A. and Stoyenko, A.D. (1990) "Comparative evaluation of high-level real-time programming languages", Real-Time Systems, Vol. 2 (4), pp. 365-382.

Hansen, P.B. (1975) "The programming language Concurrent Pascal", IEEE Transactions on Software Engineering, Vol. 1 (2), pp. 199-207.

Hohmeyer, R.E. (1968) "CDC 1700 FORTRAN for process control", IEEE Transactions on Industrial Electronics and Control Instrumentation, Vol. 15, pp. 67-70.

Holyer, I (2008) "Dictionary of Computer Science", Department of Computer Science, University of Bristol, UK, WWW website (Last accessed: November 2010) http://www.cs.bris.ac.uk/Teaching/Resources/COMS11200/jargon.html

Hughes, Z.M. and Pont, M.J. (2008) "Reducing the impact of task overruns in resource-constrained embedded systems in which a time-triggered software architecture is employed", Trans Institute of Measurement and Control.

IFIP-ICC (1966) "The IFIP-ICC Vocabulary of Information Processing", North-Holland Pub. Co., Amsterdam.

ISO (2001) "ISO 5127 Information and documentation –Vocabulary", International Organisation for Standardisation (ISO).

Jalote, P. (1997) "An integrated approach to software engineering", Springer-Verlag.

Jarvis, P.H. (1968) "Some experiences with process control languages," IEEE Transactions on Industrial Electronics and Control Instrumentation, Vol. 15, pp. 54-56.

Jones, N. (2002) "Introduction to MISRA C", Embedded.com, WWW website (Last accessed: November 2010) http://www.embedded.com/columns/beginerscorner/9900659

Kircher, O. and Turner, E.B. (1968) "On-line MISSIL", IEEE Transactions on Industrial Electronics and Control Instrumentation, Vol. 15, pp. 80-84.

Koch, B. (1999) "The Theory of Task Scheduling in Real-Time Systems: Compilation and Systematization of the Main Results", Studies thesis, University of Hamburg.

Kravetz, M. and Franke, H. (2001) "Implementation of a Multi-Queue Scheduler for Linux", IBM Linux Technology Center, Version 0.2, April 2001.

Kurian, S. and Pont, M.J. (2005) "Building reliable embedded systems using Abstract Patterns, Patterns, and Pattern Implementation Examples", In: Koelmans, A., Bystrov, A., Pont, M.J., Ong, R. and Brown, A. (Eds.), Proceedings of the Second UK Embedded Forum (Birmingham, UK, October 2005), pp. 36-59. Published by University of Newcastle upon Tyne.

Kurian, S. and Pont, M.J. (2006) "Restructuring a pattern language which supports time-triggered co-operative software architectures in resource-constrained embedded systems", Paper presented at the 11th European Conference on Pattern Languages of Programs (EuroPLoP 2006), Germany, July 2006.

Kurian, S. and Pont, M.J. (2007) "Maintenance and evolution of resource-constrained embedded systems created using design patterns", Journal of Systems and Software, Vol. 80 (1), pp. 32-41.

Labrosse, J.J. (2000) "Embedded Systems Building Blocks: Complete and Ready-to-use Modules in C", Focal Press.

Lambert, K.A. and Osborne, M. (2000) "Java: A Framework for Program Design and Data Structures", Brooks / Cole.

Laplante, P.A. (2004) "Real-time Systems Design and Analysis", Wiley-IEEE.

Liberty, J. and Jones, B. (2004) "Teach Yourself C++ in 21 Days", Sams.

Maaita, A. and Pont, M.J. (2005) "Using 'planned pre-emption' to reduce levels of task jitter in a time-triggered hybrid scheduler". In: Koelmans, A., Bystrov, A., Pont, M.J., Ong, R. and Brown, A. (Eds.), Proceedings of the Second UK Embedded Forum (Birmingham, UK, October 2005), pp. 18-35. Published by University of Newcastle upon Tyne

Martin, J. and Leben, J. (1986) "Fourth Generation Languages Volume 1: Principles", Prentice Hall.

Mensh, M. and Diehl, W. (1968) "Extended FORTRAN for process control", IEEE Transactions on Industrial Electronics and Control Instrumentation, Vol. 15, pp. 75-79.

Mitchell, J.C. (2003) "Concepts in Programming Languages", Cambridge University Press.

Mwelwa C., Pont M.J. and Ward D. (2003) "Towards a CASE Tool to Support the Development of Reliable Embedded Systems Using Design Patterns", In: Bruel, J-M [Ed.] Proceedings of the 1st International Workshop on Quality of Service in Component-Based Software Engineering, June 20th 2003, Toulouse, France, Published by Cepadues-Editions, Toulouse.

Mwelwa, C. and Pont, M.J. (2003) "Two new patterns to support the development of reliable embedded systems", Paper presented at VikingPLoP 2003 (Bergen, Norway, September 2003).

Mwelwa, C., Pont, M.J. and Ward, D. (2004) "Code generation supported by a pattern-based design methodology", In: Koelmans, A., Bystrov, A. and Pont, M.J. (Eds.)

Proceedings of the UK Embedded Forum 2004 (Birmingham, UK, October 2004), pp. 36-55. Published by University of Newcastle upon Tyne

Nahas, M. (2008) "Bridging the gap between scheduling algorithms and scheduler implementations in time-triggered embedded systems", PhD thesis, Department of Engineering, University of Leicester, UK.

Nahas, M., Pont, M.J. and Jain, A. (2004) "Reducing task jitter in shared-clock embedded systems using CAN", In: Koelmans, A., Bystrov, A. and Pont, M.J. (Eds.) Proceedings of the UK Embedded Forum 2004 (Birmingham, UK, October 2004), pp. 184-194. Published by University of Newcastle upon Tyne.

Network Dictionary (2008) "Concurrent programming", WWW website (Last accessed: November 2010) http://wiki.networkdictionary.com/index.php/Concurrent_programming

Oerter, G.W. (1968) "A new implementation of decision tables for a process control language", IEEE Transactions on Industrial Electronics and Control Instrumentation, Vol. 15, pp. 57-61.

Opler, A. (1966) "Requirements for real-time languages", Communications of the ACM, Vol. 9 (3), pp. 196-199.

O'Reilly, T. (2006) "Programming Language Trends", WWW website (Last accessed: November 2010) http://radar.oreilly.com/archives/2006/08/programming-language-trends.html

Phatrapornnant, T. (2007) "Reducing Jitter in Embedded Systems Employing a Time-Triggered Software Architecture and Dynamic Voltage Scaling", PhD thesis, Department of Engineering, University of Leicester, UK.

Phatrapornnant, T. and Pont, M.J. (2004a) "The application of dynamic voltage scaling in embedded systems employing a TTCS software architecture: A case study", Proceedings of the IEE / ACM Postgraduate Seminar on "System-On-Chip Design, Test and Technology", Loughborough, UK, 15 September 2004. Published by IEE. ISBN: 0 86341 460 5 (ISSN: 0537-9989), pp. 3-8.

Phatrapornnant, T. and Pont, M.J. (2004b) "The application of dynamic voltage scaling in embedded systems employing a TTCS software architecture: A case study", Proceedings of the IEE / ACM Postgraduate Seminar on "System-On-Chip Design, Test and Technology", Loughborough, UK, 15 September 2004. Published by IEE. ISBN: 0 86341 460 5 (ISSN: 0537-9989), pp. 3-8.

Pont, M.J. (2001) "Patterns for time-triggered embedded systems: Building reliable applications with the 8051 family of microcontrollers", ACM Press / Addison-Wesley.

Pont, M.J. (2003) "An object-oriented approach to software development for embedded systems implemented using C", Transactions of the Institute of Measurement and Control, Vol. 25 (3), pp. 217-238.

Pont, M.J. and Banner, M.P. (2004) "Designing embedded systems using patterns: A case study", Journal of Systems and Software, Vol. 71 (3), pp. 201-213.

Pont, M.J. and Mwelwa, C. (2003) "Developing reliable embedded systems using 8051 and ARM processors: Towards a new pattern language", Paper presented at Viking PLoP 2003 (Bergen, Norway, September 2003).

Pont, M.J. and Ong, H.L.R. (2003) "Using watchdog timers to improve the reliability of TTCS embedded systems", in Hruby, P. and Soressen, K. E. [Eds.]Proceedings of the First Nordic Conference on Pattern Languages of Programs, September, 2002, pp.159-200. Published by Micrsoft Business Solutions.

Pont, M.J., Kurian, S. and Bautista-Quintero, R. (2006) "Meeting real-time constraints using 'Sandwich Delays'", In: Zdun, U. and Hvatum, L. (Eds) Proceedings of the Eleventh European conference on Pattern Languages of Programs (EuroPLoP '06), Germany, July 2006: pp. 67-77. Published by Universitätsverlag Konstanz.

Pont, M.J., Norman, A.J., Mwelwa, C. and Edwards, T. (2003) "Prototyping time-triggered embedded systems using PC hardware". Paper presented at EuroPLoP 2003 (Germany, June 2003).

Rao, M.V.P, Shet, K.C, Balakrishna, R. and Roopa, K. (2008) "Development of Scheduler for Real Time and Embedded System Domain", 22nd International Conference on Advanced Information Networking and Applications - Workshops, 25-28 March 2008, AINAW, pp. 1-6.

Roberts, B.C (1968) "FORTRAN IV in a process control environment", IEEE Transactions on Industrial Electronics and Control Instrumentation, Vol. 15, pp. 61-63.

Samek, M. (2002) "Practical Statecharts in C/C++: Quantum Programming for Embedded Systems", CMP Books.

Sammet, J.E. (1969) "Programming languages: history and fundamentals", Prentice-Hall.

Sanders, J. (2007) "Simple Glossary", WWW website (Last accessed: October 2007) http://www-xray.ast.cam.ac.uk/~jss/lecture/computing/notes/out/glossary/

Schoeffler, J.D. and Temple, R.H. (1970) "A real-time language for industrial process control", Proceedings of the IEEE, Vol. 58 (1), pp. 98-111.

Schutz, H.A. (1979) "On the Design of a Language for Programming Real-Time Concurrent Processes", IEEE Transactions on Software Engineering, Vol. 5 (3), pp. 248-255.

Sickle, T.V. (1997) "Reusable Software Components: Object-Oriented Embedded Systems Programming in C", Prentice Hall.

Sommerville, I. (2007) "Software engineering", 8th edition, Harlow: Addison-Wesley.

Steusloff, H.U. (1984) "Advanced real time languages for distributed industrial process control", IEEE Computer, pp. 37-46.

Taft, S.T., Duff, R.A., Brukardt, R.L., Ploedereder, E. and Leroy, P. (2007) "Ada 2005 Reference Manual: Language and Standard Libraries", Springer.

Walls, C. (2005) "Embedded Software: The Works", Newnes.

Wang, H., Pont, M.J. and Kurian, S. (2007) "Patterns which help to avoid conflicts over shared resources in time-triggered embedded systems which employ a pre-emptive scheduler", Paper presented at the 12th European Conference on Pattern Languages of Programs (EuroPLoP 2007).

Watson, D. (1989) "High Level Languages and Their Compilers", Addison-Wesley.

Wexelblat, L. (1981) "History of Programming Languages", Academic Press.

Wilson, L.B. and Clark, R.G. (2000) "Comparative Programming Languages", Addison-Wesley.

Wirth, N (1993) "Recollections about the development of Pascal", Proceedings of the 2nd ACM SIGPLAN conference on history of programming languages, pp. 333-342.

Wirth, N. (1977) "Modula - A programming language for modular multiprogramming", Software - Practice and Experience, Vol. 7, pp. 3-35.

Wizitt (2001) "T223 – A Glossary of Terms (Block 2)", Wizard Information Technology Training (Wizitt), WWW website (Last accessed: November 2010) http://wizitt.com/t223/glossary/glossary2.htm

Zurell, K. (2000) "C programming for embedded systems", CMP Books.

Zuse, K (1995) "A Brief History of Programming Languages", Byte.com, WWW website (Last accessed: November 2010) http://www.byte.com/art/9509/sec7/art19.htm

Part 3

High-Level Synthesis, SRAM Cells, and Energy Efficiency

16

High-Level Synthesis for Embedded Systems

Michael Dossis

Technological Educational Institute of Western Macedonia,
Dept. of Informatics & Computer Technology
Greece

1. Introduction

Embedded systems comprise small-size computing platforms that are self-sufficient. This means that they contain all the software and hardware components which are "embedded" inside the system so that complete applications can be realised and executed without the aid of other means or external resources. Usually, embedded systems are found in portable computing platforms such as PDAs, mobile and smart phones as well as GPS receivers. Nevertheless, larger systems such as microwave ovens and vehicle electronics, contain embedded systems. An embedded platform can be thought of as a configuration that contains one or more general microprocessor or microprocessor core, along with a number of customized, special function co-processors or accelerators on the same electronic board or integrated inside the same System-on-Chip (Soc). Often in our days, such embedded systems are implemented using advanced Field-Programmable Gate Arrays (FPGAs) or other types of Programmable Logic Devices (PLDs). FPGAs have improved a great deal in terms of integrated area, circuit performance and low power features. FPGA implementations can be easily and rapidly prototyped, and the system can be easily reconfigured when design updates or bug fixes are present and needed.

During the last 3-4 decades, the advances on chip integration capability have increased the complexity of embedded and in general custom VLSI systems to such a level that sometimes their spec-to-product development time exceeds even their product lifetime in the market. Because of this, and in combination with the high design cost and development effort required for the delivery of such products, they often even miss their market window. This problem generates competitive disadvantages for the relevant industries that design and develop these complex computing products. The current practice in the used design and engineering flows, for the development of such systems and applications, includes to a large extent approaches which are semi-manual, add-hoc, incompatible from one level of the design flow to the next, and with a lot of design iterations caused by the discovery of functional and timing bugs, as well as specification to implementation mismatches late in the development flow. All of these issues have motivated industry and academia to invest in suitable methodologies and tools to achieve higher automation in the design of contemporary systems. Nowadays, a higher level of code abstraction is pursued as input to automated E-CAD tools. Furthermore, methodologies and tools such as High-level Synthesis (HLS) and Electronic System Level (ESL) design entry employ established techniques, which are borrowed from the computer language program compilers and mature E-CAD tools and new algorithms such as advanced scheduling, loop unrolling and code motion heuristics.

The conventional approach in designing complex digital systems is the use of Register-Transfer Level (RTL) coding in hardware description languages such as VHDL and Verilog. However, for designs that exceed an area of a hundred thousand logic gates, the use of RTL models for specification and design can result into years of design flow loops and verification simulations. Combined with the short lifetime of electronic products in the market, this constitutes a great problem for the industry. The programming style of the (hardware/software) specification code has an unavoidable impact on the quality of the synthesized system. This is deteriorated by models with hierarchical blocks, subprogram calls as well as nested control constructs (e.g. if-then-else and while loops). For these models the complexity of the transformations that are required for the synthesis tasks (compilation, algorithmic transformations, scheduling, allocation and binding) increases at an exponential rate, for a linear increase in the design size.

Usually the input code (such as ANSI-C or ADA) to HLS tool, is first transformed into a control/data flow graph (CDFG) by a front-end compilation stage. Then various synthesis transformations are applied on the CDFG to generate the final implementation. The most important HLS tasks of this process are scheduling, allocation and binding. Scheduling makes an as-much-as-possible optimal order of the operations in a number of control steps or states. Optimization at this stage includes making as many operations as possible parallel, so as to achieve shorter execution times of the generated implementation. Allocation and binding assign operations onto functional units, and variables and data structures onto registers, wires or memory positions, which are available from an implementation library.

A number of commercial HLS tools exist nowadays, which often impose their own extensions or restrictions on the programming language code that they accept as input, as well as various shortcuts and heuristics on the HLS tasks that they execute. Such tools are the CatapultC by Mentor Graphics, the Cynthesizer by Forte Design Systems, the Impulse CoDeveloper by Impulse Accelerated Technologies, the Synfony HLS by Synopsys, the C-to-silicon by Cadence, the C to Verilog Compiler by C-to-Verilog, the AutoPilot by AutoESL, the PICO by Synfora, and the CyberWorkBench by NEC System Technologies Ltd. The analysis of these tools is not the purpose of this work, but most of them are suitable for linear, dataflow dominated (e.g. stream-based) applications, such as pipelined DSP and image filtering.

An important aspect of the HLS tools is whether their transformation tasks (e.g. within the scheduler) are based on formal techniques. The latter would guarantee that the produced hardware implementations are correct-by-construction. This means that by definition of the formal process, the functionality of the implementation matches the functionality of the behavioral specification model (the source code). In this way, the design will need to be verified only at the behavioral level, without spending hours or days (or even weeks for complex designs) of simulations of the generated register-transfer level (RTL), or even worse of the netlists generated by a subsequent RTL synthesis of the implementations. Behavioral verification (at the source code level) is orders of magnitude faster than RTL or even more than gate-netlist simulations. Releasing an embedded product with bugs can be very expensive, when considering the cost of field upgrades, recalls and repairs. Something that is less measurable, but very important as well, is the damage done to the industry's reputation and the consequent loss of customer trust. However, many embedded products are indeed released without all the testing that is necessary and/or desirable. Therefore, the quality of the specification code as well as formal techniques employed during transformations ("compilations") in order to deliver the hardware and software components of the system, are receiving increasing focus in embedded application development.

This chapter reviews previous and existing work of HLS methodologies for embedded systems. It also discusses the usability and benefits using the prototype hardware compilation system which was developed by the author. Section 2 discusses related work. Section 3 presents HLS problems related to the low energy consumption which is particularly interesting for embedded system design. The hardware compilation design flow is explained in section 4. Section 5 explains the formal nature of the prototype compiler's formal logic inference rules. In section 6 the mechanism of the formal high-level synthesis transformations of the back-end compiler is presented. Section 7 outlines the structure and logic of the PARCS optimizing scheduler which is part of the back-end compiler rules. Section 8 explains the available options for target micro-architecture generation and the communication of the accelerators with their computing environment. Section 9 outlines the execution environment for the generated hardware accelerators. Sections 10 and 11 discuss experimental results, draw useful conclusions, and propose future work.

2. Background and review of ESL methodologies

2.1 The scheduling task

The scheduling problem covers two major categories: time-constrained scheduling and resource-constrained scheduling. Time-constrained scheduling attempts to achieve the lowest area or number of functional units, when the total number of control steps (states) is given (time constraint). Resource-constrained scheduling attempts to produce the fastest schedule (the fewest control states) when the amount of hardware resources or hardware area are given (resource constraint). Integer linear programming (ILP) solutions have been proposed, but their run time grows exponentially with the increase of design size, which makes them impractical. Heuristic methods have also been proposed to handle large designs and to provide sub-optimal but practical implementations. There are two heuristic scheduling techniques: constructive solutions and iterative refinement. Two constructive methods are the as-soon-as-possible (ASAP) and the as-late-as-possible (ALAP) approach.

In both ASAP and ALAP scheduling, the operations that belong to the critical path of the design are not given any special priority over other operations. Thus, excessive delay may be imposed on the critical path operations. This is not good for the quality of the produced implementation. On the contrary, list scheduling utilizes a global priority function to select the next operation to be scheduled. This global priority function can be either the mobility (Pangrle & Gajski, 1987) of the operation or its urgency (Girczyc et al., 1985). Force-directed scheduling (Paulin & Knight, 1989) calculates the range of control steps for each operation between the operation's ASAP and ALAP state assignment. It then attempts to reduce the total number of functional units of the design's implementation, in order to evenly distribute the operations of the same type into all of the available states of the range.

The problem with constructive scheduling is that there is not any lookahead into future assignment of operations into the same control step, which may lead to sub-optimal implementations. After an initial schedule is delivered by any of the above scheduling algorithms, then iterative scheduling produces new schedules, by iteratively re-scheduling sequences of operations that maximally reduce the cost functions (Park & Kyung, 1991). This method is suitable for dataflow-oriented designs with linear control. In order to schedule control-intensive designs, the use of loop pipelining (Park & Parker, 1988) and loop folding (Girczyc, 1987), have been reported in the bibliography.

2.2 Allocation and binding tasks

Allocation determines the type of resource storage and functional units, selected from the library of components, for each data object and operation of the input program. Allocation also calculates the number of resources of each type that are needed to implement every operation or data variable. Binding assigns operations, data variables, data structures and data transfers onto functional units, storage elements (registers or memory blocks) and interconnections respectively. Also binding makes sure that the design's functionality does not change by using the selected library components.

Generally, there are three kinds of solutions to the allocation problem: constructive techniques, decomposition techniques and iterative approaches. Constructive allocation techniques start with an empty implementation and progressively build the datapath and control parts of the implementation by adding more functional, storage and interconnection elements while they traverse the CDFG or any other type of internal graph/representation format. Decomposition techniques divide the allocation problem into a sequence of well-defined independent sub-tasks. Each such sub-task is a graph-based theoretical problem which is solved with any of the three well known graph methods: clique partitioning, the left-edge technique and the weighted bipartite matching technique. The task of finding the minimum cliques in the graph which is the solution for the sub-tasks, is a NP-hard problem, so heuristic approaches (Tseng & Siewiorek, 1986) are utilized for allocation.

Because the conventional sub-task of storage allocation, ignores the side-effects between the storage and interconnections allocation, when using the clique partitioning technique, graph edges are enhanced with weights that represent the effect on interconnection complexity. The left-edge algorithm is applied on the storage allocation problem, and it allocates the minimum number of registers (Kurdahi & Parker, 1987). A weighted, bipartite-matching algorithm is used to solve both the storage and functional unit allocation problems. First a bipartite graph is generated which contains two disjoint sets, e.g. one for variables and one for registers, or one for operations and one for functional units. An edge between one node of the one of the sets and one node of the other represents an allocation of e.g. a variable to a register. The bipartite-matching algorithm considers the effect of register allocation on the design's interconnection elements, since the edges of the two sets of the graph are weighted (Huang et al., 1990). In order to improve the generated datapaths iteratively, a simple assignment exchange, using the pairwise exchange of the simulated annealing, or by using a branch-and-bound approach is utilized. The latter reallocates groups of elements of different types (Tsay & Hsu, 1990).

2.3 Early high-level synthesis

HLS has been an active research field for more than two decades now. Early approaches of experimental synthesis tools that synthesized small subsets of programming constructs or proprietary modeling formats have emerged since the late 80's. As an example, an early tool that generated hardware structures from algorithmic code, written in the PASCAL-like, Digital System Specification language (DSL) is reported in (Camposano & Rosenstiel, 1989). This synthesis tool performs the circuit compilation in two steps: first step is datapath synthesis which is followed by control synthesis. Examples of other behavioral circuit specification languages of that time, apart from DSL, were DAISY (Johnson, 1984), ISPS (Barbacci et al., 1979), and MIMOLA (Marwedel, 1984).

In (Casavant et al., 1989) the circuit to be synthesized is described with a combination of algorithmic and structural level code and then the PARSIFAL tool synthesizes the code into a bit-serial DSP circuit implementation. The PARSIFAL tool is part of a larger E-CAD system called FACE and which included the FACE design representation and design manager core. FACE and PARSIFAL were suitable for DSP pipelined implementations, rather than for a more general behavioral hardware models with hierarchy and complex control.

According to (Paulin & Knight, 1989) scheduling consists of determining the propagation delay of each operation and then assigning all operations into control steps (states) of a finite state machine. List-scheduling uses a local priority function to postpone the assignment of operations into states, when resource constraints are violated. On the contrary, force-directed scheduling (FDS) tries to satisfy a global execution deadline (time constraint) while minimizing the utilized hardware resources (functional units, registers and busses). The force-directed list scheduling (FDLS) algorithm attempts to implement the fastest schedule while satisfying fixed hardware resource constraints.

The main HLS tasks in (Gajski & Ramachandran, 1994) include allocation, scheduling and binding. According to (Walker & Chaudhuri, 1995) scheduling is finding the sequence of which operations to execute in a specific order so as to produce a schedule of control steps with allocated operations in each step of the schedule; allocation is defining the required number of functional, storage and interconnect units; binding is assigning operations to functional units, variables and values to storage elements and forming the interconnections amongst them to form a complete working circuit that executes the functionality of the source behavioral model.

The V compiler (Berstis, 1989) translates sequential descriptions into RTL models using parsing, scheduling and resource allocation. The source sequential descriptions are written in the V language which includes queues, asynchronous calls and cycle blocks and it is tuned to a kind of parallel hardware RTL implementations. The V compiler utilizes percolation scheduling (Fisher, 1981) to achieve the required degree of parallelism by meeting time constraints.

A timing network is generated from the behavioral design in (Kuehlmann & Bergamaschi, 1992) and is annotated with parameters for every different scheduling approach. The scheduling approach in this work attempts to satisfy a given design cycle for a given set of resource constraints, using the timing model parameters. This approach uses an integer linear program (ILP) which minimizes a weighted sum of area and execution time of the implementation. According to the authors, their Symphony tool delivers better area and speed than ADPS (Papachristou & Konuk, 1990). This synthesis technique is suitable for data-flow designs (e.g. DSP blocks) and not for more general complex control flow designs.

The CALLAS synthesis framework (Biesenack et al., 1993), transforms algorithmic, behavioral VHDL models into VHDL RTL and gate netlists, under timing constraints. The generated circuit is implemented using a Moore-type finite state machine (FSM), which is consistent with the semantics of the VHDL subset used for the specification code. Formal verification techniques such as equivalence checking, which checks the equivalence between the original VHDL FSM and the synthesized FSM are used in the CALLAS framework by using the symbolic verifier of the Circuit Verification Environment (CVE) system (Filkorn, 1991).

The Ptolemy framework (Kalavade & Lee, 1993) allows for an integrated hardware-software co-design methodology from the specification through to synthesis of hardware and software components, simulation and evaluation of the implementation. The tools of Ptolemy can synthesize assembly code for a programmable DSP core (e.g. DSP processor), which is built for a synthesis-oriented application. In Ptolemy, an initial model of the entire system is partitioned into the software and hardware parts which are synthesized in combination with their interface synthesis.

The Cosyma hardware-software co-synthesis framework (Ernst et al., 1993) realizes an iterative partitioning process, based on a hardware extraction algorithm which is driven by a cost function. The primary target in this work is to minimize customized hardware within microcontrollers but the same time to allow for space exploration of large designs. The specialized co-processors of the embedded system can be synthesized using HLS tools. The specification language is based on C with various extensions. The generated hardware descriptions are in turn ported to the Olympus HLS tool (De Micheli et al., 1990). The presented work included tests and experimental results based on a configuration of an embedded system, which is built around the Sparc microprocessor.

Co-synthesis and hardware-software partitioning are executed in combination with control parallelism transformations in (Thomas et al., 1993). The hardware-software partition is defined by a set of application-level functions which are implemented with application-specific hardware. The control parallelism is defined by the interaction of the processes of the functional behavior of the specified system. The system behavior is modeled using a set of communicating sequential processes (Hoare, 1985). Each process is then assigned either to hardware or to software implementation.

A hardware-software co-design methodology, which employs synthesis of heterogeneous systems, is presented in (Gupta & De Micheli, 1993). The synthesis process is driven by timing constraints which drive the mapping of tasks onto hardware or software parts so that the performance requirements of the intended system are met. This method is based on using modeling and synthesis of programs written in the HardwareC language. An example application which was used to test the methodology in this work was an Ethernet-based network co-processor.

2.4 Next generation high-level synthesis tools

More advanced methodologies and tools started appearing from the late 90s and continue with improved input programming code sets as well as scheduling and other optimization algorithms. The CoWare hardware-software co-design environment (Bolsens et al., 1997) is based on a data model that allows the user to specify, simulate and produce heterogeneous implementations from heterogeneous specification source models. This synthesis approach focuses on designing telecommunication systems that contain DSP, control loops and user interfaces. The synchronous dataflow (SDF) type of algorithms found in a category of DSP applications, can easily be synthesized into hardware from languages such as SILAGE (Genin et al., 1990), DFL (Willekens et al., 1994), and LUSTRE (Halbwachs et al. 1991). In contrast to this, dynamic dataflow (DDF) algorithms consume and produce tokens that are data-dependent, and thus they allow for complex if-then-else and while loop control constructs. CAD systems that allow for specifying both SDF and DDF algorithms and perform as much as possible static scheduling are the DSP-station from Mentor Graphics (Van Canneyt, 1994), PTOLEMY (Buck et al., 1994), GRAPE-II (Lauwereins et al., 1995), COSSAP from Synopsys and SPW from the Alta group (Rafie et al., 1994).

C models that include dynamic memory allocation, pointers and the functions malloc and free are mapped onto hardware in (Semeria et al., 2001). The SpC tool which was developed in this work resolves pointer variables at compile time and thus C functional models are synthesized into Verilog hardware models. The synthesis of functions in C, and therefore the resolution of pointers and malloc/free inside of functions, is not included in this work. The different techniques and optimizations described above have been implemented using the SUIF compiler environment (Wilson et al., 1994).

A heuristic for scheduling behavioral specifications that include a lot of conditional control flow, is presented in (Kountouris & Wolinski, 2002). This heuristic is based on a powerful intermediate design representation called hierarchical conditional dependency graph (HCDG). HCDG allows chaining and multicycling, and it enables advanced techniques such as conditional resource sharing and speculative execution, which are suitable for scheduling conditional behaviors. The HLS techniques in this work were implemented in a prototype graphical interactive tool called CODESIS which used HCDG as its internal design representation. The tool generates VHDL or C code from the HCDG, but no translation of standard programming language code into HCDG are known so far.

A coordinated set of coarse-grain and fine-grain parallelizing HLS transformations on the input design model are discussed in (Gupta et al., 2004). These transformations are executed in order to deliver synthesis results that don't suffer from the negative effects of complex control constructs in the specification code. All of the HLS techniques in this work were implemented in the SPARK HLS tool, which transforms specifications in a small subset of C into RTL VHDL hardware models. SPARK utilizes both control/data flow graphs (CDFGs) as well as an encapsulation of basic design blocks inside hierarchical task graphs (HTGs), which enable coarse-grain code restructuring such as loop transformations and an efficient way to move operations across large pieces of specification code.

Typical HLS tasks such as scheduling, resource allocation, module binding, module selection, register binding and clock selection are executed simultaneously in (Wang et al., 2003) so as to achieve better optimization in design energy, power and area. The scheduling algorithm utilized in this HLS methodology applies concurrent loop optimization and multicycling and it is driven by resource constraints. The state transition graph (STG) of the design is simulated in order to generate switched capacitance matrices. These matrices are then used to estimate power/energy consumption of the design's datapath. Nevertheless, the input to the HLS tool, is not programming language code but a proprietary format representing an enhanced CDFG as well as a RTL design library and resource constraints.

An incremental floorplanner is described in (Gu et al., 2005) which is used in order to combine an incremental behavioral and physical optimization into HLS. These techniques were integrated into an existing interconnect-aware HLS tool called ISCALP (Zhong & Jha, 2002). The new combination was named IFP-HLS (incremental floorplanner high-level synthesis) tool, and it attempts to concurrently improve the design's schedule, resource binding and floorplan, by integrating high-level and physical design algorithms.

(Huang et al., 2007) discusses a HLS methodology which is suitable for the design of distributed logic and memory architectures. Beginning with a behavioral description of the system in C, the methodology starts with behavioral profiling in order to extract simulation statistics of computations and references of array data. Then array data are distributed into different partitions. An industrial tool called Cyber (Wakabayashi, 1999) was developed which generates a distributed logic/memory micro-architecture RTL model, which is synthesizable with existing RTL synthesizers, and which consists of two or more partitions, depending on the clustering of operations that was applied earlier.

A system specification containing communicating processes is synthesized in (Wang et al., 2003). The impact of the operation scheduling is considered globally in the system critical path (as opposed to the individual process critical path), in this work. It is argued by the authors in this work, that this methodology allocates the resources where they are mostly needed in the system, which is in the critical paths, and in this way it improves the overall multi-process designed system performance.

The work in (Gal et al., 2008) contributes towards incorporating memory access management within a HLS design flow. It mainly targets digital signal processing (DSP) applications but also other streaming applications can be included along with specific performance constraints. The synthesis process is performed on the extended data-flow graph (EDFG) which is based on the signal flow graph. Mutually exclusive scheduling methods (Gupta et al., 2003; Wakabayashi & Tanaka, 1992) are implemented with the EDFG. The graph which is processed by a number of annotations and improvements is then given to the GAUT HLS tool (Martin et al., 1993) to perform operator selection and allocation, scheduling and binding.

A combined execution of operation decomposition and pattern-matching techniques is targeted to reduce the total circuit area in (Molina et al., 2009). The datapath area is reduced by decomposing multicycle operations, so that they are executed on monocycle functional units (FUs that take one clock cycle to execute and deliver their results). A simple formal model that relies on a FSM-based formalism for describing and synthesizing on-chip communication protocols and protocol converters between different bus-based protocols is discussed in (Avnit, 2009). The utilized FSM-based format is at an abstraction level which is low enough so that it can be automatically translated into HDL implementations. The generated HDL models are synthesizable with commercial tools. Synchronous FSMs with bounded counters that communicate via channels are used to model communication protocols. The model devised in this work is validated with an example of communication protocol pairs which included AMBA APB and ASB. These protocols are checked regarding their compatibility, by using the formal model.

The methodology of SystemCoDesigner (Keinert et al., 2009) uses an actor-oriented approach so as to integrate HLS into electronic system level (ESL) design space exploration tools. The design starts with an executable SystemC system model. Then, commercial synthesizers such as Forte's Cynthesizer are used in order to generate hardware implementations of actors from the behavioral model. This enables the design space exploration in finding the best candidate architectures (mixtures of hardware and software modules). After deciding on the chosen solution, the suitable target platform is then synthesized with the implementations of the hardware and software parts. The final step of this methodology is to generate the FPGA-based SoC implementation from the chosen hardware/software solution. Based on the proposed methodology, it seems that SystemCoDesigner method is suitable for stream-based applications, found in areas such as DSP, image filtering and communications.

A formal approach is followed in (Kundu et al., 2010) so as to prove that every HLS translation of a source code model produces a RTL model that is functionally-equivalent to the one in the behavioral input to the HLS tools. This technique is called translation validation and it has been maturing via its use in the optimizing software compilers. The validating system in this work is called SURYA, it is using the Symplify theorem prover and it was used to validate the SPARK HLS tool. This validation work found two bugs in the SPARK compilations.

The replacement of flip-flop registers with latches is proposed in (Paik et al., 2010) in order to yield better timing in the implemented designs. The justification for this is that latches are inherently more tolerant to process variations than flip-flops. These techniques were integrated into a tool called HLS-1. HLS-1 translates behavioral VHDL code into a synthesized netlist. Nevertheless, implementing registers with latches instead of edge-triggered flip-flops is generally considered to be cumbersome due to the complicated timing behavior of latches.

3. Synthesis for low power

A number of portable and embedded computing systems and applications such as mobile (smart) phones, PDAs, etc, require low power consumption therefore synthesis for low energy is becoming very important in the whole area of VLSI and embedded system design. During the last decade, industry and academia invested on significant part of research regarding VLSI techniques and HLS for low power design. In order to achieve low energy in the results of HLS and system design, new techniques that help to estimate power consumption at the high-level description level, are needed. In (Raghunathan et al., 1996), switching activity and power consumption are estimated at the RTL description taking also into account the glitching activity on a number of signals of the datapath and the controller. The spatial locality, the regularity, the operation count and the ratio of critical path to available time are identified in (Rabaey et al., 1995) with the aim to reduce the power consumption of the interconnections. The HLS scheduling, allocation and binding tasks consider such algorithmic statistics and properties in order to reduce the fanins and fanouts of the interconnect wires. This will result into reducing the complexity and the power consumed on the capacitance of the inteconnection buses (Mehra & Rabaey, 1996).

The effect of the controller on the power consumption of the datapath is considered in (Raghunathan & Jha, 1994). Pipelining and module selection was proposed in (Goodby et al., 1994) for low power consumption. The activity of the functional units was reduced in (Musoll & Cortadella, 1995) by minimizing the transitions of the functional unit's inputs. This was utilized in a scheduling and resource binding algorithm, in order to reduce power consumption. In (Kumar et al., 1995) the DFG is simulated with profiling stimuli, provided by the user, in order to measure the activity of operations and data carriers. Then, the switching activity is reduced, by selecting a special module set and schedule. Reducing supply voltage, disabling the clock of idle elements, and architectural tradeoffs were utilized in (Martin & Knight, 1995) in order to minimize power consumption within HLS.

The energy consumption of memory subsystem and the communication lines within a multiprocessor system-on-a-chip (MPSoC) is addressed in (Issenin et al., 2008). This work targets streaming applications such as image and video processing that have regular memory access patterns. The way to realize optimal solutions for MPSoCs is to execute the memory architecture definition and the connectivity synthesis in the same step.

4. The CCC hardware synthesis method

The previous two sections reviewed related work in HLS methodologies. This section and the following six sections describe a particular, formal HLS methodology which is directly applicable on embedded system design, and it has been developed solely by the author of this chapter. The Formal Intermediate Format (FIF)[1] was invented and designed by the author of this chapter as a tool and media for the design encapsulation and the HLS transformations in the CCC (Custom Coprocessor Compilation) hardware compilation tool[2]. A near-complete analysis of FIF syntax and semantics can be found in (Dossis, 2010). The formal methodology discussed here is based on using predicate logic to describe the intermediate representations of the compilation steps,   and the resolution of a set of transformation Horn clauses (Nilsson & Maluszynski, 1995) is used, as the building blocks of the prototype HLS tool.

The front-end compiler translates the algorithmic data of the source programs into the FIF's logic statements (logic facts). The inference logic rules of the back-end compiler transform the FIF facts into the hardware implementations. There is one-to-one correspondence between the source specification's subroutines and the generated hardware modules. The source code subroutines can be hierarchical, and this hierarchy is maintained in the generated hardware implementation. Each generated hardware model is a FSM-controlled custom processor (or co-processor, or accelerator), that executes a specific task, described in the source program code. This hardware synthesis flow is depicted in Figure 1.

Essentially the front-end compilation resembles software compilation and the back-end compilation executes formal transformation tasks that are normally found in HLS tools. This whole compilation flow is a formal transformation process, which converts the source code programs into implementable RTL (Register-Transfer Level) VHDL hardware accelerator models. If there are function calls in the specification code, then each subprogram call is transformed into an interface event in the generated hardware FSM. The interface event is

---

[1] The Formal Intermediate Format is patented with patent number: 1006354, 15/4/2009, from the Greek Industrial Property Organization
[2] This hardware compiler method is patented with patent number: 1005308, 5/10/2006, from the Greek Industrial Property Organization

used so that the "calling" accelerator uses the "services" of the "called" accelerator, as it is depicted in the source code hierarchy as well.

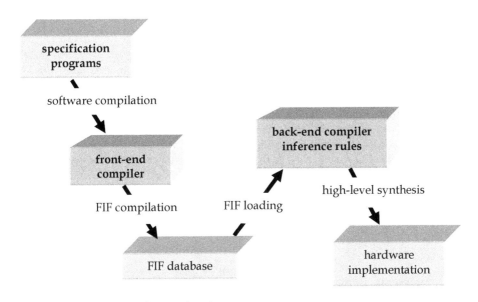

Fig. 1. Hardware synthesis flow and tools.

5. Back-end compiler inference logic rules

The back-end compiler consists of a very large number of logic rules. The back-end compiler logic rules are coded with logic programming techniques, which are used to implement the HLS algorithms of the back-end compilation phase. As an example, one of the latter algorithms reads and incorporates the FIF tables' facts into the compiler's internal inference engine of logic predicates and rules (Nilsson & Maluszynski, 1995). The back-end compiler rules are given as a great number of definite clauses of the following form:

$$A_0 \leftarrow A_1 \wedge \ldots \wedge A_n \text{ (where } n \geq 0) \hspace{2cm} \text{(form 1)}$$

where $\leftarrow$ is the logical implication symbol ($A \leftarrow B$ means that if B applies then A applies), and $A_0, \ldots, A_n$ are atomic formulas (logic facts) of the form:

$$\text{predicate\_symbol(Var\_1, Var\_2, \ldots, Var\_N)} \hspace{2cm} \text{(form 2)}$$

where the positional parameters Var_1,...,Var_N of the above predicate "predicate_symbol" are either variable names (in the case of the back-end compiler inference rules), or constants (in the case of the FIF table statements). The predicate syntax in form 2 is typical of the way that the FIF facts and other facts interact with each other, they are organized and they are used internally in the inference engine. Thus, the hardware descriptions are generated as "conclusions" of the inference engine upon the FIF "facts". This is done in a formal way from the input programs by the back-end phase, which turns the overall transformation into a provably-correct compilation process. In essence, the FIF file consists of a number of such atomic formulas, which are grouped in the FIF tables. Each such table contains a list of homogeneous facts which describe a certain aspect of the compiled program. E.g. all prog_stmt facts for a given subprogram are grouped together in the listing of the program statements table.

6. Inference logic and back-end transformations

The inference engine of the back-end compiler consists of a great number of logic rules (like the one in form 1) which conclude on a number of input logic predicate facts and produce another set of logic facts and so on. Eventually, the inference logic rules produce the logic predicates that encapsulate the writing of RTL VHDL hardware co-processor models. These hardware models are directly implementable to any hardware (e.g. ASIC or FPGA) technology, since they are technology and platform – independent. For example, generated RTL models produced in this way from the prototype compiler were synthesized successfully into hardware implementations using the Synopsys DC Ultra, the Xilinx ISE and the Mentor Graphics Precision software without the need of any manual alterations of the produced RTL VHDL code. In the following form 3 an example of such an inference rule is shown:

$$\text{dont\_schedule(Operation1, Operation2)} \leftarrow$$

$$\text{examine(Operation1, Operation2),}$$

$$\text{predecessor(Operation1, Operation2).} \qquad \text{(form 3)}$$

The meaning of this rule that combines two input logic predicate facts to produce another logic relation (dont_schedule), is that when two operations (Operation1 and Operation2) are examined and the first is a predecessor of the second (in terms of data and control dependencies), then don't schedule them in the same control step. This rule is part of a parallelizing optimizer which is called "PARCS" (meaning: Parallel, Abstract Resource – Constrained Scheduler).

The way that the inference engine rules (predicates relations-productions) work is depicted in Figure 2. The last produced (from its rule) predicate fact is the VHDL RTL writing predicate at the top of the diagram. Right bellow level 0 of predicate production rule there is a rule at the -1 level, then level -2 and so on. The first predicates that are fed into this engine of production rules belong to level –K, as shown in this figure. Level –K predicate facts include of course the FIF facts that are loaded into the inference engine along with the other predicates of this level.

In this way, the back-end compiler works with inference logic on the basis of predicate relation rules and therefore, this process is a formal transformation of the FIF source program definitions into the hardware accelerator (implementable) models. Of course in the case of the prototype compiler, there is a very large number of predicates and their relation rules that are defined inside the implementation code of the back-end compiler, but the whole concept of implementing this phase is as shown in Figure 2. The user of the back-end compiler can select certain environment command list options as well as build an external memory port parameter file as well as drive the compiler's optimizer with specific resource constraints of the available hardware operators.

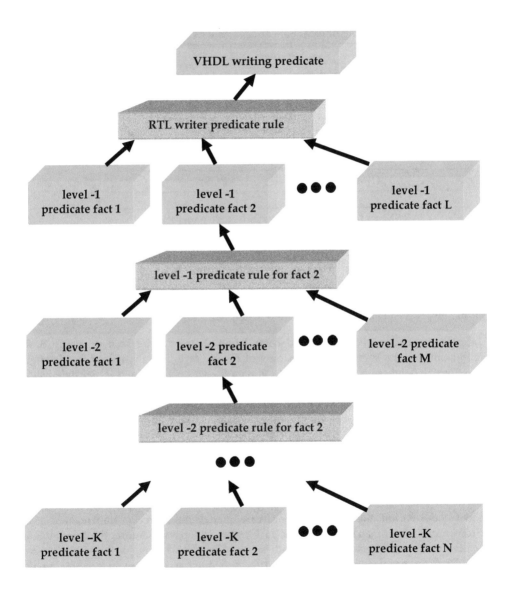

Fig. 2. The back-end inference logic rules structure.

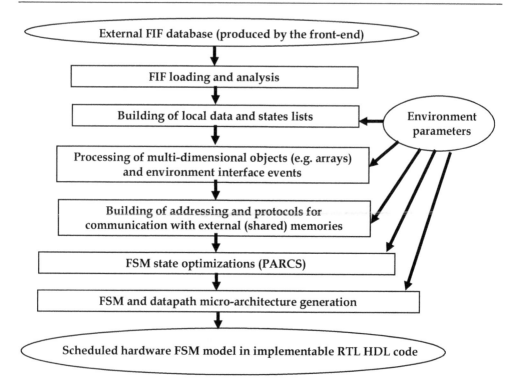

Fig. 3. The processing stages of the back-end compiler.

The most important of the back-end compilation stages can be seen in Figure 3. The compilation process starts with the loading of the FIF facts into the inference rule engine. After the FIF database is analyzed, the local data object, operation and initial state lists are built. Then the environment options are read and the temporary lists are updated with the special (communication) operations as well as the predecessor and successor dependency relation lists. After the complete initial schedule is built and concluded, the PARCS optimizer is run on it, and the optimized schedule is delivered to the micro-architecture generator. The transformation is concluded with the formation of the FSM and datapath implementation and the writing of the RTL VHDL model for each accelerator that is defined in each subprogram of the source code program.

A separate hardware accelerator model is generated from each subprogram in the system model code. All of the generated hardware models are directly implementable into hardware using commercial CAD tools, such as the Synopsys DC-ultra, the Xilinx ISE and the Mentor Graphics Precision RTL synthesizers. Also the hierarchy of the source program modules (subprograms) is maintained and the generated accelerators may be hierarchical. This means that an accelerator can invoke the services of another accelerator from within its processing states, and that other accelerator may use the services of yet another accelerator and so on. In this way, a subprogram call in the source code is translated into an external coprocessor interface event of the corresponding hardware accelerator.

7. The PARCS optimizer

PARCS aggressively attempts to schedule as many as possible operations in the same control step. The only limits to this are the data and control dependencies as well as the optional resource (operator) constraints, which are provided by the user.

1. start with the initial schedule (including the special external port operations)
2. Current PARCS state <- 1
3. Get the 1st state and make it the current state
4. Get the next state
5. Examine the next state's operations to find out if there are any dependencies with the current state
6. If there are no dependencies then absorb the next state's operations into the current PARCS state; If there are dependencies then finalize the so far absorbed operations into the current PARCS state, store the current PARCS state, PARCS state <- PARCS state + 1; make next state the current state; store the new state's operations into the current PARCS state
7. If next state is of conditional type (it is enabled by guarding conditions) then call the conditional (true/false branch) processing predicates, else continue
8. If there are more states to process then go to step 4, otherwise finalize the so far operations of the current PARCS state and terminate

Fig. 4. Pseudo-code of the PARCS scheduling algorithm.

The pseudo-code for the main procedures of the PARCS scheduler is shown in Figure 4. All of the predicate rules (like the one in form 1) of PARCS are part of the inference engine of the back-end compiler. A new design to be synthesized is loaded via its FIF into the back-end compiler's inference engine. Hence, the FIF's facts as well as the newly created predicate facts from the so far logic processing, "drive" the logic rules of the back-end compiler which generate provably-correct hardware architectures. It is worthy to note that although the HLS transformations are implemented with logic predicate rules, the PARCS optimizer is very efficient and fast. In most of benchmark cases that were run through the prototype hardware compiler flow, compilation did not exceed 1-10 minutes of run-time and the results of the compilation were very efficient as explained bellow.

8. Generated hardware architectures

The back-end stage of micro-architecture generation can be driven by command-line options. One of the options e.g. is to generate massively parallel architectures. The results of this option are shown in Figure 5. This option generates a single process – FSM VHDL description with all the data operations being dependent on different machine states. This implies that every operator is enabled by single wire activation commands that are driven by different state register values. This in turn means that there is a redundancy in the generated hardware, in a way that during part of execution time, a number of state-dedicated operators remain idle. However, this redundancy is balanced by the fact that this option achieves the fastest clock cycle, since the state command encoder, as well as the data

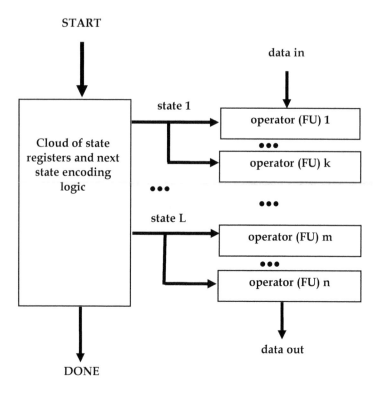

Fig. 5. Massively-parallel microarchitecture generation option.

multiplexers are replaced by single wire commands which don't exhibit any additional delay, and this option is very suitable to implement on large ASICs with plenty of resources.

Another micro-architecture option is the generation of traditional FSM + datapath based VHDL models. The results of this option are shown in Figure 6. With this option activated the generated VHDL models of the hardware accelerators include a next state process as well as signal assignments with multiplexing which correspond to the input data multiplexers of the activated operators. Although this option produces smaller hardware structures (than the massively-parallel option), it can exceed the target clock period due to larger delays through the data multiplexers that are used in the datapath of the accelerator.

Using the above micro-architecture options, the user of the CCC HLS tool can select various solutions between the fastest and larger massively-parallel micro-architecture, which may be suitable for richer technologies in terms of operators such as large ASICs, and smaller and more economic (in terms of available resources) technologies such as smaller FPGAs.

As it can be seen in Figure 5 and Figure 6, the produced co-processors (accelerators) are initiated with the input command signal START. Upon receiving this command the co-processors respond to the controlling environment using the handshake output signal BUSY

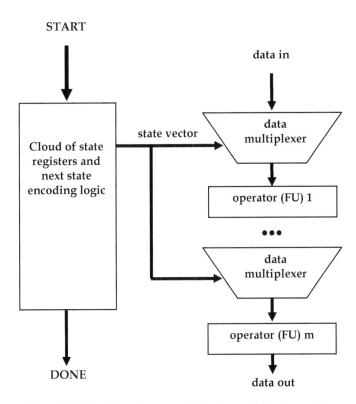

Fig. 6. The traditional FSM + datapath generated micro-architecture option.

and right after this, they start processing the input data in order to produce the results. This process may take a number of clock cycles and it is controlled by a set of states (discrete control steps). When the co-processors complete their processing, they notify their environment with the output signal DONE. In order to conclude the handshake the controlling environment (e.g. a controlling central processing unit) responds with the handshake input RESULTS_READ, to notify the accelerator that the processed result data have been read by the environment. This handshake protocol is also followed when one (higher-level) co-processor calls the services of another (lower-level) co-processor. The handshake is implemented between any number of accelerators (in pairs) using the START/BUSY and DONE/RESULTS_READ signals. Therefore, the set of executing co-processors can be also hierarchical in this way.

Other environment options, passed to the back-end compiler, control the way that the data object resources are used, such as registers and memories. Using a memory port configuration file, the user can determine that certain multi-dimensional data objects, such as arrays and array aggregates are implemented in external (e.g. central, shared) memories (e.g. system RAM). Otherwise, the default option remains that all data objects are allocated to hardware (e.g. on-chip) registers. All of the related memory communication protocols and hardware ports/signals, are automatically generated by the back-end synthesizer, and without the need for any manual editing of the RTL code by the user. Both synchronous and asynchronous memory communication protocol generation are supported.

9. Co-processor execution system

The generated accelerators can be placed inside the computing environment that they accelerate or can be executed standalone. For every subprogram in the source specification code one co-processor is generated to speed up (accelerate) the particular system task. The whole system (both hardware and software models) is modeled in algorithmic ADA code which can be compiled and executed with the host compiler and linker to run and verify the operation of the whole system at the program code level. In this way, extremely fast verification can be achieved at the algorithmic level. It is evident that such behavioral (high-level) compilation and execution is orders of magnitude faster than conventional RTL simulations.

After the required co-processors are specified, coded in ADA, generated with the prototype hardware compiler and implemented with commercial back-end tools, they can be downloaded into the target computing system (if the target system includes FPGAs) and executed to accelerate certain system tasks. This process is shown in Figure 7. The accelerators can communicate with each other and with the host computing environment using synchronous handshake signals and connections with the system's handshake logic.

10. Experimental results and evaluation of the method

In order to evaluate the efficiency of the presented HLS and ESL method, many designs from the area of hardware compilation and high-level synthesis were run through the front-end and the back-end compilers. Five selected benchmarks include a DSP FIR filter, a second order differential equation iterative solver, a well-known high-level synthesis benchmark, a RSA crypto-processor from cryptography applications, a synthetic benchmark that uses two level nested for-loops, and a large MPEG video compression engine. The fourth benchmark includes subroutines with two-dimensional data arrays stored in external memories. These data arrays are processed within the bodies of 2-level nested loops. All of the above generated accelerators were simulated and the RTL behavior matched the input source program's functionality. The state number reduction after applying the PARCS optimizer, on the various modules of the five benchmarks is shown in Table 1.

Moreover, the number of lines of RTL code is orders of magnitude more compared with the lines of the source code model for each sub-module. This indicates the gain in engineering productivity when the prototype ESL tools are used to automatically implement the computing products. It is well accepted in the engineering community that the coding & verification time at the algorithmic program level is only a small fraction of the time required for verifying designs at the RTL or the gate-netlist level. There were more than 400 states in the initial schedule of the MPEG benchmark. In addition to this, manual coding is extremely prone to errors which are very cumbersome and time-consuming to correct with (traditional) RTL simulations and debugging.

The specification (source code) model of the various benchmarks, and all of the designs using the prototype compilation flow, contains unaltered regular ADA program code,

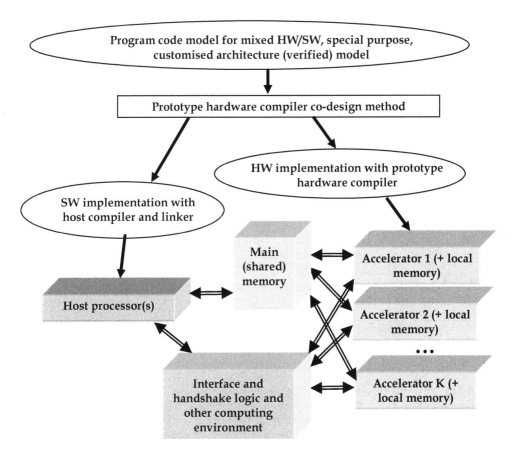

Fig. 7. Host computing environment and accelerators execution configuration.

without additional semantics and compilation directives which are usual in other synthesis tools which compile code in SystemC, HandelC, or any other modified program code with additional object class and TLM primitive libraries. This advantage of the presented methodology eliminates the need for the system designers to learn a new language, a new set of program constructs or a new set of custom libraries. Moreover, the programming constructs and semantics, that the prototype HLS compiler utilizes are the subset which is common to almost all of the imperative and procedural programming languages such as ANSI C, Pascal, Modula, Basic etc. Therefore, it is very easy for a user that is familiar with these other imperative languages, to get also familiar with the rich subset of ADA that the prototype hardware compiler processes. It is estimated that this familiarization doesn't exceed a few days, if not hours for the very experienced software/system programmer/modeler.

Module name	Initial schedule states	PARCS parallelized states	State reduction rate
FIR filter main routine	17	10	41%
Differential equation solver	20	13	35%
RSA main routine	16	11	31%
nested loops 1st subroutine	28	20	29%
nested loops 2nd subroutine (with embedded mem)	36	26	28%
nested loops 2nd subroutine (with external mem)	96	79	18%
nested loops 3rd subroutine	15	10	33%
nested loops 4th subroutine	18	12	33%
nested loops 5th subroutine	17	13	24%
MPEG 1st subroutine	88	56	36%
MPEG 2nd subroutine	88	56	36%
MPEG 3rd subroutine	37	25	32%
MPEG top subroutine (with embed. mem)	326	223	32%
MPEG top subroutine (with external mem)	462	343	26%

Table 1. State reduction statistics from the IKBS PARCS optimizer.

The following Table 2 contains the area and timing statistics of the main module of the MPEG application synthesis runs. Synthesis was executed on a Ubuntu 10.04 LTS linux server with Synopsys DC-Ultra synthesizer and the 65nm UMC technology libraries. From this table a reduction in terms of area can be observed for the FSM+datapath implementation against the massively parallel one. Nevertheless, due to the quality of the technology libraries the speed target of 2 ns clock period was achieved in all 4 cases.

Area/time statistic	massively-parallel, initial schedule	massively-parallel, PARCS schedule	FSM + datapath, initial schedule	FSM + datapath, PARCS schedule
area in square nm	117486	114579	111025	107242
equivalent number of NAND2 gates	91876	89515	86738	83783
achievable clock period	2 ns	2 ns	2 ns	2 ns
achievable clock frequency	500 MHz	500 MHz	500 MHz	500 MHz

Table 2. Area and timing statistics from UMC 65nm technology implementation.

Moreover, the area reduction for the FSM+datapath implementations of both the initial schedule and the optimized (by PARCS) one isn't dramatic and it reaches to about 6 %. This happens because the overhead of massively-parallel operators is balanced by the large amount of data and control multiplexing in the case of the FSM+datapath option.

11. Conclusions and future work

This chapter includes a discussion and survey of past and present existing ESL HLS tools and related synthesis methodologies suitable for embedded systems. Formal and heuristic techniques for the HLS tasks are discussed and more specific synthesis issues are analyzed. The conclusion from this survey is that the authors prototype ESL behavioral synthesizer is unique in terms of generality of input code constructs, the formal methodologies employed and the speed and utility of the developed hardware compiler.

One important contribution of this work is a provably-correct, ESL, and HLS method and a unified prototype tool-chain, which is based on compiler-compiler and formal logic inference techniques. The prototype tools transform a number of arbitrary input subprograms (for now coded in the ADA language) into an equivalent number of correct-by-construction and functionally-equivalent RTL VHDL hardware accelerator descriptions. Encouraging state-reduction rates of the PARCS scheduler-optimizer were observed for five benchmarks in this chapter, which exceed 30% in some cases. Using its formal flow, the prototype hardware compiler can be used to develop complex embedded systems in orders of magnitude shorter time and lower engineering effort, than that which are usually required using conventional design approaches such as RTL coding or IP encapsulation and schematic entry using custom libraries.

Existing HLS tools compile usually a small-subset of the programming language, and sometimes with severe restrictions in the type of constructs they accept (some of them don't accept while-loops for example). Furthermore, most of them are suited for linear, data-flow oriented specifications. However, a large number of applications found in embedded and telecommunication systems, mobile and other portable computing platforms involve a great deal of complex control flow with nesting and hierarchy levels. For this kind of applications most of HLS tools produce low level of quality results. The prototype ESL tool developed by the author has proved that it can deliver a better quality of results in applications with complex control such as image compression and processing standards.

Future extensions of this work include undergoing work to upgrade the front-end phase to accommodate more input programming languages (e.g. ANSI-C, C++) and the back-end HDL writer to include more back-end RTL languages (e.g. Verilog HDL), which are currently under development. Another extension could be the inclusion of more than 2 operand operations as well as multi-cycle arithmetic unit modules, such as multi-cycle operators, to be used in datapath pipelining. Moreover, there is ongoing work to extend the FIF's semantics so that it can accommodate embedding of IP blocks (such as floating-point units) into the compilation flow, and enhance further the schedule optimizer algorithm for even more reduced schedules. Furthermore, connection flows from the front-end compiler to even more front-end diagrammatic system modeling formats such as the UML formulation are currently investigated.

12. References

Avnit K., D'silva V., Sowmya A., Ramesh S. & Parameswaran S (2009) Provably correct on-chip communication: A formal approach to automatic protocol converter synthesis. *ACM Trans on Des Autom of Electr Sys (TODAES)*, ISSN: 1084-4309, Vol. 14, No. 2, article no: 19, March 2009.

Barbacci M., Barnes G., Cattell R. & Siewiorek D. (1979). *The ISPS Computer Description Language*. Report CMU-CS-79-137, dep. of Computer Science, Carnegie-Mellon University, USA.

Berstis V. (1989). The V compiler: automatic hardware design. *IEEE Des & Test of Comput*, Vol. 6, No. 2, pp. 8–17.

Biesenack J., Koster M., Langmaier A., Ledeux S., Marz S., Payer M., Pilsl M., Rumler S., Soukup H., Wehn N. & Duzy P. (1993). The Siemens high-level synthesis system CALLAS. *IEEE trans on Very Large Scale Integr (VLSI) sys*, Vol. 1, No. 3, September 1993, pp. 244-253.

Bolsens I., De Man H., Lin B., Van Rompaey K., Vercauteren S. & Verkest D. (1997). Hardware/software co-design of digital telecommunication systems. *Proceedings of the IEEE*, Vol. 85, No. 3, pp. 391-418.

Buck J., Ha S., Lee E. & Messerschmitt D. (1992). PTOLEMY: A framework for simulating and prototyping heterogeneous systems. *Invited Paper in the International Journal of Computer Simulation*, 31 August 1992. pp. 1-34.

Camposano R. & Rosenstiel W. (1989). Synthesizing circuits from behavioral descriptions. *IEEE Trans Comput-Aided Des Integr Circuits Syst*, Vol. 8, No. 2, pp. 171-180.

Casavant A., d'Abreu M., Dragomirecky M., Duff D., Jasica J., Hartman M., Hwang K. & Smith W. (1989). A synthesis environment for designing DSP systems. *IEEE Des & Test of Comput*, Vol. 6, No. 2, pp. 35–44.

De Micheli G., Ku D., Mailhot F. & Truong T. (1990). The Olympus synthesis system. *IEEE Des & Test of Comput*, Vol. 7, No. 5, October 1990, pp. 37-53.

Dossis M (2010) Intermediate Predicate Format for design automation tools. *Journal of Next Generation Information Technology (JNIT)*, Vol. 1, No. 1, pp. 100-117.

Ernst R., Henkel J. & Benner T. (1993). Hardware-software cosynthesis for microcontrollers. *IEEE Des & Test of Comput*, Vol. 10, No. 4, pp. 64-75.

Filkorn T. (1991). A method for symbolic verification of synchronous circuits, *Proceedings of the Comp Hardware Descr Lang and their Application (CHDL 91)*, pp. 229-239, Marseille, France 1991.

Fisher J (1981). Trace Scheduling: A technique for global microcode compaction. *IEEE trans. on comput*, Vol. C-30, No. 7, pp. 478-490.

Gajski D., & Ramachandran L. (1994). Introduction to high-level synthesis. *IEEE Des & Test of Comput*, Vol. 11, No. 4, pp. 44-54.

Gal B., Casseau E. & Huet S. (2008) Dynamic Memory Access Management for High-Performance DSP Applications Using High-Level Synthesis. *IEEE Trans on Very Large Scale Integr (VLSI)*, ISSN: 1063-8210, Vol. 16, No. 11, November 2008, pp. 1454-1464.

Genin D., Hilfinger P., Rabaey J., Scheers C. & De Man H. (1990). DSP specification using the SILAGE language, *Proceedings of the Int Conf on Acoust Speech Signal Process*, pp. 1056–1060, Albuquerque, NM., USA, 3-6 April 1990.

Girczyc E. (1987). Loop winding—a data flow approach to functional pipelining, *Proceedings of the International Symp on Circ and Syst*, pp. 382–385, 1987.

Girczyc E., Buhr R. & Knight J. (1985). Applicability of a subset of Ada as an algorithmic hardware description language for graph-based hardware compilation. *IEEE Trans Comput-Aided Des Integ Circuits Syst*, Vol. 4, No. 2, pp. 134-142.

Goodby L., Orailoglu A. & Chau P. (1994) Microarchitecture synthesis of performance-constrained low-power VLSI designs, *Proceedings of the Intern Conf on Comp Des (ICCD)*, ISBN: 0-8186-6565-3, Cambridge, MA , USA, 10-12 October 1994, pp. 323–326.

Gu Z., Wang J., Dick R. & Zhou H. (2005) Incremental exploration of the combined physical and behavioral design space. *Proceedings of the 42nd annual conf on des aut DAC '05*, Anaheim, CA, USA, June 13-17, 2005, pp. 208-213.

Gupta R. & De Micheli G. (1993). Hardware-software cosynthesis for digital systems. *IEEE Des & Test of Comput*, Vol. 10, No. 3, pp. 29-41.

Gupta S., Gupta R., Dutt N. & Nicolau A., (2003) Dynamically increasing the scope of code motions during the high-level synthesis of digital circuits, *Proceedings of the IEEE Conf Comput Digit Techn*, ISSN: 1350-2387, 22 Sept. 2003, Vol. 150, No. 5, pp. 330–337.

Gupta S., Gupta R., Dutt N. & Nikolau A. (2004) Coordinated Parallelizing Compiler Optimizations and High-Level Synthesis. *ACM Trans on Des Aut of Electr Sys*, Vol. 9, No. 4, September 2004, pp. 441–470.

Halbwachs N., Caspi P., Raymond P. & Pilaud D. (1991). The synchronous dataflow programming language Lustre, *Proceedings of the IEEE*, Vol. 79, No. 9, pp. 1305–1320.

Hoare C. (1985). *Communicating sequential processes*. Prentice-Hall, Englewood Cliffs, N.J., USA.

Huang C., Chen Y., Lin Y. & Hsu Y. (1990). Data path allocation based on bipartite weighted matching, *Proceedings of the Des Autom Conf (DAC)*, pp. 499–504, Orlando, Florida, USA, June, 1990.

Huang C., Ravi S., Raghunathan A. & Jha N. (2007) Generation of Heterogeneous Distributed Architectures for Memory-Intensive Applications Through High-Level Synthesis. *IEEE Trans on Very Large Scale Integr (VLSI)*, Vol. 15, No. 11, November 2007, pp. 1191-1204.

Issenin I, Brockmeyer E, Durinck B, Dutt ND (2008) Data-Reuse-Driven Energy-Aware Cosynthesis of Scratch Pad Memory and Hierarchical Bus-Based Communication Architecture for Multiprocessor Streaming Applications. *IEEE Trans on Comp-Aided Des of Integr Circ and Sys*, ISSN: 0278-0070, Vol. 27, No. 8, Aug. 2008, pp. 1439-1452.

Johnson S. (1984) *Synthesis of Digital Designs from Recursion Equations*. MA: MIT press, Cambridge.

Kalavade A. & Lee E. (1993). A hardware-software codesign methodology for DSP applications. *IEEE Des & Test of Comput*, Vol. 10, No. 3, pp. 16-28.

Keinert J., Streubuhr M., Schlichter T., Falk J., Gladigau J., Haubelt C., Teich J. & Meredith M. (2009) SystemCoDesigner—an automatic ESL synthesis approach by design space exploration and behavioral synthesis for streaming applications. *ACM Trans on Des Autom of Electr Sys (TODAES)*, ISSN: 1084-4309, Vol. 14, No. 1, article no: 1, January 2009.

Kountouris A. & Wolinski C. (2002) Efficient Scheduling of Conditional Behaviors for High-Level Synthesis. *ACM Trans. on Design Aut of Electr Sys*, Vol. 7, No. 3, July 2002, pp. 380–412.

Kuehlmann A. & Bergamaschi R. (1992). Timing analysis in high-level synthesis, *Proceedings of the 1992 IEEE/ACM international conference on Computer-aided design (ICCAD '92)*, pp. 349-354.

Kumar N., Katkoori S., Rader L. & Vemuri R. (1995) Profile-driven behavioral synthesis for low-power VLSI systems. *IEEE Des Test of Comput*, ISSN: 0740-7475, Vol. 12, No. 3, Autumn 1995, pp. 70–84.

Kundu S., Lerner S. & Gupta R. (2010) Translation Validation of High-Level Synthesis. *IEEE Trans Comput-Aided Des Integ Circuits Syst*, ISSN: 0278-0070 ,Vol. 29, No. 4, April 2010, pp. 566-579.

Kurdahi F. & Parker A. (1987). REAL: A program for register allocation, *Proceedings of the Des Autom Conf (DAC)*, pp. 210–215 , Miami Beach, Florida, USA, June, 1987.

Lauwereins R., Engels M., Ade M. & Peperstraete, J. (1995). GRAPE-II: A system level prototyping environment for DSP applications. *IEEE Computer*, Vol. 28, No. 2, February 1995, pp. 35–43.

Martin E., Santieys O. & Philippe J. (1993) GAUT, an architecture synthesis tool for dedicated signal processors, *Proceedings of the IEEE Int Eur Des Autom Conf (Euro-DAC)*, Hamburg, Germany, Sep. 1993, pp. 14–19.

Martin R. & Knight J. (1995) Power-profiler: Optimizing ASICs power consumption at the behavioral level, *Proceedings of the Des Autom Conf (DAC)*, ISBN: 0-89791-725-1, San Francisco, CA, USA, 1995, pp. 42-47.

Marwedel P. (1984). The MIMOLA design system: Tools for the design of digital processors, *Proceedings of the 21st Design Automation Conf (DAC)*, pp. 587-593.

Mehra R. & Rabaey J. (1996) Exploiting regularity for low-power design. *Dig of Techn Papers, Intern Conf on Comp-Aided Des (ICCAD)*, ISBN:0-8186-7597-7, San Jose, CA, USA, November 1996, pp. 166–172.

Molina M., Ruiz-Sautua R., Garcia-Repetto P. & Hermida R (2009) Frequent-Pattern-Guided Multilevel Decomposition of Behavioral Specifications. *IEEE Trans Comput-Aided Des Integ Circuits Syst*, ISSN: 0278-0070, Vol. 28, No. 1, January 2009, pp. 60-73.

Musoll E. & Cortadella J. (1995) Scheduling and resource binding for low power, *Proceedings of the Eighth Symp on Sys Synth*, ISBN: 0-8186-7076-2, Cannes , France, 13-15 September 1995, pp.104–109.

Nilsson U. & Maluszynski J. (1995) *Logic Programming and Prolog*. John Wiley & Sons Ltd., 2nd Edition, 1995.

Paik S., Shin I., Kim T. & Shin Y (2010) HLS-l: A High-Level Synthesis framework for latch-based architectures. *IEEE Trans Comput-Aided Des Integ Circuits Syst*, ISSN: 0278-0070, Vol. 29, No. 5, May 2010, pp. 657-670.

Pangrle B. & Gajski D. (1987). Design tools for intelligent silicon compilation. *IEEE Trans Comput-Aided Des Integ Circuits Syst*, Vol. 6, No. 6. pp. 1098–1112.

Papachristou C. & Konuk H. (1990). A Linear program driven scheduling and allocation method followed by an interconnect optimization algorithm, *Proceedings of the 27th ACM/IEEE Design Automation Conf (DAC)*, pp. 77-83.

Park I. & Kyung C. (1991). Fast and near optimal scheduling in automatic data path synthesis, *Proceedings of the Des Autom Conf (DAC)*, pp. 680–685, San Francisco, USA, 1991.

Park N. & Parker A. (1988). Sehwa: A software package for synthesis of pipelined data path from behavioral specification. *IEEE Trans Comput Aided Des Integrated Circuits Syst*, Vol. 7, No. 3, pp.356–370.

Paulin P. & Knight J. (1989). Algorithms for high-level synthesis. *IEEE Des & Test of Comput*, Vol. 6, No. 6, pp. 18-31.

Paulin P. & Knight J. (1989). Force-directed scheduling for the behavioral synthesis of ASICs. *IEEE Trans Comput-Aided Des Integ Circuits Syst*, Vol. 8, No 6, pp. 661–679.

Rabaey J., Guerra L. & Mehra R. (1995) Design guidance in the power dimension, *Proceedings of the 1995 Intern Conf on Acoustics, Speech, and Signal Proc*, ISBN: 0-7803-2431-5, Detroit, MI , USA, 9-12 May 1995, pp. 2837–2840.

Rafie M., et al. (1994) Rapid design and prototyping of a direct sequence spread-spectrum ASIC over a wireless link. *DSP and Multimedia Technol*, Vol. 3, No. 6, pp. 6–12.

Raghunathan A. & Jha N. (1994) Behavioral synthesis for low power, *Proceedings of the Intern Conf on Comp Des (ICCD)*, ISBN: 0-8186-6565-3, Cambridge, MA , USA, 10-12 October 1994 pp. 318–322.

Raghunathan A., Dey S. & Jha N. (1996) Register-transfer level estimation techniques for switching activity and power consumption, *Dig of Techn Papers, Intern Conf on Comp-Aided Des (ICCAD)*, ISBN: 0-8186-7597-7, San Jose, CA , USA, 10-14 November 1996, pp. 158–165.

Semeria L., Sato K. & De Micheli G. (2001) Synthesis of hardware models in C with pointers and complex data structures. *IEEE Trans VLSI Systems*, Vol. 9, No. 6, pp. 743–756.

Thomas D., Adams J. & Schmit H. (1993). A model and methodology for hardware-software codesign. *IEEE Des & Test of Comput*, Vol. 10, No. 3, pp. 6-15.

Tsay F., & Hsu Y. (1990). Data path construction and refinement. *Digest of Techn papers, Int Conf on Comp-Aided Des (ICCAD)*, pp. 308–311 , Santa Clara, CA, USA, November, 1990.

Tseng C. & Siewiorek D. (1986). Automatic synthesis of data path on digital systems. *IEEE Trans Comput Aided Des.Integ Circuits Syst*, Vol. 5, No. 3, pp. 379–395.

Van Canneyt M. (1994). Specification, simulation and implementation of a GSM speech codec with DSP station. *DSP and Multimedia Technol*, Vol. 3, No. 5, pp. 6–15.

Wakabayashi K. & Tanaka H. (1992) Global scheduling independent of control dependencies based on condition vectors, *Proceedings of the 29th ACM/IEEE Conf Des Autom (DAC)*, ISBN: 0-8186-2822-7, Anaheim, CA , USA, 8-12 June 1992, pp. 112-115.

Wakabayashi K. (1999) C-based synthesis experiences with a behavior synthesizer, "Cyber". *Proceedings of the Des Autom and Test in Eur Conf*, ISBN: 0-7695-0078-1, Munich, Germany, 9-12 March1999, pp. 390–393.

Walker R. & Chaudhuri S. (1995). Introduction to the scheduling problem. *IEEE Des & Test of Comput*, Vol. 12, No. 2, pp. 60–69.

Wang W., Raghunathan A., Jha N. & Dey S. (2003) High-level Synthesis of Multi-process Behavioral Descriptions, *Proceedings of the 16th IEEE International Conference on VLSI Design (VLSI'03)*, ISBN: 0-7695-1868-0, 4-8 Jan. 2003, pp. 467-473.

Wang W., Tan T., Luo J., Fei Y., Shang L., Vallerio K., Zhong L., Raghunathan A. & Jha N. (2003) A comprehensive high-level synthesis system for control-flow intensive behaviors, *Proceedings of the 13th ACM Great Lakes symp on VLSI GLSVLSI '03*, ISBN:1-58113-677-3, Washington, DC, USA, April 28-29, 2003, pp. 11-14.

Willekens P, et al (1994) Algorithm specification in DSP station using data flow language. DSP Applicat. 3(1):8–16.

Wilson R., French R., Wilson C., Amarasinghe S., Anderson J., Tjiang S., Liao S-W., Tseng C-W., Hall M., Lam M. & Hennessy J. (1994) Suif: An infrastructure for research on parallelizing and optimizing compilers. *ACM SIPLAN Notices*, Vol. 28, No. 9, December 2994, pp. 67–70.

Wilson T., Mukherjee N., Garg M. & Banerji1 D. (1995). An ILP Solution for Optimum Scheduling, Module and Register Allocation, and Operation Binding in Datapath Synthesis. *VLSI Design*, Vol. 3, No. 1, pp. 21-36.

Zhong L. & Jha N. (2002) Interconnect-aware high-level synthesis for low power. *Proceedings of the IEEE/ACM Int Conf Comp-Aided Des*, ISBN:0-7803-7607-2, November 2002, pp. 110-117.

17

SRAM Cells for Embedded Systems

Jawar Singh[1] and Balwinder Raj[2]

[1]*PDPM- Indian Institute of Information Technology, Design & Manufacturing, Jabalpur,*
[2]*ABV-Indian Institute of Information Technology and Management, Gwalior,*
India

1. Introduction

Static Random Access Memories (SRAMs) continue to be critical components across a wide range of microelectronics applications from consumer wireless to high performance server processors, multimedia and System on Chip (SoC) applications. It is also projected that the percentage of embedded SRAM in SoC products will increase further from the current 84% to as high as 94% by the year 2014 according to the International Technology Roadmap for Semiconductors (ITRS). This trend has mainly grown due to ever increased demand of performance and higher memory bandwidth requirement to minimize the latency, therefore, larger L1, L2 and even L3 caches are being integrated on-die. Hence, it may not be an exaggeration to say that the SRAM is a good technology representative and a powerful workhorse for the realization of modern SoC applications and high performance processors.

This chapter covers following SRAM aspects, basic operations of a standard 6-transistor (6T) SRAM cells and design metrics, nano-regime challenges and conflicting read-write requirements, recent trends in SRAM designs, process variation and Negative Bias Temperature Instability (NBTI), and SRAM cells for emerging devices such as Tunnel-FET (TFET) and Fin-FET. The basic operation of a SRAM cell as a storage element includes reading and writing data from/into the cell. Success of these operations is mainly gauged by two design metrics: Read Static Noise Margin (RSNM) and Write Static Noise Margin (WSNM). Apart from these metrics, an inline metric, N-curve is also used for measurement of read and write stability. The schematic diagrams and measurement process supported with HSPICE simulations results of different metrics will be presented in this chapter.

As standard 6T SRAM cell has failed to deliver the adequate read and write noise margins below 600mv for 65nm technology nodes, several new SRAM designs have been proposed in the recent past to meet the nano-regime challenges. In standard 6T, both read and write operations are performed via same pass-gate transistors, therefore, poses a conflicting sizing requirement. The recent SRAM cell designs which comprise of 7 to 10 transistor resolved the conflicting requirement by providing separate read and write ports.

SRAM cells are the first to suffer from the Process Variation (PV) induced side-effects. Because SRAM cells employ the minimum sized transistors to increase the device density into a die. PV significantly degrades the read and write noise margins and further exacerbates parametric yield when operating at low supply voltage. Furthermore, SRAM cells are particularly more susceptible to the NBTI effect because of their topologies. Since, one of the PMOS transistors is always negative bias if the cell contents are not flipped, it introduces asymmetry in the standard 6T SRAM cell due to shift in threshold voltage in either of PMOS devices, as a result poor read and write noise margin. A brief discussion on the impact of PV and NBTI on the SRAM will be covered in this chapter.

Finally, SRAM architectures for emerging devices such as TFET and Fin-FET will be discussed in this chapter. Also issues related to uni-directional devices (TFET) for realization of SRAM cell will be highlighted as uni-directional devices poses severe restriction on the implementation of SRAM cell.

2. Random-Access Memories (RAMs)

A random-access memory is a class of semiconductor memory in which the stored data can be accessed in any fashion and its access time is uniform regardless of the physical location. Random-access memories in general classified as read-only memory (ROM) and read/write memory. Read/write random-access memories are generally referred to as RAM. RAM can also be classified based on the storage mode of the memory: volatile and non-volatile memory. Volatile memory retains its data as long as power is supplied, while non-volatile memory will hold data indefinitely. RAM is referred as volatile memory, while ROM is referred as nonvolatile memory.

Memory cells used in volatile memories can be further classified into static or dynamic structures. Static RAM (SRAM) cells use feedback (or cross coupled inverters) mechanism to maintain their state, while dynamic RAM (DRAM) cells use floating capacitor to hold charge as a data. The charged stored in the floating capacitor is leaky, so dynamic cells must be refreshed periodically to retain stored data. The positive feedback mechanism, between two cross coupled inverters in SRAM provides a stable data and facilitates high speed read and write operations. However, SRAMs are faster and it requires more area per bit than DRAMs.

2.1 SRAM architecture

An SRAM cache consists of an array of memory cells along with peripheral circuitries, such as address decoder, sense amplifiers and write drivers etc. those enable reading from and writing into the array. A classic SRAM memory architecture is shown in Figure 1. The memory array consists of 2^n words of 2^m bits each. Each bit of information is stored in one memory cell. They share a common word-line (WL) in each row and a bit-line pairs (BL, complement of BL) in each column. The dimensions of each SRAM array are limited by its electrical characteristics such as capacitances and resistances of the bit lines and word lines used to access cells in the array. Therefore, large size memories may be folded into multiple blocks with limited number of rows and columns. After folding, in order to meet the bit and word line capacitance requirement each row of the memory contains 2^k words, so the array is physically organized as 2^{n-k} rows and 2^{m+k} columns. Every cell can be randomly addressed by selecting the appropriate word-line (WL) and bit-line pairs (BL, complement of BL), respectively, activated by the row and the column decoders.

The basic static RAM cell is shown in inset of Figure 1. It consists of two cross-coupled inverters (M3, M1 and M4, M2) and two access transistors (M5 and M6). The access transistors are connected to the wordline at their respective gate terminals, and the bitlines at their source/drain terminals. The wordline is used to select the cell while the bitlines are

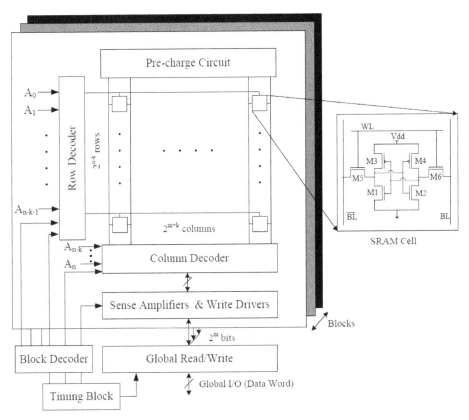

Fig. 1. SRAM architecture.

used to perform read or write operations on the cell. Internally, the cell holds the stored value on one side and its complement on the other side. The two complementary bitlines are used to improve speed and noise rejection properties [D. A. Hodges, 2003; S. M. Kang, 2003].

The voltage transfer characteristics (VTC) of cross-coupled inverters are shown in Figure 2. The VTC conveys the key cell design considerations for read and write operations. In the cross-coupled configuration, the stored values are represented by the two stable states in the VTC. The cell will retain its current state until one of the internal nodes crosses the switching threshold, V_S. When this occurs, the cell will flip its internal state. Therefore, during a read operation, we must not disturb its current state, while during the write operation we must force the internal voltage to swing past V_S to change the state.

2.2 Standard six transistor (6T) SRAM

The standard six transistor (6T) static memory cell in CMOS technology is illustrated schematically in Figure 3. The cross-coupled inverters, M_1, M_3 and M_2, M_4, act as the storage element. Major design effort is directed at minimizing the cell area and power consumption

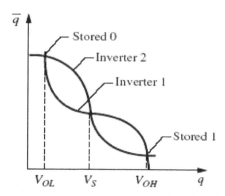

Fig. 2. Basic voltage transfer characteristics (VTC) of SRAM.

so that millions of cells can be placed on a chip. The steady state power consumption of the cell is controlled by sub-threshold leakage currents, so a larger threshold voltage is often used in memory circuits [J. Rabaey, 1999, J. P. Uyemura, 2002; A. S. Sedra 2003].

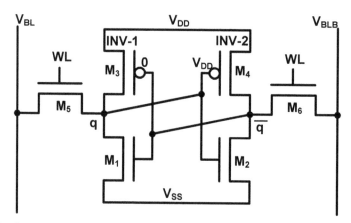

Fig. 3. Standard 6T SRAM cell.

3. Challenges in Bulk-Si SRAM scaling

Challenges for MOSFET scaling in the nanoscale regime including gate oxide leakage, control of short channel effects (SCE), contact resistance, ultra-shallow and abrupt junction technology apply to SRAM scaling as well. While it is possible to scale the classical bulk-Si MOSFET structure to sub-45 nm nodes [H. Wakabayashi *et al.*, 2003], effective control of SCE requires heavy channel doping (>5x10^{18} cm^{-3}) and heavy super-halo implants to suppress sub-surface leakage currents. As a result, carrier mobilities are severely degraded due to impurity scattering and a high transverse electric field in the ON-state. Further, more degraded SCE result in large leakage and larger subthreshold slope. Threshold voltage (V_{TH}) variability caused by random dopant fluctuations is another concern for nanoscale bulk-Si MOSFETs and is perceived as a fundamental roadblock for scaling SRAM. In addition to statistical dopant fluctuations, line-edge roughness increases the spread in transistor threshold voltage (V_{TH}) and thus the on- and off- currents and can limit the size of the cache [A. J. Bhavnagarwala *et al.*, 2001; A. Asenov *et al.*, 2001].

3.1 Process variations

The study of process variations has greatly increased due to aggressive scaling of CMOS technology. The critical sources have variation including gate length and width, random dopant fluctuation, line-edge and line-width roughness, variation associated with oxide thickness, patterning proximity effect etc. These variations result in dramatic changes in device and circuit performance and characteristics in positive and negative directions. SRAM cells are especially susceptible to process variations due to the use of minimum sized transistors within the cell to increase the SRAM density. Furthermore, the transistors within a cell must be closely matched in order to maintain good noise margins. An individual SRAM cell does not benefit from the "averaging effect" observed in multi-stage logic circuits whereby random device variations along a path tend to partially cancel one another.

The stability of a 6T SRAM cell under process variation can be verified by examining its butterfly curves obtained by voltage transfer characteristics (VTC) and inverse voltage transfer characteristics (VTC⁻¹). Under process variation the read static noise margin (SNM) of a standard 6T SRAM cell is shown in Figure. 4 (a). One can observe that the SNM window has narrowed down due to process variation and this effect becomes severe at lower V_{DD} =0.3V, as shown in Figure. 5 (a). Therefore, process variation affects the reliability and performance severely at lower voltages. However, recently different SRAM cells have been proposed to circumvent the read SNM problem in SRAM cell. The most attracting cell in this direction is referred as read SNM free 8T SRAM cell. This cell provides 2-3X times better read SNM even at lower voltages as shown in Figure. 4 (b) and 5 (b).

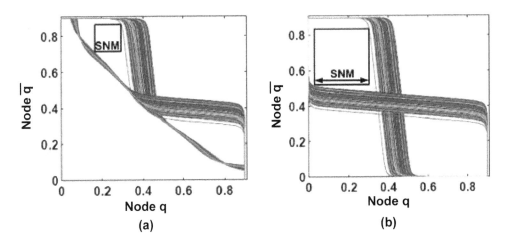

Fig. 4. Measurement of read static noise margin (SNM) at V_{DD}=0.9V for 45nm technology node (a) standard 6T SRAM cell, and (b) read SNM free 8T SRAM cell.

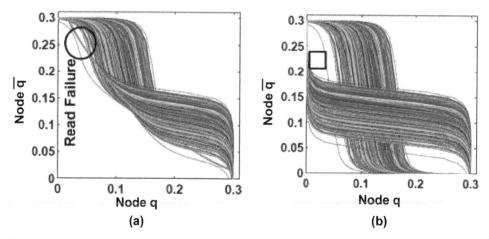

Fig. 5. Measurement of read static noise margin (SNM) at V_{DD}=0.3V for 45nm technology node (a) standard 6T SRAM cell, and (b) read SNM free 8T SRAM cell.

3.2 Device size requirements in SRAM cell

The standard 6T SRAM cell design space is continuously narrowing down due to lowering the supply voltage, shrinkage in device dimensions- attempting to achieve the high density and high performance objectives of on-chip caches. The SRAM cell stability, that is, read SNM and write-ability margins are further degraded by supply voltage scaling as shown above. The degradation in noise margins is mainly due to conflicting read and write requirements of the device size in the 6T cell. Both operations are performed via the same pass-gate (NMOS) devices, M5 and M6, as shown in Figure 3. For a better read stability (or read SNM), both pull down devices, M1 and M2 of the storage inverters must be stronger than the pass-gate devices, M5 and M6. While for write operation the opposite is desirable, that is, pass-gate devices, M5 and M6, must be stronger than pull up devices, M3 and M4, to achieve better write-ability, that is, weak storage inverters and strong pass-gate devices. Combining these constraints, yield the following relation.

strength (PMOS pull-up) < strength (NMOS access) < strength (NMOS pull-down)

The conflicting trend is also observed when read SNM and write noise margin (WNM) for different cell ratios and pull up ratios are simulated. Figure 6 shows the standard 6T SRAM cells' normalized read SNM and WNM measured for different cell ratio (CR), while the pull-up ratio is kept constant (PR=1). It can be seen from Figure 6 that the SNM is sharply increasing with increase in the cell ratio, while there is a gradual decrease in the WNM. For different pull-up ratio (PR), the normalized read SNM and WNM exhibit the similar trend. For example, there is a sharp increase in the read SNM and gradual decrease in WNM with increasing PR, while CR is kept constant to 2, as shown in Figure 7. In general, for a standard 6T cell the PR is kept to 1 while the CR is varied from 1.25 to 2.5 for a functional cell, in order to have a minimum sized cell for high density SRAM arrays. Therefore, in high density and high performance standard 6T SRAM cell, the recommended value for CR and PR are 2 and 1, respectively.

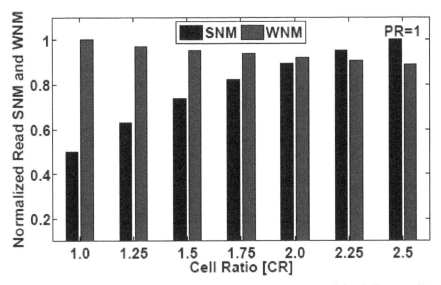

Fig. 6. Normalized read SNM and WNM of a standard 6T SRAM cell for different cell ratios (CR), while pull-up ratio (PR) was fixed to 1.

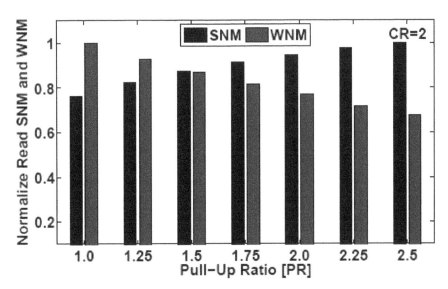

Fig. 7. Normalized read SNM and WNM of a standard 6T SRAM cell for different pull-up ratios (PR), while cell ratio (CR) is was fixed to 2.

3.3 Impact of NBTI on SRAM cells

A systematic shift in PMOS transistor parameters such as reduction in trans-conductance and drain current due to Negative Bias Temperature Instability (NBTI) over the life time of a system is becoming a significant reliability concern in nanometer regime. Particularly, sub-threshold devices and circuits which demand a high drive current for operation are hugely affected by threshold shifts and drive current losses due to NBTI. SRAM cells are particularly more susceptible to the NBTI effect because of their symmetric topologies. In other words, one of the PMOS transistor is always under stress if the SRAM cell contents are not periodically flipped. As a result, it introduces an asymmetric threshold shifts in both PMOS devices of a SRAM cell. The performance and reliability (noise margins) are significantly degraded in SRAM cells due to assymetric threshold voltage shift of PMOS devices. The degradation in read SNM of a standard 6T for different duty cycles (beta β) is shown in Figure 8. One can observe that there is a drastic reduction in read SNM of SRAM cell after five years of time span.

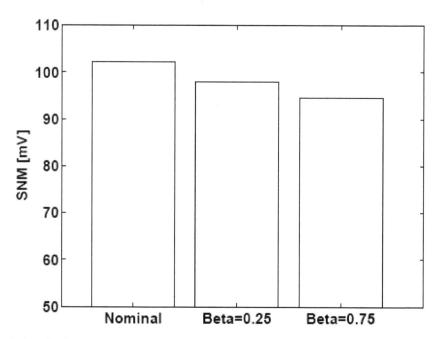

Fig. 8. Standard 6T SRAM cell read SNM degradation due to NBTI for different duty cycles.

3.4 SRAM scaling issues

Static Random Access Memory (SRAM) is by far the dominant form of embedded memory found in today's Integrated Circuits (ICs) occupying as much as 60-70% of the total chip area and about 75%-85% of the transistor count in some IC products. The most commonly used memory cell design uses Six Transistors (6-T) to store a bit, so all of the issues associated with MOSFET scaling apply to scaling of SRAM [A. Bhavnagarwala, et. al., 2005]. As memory will continue to consume a large fraction of the area in many future IC chips, scaling of memory density must continue to track the scaling trends of logic. [Z. Guo *et al.*, 2005]. Statistical dopant fluctuations, variations in oxide thickness and line-edge roughness increase the spread in transistor threshold voltage and thus on- and off- currents as the MOSFET is scaled down in the nanoscale regime [A. Bhavnagarwala *et al.*, 2005]. Increased transistor leakage and parameter variations present the biggest challenges for the scaling of 6-T SRAM memory arrays [C. H. Kim, *et. al.*, 2005, H. Qin, *et. al.*, 2004].

The functionality and density of a memory array are its most important properties. Functionality is guaranteed for large memory arrays by providing sufficiently large design margins (to be able to read without changing the state, to hold the state, to be writable and to function within a specified timeframe), which are determined by device sizing (channel widths and lengths), the supply voltage and, marginally, by the selection of transistor threshold voltages. Increase in process-induced variations results in a decrease in SRAM read and write margins, which prevents the stable operation of the memory cell and is perceived as the biggest limiter to SRAM scaling [E. J. Nowak, *et. al.*, 2003].

The 6-T SRAM cell size, thus far, has been scaled aggressively by ~0.5x every generation (Figure 9), however it remains to be seen if that trend will continue. Since the control of process variables does not track the scaling of minimum features, design margins will need to be increased to achieve large functional memory arrays. Moving to more lithography friendly regular layouts with gate lines running in one direction, has helped in gate line printability [P. Bai *et al.*, 2005], and could be the beginning of more layout regularization in the future. Also, it might become necessary to slow down the scaling of transistor dimensions to increase noise margins and ensure functionality of large arrays, i.e., tradeoff cell area for SRAM robustness. [Z. Guo *et al.*, 2005].

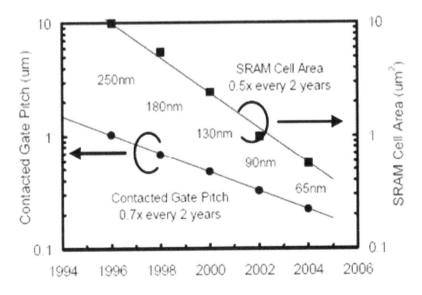

Fig. 9. SRAM cell size has been scaling at ~0.5 x per generation.

SRAM cells based on advanced transistor structures such as the planar UTB FETs and FinFETs have been demonstrated [E. J. Nowak *et al.*, 2003; T. Park *et al.*, 2003] to have excellent stability and leakage control. Some techniques to boost the SRAM cell stability, such as dynamic feedback [P. Bai *et al.*, 2005], are best implemented using FinFET technology, because there is no associated layout area or leakage penalty. FinFET-based SRAM are attractive for low-power, low voltage applications [K. Itoh, *et. al.*, 1998, M. Yamaoka, *et. al.*, 2005].

3.5 SRAM design Tradeoff's

a.   Area vs. Yield

The functionality and density of a memory array are its most important properties. The area efficiency and the reliable printing of the SRAM cell which directly impacts yield are both reliant on lithography technology. Given lithography challenges, functionality for large memory arrays is guaranteed by providing sufficiently large design margins, which are determined by device sizing (channel widths and lengths), the supply voltage and, marginally, by the selection of transistor threshold voltages. Although upsizing the transistors increases the noise margins, it increases the cell area and thus lowers the density [Z. Guo *et al.*, 2005].

b.   Hold Margin

In standby mode, when the memory is not being accessed, it still has to retain its state. The stored '1' bit is held by the PMOS load transistor (PL), which must be strong enough to compensate for the sub-threshold and gate leakage currents of all the NMOS transistors connected to the storage node V_L (Figure 8). This is becoming more of a concern due to the dramatic increase in gate leakage currents and degradation in I_{ON}/I_{OFF} ratio in recent technology nodes [H. Pilo *et al.*, 2005]. While hold stability was not of concern before, there has been a recent trend [H. Qin *et al.*, 2004] to decrease the cell supply voltage during standby to reduce static power consumption. The minimum supply voltage or the data retention voltage in standby is dictated by the hold margin. Degraded hold margins at low voltages make it increasingly more difficult to design robust low-power memory arrays. Hold stability is commonly quantified by the cell Static Noise Margin (SNM) in standby mode with the voltage on the word line $V_{WL}=0$ V. The SNM of an SRAM cell represents the minimum DC-voltage disturbance necessary to upset the cell state [E. Seevinck *et al.*, 1987], and can be quantified by the length of the side of the maximum square that can fit inside the lobes of the butterfly plot formed by the transfer characteristics of the cross-coupled inverters (Figure 10).

c.   Read Margin

During a read operation, with the bit lines (BL and CBL) in their precharged state, the Word Line (WL) is turned on (i.e., biased at V_{DD}), causing the storage node voltage, V_R, to rise above 0V, to a voltage determined by the resistive voltage divider formed by the access transistor (AXR) and the pull-down transistor (NR) between BL and ground (Figure 8). The ratio of the strengths of the NR and AXR devices (ratio of width/length of the two devices) determines how high V_R will rise, and is commonly referred to as the cell β-ratio. If V_R exceeds the trip voltage of the inverter formed by PL and NL, the cell bit will flip during the

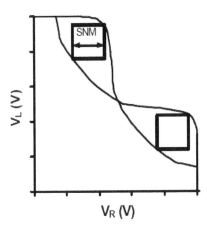

Fig. 10. Butterfly plot represents the voltage-transfer characteristics of the cross-coupled inverters in the SRAM cell.

read operation, causing a read upset. Read stability can be quantified by the cell SNM during a read access.

Since AXR operates in parallel to PR and raises V_R above 0V, the gain in the inverter transfer characteristic is decreased [A. J. Bhavnagarwala *et al.*, 2001], causing a reduction in the separation between the butterfly curves and thus in SNM. For this reason, the cell is considered most vulnerable to electrical disturbs during the read access. The read margin can be increased by upsizing the pull-down transistor, which results in an area penalty, and/or increasing the gate length of the access transistor, which increases the WL delay and also hurts the write margin. [J. M. Rabaey *et al.*, 2003] Process-induced variations result in a decrease in the SNM, which reduces the stability of the memory cell and have become a major problem for scaling SRAM. While circuit design techniques can be used to compensate for variability, it has been pointed out that these will be insufficient, and that development of new technologies, including new transistor structures, will be required [M. Yamaoka *et al.*, 2005].

d.   Write Margin

The cell is written by applying appropriate voltages to be written to the bit lines, e.g. if a '1' is to be written, the voltage on the BL is set to V_{DD} while that on the BLC is set to 0V and then the WL is pulsed to V_{DD} to store the new bit. Careful sizing of the transistors in a SRAM cell is needed to ensure proper write operation. During a write operation, with the voltage on the WL set to V_{DD}, AXL and PL form a resistive voltage divider between the BLC biased at 0V and V_{DD} (Figure 8). If the voltage divider pulls V_L below the trip voltage of the inverter formed by PR and NR, a successful write operation occurs. The write margin can be measured as the maximum BLC voltage that is able to flip the cell state while the BL voltage is kept high. The write margin can be improved by keeping the pull-up device minimum sized and upsizing the access transistor W/L, at the cost of cell area and the cell read margin [Z. Guo *et al.*, 2005].

e.   Access Time

During any read/write access, the WL voltage is raised only for a limited amount of time specified by the cell access time. If either the read or the write operation cannot be successfully carried out before the WL voltage is lowered, access failure occurs. A successful write access occurs when the voltage divider is able to pull voltage at V_L below the inverter trip voltage, after which the positive feedback in the cross-coupled inverters will cause the cell state to flip almost instantaneously. For the precharged bitline architecture that employs voltage-sensing amplifiers, a successful read access occurs if the pre-specified voltage difference, ΔV, between the bit-lines (required to trigger the sense amplifier) can be developed before the WL voltage is lowered [S. Mukhopadhyay *et al.*, 2004]. Access time is dependent on wire delays and the memory array column height. To speed up access time, segmentation of the memory into smaller blocks is commonly employed. With reductions in column height, the overhead area required for sense amplifiers can however become substantial.

4. Novel devices based SRAM design for Embedded Systems

4.1 FinFET based SRAM cell design

FinFETs have emerged as the most suitable candidate for DGFET structure as shown in figure 11 [E. Chin, et. al., 2006]. Proper optimization of the FinFET devices is necessary for reducing leakage and improving stability in FinFET based SRAM. The supply voltage (V_D), Fin height (H_{fin}) and threshold voltage (V_{th}) optimization can be used for reducing leakage in FinFET SRAMs by increasing Fin-height which allows reduction in V_D. [F. Sheikh, et. al., 2004]. However, reduction in V_D has a strong negative impact on the cell stability under parametric variations. We require a device optimization technique for FinFETs to reduce standby leakage and improve stability in an SRAM cell.

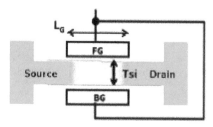

Fig. 11. Double Gate FinFET.

FinFET based SRAM cells are used to implement memories that require short access times, low power dissipation and tolerance to environmental conditions. FinFET based SRAM cells are most popular due to lowest static power dissipation among the various circuit configurations and compatibility with current logic processes. In addition, FinFET cell offers superior noise margins and switching speeds as well. Bulk MOSFET SRAM design at sub-45 nm node is challenged by increased short channel effects and sensitivity to process variations. Earlier works [Z. Guo, et. al., 2005; P. T. Su, et. al., 2006] have shown that FinFET based SRAM design shows improved performance compared to CMOS based design. Functionality and tolerance to process variation are the two important considerations for design of FinFET based SRAM at 32nm technology. Proper functionality is guaranteed by designing the SRAM cell with adequate read, write, static noise margins and lower power consumption. SRAM cells are building blocks for Random Access Memories (RAM). The cells must be sized as small as possible to achieve high densities. However, correct read operation of the FinFET based SRAM cell is dependent on careful sizing of M1 and M5 in figure 12. Correct write operation is dependent on careful sizing of M4 and M6 as shown in the figure 12. As explained [F. Sheikh, et. al., 2004], the critical operation is reading from the cell. If M5 is made of minimum-size, then M1 must be made large enough to limit the voltage rise on Q' so that the M3-M4 inverter does not inadvertently switch and accidentally write a '1' into the FinFET based SRAM cell.

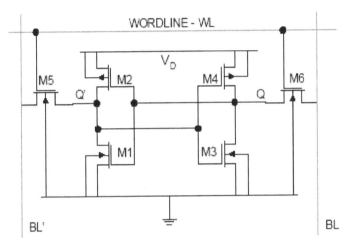

Fig. 12. 6T SRAM cell [F. Sheikh, et. al., 2004].

As explained [F. Sheikh, et. al., 2004], the sizing of the FinFET M5 and M6 is critical for correct operation once sizes for M1-M2 and M3-M4 inverters are chosen. The switching threshold for the ratioed inverter (M5-M6)-M2 must be below the switching threshold of the M3-M4 inverter to allow the flip-flop to switch from Q=0 to Q=1 state. The sizes for the FinFET can be determined through simulation, where M5 and M6 can be taken together to form a single transistor with twice the length of the individual transistors. It is well-understood that sizing affects noise margins, performance and power [Kiyoo Itoh, et. al., 1998; K. Zhang, et. al., 2005]. Therefore, sizes for pFinFET and nFinFET must be carefully selected to optimize the tradeoff between performance, reliability and power. We have studied FinFET based SRAM design issues such as: read and write cell margins, Static Noise Margin (SNM), power evaluation, performance and how they are affected by process induced variations [F. Sheikh, et. al., 2004].

4.2 Tunnel diode based SRAM cell design

As discussed in the previous sections, there is a fundamental limit to the scaling of the MOSFET threshold voltage, and hence the supply voltage. Scaling supply voltage limits the ON current (I_{ON}) and the I_{ON} - I_{OFF} ratio. This theoretical limit to threshold voltage scaling mainly arises from MOSFETs 60 mV/decade subthreshold swing at room temperature and it significantly restricts low voltage operation. Therefore, it seems that quantum transistors such as Inter-Band Tunnel Field Effect Transistors (TFETs) may be promising candidates to replace the traditional MOSFETs because the quantum tunnelling transistor has smaller dimension and steep subthreshold slope. Compared to MOSFET, TFETs have several advantages:

- Ultra-low leakage current due to the higher barrier of the reverse p-i-n junction.
- The subthreshold swing is not limited by 60mV/dec at room temperature because of its distinct working principle.
- V_t roll-off is much smaller while scaling, since threshold voltage of TFET depends on the band bending in the small tunnel region, but not in the whole channel region.
- There is no punch-through effect because of reverse biased p-i-n structure.

One key difference between TFETs and traditional MOSFETs that should be considered in the design of circuits is uni-directionality. TFETs exhibit the asymmetric behavior of conductance. For instance, in MOSFETs the source and drain are inter-changeable, with the distinction only determined by the biasing during the operation. While in TFETs, the source and drain are determined at the time of fabrication, and the flow of current I_{ON} takes place only when $V_{DS} > 0$. For $V_{DS} < 0$ a substantially less amount of current flows, referred as I_{OFF} or leakage current. Hence, TFETs can be thought to operate uni-directionally. This uni-directionality or passing a logic value only in one direction has significant implication on logic and in particularly for SRAMs design.

5. SRAM bitcell topologies

Standard 6T SRAM cell has been widely used in the implementation of high performance microprocessors and on-chip caches. However, aggressive scaling of CMOS technology presents a number of distinct challenges for embedded memory fabrics. For instance, smaller feature sizes imply a greater impact of process and design variability, including random threshold voltage (V_{TH}) variations, originating from the fluctuation in number of dopants and poly-gate edge roughness [Mahmoodi et al., 2005; Takeuchi et al., 2007]. The process and design variability leads to a greater loss of parametric yield with respect to SRAM bitcell noise margins and bitcell read currents when a large number of devices are integrated into a single die. Predictions in [A.J.Bhavnagarwala et al., 2001] suggest the variability will limit the voltage scaling because of degradation in the SNM and write margin. Furthermore, increase in device mismatch that accompanies geometrical scaling may cause data destruction at normal V_{DD} [Calhoun et al., 2005]. Therefore, a sufficiently large read Static Noise Margin (SNM) and Write-Ability Margin (WAM) in a bitcell are needed to handle the tremendous loss of parametric yield.

Recently, several SRAM bitcell topologies have been proposed to achieve different objectives such as minimum bitcell area, low static and dynamic power dissipation, improved performance and better parametric yield in terms of static noise margins (SNM) and write ability margin (WAM). The prime concern in SRAM bitcell design is a trade-off among these design metrics. For example, in sub-threshold SRAMs, noise margin (robustness) is the key design parameter and not the speed [Wang & Chandrakasan, 2004, 2005]. Some of the attracting SRAM bitcell topologies having good noise margin are as follows.

5.1 8T SRAM bitcell topology

Figure 13 shows the read SNM free 8T bitcell [Chang et al., 2005, 2008; Suzuki et al., 2008; Takeda et al., 2006; Verma & Chandrakasan, 2008], a register file type of SRAM bitcell topology, which has separate read and write ports. These separate read and write ports are controlled by read (RWL) and write (WWL) wordlines and used for accessing the bitcell during read and write cycles, respectively. In 8T bitcell topology, read and write operations of a standard 6T SRAM bitcell are de-coupled by creating an isolated read-port or read buffer (comprised of two transistors, M7 and M8). De-coupling of read and write operations yields a non-destructive read operation or SNM-free read stability. The interdependence between stability and read-current is overcome, while dependence between density and read-current remains there. An additional leakage current path is introduced by the separate read-port which increases the leakage current as compared to standard 6T bitcell. Therefore, an increased area overhead and leakage power make this design rather unattractive, since leakage power is a critical SRAM design metric, particularly for highly energy constrained applications. The read bitline leakage current problem in the 8T bitcell is similar to the problem in the standard 6T bitcell, except that the leakage currents from the un-accessed bitcells and from the accessed bitcell affect the same node, RBL. So, the leakage currents can pull down RBL regardless of the accessed bitcells state. In [Verma & Chandrakasan, 2008] the bitline leakage current from the un-accessed bitcells is managed by adding a buffer-footer, shared by the all bitcells in that word.

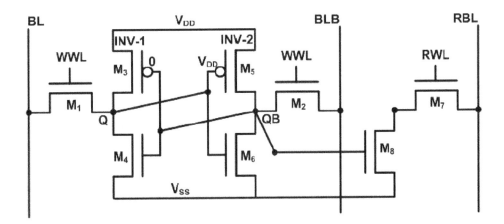

Fig. 13. Schematic diagram of read SNM free SRAM bitcell topology [Chang et al., 2005].

5.2 9T SRAM bitcell topology

Standard 6T bitcell along with three extra transistors were employed in nine-transistor (9T) SRAM bitcell [Liu & Kursun, 2008], to bypass read-current from the data storage nodes, as shown in Figure 14. This arrangement yields a non-destructive read operation or SNM-free read stability. However, it leads to 38% extra area overhead and a complex layout. Thin cell layout structure does not fit in this design and introduces jogs in the poly.

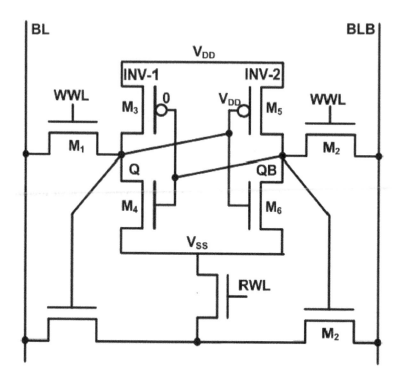

Fig. 14. Schematic diagram of 9T SRAM bitcell topology [Liu & Kursun, 2008].

5.3 10T SRAM bitcell topology

In the 10T bitcell [Calhoun & Chandrakasan, 2007], as shown in Figure 15, a separate read-port comprised of 4-transistors was used, while write access mechanism and basic data storage unit are similar to standard 6T bitcell. This bitcell also offers the same benefits as the 8T bitcell, such as a non-destructive read operation and ability to operate at ultra low voltages. But the 8T bitcell does not address the problem of read bitline leakage current, which degrades the ability to read data correctly. In particularly, the problem with the isolated read-port 8T cell is analogous to that with the standard (non-isolated read-port) 6T bitcell discussed. The only difference here is that the leakage currents from the un-accessed bitcells sharing the same read bit-line, RBL, affect the same node as the read-current from the accessed bitcell. As a result, the aggregated leakage current, which depends on the data stored in all of the unaccessed bitcells, can pull-down RBL even if the accessed bitcell based on its stored value should not do so. This problem is referred as an erroneous read. The erroneous read problem caused by the bitline leakage current from the un-accessed bitcells is managed by this 10T bitcell by providing two extra transistors in the read-port. These additional transistors help to cut-off the leakage current path from RBL when RWL is low and makes it independent of the data storage nodes content.

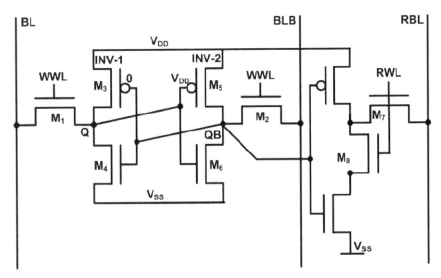

Fig. 15. Ultra-low voltage subthreshold 10T SRAM bitcell topology [Calhoun & Chandrakasan, 2007].

6. Summary

In this chapter, we have presented an existing review of bulk SRAM design and novel devices based embedded SRAM design. This literature survey has helped to identify various technical gaps in this area of research for embedded SRAM design. Through our work, we have tried to bridge these technical gaps in order to have better novel cells for low power applications in future embedded SRAM. Various research papers, books, monographic and articles have also been studied in the area of nanoscale device and memory circuits design. Articles on implementation of novel devices such as FinFET and Tunnel diode based 6T-SRAM cell for embedded system, which is having low leakage, high SNM and high speed were also incorporated.

7. References

A. Bhavnagarwala, S. Kosonocky, C. Radens, K. Stawiasz, R. Mann, and Q. Ye, "Fluctuation Limits & Scaling Opportunities for CMOS SRAM Cells," Proc. International Electron Devices Meeting, Technical Digest, Washington DC, pp. 659-662, 28.2.2005.

A. J. Bhavnagarwala, T. Xinghai, and J. D. Meindl, "The impact of intrinsic device fluctuations on CMOS SRAM cell stability," IEEE Journal of Solid-State Circuits, vol. 36, pp. 658-665, 2001.

Adel S. Sedra, Kenneth C. Smith, "Microelectronic Circuits", Fifth edition, Oxford University Press, 2003.

Calhoun, B., Daly, D., Verma, N., Finchelstein, D., Wentzloff, D., Wang, A., Cho, S.H. & Chandrakasan, A.,"Design considerations for ultra-low energy wireless microsensor nodes", Computers, IEEE Transactions on, 54, 727–740, 2005.

Calhoun, B.H. & Chandrakasan, A.P., "A 256-kb 65-nm sub-threshold sram design for ultra-low-voltage operation", Solid-State Circuits, IEEE Journal of , 42, 680–688, 2007.

Chang, L., Fried, D., Hergenrother, J., Sleight, J., Dennard, R., Montoye, R., Sekaric, L., McNab, S., Topol, A., Adams, C., Guarini, K. & Haensch, W., "Stable sram cell design for the 32 nm node and beyond", VLSI Technology, 2005. Digest of Technical Papers. 2005 Symposium on, 128–129.

Chang, L., Montoye, R., Nakamura, Y., Batson, K., Eickemeyer, R., Dennard, R., Haensch, W. & Jamsek, D., "An 8t-sram for variability tolerance and low-voltage operation in high-performance caches", Solid-State Circuits, IEEE Journal of , 43, 956–963, 2008.

Chris Hyung-il Kim, Jae-Joon Kim, "A Forward Body-Biased Low-Leakage SRAM Cache Device, Circuit and Architecture Considerations, " IEEE Transactions on Very Large Scale Integration (VLSI) Systems, vol. 13, pp. 349-357, no. 3, 2005.

David A. Hodges, "Analysis and Design of Digital Integrated Circuits", Third Edition, Tata McGraw-Hill Publishing Company Limited, 2003.

E. Chin, M. Dunga, B. Nikolic, "Design Trade-offs of a 6T FinFET SRAM Cell in the Presence of Variations," IEEE. Symp. VLSI Circuits, pp. 445- 449, 2006.

E. J. Nowak, T. Ludwig, I. Aller, J. Kedzierski, M. Leong, B. Rainey, M Breitwisch, V. Gemhoefer, J. Keinert, and D. M. Fried, "Scaling beyond the 65 nm node with FinFET-DGCMOS," Proc. CICC Custom Integrated Circuits Conference. San Jose, CA, pp.339-342, 2003

E. Seevinck, F. J. List, and J. Lohstroh, "Static-noise margin analysis of MOS SRAM cells," IEEE Journal of Solid-State Circuits, vol. SC-22, pp. 748-754, 1987.

F. Sheikh and V. Varadarajan, "The Impact of Device-Width Quantization on Digital Circuit Design Using FinFET Structures," EE 241 SPRING, pp. 1-6, 2004.

Gary Yeap, "Practical Low Power Digital VLSI Design", Kluwer Academic Publication, 1998.

H. Pilo, "SRAM Design in the Nanoscale Era," presented at International Solid- State Circuits Conference, pp. 366-367, 2005.

H. Qin, Y. Cao, D. Markovic, A. Vladimirescu, and J. Rabaey, "SRAM leakage suppression by minimizing standby supply voltage," presented at Proceedings, 5th International Symposium on Quality Electronic Design. San Jose, CA, pp. 55-60, 2004.

H. Wakabayashi, S. Yamagami, N. Ikezawa, A. Ogura, M. Narihiro, K. Arai, Y. Ochiai, K. Takeuchi, T. Yamamoto, and T. Mogami, "Sub-10-nm planar-bulk- CMOS devices using lateral junction control," presented at IEEE International Electron Devices Meeting, Washington, DC, pp. 20.7.1-20.7.4, 2003.

J. P. Uyemura, "Introduction to VLSI Circuit and Systems", Wiley, 2002. Principles of CMOS VLSI Design: A System Perspective

J. Rabaey, A. Chandrakasan, and B. Nikolic, "Digital Integrated Circuits: A Designer Perspective", Second Edition, Prentice-Hall, 2003.

Joohee Kim Marios C. Papaefthymiou, "Constant-Load Energy Recovery Memory for Efficient High-speed Operation" ISLPED'W, August 9 -1 1, 2004.

K. Zhang, U. Bhattacharya, Z. Chen, F. Hamzaoglu, D. Murray, N. Vallepalli, Y. Wang, B. Zheng, and M. Bohr, "A 3-GHz 70MB SRAM in 65nm CMOS technology with integrated column-based dynamic power supply," IEEE International Solid-State Circuits Conference. San Francisco, CA, pp.474-476, 2005.

Kaushik Roy, Sharat Prasad, "Low power CMOS VLSI Circuit Design", A Wiley Interscience Publication, 2000.

Kiyoo Itoh, "Review and Prospects of low-Power Memory Circuits", pp.313-317, 1998.

Kevin Zhang, Uddalak Bhattacharya, Zhanping Chen, "SRAM Design on 65-nm CMOS Technology With   Dynamic Sleep Transistor for Leakage Reduction," *IEEE JOURNAL OF SOLID-STATE CIRCUITS*, VOL. 40, NO. 4, APRIL 2005.

Kiyoo Itoh, "Review and Prospects of low-Power Memory Circuits", pp.313-317, 1998.

Liu, Z. & Kursun, V., "Characterization of a novel nine-transistor sram cell", Very Large Scale Integration (VLSI) Systems, IEEE Transactions on, 16, 488–492, 2008.

M. Yamaoka, R. Tsuchiya, and T. Kawahara, "SRAM Circuit with Expanded Operating Margin and Reduced Stand-by Leakage Current Using Thin-BOX FDSOI Transistors," presented at IEEE Asian Solid-State Circuits Conference, Hsinchu, Taiwan, pp. 109-112, 2005.

Mahmoodi, H., Mukhopadhyay, S. & Roy, K., "Estimation of delay variations due to random-dopant fluctuations in nanoscale cmos circuits", Solid-State Circuits, IEEE Journal of , 40, 1787–1796, 2005.

P. Bai, C. Auth, S. Balakrishnan, M. Bost, R. Brain, V. Chikarmane, R. Heussner, M. Hussein, J. Hwang, D. Ingerly, R. James, J. Jeong, C. Kenyon, E. Lee, S. H. Lee, N. Lindert, M. Liu, Z. Ma, T. Marieb, A. Murthy, R. Nagisetty, S. Natarajan, J. Neirynck, A. Ott, C. Parker, J. Sebastian, R. Shaheed, S. Sivakumar, J. Steigerwald, S. Tyagi, C. Weber, B. Woolery, A. Yeoh, K. Zhang, and M. Bohr, "A 65nm logic technology featuring 35nm gate lengths, enhanced channel strain, 8 Cu interconnect layers, low-k ILD and 0.57 μm2 SRAM cell," Proceeding International Electron Devices Meeting, San Francisco, CA, pp. 657-660, 2005

P. T. Su, C. H. Jin, C. J. Dong, H. S. Yeon, P. Donggun, K. Kinam, E. Yoon, and L. J. Ho, "Characteristics of the full CMOS SRAM cell using body tied TG MOSFETs (bulk FinFETs)," IEEE Trans. Electron Dev., vol. 53, pp. 481-487, 2006.

S. Mukhopadhyay, H. Mahmoodi-Meimand, and K. Roy, "Modeling and estimation of failure probability due to parameter variations in nano-scale SRAMs for yield enhancement," Symposium on VLSI Circuits, Digest of Technical Papers. Honolulu, HI, 2004.

Sung-Mo Kang, Yusef Leblebici, "CMOS Digital Integrated circuits-Analysis and Design", Third Edition, Tata McGraw-Hill Publishing Company Limited, 2003.

Takeda, K., Hagihara, Y., Aimoto, Y., Nomura, M., Nakazawa, Y., Ishii, T. & Kobatake, H., "A read-static-noise-margin-free sram cell for low-vdd and high-speed applications", IEEE Journal of Solid-State Circuits, 41, 113–121, 2006.

Takeuchi, K., Fukai, T., Tsunomura, T., Putra, A., Nishida, A., Kamohara, S. & Hiramoto, T., "Understanding random threshold voltage fluctuation by comparing multiple fabs and technologies", Electron Devices Meeting, IEDM 2007. IEEE International , 467–470, 2007.

Tohru Miwa, Junichi Yamada, Hiroki Koike, "A 512 Kbit low-voltage NV-SRAM with the size of a conventional SRAM", 2001 Symposium on VLSl Circuits Digest of Technical Papers.

Verma, N. & Chandrakasan, A.P., "A 256kb 65nm 8T Subthreshold SRAM Employing Sense-Amplifier Redundancy. IEEE Journal of Solid-State Circuits", 43, 141–149, 2008.

Wang, A. & Chandrakasan, A., A 180-mv subthreshold fft processor using a minimum energy design methodology. Solid-State Circuits, IEEE Journal,310–319, 2005.

Wang, A. & Chandrakasan, A., "A 180 mv fft processor using sub-threshold circuit techniques", In Proc.IEEE ISSCC Dig. Tech. Papers, 229–293, 2004.

Z. Guo, S. Balasubramanian, R. Zlatanovici, T.-J. King, and B. Nikolic', "FinFET based SRAM design," Proceeding, ISLPED, Proceedings of the International Symposium on Low Power Electronics and Design. San Diego, CA, pp. 2-7, 2005.

18

A Hierarchical C2RTL Framework for Hardware Configurable Embedded Systems

Yongpan Liu[1], Shuangchen Li[1], Huazhong Yang[1] and Pei Zhang[2]
[1]Tsinghua University, Beijing,
[2]Y Explorations, Inc., San Jose, CA
[1]P.R.China
[2]USA

1. Introduction

Embedded systems have been widely used in the mobile computing applications. The mobility requires high performance under strict power consumption, which leads to a big challenge for the traditional single-processor architecture. Hardware accelerators provide an energy efficient solution but lack the flexibility for different applications. Therefore, the hardware configurable embedded systems become the promising direction in future. For example, Intel just announced a system on chip (SoC) product, combining the ATOM processor with a FPGA in one package (Intel Inc., 2011).

The configurability puts more requirements on the hardware design productivity. It worsens the existing gap between the transistor resources and the design outcomes. To reduce the gap, design community is seeking a higher abstraction rather than the register transfer level(RTL). Compared with the manual RTL approach, the C language to RTL (C2RTL) flow provides magnitudes of improvements in productivity to better meet the new features in modern SoC designs, such as extensive use of embedded processors, huge silicon capacity, reuse of behavior IPs, extensive adoption of accelerators and more time-to-market pressure. Recently, people (Cong et al., 2011) observed a rapid rising demand for the high quality C2RTL tools.

In reality, designers have successfully developed various applications using C2RTL tools with much shorter design time, such as face detection (Schafer et al., 2010), 3G/4G wireless communication (Guo & McCain, 2006), digital video broadcasting (Rossler et al., 2009) and so on. However, the output quality of the C2RTL tools is inferior to that of the human-designed ones especially for large behavior descriptions. Recently, people proposed more scalable design architectures including different small modules connected by first-in first-out (FIFO) channels. It provides a natural way to generate a design hierarchically to solve the complexity problem.

However, there exist several major challenges of the FIFO-connected architecture in practice. First of all, the current tools leave the user to determine the FIFO capacity between modules, which is nontrivial. As shown in Section 2, the FIFO capacity has a great impact on the system performance and memory resources. Though determining the FIFO capacity via extensive

RTL-level simulations may work for several modules, the exploration space will become prohibitive large in the multiple-module case. Therefore, previous RTL-level simulating method is neither time-efficient nor optimal. Second, the processing rate among modules may bring a large mismatch, which causes a serious performance degradation. Block level parallelism should be introduced to solve the mismatches between modules. Finally, the C program partition is another challenge for the hierarchical design methodology.

This chapter proposed a novel C2RTL framework for configurable embedded systems. It supports a hierarchical way to implement complex streaming applications. The designers can determine the FIFO capacity automatically and adopt the block level parallelism. Our contributions are listed as below: 1) Unlike treating the whole algorithm as one module in the flatten design, we cut the complex streaming algorithm into modules and connect them with FIFOs. Experimental results showed that the hierarchical implementation provides up to 10.43 times speedup compared to the flatten design. 2) We formulate the parameters of modules in streaming applications and design a behavior level simulator to determine the optimal FIFO capacity very fast. Furthermore, we provide an algorithm to realize the block level parallelism under certain area requirement. 3) We demonstrate the proposed method in seven real applications with good results. Compared to the uniform FIFO capacity, our method can save memory resources by 14.46 times. Furthermore, the algorithm can optimize FIFO capacity in seconds, while extensive RTL level simulations may need hours. Finally, we show that proper block level parallelism can provide up to 22.94 times speedup in performance with reasonable area overheads.

The rest of the chapter is organized as follows. Section 2 describes the motivation of our work. We present our model framework in Section 3. The algorithm for optimal FIFO size and block level parallelism is formulated in Section 4 and 5. Section 6 presents experimental results. Section 7 illustrates the previous work in this domain. Section 8 concludes this paper.

2. Motivation

This section provides the motivation of the proposed hierarchical C2RTL framework for FIFO-connected streaming applications. We first compare the hierarchical approach with the flatten one. And then we point out the importance of the research of block level parallelism and FIFO sizing.

2.1 Hierarchical vs flatten approach

The flatten C2RTL approach automatically transforms the whole C algorithm into a large module. However, it faces two challenges in practice. 1) The translating time is unacceptable when the algorithm reaches hundreds of lines. In our experiments, compiling algorithms over one thousand lines into the hardware description language (HDL) codes may lead to several days to run or even failed. 2) The synthesized quality for larger algorithms is not so good as the small ones. Though the user may adjust the code style, unroll the loop or inline the functions, the effect is usually limited.

Unlike the flatten method, the hierarchical approach splits a large algorithm into several small ones and synthesizes them separately. Those modules are then connected by FIFOs.

It provides a flexible architecture as well as small modules with better performance. For example, we synthesized the JPEG encode algorithm into HDLs using eXCite (Y Exploration Inc., 2011) directly compared to the proposed solution. The flatten one costs 42'475'202 clock cycles with a max clock frequency of 69.74MHz to complete one computation, while the hierarchical method spends 4'070'603 clock cycles with a max clock frequency of 74.2MHz. It implies a 10.43 times performance speedup and a 7.2% clock frequency enhancement.

2.2 Performance with different block number

Among multiple blocks in a hierarchical design, there exist processing rate mismatches. It will have a great impact on the system performance. For example, Figure 1 shows the IDCT module parallelism. It is in the slowest block in the JPEG decoder. The JPEG decoder can be boosted by duplicating the IDCT module. However, block level parallelism may lead to nontrivial area overheads. It should be careful to find a balance point between the area and the performance.

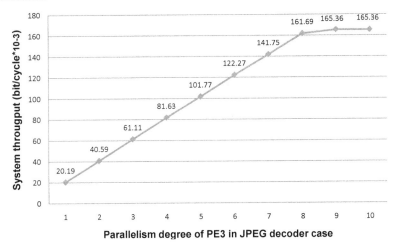

Fig. 1. System throughput under different parallelism degrees

2.3 Performance with different FIFO capacity

What's more, determining the FIFO size becomes relevant in the hierarchial method. We demonstrate the clock cycles of a JPEG encoder under different FIFO sizes in Figure 2. As we can see, the FIFO size will lead to an over 50% performance difference. It is interesting to see that the throughput cannot be boosted after a threshold. The threshold varies from several to hundreds of bits for different applications as described in Section 6. However, it is impractical to always use large enough FIFOs (several hundreds) due to the area overheads. Furthermore, designers need to decide the FIFO size in an iterative way when exploring different function partitions in the architecture level. Considering several FIFOs in a design, the optimal FIFO sizes may interact with each other. Thus, determining the proper FIFO size accurately and efficiently is important but complicated. More efficient methods are preferred.

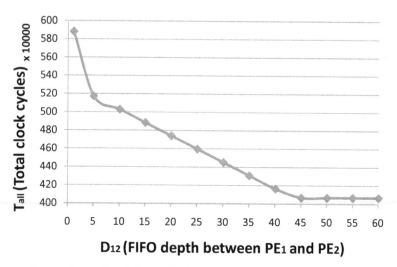

D₁₂ (FIFO depth between PE₁ and PE₂)

Fig. 2. Computing cycles under different FIFO sizes

3. Hierarchical C2RTL framework

This section first shows the diagram of the proposed hierarchical C2RTL framework. We then define four major stages: function partition, parameter extraction, block level parallelism and FIFO interconnection.

3.1 System diagram

The framework consists of four steps in Figure 3. In Step 1, we partition C codes into appropriate-size functions. In Step 2, we use C2RTL tools to transform each function into a hardware process element (PE), which has a FIFO interface. We also extract timing parameters of each PE to evaluate the partition in Step 1. If a partition violates the timing constraints, a design iteration will be done. In Step 3, we decide which PEs should be parallelized as well as the parallelism degree. In Step 4, we connect those PEs with proper sized FIFOs. Given a large-scale streaming algorithm, the framework will generate the corresponding hardware module efficiently. The synthesizing time is much shorter than that in the flatten approach. The hardware module can be encapsulated as an accelerator or a component in other designs. Its interface supports handshaking, bus, memory or FIFO. We denote several parameters for the module as below: the number of PEs in the module as N, the module's throughput as TH_{all}, the clock cycles to finish one computation as T_{all}, the clock frequency as CLK_{all} and the design area as A_{all}.

As C2RTL tools can handle the small-sized C codes synthesis (Step 2) efficiently, four main problems exist: how to partition the large-scale algorithm into proper-sized functions (Step 1), what parameters to be extracted from each PE(In Step 2), how to determine the parallelized PEs and their numbers (Step 3) and how to decide the optimal FIFO size between PEs (Step 4). We will discuss them separately.

3.2 Function partition

The C code partition greatly impacts the final performance. On one hand, the partition will affect the speed of the final hardware. For example, a very big function may lead to a very slow PE. The whole design will be slowed down, since the system's throughput is decided by the slowest PE. Therefore, we need to adjust the slowest PE's partition. The simplest method is to split it into two modules. In fact, we observe that the ideal and most efficient partition leads to an identical throughput of each PE. On the other hand, the partition will also affect the

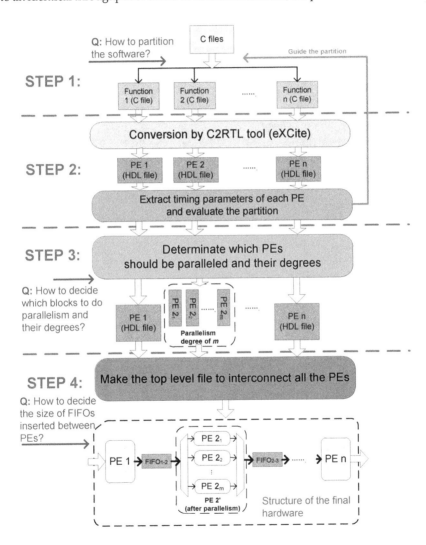

Fig. 3. Hierarchical C2RTL Flow

Name	Description	Examples[2]
Type	Interface type,I or II	II
$TH_{ni/o}$	Throughput of input or output interface	0.0755
$t_{ni/o}$	Input or output time in T_n (cycles)	128
T_n	Period of PE_n (cycles)	848
A_n	Area of PE_n (LE)	4957
f_n	$TH_{no}/TH_{ni/i}$	1
$SoP_n(m)$	State of PE_n at m^{th} cycle 0:Processing;1:Reading; 2:Writing;3:Reading and writing	

[1] m means m^{th} cycle.
[2] Output of PE_2 in the JPEG encode case, as shown in Figre 4

Table 1. The parameter of the nth PE's input/output interfaces

area. Too fine-grained partitions lead to many independent PEs, which will not only reduce the resource sharing but also increase the communication costs.

In this design flow, we use a manual partition strategy, because no timing information in C language makes the automatic partition difficult. In this framework, we introduce an iterative design flow. Based on the timing parameters[1] extracted by the PEs from the C2RTL tools, the designers can determine the C code partition. However, automatizing this partition flow is an interesting work which will be addressed in our future work.

3.3 Parameter extraction

We get the PE's timing information after the C2RTL conversion. In streaming applications, each PE has a working period T_n, under which the PE will never be stopped by overflows or underflows of an FIFO. During the period T_n, the PE will read, process, and write data. We denote the input time as t_{ni} and the output time as t_{no}. In summary, we formulate the parameters of the nth PE interface in Table 1. Based on a large number of PEs converted by *eXCite*, we have observed two types of interface parameters. Figure 4 shows the waveform of the type II. As we can see, t_n is less than T_n in this case. In type I, t_n equals to T_n, which indicates the idle time is zero.

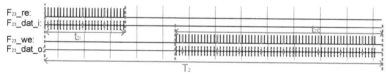

Fig. 4. Type II case: Output of PE_2 in the JPEG encoder

3.4 Block level parallelism

To implement block level parallelism, we denote the nth PE's parallelism degree as P_n.[2] Thus, $P_n=1$ means that the design does not parallelize this PE. When $P_n > 1$, we can implement block level parallelism using a MUX, a DEMUX, and a simple controller in Figure 5.

---

[1] We will define those parameters in the next section.
[2] We assume that no data dependence exists among PE_n's task.

Figure 6 illustrates the working mechanism of the nth parallelized PE. It shows a case with two-level block parallelism with $t_{ni} > t_{no}$. In this case, the input and the output of the parallelized blocks work serially. It means that the PE_{n_2} block must be delayed for t_{ni} by the controller, so as to wait for the PE_{n_1} to load its input data. However, when another work period T_n starts, the PE_{n_1} can start its work immediately without waiting for the PE_{n_2}.

As we can see, the interface of the new PE_n after parallelism remains the same as Table 1. However, the values of the input and the output parameters should be updated due to the parallelism. It will be discussed in Section 4.2.

3.5 FIFO interconnection

To deal with the FIFO interconnection, we first define the parameters of a FIFO. They will be used to analyze the performance in the next section. Figure 7 shows the signals of a FIFO. F_clk denotes the clock signal of the FIFO F. F_we and F_re denote the enable signals of writing and reading. F_dat_i and F_dat_o are the input and the output data bus. F_ful and F_emp indicate the full and empty state, which are active high. Given a FIFO, its parameters are shown in Table 2. To connect modules with FIFOs, we need to determine $D_{(n-1)n}$ and $W_{(n-1)n}$.

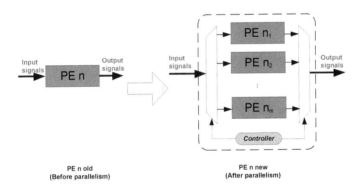

PE n old
(Before parallelism)

PE n new
(After parallelism)

Fig. 5. Realization of block level parallelism

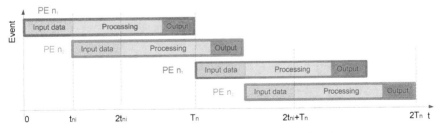

Fig. 6. Working mechanism of block level parallelism($P_n \leq \lfloor T_n / t_{ni} \rfloor$)

Name	Description	Examples[2]
$Fclk_{(n-1)n}$	Clock frequency (MHz)	50
$W_{(n-1)n}$	Data bus width	16
$A_{FIFO(n-1)n}$	Area: memory resource used (bit)	704
$D_{(n-1)n}$	FIFO depth	44
$f_{(n-1)n}(m)$ [1]	Number of data in FIFO at m^{th} cycle	
$SoF_{(n-1)n}(m)$	State of FIFO at m^{th} cycle; 1:Full; -1:Empty; 0:Other state	

[1] m means m^{th} cycle.
[2] This example comes from the FIFO between PE_1 and PE_2 in the JPEG encode case.

Table 2. The parameter of FIFO between PE_{n-1} and PE_n

4. Algorithm for block level parallelism

This section formulates the block level parallelism problem. After that, we propose an algorithm to solve the problem for multiple PEs in the system level.

4.1 Block level parallelism formulation

Given a design with N PEs, the throughput constraint TH_{ref} and the area constraint A_{ref}[3], we decide the nth PE's parallelism degree P_n. That is

$$MIN.P_n, \quad \forall n \in [1, N] \tag{1}$$

$$s.t.TH_{all} \geq TH_{ref} \quad and \quad \sum_{n=1}^{N} \widehat{A}_n \leq A_{ref} \tag{2}$$

where TH_{all} denotes the entire throughput and $\widehat{A}_n$ is the PE_n's area after the block level parallelism. Without losing generality, we assume that the capacity of all FIFOs is infinite and $A_{ref}=\infty$. We leave the FIFO sizing in the next section.

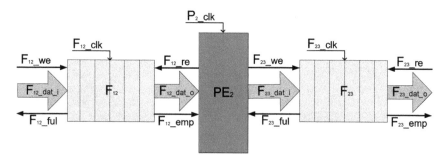

Fig. 7. Circuit diagram of FIFO blocks connecting to PE_2

---

[3] This area constraint doesn't consider the FIFO area.

4.2 Parameter extraction after block level parallelism

Before determining the parallelism degree of each PE, we first discuss how to extract new interface parameters for each PE after parallelism. That is to update the following parameters: $\widehat{TH}_{ni/o}$, $\widehat{A}_n$, $\widehat{T}_n$, $\widehat{f}_n$, and $\widehat{SoP}_n$, which are calculated based on P_n, $TH_{ni/o}$, A_n, T_n, f_n, and SoP_n.

First of all, we calculate $TH_{ni/o}$. As Figure 8 shows, larger parallelism degree won't always increase the throughput. It is limited by the input time t_{ni}. Assuming $t_{ni} > t_{no}$ and $P_n \leq \lfloor T_n/t_{ni} \rfloor$, we have

$$\widehat{TH}_{ni/o} = P_n * TH_{ni/o} \quad when \quad P_n \leq \lfloor T_n/t_{ni} \rfloor \tag{3}$$

For example, as shown in Figure 6, $\widehat{TH}_{ni/o}=2*TH_{ni/o}$ because $P_n=2< \lfloor T_n/t_{ni} \rfloor=3$. When $P_n \geq$

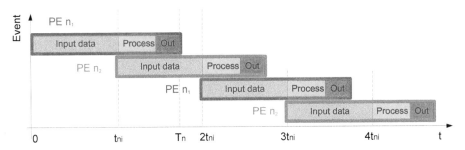

Fig. 8. Working mechanism of block level parallelism$(P_n \geq \lceil T_n/t_{ni} \rceil)$

$\lceil T_n/t_{ni} \rceil$, we have

$$\widehat{TH}_{ni/o} = T_n/t_{ni} * TH_{ni/o} \quad when \quad P_n \geq \lceil T_n/t_{ni} \rceil \tag{4}$$

where the throughput is limited by the input time t_{ni}. More parallelism degree is useless in this case. For example, as shown in Figure 8, $\widehat{TH}_{ni/o}=T_n/t_{ni}*TH_{ni/o}$, because $P_n=2=\lceil T_n/t_{ni} \rceil$. When $t_{ni}<t_{no}$ we have the similar conclusions. In summary, we have

$$\widehat{TH}_{ni/o} = \begin{cases} P_n * TH_{ni/o} & P_n < p_n \\ T_n/max\{t_{ni}, t_{no}\} * TH_{ni/o} & others \end{cases} \tag{5}$$

where

$$p_n = \lceil T_n/max\{t_{ni}, t_{no}\} \rceil \tag{6}$$

Second, we can solve $\widehat{A}_n$, $\widehat{T}_n$, and $\widehat{f}_n$. Ignoring the area of the controller, we have

$$\widehat{A}_n = P_n * A_n \tag{7}$$

Based on Figure 6 and 8, we conclude

$$\widehat{T}_n = \begin{cases} T_n + (P_n - 1) * max\{t_{ni}, t_{no}\} & P_n \leq p_n \\ P_n * max\{t_{ni}, t_{no}\} & others \end{cases} \tag{8}$$

Equation 5 shows that $\widehat{TH}_{ni}$ and $\widehat{TH}_{no}$ change at the same rate. Therefore,

$$\widehat{f}_n = \widehat{TH}_{no}/\widehat{TH}_{ni} = TH_{ni}/TH_{ni} = f_n \tag{9}$$

Furthermore, we calculate $\widehat{SoP}_n$. $\widehat{SoP}_n$ is the combination of each sub-block's SoP. Therefore

$$\widehat{SoP}_n = \begin{cases} \sum_{i=0}^{P_n} SoP_n(m - i * t_{ni}) & t_{ni} \geq t_{no} \\ \sum_{i=0}^{P_n} SoP_n(m - i * (T_n - t_{no})) & t_{ni} < t_{no} \end{cases} \tag{10}$$

Finally, we can obtain all new parameters of a PE after parallelism. We will use those parameters to decide the parallelism degree in Section 4.3 and Section 5.

4.3 Block level parallelism degree optimization

To solve the optimization question in Section 4.1, we need to understand the relationship between TH_{all} and $\widehat{TH}_{ni/o}$. When PE_n is connected to the chain from PE_1 to $PE_{(n-1)}$, we define the output interface's throughput of PE_n as TH'_{no}. This parameter is different from $\widehat{TH}_{ni/o}$ because it has considered the rate mismatch effects from previous PEs. We have

$$TH'_{no} = \begin{cases} \widehat{TH}_{no} & TH'_{(n-1)o} > \widehat{TH}_{ni} \\ \widehat{f}_n * TH'_{(n-1)o} & others \end{cases} \tag{11}$$

In fact, $TH_{all} = TH'_{No}$. Therefore, we can express TH_{all} in the following format

$$TH_{all} = \widehat{TH}_{bo} \prod_{i=b+1}^{N} f_i \tag{12}$$

where b is the index of the slowest PE_b. It is the bottleneck of the system.

To do the optimization of parallelism degrees, we purpose an algorithm shown in Algorithm 1. In the algorithm, the inputs are the number of PE N, the parameters of each PE $ParaG[N]$, each PE's maxim parallelism degree by Equation 6, and the design constraint $TH\_ref = TH_{ref}$. $ParaG[N]$ includes $TH_{ni/o}, t_{ni/o}, T_n, SoP_n$ shown in Table 1[4].

The output is each PE's optimal parallelism degree $P[N]$. Lines $1-7$ are to check if the optimization object is possible. Lines $8-14$ are the initializing process. Lines $15-20$ are the main loop. $pTH[N]$ equals to $\widehat{TH}_{ni/o}$ and TH_best denotes the best performance. Function $get\_pTH()$ returns the PE's $\widehat{TH}_{ni/o}$. Function $get\_THall()$ returns TH_now which means the TH_{all} under $\widehat{TH}_{ni/o}$ condition. Line 2 sets all the parallelism degree to its maximum value. After that, we get the fastest TH_{all} in Line 4. If the system can never approach the optimizing target, we will change the target in Line 6. In the main loop, we find the bottleneck in each step in Line 16 and add more parallelism degree to it. We will update $\widehat{TH}_{ni/o}$ in Line 18 and evaluate the system again in Line 19. We end this loop until the design constraints are satisfied.

---

[4] These parameters are initial ones got by Step 2

Algorithm 1 Block Level Parallelism Degree Optimization Algorithm

---

Input: $N, ParaG[N], p[N], TH\_ref$
Output: $P[N]$
 1: **for** $k = 1 \rightarrow N$ **do**
 2:    $pTH[k] = get\_pTH(p[k], ParaG[k], p[k]), k = k + 1$
 3: **end for**
 4: $TH\_best = get\_THall(pTH, ParaG)$
 5: **if** $TH\_best > TH\_ref$ **then**
 6:    $TH\_ref = TH\_best$
 7: **end if**
 8: **for** $k = 1 \rightarrow N$ **do**
 9:    $P[k] = 1, k = k + 1$
10: **end for**
11: **for** $k = 1 \rightarrow N$ **do**
12:    $pTH[k] = get\_pTH(P[k], ParaG[k], p[k]), k = k + 1$
13: **end for**
14: $TH\_now = get\_THall(pTH, ParaG)$
15: **while** $TH\_now \geq TH\_ref$ **do**
16:    $Bottleneck = get\_bottle(pTH, ParaG)$
17:    $P[Bottleneck] + +$
18:    $k = Bottleneck$
19:    $pTH[k] = get\_pTH(P[k], ParaG[k], p[k]), k = k + 1$
20:    $TH\_now = get\_THall(pTH, ParaG)$
21: **end while**

---

5. Algorithm for FIFO-connected blocks

This section formulates the FIFO interconnecting problem. We then demonstrate that this problem can be solved by a binary searching algorithm. Finally, we propose an algorithm to solve the FIFO interconnecting problem of multiple PEs in the system level.

5.1 FIFO interconnection formulation

Given a design consisting of N PEs, we need to determine the depth $D_{(i-1)i}$ of each FIFO[5], which maximizes the entire throughput TH_{all} and minimizes the FIFO area of $A_{FIFO_{all}}$.

$$MIN. \quad \sum_{i=2}^{N} D_{(i-1)i} \tag{13}$$

$$s.t. \quad TH_{all} \geq TH_{ref} \quad and \quad A_{FIFO_{all}} \leq A_{FIFO_{ref}} \tag{14}$$

where TH_{ref} and $A_{FIFO_{ref}}$ can be the user-specified constraints or optimal values of the design. Without losing generality, we set $TH_{ref}=(TH_{all})_{max}$ and $A_{FIFO\_ref}=\infty$. We assume that F_{01} never empties and $F_{N(N+1)}$ never fulls. That is, $\forall m, SoF_{01}(m) \neq -1$ and $SoF_{N(N+1)}(m) \neq 1$[6].

---

[5] We assume that the $W_{(i-1)i}$ is decided by the application.
[6] This means that we only consider the operating state of the design instead of the halted state.

5.2 FIFO capacity optimization

We can conclude a brief relationship between $TH_{ni/o}$ and D_i. For PE_n, we define the real throughput as $\widetilde{TH}_{ni/o}$, when connected with F_{n-1} of D_{n-1} and F_{n+1} of D_{n+1}. Then we set

$$\widetilde{TH}_{ni/o} = f(D_{n-1}, D_{n+1}) \tag{15}$$

We know that a small D_{n-1} or D_{n+1} will cause $\widetilde{TH}_{ni/o} < TH_{ni/o}$. Also, when $\widetilde{TH}_{ni/o} = TH_{ni/o}$, larger D_{n-1} or D_{n+1} will not increase performance any more. Therefore, as it is shown in Figure 2, $f(x)$ is a monotone nondecreasing function with a boundary.

With the fixed relationship between $TH_{ni/o}$ and D_i, we can solve the FIFO capacity optimization problem by a binary searching algorithm based on the system level simulations. We describe this method to determine the FIFO capacity for multiple PEs $(N > 2)$ in Algorithm 2.

Algorithm 2 FIFO Capacity Algorithm for $N \geq 2$

Input: $N, ParaG[N], Inital\_D[N]$
Output: $D[N]$
1: $k = 1, n = 1$
2: **while** $k < N$ **do**
3:    $D[k] = Initial\_D[k]$
4: **end while**
5: $TH\_obj = get\_TH(D, ParaG)$
6: $TH\_new = TH\_obj, Upper = D[1], Mid = D[1], Lower = 1$
7: **while** $n < N$ **do**
8:    **if** $TH\_new = TH\_obj$ **then**
9:       $D[n] = ceil((Mid - Lower)/2)$
10:       $Upper = Mid, Mid = D[n]$
11:    **else**
12:       $D[n] = ceil((Upper - Mid)/2)$
13:       $Lower = Mid, Mid = D[n]$
14:    **end if**
15:    $TH\_new = get\_TH(D, ParaG)$
16:    **if** $Upper = Lower$ **then**
17:       $n = n + 1$
18:       $Upper = D[n], Mid = D[n], Lower = 1$
19:    **end if**
20: **end while**

The inputs are the number of PE N, the parameters of each PE $ParaG[N]$ and each FIFO's initial capacity $Initial\_D[N]$. $ParaG[N]$ includes $TH_{ni/o}, t_{ni/o}, T_n, SoP_n$ shown in Table 1[7]. $Initial\_D[n]$ means the initial searching value of $D_{n(n+1)}$, which is big enough to ensure $\widetilde{TH}_{ni/o} = TH_{all}$. The output is each FIFO's optimal depth $D[N]$. Lines $1 - 6$ are the initializing process. Lines $7 - 20$ are the main loop. Function $get\_TH()$ in line 5 and 15 can return the entire throughput under different $D[N]$ settings. Variable TH_obj is the searching object calculated by $Initial\_D[N]$. $Initial\_D[N]$ equals to TH_{all} and TH_new is the current throughput calculated based on $D[N]$. $Upper$, Mid, and $Lower$ decide the binary searching range. In each loop, n means that the capacity of $F_{n(n+1)}$ is processed. We get the searching

---

[7] These parameters are updated by Block Level Parallelism step

point and the range according to TH_new in lines $8 - 14$. We update TH_new in line 15. The end condition is checked in line 16. When $n = N$, it means that all FIFOs have their optimal capacity. As we can see, the most time-consuming part of the algorithm is the $getTH()$ function. It calls for an entire simulation of the hardware. Therefore, we build a system level simulator instead of a RTL level one. It can shorten the optimization greatly. The system level simulator adopts the parameters extracted in Step 2. The C-based system level simulator will be released on our website soon.

6. Experiments

In this section, we first explain our experimental configurations. Then, we compare the flatten approach, the hierarchical method without block level parallelism (BLP) and with BLP under several real benchmarks. After that, we break down the advantages by two aspects: the block level parallelism and the FIFO sizing. We then show the effectiveness of the proposed algorithm to optimize the parallel degree. Finally, we demonstrate the advantages from the FIFO sizing method.

6.1 Experimental configurations

In our experiments, we use a C2RTL tool called *eXCite* (Y Exploration Inc., 2011). The HDL files are simulated by Mentor Graphics' ModelSim to get the timing information. The area and clock information is obtained by Quartus II from Altera. *Cyclone* II FPGAs are selected as the target hardware. We derive seven large streaming applications from the high-level synthesis benchmark suits CHstone(Hara et al. (2008)). They come from real applications and consist of programs from the areas of image processing, security, telecommunication and digital signal processing.

- *JPEG encode/decode*: JPEG transforms image between JPEG and BMP format.
- *AES encryption/decryption*: AES (Advanced Encryption Standard) is a symmetric key crypto system.
- *GSM*: LPC (Linear Predictive Coding) analysis of GSM (Global System for Mobile Communications).
- *ADPCM*: Adaptive Differential Pulse Code Modulation is an algorithm for voice compression.
- *Filter Group*: The group includes two FIR filters, a FFT and an IFFT block.

6.2 System optimization for real cases

We show the synthesized results for seven benchmarks and compare the flatten approach, the hierarchical approach without and with BLP. Table 3 shows the clock cycles saved by the hierarchical method without and with BLP. The last column in Table 3 shows the BLP vector for each PE. The ith element in the vector denotes the parallel degree of the PE_i. The total speedup represents the clock cycle reductions from the hierarchical approach with BLP. As we can see, the hierarchical method without BLP achieves up to 10.43 times speedup compared with the flatten approach. However, the BLP can provide considerable extra up to another 5 times speedup compared with the hierarchial method without BLP. It should be noted that

Benchmark	Flatten approach	Hierarchical W.O. BLP(speedup)	Hierarchical W. BLP(speedup)	BLP degree $(P_1..P_n)$
JPEG encode	42,475,202	4,070,603 (x10.43)	1,850,907 (x22.94)	(1,3,1)
JPEG decode	623,090	456,821 (x1.364)	115,622 (x5.389)	(1,1,4,1)
AES encryption	1,904,802	719,263 (x2.648)	216,393 (x8.803)	(4,2,3,2)
AES decryption	2,185,802	867,306 (x2.388)	229,570 (x9.521)	(4,2,4,2)
GSM	620,802	204,356 (x3.038)	55,306 (x11.22)	(4,4,4,1,1,1)
ADPCM	35,691	12,464 (x2.864)	3,762 (x9.487)	(4,2,2,2,3)
Filter groups	6,537,416	1,702,406 (x3.84)	511,853 (x12.77)	(2,1,1,4,1,2)

Min T_{all} (cycles)

BLP: Block level parallelism.

Table 3. System optimization result of minimal clock cycles

Benchmark	Flatten approach	Hierarchical W.O. BLP	Hierarchical W. BLP	Total Speedup
JPEG encode	69.74	74.2	74.2	x1.064
JPEG decode	71.15	71.3	71.3	x1.002
AES encode	71.24	91.06	91.06	x1.278
AES decode	75.56	87.35	87.35	x1.156
GSM	55.73	59.16	59.16	x1.062
ADPCM	53.29	68.32	68.32	x1.282
Filter groupe	93.41	96.69	96.69	x1.035

Max Clk_{all} (MHz)

BLP: Block level parallelism.

Table 4. System optimization result of maximal clock frequency

the BLP will lead to area overheads in some extents. We will discuss those challenges in the following experiments. Furthermore, Table 4 shows the maximum clock frequency of three approaches. As we can see, the BLP does not introduce extra delay compared with the pure hierarchical method.

6.3 Block level parallelism

The previous experimental results show the total advantages from the hierarchial method with BLP. This section will discuss the performance and the area overheads of BLP alone. We show the throughput improvement and the area costs in the GSM benchmark in Figure 9[8]. We list the BLP vector as the horizontal axis. As we can see, parallelizing some PEs will increase the throughput. For the BLP vector $(1, 2, 1, 1, 1, 1)$, we duplicate the second PE_2 by two. It will improve the performance by 4% with 48% area overheads. The result comes from the rate mismatch between PEs. It indicates that duplicating single PE may not increase the throughput effectively and the area overheads may be quite large. Therefore, we should develop an algorithm to find the optimal BLP vector to boost the performance without introducing too many overheads. For example, the BLP vector $(4, 4, 4, 1, 1, 1)$ leads to over 4 times performance speedup while with only less than 3 times area overheads.

Furthermore, we evaluate the proposed BLP algorithm with the approach duplicating the entire hardware. Figure 10 demonstrates that our algorithm can increase the throughput with less area. It is because the BLP algorithm does not parallelize every PE and can explore more fine-grained design space. Obviously, the BLP method provides a solution to trade off

---

[8] We observe similar trends in other cases.

performance with area more flexibly and efficiently. In fact, as the modern FPGA can provide more and more logic elements, it makes the area not so urgent as the performance, which is the first-priority metric in most cases.

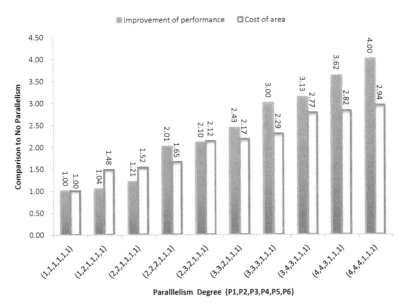

Fig. 9. Speedup and Area cost in GSM case

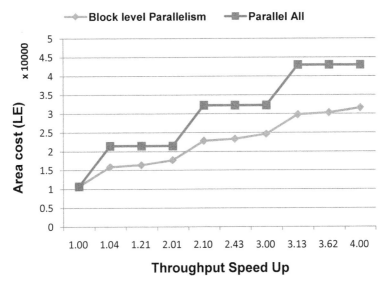

Fig. 10. Advantage of Block Level Parallelism algorithm

Benchmark		D_{12}	D_{23}	D_{34}	D_{45}	D_{56}	T_{all}
JPEG encode	System Level	43	2	-	-	-	4080201
	RTL Level	44	2	-	-	-	4070603
JPEG decode	System Level	2	33	17	2	-	456964
	RTL Level	2	33	18	2	-	456821
AES encryption	System Level	2	2	2	-	-	719364
	RTL Level	3	2	3	-	-	719263
AES decryption	System Level	2	257	2	-	-	867407
	RTL Level	3	249	3	-	-	867306
GSM	System Level	54	2	2	2	2	204554
	RTL Level	55	2	2	2	2	204356
ADPCM	System Level	2	2	2	2	2	12464
	RTL Level	2	2	2	2	1	12464
Filter group	System Level	2	2	86	2	2	1701896
	RTL Level	2	2	87	2	2	1701846

Table 5. Optimal FIFO capacity algorithm experiment result in 7 real cases

6.4 Optimal FIFO capacity

We show the simulated results for real designs with multiple PEs. First of all, we show the relationship between the FIFO size and the running time T_{all}. Figure 11 shows the JPEG encoding case. As we can see, the FIFO size has a great impact on the performance of the design. In this case, the optimal FIFO capacity should be $D_{12}=44$, $D_{23}=2$.

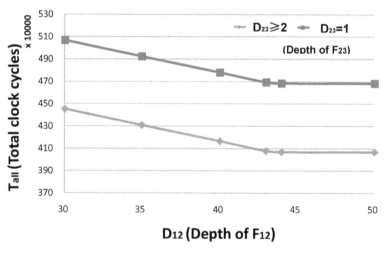

Fig. 11. FIFO capacity in JPEG encode case

Table 5 lists both the system level simulation results and the RTL level experimental ones on FIFO size in seven cases. It shows that our approach is accurate enough for those real cases. Though little mismatch exists, the difference is very small. Compared to the magnitudes of speedup to determine the FIFO size, our approach is quite promising to be used in architecture level design space exploration.

Benchmark	Memory resource used(bit)		Savings
	FIFOs with [1] enough size	FIFOs with optimized size	
JPEG encode	10,048	2,624	x3.83
JPEG decode	38,776	8,376	x4.63
AES encode	92,160	67,968	x1.36
AES decode[2]	92,160	75,808	x1.22
GSM	36,028	8,602	x4.19
ADPCM	54,040	3,736	x14.46
Filter groupe	114,400	76,736	x1.49

[1] We set each FIFO depth as 128.
[2] In this case we set each FIFO depth as 256.

Table 6. Area saved

The memory resource savings by well designing FIFO are listed in Table 6. Compared to the large enough design strategy, the memory savings are significant. Moreover, compared to the method using RTL level simulator to decide FIFO capacity, our work is extremely time efficient. Considering a hardware with N FIFO to design, each FIFO size is fixed using a binary searching algorithm. It will request $log_2(p)$ times simulations with the initial FIFO depth value $D_{(n-1)n} = p$. Assuming that the average time cost by *ModelSim* RTL level simulation is C, the entire exploration time is $N * log_2(p) * C$. Considering the *FilerGroup* case with $N = 5$, $p = 128$ and $C = 170$ seconds, which are typical values on a normal PC, we have to wait 100 minutes to find the optimal FIFO size. However, our system level solution can finish the exploration in seconds.

7. Related works

Many C2RTL tools (Gokhale et al., 2000; Lhairech-Lebreton et al., 2010; Mencer, 2006; Villarreal et al., 2010) are focusing on streaming applications. They create design architectures including different modules connected by first-in first-out (FIFO) channels. There are some other tools focusing on general purpose applications. For example, Catapult C (Mentor Graphics, 2011) takes different timing and area constraints to generate Pareto-optimal solutions from common C algorithms. However, little control on the architecture leads to suboptimal results. As (Agarwal, 2009) has shown, FIFO-connected architecture can generate much faster and smaller results in streaming applications.

Among C2RTL tools for streaming applications, GAUT (Lhairech-Lebreton et al., 2010) transforms C functions into pipelined modules consisting of processing units, memory units and communication units. Global asynchronous local synchronous interconnections are adopted to connect different modules with multiple clocks. ROCCC (Villarreal et al., 2010) can create efficient pipelined circuits from C to be re-used in other modules or system codes. Impulse C (Gokhale et al., 2000) provides a C language extension to define parallel processes and communication channels among modules. ASC (Mencer, 2006) provides a design environment for users to optimize systems from algorithm level to gate level, all within the same C++ program. However, previous works keep how to determine the FIFO capacity efficiently unsolved. Most recently, (Li et al., 2012) presented a hierarchical C2RTL framework with analytical formulas to determine the FIFO capacity. However, block level parallelism

Concepts and Design of Embedded Systems

is not supported and their FIFO sizing method is limited to PEs with certain input/output interfaces.

During the hierarchical C2RTL flow, a key step is to partition a large C program into several functions. Plenty of works have been done in this field. Many C-based high level synthesis tools, such as SPARK (Gupta et al., 2004), eXcite (Y Exploration Inc., 2011), Cyber (NEC Inc., 2011) and CCAP (Nishimura et al., 2006), can partition the input code into several functions. Each function has a corresponding hardware module. However, it leads to a nontrivial datapath area overhead because it eliminates the resource sharing among modules. On the contrary, function inline technique can reduce the datapath area via resource sharing. The fast increasing complexity of the controller makes the method inefficient. Appropriate function clustering (Okada et al., 2002) in a sub module provides a more elegant way to solve the partition problem. But it is hard to find a proper clustering rule. For example, too many functions in one cluster will also lead to a prohibitive complexity in controllers. In practise, architects often help the partition program to divide the C algorithms manually.

Similar to the hierarchical C2RTL, multiple FIFO-connected processing elements (PE) are used to process audio and video streams in the mobile embedded devices. Researchers had investigated on the input streaming rates to make sure that the FIFO between PEs will not overflow, while the real-time processing requirements are met. On-chip traffic analysis of the SoC architecture (Lahiri et al., 2001) had been explored. However, their simulation-based approaches suffer from a long executing time and fail in exploring large design space. A mathematical framework of rate analysis for streaming applications have been proposed in reference (Cruz, 1995). Based on the network calculus, reference (Maxiaguine et al., 2004) extended the service curves to show how to shape an input stream to meet buffer constraints. Furthermore, reference (Liu et al., 2006) discussed the generalized rate analysis for multimedia processing platforms. However, all of them adopts a more complicated behavior model for PE streams, which is not necessary in the hierarchical C2RTL framework.

8. Conclusion

Improving the booming design methodology of C2RTL to make it more widely used is the goal of many researchers. Our work of the framework does have achieved the improvement. We first propose a hierarchical C2RTL design flow to increase the performance of a traditional flatten one. Moreover, we propose a method to increase throughput by making block level parallelism and an algorithm to decide the degree. Finally, we develop an heuristic algorithm to find the optimal FIFO capacity in a multiple-module design. Experimental results show that hierarchical approach can improve performance by up to 10.43 times speedup, and block level parallelism can make extra 4 times speedup with 194% area overhead. What's more, it determines the optimal FIFO capacity accurately and fast. The future work includes automatical C code partition in the hierarchical C2RTL framework and adopting our optimizing algorithm in more complex architectures with feedback and branches.

9. Acknowledgement

The authors would like to thank reviewers for their helpful suggestions to improve the chapter. This work was supported in part by the NSFC under grant 60976032 and 61021001,

National Science and Technology Major Project under contract 2010ZX03006-003-01, and High-Tech Research and Development (863) Program under contract 2009AA01Z130.

10. References

Agarwal, A. (2009). *Comparison of high level design methodologies for algorithmic IPs: Bluespec and C-based synthesis*, PhD thesis, Massachusetts Institute of Technology.

Cong, J., Liu, B., Neuendorffer, S., Noguera, J., Vissers, K. & Zhang, Z. (2011). High-level synthesis for fpgas: From prototyping to deployment, *Computer-Aided Design of Integrated Circuits and Systems, IEEE Transactions on* 30(4): 473–491.

Cruz, R. (1995). Quality of service guarantees in virtual circuit switched networks, *Selected Areas in Communications, IEEE Journal on* 13(6): 1048–1056.

Gokhale, M., Stone, J., Arnold, J. & Kalinowski, M. (2000). Stream-oriented fpga computing in the streams-c high level language, *fccm*, IEEE, p. 49.

Guo, Y. & McCain, D. (2006). Rapid prototyping and vlsi exploration for 3g/4g mimo wireless systems using integrated catapult-c methodology, *Wireless Communications and Networking Conference, 2006. WCNC 2006. IEEE*, Vol. 2, IEEE, pp. 958–963.

Gupta, S., Gupta, R. & Dutt, N. (2004). *SPARK: a parallelizing approach to the high-level synthesis of digital circuits*, Vol. 1, Kluwer Academic Pub.

Hara, Y., Tomiyama, H., Honda, S., Takada, H. & Ishii, K. (2008). Chstone: A benchmark program suite for practical c-based high-level synthesis, *IEEE International Symposium on Circuits and Systems*, IEEE, pp. 1192–1195.

Intel Inc. (2011). Stellarton atom processor, *Website: http://www. intel. com* .

Lahiri, K., Raghunathan, A. & Dey, S. (2001). System-level performance analysis for designing on-chip communication architectures, *Computer-Aided Design of Integrated Circuits and Systems, IEEE Transactions on* 20(6): 768–783.

Lhairech-Lebreton, G., Coussy, P. & Martin, E. (2010). Hierarchical and multiple-clock domain high-level synthesis for low-power design on fpga, *2010 International Conference on Field Programmable Logic and Applications*, IEEE, pp. 464–468.

Li, S., Liu, Y., Zhang, D., He, X., Zhang, P. & Yang, H. (2012). A hierarchical c2rtl framework for fifo-connected stream applications, *Proceedings of the 2012 Asia and South Pacific Design Automation Conference*, IEEE Press, pp. 1–4.

Liu, Y., Chakraborty, S. & Marculescu, R. (2006). Generalized rate analysis for media-processing platforms, *Proceedings of the 12th IEEE International Conference on Embedded and Real-Time Computing Systems and Applications, RTCSA*, Vol. 6, Citeseer, pp. 305–314.

Maxiaguine, A., Künzli, S., Chakraborty, S. & Thiele, L. (2004). Rate analysis for streaming applications with on-chip buffer constraints, *Proceedings of the 2004 Asia and South Pacific Design Automation Conference*, IEEE Press, pp. 131–136.

Mencer, O. (2006). Asc: a stream compiler for computing with fpgas, *Computer-Aided Design of Integrated Circuits and Systems, IEEE Transactions on* 25(9): 1603–1617.

Mentor Graphics, M. (2011). Catapult c synthesis, *Website: http://www. mentor. com* .

NEC Inc. (2011). CyberWorkBench, *Website: http://www.nec.com/global/prod/cwb/* .

Nishimura, M., Nishiguchi, K., Ishiura, N., Kanbara, H., Tomiyama, H., Takatsukasa, Y. & Kotani, M. (2006). High-level synthesis of variable accesses and function calls in software compatible hardware synthesizer ccap, *Proc. Synthesis And System Integration of Mixed Information technologies (SASIMI)* pp. 29–34.

Okada, K., Yamada, A. & Kambe, T. (2002). Hardware algorithm optimization using bach c, *IEICE Transactions on Fundamentals of Electronics, Communications and Computer Sciences* 85(4): 835–841.

Rossler, M., Wang, H., Heinkel, U., Engin, N. & Drescher, W. (2009). Rapid prototyping of a dvb-sh turbo decoder using high-level-synthesis, *Forum on Specification & Design Languages, 2009.*, IEEE, pp. 1–6.

Schafer, B., Trambadia, A. & Wakabayashi, K. (2010). Design of complex image processing systems in esl, *Proceedings of the 2010 Asia and South Pacific Design Automation Conference*, IEEE Press, pp. 809–814.

Villarreal, J., Park, A., Najjar, W. & Halstead, R. (2010). Designing modular hardware accelerators in c with roccc 2.0, *2010 18th IEEE Annual International Symposium on Field-Programmable Custom Computing Machines*, IEEE, pp. 127–134.

Y Exploration Inc. (2011). eXCite, *Website: http://www.yxi.com* .

19

Development of Energy Efficiency Aware Applications Using Commercial Low Power Embedded Systems

Konstantin Mikhaylov[1], Jouni Tervonen[1] and Dmitry Fadeev[2]
[1]*Oulu Southern Institute, University of Oulu*
[2]*Saint-Petersburg State Polytechnical University*
[1]*Finland*
[2]*Russian Federation*

1. Introduction

In recent years, different devices that encapsulate different types of embedded system processors (ESPs) are becoming increasingly commonplace in everyday life. The number of machines built around embedded systems (ESs) that are now being used in households and industry is growing rapidly every year. Accordingly, the amount of energy required for their operation is also increasing. The United States (U.S.) Energy Information Administration (EIA) estimates that the share of residential electricity used by appliances and electronics in U.S. homes has nearly doubled over the last three decades. In 2005, this accounted for an increase of around 31% in the overall household energy consumption or 3.4 exajoule (EJ) of energy across the entire country(USEIA, 2011).

Portable devices built around different ESs are often supplied using different primary or secondary batteries. According to (FreedoniaGroup, 2011), the battery market in 2012 in the U.S. alone will exceed $16.4 billion and will be over $50 billion worldwide (Munsey, 2011). Based on the previous year's consumption data analysis (e.g., (Munsey, 2011)), a significant percentage of batteries will be used by different communication, computer, medical and other devices containing ES chips. Therefore, improvement in the energy efficiency of ESs, which would also result in reduction of energy consumption of the services provided, becomes one of the most critical problems today, both for the research community and the industry. The problem of energy efficiency of ESs has recently become the focus of governmental research programs such as the European FP7 and ARTEMIS and CISE/ENG in the U.S., etc. Resolution of this problem would have additional value due to recent CO_2 reduction initiatives, as the increase in energy efficiency for the upcoming systems would allow reduction of the energy consumption and corresponding CO_2 emissions arising during energy production (Earth, 2011).

The problem of ES energy efficiency can be divided into two major components:

- the development of an ES chip that would consume the minimum amount of energy during its operation and during its manufacturing;

- the development of applications based on existing ES chips, so that the minimum amount of energy would be consumed during fulfilment of the specified tasks.

The first part of the problem is currently under intensive investigation by the leading ESP manufacturers and research laboratories, which are bringing more energy efficient ESPs to the market every year. The development of a novel ESP is quite a complicated task and requires special skills and knowledge in various disciplines, special equipment and substantial resources.

Unlike the development of the energy efficient ESP itself, the development of energy efficient applications that use existing commercial ESPs is quite a common task faced by today's engineers and researchers. An efficient solution to this problem requires knowledge of ESP parameters and how they influence power consumption, as well as knowing how the power consumption affects the device's efficiency with different power supply options. This chapter will answer these questions and provide the readers with references that describe the most widespread ES power supply options and their features, the effect of the different ES parameters on the overall device power consumption and the existing methods for increasing energy efficiency. Although the main focus of this chapter will be on low-power ESs - and low-power microcontrollers in particular - we will also provide some hints concerning the energy efficient use of other ESs.

Most of the general-purpose ES-based devices in use today have a structure similar to that shown in Fig. 1. Therefore, all of the components of these devices can be attributed to three major groups: 1) the power supply system, which provides the required power for device operation, 2) the ES with the compulsory peripherals that execute the application program and 3) the application specific peripherals that are used by the ES. As the number of the possible application specific peripherals is extremely large at present, we will not consider these in this chapter and will focus mainly on the basic parameters of the ES, the ES compulsory peripherals and the power system parameters. To provide a comprehensive approach for the stated problem, the remainder of this chapter is organized as follows. Section 2 reviews the details of possible power supply options that can be used for the ESs. Section 3 describes the effect of the different ES parameters and features on its power consumption. Section 4 shows how the parameters and features discussed in Sections 2 and 3 could be used to increase the energy efficiency of a real ES-based device. Finally, Section 5 gives a short summary and discusses some of the existing research problems.

2. Embedded system power supply options

Three possible options are presently available for providing ESs with the required energy for operation:

- mains;
- primary or secondary batteries;
- energy from environment harvesting system.

Each of these options has specific features that are described in more detail in Subsections 2.1-2.3.

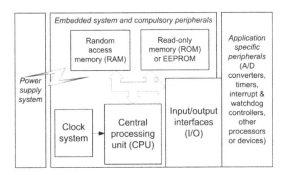

Fig. 1. Architecture of typical embedded system-based devices

2.1 Embedded systems power supply from mains

The power supply of the ESP from mains is the most universal method and is applicable for the devices that utilize low-power microcontrollers and high-end Application-Specific Instruction-Set Processors (ASIPs) or Field-Programmable Gate Arrays (FPGAs). The utilization of mains for ES power supply is usually capable of providing the attached ES with any required amount of energy, thereby reducing the importance of energy efficiency for these applications. Nevertheless, the energy efficiency increase for mains supplied devices allows reduction of their exploitation costs and can produce a positive environmental impact.

One of the major considerations while using mains for ES power supply is the necessity of converting the Alternating Current (AC) into the required Direct Current (DC) supply voltage for the given ESP (for examples, see Table 3). This conversion causes some energy losses that depend on the parameters of the AC/DC converter used and usually account for about 5-10% of the overall energy for high loads and high power, and increase dramatically for lower loads (Jang & Jovanovic, 2010). The typical curves for conversion efficiency dependance on the output current for the low power and high-power AC/DC converters available on the market are presented on Fig. 2. This Figure also shows the conversion efficiency curves for the low-power DC/DC converter with adjustable output voltage (V_{out}).

The data in Fig. 2 allow prediction that the use of extremely low-power modes for mains-supplied devices will not often result in any significant reduction in overall device energy consumption due to the low AC/DC conversion efficiency at low loads.

2.2 Embedded system power supply from primary and secondary batteries

The non-rechargeable (primary) and rechargeable (secondary) batteries are often used as power supply sources for various portable devices utilizing ESs. Unlike the mains, batteries are capable of providing the attached ESs only with *a limited* amount of energy, which depends as well on the battery characteristics and *the attached ES operation mode*. This fact makes the problem of energy efficiency for battery supplied ESs very real, as higher energy efficiency allows extension of the period of time during which the device is able to fulfil its function; i.e.,

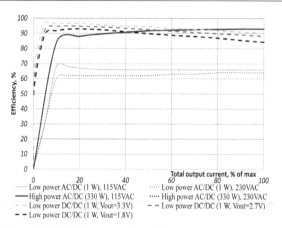

Fig. 2. Typical AC/DC and DC/DC conversion efficiency curves

the device's lifetime. The nominal characteristics of the most widely used batteries for power supplies for ES-based devices are presented in Table 1.

As Table 1 reveals, the nominal DC voltages provided by the batteries depend on the battery chemistry and are in the range of 1.2 to 12 Volts. Therefore, as can be noted from Table 3, for the battery-supplied ESs, voltage conversion is often not required, although this can allow extension of the overall operation time in some cases (see Section 4).

As can be seen in Table 1 and Fig. 3, compared to primary batteries, secondary batteries usually (Crompton, 2000; Linden & Reddy, 2002):

- have lower overall capacity;
- have better performance on discharges at higher current drains;
- have better performance on discharges at lower temperatures;
- have flatter discharge profiles;
- have much lower charge retention and shelf life.

Therefore, based on the presented data, the conclusion can be drawn that the use of the primary batteries is most convenient for those applications with low-power consumption, where a long service life is required, or in the applications with low duty cycles. Secondary batteries should be used in applications where they will operate as the energy storage buffer that is charged by the main energy source and will provide the energy when the main energy source is not available. Secondary batteries can also be convenient for applications where the battery can be recharged after use to provide higher overall cost efficiency.

According to recent battery market analyses (FreedoniaGroup, 2011; INOBAT, 2009; Munsey, 2011), the most widely used batteries today are alkaline, lithium and zinc-air primary batteries and lead-acid, rechargeable lithium-ion and nickel-metal hydride secondary batteries.

Alkaline primary batteries are currently the most widely used primary battery type (FreedoniaGroup, 2011; Linden & Reddy, 2002; Munsey, 2011). These batteries are capable of providing good performance at rather high current drains and low temperatures, have long shelf lives and are readily available at moderate cost per unit (Linden & Reddy, 2002).

Battery envelope	Common battery names	Battery chemistry	Dimensions: diameter × height, mm	Weight, g	Nominal voltage, V	Cost, USD [a]	Typical capacity, mAh [b]	Charge retention, months	Recharge cycles
9-Volt	6LR61/1604A	alkaline	48.5 × 26.5 × 17.5 [c]	45.9	9	1.71	500-600	5-7	0
	6HR61/7.2H5	nickel-metal hydride	48.5 × 26.5 × 17.5 [c]	41	7.2-9.6	10	300-400	0.25-0.5	400-500
D	LR20/13A	alkaline	34.2 × 61.5	134	1.5	2.34	12000-17000	5-7	0
C	LR14/14A	alkaline	26.2 × 50	65.8	1.5	1.4	6000-8000	5-7	0
AA	LR6/24A	alkaline	14.5 × 50.5	22.7	1.5	0.11	1500-3000	5-7	0
	R6/15D	carbon-zinc	14.5 × 50.5	15	1.5	0.05	500-1100	5-7	0
	HR6/1.2H2	nickel-metal hydride	14.5 × 50.5	27	1.2	0.42	1300-3000	0.25-0.5	400-500
	14500	lithium - ion	14.5 × 50.5	17	3	1.16	800-2000	0.75-1	1000
AAA	LR03/24A	alkaline	10.5 × 44.5	10.8	1.5	0.09	600-1200	5-7	0
	R03/24D	carbon-zinc	10.5 × 44.5	9.7	1.5	0.05	300-600	3-5	0
	HR03	nickel-metal hydride	10.5 × 44.5	12	1.2	0.21	300-1200	0.25-0.5	400-500
CR123A	CR17345	lithium	17 × 34.5	17	3	0.87	1000-1500	5-10	0
	16340	lithium - ion	17 × 34.5	17	3	1.54	750-1000	0.75-1	1000
A27	GP27A/L828	alkaline	8 × 28	4.4	12	0.2	18-22	5-7	0
CR2032	5004LC	lithium	20 × 3.2	6.6	3	0.04	200-225	5-10	0
XR44	LR44/AG13	alkaline	11.6 × 5.4	2	1.5	0.01	100-150	5-7	0
	PR44/A675	zinc-air	11.6 × 5.4	1.82	1.4	0.29	600-650	1-5	0
CR1025	5033LC	lithium	10 × 2.5	0.6	3	0.1	30	5-10	0
LR66	AG4	alkaline	6.8 × 2.6	0.3	1.5	0.01	12-18	5-7	0
A10	PR70	zinc-air	5.8 × 3.6	0.4	1.4	1.34	90-100	1-5	0

[a] minimum single unit price, estimated using the price lists from battery distributors

[b] depending on the discharge profile, the presented values are for each battery's most common usage scenarios

[c] height, mm x width, mm x length, mm

Table 1. Nominal parameters for the most widely used primary and secondary batteries[1]

[1] The table summarizes the characteristics of the typical batteries, which have been obtained from different open sources and battery specifications from different manufacturers

The average voltage supplied by an alkaline battery over its lifetime is usually around 1.3 V, which requires some ESPs to use two alkaline batteries as a power supply.

Lithium primary batteries have the advantage of a high specific energy (the amount of energy per unit mass), as well as the ability to operate over a very wide temperature range. They also have a long shelf life and are often manufactured in button or coin form. The voltage supplied by these batteries is usually around 3 Volts, which allows powering of the attached ES-based device with a single lithium battery. The cost is usually higher for lithium than for alkaline batteries.

Zinc-air primary batteries have very high specific energy, which determines their use in battery-sized critical applications with low current consumption, such as hearing aids. The main disadvantages of zinc-air batteries are their sensitivity to environmental factors and their short lifetime once exposed to air.

Although lead-acid batteries currently represent a significant part of the secondary battery market, most of these are used as the automobile Starting, Lighting and Ignition (SLI) batteries, industrial storage batteries or backup power supplies. Lead-acid batteries have very low cost but also have relatively low specific energy compared to other secondary batteries.

The rechargeable lithium-ion batteries have high specific energy as well as long cycle and shelf lifetimes, and unlike the other batteries, have high efficiency even at high loads (see Fig. 3). These features make lithium-ion batteries very popular for powering portable consumer electronic devices such as laptop computers, cell phones and camcorders. The disadvantage of the rechargeable lithium-ion batteries is their higher cost compared to lead-acid or nickel-metal hydride batteries.

Nickel-metal hydride secondary batteries are often used when common AA or AAA primary batteries are replaced with rechargeable ones. Although nickel-metal hydride batteries have a lower fully-charged voltage (1.4 V comparing to, e.g., 1.6-1.8 V for primary alkaline batteries), they have a flatter discharge curve (see Fig. 3), which allows them to generate around 1.2 V constant voltage for most of the discharge cycle. The nickel-metal hydride batteries have average specific energy, but also have lower charge retention compared to lithium and lead-acid batteries.

As revealed in Fig. 3, temperature is one parameter that influences the amount of energy obtainable from the battery. Two other critical parameters that define the amount of energy available from the battery are the battery load and duty cycle. The charts in Fig. 4 show the discharge curves for different loads and energy consumption profiles for the real-life common Commercially-available Off-The-Shelf (COTS) alkaline AAA batteries with nominal capacity of 1000 mAh. Note that the amount of the energy available from the battery decreases with the increase in load and that for a 680 Ohm load (2.2 mA @ 1.5 Volts), the alkaline AAA battery can provide over 1.95 Watt hours (Wh) of energy, whereas a 330 Ohm load (4.5 mA @ 1.5 Volts) from the same battery would get less than 1.75 Wh. At higher loads, as Fig. 3 reveals, the amount of available energy will decrease even at a higher rate. For batteries under intermittent discharge, the longer relaxation period between load connection (OFF time on Fig. 4), as noted in Fig. 4, also allows an increase in the amount of energy obtainable from the battery.

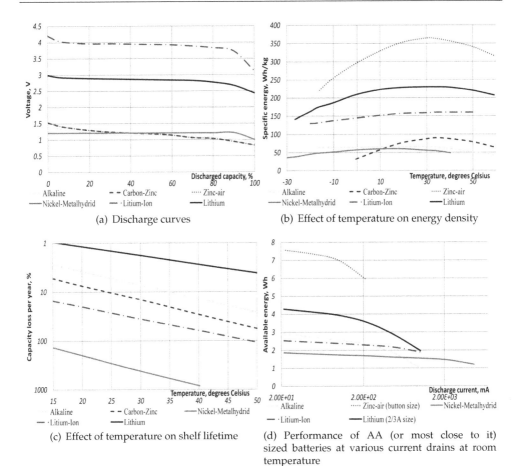

(a) Discharge curves

(b) Effect of temperature on energy density

(c) Effect of temperature on shelf lifetime

(d) Performance of AA (or most close to it) sized batteries at various current drains at room temperature

Fig. 3. Effect of the chemistry on battery performance[2]

2.3 Embedded systems power supply using energy scavenging systems

The final-and a very promising-ES power supply option that became possible due to recent technological advances, and that is currently gaining popularity, is the use of energy harvested from the environment as an ES power supply. Numerous demonstrations have now been reported for powering ESs utilizing the energy from such environment elements as:

- light (Hande et al., 2007; Knight et al., 2008; Morais et al., 2008; Valenzuela, 2008);
- temperature difference (Knight et al., 2008; Mathuna et al., 2008);
- vibration or movement (Knight et al., 2008; Mathuna et al., 2008; Mitcheson et al., 2008);
- water, air or gas flow (Hande et al., 2007; Mitcheson et al., 2008; Morais et al., 2008);

---

[2] The presented charts compile the results of (Crompton, 2000; Linden & Reddy, 2002) and different open sources

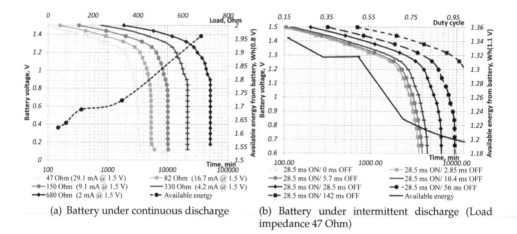

(a) Battery under continuous discharge

(b) Battery under intermittent discharge (Load impedance 47 Ohm)

Fig. 4. Typical discharge curves and available energy for alkaline AAA batteries[3]

Source	Conditions	Power density	Reference
Acoustic	75dB	$0.003~\mu W/cm^3$	(Yildiz, 2009)
	100dB	$0.96~\mu W/cm^3$	(Hande et al., 2007)
Air flow		$1\text{-}800~\mu W/cm^3$	(Knight et al., 2008; Yildiz, 2009)
Radio	GSM	$0.1~\mu W/cm^2$	(Raju, 2008)
	WiFi	$1~\mu W/cm^2$	(Raju, 2008; Yildiz, 2009)
Solar	Outdoors	up to $15000~\mu W/cm^2$	(Hande et al., 2007; Knight et al., 2008)
	Indoors	$100~\mu W/cm^2$	(Mathuna et al., 2008)
Thermal		$5\text{-}40~\mu W/cm^2$	(Hande et al., 2007; Knight et al., 2008)
Vibration		$4\text{-}800~\mu W/cm^3$	(Knight et al., 2008)
Water flow		up to $500000~\mu W/cm^3$	(Knight et al., 2008)

Table 2. Available energy harvesting technologies and their efficiency (based on (Hande et al., 2007; Knight et al., 2008; Mathuna et al., 2008; Raju, 2008; Yildiz, 2009))

- electrical or magnetic fields (Arnold, 2007; Knight et al., 2008; Mathuna et al., 2008);
- and biochemical reactions (e.g. Thomson (2008); Valenzuela (2008)).

Regardless of the energy harvesting method used, the energy should be initially harvested from the environment, converted to electric energy and buffered within a special storage system, which will later supply it to the attached ES. Usually, the amount of the energy that can be collected from the environment at any period of time is rather small (see Table 2). Therefore, the accumulation of energy over relatively long period of time is often required before the attached ES would be able to start operating. In real-life implementations (see Fig. 5(a)), thin film capacitors or super-capacitors are usually used for collected energy storage. Although supporting multiple charge/discharge cycles, these capacitors have very limited capacity and self-discharge rapidly (Mikhaylov & Tervonen, 2010b; Valenzuela, 2008). Energy storage over a long period of time is not possible without harvested energy being available. The

---

[3] The charts present the real-life measurement results for commercially available off-the-shelf alkaline AAA batteries

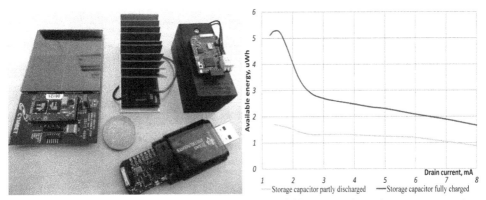

(a) Examples of COTS energy-harvesting hardware implementations: eZ430-RF2500SEH(Light), Micropelt TE-Power NODE(Temperature) and AdaptivEnergy Joule-Thief(Vibration)

(b) Available energy from the storage capacitor Cymbet-TI depending on the load for the real-life energy scavenging system

Fig. 5. Real-life energy harvesting applications

devices that are supplied with energy harvested from the environment can therefore suffer from frequent restarts due to energy unavailability and they must have very energy-efficient applications with low duty cycles and the appropriate mechanisms for recovery after energy exhaustion (Mikhaylov & Tervonen, 2011).

The parameters of the energy storage system used in energy scavenging devices have much in common with the secondary batteries discussed in Section 2.2. Thus, like the secondary batteries, the amount of energy obtainable from a harvested energy storage capacitor will decrease with increasing load (see Fig. 5(b))(Mikhaylov & Tervonen, 2010b).

3. Effect of the embedded system processor working mode and compulsory peripherals on the power consumption

3.1 Contemporary embedded systems

The market today offers a broad choice of commercial ESs, each having different purposes and characteristics. Table 3 provides a brief summary of the main parameters and required power supplied for the four main types of commercial ESPs.

Microcontrollers are the most commonly used ESPs (Emitt, 2008). Contemporary microcontrollers usually have an architecture based on a lightweight Central Processing Unit (CPU) with sequential command execution. The existing microcontrollers often have on chip all of the peripherals required for operation, such as volatile (e.g., Random Access Memory RAM) and non-volatile (e.g., Read Only Memory -ROM) memories, controllers for the digital communication interfaces (e.g., I2C, SPI, UART), analogue-to-digital converters (ADC), timers and clock generators. The microcontrollers have rather low cost, size and power consumption, which defines their wide usage in the wide range of the simple single task applications. The latest microcontroller generations, such as Texas Instruments (TI) MSP430L092 low-voltage microcontrollers, are capable of working using as low as 0.9 V power supply. Some of the recently developed microcontrollers already include such application-specific components as radio communication devices (e.g., TI CC2530 or CC430, Atmel ATmega128RFA1) or operational amplifiers (e.g., TI MSP430F2274).

Embedded system processor	Clock frequency, MHz	Supply voltage, V	Power consumption, W
microcontroller	0.032-30	0.9-3.6	0.00005-0.05
microprocessor	50-4000	1-3	0.5-150
ASIP	20-1200	1-5	0.2-10
FPGA	1500-8000000[a]	0.9-3	0.001-5

[a] number of gates

Table 3. Typical parameters of the contemporary embedded system's processors [4]

Contemporary microprocessors usually do not include any compulsory peripherals, thus implementing a standalone general purpose CPU. These microprocessors usually work at higher clock frequencies than the microcontrollers and are often used for different multi-task applications. The power consumption and the cost are usually higher for the processors than for the microcontrollers. The microprocessors nowadays can have multiple cores for implementing parallel data processing.

The Application-Specific Instruction-Set Processors (ASIPs) are the specially designed processors aimed for specific tasks such as Digital Signal Processors (DSPs), which are intended for efficient digital signal processing implementation, or Network Processors that can optimize packet processing during the communication within a network. Today, ASIPs are mostly used in applications that implement one specific task that requires significant processing capabilities, such as audio/video or communication processing.

The Field-Programmable Gate Arrays (FPGAs) contain reconfigurable logic elements (LEs) with interconnections that can be changed to implement the required functionality. This allows the use of FPGAs for implementing efficient high-speed parallel data processing, which is often required for high-speed video and signal processing. The contemporary FPGAs are often capable of using reconfigurable LEs to implement the software processors (e.g., MicroBlaze for Xilinx or NIOS II for Altera). The power consumption of FPGAs depends on the number of actually used LEs, the maximum number of which can vary from several thousands and up to 8 million.

In Section 3.2, the different parameters that influence the power consumption of ESs and the mechanisms underlying their effects are discussed.

3.2 Parameters influencing the power consumption for contemporary embedded system's processors

The energy consumed by a device at a given period of time (the power) is one of the parameters that defines the energy efficiency of every electrical device. In this subsection, we will focus the different parameters that influence the power consumption of ESs. For the sake of simplicity, we will assume that the ESs are supplied by an ideal source of power, which can be controlled by the ES.

---

[4] Based on the analysis of the data sheets and information from the main ESP manufacturers and open sources, data are presented for the most typical use case scenarios for each processor type.

The most widely used technology for implementing the different digital circuits today is the Complementary Metal-Oxide-Semiconductor (CMOS) technology (Benini et al., 2001; Hwang, 2006). The power consumption for a device built according to CMOS can be approximated using Equation 1 (Chandrakasan & Brodersen, 1995; SiLabs, 2003; Starzyk & He, 2007).

$$P = \alpha_{0\rightarrow1} \cdot C \cdot V^2 \cdot f + I_{peak} \cdot V \cdot t_{sc} \cdot f + I_l \cdot V \qquad (1)$$

In this equation, the first term represents the switching or dynamic capacitive power consumption due to charging of the CMOS circuit capacitive load through P-type Metal-Oxide-Semiconductor (PMOS) transistors, to make a voltage transition from the low to the high voltage level. The switching power depends on the average number of power consuming transitions made by the device over one clock period $\alpha_{0\rightarrow1}$, the CMOS device load capacitance C, the supply voltage level V and the clock frequency f. The second term represents the short circuit power consumed due to the appearance of the direct short current I_{peak} from the supply voltage to the ground, while PMOS and N-type Metal-Oxide-Semiconductor (NMOS) transistors are switched on simultaneously for a very short period of time t_{sc} during switching. The third term represents the static power consumed due to the leakage current I_l and does not depend on the clock frequency.

Of the three components that influence the circuit power consumption, the dynamic capacitive power is usually the dominant one when the circuit is in operational mode (Starzyk & He, 2007). In practice, the power consumed by the short-circuit current is typically less than 10% of the total dynamic power and the leakage currents cause significant consumption only if the circuit spends most of the time in standby mode(Chandrakasan & Brodersen, 1995)[5].

For a real-life ES-based device, apart from the power consumption of the ESP itself, which is described by Equation 1, the effect of other ESP compulsory peripherals (e.g., clock generator or used memory) need also to be considered.

3.2.1 Clock frequency

The clock frequency is one of the fundamental parameters for any synchronous circuit, including all of the CPU-based embedded systems (microcontrollers and microprocessors). The clock frequency is one of the parameters that - together with the processor architecture, command set and available peripherals used - would define the performance of the CPU.

Equation 1 reveals that the dynamic power consumed by the ESP for the particular supply voltage level should linearly increase with the increase of clock frequency. Note also that the most efficient strategy from the perspective of the consumed power per single operation, for the case when the third term in Equation 1 is above zero, would be to use, for any particular voltage, the *maximum clock frequency supportable at that supply voltage level*. The measurements for the real-life ESP presented in Fig. 6 confirm these statements (Dudacek & Vavricka, 2007; Mikhaylov & Tervonen, 2010b).

Fig. 6 reveals that the maximum achievable ESP clock frequency is influenced by the level of the supply voltage. For most ESPs, obtaining a high clock frequency is impossible while

---

[5] As revealed in (Ekekwe & Etienne-Cummings, 2006; Roy et al., 2003) the leakage current increases as technology scales down and can become the major contributor to the total power consumption in the future

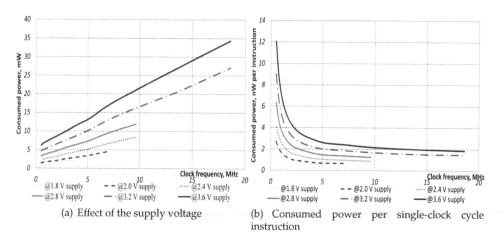

(a) Effect of the supply voltage   (b) Consumed power per single-clock cycle instruction

Fig. 6. Effect of the clock frequency on power consumption for the TI MSP430F2274 low-power microcontroller

maintaining a minimum supply voltage. The maximum allowable clock frequency for a particular supply voltage level can be estimated using Equation 2 (Chandrakasan et al., 1995; Cho & Chang, 2006). In Equation 2, V is the level of supply voltage, V_{th} is the threshold voltage and k and a are constants for a given technology process, which should be determined experimentally.

$$f = \frac{(V - V_{th})^a}{k \cdot V} \tag{2}$$

As previously noted (e.g., (Mikhaylov & Tervonen, 2010b)), a hysteresis exists for real-life ESPs for switch-on and switch-off threshold voltages (e.g., the MSP430 microcontroller using nominal clock frequency of 1 MHz will start operating with a supply voltage above 1.5 V and will continue working until the supply voltage drops to below 1.38 V).

Other research (e.g., (Dighe et al., 2007)) show that, for CPU-based ESPs other than microcontrollers, the power-frequency dependencies are similar to those presented in Fig. 6.

3.2.2 Supply voltage

As already noted in Subsection 3.2.1, the maximum possible clock frequency for the CPUs depends on the available supply voltage level. A further analysis of Equation 1 reveals that the supply voltage has a strong effect on the power components of both the dynamic and static systems. The charts showing the effect of the supply voltage on the overall power consumed by the system and the required power per single clock instruction execution for a real-life device are presented in Fig. 7. Equation 1 allows prediction that the most power efficient of any particular clock frequency would be one obtained using the minimum possible supply voltage. Equation 1 also reveals that, from the point of view of power consumption per operation, the most efficient strategy would be to use the *maximum clock frequency* at *the minimum possible supply voltage level*. Taking into account the clock frequency hysteresis for switch-on and switch-off voltage, further power efficiency can be obtained by first switching

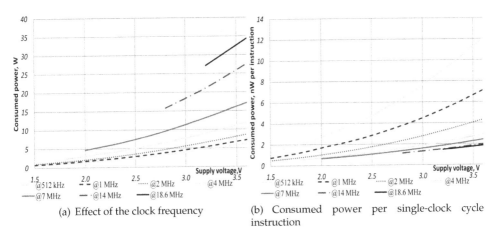

(a) Effect of the clock frequency

(b) Consumed power per single-clock cycle instruction

Fig. 7. Effect of the supply voltage on power consumption for the TI MSP430F2274 low-power microcontroller

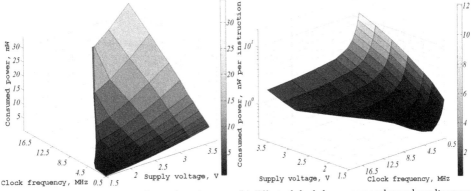

(a) Effect of clock frequency and supply voltage on the power consumption

(b) Effect of clock frequency and supply voltage on the power consumption per instruction

Fig. 8. Effect of clock frequency and supply voltage on the power consumption for the TI MSP430F2274 low-power microcontroller

the required clock frequency using a higher supply voltage level and later reducing the supply voltage up to a level slightly above the switch-off threshold (Mikhaylov & Tervonen, 2010b).

To summarize the effect of clock frequency and supply voltage for a real system, Fig. 8 presents the 3-D charts showing the overall consumed power and single-clock instruction power efficiency for the TI MSP430 microcontroller for different working modes. As expected, Fig. 8 reveals that the most efficient strategy from the perspective of power consumption per instruction would be to use the maximum supported clock frequency at a minimum possible supply voltage level. Similar results can be seen from other work (e.g.,(Luo et al., 2003)) and multiple desktop processor tests could be also obtained for the other types of ESPs and even FPGAs (Thatte & Blaine, 2002).

Nowadays, the dynamic tuning of the supply voltage level (dynamic voltage scaling) and clock frequency (dynamic frequency scaling) depending on the required system performance are the most widely used and the most effective techniques for improving ESP energy efficiency. Nonetheless, the practical implementation of voltage scaling has some pitfalls, the main one being that the efficiency of the DC/DC voltage converter, which will implement the voltage scaling, is usually on the order of 90-95% and will significantly decrease for the low load case, as also happens for the AC/DC converters discussed in 2.1.

3.2.3 CPU utilization

The CPU utilization, or time-loading factor, is the parameter that is often used for different general-purpose processors to measure their real time performance. The CPU utilization can be defined as the percentage of non-idle processing relative to the overall processing time (Laplante, 2004). Indeed, depending on the application, ESPs are required to fulfil a specified number of instructions at a specified period of time. After that, the ESP can switch to other tasks, execute no-ops, or move to a low-power mode (if it has the appropriate "waking-up" system).

Sections 3.2.1-3.2.2 have already shown that the most power efficient strategy for contemporary ESPs would be to use higher clock frequencies than to use lower clock frequencies at a particular supply voltage level and to use lower supply voltages, rather than higher ones. These statements indicate that, from the perspective of power efficiency, it would be optimal to have the CPU operating at a *minimum possible supply voltage that would support the clock frequency, which would allow fulfilment of the required number of instructions within the specified period of time.*

The problem of CPU utilization effects on processor power consumption has been described details e.g. in (Li et al., 2009; Uhrig & Ungerer, 2005), where appropriate real-life applications and measurements results are discussed.

3.3 Effect of the embedded system processor's compulsory peripherals on power consumption

The power consumption of a contemporary embedded system-based device is defined not only by the consumption of the actual ESP, but also by the cumulative power consumption of the all peripherals that are used by the application.   Apart from the actual ESP, the end-device will typically include a clock generation system, RAM, ROM, different input/output interfaces and some other peripherals (see Fig.1). As shown in Section 3.1, certain ES types can have some of the peripherals already integrated with the CPU. The actual set of peripherals used will clearly be defined by each particular application requirement; therefore, the most critical ones will be discussed in a Sections 3.3.1-3.3.5.

3.3.1 The clock generator

The clock generator is intended to provide the ESP and other peripherals with the required clock signal reference. Most present-day ESPs have the possibility either to use the external clock generator or to generate the clock signal using an internal clock crystal.   Most contemporary ESPs have inbuilt clock management systems, which can generate the required number of internal clock signals by multiplying or dividing the input one. Note, however, that higher power consumption occurs with the generation of a high clock frequency than with lower clock frequencies. Further clock conversions in ESPs would cause additional power consumption. Therefore, as has been shown previously (e.g., (Schmid et al., 2010; SiLabs, 2003)), from the point of view of power consumption, using the external low-frequency clock crystal is often much more convenient than using a high-frequency internal crystal and later dividing the frequency.

3.3.2 Random access memory

RAM is the memory type that is usually used for storing temporary data with critical access latency. The advantage of the RAM is that the data stored in it can be accessed both for reading and writing as single bytes (or small data blocks for recent chips) having the fixed access time regardless of the accessed location (Chen, 2004). As previously noted (e.g., (Mikhaylov & Tervonen, 2010a; Ou et al., 2011)), the RAM is usually the most efficient memory type from the point of view of power consumption. The disadvantage of RAM is that it is usually a volatile type of memory, meaning that the stored information is lost once the power supply is removed. Nonetheless, as has been shown previously (e.g., (Halderman et al., 2008; Mikhaylov & Tervonen, 2011)), the information in RAM remains undamaged for some time (5-60 seconds, depending on the RAM type and its working mode). This can be used to reduce the overall system power consumption through periodic power on/off switching of RAM memory when it is not being used.

The power consumption of RAM, similarly to the power consumption of the other already discussed CMOS systems (see Section 3.2), is influenced by the level of the supply voltage and the clock frequency (Cho & Chang, 2004; Fan et al., 2003). Quite often, the levels of supply voltage and clock frequency that minimize the power consumption for the RAM differ from the ones minimizing the consumption of the CPU, which requires resolution of the joint optimization problem for combined system (Cho & Chang, 2004; Fan et al., 2003).

3.3.3 Read-only and electrically erasable programmable read-only memory

ROM memory is a type of memory that is used for permanent data storage. The data in ROM either cannot be modified at all (e.g., masked ROM), or requires significant effort and time for data changing (e.g., electrically erasable programmable read-only memory (EEPROM) or Flash ROM). The advantage of ROM is that it is a non-volatile type of memory and retains the stored data even if no power supplied. The common disadvantages of ROM compared to RAM are the higher data access time and power consumption (Chen, 2004; Mikhaylov & Tervonen, 2011; Ou et al., 2011). Another common feature of ROM and especially EEPROM, which is currently mostly often used in the ES, is that writing to the memory should be done by so-called pages; i.e., data blocks with the sizes in the range of 64 and 512 bytes depending on the memory chip architecture. Therefore, changing the data in EEPROM first requires erasing the entire page containing the data to be changed. After that, the new values for the bytes within the erased page can be written either byte-wise or in burst mode. Rather often, especially for the EEPROM integrated into microcontrollers, the cleaning and writing to EEPROM requires a higher supply voltage level than the one required for normal CPU operation. This complicated rewrite process causes the Flash memory to have very significant power consumption during data rewritings, which can be several orders of magnitude higher than while writing to RAM. The number of rewrite cycles for contemporary EEPROMs can reach 10.000 to 10.000.000, but it is by no means infinite.

Although ROM is now often used for storing the executable application program codes for different ESPs, as shown previously (e.g., (Mikhaylov & Tervonen, 2010b)), the running of ESP programs stored in RAM allows a reduction of the overall power consumption by 5% to 10%.

3.3.4 Input/output interfaces

The input/output (I/O) interfaces are the essential ESP peripherals that allow ESPs to interact with the external world. Since the I/O interfaces are implemented using the same CMOS blocks as the rest of ESP, the conclusions made within Section 3.2 are also applicable for the I/O interfaces (Dake & Svensson, 1994). In addition to the actual power consumption of the I/O interfaces, the wire propagation effects, such as attenuation, distortion, noise and interferences, must also be considered. Therefore, the conclusion can be made that implementation of power efficient communication over a particular I/O interface should use the lowest possible level of the supply voltage together with highest data rate that allows provision of reliable communication with the required throughput.

Quite often, the developed ES-based application does not use all of the available ESP's digital pins. To reduce the overall system power consumption, these pins should be configured as outputs. Whether initialized as high or low, the output voltage will not subject the enabled digital input circuitry to a leakage-current-inducing voltage in the middle range (Peatman, 2008).

3.3.5 Other peripherals

Depending on the application, the ESPs can require a wide range of other peripherals. The two basic rules for power effective peripheral usage are:

- the peripherals should be provided with the minimum level of supply voltage that allow their reliable operation;
- the peripherals should be powered off when not in use.

As previously shown (e.g., (Curd, 2007)), the use of embedded blocks for special function implementing in FPGAs dramatically reduces the dynamic power consumption when compared to implementing these functions in general purpose FPGA logic. This is also valid for other types of ESPs.

4. Energy efficiency-aware low-power embedded systems utilization

The two previous sections discussed the different power supply options that can be used for existing ESs (Section 2) and the parameters influencing the power consumption for the standalone ES (Section 3). These discussions confirm that *the real energy efficiency maximization for an ES-based application requires a joint consideration of the power supply system and the ES itself*. The current section will show how ES parameters influence the power consumption of a real-life device supplied using different power supply sources. It will also discuss the efficiency of the methods that can be used to improve the system's overall power efficiency.

4.1 Energy efficiency for mains-supplied low-power embedded systems

Fig. 9 shows the power consumption for a low-power microcontroller-based device supplied from mains via an AC/DC converter, with (Fig.9(a)) and without (Fig. 9(b)) a voltage

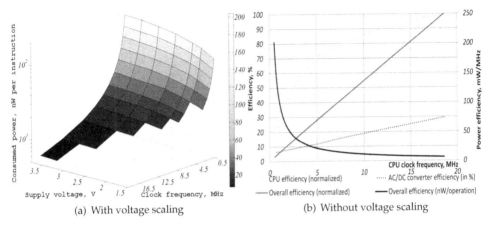

(a) With voltage scaling   (b) Without voltage scaling

Fig. 9. Power efficiency for a MSP430-based system supplied from mains via an AC/DC converter[6]

scaling system. Comparing the results in Fig. 9 with the standalone microcontroller power consumption (see Fig. 8) shows that the situation changed dramatically. For the standalone microcontroller, the most efficient strategy from the point of system power consumption per instruction was to operate at the *maximum clock frequency supported, using the minimum supply voltage level* (see Section 3.2.2), while for the mains-supplied system, the most effective strategy is to use the *minimum supply voltage level that supported the maximum possible clock frequency*. At first glance, these results seem contradictory, but they can be easily explained if the conversion efficiency curves for the real-life AC/DC and DC/DC converters, which are presented in Figs. 2 and 9(b), are also taken into account. As shown in Fig. 9(a), the use of voltage scaling for the low-power ES does not significantly increase the overall power efficiency due to the very low AC/DC conversion efficiency for the microcontroller low-power modes.

Nonetheless, as Fig. 2 reveals, the efficiency of AC/DC and DC/DC converters under the higher loads increases to more than 90% and becomes consistent, which allows efficient use of the dynamic voltage and frequency scaling techniques for improving the power consumption of high-power ESPs supplied from mains (as shown previously by e.g., (Cho & Chang, 2006; Simunic et al., 2001)).

4.2 Energy efficiency for battery-supplied low power embedded systems

To illustrate the effect of the ESP parameters on a battery-supplied system, we investigated the operation of the same low-power microcontroller-based system discussed in Subsection 4.1, but now supplying power from two alkaline batteries. The charts summarizing the results are presented in Fig. 10 for AAA batteries and in Fig. 11 for AG3 button batteries. The presented charts has been built using the battery capacity models (Equation 3, with the parameters from Table 4), which are based on the real-life battery capacity measurements (see, e.g., Fig. 4). The presented charts illustrate the system efficiency (measured as the number of single clock instructions computed over the system lifetime) for the system built around a low-power ESP,

---

[6] The presented charts have been obtained through simulations based on the real AC/DC and DC/DC converters characteristics.

with (Figs. 10(a) and 11(a)) and without (Figs. 10(b) and 11(b)) the voltage scaling mechanism. For the sake of simplicity, in the used model, we assume that the ESP is working with 100% CPU utilization and that it switches off when the voltage acquired from the battery supply falls below the minimum supply voltage required to support the ESP operation at a defined clock frequency (see Section 3.2.1).

$$E = C_1 \cdot (P_{avg})^{C_2}$$

(3)

The charts for the battery-supplied ESP-and likewise for the standalone ESP-show that an optimal working mode exists that allows maximizing of the system efficiency within the used metrics. Figs. 10(b) and 11(b) show that the number of operations executed by the battery-supplied ESP over its lifetime strongly depend on the clock frequency used; e.g., for AAA batteries for clock frequencies 2.5 times higher and lower than the optimal one, the number of possible operations decreases 2 times. Nonetheless, the optimal working mode for the system supplied from the battery is slightly different from the one for the standalone system. For the standalone system, as shown in Fig. 8, use of a 3 MHz clock frequency with 1.5 V supply voltage level was optimal, while for battery supplied system, use of a 4.4 MHz clock frequency with 1.8 V supply was optimal. The main reasons for this observation are: the lower efficiency of DC/DC conversion of the voltage controlling system for lower loads (see Fig. 2), and the different amounts of energy available from the battery for various loads (see Figs. 4, 10(b) and 11(b)).

As Figs. 10(a) and 11(a) reveal, the voltage scaling possibility allows an increase in the number of executable operations by the ESP by more than 2.5 times compared to the system without voltage control. The optimum working mode for the battery supplied ESP with the voltage control possibility appears to be the same as for the standalone system (3 MHz at 1.5 V supply) and differs from the battery supplied system without voltage conversion. Nonetheless, the use of voltage conversion circuits would have one significant drawback for the devices working at low duty cycle: the typical DC/DC voltage converter chips have a standby current on the order of dozens μA, while the standby current of contemporary microcontrollers in the low-power mode is below 1 μA. Therefore, the use of a voltage controlling system for a low duty cycle system can dramatically increase the sleep-mode power consumption, thereby reducing the overall system lifetime.

As can be noted comparing Figs. 10(b) and 11(b), the small sized AG3 alkaline batteries have a much lower capacity and lower performance while using higher load. These figures also reveal that the optimal clock frequency for both batteries is slightly different: the optimal clock frequency for an AAA battery appears to be slightly higher than for the button style.

Threshold,V	AAA battery			AG3 button battery		
	C_1	C_2	R^2 [a]	C_1	C_2	R^2 [a]
0.75	1.063681	-0.08033	>0.95	0.004009	-0.36878	>0.98
0.9	0.995933	-0.08998	>0.95	0.003345	-0.39978	>0.98
1	0.996802	-0.07764	>0.99	0.001494	-0.53116	>0.98
1.2	0.888353	-0.06021	>0.98	0.000104	-0.92647	>0.98
1.4	0.15627	-0.21778	>0.97	0.000153	-0.89025	>0.99

[a] The coefficient of determination for model

Table 4. Parameters of the used battery discharge models

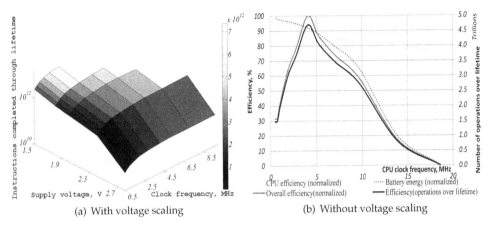

(a) With voltage scaling (b) Without voltage scaling

Fig. 10. Energy efficiency for a MSP430-based system supplied from AAA alkaline batteries

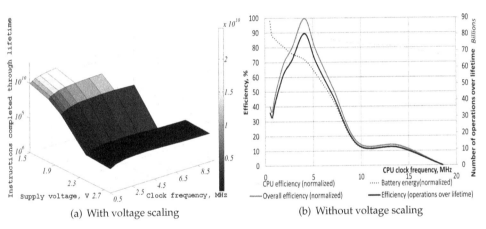

(a) With voltage scaling (b) Without voltage scaling

Fig. 11. Energy efficiency for a MSP430-based system supplied from AG3 alkaline batteries

In the current section, we have focused on the Alkaline batteries, as they are most commonly used today. It has been shown, that for the batteries of the same chemistry but different form-factor the ESPs optimal parameters are slightly different. For the batteries that use other chemistries, as suggested by the data in Fig. 3, the optimal energy work mode parameters will differ significantly (see e.g., (Raskovic & Giessel, 2009)). The system lifetime for the other types of ESPs supplied from batteries would follow the same general trends.

4.3 Energy efficiency for low-power embedded systems supplied by energy harvesting

Fig. 12 illustrates the effects of the ESP parameters on the operation of the system supplied using an energy harvesting system. The charts show results of practical measurements for a real system utilizing the MSP430F2274 microcontroller board and a light-energy harvesting system using a thin-film rechargeable EnerChips energy storage system (Texas, 2010). The

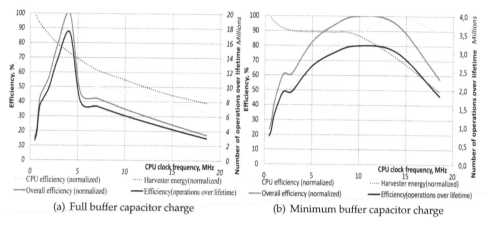

(a) Full buffer capacitor charge   (b) Minimum buffer capacitor charge

Fig. 12. Energy efficiency for a MSP430-based system supplied from an energy harvesting system with a thin-film rechargeable EnerChips storage system

presented charts illustrate the system operation for the cases when the storage system has been initially fully charged (Fig. 12(a)) and when the storage system had only minimum amount of energy [7] (Fig. 12(b)). During the measurements, the system was located indoors under the light with intensity of around 275 Lux. For evaluating the energy efficiency for the system supplied using energy harvested from the environment, we have used the same metrics as described for the battery supplied system; namely, the number of single clock instructions which the ESP is able to execute until energy storage system is discharged.

Figs. 12(a) and 12(b) reveal that the optimal work mode parameters for the ESP for an energy harvesting supplied system are different for various energy storage system initial states. Fig. 12(a) shows that a well-defined clock frequency exists for the fully charged storage system, which allows the execution of the maximum number of instructions to be achieved. For a system with minimum storage system initial charge, the optimum clock frequency that will maximize the number of ESP operations is shifted to higher clock frequencies.

Due to the already discussed high standby current for the DC/DC converters, the use of the voltage control circuits within the system supplied by energy harvesting appeared to be ineffective.

Table 2 shows that the amount of energy that the small sized energy harvesting systems can collect from environment is rather small. This means that energy scavenging applications using high-power or high-duty cycle ESPs will need to have rather volumetric supply systems. Therefore, this power supply options is now mostly often used with low-power ESPs in Wireless Sensor Networks (WSN), toys and consumer electronics applications.

5. Conclusions and further research

In this chapter, we have discussed the different aspects of the energy efficient operation of the commercial low-power embedded systems. The possible supply sources that can be

---

[7] The energy storage system is connected to the load only once the amount of available energy exceeds the threshold - see (Texas, 2010)

used in ES-based applications, the ES parameters that influence the energy consumption and the mechanisms underlying their effect have been discussed in detail. Finally, real-life examples were used to show that real energy efficiency for ES-based applications is possible *only* when the characteristics of the used supply system and the embedded system itself are considered as a whole. The results presented in the chapter have been obtained by the authors through multiple years of practical research and development experience within the field of low power embedded systems applications, and they could be valuable for both engineers and researchers working in this field.

The problem of energy efficiency is a versatile one, and many open questions still remain. For the energy efficiency optimization, one needs to have full information on the source of power characteristics, the characteristics of the embedded system itself and the user application requirements. This requires a standardized way to store this type of information and mechanisms that would allow identification of the source of power and peripherals attached to the embedded system and that would obtain the information required for operation optimization. Once all of the required information was available, this would advance the possibility of developing the algorithms needed to allow the embedded system to adapt its operation to the available resources and to the application requirements. The other open problem currently limiting the possibility of developing automated power optimization algorithms is that most of the currently existing embedded systems do not implement any mechanism for measuring their power consumption.

6. References

Arnold, D. (2007). Review of microscale magnetic power generation, *IEEE Transactions on Magnetics* Vol. 43(No. 11): 3940 – 3951.

Benini, L., Micheli, G. D. & Macii, E. (2001). Designing low-power circuits: practical recipes, *IEEE Circuits and Systems Magazine* Vol. 1(No. 1): 6 – 25.

Chandrakasan, A. & Brodersen, R. (1995). Minimizing power consumption in digital CMOS circuits, *Proceedings of the IEEE* Vol. 83(No. 4): 498 – 523.

Chandrakasan, A., Potkonjak, M., Mehra, R., Rabaey, J. & Brodersen, R. (1995). Optimizing power using transformations, *IEEE Transactions on Computer-Aided Design of Integrated Circuits and Systems* Vol. 14: 12 – 31.

Chen, W. (2004). *The Electrical Engineering Handbook*, Academic press.

Cho, Y. & Chang, N. (2004). Memory-aware energy-optimal frequency assignment for dynamic supply voltage scaling, *Proceedings of ISLPED '04*, pp. 387–392.

Cho, Y. & Chang, N. (2006). Energy-aware clock-frequency assignment in microprocessors and memory devices for dynamic voltage scaling, *IEEE Transactions on Computer-Aided Design of Integrated Circuits and Systems* Vol. 26(No. 6): 1030 – 1040.

Crompton, T. (2000). *Battery Reference Book*, Newnes.

Curd, D. (2007). Power consumption in 65 nm FPGAs.
    URL: *http://www.xilinx.com/support/documentation/white%5Fpapers/wp246.pdf*

Dake, L. & Svensson, C. (1994). Power consumption estimation in CMOS VLSI chips, *IEEE Journal of Solid-State Circuits* Vol. 29(No. 6): 663 – 670.

Dighe, S., Vangal, S., Aseron, R., Kumar, S., Jacob, T., Bowman, K., Howard, J., Tschanz, J., Erraguntla, V., Borkar, N., De, V. & Borkar, S. (2007). Within-die variation-aware dynamic-voltage-frequency-scaling with optimal core allocation and thread hopping for the 80-core TeraFLOPS processor, *IEEE Journal of Solid-State Circuits* Vol. 46(No. 1): 184 – 193.

Dudacek, K. & Vavricka, V. (2007). Experimental evaluation of the MSP430 microcontroller power requirements, *Proceedings of EUROCON'07*, pp. 400–404.

Earth (2011). INFSO-ICT-247733 EARTH: Deliverable D2.1: Economic and ecological impact of ICT.
    URL: *https://bscw.ict-earth.eu/pub/bscw.cgi/d38532/EARTH%5FWP2%5FD2.1%5Fv2.pdf*

Ekekwe, N. & Etienne-Cummings, R. (2006). Power dissipation sources and possible control techniques in ultra deep submicron CMOS technologies, *Elsevier Microelectronics Journal* Vol. 37: 851 – 860.

Emitt (2008). Microcontroller market and technology analysis report - 2008.
    URL: *http://www.emittsolutions.com/images/microcontroller%5Fmarket%5Fanalysis%5F2008.pdf*

Fan, X., Ellis, C. & Lebeck, A. (2003). Interactions of power-aware memory systems and processor voltage scaling, *Proceedings of PACS'03*, pp. 1–12.

FreedoniaGroup (2011). Study 2449: Batteries.
    URL: *http://www.freedoniagroup.com/brochure/24xx/2449smwe.pdf*

Halderman, J., Schoen, S., Heninger, N., Clarkson, W., Paul, W., Calandrino, J., Feldman, A., Appelbaum, J. & Felten, E. (2008). Lest we remember: Cold boot attacks on encryption keys, *Proceedings of USENIX Security '08*, pp. 1–16.

Hande, A., T.Polk, Walker, W. & Bhatia, D. (2007). Indoor solar energy harvesting for sensor network router nodes, *Future beyond Science* Vol. 31(No. 6): 420 – 432.

Hwang, E. (2006). *Digital Logic and Microprocessor Design with VHDL*, Thomson.

INOBAT (2009). Absatzzahlen 2008.
    URL: *http://www.inobat.ch/fileadmin/user%5Fupload/pdf%5F09/Absatz%5FStatistik%5F2008.pdf*

Jang, Y. & Jovanovic, M. (2010). Light-load efficiency optimization method, *IEEE Transactions on Power Electronics* Vol. 25(No. 1): 67 – 74.

Knight, C., Davidson, J. & Behrens, S. (2008). Energy options for wireless sensor nodes, *Sensors* Vol. 8: 8037 – 8066.

Laplante, P. (2004). *Real-time systems design and analysis*, Wiley-IEEE.

Li, L., RuiXiong, T., Bo, Y. & ZhiGuo, G. (2009). A model of web server's performance-power relationship, *Proceedings of ICCSN'09*, pp. 260–264.

Linden, D. & Reddy, T. (2002). *Handbook of batteries*, McGraw-Hill.

Luo, J., Peh, L. & Jha, N. (2003). Simultaneous dynamic voltage scaling of processors and communication links in real-time distributed embedded systems, *Proceedings of DATE'03*, pp. 1150–1151.

Mathuna, C., O'Donnell, T., Martinez-Catala, R., Rohan, J. & B.O'Flynn (2008). Energy scavenging for long-term deployable wireless sensor networks, *Future beyond Science* Vol. 75(No. 3): 613 – 623.

Mikhaylov, K. & Tervonen, J. (2010a). Improvement of energy consumption for over-the-air reprogramming in wireless sensor networks, *Proceedings of ISWPC'10*, pp. 86–92.

Mikhaylov, K. & Tervonen, J. (2010b). Optimization of microcontroller hardware parameters for wireless sensor network node power consumption and lifetime improvement, *Proceedings of ICUMT'10*, pp. 1150–1156.

Mikhaylov, K. & Tervonen, J. (2011). Energy efficient data restoring after power-downs for wireless sensor networks nodes with energy scavenging, *Proceedings of NTMS'11*, pp. 1–5.

Mitcheson, P., Yeatman, E., Rao, G., Holmes, A. & Green, T. (2008). Energy harvesting from human and machine motion for wireless electronic devices, *Proceedings of the IEEE* Vol. 96(No. 9): 1457 – 1486.

Morais, R., Matos, S., Fernandes, M., Valentea, A., Soares, S., Ferreira, P. & Reis, M. (2008). Sun, wind and water flow as energy supply for small stationary data acquisition platforms, *Computers and Electronics in Agriculture* Vol. 6(No. 2): 120 – 132.

Munsey, B. (2011). New developments in battery design and trends.
URL: *http://www.houseofbatteries.com/documents/New%20Chemistries%20April%202010%20V2.pdf*

Ou, Y., & Harder, T. (2011). Trading memory for performance and energy, *Proceedings of DASFAA'11*, pp. 1–5.

Peatman, J. (2008). *Coin-Cell-Powered Embedded Design*, Qwik&Low Books.

Raju, M. (2008). Energy harvesting.
URL: *http://www.ti.com/corp/docs/landing/cc430/graphics/slyy018%5F20081031.pdf*

Raskovic, D. & Giessel, D. (2009). Dynamic voltage and frequency scaling for on-demand performance and availability of biomedical embedded systems, *IEEE Transactions on Information Technology in Biomedicine* Vol.13(No. 6): 903 – 909.

Roy, K., Mukhopadhyay, S. & Mahmoodi-Meimand, H. (2003). Leakage current mechanisms and leakage reduction techniques in deep-submicrometer CMOS circuits, *Proceedings of the IEEE* Vol. 91(2): 305 – 327.

Schmid, T., Friedman, J., Charbiwala, Z., Cho, Y. & Srivastava, M. (2010). Low-power high-accuracy timing systems for efficient duty cycling, *Proceedings of ISLPED '08*, pp. 75–80.

SiLabs (2003). AN116: Power management techniques and calculation.
URL: *http://www.silabs.com/Support%20Documents/TechnicalDocs/an116.pdf*

Simunic, T., Benini, L., Acquaviva, A., Glynn, P. & De Micheli, G. (2001). Dynamic voltage scaling and power management for portable systems, *Proceedings of DAC'01*, pp. 524–529.

Starzyk, J. & He, H. (2007). A novel low-power logic circuit design scheme, *IEEE Transactions on Circuits and Systems* Vol. 54(No. 2): 176 – 180.

Texas (2010). eZ430-RF2500-SEH solar energy harvesting development tool (SLAU273C).
URL: *http://www.ti.com/lit/ug/slau273c/slau273c.pdf*

Thatte, S. & Blaine, J. (2002). Power consumption in advanced FPGAs, *Xcell Journal* .
URL: *http://cdserv1.wbut.ac.in/81-312-0257-7/Xilinx/files/Xcell%20Journal%20Articles/xcell%5Fpdfs/xc%5Fsynplicity44.pdf*

Thomson, E. (2008). Preventing forest fires with tree power, *MIT Tech Talk* Vol. 53(No. 3): 4 – 4.
URL: *http://web.mit.edu/newsoffice/2008/techtalk53-3.pdf*

Uhrig, S. & Ungerer, T. (2005). Energy management for embedded multithreaded processors with integrated EDF scheduling, *Proceedings of ARCS'05*, pp. 1–17.

USEIA (2011). RECS 2009.
URL: *http://www.eia.gov/consumption/residential/reports/electronics.cfm*

Valenzuela, A. (2008). Energy harvesting for no-power embedded systems.
    URL: *http://focus.ti.com/graphics/mcu/ulp/energy%5Fharvesting%5Fembedded%5Fsystems
    %5Fusing%5Fmsp430.pdf*
Yildiz, F. (2009). Potential ambient energy-harvesting sources and techniques, *The Journal of
    Technology Studies* Vol. 35(No. 1): 40 – 48.

Permissions

The contributors of this book come from diverse backgrounds, making this book a truly international effort. This book will bring forth new frontiers with its revolutionizing research information and detailed analysis of the nascent developments around the world.

We would like to thank Kiyofumi Tanaka, for lending his expertise to make the book truly unique. He has played a crucial role in the development of this book. Without his invaluable contribution this book wouldn't have been possible. He has made vital efforts to compile up to date information on the varied aspects of this subject to make this book a valuable addition to the collection of many professionals and students.

This book was conceptualized with the vision of imparting up-to-date information and advanced data in this field. To ensure the same, a matchless editorial board was set up. Every individual on the board went through rigorous rounds of assessment to prove their worth. After which they invested a large part of their time researching and compiling the most relevant data for our readers. Conferences and sessions were held from time to time between the editorial board and the contributing authors to present the data in the most comprehensible form. The editorial team has worked tirelessly to provide valuable and valid information to help people across the globe.

Every chapter published in this book has been scrutinized by our experts. Their significance has been extensively debated. The topics covered herein carry significant findings which will fuel the growth of the discipline. They may even be implemented as practical applications or may be referred to as a beginning point for another development. Chapters in this book were first published by InTech; hereby published with permission under the Creative Commons Attribution License or equivalent.

The editorial board has been involved in producing this book since its inception. They have spent rigorous hours researching and exploring the diverse topics which have resulted in the successful publishing of this book. They have passed on their knowledge of decades through this book. To expedite this challenging task, the publisher supported the team at every step. A small team of assistant editors was also appointed to further simplify the editing procedure and attain best results for the readers.

Our editorial team has been hand-picked from every corner of the world. Their multi-ethnicity adds dynamic inputs to the discussions which result in innovative outcomes. These outcomes are then further discussed with the researchers and contributors who give their valuable feedback and opinion regarding the same. The feedback is then collaborated with the researches and they are edited in a comprehensive manner to aid the understanding of the subject.

Apart from the editorial board, the designing team has also invested a significant amount of their time in understanding the subject and creating the most relevant covers. They scrutinized every image to scout for the most suitable representation of the subject and create an appropriate cover for the book.

The publishing team has been involved in this book since its early stages. They were actively engaged in every process, be it collecting the data, connecting with the contributors or procuring relevant information. The team has been an ardent support to the editorial, designing and production team. Their endless efforts to recruit the best for this project, has resulted in the accomplishment of this book. They are a veteran in the field of academics and their pool of knowledge is as vast as their experience in printing. Their expertise and guidance has proved useful at every step. Their uncompromising quality standards have made this book an exceptional effort. Their encouragement from time to time has been an inspiration for everyone.

The publisher and the editorial board hope that this book will prove to be a valuable piece of knowledge for researchers, students, practitioners and scholars across the globe.

List of Contributors

Mouaaz Nahas and Ahmed M. Nahhas
Department of Electrical Engineering, College of Engineering and Islamic Architecture, Umm Al-Qura University, Makkah, Saudi Arabia

Javier D. Orozco and Rodrigo M. Santos
Universidad Nacional Del Sur – CONICET, Argentina

Juergen Mottok
Regensburg University of Applied Sciences, Germany

Frank Schiller
Beckhoff Automation GmbH, Germany

Thomas Zeitler
Continental Automotive GmbH, Germany

Yung-Yuan Chen and Tong-Ying Juang
National Taipei University, Taiwan

Makoto Sugihara
Kyushu University, Japan

Susanna Pantsar-Syväniemi
VTT Technical Research Centre of Finland, Finland

Nikolaos Kavvadias, Vasiliki Giannakopoulou and Kostas Masselos
Department of Computer Science and Technology, University of Peloponnese, Tripoli, Greece

Takaaki Goto and Tetsuro Nishino
The University of Electro-Communications, Japan

Yasunori Shiono, Tomoo Sumida and Kensei Tsuchida
Toyo University, Japan

Takeo Yaku
Nihon University, Japan

Philippe Dhaussy and Jean-Charles Roger
Ensta-Bretagne, France

Frédéric Boniol
ONERA, France

Pablo Peñil, Fernando Herrera and Eugenio Villar
Microelectronics Engineering Group of the University of Cantabria, Spain

F. Herrera and I. Ugarte
University of Cantabria, Spain

Meng Shao
Computer Centre, Hangzhou First People's Hospital, Hangzhou, China

Zhe Peng and Longhua Ma
School of Aeronautics and Astronautics, Zhejiang University, Hangzhou, China

Héctor Posadas, Álvaro Díaz and Eugenio Villar
Microelectronics Engineering Group of the University of Cantabria, Spain

Adi Maaita
Software Engineering Department, Faculty of Information Technology, Isra University, Amman, Jordan

Michael Dossis
Technological Educational Institute of Western Macedonia, Dept. of Informatics & Computer Technology, Greece

Jawar Singh
PDPM- Indian Institute of Information Technology, Design & Manufacturing, Jabalpur, India

Balwinder Raj
ABV-Indian Institute of Information Technology and Management, Gwalior, India

Yongpan Liu, Shuangchen Li and Huazhong Yang
Tsinghua University, Beijing, P.R. China

Pei Zhang
Y Explorations, Inc., San Jose, CA USA

Konstantin Mikhaylov and Jouni Tervonen
Oulu Southern Institute, University of Oulu, Finland

Dmitry Fadeev
Saint-Petersburg State Polytechnic University, Russian Federation

9 781632 401168